AF864680

Springer Handbook of Enzymes · Volume 13

Springer
Berlin
Heidelberg
New York
Hong Kong
London
Milan
Paris
Tokyo

Dietmar Schomburg and
Ida Schomburg (Eds.)

Springer Handbook of Enzymes

Volume 13

Class 3.2 · Hydrolases VIII

EC 3.2.1.48–3.2.1.149

coedited by Antje Chang

Second Edition

Springer

Professor DIETMAR SCHOMBURG
e-mail: d.schomburg@uni-koeln.de

Dr. IDA SCHOMBURG
e-mail: i.schomburg@uni-koeln.de

Dr. Antje Chang
e-mail: a.chang@uni-koeln.de

University to Cologne
Institute for Biochemistry
Zülpicher Strasse 47
50674 Cologne
Germany

ISBN 3-540-00906-X 2nd Edition Springer-Verlag Berlin Heidelberg New York

The first edition was published as Volume 4 (ISBN 3-540-53730-9) and supplemented in Vol. 15 (ISBN 3-540-64116-5)of the "Enzyme Handbook".

Library of Congress Cataloging-in-Publication Data

Springer handbook of enzymes /
Dietmar Schomburg and Ida Schomburg (ed.) –
Berlin ; Heidelberg ; New York ; Barcelona ; Hong Kong ;
London ; Milan ; Paris ; Tokyo : Springer

Vol. 13 Class 3.2, hydrolases VIII: EC 3.2.1.48–3.2.1.149 / coed. by Antje Chang. – 2. ed. – 2003
ISBN 3-540-00906-X

Springer-Verlag Berlin Heidelberg New York
a member of BertelsmannSpringer Science+Business Media GmbH

http://www.springer.de

© Springer-Verlag Berlin Heidelberg 2003
Printed in Germany

The use of general descriptive names, registered names, etc. in this publication does not imply, even in the absence of a specific statement, that such names are exempt from the relevant protective laws and regulations and free for general use.

The publisher cannot assume any legal responsibility for given data, especially as far as directions for the use and the handling of chemicals and biological material are concerned. This information can be obtained from the instructions on safe laboratory practice and from the manufacturers of chemicals and laboratory equipment.

Cover design: Erich Kirchner, Heidelberg
Typesetting: medio Technologies AG, Berlin

2/3020xv-5 4 3 2 1 0 – printed on acid-free paper

Attention all Users of the "Springer Handbook of Enzymes"

Information on this handbook can be found on the internet at http://www.springer.de/enzymes/

A complete list of all enzyme entries either as an alphabetical Name Index or as the EC-Number Index is available at the above mentioned URL. You can download and print them free of charge.

A complete list of all synonyms (> 25,000 entries) used for the enzymes is available in print form (ISBN 3-540-41830-X).

Save 15 %

We recommend a standing order for the series to ensure you automatically receive all volumes and all supplements and save 15 % on the list price.

Preface

Today, as the full information about the genome is becoming available for a rapidly increasing number of organisms and transcriptome and proteome analyses are beginning to provide us with a much wider image of protein regulation and function, it is obvious that there are limitations to our ability to access functional data for the gene products – the proteins and, in particular, for enzymes. Those data are inherently very difficult to collect, interpret and standardize as they are widely distributed among journals from different fields and are often subject to experimental conditions. Nevertheless a systematic collection is essential for our interpretation of genome information and more so for applications of this knowledge in the fields of medicine, agriculture, etc. Progress on enzyme immobilisation, enzyme production, enzyme inhibition, coenzyme regeneration and enzyme engineering has opened up fascinating new fields for the potential application of enzymes in a wide range of different areas.
The development of the enzyme data information system BRENDA was started in 1987 at the German National Research Centre for Biotechnology in Braunschweig (GBF) and is now continuing at the University at Cologne, Institute of Biochemistry. The present book "Springer Handbook of Enzymes" represents the printed version of this data bank. The information system has been developed into a full metabolic database.
The enzymes in this Handbook are arranged according to the Enzyme Commission list of enzymes. Some 4,000 "different" enzymes are covered. Frequently enzymes with very different properties are included under the same EC- number. Although we intend to give a representative overview on the characteristics and variability of each enzyme, the Handbook is not a compendium. The reader will have to go to the primary literature for more detailed information. Naturally it is not possible to cover all the numerous literature references for each enzyme (for some enzymes up to 40,000) if the data representation is to be concise as is intended.
It should be mentioned here that the data have been extracted from the literature and critically evaluated by qualified scientists. On the other hand, the original authors' nomenclature for enzyme forms and subunits is retained. In order to keep the tables concise, redundant information is avoided as far as possible (e.g. if K_m values are measured in the presence of an obvious cosubstrate, only the name of the cosubstrate is given in parentheses as a commentary without reference to its specific role).
The authors are grateful to the following biologists and chemists for invaluable help in the compilation of data: Cornelia Munaretto and Dr. Antje Chang.

Cologne
Spring 2003

Dietmar Schomburg, Ida Schomburg

List of Abbreviations

A	adenine
Ac	acetyl
ADP	adenosine 5'-diphosphate
Ala	alanine
All	allose
Alt	altrose
AMP	adenosine 5'-monophosphate
Ara	arabinose
Arg	arginine
Asn	asparagine
Asp	aspartic acid
ATP	adenosine 5'-triphosphate
Bicine	N,N'-bis(2-hydroxyethyl)glycine
C	cytosine
cal	calorie
CDP	cytidine 5'-diphosphate
CDTA	trans-1,2-diaminocyclohexane-N,N,N,N-tetraacetic acid
CMP	cytidine 5'-monophosphate
CoA	coenzyme A
CTP	cytidine 5'-triphosphate
Cys	cysteine
d	deoxy-
D-	(and L-) prefixes indicating configuration
DFP	diisopropyl fluorophosphate
DNA	deoxyribonucleic acid
DPN	diphosphopyridinium nucleotide (now NAD)
DTNB	5,5'-dithiobis(2-nitrobenzoate)
DTT	dithiothreitol (i.e. Cleland's reagent)
EC	number of enzyme in Enzyme Commission's system
E. coli	Escherichia coli
EDTA	ethylene diaminetetraacetate
EGTA	ethylene glycol bis(-aminoethyl ether) tetraacetate
ER	endoplasmic reticulum
Et	ethyl
EXAFS	extended X-ray absorption fine structure
FAD	flavin-adenine dinucleotide
FMN	flavin mononucleotide (riboflavin 5'-monophosphate)
Fru	fructose
Fuc	fucose
G	guanine
Gal	galactose

GDP	guanosine 5'-diphosphate
Glc	glucose
GlcN	glucosamine
GlcNAc	N-acetylglucosamine
Gln	glutamine
Glu	glutamic acid
Gly	glycine
GMP	guanosine 5'-monophosphate
GSH	glutathione
GSSG	oxidized glutathione
GTP	guanosine 5'-triphosphate
Gul	gulose
h	hour
H4	tetrahydro
HEPES	4-(2-hydroxyethyl)-1-piperazineethane sulfonic acid
His	histidine
HPLC	high performance liquid chromatography
Hyl	hydroxylysine
Hyp	hydroxyproline
IAA	iodoacetamide
Ig	immunoglobulin
Ile	isoleucine
Ido	idose
IDP	inosine 5'-diphosphate
IMP	inosine 5'-monophosphate
ITP	inosine 5'-triphosphate
K_m	Michaelis constant
L-	(and D-) prefixes indicating configuration
Leu	leucine
Lys	lysine
Lyx	lyxose
M	mol/l
mM	millimol/l
m-	*meta*-
Man	mannose
MES	2-(N-morpholino)ethane sulfonate
Met	methionine
min	minute
MOPS	3-(N-morpholino)propane sulfonate
Mur	muramic acid
MW	molecular weight
NAD^+	nicotinamide-adenine dinucleotide
NADH	reduced NAD
$NADP^+$	NAD phosphate
NADPH	reduced NADP
NAD(P)H	indicates either NADH or NADPH
NBS	N-bromosuccinimide

NDP	nucleoside 5'-diphosphate
NEM	N-ethylmaleimide
Neu	neuraminic acid
NMN	nicotinamide mononucleotide
NMP	nucleoside 5'-monophosphate
NTP	nucleoside 5'-triphosphate
o-	*ortho-*
Orn	ornithine
p-	*para-*
PBS	phosphate-buffered saline
PCMB	*p*-chloromercuribenzoate
PEP	phosphoenolpyruvate
pH	$-\log10[H^+]$
Ph	phenyl
Phe	phenylalanine
PHMB	*p*-hydroxymercuribenzoate
PIXE	proton-induced X-ray emission
PMSF	phenylmethane-sulfonylfluoride
p-NPP	*p*-nitrophenyl phosphate
Pro	proline
Q_{10}	factor for the change in reaction rate for a 10 °C temperature increase
Rha	rhamnose
Rib	ribose
RNA	ribonucleic acid
mRNA	messenger RNA
rRNA	ribosomal RNA
tRNA	transfer RNA
Sar	N-methylglycine (sarcosine)
SDS-PAGE	sodium dodecyl sulfate polyacrylamide gel electrophoresis
Ser	serine
T	thymine
t_H	time for half-completion of reaction
Tal	talose
TDP	thymidine 5'-diphosphate
TEA	triethanolamine
Thr	threonine
TLCK	N^α-*p*-tosyl-L-lysine chloromethyl ketone
T_m	melting temperature
TMP	thymidine 5'-monophosphate
Tos-	tosyl-(*p*-toluenesulfonyl-)
TPN	triphosphopyridinium nucleotide (now $NADP^+$)
Tris	tris(hydroxymethyl)-aminomethane
Trp	tryptophan
TTP	thymidine 5'-triphosphate
Tyr	tyrosine
U	uridine
U/mg	µmol/(mg*min)

UDP	uridine 5'-diphosphate
UMP	uridine 5'-monophosphate
UTP	uridine 5'-triphosphate
Val	valine
Xaa	symbol for an amino acid of unknown constitution in peptide formula
XAS	X-ray absorption spectroscopy
Xyl	xylose

List of Deleted and Transferred Enzymes

Since its foundation in 1956 the Nomenclature Committee of the International Union of Biochemistry and Molecular Biology (NC-IUBMB) has continually revised and updated the list of enzymes. Entries for new enzymes have been added, others have been deleted completely, or transferred to another EC number in the original class or to different EC classes, catalyzing other types of chemical reactions. The old numbers have not been allotted to new enzymes; instead the place has been left vacant or cross-references given to the changes in nomenclature.

Deleted Enzymes

For EC class 3.2.1.48–3.2.1.149 these changes are:

Recommended name	Old EC number	Alteration
amylopectin 6-glucanohydrolase	3.2.1.69	deleted, included in EC 3.2.1.41
α-L-arabinofuranoside hydrolase	3.2.1.79	deleted, included in EC 3.2.1.55
arabinogalactan endo-1,3-β-galactosidase	3.2.1.90	deleted, not sufficiently characterized

Index of Recommended Enzyme Names

Description of Data Fields

All information except the nomenclature of the enzymes (which is based on the recommendations of the Nomenclature Committee of IUBMB (International Union of Biochemistry and Molecular Biology) and IUPAC (International Union of Pure and Applied Chemistry)) is extracted from original literature (or reviews for very well characterized enzymes). The quality and reliability of the data depends on the method of determination, and for older literature on the techniques available at that time. This is especially true for the fields Molecular Weight and Subunits.
The general structure of the fields is: **Information – Organism – Commentary – Literature**
The information can be found in the form of numerical values (temperature, pH, K_m etc.) or as text (cofactors, inhibitors etc.).
Sometimes data are classified as Additional Information. Here you may find data that cannot be recalculated to the units required for a field or also general information being valid for all values. For example, for Inhibitors, Additional Information may contain a list of compounds that are not inhibitory.
The detailed structure and contents of each field is described below. If one of these fields is missing for a particular enzyme, this means that for this field, no data are available.

1 Nomenclature

EC number

The number is as given by the IUBMB, classes of enzymes and subclasses defined according to the reaction catalyzed.

Systematic name

This is the name as given by the IUBMB/IUPAC Nomenclature Committee

Recommended name

This is the name as given by the IUBMB/IUPAC Nomenclature Committee

Synonyms

Synonyms which are found in other databases or in the literature, abbreviations, names of commercially available products. If identical names are frequently used for different enzymes, these will be mentioned here, cross references are given. If another EC number has been included in this entry, it is mentioned here.

CAS registry number

The majority of enzymes have a single chemical abstract (CAS) number. Some have no number at all, some have two or more numbers. Sometimes two enzymes share a common number. When this occurs, it is mentioned in the commentary.

2 Source Organism

For listing organisms their systematic name is preferred. If these are not mentioned in the literature, the names from the respective literature are used. For example if an enzyme from yeast is described without being specified further, yeast will be the entry. This field defines the code numbers for the organisms in which the enzyme with the respective EC number is found. These code numbers (form <_>) are displayed together with each entry in all fields of Brenda where organism-specific information is given.

3 Reaction and Specificity

Catalyzed reaction

The reaction as defined by the IUBMB. The commentary gives information on the mechanism, the stereochemistry, or on thermodynamic data of the reaction.

Reaction type

According to the enzyme class a type can be attributed. These can be oxidation, reduction, elimination, addition, or a name (e.g. Knorr reaction)

Natural substrates and products

These are substrates and products which are metabolized in vivo. A natural substrate is only given if it is mentioned in the literature. The commentary gives information on the pathways for which this enzyme is important. If the enzyme is induced by a specific compound or growth conditions, this will be included in the commentary. In *Additional information* you will find comments on the metabolic role, sometimes only assumptions can be found in the references ot hte natural substrates are unknown.

In the listings, each natural substrate (indicated by a bold **S**) is followed by its respective product (indicated by a bold **P**). Products are given with organisms and references included only if the respective authors were able to demonstrate the formation of the specific product. If only the disappearance of the substrate was observed, the product is included without organisms of references. In cases with unclear product formation only a ? as a dummy is given.

Substrates and products

All natural or synthetic substrates are listed (not in stoichiometric quantities) The commentary gives information on the reversibility of the reaction, on isomers accepted as substrates it and compares the efficiency of substrates. If a specific substrate is accepted by only one of several isozymes, this will be stated here.

The field *Additional Information* summarizes compounds that are not accepted as substrates or general comments which are valid for all substrates. In the listings, each substrate (indicated by a bold **S**) is followed by its respective product (indicated by a bold **P**). Products are given with organisms and references included if the respective authors demonstrated the formation of the specific product. If only the disappearance of the substrate was observed, the product will be included without organisms or references. In cases with unclear product formation only a ? as a dummy is given.

Inhibitors

Compounds found to be inhibitory are listed. The commentary may explain experimental conditions, the concentration yielding a specific degree of inhibition or the inhibition constant. If a substance is activating at a specific concentration but inhibiting at a higher or lower value, the commentary will explain this.

Cofactors, prosthetic groups

This field contains cofactors which participate in the reaction but are not bound to the enzyme, and prosthetic groups being tightly bound. The commentary explains the function or, if known, the stereochemistry, or whether the cofactor can be replaced by a similar compound with higher or lower efficiency.

Activating Compounds

This field lists compounds with a positive effect on the activity. The enzyme may be inactive in the absence of certain compounds or may require activating molecules like sulfhydryl compounds, chelating agents, or lipids. If a substance is activating at a specific concentration but inhibiting at a higher or lower value, the commentary will explain this.

Metals, ions

This field lists all metals or ions that have activating effects. The commentary explains the role each of the cited metal has, being either bound e.g. as Fe-S centers or being required in solution. If an ion plays a dual role, activating at a certain concentration but inhibiting at a higher or lower concentration, this will be given in the commentary.

Turnover number (min^{-1})

The k_{cat} is given in the unit min^{-1}. The commentary lists the names of the substrates, sometimes with information on the reaction conditions or the type of reaction if the enzyme is capable of catalyzing different reactions with a single substrate. For cases where it is impossible to give the turnover number in the defined unit (e.g., substrates without a defined molecular weight, or an undefined amount of protein) this is summarized in Additional Information.

Specific activity (U/mg)

The unit is micromole/minute/milligram of protein. The commentary may contain information on specific assay conditions or if another than the natural substrate was used in the assay. Entries in Additional Information are included if the units of the activity are missing in the literature or are not calculable to the obligatory unit. Information on literature with a detailed description of the assay method may also be found.

K_m-Value (mM)

The unit is mM. Each value is connected to a substrate name. The commentary gives, if available, information on specific reaction condition, isozymes or presence of activators. The references for values which cannot be expressed in mM (e.g. for macromolecular, not precisely defined substrates) are given in Additional Information. In this field we also cite literature with detailed kinetic analyses.

pH-Optimum

The value is given to one decimal place. The commentary may contain information on specific assay conditions, such as temperature, presence of activators or if this optimum is valid for only one of several isozymes. If the enzyme has a second optimum, this will be mentioned here.

pH-Range

Mostly given as a range e.g. 4.0–7.0 with an added commentary explaining the activity in this range. Sometimes, not a range but a single value indicating the upper or lower limit of enzyme activity is given. In this case, the commentary is obligatory.

Temperature optimum (°C)

Sometimes, if no temperature optimum is found in the literature, the temperature of the assay is given instead. This is always mentioned in the commentary.

Temperature range (°C)

This is the range over which the enzyme is active. The commentary may give the percentage of activity at the outer limits. Also commentaries on specific assay conditions, additives etc.

4 Enzyme Structure

Molecular weight

This field gives the molecular weight of the holoenzyme. For monomeric enzymes it is identical to the value given for subunits. As the accuracy depends on the method of determination this is given in the commentary if provided in the literature. Some enzymes are only active as multienzyme complexes for which the names and/or EC numbers of all participating enzymes are given in the commentary.

Subunits

The tertiary structure of the active species is described. The enzyme can be active as a monomer a dimer, trimer and so on. The stoichiometry of subunit composition is given. Some enzymes can be active in more than one state of complexation with differing effectivities. The analytical method is included.

Posttranslational modifications

The main entries in this field may be proteolytic modification, or side-chain modification, or no modification. The commentary will give details of the modifications e.g.:

- proteolytic modification <1> (<1>, propeptide Name) [1];
- side-chain modification <2> (<2>, N-glycosylated, 12% mannose) [2];
- no modification <3> [3]

5 Isolation / Preparation / Mutation / Application

Source / tissue

For multicellular organisms, the tissue used for isolation of the enzyme or the tissue in which the enzyme is present is given. Cell-lines may also be a source of enzymes.

Localization

The subcellular localization is described. Typical entries are: cytoplasm, nucleus, extracellular, membrane.

Purification

The field consists of an organism and a reference. Only references with a detailed description of the purification procedure are cited.

Renaturation

Commentary on denaturant or renaturation procedure.

Crystallization

The literature is cited which describes the procedure of crystallization, or the X-ray structure.

Cloning

Lists of organisms and references, sometimes a commentary about expression or gene structure.

Engineering

The properties of modified proteins are described.

Application

Actual or possible applications in the fields of pharmacology, medicine, synthesis, analysis, agriculture, nutrition are described.

6 Stability

pH-Stability

This field can either give a range in which the enzyme is stable or a single value. In the latter case the commentary is obligatory and explains the conditions and stability at this value.

Temperature stability

This field can either give a range in which the enzyme is stable or a single value. In the latter case the commentary is obligatory and explains the conditions and stability at this value.

Oxidation stability

Stability in the presence of oxidizing agents, e.g. O_2, H_2O_2, especially important for enzymes which are only active under anaerobic conditions.

Organic solvent stability

The stability in the presence of organic solvents is described.

General stability information

This field summarizes general information on stability, e.g., increased stability of immobilized enzymes, stabilization by SH-reagents, detergents, glycerol or albumins etc.

Storage stability

Storage conditions and reported stability or loss of activity during storage.

References

Authors, Title, Journal, Volume, Pages, Year.

Sucrose α-glucosidase 3.2.1.48

1 Nomenclature

EC number
3.2.1.48

Systematic name
sucrose α-D-glucohydrolase

Recommended name
sucrose α-glucosidase

Synonyms
glucosidase, sucrose α-
intestinal sucrase
sucrase
sucrase-invertase
sucrase-isomaltase
sucrose α-glucohydrolase
Additional information (enzyme isolated from intestinal mucosa is a single polypeptide chain also displaying activity towards isomaltose, oligo-1;6-glucosidase, EC 3.2.1.10)

CAS registry number
37288-39-4

2 Source Organism

<1> *Sus domestica* (pig [12]) [12]
<2> *Cuniculus sp.* (rabbit [4, 6, 8, 11, 15]) [4, 6, 8, 11, 15]
<3> *Homo sapiens* (human, sucrase/isomaltase molecule is an hybrid consisting of 2 distinct α-glucosidases [9]; sucrase-isomaltase in adult human colon and colonic polyps [17]) [5, 9, 17, 21, 24]
<4> *Streptococcus salivarius* [1]
<5> *Rattus rattus* (rat, single protein with sucrase and isomaltase activity [7]; sucrase-isomaltase synthesized as a one-chain polypeptide, precursor that is split into the subunits after its transfer to the microvillus membrane [10]; sucrase-isomaltase [13]) [2, 7, 10, 13, 14, 19]
<6> *Gallus gallus domesticus* (chicken [3]) [3]
<7> *Anser anser* (goose [16]) [16]
<8> *Perna viridis* (green mussel *Perna viridis* L. [18]) [18]
<9> *monkey* [20]

<10> *Mus musculus* (mouse, inbred strains CBA/J, CBA/Ca, CBA/CaH, CBA/CaHN, C57B1/6 [22]) [22]
<11> *Actinomyces naeslundii* (T14V [23]) [23]
<12> *Leuconostoc mesenteroides* [25]
<13> *Bacillus subtilis* (maltose-inducible α-glucosidase MalL, sucrase-isomaltase-maltase [26]) [26]
<14> *Arion ater* (slug [18]) [18]
<15> *Achatina fulica* (snail [18]) [18]

3 Reaction and Specificity

Catalyzed reaction
hydrolysis of sucrose and maltose by an α-D-glucosidase-type action

Reaction type
hydrolysis of O-glycosyl bond

Natural substrates and products
S isomaltose + H_2O <5, 6, 9> [3, 13, 20]
P glucose <5, 6> [3, 13]
S maltose + H_2O <1-15> [1-26]
P glucose <1-15> [1-26]
S sucrose + H_2O <1-15> [1-26]
P glucose <1-15> [1-26]

Substrates and products
S 6-bromo-2-naphthyl-α-D-glucoside + H_2O <5> [13]
P 6-bromonaphthol + α-D-glucose <5> [13]
S L-ascorbic acid α-glucoside + H_2O <5> [14]
P ?
S dextran + H_2O <5> [13]
P D-glucose <5> [13]
S inulin + H_2O <11> [23]
P ?
S isomaltose + H_2O <5, 6, 13> [3, 13, 26]
P D-glucose <5, 6, 13> [3, 13, 26]
S levan + H_2O <11> [23]
P ?
S maltopentaose + H_2O <9> [20]
P ?
S maltose + H_2O <5> [14]
P L-ascorbic acid α-glucoside <5> (<5> enzyme form L-ascorbic acid α-glucoside by splitting maltose among the disaccharides [14]) [14]
S maltose + H_2O <5, 6, 13> [3, 13, 26]
P D-glucose <5> [13]
S maltotetraose + H_2O <9> [20]
P ?

S maltotriose + H_2O <9> [20]
P ?
S *p*-nitrophenyl-α-glucoside + H_2O <5, 13> [13, 26]
P *p*-nitrophenol + D-glucose <5, 13> [13, 26]
S raffinose + H_2O <11> [23]
P ?
S sucrose + H_2O <5, 11-13> [13, 23, 25, 26]
P glucose + fructose <5, 11-13> [13, 23, 25, 26]

Inhibitors

2-amino-2-ethyl-1,3-propanediol <3> [24]
2-amino-2-methyl-1,3-propanediol <3> [24]
2-amino-2-methyl-1-propanol <3> [24]
6-bromo-2-naphthyl-α-glucoside <5> [13]
bis-Tris <3> [24]
$CdCl_2$ <8> (<8> 0.0056 mM 50% inactivation [18]) [18]
$CuSO_4$ <8> (<8> 0.0028 mM 50% inactivation [18]) [18]
H^+ <2> (<2> competitive inhibitor in the interaction of sucrase with sucrose and sodium [4, 6]) [4, 6]
$HgCl_2$ <8> (<8> 0.0022 mM 50% inactivation [18]) [18]
Li^+ <2> (<2> high concentration, 300 mM Li^+, pH dependent inhibition, complete at pH 8 [4, 6, 15]) [4, 6, 15]
Na^+ <2, 6> (<2, 6> high concentration [3, 4]) [3, 4]
PCMB <9> [20]
$PbCl_2$ <8> (<8> 0.0079 mM 50% inactivation [18]) [18]
Tris <2, 3, 6> (<2, 3, 6> concentration-dependent biphasic effect, first causing activation, fully competitive inhibition above pH 6.8, inhibition is stronger at alkaline pH values [3, 11, 15, 24]) [3, 11, 15, 24]
$ZnCl_2$ <8> (<8> 0.0133 mM 50% inactivation [18]) [18]
acarbose <2> (<2> BAY g 5421 [8]) [8]
castanospermine <3> (<3> no effect on the biosynthesis of sucrase [5]) [5]
deoxynojirimycin <2> [8]
diethanolamine <3> [24]
isomaltose <9> (<9> competitive inhibition [20]) [20]
maltose <5> [13]
monoethanolamine <3> [24]
nojirimycin <2> [8]
p-nitrophenyl-α-glucoside <5> (<5> competitive inhibitor, inhibits both sucrase and isomaltase [13]) [13]
sucrose <6> (<6> high concentration [3]) [3]
triethanolamine <3> [24]

Activating compounds

2-amino-2-methyl-1-propanol <2> (<2> causes activation at pH 5.2 [15]) [15]
maltose <13> [26]

Metals, ions

Li^+ <2> (<2> low concentration: activation [4]) [4]
Na^+ <2, 5> (<2,6> low concentration: activation [3, 4, 6, 11]) [4, 6, 11, 13]

Specific activity (U/mg)

0.05 <9> [20]
0.1 <2> [11]
0.51 <5> [7]
1.46 <5> (<5> adult rats [7]) [7]
1.47 <13> (<13> maltose [26]) [26]
2.02 <13> (<13> isomaltose [26]) [26]
2.32 <3> [24]
2.95 <7> [16]
4.47 <13> (<13> sucrose [26]) [26]
9.5 <5> (<5> maltose to L-ascorbic acid α-glucoside forming activity [14]) [14]
17.5 <2, 10> [11, 22]
47 <3> [9]
Additional information <1> [12]

K_m-Value (mM)

0.12 <5> (6-bromo-2-naphthyl-α-D-glucoside) [13]
0.135 <13> (maltose) [26]
0.184 <8> (sucrose) [18]
0.21 <13> (*p*-nitrophenyl-α-D-glucoside) [26]
0.455 <13> (isomaltose) [26]
1 <9> (maltose) [20]
1.5 <5> (*p*-nitrophenyl-α-D-glucoside) [13]
3.1 <5> (*p*-nitrophenyl-α-D-glucoside) [13]
3.2 <9> (isomaltose) [20]
3.8 <9> (maltotriose) [20]
7.2 <5> (maltose) [13]
8.4 <2> (sucrose) [11]
9 <3> (sucrose) [5]
10.2 <13> (sucrose) [26]
10.5 <6> (sucrose) [3]
11.1 <9> (maltotetraose) [20]
11.5 <5> (isomaltose) [13]
13.3 <9> (maltopentaose) [20]
13.5 <5> (maltose, <5> the existence of 2 K_m values supports the notion of substrate hydrolysis at 2 active sites on the sucrase-isomaltase complex [13]) [13]
15.9 <2> (sucrose, <2> pH 7.6 [15]) [15]
16 <5> (sucrose) [2]
19.7 <5> (sucrose) [7]
20 <3, 5, 9> (sucrose) [7, 9, 20]
20.2-21.2 <10> (sucrose, <10> inbred strains of mice with genetic and post-translational variations in SI complex structure, incubation at 37°C [22]) [22]

22 <2> (sucrose, <2> value remains constant at Li^+ concentration from 20-120 mM [4]) [4]
30.8 <5> (sucrose) [13]
48 <4> (sucrose) [1]

pH-Optimum
4.5-5 <5> (<5> L-ascorbic acid α-glucoside formation [14]) [14]
5.6 <15> [18]
5.8 <8> [18]
5.9 <5, 15> (<5> in the presence of 50 mM Na^+ [13]) [13, 18]
6-6.5 <3> [9]
6.2 <9> [20]
6.5 <4, 6> [1, 3]
6.7 <5> [13]
6.8 <2> [8]
7 <13> (<13> with PNPG as substrate, at pH 5.75 and 8.0 activity reduced to 50% [26]) [26]

pH-Range
4-7.8 <3> (<3> pH 4, 35% of maximal activity, pH 7.8, 50% of maximal activity [9]) [9]
4.1-8.8 <6> (<6> activity almost neglegible at pH 4.1 and pH 8.8 [3]) [3]
4.5-7.4 <2> [6]
4.8-9.2 <2> (<2> activating effect of sodium in the pH-range 4.8-7.2 [6]) [6]

Temperature optimum (°C)
37 <3, 6> (<3,6> assay at [3, 9]) [3, 9]
40 <8> [18]
45 <15> [18]
60 <5> (<5> maximal activity of L-ascorbic acid α-glucoside formation and maltose hydrolysis [14]) [14]

Temperature range (°C)
17-37 <2> [6]
20-65 <8> (<8> from 20°C to 40°C linear increase of sucrase activity, at 65°C, 78% activity retained [18]) [18]

4 Enzyme Structure

Molecular weight
99320 <11> (<11> gene coding sucrase cloned in pPNG102 contains a ORF which defines a polypeptide of 99319 Da, the sequence shows similarity to previously sequenced sucrose-hydrolyzing enzymes [23]) [23]
130000 <3> (<3> gel filtration, density gradient, equilibrium centrifugation, sucrase subunit [9, 24]) [9, 24]
140000 <1, 2, 5> (<1> SDS-PAGE, intact pancreatic duct, subunit I [12]; <2> SDS-PAGE, sucrase-isomaltase [11]; <5> gel filtration, luminal [2]) [2, 11, 12]
150000 <1> (<1> SDS-PAGE, intact pancreatic duct, subunit II [12]) [12]

220000 <2, 3, 10> (<3> ultracentrifugation analysis, mature complex precursor [17]) [11, 17, 22]
245000 <3> (<3> fully glycosylated enzyme in the Golgi apparatus [21]) [21]
260000 <1, 9> (<1> SDS-PAGE, disconnected pancreatic duct [12]; <9> gel filtration [20]) [12, 20]
270000 <6> (<6> SDS-PAGE [3]) [3]
280000 <3, 7> (<3, 7> gel filtration, density gradient equilibrium centrifugation, human, sucrase-isomaltase complex [9, 16]) [9, 16]
290000 <5> (<5> gel filtration, sucrase-isomaltase complex, intestinal mucosa [2]) [2]
380000 <7> (<7> 380000 sucrase + 543000 isomaltase, enzyme complex after solubilisation by emulphogen and papain [16]) [16]

Subunits
dimer <1, 3, 5, 9, 10> (<1> 1 * 140000 + 1 * 150000, intact pancreatic duct, SDS -PAGE [12]; <5> 1 * 140000 - 160000 + 1 * 140000 -160000 2 large subunits, EC 3.2.1.48 and EC 3.2.1.10, functional dimer, SDS-PAGE [7, 10]; <9> 1 * 112000 + 1* 100000, hetero-dimer, 2 catalytically active sites for maltase and sucrase [20]; <3> 1 * 145000, sucrase + 1* 151000, isomaltase [21]) [7, 12, 17, 20, 21, 22]
monomer <1> (<1> disconnected pancreatic duct, SDS-PAGE [12]) [12]
Additional information <6> [3]

5 Isolation/Preparation/Mutation/Application

Purification
<1> [12]
<2> [15]
<3> [9, 17, 24]
<5> [13, 14]
<6> [3]
<7> [16]
<9> [20]
<13> (produced and purified in E. coli RB791/pMalL [26]) [26]

Cloning
<4> (expression in E. coli [1]) [1]
<10> (mouse SI mRNA synthesized from mouse SI cDNA, pMSI-A1 cloned into pBluescript KS, cDNA cloned from mouse intestinal RNA with anchored polymerase chain reaction method [22]) [22]
<11> (gene levJ from a library of T14V DNA constructed in plasmid pUC18, pPNG102 contains the ORF for the sucrase polypeptide [23]) [23]
<12> (expression in E. coli, IPTG induction does not increase sucrase activity in E. coli, the cloned gene may be transcribed from ist own promoter, the first sucrase cloned from Leuconostoc mesenteroides with invertase activity [25]) [25]

<13> (expression in E. coli, the gene coding the enzyme has been identified in the ORF yvdL/malL [26]) [26]

Application

agriculture <4> (<4> sucrose-metabolizing enzymes are tools in plant science, genes enconding various sucrose-metabolizing enzymes are valuable in genetic engineering of higher plants [1]) [1]
medicine <3> (<3> sucrase-isomaltase SI is a marker enzyme of the absorptive villose cells in adult small intestine; human intestinal sucrase-isomaltase help develop a specific inhibitor of sucrase, to retard the absorption of sucrose in the intestinal tract of diabetic patients [17, 24]) [17, 24]

6 Stability

Temperature stability

50 <5, 10> (<5,10> pH 5.9, inactivation, soluble enzyme activity is stable to heat treatment, membrane associated enzyme was heat-labile under these conditions [13, 19, 22]) [13, 19, 22]
60 <5, 7> (<5, 7> soluble enzyme is stable at 60°C for 30 min in cortisone and thyroxine treated animals, heat-labile in control and insulin given animals, 3% activity less, incubation for 40 min [16, 19]) [16, 19]
65 <8> (<8> thermostable, 78% activity [18]) [18]

Storage stability

<2>, 4°C or -20°C, 0.02% sodium azide, Triton X-100, several weeks, sucrase-isomaltase [11, 15]
<3>, -20°C [24]

References

[1] Houck, C.M.; Pear, J.R.; Elliott, R.; Perchorowicz, J. T.: Isolation of DNA encoding sucrase genes from Streptococcus salivarius and partial characterization of the enzymes expressed in Escherichia coli. J. Bacteriol., **169**, 3679-3684 (1987)
[2] Abe, M.; Yamada, K.; Hosoya, N.; Moriuchi, S.: Some properties of luminal sucrase and sucrase-isomaltase complex in rat small intestine. J. Nutr. Sci. Vitaminol., **31**, 243-252 (1985)
[3] Matsushita, S.: Purification and partial characterization of chick intestinal sucrase. Comp. Biochem. Physiol. B, **76**, 465-470 (1983)
[4] Vasseur, M.; Van Melle, G.; Frangne, R.; Alvarado, F.: Alkali-metal-ion- and H^+-dependent activation and/or inhibition of intestinal brush-border sucrase. A model involving three functionally distinct key prototropic groups. Biochem. J., **251**, 667-675 (1988)
[5] Trugnan, G.; Rousset, M.; Zweibaum, A.: Castanospermine: a potent inhibitor of sucrase from the human enterocyte-like cell line Caco-2. FEBS Lett., **195**, 28-32 (1986)

[6] Vasseur, M.; Tellier, Ch.; Alvarado, F.: Sodium-dependent activation of intestinal brush-border sucrase: correlation with activation by deprotonation from pH 5 to 7. Arch. Biochem. Biophys., **218**, 263-274 (1982)
[7] Montgomery, R.K.; Sybicki, M.A.; Forcier, A.G.; Grand, R.J.: Rat intestinal microvillus membrane sucrase-isomaltase is a single high molecular weight protein and fully active enzyme in the absence of luminal factors. Biochim. Biophys. Acta, **661**, 346-349 (1981)
[8] Hanozet, G.; Pircher, H.-P.; Vanni, P.; Oesch, B.; Semenza, G.: An example of enzyme hysteresis. The slow and tight interaction of some fully competitive inhibitors with small intestinal sucrase. J. Biol. Chem., **256**, 3703-3711 (1981)
[9] Conklin, K.A.; Yamashiro, K.M.; Gray, G.M.: Human intestinal sucrase-isomaltase. Identification of free sucrase and isomaltase and cleavage of the hybrid into active distinct subunits. J. Biol. Chem., **250**, 5735-5741 (1975)
[10] Hauri, H.-P.; Quaroni, A.; Isselbacher, K.J.: Biogenesis of intestinal plasma membrane: posttranslational route and cleavage of sucrase-isomaltase. Proc. Natl. Acad. Sci. USA, **76**, 5183-5186 (1979)
[11] Sigrist, H.; Ronner, P.; Semenza, G.: A hydrophobic form of the small-intestinal sucrase-isomaltase complex. Biochim. Biophys. Acta, **406**, 433-446 (1975)
[12] Sjöström, H.; Noren, O.; Christiansen, L.; Wacker, H.; Semenza, G.: A fully active, two-active-site, single-chain sucrase. Isomaltase from pig small intestine. Implications for the biosynthesis of a mammalian integral stalked membrane protein. J. Biol. Chem., **255**, 11332-11338 (1980)
[13] Kolinska, J.; Kraml, J.: Separation and characterization of sucrose-isomaltase and of glucoamylase of rat intestine. Biochim. Biophys. Acta, **284**, 235-247 (1972)
[14] Muto, N.; Nakamura, T.; Yamamoto, I.: Enzymatic formation of a nonreducing L-ascorbic acid α-glucoside: purification and properties of α-glucosidases catalyzing site-specific transglucosylation from rat small intestine. J.Biochem., **107**, 222-227 (1990)
[15] Vasseur, M.,Frangne, R.; Caüzac, M.; Mahmood, A.; Alvarado, F.: pH-dependent inhibitory effects of tris and lithium ion on intestinal brush-border sucrase. J. Enzyme Inhib., **4**, 15-26 (1990)
[16] Galand, G.: First purification and characterization of a sucrase-isomaltase from goose kidney microvillous membrane. Biochim. Biophys. Acta, **1033**, 35-40 (1990)
[17] Beaulieu, J.-F.; Weiser, M.M.; Herrera, L.; Quaroni, A.: Detection and characterization of sucrase-isomaltase in adult human colon and in colonic polyps. Gastroenterology, **98**, 1467-1477 (1990)
[18] Teo, L.H.; Lateef, Z.; Ip, Y.K.: Some properties of the sucrase from the digestive gland of the green mussel Perna viridis L.. Comp.Biochem.Physiol., **96**, 47-51 (1990)
[19] Sandhu, M.; Mahmood, A.: Kinetic characteristics of soluble and brush border alkaline phosphatase and sucrase activities in developing rat intestine: Effect of hormones. Indian J. Biochem. Biophys., **27**, 88-92 (1990)

[20] Kapadia, H. A.; Sivakami, S.: Some properties of monkey intestinal sucrase. Indian J. Biochem. Biophys., **27**, 93-97 (1990)
[21] Fransen, J.A.M.; Hauri, H.-P.; Ginsel, L.A.; Naim, H.Y.: Naturally occuring mutations in intestinal sucrase-isomaltase provide evidence for the existence of an intracellular sorting signal in the isomaltase subunit. J. Cell Biol., **115**, 45-47 (1991)
[22] Quezada-Calvillo, R.; Markowitz, A.J.; Traber, P.G.; Underdown, B.J.: Murine intestinal disaccharidases: identification of structural variants of sucrase-isomaltose complex. Am.J. Physiol., **265**, 1142-1149 (1993)
[23] Norman, J.M.; Bunny, K.L.; Giffard, P.M.: Characterization of levJ, a sucrase/fructanase-encoding gene from Actinomyces naeslundi T14V, and comparison of its product with other sucrose-cleaving enzymes. Gene, **152**, 93-98 (1995)
[24] Kano, T.; Usami, Y.; Adachi, T.; Tatematsu, M.; Hirano, K.: Inhibition of purified human sucrase and isomaltase by ethanolamine derivatives. Biol. Pharm. Bull., **19**, 341-344 (1996)
[25] Holt, S.M.; Cote, G.L.: Cloning and characterization of a sucrase from Leuconostoc mesenteroides. Biotechnol.Lett., **19.9**, 903-907 (1997)
[26] Schönert, S.; Buder, T.; Dahl, M.K.: Identification and enzymatic characterization of the maltose-inducible α-glucosidase MalL (sucrase-isomaltase-maltase) of Bacillus subtilis. J. Bacteriol., **5**, 2574-2578 (1998)

α-N-Acetylgalactosaminidase 3.2.1.49

1 Nomenclature

EC number
3.2.1.49

Systematic name
α-N-acetyl-D-galactosaminide N-acetylgalactosaminohydrolase

Recommended name
α-N-acetylgalactosaminidase

Synonyms
2-acetamido-2-deoxy-α-D-galactoside acetamidodeoxygalactohydrolase
4-nitrophenyl-α-N-acetylgalactosaminidase
N-acetyl-α-D-galactosaminidase
N-acetyl-α-galactosaminidase
α-acetylgalactosaminidase
α-galactosidase B
exo-α-N-acetylgalactosaminidase

CAS registry number
9075-63-2

2 Source Organism

<1> *Acremonium sp.* [1]
<2> *Katsuwonus pelamis* (skipjack tuna) [2]
<3> *Solenocera melantho* (shrimp) [3]
<4> *Entodinium caudatum* [4]
<5> *Todarodes pacificus* (squid) [5]
<6> *Homo sapiens* [6, 7, 10-13, 15, 16, 21, 25]
<7> *Clostridium perfringens* [8]
<8> *Sus scrofa* [9, 18]
<9> *Patella vulgata* [14]
<10> *Bos taurus* [17-19]
<11> *Helix pomatia* [20]
<12> *Gallus domesticus* (chicken) [22, 23, 24]
<13> *Ruminococcus torques* [26]
<14> *Aspergillus niger* [27]
<15> *Mus musculus* [28, 29]
<16> *Bacteroides succinogenes* [4]

<17> *Bacteroides amylophilus* [4]
<18> *Bacteroides fragilis* [4]
<19> *Ruminococcus albus* [4]
<20> *Ruminococcus flavefaciens* [4]
<21> *Succinivibrio dextrinosolvens* [4]
<22> *Selenomonas ruminantium* [4]
<23> *Lachnospira multiparus* [4]
<24> *Butyrivibrio fibrisolvens* [4]
<25> *Eubacterium ruminantium* [4]
<26> *Streptococcus bovis* [4]
<27> *Megasphaera elsdenii* [4]
<28> *Epidinium ecaudatum caudatum* [4]
<29> *Eremoplastron bovis* [4]
<30> *Eudiplodinium maggii* [4]
<31> *Entodinium simplex* [4]

3 Reaction and Specificity

Catalyzed reaction
hydrolysis of terminal non-reducing N-acetyl-D-galactosamine residues in N-acetyl-α-D-galactosaminides

Reaction type
hydrolysis of O-glycosyl bond

Natural substrates and products
S N-acetyl-α-D-galactosaminide + H_2O <1-11, 16-31> [1-20]
P N-acetyl-D-galactosamine <1-11, 16-31> [1-20]

Substrates and products
S 4-methylumbelliferyl-α-D-galactopyranoside + H_2O <6> [11]
P 4-methylumbelliferone + α-D-galactopyranose
S N-acetyl-α-D-galactosaminides + H_2O <1-11, 16-31> [1-20]
P N-acetyl-D-galactosamine <1-11, 16-31> [1-20]
S UDP-α-N-acetylglucosamine + H_2O <10> [19]
P UDP + α-N-acetylglucosamine
S Forssman hapten + H_2O <8, 6, 13> [9, 13, 26]
P ?
S *o*-nitrophenyl-α-D-fucopyranoside + H_2O <6> [10]
P *o*-nitrophenol + α-D-fucopyranose
S *o*-nitrophenyl-α-N-acetylgalactosaminide + H_2O <6, 12> [13, 22]
P *o*-nitrophenol + α-N-acetylgalactosamine
S ovine submaxillary asialoglycoprotein + H_2O <2> [2]
P ?
S *p*-nitrophenyl-2-deoxy-α-D-galactopyranoside + H_2O <6> [10]
P *p*-nitrophenol + 2-deoxy-α-D-galactopyranose

S *p*-nitrophenyl-α-N-acetylgalactosaminide + H_2O <1-3, 5-9, 12-14> [1-3, 5, 7-9, 14, 15, 18, 24, 26, 27]
P *p*-nitrophenol + α-N-acetylgalactosamine
S phenyl-α-N-acetylglucosaminide + H_2O <10> [19]
P N-acetylgalactosamine + phenol <10> [19]

Inhibitors

Ag^+ <3, 5, 6, 9> (<5> isoenzyme II [5]) [3, 5, 12, 14]
Cu^{2+} <5, 6, 13> (<5> isoenzyme II [5]) [5, 12, 26]
D-galactose <3, 5, 9, 6> (<5> isoenzyme I [5]) [3, 5, 14, 25]
Fe^{2+} <8> [18]
Hg^{2+} <1, 5, 7, 9, 13> (<5> isoenzyme II [5]) [1, 5, 8, 14, 26]
N-acetylgalactosamine <3, 5, 9, 8> [3, 5, 14, 18]
SDS <6> [12]

Specific activity (U/mg)

51.2 <12> (<12> recombinant enzyme [24]) [24]
56.4 <12> (<12> native enzyme [24]) [24]
Additional information <4, 16-31> (<4> list of rumen bacteria and protozoa [4]) [4]

K_m-Value (mM)

0.036 <9> (forssman hapten) [14]
0.05 <2> (ovine submaxillary asialoglycoprotein) [2]
0.23 <14> (*p*-nitrophenyl-α-N-acetylgalactosaminide) [27]
0.59 <6> (forssman hapten) [13]
0.6 <9> (*p*-nitrophenyl-α-N-acetylgalactosaminide) [14]
0.77 <7> (*p*-nitrophenyl-α-N-acetylgalactosaminide) [8]
0.8 <3, 12> (*p*-nitrophenyl-α-N-acetylgalactosaminide) [3, 24]
1.07 <6> (*p*-nitrophenyl-N-acetyl-α-galactosaminide) [15]
1.3 <6> (*o*-nitrophenyl-α-N-acetylgalactosaminide) [13]
1.3 <1> (*p*-nitrophenyl-α-N-acetylgalactosaminide) [1]
1.55 <5> (*p*-nitrophenyl-α-N-acetylgalactosaminide, <5> isoenzyme II [5]) [5]
2.2-5.8 <13> (*p*-nitrophenyl-α-N-acetylgalactosaminide, <13> three isoforms [26]) [26]
2.4 <12> (*o*-nitrophenyl-α-N-acetylgalactosaminide) [22]
2.4 <6> (*p*-nitrophenyl-α-N-acetylgalactosaminide) [7]
2.5 <8> (*p*-nitrophenyl-α-N-acetylgalactosaminide) [9]
3.33 <5> (*p*-nitrophenyl-α-N-acetylgalactosaminide, <5> isoenzyme I [5]) [5]
4.2 <2> (*p*-nitrophenyl-α-N-acetylgalactosaminide) [2]
6.6 <8> (*p*-nitrophenyl-α-N-acetylgalactosaminide) [18]
6.8 <6> (4-methylumbelliferyl-α-D-galactopyranoside) [11, 13]
7 <6> (N-acetyl-α-D-galactosaminide) [25]
8.8 <6> (*o*-nitrophenyl-α-D-fucopyranoside) [10]
14.7 <6> (*p*-nitrophenyl-2-deoxy-α-D-galactopyranoside) [10]
19.6 <6> (*p*-nitrophenyl-N-acetyl-α-galactosaminide) [15]

50 <10> (phenyl-α-N-acetylglucosaminide) [19]
Additional information <6, 8, 12> [9, 11, 22, 21]

pH-Optimum

2.6-3.2 <14> [27]
2.7-3.5 <5> (<5> isoenzyme I, *p*-nitrophenyl-α-N-acetylgalactosaminide [5]) [5]
3.7 <12> [22, 24]
3.8 <3, 9> (<3, 9> *p*-nitrophenyl-α-N-acetylgalactosaminide [3, 14]) [3, 14]
3.9 <8> (<8> forssman hapten [9]) [9]
4 <2> (<2> *p*-nitrophenyl-α-N-acetylgalactosaminide [2]) [2]
4-4.5 <1> (<1> *p*-nitrophenyl-α-N-acetylgalactosaminide [1]) [1]
4.2 <5> (<5> isoenzyme II, *p*-nitrophenyl-α-N-acetylgalactosaminide [5]) [5]
4.3-4.7 <8, 10> (<8, 10> *p*-nitrophenyl-α-N-acetylgalactosaminide [9, 18]) [9, 18]
4.5 <6> (<6> 4-methylumbelliferyl-α-galactopyranoside [12]; <6> N-acetyl-α-D-galactosaminide [25]) [12, 25]
4.8 <6> (<6> 4-methylumbelliferyl-α-galactopyranoside [11]) [11]
5.8-6.8 <13> (<13> three isoforms [26]) [26]
6 <7> (<7> *p*-nitrophenyl-α-N-acetylgalactosaminide [8]) [8]
Additional information <6, 8> (<6> other substrates [11]; <8> more values of isoenzymes [18]) [11, 18]

pH-Range

2.5-8 <1> (<1> *p*-nitrophenyl-α-N-acetylgalactosaminide [1]) [1]
4-7 <2> (<2> not active above, *p*-nitrophenyl-α-N-acetylgalactosaminide [2]) [2]
5-9.5 <13> [26]

Temperature optimum (°C)

37 <7, 6> [8, 25]

4 Enzyme Structure

Molecular weight

40000 <2> (<2> pH 7.2, gel filtration [2]) [2]
48000 <6> (<6> gel filtration [6]) [6]
55000 <1, 3> (<1> gel filtration [1,3]) [1, 3]
80000 <2> (<2> pH 4.2, gel filtration [2]) [2]
102000 <8> (<8> gel filtration [9]) [9]
115000 <14> (<14> non-denaturating polyacrylamide gel electrophoresis [27]) [27]

Subunits

dimer <2> (<2> 2 * 35000, monomers associate to dimers at pH 4.2, SDS-PAGE [2]) [2]

dimer <8> (<8> 2 * 52000, probably two subunits with identical molecular weight, SDS-PAGE [9]) [9]
monomer <1> (<1> 57000, SDS-PAGE [1]) [1]
monomer <12> (<12> 43000, SDS-PAGE [24]) [24]
multimer <12> (<12> x * 49100, SDS-PAGE [22]) [22]
multimer <13> (<13> 160000-200000, SDS-PAGE, three isoforms [26]) [26]
tetramer <9> (4 * 50000, SDS-PAGE [14]) [14]

Posttranslational modification
glycoprotein <6, 8> (<6> precursor is processed to the mature protein [6]; <8> 7.0% neutral sugars [9]) [6, 9]

5 Isolation/Preparation/Mutation/Application

Source/tissue
fibroblast <6> [6]
hepatopancreas <3, 11> [3, 20]
leukocyte <6> [7, 12]
liver <2, 5, 6, 8, 9, 12> [2, 5, 9, 10, 11, 13, 14, 16, 23]
lymphocyte <6> [7]
lymphoid <6> [7]
spleen <10> [19]

Localization
lysosome <8> [18]

Purification
<1> [1]
<2> [2]
<3> [3]
<5> [5]
<6> [6, 11, 13]
<7> [8]
<8> (partially [18]) [9, 18]
<9> [14]
<10> (partially [18]) [18, 19]
<11> [20]

Cloning
<12> (expression in bacteria, rabbit reticulocyte lysate and yeast [23]; expression in yeast Pichia pastoris [24]) [23, 24]
<15> (expression in B-lymphoblast cell line [28]) [28]
<15> (expression in COS-1 cells [29]) [29]

6 Stability

pH-Stability

2.5-8 <14> [27]
3-7.5 <3> [3]
3.6-6.8 <9> [14]
4-6 <2> [2]
4.5-5.5 <5> (<5> isoenzyme I [5]) [5]
4.5-6.5 <5> (<5> isoenzyme II [5]) [5]
4.5-9.5 <13> [26]
6-7.5 <1> [1]

Temperature stability

40 <1-3, 9> (<1> not stable above [1]) [1-3, 14]
50 <14> (<14> not stable above [27]) [27]
50 <5> (<5> isoenzyme I stable for 30 min, isoenzyme II loses activity [5]) [5]
55 <13> (<13> not stable above [26]) [26]

Storage stability

<1>, 0°C, 1 month [1]
<2>, 0°C, sodium citrate buffer, bovine serum albumin, 2 weeks, pH 4.2 [2]
<6>, -20°C, acidified plasma, at least one year [25]
<13>, 4°C, enzyme separation buffer containing 3 mM sodium azide, 25-31 months [26]
<2, 6, 7>, -20°C, lyophilized, several years [2, 6, 8]
<8, 11>, -20°C, 6 months [9, 20]

References

[1] Kadowaki, S.; Ueda, T.; Yamamoto, K.; Kumagai, H.; Tochikura, T.: Isolation and characterization of a blood group A substance-degrading α-N-acetylgalactosaminidase from an Acremonium sp.. Agric. Biol. Chem., **53**, 111-120 (1989)

[2] Nakagawa, H.; Asakawa, M.; Enomoto, N.: Purification and characterization of α-N-acetylgalactosaminidase from skipjack liver. J. Biochem., **101**, 855-862 (1987)

[3] Chien, S.F.: Purification and characterization of α-N-acetylgalactosaminidase from shrimp hepatopancreas. J. Chin. Biochem. Soc., **15**, 86-96 (1986)

[4] Williams, A.G.; Withers, S.E.; Coleman, G.S.: Glycoside hydrolase of rumen bacteria and protozoa. Curr. Microbiol., **10**, 287-294 (1984)

[5] Itoh, T.; Uda, Y.: α-N-Acetylgalactosaminidase from squid liver: purification and characterization of two enzymes. J. Biochem., **95**, 959-970 (1984)

[6] Sweeley, C.C.; Ledonne, N.C.; Robbins, P.W.: Post-translational processing reactions involved in the biosynthesis of lysosomal α-N-acetylgalactosami-

nidase in cultured human fibroblasts. Arch. Biochem. Biophys., **223**, 158-165 (1983)
[7] Salvayre, R.; Negre, A.; Maret, A.; Lenoir, G.; Douste-Blazy, L.: Separation and properties of molecular forms of α-galactosidase and α-N-acetylgalactosaminidase from blood lymphocytes and lymphoid cell lines transformed by Epstein-Barr virus. Biochim. Biophys. Acta, **659**, 445-456 (1981)
[8] Levy, G.N.; Aminoff, D.: Purification and properties of α-N-acetylgalactosaminidase from Clostridium perfringens. J. Biol. Chem., **255**, 11737-11742 (1980)
[9] Sung, S.S.J.; Sweely, C.C.: Purification and partial characterization of porcine liver α-N-acetylgalactosaminidase. J. Biol. Chem., **255**, 6589-6594 (1980)
[10] Dean, K.J.; Sweeley, C.C.: Studies on human liver α-galactosidases. III. Partial characterization of carbohydrate-binding specificities. J. Biol. Chem., **254**, 10006-10010 (1979)
[11] Dean, K.J.; Sweeley, C.C.: Studies on human liver α-galactosidases. II. Purification and enzymatic properties of α-galactosidase B (α-N-acetylgalactosaminidase). J. Biol. Chem., **254**, 10001-10005 (1979)
[12] Salvayre, R.; Maret, A.; Negre, A.; Douste-Blazy, L.: Properties of multiple molecular forms of α-galactosidase and α-N-acetylgalactosaminidase from normal and Fabry leukocytes. Eur. J. Biochem., **100**, 377-383 (1979)
[13] Dean, K.J.; Sung, S.S.J.; Sweeley, C.C.: The identification of α-galactosidase B from human liver as an α-N-acetylgalactosaminidase. Biochem. Biophys. Res. Commun., **77**, 1411-1417 (1977)
[14] Uda, Y.; Li, S.C.; Li, Y.T.: α-N-Acetylgalactosaminidase from the limpet, Patella vulgata. J. Biol. Chem., **252**, 5194-5200 (1977)
[15] Schram, A.W.; Hamers, M.N.; Tager, J.M.: The identity of α-galactosidase B from human liver. Biochim. Biophys. Acta, **482**, 138-144 (1977)
[16] Callahan, J.W.; Lassila, E.L.; den Tandt, W.; Philippart, M.: α-N-Acetylgalactosaminidase: isolation, properties and distribution of the human enzyme. Biochem. Med., **7**, 424-431 (1973)
[17] Wang, C.T.; Weissmann, B.: Association-dissociation and abnormal kinetics of bovine α-acetylgalactosaminidase. Biochemistry, **10**, 1067-1072 (1971)
[18] Weissmann, B.; Hinrichsen, D.F.: Mammalian α-acetylgalactosaminidase. Occurrence, partial purification, and action on linkages in submaxillary mucins. Biochemistry, **8**, 2034-2043 (1969)
[19] Werries, E.; Wollek, E.; Gottschalk, A.; Buddecke, E.: Separation of N-acetyl-α-glucosaminidase and N-acetyl-α-galactosaminidase from ox spleen. Cleavage of the O-glycosidic linkage between carbohydrate and polypeptide in ovine and bovine submaxillary glycoprotein by N-acetyl-α-galactosaminidase. Eur. J. Biochem., **10**, 445-449 (1969)
[20] Tuppy, H.; Staudenbauer, W.L.: The action on soluble blood group A substances of an α-N-acetylgalactosaminidase from Helix pomatia. Biochemistry, **5**, 1742-1747 (1966)
[21] Yagi, F.; Eckhardt, A.E.; Goldstein, I.J.: Gycosidases of Ehrlich ascites tumor cells and ascitic fluid-purification and substrate specificity of α-N-acetylga-

lactosaminidase and α-galactosidase: comparison with coffee bean α galactosidase. Arch. Biochem. Biophys., **280**, 61-67 (1990)
[22] Hata, J.; Dhar, M.; Mitra, M.; Harmata, M.; Haibach, P.; Sun, P.; Smith, D.: Purification and characterization of N-acetyl-α-D-galactosaminidase from Gallus domesticus. Biochem. Int., **28**, 77-86 (1992)
[23] Zhu, A.; Goldstein, J.: Cloning and characterization of a cDNA encoding chicken liver α-N-acetylgalactosaminidase. Gene, **137**, 309-314 (1993)
[24] Zhu, A.; Monahan c.; Wang Z.; Goldstein J.: Expression, purification and characterization of recombinant α-N-acetylgalactosaminidase produced in the yeast Pichia pastoris. Protein Expr. Purif., **8**, 456-462 (1996)
[25] den Tandt, W.R.; Scharpé, S.: Micromethod for the fluorimetric determination of plasma N-acetyl-α-galactosaminidase and study of some of its characteristics. Enzyme Protein, **49**, 273-280 (1996)
[26] Hoskins, L.C.; Boulding, E.T.; Larson G.: Purification and characterization of blood group A-degrading isoforms of α-N-acetylgalactosaminidase from Ruminococcus torques strain IX-70. J. Biol. Chem., **272**, 7932-7939 (1997)
[27] Naundorf, A.; Ajisaka, K.: Purification of α-N-acetylgalactosaminidase from Aspergillus niger and its use in the synthesis of GalNAc-α-(1-O)-serine. Enzyme Microb. Technol., **25**, 483-488 (1999)
[28] Hermann, T.; Schindler, D.; Tabe, H.; Onadera, O.; Igarashi, S.; Polack, A.; Zehnpfennig, D.; Tsuji S.: Molecular cloning, structural organization, sequence, chromosomal assignment, and expression of the mouse α-N-acetylgalactosaminidase gene. Gene, **211**, 205-214 (1998)
[29] Wang, A.M.; Ioannou, Y.A.; Zeidner, K.M.; Desnick, R.J.: Murine α-N-acetylgalactosaminidase: Isolation and expression of a full-lenght cDNA and genomic organization: further evidence of an α galactosidase gene family. Mol. Genet. Metab., **65**, 165-173 (1998)

α-N-Acetylglucosaminidase 3.2.1.50

1 Nomenclature

EC number
3.2.1.50

Systematic name
α-N-acetyl-D-glucosaminide N-acetylglucosaminohydrolase

Recommended name
α-N-acetylglucosaminidase

Synonyms
2-acetamido-2-deoxy-α-D-glucoside acetamidodeoxyglucohydrolase <1, 2, 4> [6, 8, 11]
N-acetyl-α-D-glucosaminidase <1> [14]
N-acetyl-α-glucosaminidase <1, 2> [6, 10]
Sanfilippo b corrective factor <1> (<1> previously [1]) [1]
α-acetylglucosaminidase <4> [11]

CAS registry number
37288-40-7

2 Source Organism

<1> *Homo sapiens* [1, 3-7, 12-14, 16, 17]
<2> *Bos taurus* [8, 10]
<3> *Patella vulgata* [9]
<4> *Sus scrofa* [11]
<5> *Oryctolagus cuniculus* [2]
<6> *Pomacea sp.* [15]
<7> *Rattus norvegicus* [11]

3 Reaction and Specificity

Catalyzed reaction
hydrolysis of terminal non-reducing N-acetyl-D-glucosamine residues in N-acetyl-α-D-glucosaminides

Reaction type
hydrolysis of O-glycosyl bond

Natural substrates and products

S N-acetyl-α-D-glucosaminide + H_2O <1-5> (<1, 2, 6> involved in the degradation of heparin and heparan sulfate [3, 5, 8, 12, 15, 17]) [1-11]
P N-acetyl-α-D-glucosamine <1-5> [1-11]
S UDP-N-acetyl-α-D-glucosamine + H_2O <2> [10]
P uridine-5'-diphosphate + N-acetyl-α-D-glucosamine <2> [10]

Substrates and products

S heparan sulfate + H_2O <1, 6> [5, 15]
P N-acetyl-α-D-glucosamine <1, 6> [5, 15]
S heparin + H_2O <1> [5]
P N-acetyl-α-D-glucosamine <1> [5]
S methylumbelliferyl-N-acetyl-α-D-glucosaminide + H_2O <1> [14, 17]
P methylumbelliferone + N-acetyl-α-D-glucosamine <1> [14, 17]
S *o*-nitrophenyl-N-acetyl-α-D-glucosaminide + H_2O <1, 4> [3, 4, 11]
P *o*-nitrophenol + N-acetyl-α-D-glucosamine <1, 4> [3, 4, 11]
S *p*-nitrophenyl-N-acetyl-α-D-glucosaminide + H_2O <1, 3-5> [1-4, 7, 9, 11-13]
P *p*-nitrophenol + N-acetyl-α-D-glucosamine <1, 3-5> [1-4, 7, 9, 11-13]
S phenyl-N-acetyl-α-D-glucosaminide + H_2O <1, 2, 4> [3, 4, 6, 8, 10, 11]
P phenol + N-acetyl-α-D-glucosamine <1, 2, 4> [3, 4, 6, 8, 10, 11]
S uridine-5'-diphospho-N-acetyl-α-D-glucosamine <1, 2> [3, 4, 10, 17]
P uridine-5'-diphosphate + N-acetyl-α-D-glucosamine <1, 2> [3, 4, 10, 17]

Inhibitors

2-acetamido-1,2-dideoxynojirimycin <1> (<1> competitive inhibition [17]) [17]
6-acetamido-6-deoxycastanospermine <1> (<1> competitive inhibition [17]) [17]
N-acetylglucosamine <4> (<4> competitive inhibition [11]) [11]
N-acetylglucosaminolactone <4> (competitive inhibition [11]) [11]
N-acetylmannosamine <4> (competitive inhibition [11]) [11]
dermatan sulfate <1> (<1> competitive inhibition at 1 mM with *p*-nitrophenyl-α-N-acetylglucosaminide and UDP-N-acetylglucosamine as substrates [4]) [4]
iodoacetate <1> (<1> complete inhibition at 2.5 mM [17]) [17]
mouse anti-serum <1> (<1> raised in BALB/c mice, complete inhibition at pH 7.5 [12, 13]) [12, 13]
p-chloromercuribenzoate <1> (<1> complete inhibition at 0.005 mM [17]) [17]

Activating compounds

bovine serum albumin <1> (<1> stabilization and acceleration of activity betweeen 0.05-0.1%, 2-3fold activity at 0.025% [3, 4]) [3-5]

Metals, ions

Cu^{2+} <1, 4> (<1> complete inhibition at 10 mM [4]; <4> 82% inhibition by 1 mM $CuSO_4$ [11]) [4, 11]
Fe^{3+} <4> (<4> 76% inhibition by 1 mM ferric chloride [11]) [11]

Hg^{2+} <1, 2> (<1> complete inhibition at 10 mM [4]; <2> complete inactivation at 20 mM $HgCl_2$ [8]; <1> complete inhibition at 0.0015 mM [17]) [4, 8, 17]

Specific activity (U/mg)

0.000009 <1> (<1> methylumbelliferyl-N-acetyl-α-D-glucosaminide as substrate [14]) [14]
0.00012 <1> (<1> phenyl-N-acetyl-α-D-glucosaminide as substrate [6]) [6]
0.00015-0.0003 <2> (<2> homogenates of various organes [10]) [10]
0.001 <5> (<5> in lamellar bodies [2]) [2]
0.0012 <5> (<5> in lysosomes [2]) [2]
0.0075 <2> (<2> purified enzyme [8]) [8]
0.027 <4> (<4> partially purified enzyme [11]) [11]
0.115-1.2 <1> (<1> purified enzyme, *p*-nitrophenyl-N-acetyl-α-D-glucosaminide as substrate [1,3,4]) [1, 3, 4]
0.58 <1> (<1> purified enzyme, methylumbelliferyl-N-acetyl-α-D-glucosaminide as substrate [17]) [17]
1.35 <3> (<3> purified enzyme [9]) [9]

K_m-Value (mM)

0.043-0.61 <1, 2> (*o*-nitrophenyl-α-N-acetylglucosaminide, <1> 0.23-0.25 in the presence of 0.025% bovine serum albumin [3, 5]; <2> form I 0.067, form II 0.043 [8]) [3, 5, 8]
0.08 <4> (*o*-nitrophenyl-α-N-acetylglucosaminide, <4> 0.08 partially purified enzyme [11]) [11]
0.12-1.6 <1, 2> (phenyl-α-N-acetylglucosaminide, <1> 0.46-0.66 in the presence of 0.025% bovine serum albumin [3, 5]; <1> 1 M citrate buffer, pH 4.5 [6]; <2> form I 0.32, form II 0.29 [8]) [3, 5, 6, 8, 10]
0.13-0.2 <1> (methylumbelliferyl-N-acetyl-α-D-glucosaminide) [14]
0.13-0.2 <1> (*p*-nitrophenyl-α-N-acetylglucosaminide, <1> 0.21-0.55 in the presence of 0.025% bovine serum albumin [3, 5]) [3, 5, 12, 13]
0.22 <1> (methylumbelliferyl-N-acetyl-α-D-glucosaminide, <1> purified enzyme [17]) [17]
0.3 <4> (*p*-nitrophenyl-α-N-acetylglucosaminide, <4> partially purified enzyme [11]) [11]
0.39 <1> (UDP-N-acetylglucosamine, <1> 0.58 in the presence of 0.025% bovine serum albumin [4]) [4]
1.4 <4> (phenyl-α-N-acetylglucosaminide, <4> partially purified enzyme [11]) [11]

pH-Optimum

4 <1> (<1> methylumbelliferyl-N-acetyl-α-D-glucosaminide [14]) [14]
4.2 <1> (<1> UDP-N-acetylglucosamine [4]) [4]
4.4 <1> (<1> *p*-nitrophenyl-α-N-acetylglucosaminide [3]; <1> heparin [5]; <1> heparan sulfate [5]) [3, 5]
4.5 <1, 2, 6> (<1> *p*-nitrophenyl-α-N-acetylglucosaminide [12,13]; <1> *o*-nitrophenyl-α-N-acetylglucosaminide [3]; <1, 2> phenyl-α-N-acetylglucosaminide [6, 10]; <6> heparan sulfate [15]) [3, 6, 10, 12, 13, 15]

4.6 <1, 4> (<1,4> phenyl-α-N-acetylglucosaminide [4, 11]) [4, 11]
4.7 <2, 4> (<2> phenyl-α-N-acetylglucosaminide [8]; <4> *p*-nitrophenyl-α-N-acetylglucosaminide [11]) [8, 11]
4.8 <1> (<1> phenyl-α-N-acetylglucosaminide [3]; <1> *p*-nitrophenyl-α-N-acetylglucosaminide [4]) [3, 4]
5.7 <1> (<1> sucrose-stabilized pH-gradient, after treatment with neuraminidase [3]) [3]

pH-Range
3.5-6.5 <1> [5]
4.3-6.5 <1> [3]

Temperature optimum (°C)
45 <6> [15]

4 Enzyme Structure

Molecular weight
64500 <2> (<2> form II after gel filtration or disc electrophoresis [8]) [8]
80000 <1> (<1> SDS-PAGE [12, 13]) [12, 13]
83000 <1> (<1> SDS-PAGE [17]) [17]
86000 <1> (<1> SDS-PAGE after reduction and alkylation of native enzyme [1]) [1]
90000-180000 <1> (<1> monomer of 90000 and dimer of 180000 in the ratio 1:2, gel filtration [17]) [17]
127500 <2> (<2> form I after gel filtration or disc electrophoresis [8]) [8]
217000 <3> [9]
240000 <1> (<1> native enzyme, sucrose density gradient centrifugation [1]) [1]
300000-307000 <1> (<1> gel filtration [3, 4, 12, 13]) [3, 4, 12, 13]

Subunits
dimer <1> (<1> α, β, 1 * 86500 + 1 * 81000, SDS-PAGE [3]; 2 * 90000, monomer of 90000 and dimer in a ratio 1:2, gel filtration [17]) [3, 17]
multimer <1> [1]

Posttranslational modification
glycoprotein <1> (<1> 23.4% carbohydrates, e.g. glucose, mannose, galactose [3]; <1> naturally occuring enzyme contains mannose-6-phosphate [17]) [3, 17]

5 Isolation/Preparation/Mutation/Application

Source/tissue
adrenal <7> [11]
amnion <1> [6]
amniotic fluid <1> [6]

brain <7> [11]
digestive gland <3> [9]
duodenum <7> [11]
egg <6> (<6> from day 1-15 of oviposition [15]) [15]
epididymis <2, 7> [10, 11]
heart <7> [11]
jejunum <7> [11]
kidney <2, 7> [10, 11]
leukocyte <1> [14]
liver <1, 2, 4, 7> [10-13]
lung <5, 7> [2, 11]
ovary cell <1> [17]
pancreas <7> [11]
placenta <1> [3]
skeletal muscle <7> [11]
skin <7> [11]
spleen <2, 7> [8, 10, 11]
testis <2, 7> [10, 11]
urine <1> [1, 4, 5, 7]

Localization
lamellar body <5> [2]
lysosome <1, 4, 5> [1, 2, 11]

Purification
<1> [1, 3, 4, 7, 12, 13, 17]
<2> [8, 10]
<3> [9]
<4> (partial [11]) [11]

Cloning
<1> [16]
<1> (<1> expressed in chinese hamster ovary cells [17]) [17]

Application
medicine <1> (<1> antenatal diagnosis of heritable lysosomal storage diseases [6]) [6]

6 Stability

pH-Stability
3-8 <1> (<1> stable at 0°C and 23°C for 8 h [14]) [14]
4.6-8.1 <1> [17]
5 <4> (<4> not stable below 4°C [11]) [11]
5-9 <1> (<1> no loss of activity by dialysis against 100 mM NaCl for 16 h, rapid loss of activity below pH 4 and above pH 10 [3]) [3, 4]
5-10.5 <2> (<2> rapid loss of activity above and below [8]) [8]
6.5-8.5 <1> [12]

Temperature stability

0-50 <1> (<1> stable at 0°C and 23°C for 24 h, at 37°C loss of 45%, at 50°C loss of 85% of activity [14]) [14]

50-70 <1, 2> (<1> no effect by 50°C heating, 50% inactivation after 12 min at 60°C and 2 min at 70°C [3]; <1> thermal stability increased after complex formation with rabbit antiserum [7]; <1> 90% loss of activity after heating at 60°C for 10 min in the presence of 0.5% bovine serum albumin [4]; <1> stable at 50°C for 1 h, at 55°C loss of 17%, at 60°C loss of 77% of activity [12]; <2> in citrate buffer, pH 6.0, no inactivation by 50°C heating, 90% inactivation at 70°C for 30 min [8]; <1> complete loss of activity by heating at 70°C for 4 min [17]) [3, 4, 8, 12, 17]

Oxidation stability

<1>, oxidation by periodate causes loss of 28% of activity after treatment with 10 mM at 4°C for 240 min [5]

General stability information

<1>, no proteolysis with pronase, bromelain and trypsin, 15% loss of activity after 30 min incubation with papain [4]

<2>, entire inactivation by gel electrophoresis in the presence of sodium dodecylsulfate [8]

Storage stability

<1>, -20°C, 0.9% NaCl, cultivated cells centrifuged at 1500 g, 2 weeks, no loss of activity [6]

<1>, 4°C, 10 mM sodium phosphate buffer, pH 7.2, 150 mM NaCl, 1 year, no significant loss of activity [17]

<2>, 37°C, 4 days, no loss of activity, but complete dissociation of form I into form II [8]

<4>, 4°C, pH 6, crude enzyme, several months, no loss of activity [11]

References

[1] Salvatore, D.; Bonatti, S.; Di Natale, P.: Lysosomal α-N-acetylglucosamidase: purification and characterization of the human urinary enzyme. Bull. Mol. Biol. Med., **9**, 111-121 (1984)

[2] Hook, G.E.R.; Gilmore, L.B.: Hydrolases of pulmonary lysosomes and lamellar bodies. J. Biol. Chem., **257**, 9211-9220 (1982)

[3] Roehrborn, W.; von Figura, K.: Human placenta α-N-acetylglucosaminidase: purification, characterization and demonstration of multiple recognition forms. Hoppe-Seyler's Z. Physiol. Chem., **359**, 1353-1362 (1978)

[4] von Figura, K.: Human α-N-acetylglucosaminidase. 1. Purification and properties. Eur. J. Biochem., **80**, 525-533 (1977)

[5] von Figura, K.: Human α-N-acetylglucosaminidase. 2. Activity towards natural substrates and multiple recognition forms. Eur. J. Biochem., **80**, 535-542 (1977)

[6] Hultberg, B.; Lindsten, J.; Sjöblad, S.: Molecular forms and activities of glycosidases in cultures of amniotic-fluid cells. Biochem. J., **155**, 599-605 (1976)
[7] von Figura, K.; Kresse, H.: Sanfilippo disease type B: presence of material cross reacting with antibodies against α-N-acetylglucosaminidase. Eur. J. Biochem., **61**, 581-588 (1976)
[8] Mersman, G.; von Figura, K.; Buddecke, E.: Physical properties and biological activities of two forms of α-N-acetylglucosaminidase from bovine spleen. Biochim. Biophys. Acta, **364**, 88-96 (1974)
[9] Bannister, J.V.; Phizackerley, P.J.R.: Purification of an N-Acetylglucosaminidase from the limpet Patella vulgata L.. FEBS Lett., **29**, 313-317 (1973)
[10] Werries, E.; Wollek, E.; Gottschalk, A.; Buddecke, E.: Separation of N-acetyl-α-glucosaminidase and N-acetyl-α-galactosaminidase from ox spleen. Cleavage of the O-glycosidic linkage between carbohydrate and polypeptide in ovine and bovine submaxillary glycoprotein by N-acetyl-α-galactosaminidase. Eur. J. Biochem., **10**, 445-449 (1969)
[11] Weissmann, B.; Rowin, G.; Marshall, J.; Friederici, D.: : Mammalian α-acetylglucosaminidase. Enzymic properties, tissue distribution, and intracellular localization. Biochemistry, **6**, 207-214 (1967)
[12] Sasaki, T.; Sukegawa, K.; Masue, M.; Fukuda, S.; Tomatsu, S.; Orii, T.: Purification and partial characterization of α-N-acetylglucosaminidase from human liver. J. Biochem., **110**, 842-846 (1991)
[13] Sasaki, T.; Sukegawa, K.; Masue, M.; Fukuda, S.; Tomatsu, S.; Inoue, N.; Orii, T.: Mucopolysaccharidosis type III B: Purification of α-N-acetylglucosaminidase from human liver. Connect. Tissue Res., **23**, 156-157 (1992)
[14] Den Tandt, W.R.; Scharpé, S.: Micromethod determination of N-acetyl-α-D-glucosaminidase in human leukocytes and study of some of its characteristics. Int. J. Biochem., **25**, 209-212 (1993)
[15] Oliveira, F.W.; Chavante, S.F.; Santos, E.A.; Dietrich, C.P.; Nader, H.B.: Appearance and fate of a β-galactanase, α,β-galactosidases, heparan sulfate and chondroitin sulfate degrading enzymes during embryonic development of the mollusc Pomacea sp.. Biochim. Biophys. Acta, **1200**, 241-246 (1994)
[16] Zhao, Z.; Yazdani, A.; Shen, Y.; Sun, Z.-S.; Bailey, J.; Caskey, C.T.; Lee, C.-C.: Molecular dissection of a cosmid gene from a gene-rich region in 17q21 and characterization of a candidate gene for α-N-acetylglucosaminidase with two cDNA isoforms. Mamm. Genome, **7**, 686-690 (1996)
[17] Zhao, K.-W.; Neufeld, E.F.: Purification and characterization of recombinant human α-N-acetylglucosaminidase secreted by chinese hamster ovary cells. Protein Expr. Purif., **19**, 202-211 (2000)

α-L-Fucosidase 3.2.1.51

1 Nomenclature

EC number
3.2.1.51

Systematic name
α-L-fucoside fucohydrolase

Recommended name
α-L-fucosidase

Synonyms
α-fucosidase

CAS registry number
9037-65-4

2 Source Organism

<-1> no activity in *Trichomonas foetus* [49]
<1> *Canis lupus lupus* (canine) [1]
<2> *Pomacea canaliculata* [2, 7, 9]
<3> *Rattus sp.* [3, 6, 10, 16, 36, 38, 40, 47, 49]
<4> *Penicillium multicolor* [4]
<5> *Aspergillus niger* [4]
<6> *Unio elongatulus* [5]
<7> *Cricetulus griseus* (Chinese hamster ovary cells) [8]
<8> *Mesocricetus auratus* (Syrian hamster) [10, 16]
<9> *Mus musculus* [10, 16, 17, 22, 49]
<10> *Cavia porcellus* (guinea pig) [10, 16]
<11> *Dictyostelium discoideum* [11]
<12> *Homo sapiens* [12, 14, 22, 23, 26, 28, 30, 31, 34, 42-45]
<13> *Bos taurus* (bovine) [13, 19, 20, 49]
<14> *Penaeus monodon* [14]
<15> *Bacillus circulans* [15]
<16> *Fusarium oxysporum* [18, 21]
<17> *Prunus dulcis* (almond) [24, 35, 39]
<18> *Octopus vulgaris* [25]
<19> *Sus scrofa* [27, 49]
<20> *Bacteroides fragilis* [29]
<21> *Venus mercenaria* [32]

<22> *Chamelea gallina* [33, 41]
<23> *Tapes rhomboideus* [33]
<24> *Littorina littorea* [33]
<25> *Macaca radiata* [37]
<26> *Clostridium perfringens* [38]
<27> *Limulus polyphemus* [38]
<28> *Bacillus fulminans* [46]
<29> *Haliotis gigantea* (abalone) [48]
<30> *Patella vulgata* [49]

3 Reaction and Specificity

Catalyzed reaction

an α-fucoside + H_2O = L-fucose + an alcohol

Reaction type

hydrolysis of O-glycosyl bond

Natural substrates and products

S 2'-fucosyllactitol + H_2O <2, 11, 17> (<11> low activity [11]; <17> α-fucosidase I, cleavage of α-(1-3)- and α-(1-4)-linkages in oligosaccharides and glycopeptides, doesn't cleave Fuc-α-(1-2)-Gal- and Fuc-α-(1-6)-GlcNAc-linkages [24]; <17> α-fucosidase II highly active [39]) [9, 11, 24, 39]

P α-L-fucose + lactitol <2, 11, 17> [9, 11, 24, 39]

S 3-fucosyllactitol + H_2O <11, 17> (<11> very low activity [11]; <17> low activity [39]) [11, 39]

P α-L-fucose + lactitol <11, 17> [11, 39]

S 6-O-α-L-fucopyranosyl-N-acetylglucosamine + H_2O <11> (<11> Fuc-α-1,6-GlcNAc-linkage most susceptible [11]) [11]

P α-L-fucopyranose + N-acetyl-glucosamine <11> [11]

S Fuc-α-(1-2)-Gal + H_2O <12, 22> [26, 41]

P α-L-fucose + D-galactose <12, 22> [26, 41]

S Fuc-α-(1-2)-Gal-β-(1-3)-GlcNAc-β-(1-3)-Gal-β-(1-4)-Glc + H_2O <12, 15, 17, 19, 21, 28> (<12, 15, 17, 19, 21, 28> lacto-N-fucopentaose I [15, 27, 28, 32, 35, 43, 46]; <12, 21> low activity [32, 43]; <17> α-fucosidase I [35]) [15, 27, 28, 32, 35, 43, 46]

P α-L-fucose + Gal-β-(1-3)-GlcNAc-β-(1-3)-Gal-β-(1-4)-Glc <12, 15, 17, 19, 21, 28> [15, 27, 28, 32, 35, 43, 46]

S Fuc-α-(1-2)-Gal-β-(1-3)-GlcNAc-β-(1-4)-GlcOT + H_2O <2, 17> (<2, 17> lacto-N-fucopentaitol I, subscript OT is used to indicate $NaBH_4$-reduced oligosaccharide [9, 39]; <17> only hydrolysed by α-L-fucosidase II [39]) [9, 39]

P Gal-β-(1-3)-GlcNAc-β-(1-3)-Gal-β-(1-4)-GlcOT + α-L-fucose <2, 17> [9, 39]

S Fuc-α-(1-2)-Gal-β-(1-3)-[Fuc-α-(1-4)]-GlcNAc-β-(1-3)-Gal-β-(1-4)-GlcOT + H_2O <2> (<2> lacto-N-difucohexaitol I [9]) [9]

P Gal-β-(1-3)-GlcNAc-β-(1-3)-Gal-β-(1-4)-GlcOT + α-L-fucose <2> [9]
S Fuc-α-(1-2)-Gal-β-(1-4)-Glc + H_2O <7, 12, 15, 17, 22, 28> (<7, 12, 15, 17, 22, 28> 2'-fucosyllactose [8, 12, 15, 26, 28, 39, 41, 43, 46]; <17> only hydrolysed by α-fucosidase II [39]) [8, 12, 15, 26, 28, 39, 41, 43, 46]
P α-L-fucose + lactose <7, 12, 15, 17, 22, 28> [8, 12, 15, 26, 28, 39, 41, 43, 46]
S Fuc-α-(1-2)-Gal-β-(1-4)-[Fuc-α-(1-3)]-GlcNAc-β-(1-3)-Gal-β-(1-4)-GlcOT + H_2O <2> (<2> hydrolysis of Fuc-α-(1-2)-Gal- and Fuc-α-(1-4)-GlcNAc- and Fuc-α-(1-6)-GlcNAc- linkages but hardly the Fuc-α-(1-3)-GlcNAc-linkage [9]) [9]
P Gal-β-(1-4)-[Fuc-α-(1-3)]-GlcNAc-β-(1-3)-Gal-β-(1-4)-GlcOT + α-L-fucose <2> [9]
S Fuc-α-(1-2)-Gal-β-(1-4)-sorbitol + H_2O <17> (<17> α-fucosidase II, does not cleave α-(1-4)-GlcNAc-, α-(1-3)-GlcNAc- or α-(1-6)-GlcNAc-linkages and does not act on lacto-N-fucopentaoitol I, which has a Fuc-α-(1-2)-Gal-group [24]) [24]
P α-L-fucose + Gal-β-(1-4)-sorbitol <17> [24]
S Fuc-α-(1-2)-Glc + H_2O <12> [26]
P α-L-fucose + D-glucose <12> [26]
S Fuc-α-(1-2)-Xyl + H_2O <12> [26]
P α-L-fucose + D-xylose <12> [26]
S Fuc-α-(1-3)-Gal + H_2O <12> [26]
P α-L-fucose + D-galactose <12> [26]
S Fuc-α-(1-3)-Gal-β-(1-4)-Glc + H_2O <7, 15, 22> (<7, 15, 22> 3-fucosyllactose [8, 15, 41]; <7> low activity [8]) [8, 15, 41]
P α-D-fucose + lactose <7, 15, 22> [8, 15, 39, 41]
S Fuc-α-(1-3)-Glc + H_2O <12> [26]
P α-L-fucose + D-glucose <12> [26]
S Fuc-α-(1-3)-GlcNAc + H_2O <4, 5, 12> (<4,5> slow hydrolysis rate [4]) [4, 26]
P α-L-fucose + N-acetyl-D-glucosamine <4, 5, 12> [4, 26]
S Fuc-α-(1-4)-GlcNAc + H_2O <12> [26]
P α-L-fucose + N-acetyl-D-glucosamine <12> [26]
S Fuc-α-(1-6)-Gal + H_2O <12> [26]
P α-L-fucose + D-galactose <12> [26]
S Fuc-α-(1-6)-GlcNAc + H_2O <12> [26, 28]
P α-L-fucose + N-acetyl-D-glucosamine <12> [26, 28]
S Fuc-α-(1-6)-GlcNAc + H_2O <7> ([<7> low activity [8]) [8]
P α-L-fucose + N-acteyl-D-glucosamine <7> [8]
S Fuc-α-(1-6)-GlcNAc-β-Asn + H_2O <12> [26]
P α-L-fucose + N-actyl-D-glucosaminyl-Asn <12> [26]
S Gal-β-(1-3)-GlcNAc-β-(1-3)-[Fuc-α-(1-4)]-Gal-β-(1-4)-GlcOT + H_2O <2, 17> (<2, 17> lacto-N-fucopentaitol II, subscript OT is used to indicate $NaBH_4$-reduced oligosaccharide [9, 24, 39]; <17> low activity for α-L-fucosidase II [39]) [9, 24, 39]
P α-L-fucose + Gal-β-(1-3)-GlcNAc-β-(1-3)-Gal-β-(1-4)-GlcOT <2, 17> [9, 24, 39]

S Gal-β-(1-4)-GlcNAc-[Fuc-α-(1-4)]-β-(1-3)-Gal-β-(1-4)-Glc + H_2O <15, 22> (<15, 22> lacto-N-fucopentaose III [15, 41]) [15, 41]
P α-L-fucose + Gal-β-(1-4)-GlcNAc-β-(1-3)-Gal-β-(1-4)-Glc <15, 22> [15, 41]
S Gal-β-(1-4)[Fuc-α-(1-3)]-GlcOT + H_2O <2> (<2> very slow hydrolysis [9]) [9]
P Gal-β-(1-4)-GlcOT + α-L-fucose <2> [9]
S GalNAc-α-(1-3)-Gal-[Fuc-α-(1-2)] + H_2O <12, 22> (<12> low activity [12]) [12, 41]
P α-L-fucose + GalNAc-α-(1-3)-Gal <12, 22> [12, 41]
S IgG Fc peptide + H_2O <3> (<3> human, rabbit, horse [47]) [47]
P α-L-fucose + ? <3> [47]
S IgG L peptide + H_2O <3> [47]
P α-L-fucose + ? <3> [47]
S LH-β peptic glycopeptide + H_2O <3> [47]
P α-L-fucose + ? <3> [47]
S [Man-β-(1-6)-Man-α-(1-3)]-Man-β-(1-4)-GlcNAc-β-(1-4)-GlcNAcOT-[Fuc-α-(1-6)] + H_2O <2> [9]
P [Man-β-(1-6)-Man-α-(1-3)]-Man-β-(1-4)-GlcNAc-β-(1-4)-GlcNAcOT + α-L-fucose <2> [9]
S α-L-Fuc-(1-2)-D-Gal + N-acetyl-D-glucosamine + H_2O <4> (<4> same yield of transglycosylation and hydrolysis products [4]) [4]
P α-L-Fuc-(1-2)-D-GlcNAc + α-L-fucose + D-galactose <4> [4]
S α-L-Fuc-(1-3)-D-Gal + N-acetyl-D-glucosamine + H_2O <4> (<4> slower hydrolysis than for the α-(1-6)- or α-(1-2)-isomers, transglycosylation activity [4]) [4]
P α-L-Fuc-(1-3)-D-GlcNAc + α-L-fucose + D-galactose <4> [4]
S α-L-Fuc-(1-6)-D-Gal + N-acetyl-D-glucosamine + H_2O <4> (<4> same yield of transglycosylation and hydrolysis products [4]) [4]
P α-L-Fuc-(1-6)-D-GlcNAc + α-L-fucose + D-galactose <4> [4]
S fucoidan + H_2O <16> [18]
P α-L-fucose + ? <16> [18]
S gastric mucin + H_2O <15, 16, 26> (<15, 16, 26> porcine [15, 18, 20, 38]) [15, 18, 20, 38]
P α-L-fucose + ? <15, 16, 26> [15, 18, 20, 38]
S glycopeptide + H_2O <3, 18, 28> (<18> cellular source of substrate [25]; <3> not active on native glycoproteins [40]; <28> desialyzed porcine submaxillary mucin [46]; <28> meconium H-1, H-2 and X-1 [46]) [25, 40, 46]
P α-L-fucose + ? <3, 18, 28> [25, 40, 46]
S glycoprotein + H_2O <21, 28> (<21> salivary human type H [32]; <28> specifically hydrolyses terminal α-(1-2)-fucosidic bonds of glycoproteins, glycopeptides and oligosaccharides [46]) [32, 46]
P α-L-fucose + ? <21, 28> [32, 46]
S glycosphingolipid + H_2O <21> (<21> blood-group active types B and H [32]) [32]
P α-L-fucose + ? <21> [32]

S gp120 oligosaccharides + H_2O <7> (<7> released from recombinant CHO cell-produced gp120, glycoproteins containing fucose in α-1,3- and α-1,6-linkages are no substrates [8]) [8]
P α-L-fucose + ? <7> [8]
S lacto-N-difucohexaitol I + H_2O <17> (<17> low activity [39]) [39]
P α-L-fucose + ? <17> [39]
S lacto-N-difucohexaose I + H_2O <28> [46]
P α-L-fucose + ? <28> [46]
S lacto-N-fucopentaitol III + H_2O <17> (<17> low activity for α-L-fucosidase II [39]) [24, 39]
P α-L-fucose + ? <17> [24, 39]
S lacto-N-fucopentaose II + H_2O <12, 15, 17, 19, 21, 22> (<12> low activity [12]; <17> α-fucosidase II [24]; <12> preferentially hydrolysed by α-fucosidase II [28]; <17> α-fucosidase II [35]) [12, 15, 24, 27, 28, 32, 35, 41]
P α-L-fucose + Gal-β-(1-3)-GlcNAc-β-(1-3)-Gal-β-(1-4)-Glc <12, 15, 17, 19, 21, 22> [12, 15, 24, 27, 28, 32, 35, 41]
S lactoferrin + H_2O <15> (<15> low activity [15]) [15]
P α-L-fucose + ? <15> [15]
S meconium H-1 + H_2O <28> [46]
P α-L-fucose + ? <28> [46]
S meconium H-2 + H_2O <28> [46]
P α-L-fucose + ? <28> [46]
S meconium X-1 + H_2O <28> (<28> low activity [46]) [46]
P α-L-fucose + ? <28> [46]
S milk + H_2O <15> (<15> human [15]) [15]
P α-L-fucose + ? <15> [15]
S oligosaccharide + H_2O <22, 24> (<22, 24> di-, tri- and pentasaacharides having α-1-2, α-1-3, and α-1-4 bonds [33, 41]) [33, 41]
P α-L-fucose + ? <22, 24> [33]
S orosomucoid + H_2O <15> (<15> low activity, higher activity if desialyzed [15]) [15]
P α-L-fucose + ? <15> [15]
S saliva + H_2O <15, 16> (<15> human [15]; <16> Ulex europaeus [21]) [15, 21]
P α-L-fucose + ? <15, 16> [15, 21]
S submaxillary mucin + H_2O <15, 22, 24, 28, 29> (<15, 22, 24, 29> A-minus type and desialyzed, porcine [15, 33, 48]; <28> desialyzed and not, desialyzed mucine best substrate tested [46]; <29> one form, optimally active at pH 5, not able to hydrolyse *p*-nitrophenyl-α-L-fucopyranoside [48]) [15, 33, 41, 46, 48]
P α-L-fucose + ? <15, 22, 24, 28> [15, 33, 41, 46, 48]
S submaxillary porcine trisaccharide + H_2O <19> [27]
P α-L-fucose + disaccharide <19> [27]
S thyroglobulin + H_2O <19, 21, 22, 24> (<19> hydrolysis only after pronase or neuraminidase treatment [27]; <21> bovine [32]) [27, 32, 33, 41]
P α-L-fucose + ? <19, 21, 22, 24> [27, 32, 33, 41]

Substrates and products

S 2'-fucosyllactitol + H_2O <2, 11, 17> (<11> low activity [11]; <17> α-fucosidase I, cleavage of α-(1-3)- and α-(1-4)-linkages in oligosaccharides and glycopeptides, doesn't cleave Fuc-α-(1-2)-Gal- and Fuc-α-(1-6)-GlcNAc-linkages [24]; <17> α-fucosidase II highly active [39]) [9, 11, 24, 39]

P α-L-fucose + lactitol <2, 11, 17> [9, 11, 24, 39]

S 3-fucosyllactitol + H_2O <11, 17> (<11> very low activity [11]; <17> low activity [39]) [11, 39]

P α-L-fucose + lactitol <11, 17> [11, 39]

S 4-methylumbelliferyl-α-L-fucopyranoside + H_2O <1, 3, 7-10, 12-14, 19> (<3> involved in sperm-egg interactions [6]; <13> induces acrosome reaction of guinea pig spermatozoa [19]; <12> responsible for the intralysosomal hydrolysis of fucose-containing glypeptide and glycolipids [34]) [1, 3, 6, 8, 10, 14, 16, 17, 19, 22, 23, 27, 28, 30, 31, 34, 42, 43]

P 4-methylumbelliferone + α-L-fucose <1, 3, 7-10, 12-14, 19> [1, 3, 7, 8, 10, 14, 16, 17, 19, 22, 23, 27, 28, 30, 31, 34, 42, 43]

S 4-nitrophenyl-α-L-fucopyranoside + H_2O <2, 3, 6, 9, 11-13, 16, 18-25, 27, 29, 30> (<3,22> can cleave α-(1-2)-, α-(1-3)- and α-(1-4)- fucosyl linkages [36, 40, 41]) [2, 5, 7, 9, 11-13, 20, 21, 25-27, 29, 32, 33, 36-38, 40-45, 47-49]

P 4-nitrophenol + α-L-fucose <2, 3, 6, 9, 11-13, 16, 18-25, 27, 29, 30> [2, 5, 7, 9, 11-13, 18, 20, 21, 25-27, 29, 32, 33, 36-38, 40-45, 47-49]

S 4-nitrophenyl-α-L-fucopyranoside + N-acetyl-D-glucosamine + H_2O <4, 5> (<4, 5> transglycosylation activity [4]) [4]

P 4-nitrophenol + α-L-Fuc-(1-3)-D-GlcNAc + <4, 5> (<4, 5> no other isomer [4]) [4]

S 6-O-α-L-fucopyranosyl-N-acetylglucosamine + H_2O <11> (<11> Fuc-α-1,6-GlcNAc-linkage most susceptible [11]) [11]

P α-L-fucopyranose + N-acetyl-glucosamine <11> [11]

S Fuc-α-(1-2)-Fuc-β-4-nitrophenol + H_2O <12> [26]

P α-L-fucose + Fuc-β-4-nitrophenol <12> [26]

S Fuc-α-(1-2)-Gal + H_2O <12, 22> [26, 41]

P α-L-fucose + D-galactose <12, 22> [26, 41]

S Fuc-α-(1-2)-Gal-β-(1-3)-GalNAc-α-phenol + H_2O <12> [26]

P α-L-fucose + Gal-β-(1-3)-N-acetyl-galactosaminyl-phenol <12> [26]

S Fuc-α-(1-2)-Gal-β-(1-3)-GlcNAc-β-(1-3)-Gal-β-(1-4)-Glc + H_2O <12, 15, 17, 19, 21, 28> (<12, 15, 17, 19, 21, 28> lacto-N-fucopentaose I [15, 27, 28, 32, 35, 43, 46]; <12, 21> low activity [32, 43]; <17> α-fucosidase I [35]) [15, 27, 28, 32, 35, 43, 46]

P α-L-fucose + Gal-β-(1-3)-GlcNAc-β-(1-3)-Gal-β-(1-4)-Glc <12, 15, 17, 19, 21, 28> [15, 27, 28, 32, 35, 43, 46]

S Fuc-α-(1-2)-Gal-β-(1-3)-GlcNAc-β-(1-4)-GlcOT + H_2O <2, 17> (<2, 17> lacto-N-fucopentaitol I, subscript OT is used to indicate $NaBH_4$-reduced oligosaccharide [9, 39]; <17> only hydrolysed by α-L-fucosidase II [39]) [9, 39]

P Gal-β-(1-3)-GlcNAc-β-(1-3)-Gal-β-(1-4)-GlcOT + α-L-fucose <2, 17> [9, 39]
S Fuc-α-(1-2)-Gal-β-(1-3)-[Fuc-α-(1-4)]-GlcNAc-β-(1-3)-Gal-β-(1-4)-GlcOT + H_2O <2> (<2> lacto-N-difucohexaitol I, subscript OT is used to indicate $NaBH_4$-reduced oligosaccharide [9]) [9]
P Gal-β-(1-3)-GlcNAc-β-(1-3)-Gal-β-(1-4)-GlcOT + α-L-fucose <2> [9]
S Fuc-α-(1-2)-Gal-β-(1-4)-Glc + H_2O <7, 12, 15, 17, 22, 28> (<7, 12, 15, 17, 22, 28> 2'-fucosyllactose [8, 12, 15, 26, 28, 39, 41, 43, 46]; <17> only hydrolysed by α-fucosidase II [39]) [8, 12, 15, 26, 28, 39, 41, 43, 46]
P α-L-fucose + lactose <7, 12, 15, 17, 22, 28> [8, 12, 15, 26, 28, 39, 41, 43, 46]
S Fuc-α-(1-2)-Gal-β-(1-4)-[Fuc-α-(1-3)]-GlcNAc-β-(1-3)-Gal-β-(1-4)-GlcOT + H_2O <2> (<2> subscript OT is used to indicate $NaBH_4$-reduced oligosaccharide, hydrolysis of Fuc-α-(1-2)-Gal- and Fuc-α-(1-4)-GlcNAc- and Fuc-α-(1-6)-GlcNAc- linkages but hardly the Fuc-α-(1-3)-GlcNAc-linkage [9]) [9]
P Gal-β-(1-4)-[Fuc-α-(1-3)]-GlcNAc-β-(1-3)-Gal-β-(1-4)-GlcOT + α-L-fucose <2> [9]
S Fuc-α-(1-2)-Gal-β-(1-4)-sorbitol + H_2O <17> (<17> α-fucosidase II, does not cleave α-(1-4)-GlcNAc, α-(1-3)-GlcNAc or α-(1-6)-GlcNAc-linkages and does not act on lacto-N-fucopentaoitol I which has a Fuc-α-(1-2)-Gal-group [24]) [24]
P α-L-fucose + Gal-β-(1-4)-sorbitol <17> [24]
S Fuc-α-(1-2)-Gal-β-4-nitrophenol + H_2O <12> [26]
P α-L-fucose + β-galactosyl-4-nitrophenol <12> [26]
S Fuc-α-(1-2)-Glc + H_2O <12> [26]
P α-L-fucose + D-glucose <12> [26]
S Fuc-α-(1-2)-Xyl + H_2O <12> [26]
P α-L-fucose + D-xylose <12> [26]
S Fuc-α-(1-2)-[Gal-α-(1-3)]-Gal-β-phenol + H_2O <12> [26]
P α-L-fucose + Gal-α-(1-3)-Gal-β-phenol <12> [26]
S Fuc-α-(1-3)-Gal + H_2O <12> [26]
P α-L-fucose + D-galactose <12> [26]
S Fuc-α-(1-3)-Gal-β-(1-4)-Glc + H_2O <7, 15, 22> (<7, 15, 22> 3-fucosyllactose [8, 15, 41]; <7> low activity [8]) [8, 15, 41]
P α-L-fucose + lactose <7, 15, 22> [8, 15, 39, 41]
S Fuc-α-(1-3)-Glc + H_2O <12> [26]
P α-L-fucose + D-glucose <12> [26]
S Fuc-α-(1-3)-GlcNAc + H_2O <4, 5, 12> (<4, 5> slow hydrolysis rate [4]) [4, 26]
P α-L-fucose + N-acetyl-D-glucosamine <4, 5, 12> [4, 26]
S Fuc-α-(1-3)-GlcNAc-β-$O(CH_2)_6NHCOCF_3$ + H_2O <12> [26]
P α-L-fucose + GlcNAc-β-$O(CH_2)_6NHCOCF_3$ <12> [26]
S Fuc-α-(1-3)-GlcNAc-β-phenol + H_2O <12> [26]
P α-L-fucose + N-acetyl-glucosaminyl-β-phenol <12> [26]
S Fuc-α-(1-4)-GlcNAc + H_2O <12> [26]
P α-L-fucose + N-acetyl-D-glucosamine <12> [26]

S Fuc-α-(1-4)-GlcNAc-β-phenol + H_2O <12> [26]
P α-L-fucose + β-N-acetyl-glucosaminyl-phenol <12> [26]
S Fuc-α-(1-4)-[Gal-β-(1-3)]-GlcNAc-β-phenol + H_2O <12> [26]
P α-L-fucose + Gal-β-(1-3)-GlcNAc-β-phenol <12> [26]
S Fuc-α-(1-6)-Gal + H_2O <12> [26]
P α-L-fucose + D-galactose <12> [26]
S Fuc-α-(1-6)-GlcNAc + H_2O <12> [26, 28]
P α-L-fucose + N-acetyl-D-glucosamine <12> [26, 28]
S Fuc-α-(1-6)-GlcNAc + H_2O <7> ([<7> low activity [8]) [8]
P α-L-fucose + N-acteyl-D-glucosamine <7> [8]
S Fuc-α-(1-6)-GlcNAc-4-nitrophenol + H_2O <12> [26]
P α-L-fucose + GlcNAc-4-nitrophenol <12> [26]
S Fuc-α-(1-6)-GlcNAc-β-Asn + H_2O <12> [26]
P α-L-fucose + N-acetyl-D-glucosaminyl-Asn <12> [26]
S Fuc-α-(1-6)-[Gal-β-(1-3)]-GlcNAc-β-OCH_3 + H_2O <12> [26]
P α-L-fucose + Gal-β-(1-3)-GlcNAc-β-OCH_3 <12> [26]
S Gal-β-(1-3)-GlcNAc-β-(1-3)-[Fuc-α-(1-4)]-Gal-β-(1-4)-GlcOT + H_2O <2, 17> (<2, 17> lacto-N-fucopentaitol II, subscript OT is used to indicate $NaBH_4$-reduced oligosaccharide [9, 24, 39]; <17> low activity for α-L-fucosidase II [39]) [9, 24, 39]
P α-L-fucose + Gal-β-(1-3)-GlcNAc-β-(1-3)-Gal-β-(1-4)-GlcOT <2, 17> [9, 24, 39]
S Gal-β-(1-4)-GlcNAc-[Fuc-α-(1-4)]-β-(1-3)-Gal-β-(1-4)-Glc + H_2O <15, 22> (<15, 22> lacto-N-fucopentaose III [15, 41]) [15, 41]
P α-L-fucose + Gal-β-(1-4)-GlcNAc-β-(1-3)-Gal-β-(1-4)-Glc <15, 22> [15, 41]
S Gal-β-(1-4)[Fuc-α-(1-3)]-GlcOT + H_2O <2> (<2> subscript OT is used to indicate $NaBH_4$-reduced oligosaccharide, very slow hydrolysis [9]) [9]
P Gal-β-(1-4)-GlcOT + α-L-fucose <2> [9]
S GalNAc-α-(1-3)-Gal-[Fuc-α-(1-2)] + H_2O <12, 22> (<12> low activity [12]) [12, 41]
P α-L-fucose + GalNAc-α-(1-3)-Gal <12, 22> [12, 41]
S GlcNAc-β-(1-2)-Man-α-(1-6)-[GlcNAc-β-(1-2)-Man-α-(1-3)]-[Xyl-β-(1-2)]-Man-β-(1-4)-GlcNAc-β-(1-4)-[Fuc-α-(1-3)]-GlcNAc-pyridylaminoside + H_2O <2> (<2> not completely hydrolysed due to steric hinderance by β-(1-2)-GlcNAc residues [7]) [7]
P α-L-fucose + [GlcNAc-β-(1-2)-Man-α-(1-6)]-[GlcNAc-β-(1-2)-Man-α-(1-3)]-[Xyl-β-(1-2)]-Man-β-(1-4)-GlcNAc-β-(1-4)-GlcNAc-pyridylaminoside <2> [7]
S IgG Fc peptide + H_2O <3> (<3> from human, rabbit, horse [47]) [47]
P α-L-fucose + ? <3> [47]
S IgG L peptide + H_2O <3> [47]
P α-L-fucose + ? <3> [47]
S LH-β peptic glycopeptide + H_2O <3> [47]
P α-L-fucose + ? <3> [47]
S Man-β-(1-6)-[Man-α-(1-3)-Xyl-β-(1-2)]-Man-β-(1-4)-GlcNAc-β-(1-4)-[Fuc-α-(1-3)]-GlcNAc-pyridylaminoside + H_2O <2> [7]

P α-L-fucose + Man-β-(1-6)-[Man-α-(1-3)-Xyl-β-(1-2)]-Man-β-(1-4)-GlcNAc-β-(1-4)-GlcNAc-pyridylaminoside <2> [7]
S Xyl-β-(1-2)-Man-β-(1-4)-GlcNAc-β-(1-4)-[Fuc-α-(1-3)]-GlcNAc-pyridylaminoside + H_2O <2> [7]
P α-L-fucose + Xyl-β-(1-2)-Man-β-(1-4)-GlcNAc-β-pyridylaminoside <2> [7]
S [Man-β-(1-6)-Man-α-(1-3)]-Man-β-(1-4)-GlcNAc-β-(1-4)-GlcNAcOT-[Fuc-α-(1-6)] + H_2O <2> (<2> subscript OT is used to indicate $NaBH_4$-reduced oligosaccharide [9]) [9]
P [Man-β-(1-6)-Man-α-(1-3)]-Man-β-(1-4)-GlcNAc-β-(1-4)-GlcNAcOT + α-L-fucose <2> [9]
S α-L-Fuc-(1-2)-D-Gal + N-acetyl-D-glucosamine + H_2O <4> (<4> same yield of transglycosylation and hydrolysis products [4]) [4]
P α-L-Fuc-(1-2)-D-GlcNAc + α-L-fucose + D-galactose <4> [4]
S α-L-Fuc-(1-3)-D-Gal + N-acetyl-D-glucosamine + H_2O <4> (<4> slower hydrolysis than for the α-(1-6)- or α-(1-2)-isomers, transglycosylation activity [4]) [4]
P α-L-Fuc-(1-3)-D-GlcNAc + α-L-fucose + D-galactose <4> [4]
S α-L-Fuc-(1-6)-D-Gal + N-acetyl-D-glucosamine + H_2O <4> (<4> same yield of transglycosylation and hydrolysis products [4]) [4]
P α-L-Fuc-(1-6)-D-GlcNAc + α-L-fucose + D-galactose <4> [4]
S fucoidan + H_2O <16> [18]
P α-L-fucose + ? <16> [18]
S gastric mucin + H_2O <15, 16, 26> (<15, 16, 26> porcine [15, 18, 20, 38]) [15, 18, 20, 38]
P α-L-fucose + ? <15, 16, 26> [15, 18, 20, 38]
S glycopeptide + H_2O <3, 18, 28> (<18> cellular source of substrate [25]; <3> not active on native glycoproteins [40]; <28> desialyzed porcine submaxillry mucin [46]; <28> from meconium H-1, H-2 and X-1 [46]) [25, 40, 46]
P α-L-fucose + ? <3, 18, 28> [25, 40, 46]
S glycoprotein + H_2O <21, 28> (<21> salivary human type H [32]; <28> specifically hydrolyses terminal α-(1-2)-fucosidic bonds of glycoproteins, glycopeptides and oligosaccharides [46]) [32, 46]
P α-L-fucose + ? <21, 28> [32, 46]
S glycosphingolipid + H_2O <21> (<21> blood-group active types B and H [32]) [32]
P α-L-fucose + ? <21> [32]
S gp120 oligosaccharides + H_2O <7> (<7> released from recombinant CHO cell-produced gp120, glycoproteins containing fucose in α-1,3- and α-1,6-linkages are no substrates [8]) [8]
P α-L-fucose + ? <7> [8]
S lacto-N-difucohexaitol I + H_2O <17> (<17> low activity [39]) [39]
P α-L-fucose + ? <17> [39]
S lacto-N-difucohexaose I + H_2O <28> [46]
P α-L-fucose + ? <28> [46]

S lacto-N-fucopentaitol III + H_2O <17> (<17> low activity for α-L-fucosidase II [39]) [24, 39]
P α-L-fucose + ? <17> [24, 39]
S lacto-N-fucopentaose II + H_2O <12, 15, 17, 19, 21, 22> (<12> low activity [12]; <17> α-fucosidase II [24]; <12> preferentially hydrolysed by α-fucosidase II [28]; <17> α-fucosidase II [35]) [12, 15, 24, 27, 28, 32, 35, 41]
P α-L-fucose + Gal-β-(1-3)-GlcNAc-β-(1-3)-Gal-β-(1-4)-Glc <12, 15, 17, 19, 21, 22> [12, 15, 24, 27, 28, 32, 35, 41]
S lactoferrin + H_2O <15> (<15> low activity [15]) [15]
P α-L-fucose + ? <15> [15]
S meconium H-1 + H_2O <28> [46]
P α-L-fucose + ? <28> [46]
S meconium H-2 + H_2O <28> [46]
P α-L-fucose + ? <28> [46]
S meconium X-1 + H_2O <28> (<28> low activity [46]) [46]
P α-L-fucose + ? <28> [46]
S milk + H_2O <15> (<15> human [15]) [15]
P α-L-fucose + ? <15> [15]
S oligosaccharides + H_2O <22, 24> (<22, 24> di-, tri- and pentasaacharides having α-1-2, α-1-3, and α-1-4 bonds [33, 41]) [33, 41]
P α-L-fucose + ? <22, 24> [33]
S orosomucoid + H_2O <15> (<15> low activity, higher activity if desialyzed [15]) [15]
P α-L-fucose + ? <15> [15]
S saliva + H_2O <15, 16> (<15> human [15]; <16> Ulex europaeus [21]) [15, 21]
P α-L-fucose + ? <15, 16> [15, 21]
S stomach + H_2O <28> [46]
P α-L-fucose + ? <28> [46]
S submaxillary mucin + H_2O <15, 22, 24, 28, 29> (<15, 22, 24, 29> A-minus type and desialyzed, porcine [15, 33, 48]; <28> desialyzed and not, desialyzed mucine best substrate tested [46]; <29> one form, optimally active at pH 5, not able to hydrolyse *p*-nitrophenyl-α-L-fucopyranoside [48]) [15, 33, 41, 46, 48]
P α-L-fucose + ? <15, 22, 24, 28> [15, 33, 41, 46, 48]
S submaxilary mucin trisaccharide + H_2O <19> [27]
P α-L-fucose + disaccharide <19> [27]
S thyroglobulin + H_2O <19, 21, 22, 24> (<19> hydrolysis only after pronase or neuraminidase treatment [27]; <21> from bovine [32]) [27, 32, 33, 41]
P α-L-fucose + ? <19, 21, 22, 24> [27, 32, 33, 41]

Inhibitors

1-cyclohexyl-3-(2-morpholinoethyl)-carbodiimide metho-*p*-toluenesulfonate <19> (<19> complete inactivation at 22 mM [27]) [27]

2-(hydroxymethyl)-5-methylpyrrolidine-3,4-diol <13> (<13> competitive or uncompetitive inhibition, K_i: 0.004-26 mM, depending on conformation [13]) [13]

2-methyltetrahydro-2H-pyran-3,4-diol <13> (<13> uncompetitive inhibition, K_i: 23 mM [13]) [13]
2-phenyl-tetrahydro-2H-pyran-3,4-diol <13> (<13> uncompetitive inhibition, K_i: 22-83 mM, depending on conformation [13]) [13]
4-aminophenyl 1-thio α-L-fucoside <3> (<3> weak competitive inhibition, K_i: 1.25 mM [38]) [38]
4-chloromercuribenzoate <2, 12, 25> (<2> 71% inhibition at 1 mM [7]; <25> 50% inhibition at 0.0015 mM, totally reversal by addition of 4 mM DTT [37]; <12> 92% inhibition at 0.1 mM [44]) [7, 37, 44]
4-chloromercuriphenylsulfonic acid <3, 17> (<3> complete inhibition at 0.002 mM [36]; <17> slight inhbition at 1 mM [39]) [36, 39]
4-hydroxy-2-methyltetrahydro-2H-pyran-3-yl acetate <13> (<13> competitive inhibition, K_i: 0.09 mM [13]) [13]
4-mercuribenzoate <19> (<19> complete inhibition at 0.25 mM [27]) [27]
5-(hydroxymethyl)pyrrolidine-2,3,4-triol <13> (<13> competitive inhibition, K_i: 0.91 mM [13]) [13]
Ag^+ <3, 12> (<12> complete inactivation at 0.1 mM, completely reversal by addition of 1.1 mM DTT [43]; <12> 59% inhibition at 3 mM [44]; <3> effective inhibition at 5 mM [47]) [43, 44, 47]
Ag^{2+} <25> (<25> complete inactivation at 1 mM [37]) [37]
Ca^{2+} <17> (<17> α-fucosidase I, slight inhibition at 10 mM [39]) [39]
Cd^{2+} <3> (<3> effective inhibition at 5 mM [47]) [47]
Cu^{2+} <3, 12, 25> (<25> complete inactivation at 1 mM [37]; <12> 88% inhibition at 2 mM [44]; <3> effective inhibition at 5 mM [47]) [36, 37, 44, 47]
D-1-deoxymannojirimycin <13> (<13> competitive inhibition, K_i: 0.03 mM [13]) [13]
D-1-deoxyrhamnojirimycin <13> (<13> competitive inhibition, K_i: 0.011 mM [13]) [13]
D-1-deoxytalonojirimycin <13> (<13> competitive inhibition, K_i: 0.053 mM [13]) [13]
D-arabinose <16, 19> (<16> 95% inhibition at 2.5 mM [21]; <19> competitive inhibition, K_i: 50 mM [27]) [21, 27]
D-fucal <13> (<13> uncompetitive inhibition, K_i: 35 mM [13]) [13]
D-fucose <22> (<22> 15% inhibition at 50 mM [41]) [41]
D-galactono-1,4-lactone <22> (<22> 15% inhibition at 50 mM [41]) [41]
D-galactose <19, 22> (<19> competitive inhibition, K_i: 1.7 mM [27]; <22> 15% inhibition at 50 mM [41]) [27, 41]
D-glucono-1,5-lactone <22> (<22> 30% inhibition at 50 mM [41]) [41]
D-glucose <12, 22> (<22> 15% inhibition at 50 mM [41]; <12> 21% inhibition at 6 mM [44]) [41, 44]
Fe^{3+} <12> (<12> 40% inhibition at 2 mM [44]) [44]
Hg^{2+} <3, 12, 15, 16, 22-25, 28> (<15> potent inhibitor at 1 mM [15]; <16, 25> complete inhibition at 1 mM [18, 37]; <16> marked inhibition at 20 mM [21]; <22, 23, 24> effective non-competitive inhibition at 2 mM [33, 41]; <12> complete inactivation at 0.1 mM, complete reversal by addition of 1.1 mM DTT [43]; <12> 92% inhibition at 0.1 mM [44]; <28> marked inhibition at

0.1 mM [46]; <3> effective inhibition at 5 mM [47]) [15, 18, 21, 33, 36, 37, 41, 43, 44, 46, 47]
L-1-deoxyfuconojirimycin <13> (<13> competitive inhibition, K_i: 0.00003 mM [13]) [13]
L-1-deoxymannojirimycin <13> (<13> competitive inhibition, K_i: 90 mM [13]) [13]
L-1-deoxyrhamnojirimycin <13> (<13> competitive inhibition, K_i: 1 mM [13]) [13]
L-cysteine <12, 17> (<17> α-fucosidase I, slight inhibition at 10 mM [39]; <12> 20% inhibition at 30 mM [44]) [39, 44]
L-fucal <13> (<13> competitive inhibition, best inhibitor of glycal compounds tested, K_i: 0.05 mM [13]) [13]
L-galactono-1,4-lactone <22> (<22> 30% inhibition at 50 mM [41]) [41]
L-rhamnose <19> (<19> competitive inhibition, K_i: 140 mM [27]) [27]
Mg^{2+} <17> (<17> slight inhibition at 10 mM [39]) [39]
Mn^{2+} <12, 28> (<12> 26% inhibition at 2 mM [44]; <28> slight inhibition at 0.1 mM [46]) [44, 46]
N-ethylmaleimide <25> (<25> 30% inhibition at 1 mM [37]) [37]
Ni^{2+} <3> (<3> effective inhibition at 5 mM [47]) [36, 47]
Pb^{2+} <3> (<3> effective inhibition at 5 mM [47]) [47]
SDS <28> (<28> 42% inhibition at 0.02%, w/v [46]) [46]
Sn^{2+} <12> (<12> 24% inhibition at 2 mM [44]) [44]
Zn-acetate <3> (<3> more than 90% inhibition at 5 mM [47]) [47]
α-L-fucose <3, 7, 13, 16, 19, 22-25> (<7> competitive inhibition, K_i: 0.065 mM [8]; <13> competitive inhibition, K_i: 0.3 mM [13]; <16> 90% competitive inhibition at 5 mM, K_i: 0.12 mM [18]; <16> 85% inhibition at 2.5 mM [21]; <19> competitive inhibition, K_i: 0.13 mM [27]; <22, 23, 24> competitive inhibition at 50 mM [33, 41]; <3> competitive inhibition, K_i: 1.6 mM [36, 40]; <25> competitive inhibition, K_i: 0.275 mM [37]; <3> competitive inhibition, K_i: 0.285 mM [38]) [8, 13, 18, 21, 27, 33, 36-38, 40, 41]
deoxycholate <12, 28> (<12> 36% inhibition at 0.012%, w/v, [44]; <28> 34% inhibition at 0.2%, w/v [46]) [44, 46]
deoxyfuconojirimycin <7> (<7> competitive inhibition, K_i: 0.000024 mM [8]) [8]
ethanolamine <25> (<25> strong inhibition [37]) [37]
ethylenediaminetetraacetic acid <28> (<28> 68% inhibition at 0.1 mM [46]) [46]
formaldehyde <12> (<12> 57% inhibition at 160 mM [44]) [44]
fucosylamine <7> (<7> competitive inhibition, K_i: 0.014 mM [8]) [8]
iodoacetamide <25> (<25> 30% inhibition at 1 mM [37]) [37]
phenol <13> (<13> noncompetitive inhibition, K_i: 112 mM [13]) [13]

Activating compounds

CN^- <22> (<22> 30% activation at 10 mM [41]) [41]
KCl <12> (<12> 17% activation at 300 mM [44]) [44]
bovine serum albumin <12, 25> (<12, 25> increases acitivity up to 100% [37,42]) [37, 42]

dithiothreitol <12> (<12> 13% activation at 0.2 mM [43]) [43]
ethylenediaminetetraacetic acid <22> (<22> 5% activation at 10 mM [41]) [41]

Metals, ions

Ca^{2+} <17, 22> (<17> α-fucosidase II, 21% activation at 10 mM [39]; <22> 10% activation at 10 mM [41]) [39, 41]
Fe^{2+} <12> (<12> 12% activation at 2 mM [44]) [44]
Mn^{2+} <12> (<12> 22% stimulation at 5 mM [43]) [43]
Zn^{2+} <22> (<22> 10% activation at 10 mM [41]) [41]

Specific activity (U/mg)

0.01 <13> (<13> in corpus epididymidis [20]) [20]
0.013 <13> (<13> in spermatozoa [20]) [20]
0.015 <13> (<13> in caput epididymidis [20]) [20]
0.018 <13> (<13> in cauda epididymidis [20]) [20]
0.025 <8> [10]
0.04 <4> [4]
0.09 <20> (<20> at pI 5.9 [29]) [29]
0.12 <28> [46]
0.19 <17> (<17> α-L-fucosidase II [39]) [39]
0.37 <13> [19]
0.39 <27> [38]
1.46 <17> (<17> α-L-fucosidase I [39]) [39]
1.69 <20> (<20> at pI 6.3 [29]) [29]
1.8 <12> (<12> α-L-fucosidase B [42]) [42]
1.93 <12> [45]
2.17 <20> (<20> at pI 8.2 [29]) [29]
2.58 <16> [21]
2.61 <26> [38]
3.81 <12> (<12> α-L-fucosidase A [42]) [42]
4.76 <12> [34]
5 <11> [11]
5.3 <2> [9]
5.69 <2> (<2> isoform F2 [2]) [2]
5.84 <9> [17]
5.89 <2> (<2> isoform F1 [2]) [2]
7.8 <19> [27]
10.66 <12> [26]
12.8 <3> [47]
14.9-17.1 <1> (<1> recombinant enzyme, value depending on expression systems and buffer [1]) [1]
15.6 <3> [38]
20 <15> [15]
24.2 <12> [43]
27-28 <3> [36, 40]
38.9 <18> [25]
39.7 <12> [31]

64.8 <7> [8]
110 <16> [18]
1620 <22> [41]
31250 <14> [14]

K_m-Value (mM)

0.022 <14> (4-methylumbelliferyl-α-L-fucopyranoside) [14]
0.031 <7> (4-methylumbelliferyl-α-L-fucopyranoside, <7> at pH 7.0 [8]) [8]
0.05 <3> (4-methylumbelliferyl-α-L-fucopyranoside) [16]
0.05-0.06 <9, 12> (4-methylumbelliferyl-α-L-fucopyranoside) [22, 31]
0.07 <8> (4-methylumbelliferyl-α-L-fucopyranoside) [16]
0.07 <22> (4-nitrophenyl-α-L-fucopyranoside) [33, 41]
0.07-0.08 <12> (4-methylumbelliferyl-α-L-fucopyranoside) [23]
0.072 <7> (4-methylumbelliferyl-α-L-fucopyranoside, <7> at pH 5.0 [8]) [8]
0.08 <9> (4-methylumbelliferyl-α-L-fucopyranoside) [16]
0.083 <12> (4-methylumbelliferyl-α-L-fucopyranoside) [28]
0.095 <17> (lacto-N-fucopentaitol III, <17> α-fucosidase I [24, 39]) [24, 39]
0.1 <17> (lacto-N-fucopentaitol II, <17> α-fucosidase I [24, 39]) [24, 39]
0.12 <10> (4-methylumbelliferyl-α-L-fucopyranoside) [16]
0.125 <12> (4-methylumbelliferyl-α-L-fucopyranoside, <12> α-fucosidase I [28]) [28]
0.14 <1> (4-methylumbelliferyl-α-L-fucopyranoside) [1]
0.14 <12> (4-methylumbelliferyl-α-L-fucopyranoside, <12> 4-5fold increase in fucosidosis patients [30]) [30]
0.14 <23> (4-nitrophenyl-α-L-fucopyransoide) [33]
0.168 <12> (4-nitrophenyl-α-L-fucopyranoside) [44]
0.18 <12> (4-nitrophenyl-α-L-fucopyranoside) [26]
0.19 <3> (4-nitrophenyl-α-L-fucopyranoside) [36, 40]
0.21 <3> (4-nitrophenyl-α-L-fucopyranoside) [49]
0.22 <12> (4-methylumbelliferyl-α-L-fucopyranoside) [43]
0.22 <13, 25> (4-nitrophenyl-α-L-fucopyranoside) [37, 49]
0.23 <2> (4-nitrophenyl-α-L-fucopyranoside, <2> isoform F2 [2]) [2]
0.242 <2> (4-nitrophenyl-α-L-fucopyranoside, <2> isoform F1 [2]) [2]
0.26 <12> (4-nitrophenyl-α-L-fucopyranoside) [12]
0.277 <3> (4-nitrophenyl-α-L-fucopyranoside) [38]
0.3 <12> (4-methylumbelliferyl-α-L-fucopyranoside) [42]
0.3 <19, 24> (4-nitrophenyl-α-L-fucopyranoside) [27, 33]
0.33 <28> (porcine submaxillary mucine, desialyzed) [46]
0.358 <18> (4-nitrophenyl-α-L-fucopyranoside) [25]
0.38 <12> (4-nitrophenyl-α-L-fucopyranoside) [42]
0.4 <3> (4-methylumbelliferyl-α-L-fucopyranoside) [3]
0.43 <12> (4-nitrophenyl-α-L-fucopyranoside) [43]
0.45 <2> (4-nitrophenyl-α-L-fucopyranoside) [9]
0.48 <16> (4-nitrophenyl-α-L-fucopyranoside) [18]
0.52 <12> (Fuc-α-(1-2)-Gal-β-4-nitrophenol) [26]
0.6 <21> (4-nitrophenyl-α-L-fucopyranoside) [32]
0.63 <15> (2'-fucosyllactose) [15]

0.67 <17> (2'-fucosyllactitol, <17> α-fucosidase II [24, 39]) [24, 39]
0.71 <6> (4-nitrophenyl-α-L-fucopyranoside, <6> sperm-bound enzyme [5]) [5]
0.87 <16> (4-nitrophenyl-α-L-fucopyranoside) [21]
0.88 <15> (gastric mucin) [15]
0.91 <6> (4-nitrophenyl-α-L-fucopyranoside, <6> gonad fluid enzyme [5]) [5]
1.02 <7> (gp120 oligosaccharide, <7> a7 pH 5.0 [8]) [8]
1.18 <2> (2'-fucosyllactitol) [9]
1.2 <11> (4-nitrophenyl-α-L-fucopyranoside) [11]
1.25 <15> (3-fucosyllactose) [15]
1.56 <12> (Fuc-α-(1-2)-Gal-β-(1-4)-Glc) [26]
1.6 <28> (glycopeptide from desialyzed porcine submaxillary mucin) [46]
1.63 <12> (Fuc-α-(1-3)-GlcNAc-phenol) [26]
1.75 <7> (Fuc-α-(1-6)-GlcNAc, <7> at pH 5.0 [8]) [8]
1.79 <7> (2'-fucosyllactose, <7> at pH 5.0 [8]) [8]
1.9 <7> (3-fucosyllactose, <7> at pH 5.0 [8]) [8]
2 <28> (lacto-N-fucopentaose) [46]
2.01 <12> (Fuc-α-(1-4)-GlcNAC-β-phenol) [26]
2.5 <28> (2'-fucosyllactose) [46]
2.9 <12> (Fuc-α-(1-2)-Gal) [26]
10.6 <7> (2'-fucosyllactose, <7> at pH 7.0 [8]) [8]

pH-Optimum

2 <29> (<29> form S3 [48]) [48]
2.5 <2> (<2> two pH-optima [9]) [9]
3 <18> (<18> one of two pH-optima [25]) [25]
3.2-5.2 <2> [7]
3.5-5.5 <2> (<2> isoforms F1 [2]) [2]
3.7 <11> [11]
3.9-4 <12> (<12> major pH-optimum [31]) [31]
4-6 <2> (<2> isoforms F2 [2]) [2]
4.3 <12> (<12> one of two pH-optima [22]) [22]
4.5-5.5 <16> [21]
4.6 <12> (<12> one of two pH-optima [43]) [43]
5 <4, 6, 7, 12, 25, 29> (<6> gonad fluid enzyme sharper pH dependence curve than the sperm-bound enzyme [5]; <7> optimal hydrolysis rate for 2'-fucosyllactose and Fuc-α-(1-6)-GlcNAc [8]; <2> two pH-optima [9]; <12> α-fucosidase II [28]; <25> slight different pH-optima for monomeric, dimeric and tetrameric forms [37]; <29> form S1 [48]) [4, 5, 8, 9, 12, 26, 28, 37, 48]
5-6.5 <17> (<17> α-fucosidase II [24, 39]) [24, 39]
5.1 <19> [27]
5.2 <22> [33, 41]
5.3 <12, 13, 23> (<13> isoform GF-1 [20]; <12> α-fucosidase I [28]) [20, 28, 33]
5.3-5.5 <13> (<13> isoform GF-2 [20]) [20]
5.4 <12> (<12> deglycosylated enzyme [12]) [12]

5.4-5.6 <12> [44]
5.5 <9, 12, 14, 15, 17, 18, 21> (<15> in citrate buffer [15]; <17> α-fucosidase I [24, 39]; <18> one of two pH-optima [25]; <12> for 4-nitrophenyl-α-L-fucopyranoside [42]) [14, 15, 22-25, 32, 39, 42]
5.5-5.8 <3> [36]
5.7 <2> [7]
5.7-5.9 <3> [40]
5.8 <9, 10> (<9> broad optimum [16]) [16]
6 <3, 8, 12, 13> (<8> extremely broad optimum [16]; <12> for 4-methylumbelliferyl-α-L-fucopyranoside [42]) [16, 42, 49]
6-6.5 <12, 15> (<15> in potassium phosphate buffer [15]) [15, 22]
6.1 <20> (<20> neutral α-fucosidase form [29]) [29]
6.3 <3, 12, 20> (<12> second pH-optimum for native and deglycosylated enzymes [12]) [12, 29, 47]
6.5 <12, 24> (<12> second pH-optimum [43]) [33, 43]
6.8 <8, 12> (<12> one of two pH-optima [22]) [10, 22]
6.9 <3, 12> (<12> second minor pH-optimum [31]) [16, 31]
6.9-7.1 <3> [6]
7 <1, 3, 7> (<1> fluorogenic substrate, natural substrates, isolated from dog brain, more readily digested at acidic pH [1]; <7> optimal hydrolysis rate for 4-methylumbelliferyl-α-L-fucopyranoside [8]) [1, 3, 8]

pH-Range

3-5 <3, 8, 9, 10> (<3, 9, 10> low activity between [6, 16]; <8> high activity between [16]; <3> appreciable activity between [16]) [6, 16]
4.5-6 <16> (<16> optimal activity within [18]) [18]
5-9 <1> [1]
5.5-7.5 <8> (<8> high activity between [10]) [10]

Temperature optimum (°C)

45 <12> [44]
50 <15, 16> [15, 21]
50-60 <6> [5]
55 <29> (<29> form S1 [48]) [48]
60 <16> [18]
70 <18> [25]

4 Enzyme Structure

Molecular weight

50000 <12> (<12> α-fucosidase II, gel filtration [28]) [28]
51000 <7> (<7> SDS-PAGE [8]) [8]
55970 <6> (<6> gonad fluid enzyme, mass spectrometry [5]) [5]
62000 <11> (<11> gel filtration [11]) [11]
67850 <6> (<6> sperm-bound enzyme, mass spectrometry [5]) [5]
68000 <20> (<20> gel filtration, two acid forms [29]) [29]
70000-75000 <18> (<18> gel filtration [25]) [25]

70000-80000 <28> (<28> gel filtration [46]) [46]
75000 <16> (<16> gel filtration [18]) [18]
80000 <16> (<16> gel filtration [21]) [21]
85000 <20> (<20> gel filtration, neutral form [29]) [29]
93800 <12> (<12> membrane-bound form [23]) [23]
100000 <12> (<12> gel filtration, α-fucosidase I [28]) [28]
109000-116000 <12> (<12> gel filtration, soluble form [23]) [23]
156000 <1> (<1> gel filtration [1]) [1]
180000 <4> (<4> gel filtration [4]) [4]
192000 <19> (<19> gel filtration [27]) [27]
200000 <12, 13, 22, 23> (isoform GF-2, gel filtration [20]; <22, 23> gel filtration [33, 41]; <12> gel filtration [42]) [20, 33, 41, 42]
210000-220000 <3> (<3> sedimentation equilibrium centrifugation [47]) [47]
217000 <3> (<3> sedimentation equilibrium centrifugation [36, 40]) [36, 40]
220000 <24> (gel filtration [33]) [33]
230000 <2, 12> (<2, 12> sedimentation equilibrium centrifugation [7, 43]) [7, 43]
233000 <14> (<14> gel filtration [14]) [14]
260000 <2> (<2> gel filtration [2, 9]) [2, 9]
270000 <12> (<12> gel filtration at pH 7.0 [26]) [26]
285000 <15, 25> (<15, 25> gel filtration [15, 37]) [15, 37]
305000 <12> (<12> gel filtration at pH 5.0 or 6.0, molecular weight depends on pH-value [34]) [34]
330000 <13> (<13> isoform GF-1, gel filtration [20]) [20]
390000 <12> (<12> gel filtration at pH 5.0 [26]) [26]

Subunits

dimer <1, 12, 15> (<1> 2 * 50000, SDS-PAGE [1]; <15> 2 * 135000, SDS-PAGE [15] ; <12> α-fucosidase I, 2 * 50000, SDS-PAGE, can aggregate to tetra- and hexamers [28]) [1, 15, 28]
monomer <6, 7, 11, 12, 16> (<6> 1 * 55971, gonad fluid enzyme, 1 * 67847, sperm-bound enzyme, mass spectrometry [5]; <7> 1 * 51000, SDS-PAGE [8]; <11> 1 * 62000, SDS-PAGE [11]; <16> 1 * 75000, SDS-PAGE [18]; <12> α-fucosidase II, 1 * 50000, SDS-PAGE, less-biologically active form, can aggregate to tetra- and hexamers [28]) [5, 8, 11, 18, 28]
oligomer <2, 12> (<2> contains subunits of 55 kDa and 52 kDa, SDS-PAGE [2]; <12> contains subunits of 63 kDa, SDS-PAGE [26]; <12> subunits of 50.1 kDa each, SDS-PAGE [43]) [2, 26, 43]
tetramer <2, 3, 12, 14, 19, 25> (<2> 4 * 54000, SDS-PAGE [7] ; <2> 4 * 60000, SDS-PAGE [9] ; <14> 4 * 63000, SDS-PAGE [14] ; <19> 4 * 55000, SDS-PAGE [27] ; <3> 4 * 55000, SDS-PAGE [36, 40] ; <25> 4 * 73500, SDS-PAGE, monomers and dimers also active [37] ; <12> 4 * 50000, gel filtration [42]; <3> 2 * 60000 + 2 * 2 * 47000, SDS-PAGE [47]) [7, 9, 14, 27, 36, 37, 40, 42, 47]

Posttranslational modification

glycoprotein <1-3, 6-12, 14, 16, 19, 24> (<1> 10%, w/w, carbohydrate content [1]; <2> 4.7%, w/w, mannose content for isoform F1, 5%, w/w, mannose content for isoforms F2 [2]; <3> isoform A derived by sialylation of isoform B near the end of epididymal maturation [3]; <6> N-linked oligosaccharides only for the gonad fluid enzyme, sperm-bound enzyme not N- or O-linked glycosylated [5]; <9, 14> sialylated [14, 22]; <12> α-fucosidase I: sialoglycoprotein, mannose-rich: α-fucosidase II [28]; <12> altered carbohydrate composition from enzymes of liver I-cell disease or pseudo-Hurler polydystrophy urine samples [31]; <24> contains 7.4% glucose, 2.2% galactose, 1.8% mannose, 2.2% hexosamine [33]; <12> contains 0.9% galactose, 1.9% mannose, 1.9% N-acetylglucosamine, 1.9% N-acetylneuraminic acid, w/w, [34]; <3> residues per tetramer: fucose 3.5, mannose 32, galactose 8, glucose 9, glucosamine 32, sialic acid 8 [36, 40]; <12> 1% carbohydrate content per molecule [43]; <3> 0.6% carbohydrate per native molecule [47]) [1-3, 5, 7, 8, 11, 12, 14, 16, 18, 22, 27, 28, 31, 33, 34, 36, 40, 42, 43, 45, 47]

5 Isolation/Preparation/Mutation/Application

Source/tissue

brain <3, 8-10, 12, 25> (<8,10> activities below and above pI 7.0 [10]) [10, 23, 28, 37, 42]
cauda epididymidis <13> [20]
cell culture <7, 11, 15, 16, 20, 26, 28> [8, 11, 15, 18, 21, 29, 38, 46]
commercial preparation <4> (<4> lactase-P [4]) [4]
emulsin <17> (<17> two forms, α-fucosidase I and II [24, 35, 39]) [24, 35, 39]
epididymis <3, 9, 13, 19> (<13> partly secretory and partly non-secretory bound form [20]) [20, 38, 47, 49]
foetal tissue <9> [22]
gonad fluid <6> [5]
heart <12> [42]
hepatopancreas <2, 14, 18, 21-24> [9, 14, 25, 32, 33, 41]
kidney <3, 9, 12, 13, 19> [13, 28, 42, 49]
leukocyte <12> [44]
liver <3, 8-10, 12, 13, 19, 29> (<8, 9> activities below and above pI 7.0 [10, 16]; <3, 10> no activity above pI 7.0 [16]; <12> characterization of α-fucosidase from fucosidosis patients [30]; <12> normal and I-cell disease samples [31]; <29> two forms: S1 and S3 [48]) [10, 12, 16, 17, 22, 28, 30, 31, 36, 40, 43, 48, 49]
ovary <3, 9, 13, 19> [49]
placenta <9, 12> [14, 22, 34, 42, 45]
preputial parts <3, 9> [49]
seminal plasma <13> [19, 20]
serum <12, 27> [26, 38, 42]
sperm <3, 6> (<3> two isoforms: A, B, characterized during epididymal maturation [3]; <6> one form tightly bound to the sperm plasma membrane, one form present in the gonad fluid [5]) [3, 5, 6]
spermatozoa <13> [20]
spleen <3, 8-10, 12> (<8, 10> activities below and above pI 7.0 [10]) [10, 28, 49]

testis <3, 9, 13, 19> [49]
thyroid <19> [27]
urine <12> (<12> samples from pseudo-Hurler polydystrophy [31]) [28, 31]
uterus <3, 9, 13, 19> [49]
viscera <2, 30> [2, 7, 49]

Localization

Golgi apparatus <3> [6]
cytosol <12> [23]
extracellular <4, 5, 7, 11, 13, 15, 16, 20, 26> (<7> normally a lysosomal glycoprotein, released by secretion and/or cell lysis [8]; <11> increasing activity up to aggregation stage, not detectable in the growth phase [11]; <15> maximal enzyme synthesis by growth on porcine gastric mucin containing media [15]; <16> induction of enzyme synthesis by addition of L-fucose to the medium [18]; <13> partly secretory from cauda epididymidis [20]; <16> most induction of enzyme synthesis if cells grown in 2% active sludge hydrolyzate medium [21]) [4, 8, 11, 15, 18-21, 29, 38]
intracellular <28> [46]
lysosome <3, 7, 11, 12> [8, 11, 12, 36, 40, 44]
membrane-bound <12> [23]
plasma membrane <3, 6> (<3> specifically localized to the convex region of the prinicipal segment of testicular and cauda epididymal sperm heads [6]) [3, 5, 6]

Purification

<1> (to homogeneity, recombinant enzyme [1]) [1]
<2> (to homogeneity, chromatography techniques, two isoforms: F1, F2 [2]; to homogeneity, chromatography steps [7, 9]) [2, 7, 9]
<3> (to homogeneity, affinity chromatography [36, 38, 40]; near homogeneity [47]) [36, 38, 40, 47]
<4> (partial [4]) [4]
<6> (to homogeneity, two forms, affinity chromatography [5]) [5]
<7> (to homogeneity, affinity chromatography [8]) [8]
<8> (near homogeneity [10]) [10]
<9> (near homogeneity, antibody affinity chromatography [17]) [17]
<11> (to homogeneity, affinity chromatography [11]) [11]
<12> (to homogeneity, affinity chromatography [12, 26, 31, 34]; to homogeneity, two forms: α-fucosidase I and II, affinity chromatography [28]; to homogeneity, affinity chromatography, isoelectric focusing, 2 forms, α-L-fucosidase A and B [42]; to homogeneity, 2step affinity chromatography [43]; near homogeneity, 1step affinity chromatography [45]) [12, 26, 28, 31, 34, 42, 43, 45]
<13> (homogeneous, chromatography techniques [19]) [19]
<14> (to homogeneity, chromatography techniques [14]) [14]
<15> (to homogeneity, chromatography techniques [15]) [15]
<16> (homogeneous, chromatography steps [18, 21]) [18, 21]
<17> (partial, two forms [24, 39]; to homogeneity, affinity chromatography, α-fucosidase I [35]) [24, 35, 39]

<18> (to homogeneity, affinity chromatography [25]) [25]
<19> (to homogeneity, affinity chromatography [27]) [27]
<20> (partial, three forms at different pI-values [29]) [29]
<21> (to homogeneity, affinity chromatography, two isoforms [32]) [32]
<22> (to homogeneity, isoelectric focusing [33, 41]) [33, 41]
<23> (to homogeneity, isoelectric focusing [33]) [33]
<24> (to homogeneity, isoelectric focusing [33]) [33]
<25> (to homogeneity, chromatography steps [37]) [37]
<26> (to homogeneity, affinity chromatography [38]) [38]
<27> (to homogeneity, affinity chromatography [38]) [38]
<28> (near homogeneity, chromatography techniques [46]) [46]
<29> (partial [48]) [48]

Cloning

<1> (expression in CHO cells [1]; expression in MDCK cells [1]) [1]

Application

analysis <2> (<2> studies of structures and functions of glycoconjugates and oligosaccharides due to unique substrate specificity [9]) [9]
medicine <1> (<1> recombinant enzyme will extend the proven efficiacy of enzyme replacement therapy in lysosomal storage disorders e.g. fucosidosis [1]) [1]

6 Stability

pH-Stability

3-5 <29> (<29> form S1, most stable between [48]) [48]
3-7 <4> (<4> stable within [4]) [4]
4-6 <22, 23> (<22, 23> stable between [33, 41]) [33, 41]
4-7 <2> (<2> isoform F1: stable within [2]) [2]
4-8 <3, 16> (<16> stable within [21]; <3> stable in the presence of 0.05 mg/mg bovine serum albumin, 1 mM 2-mercaptoethanol [36]) [21, 36]
4-10 <2> (<2> fairly stable within [7]) [7]
4.5-6 <3, 13> (<3, 13> stable within [49]) [49]
4.5-7 <2> (<2> isoform F2: stable within [2]) [2]
4.5-9 <15> (<15> stable within, kept at 4°C [15]) [15]
5-6 <29> (<29> most stable within, form S3 [48]) [48]
5-8 <24> (<24> stable between [33]) [33]
5.5 <25> (<25> instability below [37]) [37]
5.5-10 <16> (<16> stable within [18]) [18]

Temperature stability

-20 <3> (<3> loss of activity within 1 week [6]) [6]
4-55 <12> (<12> highly stable [44]) [44]
25 <3> (<3> 50% remaining activity after 2 h [36]) [36]
37 <7, 12> (<7> half-life: 1 h [8]; <12> gradual loss of activity after 6 min for the purified enzyme, in crude state: stable for 4 h [43]) [8, 43]

45 <15, 16> (<15> stable up to [15]; <16> stable below [18]; <16> stable up to [21]) [15, 18, 21]
50 <3, 4, 5, 9, 12, 13, 28> (<3> isoform A: 50% inactivation in 1 h, isoform B: 50% inactivation in 10 min [3]; <4,5> stable up to [4]; <12> native enzyme: 80% residual activity after 2 h, deglycosylated enzyme: 38% residual activity after 2 h [12]; <13> inactivation above, isoform GF-1 more heat stable than GF-2 [20]; <9> from mouse liver: more labile against inactivation than from human liver or mouse placental enzymes [22]; <12> 30% loss of activity after 3 h [23]; <12> α-fucosidase I: 50% loss of activity in 40 min, α-fucosidase II: 20% loss of activity in 40 min [28]; <28> 7% loss of activity after 5 min [46]) [3, 4, 12, 20, 22, 23, 28, 46]
50-55 <12, 25> (<12> greatly decreased thermostability in fucosidosis patients [30]; <25> triphasic heat inactivation due to 3 different heat sensitive forms, forms I and II: less heat sensitive, form III: most heat sensitive [37]) [30, 37]
51 <12> (<12> rapid inactivation for α-L-fucosidase A, form B more heat-stable, stabilizing effect of bovine serum albumin, addition of 0.5 M potassium thiocyanate leads to more heat sensitivity for both forms [42]) [42]
60 <22-24> (<22-24> fully active for 5 min [33]) [33]
60-65 <2> (<2> isoform F2: stable up to [2]) [2]
65 <2, 12, 14, 22> (<2> isoform F1: stable up to [2]; <2> fairly stable up to [7]; <14> stable for 10 min [14]; <12> completely abolished [14]; <22> fully active after 5 min [41]) [2, 7, 14, 41]
70 <12, 18, 28> (<18> 50% loss of activity after 5 min [25]; <12> 60% loss of activity [44]; <28> 77% loss of activity after 5 min [46]) [25, 44, 46]

Organic solvent stability

dimethyl formamide <4, 5> (<4> more stable than the Aspergillus niger enzyme, no loss of activity up to 10%, v/v [4]) [4]
dimethyl sulfoxide <4, 5> (<4> more stable than the Aspergillus niger enzyme, no loss of activity up to 20%, v/v [4]) [4]
dioxane <4, 5> (<4> more stable than the Aspergillus niger enzyme, no loss of activity up to 10%, v/v [4]) [4]

General stability information

<3>, stabilizing effect by addition of 0.05 mM/ml bovine serum albumin, 1 mM 2-mercaptoethanol, 1 mM $MgSO_4$, high protein concentrations required [36, 40]
<7>, increased half-life to 23 h at 37°C and stability to repeated freezing and thawing by addition of 0.5 mg/ml bovine serum albumin [8]
<17>, 60% loss of activity after lyophilization [24]
<22>, sensitive to repeated freezing and thawing [41]

Storage stability

<3>, -20°C, acetate buffer, pH 6.0, 100 mM L-fucose, several months [36]
<3>, 2-4°C, stable for one month [6]
<6>, 4°C, 10 mM sodium phosphate, pH 5.5, 1 mM EDTA, 1 mM PMSF, one month [5]

<7>, -20°C, 0.5 mg/ml bovine serum albumin, 6 months [8]
<9>, 2-4°C, 29 days [22]
<12>, -10°C, 3-6% loss of activity after 3 days [44]
<12>, -20°C, fully active after 29 days, 0-4°C, 10 mM Na_2HPO_4, pH 5.5, 0.02% NaN_3, w/v, 100 mM L-fucose, 163 days [23, 43]
<15>, frozen, 1 year [15]
<17>, frozen, more than 90% of its initial activity after 3 months [24]
<3, 26, 27>, 4°C, 0.2 M sodium citrate, pH 6.0, 20% polyethylene glycol, 2 months [38]

References

[1] Bielicki, J.; Muller, V.; Fuller, M.; Hopwood, J.J.; Anson, D.S.: Recombinant canine α-L-fucosidase: Expression, purification, and characterization. Mol. Genet. Metab., **69**, 24-32 (2000)
[2] Hirata, K.; Aso, Y.; Yasuda, S.; Ishiguro, M.: Purification of Pomacea canaliculata α-fucosidase isoforms with different thermostabilities. Biosci. Biotechnol. Biochem., **62**, 807-810 (1998)
[3] Abascal, I.; Skalaban, S.R.; Grimm, K.M.; Aviles, M.; Martinez-Menarguez, J.A.: Alteration of the isoform composition of plasma-membrane-associated rat sperm α-L-fucosidase during late epididymal maturation: Comparative characterization of the acidic and neutral isoforms. Biochem. J., **333**, 201-207 (1998)
[4] Ajisaka, K.; Fujimoto, H.; Miyasato, M.: An α-L-fucosidase from Penicillium multicolor as a candidate enzyme for the synthesis of α (1-3)-linked fucosyl oligosaccharides by transglycosylation. Carbohydr. Res., **309**, 125-129 (1998)
[5] Focarelli, R.; Cacace, M.G.; Seraglia, R.; Rosati, F.: A nonglycosylated, 68-kDa α-L-fucosidase is bound to the mollusc bivalve Unio elongatulus sperm plasma membrane and differs from a glycosylated 56-kDa form present in the seminal fluid. Biochem. Biophys. Res. Commun., **234**, 54-58 (1997)
[6] Aviles, M.; Bascal, I.; Martinez-Menarguez, J.A.; Castells, M.T.; Skalaban, S.R.; Ballesta, J.; Alhadeff, J.A.: Immunocytochemical localization and biochemical characterization of a novel plasma membrane-associated, neutral pH optimum α-L-fucosidase from rat testis and epididymal spermatozoa. Biochem. J., **318**, 821-831 (1996)
[7] Hirata, K.; Nkahara, Y.; Kimura, Y.; Funatsu, G.: Purification and some properties of a β-xylosidase and an α-fucosidase from apple snails (Pomacea canaliculata). Biosci. Biotechnol. Biochem., **60**, 249-254 (1996)
[8] Gramer, M.J.; Schaffer, D.V.; Sliwkowski, M.B.; Goochee, C.F.: Purification and characterization of α-L-fucosidase from Chinese hamster ovary cell culture supernatant. Glycobiology, **4**, 611-616 (1994)
[9] Endo, T.; Tsukada, T.; Hiraiwa, M.; Uda, Y.; Kobata, A.: Purification and characterization of an α-L-fucosidase from Pomacea canaliculata. Arch. Biochem. Biophys., **302**, 152-160 (1993)

[10] Shoarinejad, F.; Alhadeff, J.A.: Rodent tissue α-L-fucosidases: analysis of brain and spleen isoforms and characterization of the purified hamster liver enzyme. Comp. Biochem. Physiol. B, **105**, 523-528 (1993)
[11] Schopohl, D.; Müller-Taubenberger, A.; Orthen, B.; Hess, H.; Reutter, W.: Purification and properties of a secreted and developmentally regulated α-L-fucosidase from Dictyostelium discoideum. J. Biol. Chem., **267**, 2400-2405 (1992)
[12] Piesecki, S.; Alhadeff, J.A.: The effect of carbohydrate removal on the properties of human liver α-L-fucosidase. Biochim. Biophys. Acta, **1119**, 194-200 (1992)
[13] Dumas, D.P.; Kajimoto, T.; Liu, K.K.C.; Wong, C.H.: Azasugars and glycal inhibitors of α-L-fucosidase. Bioorg. Med. Chem. Lett., **2**, 33-36 (1992)
[14] Chuang, N.N.; Yeh, C.C.; Lin, K.S.: The basic isoelectric form of α-L-fucosidase from the hepatopancreas of the shrimp Penaeus monodon (crustacea: decapoda). Comp. Biochem. Physiol. B, **99**, 395-398 (1991)
[15] Tsuji, Y.; Yamamoto, K.; Tochikura, T.; Seno, T.; Ohkubo, Y.; Yamaguchi, H.: Purification and characterization of α-L-fucosidase from Bacillus circulans grown on porcine gastric mucin. J. Biochem., **107**, 324-330 (1990)
[16] Johnson, S.W.; Alhadeff, J.A.: Isoform, kinetic and immunochemical characterization of rodent liver α-L-fucosidases. Comp. Biochem. Physiol. B, **97**, 713-717 (1990)
[17] Laury-Kleintop, L.D.; Damjanov, I.; Alhadeff, J.A.: Antibody-affinity purification of novel α-L-fucosidase from mouse liver. Biochem. J., **245**, 589-593 (1987)
[18] Yamamoto, K.; Tsuji, Y.; Kumagai, H.; Tochikura, T.: Induction and purification of α-L-fucosidase from Fusarium oxysporum. Agric. Biol. Chem., **50**, 1689-1695 (1986)
[19] Srivastava, P.N.; Arbtan, K.; Takei, G.H.; Huang, T.T.F.; Yanagimachi, R.: α-L-Fucosidase from bull seminal plasma: Its purification and acrosome reaction promoting property. Biochem. Biophys. Res. Commun., **137**, 1061-1068 (1986)
[20] Jauhiainen, A.; Vanha-Perttula, T.: α-L-Fucosidase in the reproductive organs and seminal plasma of the bull. Biochim. Biophys. Acta, **880**, 91-95 (1986)
[21] Yano, T.; Yamamoto, K.; Kumagai, H.; Tochikura, T.; Yokoyama, T.; Seno, T.; Yamaguchi, H.: Purification and characterization of a novel α-L-fucosidase from Fusarium oxysporum grown on sludge. Agric. Biol. Chem., **49**, 3179-3187 (1985)
[22] Laury-Kleintop, L.D.; Damjanov, I.; Alhadeff, J.A.: Characterization of mouse liver α-L-fucosidase. Demonstration of unusual basic isoelectric forms of the enzyme that appear to be developmentally regulated. Biochem. J., **230**, 75-82 (1985)
[23] Hopfer, R.L.; Alhadeff, J.A.: Solubilization and characterization of pellet-associated human brain α-L-fucosidase activity. Biochem. J., **229**, 679-685 (1985)
[24] Kobata, A.: α-L-Fucosidases from almond emulsin. Methods Enzymol., **83**, 625-631 (1982)

[25] DÁniello, A.; Hakimi, J.; Cacace, G.M.; Ceccarini, C.: The purification and characterization of α-L-fucosidase from hepatopancreas of Octopus vulgaris. J. Biochem., **91**, 1073-1080 (1982)
[26] DiCioccio, R.A.; Barlow, J.J.; Matta, K.L.: Substrate specificity and other properties of α-L-fucosidase from human serum. J. Biol. Chem., **257**, 714-718 (1982)
[27] Grove, D.S.; Serif, G.S.: Porcine thyroid fucosidase. Biochim. Biophys. Acta, **662**, 246-255 (1981)
[28] Chien, S.-F.; Dawson, G.: Purification and properties of two forms of human α-L-fucosidase. Biochim. Biophys. Acta, **614**, 476-488 (1980)
[29] Berg, J.-O.; Lindqvist, L.; Nord, C.E.: Purification of glycoside hydrolases from Bacteroides fragilis. Appl. Environ. Microbiol., **40**, 40-47 (1980)
[30] Ahadeff, J.A; Andrews-Smith, G.L.: Purification and characterization of α-L-fucosidase from the liver of a fucosidosis patient. Biochem. J., **187**, 45-51 (1980)
[31] Kress, B.C.; Freeze, H.H.; Herd, J.K.; Alhadeff, J.A.; Miller, A.L.: Purification and characterization of I-cell disease α-L-fucosidase. J. Biol. Chem., **255**, 955-961 (1980)
[32] Concha-Slebe, I.; Presper, K.A.; Basu, S.: Purification and properties of α-L-fucosidase from Venus mercenaria. Glycoconjugate Res., **2**, 873-875 (1979)
[33] Cabezas, J.A.; Reglero, A.; Calvo, P.; Pedro de, M.A.: Comparative stuy of α-L-fucosidase from three species os marine molluscs: Purification and properties. Glycoconjugate Res., **2**, 867-871 (1979)
[34] Turner, B.M.: Purification and characterization of α-L-fucosidase from human placenta. pH-dependent changes in molecular size. Biochim. Biophys. Acta, **578**, 325-336 (1979)
[35] Yoshima, H.; Takasaki, S.; Ito-Mega, S.; Kobata, A.: Purification of almond emulsin α-L-fucosidase I by affinity chromatography. Arch. Biochem. Biophys., **194**, 394-398 (1979)
[36] Opheim, D.J.; Touster, O.: α-L-Fucosidase from rat liver lysosomes. Methods Enzymol., **50**, 505-510 (1978)
[37] Alam, T.; Balasubramanian, A.S.: The purification, properties and characterization of three forms of α-L-fucosidase from monkey brain. Biochim. Biophys. Acta, **524**, 373-384 (1978)
[38] Jain, R.S.; Binder, R.L.; Levy-Benshimol, A.; Buck, C.A.; Warren, L.: Purification of α-L-fucosidase from various sources by affinity chromatography. J. Chromatogr., **139**, 283-290 (1977)
[39] Ogata-Arakawa, M.; Muramatsu, T.; Kobata, A.: α-L-Fucosidases from almond emulsin: Characterization of the two enzymes with different specificities. Arch. Biochem. Biophys., **181**, 353-358 (1977)
[40] Opheim, D.J.; Touster, O.: The purification and characterization of rat liver lysosomal α-L-fucosidase. J. Biol. Chem., **252**, 739-743 (1977)
[41] Reglero, A.; Cabezas, J.A.: Glycosidases of molluscs. Purification and properties of α-L-fucosidase from Chamelea gallina L.. Eur. J. Biochem., **66**, 379-387 (1976)
[42] Di Matteo, G.; Orfeo, M.A.; Romeo, G.: Human α-fucosidase. Purification and properties. Biochim. Biophys. Acta, **429**, 527-537 (1976)

[43] Alhadeff, J.A.; Miller, A.L.; Wenaas, H.; Vedvick, T.; O'Brien, J.S.: Human liver α-L-fucosidase. Purification, characterization, and immunochemical studies. J. Biol. Chem., **250**, 7106-7113 (1975)
[44] Avila, J.L.; Convit, J.: Studies on human polymorphonuclear leukocyte enzymes IV. Intracellular distribution and properties of α-L-fucosidase. Biochim. Biophys. Acta, **358**, 308-318 (1974)
[45] Alhadeff, J.A.; Miller, A.L.; O'Brien, J.S.: Purification of human placental α-L-fucosidase by affinity chromatography. Anal. Biochem., **60**, 424-430 (1974)
[46] Kochibe, N.: Purification and properties of α-L-fucosidase from Bacillus fulminans. J. Biochem., **74**, 1141-1149 (1973)
[47] Carlsen, R.B.; Pierce, J.G.: Purification and properties of an α-L-fucosidase from rat epididymis. J. Biol. Chem., **247**, 23-32 (1972)
[48] Tanakai, K.; Nakano, T.; Noguchi, S.; Pigman, W.: Purification of α-L-fucosidase of abalone livers. Arch. Biochem. Biophys., **126**, 624-633 (1968)
[49] Levvy, G.A.; McAllan, A.: Mammalian fucosidases. 2. α-L-fucosidase. Biochem. J., **80**, 435-439 (1961)

β-N-Acetylhexosaminidase 3.2.1.52

1 Nomenclature

EC number
3.2.1.52

Systematic name
β-N-acetyl-β-D-hexosaminide N-acetylhexosaminohydrolase

Recommended name
β-N-acetylhexosaminidase

Synonyms
2-acetamido-2-deoxy-β-D-glucoside acetamidodeoxyglucohydrolase
EC 3.2.1.30 (formerly)
Hex
N-acetyl-β-D-hexosaminidase
N-acetyl-β-glucosaminidase
N-acetyl-β-hexosaminidase
N-acetylhexosaminidase
β-D-N-acetylhexosaminidase
β-D-hexosaminidase
β-N-acetyl-D-hexosaminidase
β-N-acetyl-hexosaminidase
β-N-acetylgalactosaminidase
β-N-acetylglucosaminidase
β-NAHA
β-acetylaminodeoxyhexosidase
β-acetylhexosaminidase
β-acetylhexosaminidinase
β-hexosaminidase
chitobiase
endo-β-N-acetylglucosaminidase (<22>, Endo-BI and -BII [32]; <25>, endo-β-N-acetylglucosaminidase L [79]) [32, 79]
hexosaminidase
hexosaminidase A

CAS registry number
9027-52-5

2 Source Organism

<1> *Triticum aestivum* (wheat [1]) [1]

<2> *Pycnoporus cinnabarinus* (IFO 6139 [17]) [2, 17]

<3> *Glycine max* (soybean [3]) [3]

<4> *Homo sapiens* (healthy persons and mucolipidosis patients [15]; patient with Sandhoffs disease [19]; patient with GM2-gangliosidosis B1 variant [31]; patients with GM2-gangliosidosis B1 variant from the north of Portugal [40]; patients with Tay-Sachs-disease [41]; newborns [42]; patients with I-cell disease and healthy persons [45]; pregnant women [77]) [4, 9, 15, 19, 28, 31, 40, 41, 42, 44, 45, 51, 52, 53, 72, 77]

<5> *Mugil cephalus* (striped mullet [5]) [5]

<6> *Citrullus lanatus* (watermelon [6]) [6]

<7> *Pisum sativum* (pea [7]) [7]

<8> *Mucor fragilis* [8]

<9> *Prunus serotina* (black cherry [10]) [10]

<10> *Penicillium oxalicum* (IFO 5748 [11]) [11]

<11> *Gallus gallus* [12]

<12> *Rattus norvegicus* [13, 21, 30, 39]

<13> *Halocynthia roretzi* [14]

<14> *Styela plicata* (sea-squirt [16]) [16]

<15> *Oryctolagus cuniculus* [18]

<16> *Paecilomyces persicinus* (P-10 [20]) [20]

<17> *Helicella ericetorum* (Müller [22]) [22]

<18> *Bos taurus* (calf [30]) [23, 27, 30, 58, 66, 76]

<19> *Trigonella foenum-graecum* [24, 25]

<20> *Trichomonas foetus* [26]

<21> *Canavalia ensiformis* (Jack bean [29]) [29]

<22> *Bacillus sp.* (B. subtilis [73, 74]) [32, 73, 74]

<23> *Boophilus microplus* (bovine tick, ixodide [33]) [33]

<24> *Trichoderma harzianum* [34]

<25> *Streptomyces sp.* (S. plicatus [35, 79]; strain L-13 [54]; ATCC 27800 [79]) [35, 54, 79]

<26> *Vibrio furnissii* [36]

<27> *Sus scrofa* [37, 60]

<28> *Brassica oleracea* [38]

<29> *Bipolaris sorokiniana* [43]

<30> *Patinopecten yessoensis* [46]

<31> *Turbo cornutus* [46]

<32> *Bacillus stearothermophilus* (CH-4 [47]) [47]

<33> *Penaeus japonicus* (prawn [48]) [48]

<34> *Lupinus luteus* [49, 50]

<35> *Bombyx mori* (silkworm, pupae and fifth-instar larvae at the SP3 stage [55]; 5-day old pupae [56]; full-grown fifth-instar larvae [68]) [55, 56, 68]

<36> *Phycomyces blakesleeanus* (NRRL1555(-) [57]) [57]

<37> *Turbatrix aceti* [59]

<38> *Octopus vulgaris* [61]
<39> *Conidiobolus lamprauges* [62]
<40> *Mycobacterium leprae* (from armadillo liver [63]) [63]
<41> *Mycobacterium phlei* [63]
<42> *Mycobacterium microti* [63]
<43> *Armadillo officinalis* [63]
<44> *Aspergillus niger* [64]
<45> *Cupiennius salei* [65]
<46> *Limulus polyphemus* (horseshoe crab [67]) [67]
<47> *Escherichia coli* (strain K-12 [69]) [69]
<48> *Dictyostelium discoideum* (strain A3 [70]) [70]
<49> *Patella vulgata* [71, 80]
<50> *Sclerotinia fructigena* [75]
<51> *Hordeum vulgare* (barley, malted [78]) [78]
<52> *Candida albicans* (ATCC 10261 [81]) [81]
<53> *Tremella fuciformis* (strain T-7 [82]) [82]
<54> *Ovis aries* (male [83]) [83]

3 Reaction and Specificity

Catalyzed reaction

hydrolysis of terminal non-reducing N-acetyl-D-hexosamine residues in N-acetyl-β-D-hexosaminides (<1, 9, 12, 17, 18, 19, 20, 21, 27, 34, 54>, same active site for N-glucosaminidase and N-galactosaminidase activities [1, 10, 22, 25, 26, 29, 37, 39, 49, 50, 76, 83]; <2, 10, 18, 22, 27, 28, 33, 35, 37, 38, 44, 47>, exo-type reaction [2, 11, 27, 37, 38, 48, 55, 56, 59, 61, 64, 68, 69, 73]; <12, 18>, seperate active sites for glucosaminidase and galactosaminidase activities [21, 66]; <17, 22, 25>, endo-type reaction [22, 32, 79])

Reaction type

hydrolysis of O-glycosyl bond

Natural substrates and products

S N-acetyl-β-D-hexosaminides + H_2O <1-54> [1-83]
P N-acetyl-D-hexosamines <1-54> [1-83]

Inhibitors

(N-acetyl-D-glucosamine)$_2$ <26> (<26>, non-competitive [36]) [36]
(N-acetyl-D-glucosamine)$_{3-6}$ <26> (<26>, competitive inhibition [36]) [36]
2-acetamido-2-deoxy-D-fucono(1-4)lactone <34> [50]
2-acetamido-2-deoxy-D-galactono(1-4)lactone <17, 18, 27, 34> (<17>, complete inhibition with 100 mM [22]; <27>, with 10 mM 75% inhibition [37]) [22, 27, 37, 50]
2-acetamido-2-deoxy-D-galactonolactone <54> [83]
2-acetamido-2-deoxy-D-galactose <20> [26]
2-acetamido-2-deoxy-D-glucose <20> [26]
2-acetamido-2-deoxy-D-mannono-1,4-lactone <18> [27]

2-acetamido-2-deoxy-L-arabinono(1-4)lactone <34> [50]
2-acetamido-2-deoxyglucono-1,4-lactone <1, 18> [1, 27]
2-acetamido-6-azido-2,6-dideoxy-D-galactono(1-4)lactone <34> [50]
2-deoxy-2-acetamido-D-glucono(1-5)lactone <22, 45> [65, 73]
2-mercaptoethanol <16> (<16>, with 2 mM hydrolysis of *p*-nitrophenyl-β-2-acetamido-2-deoxy-D-glucopyranoside is reduced to 22%, hydrolysis of *p*-nitrophenyl-β-2-acetamido-2-deoxy-D-galactopyranoside to 20% [20]) [20]
7-diazonaphthalene-1,3-disulfonic acid <40> (<40>, extract: 45% of activity remains [63]) [63]
Ag^{+} <1, 3, 4, 12, 13, 14, 15, 18, 19, 21, 22, 24, 30, 44, 48> (<13>, with 0.1 mM 63% of activity remains [14]; <14>, with 7 mM 5% of activity remains [16]; <15>, 1 and 2 mM cause 70 and 92% inhibition respectively [18]; <12>, completely inactivates all 3 enzyme forms in colonic mucosa [21]; <19>, all 4 forms [24]; <21>, complete inactivation with 0.1 mM [29]; <12,18>, inhibits N-acetylhexosaminidase, -galactosaminidase and -glucosaminidase [30]; <24>, with 1 mM 43% of N-acetylgalactosaminidase and 59% of N-acetylglucosaminidase activity remains [34]; <30>, concentration of 2 mM causes 70.9% inhibition [46]; <44>, 100 mM [64]; <48>, minor isozyme: 2% of activity remains with 5 mM, major isozyme: 38% of activity remains with 2 mM [70]; <4>, enzyme incubated before addition of substrate: with 0.0025 mM 18% of activity remains [72]; <22>, 0.01 mM, 30% inhibition, 2 mM, 100% inhibition [73]) [1, 3, 14, 16, 18, 21, 24, 29, 30, 34, 46, 64, 66, 70, 72, 73, 76]
Ba^{2+} <21> (<21>, with 10 mM 85% of activity remains [29]) [29]
Ca^{2+} <22, 35, 44> (<35>, with 0.03 mM 42.9% of activity remains [56]; <44>, 100 mM [64]; <22>, 1 M and more cause 100% inhibition [74]) [56, 64, 73, 74]
Cd^{2+} <16, 22, 53> (<16>, with 2 mM hydrolysis of *p*-nitrophenyl-β-2-acetamido-2-deoxy-D-glucopyranoside is reduced to 41%, hydrolysis of *p*-nitrophenyl-β-2-acetamido-2-deoxy-D-galactopyranoside to 38% [20]; <22>, 2 mM cause 41% inhibition, 10 mM cause 59% inhibition [73]; <53>, 14%, 20% and 67% inhibition with 0.01 mM, 2 mM and 10 mM [82]) [20, 73, 82]
Co^{2+} <22> (<22>, 10 mM cause 37% inhibition [73]) [73]
Cu^{2+} <1, 4, 7, 13, 21, 44, 48> (<7>, all 4 isozymes [7]; <13>, with 10 mM 47% of activity remains [14]; <21>, with 10 mM 78% of activity remains [29]; <44>, 100 mM [64]; <48>, minor isozyme: 19% of activity remains with 10 mM, major isozyme: 47% of activity remains with 10 mM [70]; <4>, only in presence of ascorbate [72]) [1, 7, 14, 29, 64, 70, 72]
D-mannose <17, 27, 32> (<17>, 75% inhibition with 100 mM [22]; <27>, with 10 mM 30% inhibition [37]; <32>, concentration of 10 mM causes 33% inhibition [47]) [22, 37, 47]
EDTA <18, 37> (<18>, with 50 mM 54% inhibition [23]; <37>, concentration of 1 mM causes 40% inhibition [59]) [23, 59]
Fe^{2+} <4, 7, 9, 14, 18, 19, 30> (<7>, 2 mM causes 30-35% loss of activity of β-NAHA-IA, 10 mM causes 30-35% loss of activity of β-NAHA-IB [7]; <9>, 1 mM reduces activity by 59% [10]; <14>, with 7 mM 16% of activity remains [16]; <19>, both hexosaminidase activities of forms II, III and IV inhibited

[24]; <30>, concentration of 2 mM causes 7.8% inhibition [46]; <4>, with 0.2 mM 28% of activity remains, citrate buffer, not inhibitory in succinate or acetate buffer [72]) [7, 10, 16, 24, 46, 72, 76]
Fe^{3+} <9, 17, 18, 19, 21, 30, 37> (<9>, concentration of 10 mM reduces activity by 23% [10]; <17>, concentration of 10 mM results in 85% inhibition [22]; <19>, glucosaminidase-activity of forms II, III and IV are depressed, galactosaminidase-activity of forms III and IV are inhibited [24]; <21>, with 10 mM 5% of activity remains [29]; <30>, concentration of 2 mM causes 53.1% inhibition [46]; <37>, concentration of 1 mM causes 70-80% inhibition [59]) [10, 22, 24, 29, 46, 59, 76]
Hg^{2+} <1, 2, 4, 7, 8, 9, 10, 12, 13, 14, 15, 16, 17, 18, 19, 21, 22, 23, 26, 29, 30, 32, 33, 35, 37, 44, 48, 53> (<2>, slight inhibition at 10 mM [2,17]; <7>, all 4 isozymes, at 1 mM [7]; <8>, low molecular weight form EII, at 0.25 mM [8]; <9>, 1 mM reduces activity by 89% [10]; <10>, 0.05 mM results in significant inhibition [11]; <13>, concentration of 0.1 mM results in total loss of activity [14]; <14>, concentration of 7 mM results in total loss of activity [16]; <15>, 0.05 and 0.1 mM cause 45 and 70% inhibition [18]; <16>, with 2 mM hydrolysis of *p*-nitrophenyl-β-2-acetamido-2-deoxy-D-glucopyranoside is reduced to 24%, hydrolysis of *p*-nitrophenyl-β-2-acetamido-2-deoxy-D-galactopyranoside to 20% [20]; <12>, completely inactivates all 3 enzyme forms in colonic mucosa [21]; <17>, concentration of 10 mM results in 97% inhibition [22]; <19>, all 4 forms [24]; <21>, complete inactivation with 0.1 mM [29]; <23>, with 1 mM 13% of activity remains [33]; <24>, with 1 mM 84% of N-acetylgalactosaminidase and 90% of N-acetylglucosaminidase activity remains [34]; <26>, with 1 mM 50% of activity remains, with 5 mM 10% of activity remains [36]; <30>, concentration of 2 mM causes 98.9% inhibition [46]; <33>, concentration of 10 mM causes 39% inhibition [48]; <4>, with 0.75 M 6% of activity remains [51]; <35>, Ex 1, Ex 2 and alimentary canal enzyme [55]; <35>, with 0.03 mM 7.1% of activity remains [56]; <37>, concentration of 1 mM causes complete inhibition [59]; <44>, 100 mM [64]; <48>, minor isozyme: 5% of activity remains with 0.01 mM, major isozyme: 53% of activity remains with 0.01 mM [70]; <4>, added with substrate: with 0.0025 mM 45% of activity remains, enzyme incubated before addition of substrate: with 0.0025 mM 25% of activity remains [72]; <22>, 0.01 mM cause 19% inhibition, 2 mM cause 97% inhibition [73]; <53>, 37% inhibition with 0.01 mM, complete inhibition with 2 mM [82]) [1, 2, 7, 8, 10, 11, 14, 16, 17, 18, 20, 21, 22, 24, 29, 33, 36, 43, 46, 47, 48, 55, 56, 59, 64, 66, 70, 72, 73, 76, 82]
KCl <26> (<26>, with 1 mM 80% of activity remains, with 2 mM 50% of activity remains [36]) [36]
Mg^{2+} <21, 22, 35> (<21>, with 10 mM 85% of activity remains [29]; <35>, with 0.03 mM 71.4% of activity remains [56]; <22>, 0.5 M causes 78% inhibition, 1 M causes 82% inhibition, 2 M and more cause 100% inhibition [74]) [29, 56, 73, 74]
Mn^{2+} <21, 32> (<21>, with 10 mM 82% of activity remains [29]; <32>, slight inhibition [47]) [29, 47]
N,N,N,N-tetraacetylchitotetraose <2> (<2>, inhibits hydrolysis of *p*-nitrophenyl-β-2-acetamido-2-deoxy-D-galactopyranoside stronger than hydrolysis

of *p*-nitrophenyl-β-2-acetamido-2-deoxy-D-glucopyranoside [2, 17]; <2>, with 1 mM 45% of N-glucosaminidase- but only 16% of N-galactosaminidase-activity remains [17]) [2, 17]
N,N,N-triacetylchitotriose <2> (<2>, inhibits hydrolysis of *p*-nitrophenyl-β-2-acetamido-2-deoxy-D-galactopyranoside stronger than hydrolysis of *p*-nitrophenyl-β-2-acetamido-2-deoxy-D-glucopyranoside [2, 17]; <2>, with 1 mM 45% of N-glucosaminidase- but only 19% of N-galactosaminidase-activity remains [17]) [2, 17]
N,N-diacetylchitobiose <2, 36, 45> (<2>, inhibits hydrolysis of *p*-nitrophenyl-β-2-acetamido-2-deoxy-D-galactopyranoside stronger than hydrolysis of *p*-nitrophenyl-β-2-acetamido-2-deoxy-D-glucopyranoside [2, 17]; <2>, with 1 mM 54% of N-glucosaminidase- but only 23% of N-galactosaminidase-activity remains [17]; <36>, with 0.1 mM 81% and with 2 mM 32% of activity remains [57]) [2, 17, 57, 65]
N-acetyl-D-galactosamine <1, 3, 4, 9, 12, 16, 17, 18, 20, 21, 22, 25, 27, 29, 30, 32, 45, 54> (<1, 3, 20>, competitive inhibitor [1, 3, 26]; <3>, more potent than N-acetyl-D-glucosamine [3]; <9>, 45% inhibition at 20 mM [10]; <12>, inhibits N-galactosaminidase [13]; <4>, with 30 mM 12.1% activity of liver enzyme type S, 13.6% of brain enzyme type S, 3.2-8.6% of liver enzyme type A and 2.9-9.9% of liver enzyme type B remains [19]; <16>, with 10 mM the hydrolysis of *p*-nitrophenyl-β-2-acetamido-2-deoxy-D-glucopyranoside is reduced to 26% [20]; <17>, 80% inhibition with 100 mM [22]; <21>, with 10 mM 20% of activity remains [29]; <12, 18>, inhibits N-acetylhexosaminidase and -galactosaminidase [30]; <27>, concentration of 10 mM causes 80% inhibition [37]; <29>, with 60 mM 45% of activity remains [43]; <32>, concentration of 10 mM causes 28% inhibition [47]; <25>, with 10 mM 52.6% of activity remains [54]; <27>, PI: concentration of 30 mM causes 69% inhibition, PII: concentration of 30 mM causes 80% inhibition [60]; <4>, 60% inhibition with 1 mM [77]) [1, 3, 10, 13, 19, 20, 21, 22, 26, 29, 30, 37, 43, 46, 47, 51, 54, 60, 65, 66, 73, 77, 83]
N-acetyl-D-glucosamine <1, 2, 3, 4, 12, 16, 17, 18, 20, 21, 22, 25, 27, 29, 30, 32, 36, 44, 45, 53, 54> (<1, 2, 3, 20, 32>, competitive inhibitor [1, 2, 3, 26, 47]; <3>, less potent than N-acetyl-D-galactosamine [3]; <12>, inhibits N-glucosaminidase [13]; <2>, with 10 mM 33-35% of activity remains [17]; <4>, with 30 mM 40.7% activity of liver enzyme type S, 19.9% of brain enzyme type S, 19.9-24.6% of liver enzyme type A and 20.9-30.5% of liver enzyme type B remains [19]; <16>, with 10 mM the hydrolysis of *p*-nitrophenyl-β-2-acetamido-2-deoxy-D-galactopyranoside is reduced to 32% [20]; <17>, 80% inhibition with 100 mM [22]; <21>, with 10 mM 88% of activity remains [29]; <12, 18>, inhibits N-acetylhexosaminidase and -glucosaminidase [30]; <24>, inhibits N-acetylgalactosaminidase activity [34]; <27>, concentration of 10 mM causes 60% inhibition [37]; <29>, with 60 mM 24% of activity remains [43]; <32>, concentration of 10 mM causes 80% inhibition [47]; <4>, with 0.75 M 34% of activity remains [51]; <25>, with 10 mM 64.2% of activity remains [54]; <36>, with 2 mM 85%, with 4 mM 75%, with 8 mM 65% and with 12 mM 59% of activity remains [57]) [1, 2, 3, 13, 17, 19, 20, 21, 22, 26, 29, 30, 34, 37, 43, 46, 47, 51, 54, 57, 64, 65, 66, 73, 82, 83]

N-acetyl-β-D-galactosaminolactone <12, 19> [21, 25]
N-acetyl-β-D-glucosaminolactone <12, 17, 19, 21, 25, 27, 40, 43, 44, 50, 54> (<21>, with 0.002 mM 65% of activity remains [29]; <12>, both isozymes [39]; <25>, with 1 mM 4.7% of activity and with 0.1 mM 6.6% of activity remains [54]; <17>, complete inhibition with 100 mM [22]; <27>, with 10 mM 75% inhibition [37]; <40>, untreated: with 0.02 mM 65% of activity remains, treated with 7-diazonaphtalene-1,3-disulfonic acid: with 0.02 mM 70% of activity remains [63]; <43>, with 0.02 30% of activity remains [63]) [21, 22, 25, 29, 37, 39, 54, 63, 64, 75, 83]
N-acetylglucosamine(1-5)lactone <45> [65]
N-acetylglucosamine-6-sulfate <29> (<29>, with 10 mM 86% of activity remains [43]) [43]
N-acetylglucosamine-isothiocyanate <29> (<29>, with 10 mM 46% of activity remains [43]) [43]
N-acetylleucine <45> [65]
N-acetylmannosamine <29, 45> (<29>, with 10 mM 83% of activity remains [43]) [43, 65]
N-bromosuccinimide <28> (<28>, with 0.025 mM 4.4% of activity remains [38]) [38]
N-ethyl-5-phenyl-isoxazolium-3'-sulfonate <28> (<28>, with 30 mM 23.3% of activity remains [38]) [38]
N-ethylmaleimide <12, 18> (<12,18>, N-acetylglucosaminidase [30]) [30]
NH_4^+ <22> (<22>, 20% inhibition with 2 M, 5% inhibition with 2 M [74]) [74]
NaCl <26> (<26>, with 1 mM 80% of activity remains, with 2 mM 50% of activity remains [36]) [36]
NaOH <40> (<40>, unbroken bacteria: total inhibition with 1 M, extract: with 1 M 9% of activity remains [63]) [63]
O-2-acetamido-2-deoxy-β-D-glucopyranosyl-(1-4)-2-acetamido-3-O-(D-I-carboxyethyl)-2-deoxy-D-glucose <22> (<22>, substrate inhibition above 0.2 mM [73]) [73]
Pb^{2+} <29> [43]
Zn^{2+} <3, 7, 22, 35, 37, 44, 48, 53> (<7>, all 4 isozymes, at 10 mM inhibition ranges from 20% for β-NAHA-IC to 45% for β-NAHA-II [7]; <35>, with 0.03 mM 88.6% of activity remains [56]; <37>, concentration of 1 mM causes 70-80% inhibition [59]; <44>, 100 mM [64]; <48>, minor isozyme: 8% of activity remains with 5 mM, major isozyme: 40% of activity remains with 2 mM [70]; <22>, 2 mM cause 57% inhibition, 10 mM cause 78% inhibition [73]; <53>, 30%, 58% and 73% inhibition with 0.01 mM, 2 mM and 10 mM [82]) [3, 7, 56, 59, 64, 70, 73, 82]
acetamide <1, 21> (<1>, inhibits hydrolysis of *p*-nitrophenyl-β-2-acetamido-2-deoxy-D-galactopyranoside [1]; <21>, with 10 mM 70% of activity remains [29]) [1, 29]
acetate <1, 4, 12, 18, 29, 37, 43> (<1>, inhibits hydrolysis of *p*-nitrophenyl-β-2-acetamido-2-deoxy-D-galactopyranoside [1]; <4>, with 30 mM 39.8% of liver enzyme type S, 71.4% of brain enzyme type S, 33.3-39.4% of liver enzyme type A and 30.6-38.2% of liver enzyme type B remains [19]; <12, 18>, inhibits

N-acetylhexosaminidase [30]; <37>, concentration of 1 mM causes 30% inhibition [59]; <43>, with 50 mM 45% of activity remains [63]; <4>, with 0.02 M 49% of activity remains [72]; <4>, 80% inhibition with 66 mM [77]) [1, 19, 30, 43, 59, 63, 66, 72, 77]
antibodies <23> (<23>, inhibit about 90% of in vitro activity [33]) [33]
ascorbate <4> (<4>, only in presence of Cu^{2+} [72]) [72]
cellobiose <26> [36]
chitobiose <32> (<32>, concentration of 10 mM causes 25% inhibition [47]) [47]
diethyl dicarbonate <28> (<28>, with 30 mM 1.1% of activity remains [38]) [38]
dithioerythritol <12> (<12>, concentration of 2.5 mM decreases activity of the 3 enzyme forms in colonic mucosa by 20% at pH 4.2 and by 25% at pH 7.0, similar results with the 2 forms from colon tumors [21]) [21]
fructose <32> (<32>, concentration of 10 mM causes 38% inhibition [47]) [47]
galactonolactone <18> [66]
galactosamine <3, 4, 12, 18, 25> (<3>, competitive inhibitor, less potent than glucosamine [3]; <4>, with 30 mM 84.3-89.3% of liver enzyme type A and 79.6-87.8% of liver enzyme type B remains [19]; <12,18>, inhibits N-acetylhexosaminidase [30]; <25>, with 10 mM 84.0% of activity remains [54]) [3, 19, 30, 54, 66]
galactose <3, 32> (<3>, very poor [3]; <32>, concentration of 10 mM causes 15% inhibition [47]) [3, 47]
gluconolactone <18> [66]
glucosamine <3, 4, 12, 18, 25> (<3>, competitive inhibitor, more potent than galactosamine [3]; <4>, with 30 mM 72.2-76.3% of liver enzyme type A and 65.7-73.3% of liver enzyme type B remains [19]; <12,18>, inhibits N-acetylhexosaminidase [30]; <25>, with 10 mM 29.3% of activity remains [54]) [3, 19, 30, 54, 66]
glucosamine-HCl <45> [65]
glucose <3, 32> (<3>, very poor [3]; <32>, concentration of 10 mM causes 30% inhibition [47]) [3, 47]
high ionic strength <5> (<5>, above 0.1 no cleavage of GM2 [5]) [5]
iodoacetamide <18> [66]
lactose <26, 32> (<32>, concentration of 10 mM causes 25% inhibition [47]) [36, 47]
maltose <26> [36]
mannosamine <12, 27> (<27>, with 10 mM 80% inhibition [37]) [21, 37]
methyl-α-D-mannoside <27> (<27>, concentration of 10 mM causes 25% inhibition [37]) [37]
monoiodoacetic acid <14> (<14>, with 1 mM 73% of activity remains [16]) [16]
o-phenanthroline <37> (<37>, concentration of 1 mM causes 25% inhibition [59]) [59]
p-aminophenyl 1-thio-β-L-fucopyranoside <46> [67]

p-chloromercuribenzoate <4, 14, 15, 18, 19, 22, 30, 53> (<14>, with 1mM 20% of activity remains [16]; <15>, 0.025 and 0.05 mM cause 70 and 83% inhibition [18]; <19>, inhibits forms II, III and IV, competitive inhibition [24]; <30>, concentration of 2 mM causes 28.5% inhibition [46]; <22>, 0.1 mM cause 16% inhibition [73]; <4>, form A: 93% inhibition with 0.1 mM, form P: 98% inhibition with 0.1 mM [77]; <53>, 43% inhibition with 0.5 mM [82]) [16, 18, 24, 46, 66, 73, 76, 77, 82]
p-chloromercuriphenylsulfonate <4, 9> (<9>, concentration of 10 mM inhibited activity by 46% [10]; <4>, with 1 mM 38.3% of liver enzyme type S, 14.2% of brain enzyme type S, 1.9-2.2% of liver enzyme type A and 2.6-9.9% of liver enzyme type B remains [19]) [10, 19]
p-hydroxymercuribenzoate <12, 18> (<12>, forms A and I retain 50% of glucosaminidase-activity if treated with 0.005 mM at pH 4.2, 5.6 or 7.0 for 15 min, glucosaminidase-activity of form B is stable at pH 4.2 and 7.0 in presence of 0.005 mM and inactivated by less than 5% at pH 5.6, galactosaminidase-activity is decreased by 50% if form A and I was treated with 0.1 mM at pH 4.2, 5.6 or 7.0, galactosaminidase-activity of form B is stable [21]; <12, 18>, inhibits N-acetylhexosaminidase, -galactosaminidase and -glucosaminidase [30]) [21, 30]
p-nitrophenyl 2-deoxy-2-thioacetamido-β-D-galactopyranoside <37> (<37>, competitive inhibition [59]) [59]
p-nitrophenyl 2-deoxy-2-thioacetamido-β-D-glucopyranoside <37> (<37>, competitive inhibition [59]) [59]
p-nitrophenyl-β-2-acetamido-2-deoxy-D-glucopyranoside <36> [57]
p-nitrophenyl-β-2-acetamido-2-deoxy-D-glucopyranoside <50> (<50>, inhibition by excess of substrate [75]) [75]
p-nitrophenyl-β-2-acetamido-2-deoxy-D-glucopyranoside <7, 10, 34> (<7>, only β-NAHA-IA, above 3.5 mM [7]; <10>, at higher concentrations [11]) [7, 11, 50]
p-thiophenyl-1-thio-N-acetyl-β-D-glucosaminide <44> [64]
phenyl N-acetylgalactosaminide <54> [83]
phenyl N-acetylglucosaminide <54> [83]
sodium acetate <2, 17, 18, 21, 30> (<2>, inhibits only N-acetylglucosaminidase-activity, concentration of 10 mM causes 30% inhibition [17]; <17>, concentration of 10 mM results in 20% inhibition [22]; <18>, with 50 mM 27% inhibition [23]; <21>, with 10 mM 50% of activity remains [29]; <30>, concentration of 60 mM causes 80% inhibition [46]) [17, 22, 23, 29, 46]
sodium taurodeoxycholate <13> [14]
sucrose <32> (<32>, concentration of 10 mM causes 9% inhibition [47]) [47]
thimerosal <12> (<12>, inhibits activity in mucosal and tumor enzymes, 50% of activity remains after 1 h if form A is treated with 0.20 mM, form B with 1.75 mM and form I with 0.65 mM [21]) [21]
thiophenyl-N-acetyl-β-D-glucosaminide <34> [50]
urea <16, 22> (<16>, with 2 mM hydrolysis of *p*-nitrophenyl-β-2-acetamido-2-deoxy-D-glucopyranoside is reduced to 37%, hydrolysis of *p*-nitrophenyl-β-2-acetamido-2-deoxy-D-galactopyranoside to 34% [20]; <22>, 19%, 33% and 40% inhibition with 4, 6 and 8 M [74]) [20, 74]

Additional information <2, 26, 27, 35> (<2>, not: N-acetyl-D-galactosamine [2]; <35>, high substrate concentrations, i.e. concentrations above 0.2 mM [56]) [2, 36, 37, 56]

Activating compounds

2-mercaptoethanol <7> (<7>, blocks inhibition by Hg^{2+}-ions [7]) [7]
D-galactosamine <4, 32> (<4>, with 30 mM the activity of liver and brain enzymes reaches 135% and 114.3% [19]; <32>, concentration of 100 mM: activity reaches 140% [47]) [19, 47]
D-glucosamine <4, 32> (<4>, with 30 mM the activity of liver and brain enzymes reaches 135% and 110.7% [19]; <32>, concentration of 30 mM: activity reaches 140% [47]) [19, 47]
EDTA <22, 35> (<35>, 0.0075 mM: activity reaches 110.0% [56]; <22>, 1 and 20 mM cause 10% stimulation [73]) [56, 73]
L-cysteine <12, 19> (<12>, presence decreases extent of inactivation by Hg^{2+}, Ag^{2+}, thimerosal or *p*-hydroxymercuribenzoate [21]; <19>, concentrations of 0.05, 0.1, 0.4 and 0.7 mM are needed to reactivate both hexosaminidase activities of forms I, II, III and IV [24]) [21, 24]
N-ethylmaleimide <12, 18> (<12,18>, N-acetylgalactosaminidase [30]) [30]
Triton X-100 <54> (<54>, prevents loss of activity caused by dilution [83]) [83]
acetate <40> (<40>, untreated, concentration of 50 mM: activity reaches 170% [63]) [63]
bovine serum albumin <4, 12, 18, 19, 54> (<12>, N-acetylgalactosaminidase, stimulation up to 4fold, stimulation of GM2-hydrolysis, effect on N-acetylglucosaminidase only slight [13]; <19>, at 5% both hexosaminidase activities of forms II and I are increased 2- and 1.5fold [24]; <54>, prevents loss of activity caused by dilution [83]) [13, 24, 72, 76, 83]
cysteine <4, 7, 18> (<4,7>, prevents inhibition by Hg^{2+}-ions [7,66,72]; <4>, prevents inhibition by Ag^{2+}-ions [72]) [7, 66, 72]
dithioerythritol <12> (<12>, presence decreases extent of inactivation by Hg^{2+}, Ag^{2+}, thimerosal or *p*-hydroxymercuribenzoate [21]) [21]
dithiothreitol <12> (<12>, stimulates N-acetylglucosaminidase even at low concentrations [13]) [13]
glutathione <12, 15> (<15>, 1 mM leads to moderate stimulation of *p*-chloromercuribenzoate-treated enzyme, no effect alone [18]; <12>, presence decreases extent of inactivation by Hg^{2+}, Ag^{2+}, thimerosal or *p*-hydroxymercuribenzoate [21]) [18, 21]
glycopeptide <47> (<47> dimeric [69]) [69]
human serum albumin <4> [72]
low molecular weight protein <5> (<5>, restores the enzyme's ability to hydrolyze GM2 in high ionic strength [5]) [5]
neutral N-acetylglucosaminidase <12> (<12>, stimulates GM2-hydrolysis by N-acetylgalactosaminidase [13]) [13]
p-nitrophenyl-N-acetylgalactosaminide <12> (<12>, presence at a concentration 10 times that of *p*-hydroxymercuribenzoate lowers loss of galactosaminidase-activity by 25.5% [21]) [21]

p-nitrophenyl-N-acetylglucosaminide <12> (<12>, presence at a concentration 50 times that of Hg^{2+}, Ag^{2+}, thimerosal or *p*-hydroxymercuribenzoate decreases loss of glucosaminidase-activity by 20.5% [21]) [21]
taurocholate <4> (<4>, concentrations of 0.05 M and 0.1 M cause 140% and 160% of activity [51]) [51]

Metals, ions

Ca^{2+} <7, 32> (<7>, β-NAHA-II [7]; <32>, concentration of 4.0 mM: activity reaches 171% [47]) [7, 47]
Fe^{2+} <19, 32> (<19>, stimulates both hexosaminidase activities of form I [24]; <32>, concentration of 0.6 mM: activity reaches 204% [47]) [24, 47]
Fe^{3+} <19> (<19>, stimulates galactosaminidase activitiy of form I [24]) [24]
Mg^{2+} <7, 23> (<7>, β-NAHA-II [7]) [7, 33]
Mn^{2+} <7, 23, 29> (<7>, β-NAHA-II [7]) [7, 33, 43]
NaCl <30, 31> (<30>, concentration of 0.2-1.5 M activates hydrolysis of *p*-nitrophenyl-β-2-acetamido-2-deoxy-D-glucopyranoside and -galactosaminide about 40% and 80% [46]; <31>, 0.3-1.0 M causes activation of about 140% [46]) [46]
Ni^{2+} <29> [43]
Zn^{2+} <32> (<32>, concentration of 0.4 mM: activity reaches 142% [47]) [47]

Turnover number (min^{-1})

1 <44> (*p*-nitrophenyl-β-2-acetamido-2-deoxy-D-glucopyranoside, <44>, in D_2O, pD 7.0 [64]) [64]
1.27 <44> (*p*-nitrophenyl-β-2-acetamido-2-deoxy-D-glucopyranoside, <44>, in D_2O, pD 4.6 [64]) [64]
9.8 <12> (*p*-nitrophenyl-β-2-acetamido-2-deoxy-D-galactopyranoside, <12>, caput, Hex B [39]) [39]
12.5 <12> (*p*-nitrophenyl-β-2-acetamido-2-deoxy-D-galactopyranoside, <12>, caput, Hex A [39]) [39]
24.5 <12> (*p*-nitrophenyl-β-2-acetamido-2-deoxy-D-glucopyranoside, <12>, caput, Hex B [39]) [39]
27 <12> (*p*-nitrophenyl-β-2-acetamido-2-deoxy-D-glucopyranoside, <12>, caput, Hex A [39]) [39]
66.9 <12> (*p*-nitrophenyl-β-2-acetamido-2-deoxy-D-galactopyranoside, <12>, caudal, Hex B [39]) [39]
74.6 <12> (*p*-nitrophenyl-β-2-acetamido-2-deoxy-D-galactopyranoside, <12>, caudal, Hex A [39]) [39]
82.9 <12> (*p*-nitrophenyl-β-2-acetamido-2-deoxy-D-galactopyranoside, <12>, corpus, Hex B [39]) [39]
94.6 <12> (*p*-nitrophenyl-β-2-acetamido-2-deoxy-D-galactopyranoside, <12>, corpus, Hex A [39]) [39]
97.9 <12> (*p*-nitrophenyl-β-2-acetamido-2-deoxy-D-glucopyranoside, <12>, caudal, Hex B [39]) [39]
115.6 <12> (*p*-nitrophenyl-β-2-acetamido-2-deoxy-D-glucopyranoside, <12>, caudal, Hex A [39]) [39]
125.1 <12> (*p*-nitrophenyl-β-2-acetamido-2-deoxy-D-glucopyranoside, <12>, corpus, Hex B [39]) [39]

159.7 <12> (*p*-nitrophenyl-*β*-2-acetamido-2-deoxy-D-glucopyranoside, <12>, corpus, Hex A [39]) [39]
168 <45> (*p*-nitrophenyl-*β*-2-acetamido-2-deoxy-D-galactopyranoside) [65]
170 <44> (*p*-nitrophenyl-thio-*β*-N-acetyl-D-glucosaminide) [64]
498 <33> ((N-acetyl-D-glucosamine)$_6$) [48]
882 <33> ((N-acetyl-D-glucosamine)$_5$) [48]
1710 <45> (N,N-diacetylchitobiose) [65]
1740 <35> ((N-acetyl-D-glucosamine)$_6$, <35>, Ex 2 [55]) [55]
1806 <33> ((N-acetyl-D-glucosamine)$_4$) [48]
1812 <45> (*p*-nitrophenyl-*β*-2-acetamido-2-deoxy-D-glucopyranoside) [65]
2300 <44> (2,4-dinitrophenyl-N-acetyl-*β*-D-glucosaminide) [64]
2520 <35> ((N-acetyl-D-glucosamine)$_4$, <35>, Ex 2 [55]) [55]
3114 <35> ((N-acetyl-D-glucosamine)$_6$) [56]
3300 <35> ((N-acetyl-D-glucosamine)$_6$, <35>, Ex 1 [55]) [55]
3420 <35> ((N-acetyl-D-glucosamine)$_5$, <35>, Ex 1 [55]) [55]
4536 <35> ((N-acetyl-D-glucosamine)$_5$) [56]
4560 <35> ((N-acetyl-D-glucosamine)$_5$, <35>, Ex 2 [55]) [55]
4920 <35> ((N-acetyl-D-glucosamine)$_3$, <35>, Ex 2 [55]) [55]
5784 <33> ((N-acetyl-D-glucosamine)$_3$) [48]
6600 <35> ((N-acetyl-D-glucosamine)$_4$, <35>, Ex 1 [55]) [55]
6620 <44> (*p*-nitrophenyl-*β*-2-acetamido-2-deoxy-D-galactopyranoside) [64]
7746 <35> ((N-acetyl-D-glucosamine)$_4$) [56]
7800 <35> ((N-acetyl-D-glucosamine)$_3$, <35>, Ex 1 [55]) [55]
11200 <44> (*p*-nitrophenyl-*β*-2-acetamido-2-deoxy-D-glucopyranoside) [64]
12560 <35> ((N-acetyl-D-glucosamine)$_3$) [56]
17880 <33> ((N-acetyl-D-glucosamine)$_2$) [48]
19200 <35> (*p*-nitrophenyl-*β*-2-acetamido-2-deoxy-D-galactopyranoside, <35>, Ex 2 [55]) [55]
34200 <35> (*p*-nitrophenyl-*β*-2-acetamido-2-deoxy-D-galactopyranoside, <35>, Ex 1 [55]) [55]
34800 <35> ((N-acetyl-D-glucosamine)$_2$, <35>, Ex 2 [55]) [55]
35880 <33> (*p*-nitrophenyl-*β*-2-acetamido-2-deoxy-D-glucopyranoside) [48]
37200 <35> ((N-acetyl-D-glucosamine)$_2$, <35>, Ex 1 [55]) [55]
43200 <35> (*p*-nitrophenyl-*β*-2-acetamido-2-deoxy-D-glucopyranoside, <35>, Ex 2 [55]) [55]
49210 <35> (*p*-nitrophenyl-*β*-2-acetamido-2-deoxy-D-glucopyranoside) [56]
53400 <35> (*p*-nitrophenyl-*β*-2-acetamido-2-deoxy-D-glucopyranoside, <35>, Ex 1 [55]) [55]
60800 <35> ((N-acetyl-D-glucosamine)$_2$) [56]
68910 <35> (*p*-nitrophenyl-*β*-2-acetamido-2-deoxy-D-galactopyranoside) [56]
Additional information <44> [64]

Specific activity (U/mg)

8.5e-006 <24> (<24>, substrate: pyridylamino-glycoside A [34]) [34]
0.0012 <22> (<22>, substrate: dansyl-ovomucoid, Endo-BI [32]) [32]
0.0012-0.0025 <4> (<4>, substrate: 4-methylumbelliferyl-2-acetamido-2-deoxy-*β*-D-glucopyranoside-6-sulfate [51]) [51]

0.0014 <24> (<24>, substrate: pyridylamino-glycoside C [34]) [34]
0.0024-0.01 <4> (<4>, light lysosomal fraction, I-cell [45]) [45]
0.0032-0.008 <4> (<4>, light lysosomal fraction, normal [45]) [45]
0.0033-0.0066 <4> (<4>, light lysosomal fraction, Pseudo-Hurler [45]) [45]
0.0051-0.0079 <4> (<4>, heavy lysosomal fraction, I-cell [45]) [45]
0.0061-0.0075 <4> (<4>, heavy lysosomal fraction, Pseudo-Hurler [45]) [45]
0.0061-0.0211 <4> (<4>, heavy lysosomal fraction, normal [45]) [45]
0.0216 <22> (<22>, substrate: dansyl-ovomucoid, Endo-BI [32]) [32]
0.03 <12, 18> (<12,18>, N-acetylhexosaminidase, substrate: *p*-nitrophenyl-β-2-acetamido-2-deoxy-D-galactopyranoside [30]) [30]
0.067-0.117 <46> (<46>, substrate: *p*-nitrophenyl-β-N-acetyl-D-glucosaminide [67]) [67]
0.123 <4> (<4>, substrate: 4-methylumbelliferyl-β-N-acetylglucosaminide, brain enzyme type S [19]) [19]
0.22 <12, 18> (<12, 18>, N-acetylhexosaminidase, substrate: *p*-nitrophenyl-β-2-acetamido-2-deoxy-D-glucopyranoside [30]) [30]
0.447 <4> (<4>, substrate: 4-methylumbelliferyl-β-N-acetylglucosaminide, liver enzyme type S [19]) [19]
0.63 <18> [66]
0.75 <12> (<12>, substrate: *p*-nitrophenyl-β-2-acetamido-2-deoxy-D-glucopyranoside, colonic mucosa, form I [21]) [21]
0.88 <12> (<12>, substrate: *p*-nitrophenyl-β-2-acetamido-2-deoxy-D-glucopyranoside, colonic mucosa, form A [21]) [21]
0.91 <24> (<24>, substrate: pyridylamino-glycoside B [34]) [34]
0.926 <16> (<16>, substrate: *p*-nitrophenyl-β-2-acetamido-2-deoxy-D-glucopyranoside [20]) [20]
0.94 <12> (<12>, substrate: *p*-nitrophenyl-β-2-acetamido-2-deoxy-D-glucopyranoside, colonic mucosa, form B [21]) [21]
0.97 <25> (<25>, pool III [79]) [79]
1 <25> (<25>, pool I [79]) [79]
1.05 <25> (<25>, pool II [79]) [79]
1.4 <48> (<48>, vegetative enzyme [70]) [70]
1.43 <18> (<18>, substrate: *p*-nitrophenyl-β-2-acetamido-2-deoxy-D-glucopyranoside, form A [58]) [58]
1.78 <12> (<12>, substrate: *p*-nitrophenyl-β-2-acetamido-2-deoxy-D-glucopyranoside, colonic tumor, form A [21]) [21]
1.97 <12> (<12>, substrate: *p*-nitrophenyl-β-2-acetamido-2-deoxy-D-glucopyranoside, colonic tumor, form B [21]) [21]
2.08 <7, 18> (<7>, β-NAHA-IC, substrate: *p*-nitrophenyl-β-2-acetamido-2-deoxy-D-glucopyranoside [7]; <18>, substrate: *p*-nitrophenyl-β-2-acetamido-2-deoxy-D-glucopyranoside, form B [58]) [7, 58]
2.4 <12> (<12>, Hex B [39]) [39]
2.64 <19> (<19>, form II [24]) [24]
2.8 <12> (<12>, Hex A [39]) [39]
2.82 <19> (<19>, form I [24]) [24]
3.04 <53> (<53>, substrate: *p*-nitrophenyl-β-2-acetamido-2-deoxy-D-glucopyranoside [82]) [82]

3.17 <7> (<7>, β-NAHA-IA, substrate: *p*-nitrophenyl-β-2-acetamido-2-deoxy-D-glucopyranoside [7]) [7]

5.1 <27> (<27>, N-acetylgalactosaminidase activity [37]) [37]

5.64 <16> (<16>, substrate: *p*-nitrophenyl-β-2-acetamido-2-deoxy-D-galactopyranoside [20]) [20]

6.35 <19> (<19>, form IV [24]) [24]

7.25 <7> (<7>, β-NAHA-IB, substrate: *p*-nitrophenyl-β-2-acetamido-2-deoxy-D-glucopyranoside [7]) [7]

7.8 <17> (<17>, substrate: *p*-nitrophenyl-β-2-acetamido-2-deoxy-D-galactopyranoside [22]) [22]

9.737 <28> (<28>, substrate: *p*-nitrophenyl-β-2-acetamido-2-deoxy-D-glucopyranoside [38]) [38]

11.7 <32> (<32>, substrate: *p*-nitrophenyl-β-2-acetamido-2-deoxy-D-glucopyranoside [47]) [47]

12 <25> (<25>, substrate: *p*-nitrophenyl-β-2-acetamido-2-deoxy-D-glucopyranoside [54]) [54]

13.4 <7> (<7>, β-NAHA-II, substrate: *p*-nitrophenyl-β-2-acetamido-2-deoxy-D-glucopyranoside [7]) [7]

14.2 <8> (<8>, low molecular weight form EII, substrate: *p*-nitrophenyl-β-2-acetamido-2-deoxy-D-galactopyranoside [8]) [8]

15.85 <45> (<45>, substrate: N,N-diacetylchitobiose [65]) [65]

18.1 <18> (<18>, substrate: *p*-nitrophenyl-β-2-acetamido-2-deoxy-D-glucopyranoside, form B [76]) [76]

21.1 <18> (<18>, substrate: *p*-nitrophenyl-β-2-acetamido-2-deoxy-D-glucopyranoside, form A [76]) [76]

21.3 <5> [5]

23.3 <22> (<22>, substrate: *p*-nirophenyl-2-acetamido-2-deoxyglucose [74]) [74]

25.68 <19> (<19>, form III [24]) [24]

33 <17> (<17>, substrate: *p*-nitrophenyl-β-2-acetamido-2-deoxy-D-glucopyranoside [22]) [22]

35.2 <8> (<8>, low molecular weight form EII, substrate: *p*-nitrophenyl-β-2-acetamido-2-deoxy-D-glucopyranoside [8]) [8]

42.5 <30> (<30>, substrate: *p*-nitrophenyl-β-2-acetamido-2-deoxy-D-glucopyranoside [46]) [46]

46.7 <27> (<27>, N-acetylglucosaminidase activity [37]) [37]

53 <20> (<20>, substrate: *p*-nitrophenyl-β-2-acetamido-2-deoxy-D-glucopyranoside [26]) [26]

56 <15> (<15>, substrate: *p*-nitrophenyl-β-2-acetamido-2-deoxy-D-glucopyranoside [18]) [18]

60.6 <35> (<35>, substrate: *p*-nitrophenyl-β-2-acetamido-2-deoxy-D-glucopyranoside [56]) [56]

62 <11> (<11>, substrate: *p*-nitrophenyl-β-2-acetamido-2-deoxy-D-glucopyranoside, microsomal enzyme [12]) [12]

62.9 <38> (<38>, substrate: phenyl-N-acetyl-β-D-glucosaminide [61]) [61]

67 <2, 14> (<2, 14>, substrate: *p*-nitrophenyl-β-2-acetamido-2-deoxy-D-glucopyranoside [2, 16, 17]) [2, 16, 17]
68.4 <11> (<11>, substrate: *p*-nitrophenyl-β-2-acetamido-2-deoxy-D-glucopyranoside, egg white-enzyme [12]) [12]
77.84 <4> (<4>, substrate: 4-methylumbelliferyl-2-acetamido-2-deoxy-β-D-glucopyranoside, normal enzyme [15]) [15]
84.7 <11> (<11>, substrate: *p*-nitrophenyl-β-2-acetamido-2-deoxy-D-glucopyranoside, lysosomal enzyme [12]) [12]
90 <37> (<37>, substrate: phenyl-N-acetyl-β-D-glucosaminide [59]) [59]
90.63 <4> (<4>, substrate: 4-methylumbelliferyl-2-acetamido-2-deoxy-β-D-glucopyranoside, I-cell enzyme [15]) [15]
92 <6> [6]
101.9 <21> (<21>, substrate: *p*-nitrophenyl-2-acetamido-2-deoxy-β-D-glucopyranoside [29]) [29]
102 <24> [34]
126 <10> (<10>, substrate: *p*-nitrophenyl-β-2-acetamido-2-deoxy-D-glucopyranoside [11]) [11]
130 <4> (<4>, hexosaminidase I2, substrate: 4-methylumbelliferyl-2-acetamido-2-deoxy-β-D-glucopyranoside [9]) [9]
150 <4> (<4>, hexosaminidase A, substrate: 4-methylumbelliferyl-2-acetamido-2-deoxy-β-D-glucopyranoside [9]) [9]
162.3 <47> (<47>, substrate: *p*-nitrophenyl-β-2-acetamido-2-deoxy-D-glucopyranoside [69]) [69]
184.2 <9> (<9>, substrate: *p*-nitrophenyl-β-2-acetamido-2-deoxy-D-glucopyranoside [10]) [10]
186 <10> (<10>, substrate: *p*-nitrophenyl-β-2-acetamido-2-deoxy-D-galactopyranoside [11]) [11]
200 <26> (<26>, substrate: *p*-nitrophenyl-β-2-acetamido-2-deoxy-D-glucopyranoside [36]) [36]
214.1 <13> [14]
262 <48> (<48>, stage-specific enzyme [70]) [70]
310 <52> (<52>, periplasmic enzyme [81]) [81]
320 <4> (<4>, hexosaminidase B, substrate: 4-methylumbelliferyl-2-acetamido-2-deoxy-β-D-glucopyranoside [9]) [9]
362.9 <35> [68]
817.8 <29> (<29>, substrate: *p*-nitrophenyl-β-2-acetamido-2-deoxy-D-glucopyranoside [43]) [43]
4100 <52> (<52>, extracellular enzyme [81]) [81]
77290 <23> (<23>, substrate: *p*-nitrophenyl-β-2-acetamido-2-deoxy-D-glucopyranoside [33]) [33]
123000 <33> (<33>, substrate: *p*-nitrophenyl-β-2-acetamido-2-deoxy-D-glucopyranoside [48]) [48]
Additional information <1, 3, 4, 19, 34, 39, 40, 41, 42, 43, 44, 45, 50, 51> [1, 3, 25, 40, 49, 62, 63, 64, 65, 75, 77, 78]

K_m-Value (mM)

0.018 <22> (O-2-acetamido-2-deoxy-β-D-glucopyranosyl-(1-4)-2-acetamido-3-O-(D-carboxyethyl)-2-deoxy-D-glucose) [73]
0.02 <25> (4-methylumbelliferyl-β-N-acetylglucosaminide, <25>, E314Q [35]) [35]
0.03 <35> ((N-acetyl-D-glucosamine)$_4$, <35>, Ex 2 [55]) [55]
0.04 <19> (*p*-nitrophenyl-β-2-acetamido-2-deoxy-D-galactopyranoside, <19>, form IV [24,25]) [24, 25]
0.04 <19> (*p*-nitrophenyl-β-2-acetamido-2-deoxy-D-glucopyranoside, <19>, form I [24,25]) [24, 25]
0.044 <33> ((N-acetyl-D-glucosamine)$_4$) [48]
0.045 <33> ((N-acetyl-D-glucosamine)$_5$) [48]
0.047 <33> ((N-acetyl-D-glucosamine)$_6$) [48]
0.05 <35> ((N-acetyl-D-glucosamine)$_3$, <35>, Ex 2 [55]) [55]
0.055 <33> ((N-acetyl-D-glucosamine)$_3$) [48]
0.056 <35> ((N-acetyl-D-glucosamine)$_6$) [56]
0.06 <35> ((N-acetyl-D-glucosamine)$_6$, <35>, Ex 2 [55]) [55]
0.064 <12> (*p*-nitrophenyl-β-2-acetamido-2-deoxy-D-galactopyranoside, <12>, colonic tumor, form B [21]) [21]
0.07 <35> ((N-acetyl-D-glucosamine)$_5$, <35>, Ex 1 [55]) [55]
0.08 <18> (*p*-nitrophenyl-β-2-acetamido-2-deoxy-D-galactopyranoside, <18>, form B, pH 4.55 [76]) [76]
0.08 <19> (*p*-nitrophenyl-β-2-acetamido-2-deoxy-D-glucopyranoside, <19>, form III [25]) [25]
0.081 <29> (*p*-nitrophenyl-β-2-acetamido-2-deoxy-D-glucopyranoside) [43]
0.087 <19> (*p*-nitrophenyl-β-2-acetamido-2-deoxy-D-glucopyranoside, <19>, form III [24]) [24]
0.09 <35> ((N-acetyl-D-glucosamine)$_5$) [56]
0.09 <35> ((N-acetyl-D-glucosamine)$_5$, <35>, Ex 2 [55]) [55]
0.09 <19> (*p*-nitrophenyl-β-2-acetamido-2-deoxy-D-galactopyranoside, <19>, form III [24, 25]) [24, 25]
0.09 <26> (*p*-nitrophenyl-β-2-acetamido-2-deoxy-D-glucopyranoside) [36]
0.095 <12> (*p*-nitrophenyl-β-2-acetamido-2-deoxy-D-galactopyranoside, <12>, colonic mucosa, form B [21]) [21]
0.1 <35> ((N-acetyl-D-glucosamine)$_6$, <35>, Ex 1 [55]) [55]
0.108 <4> (4-methylumbelliferyl-2-acetamido-2-deoxy-β-D-galactopyranoside) [72]
0.11 <22> (4-methylumbelliferyl-2-acetamido-2-deoxy-β-D-glucopyranoside) [73]
0.117 <33> ((N-acetyl-D-glucosamine)$_2$) [48]
0.118 <35> ((N-acetyl-D-glucosamine)$_4$) [56]
0.122 <35> ((N-acetyl-D-glucosamine)$_3$) [56]
0.13 <32> (*p*-nitrophenyl-acetylchitobioside) [47]
0.137 <33> (*p*-nitrophenyl-β-2-acetamido-2-deoxy-D-glucopyranoside) [48]
0.14 <25> (4-methylumbelliferyl-β-N-acetylglucosaminide, <25>, wild type [35]) [35]

0.15 <18> (*p*-nitrophenyl-β-2-acetamido-2-deoxy-D-galactopyranoside, <18>, form A, pH 4.55 [76]) [76]
0.15 <19, 30> (*p*-nitrophenyl-β-2-acetamido-2-deoxy-D-galactopyranoside, <19>, form I [24,25]) [24, 25, 46]
0.15 <22> (*p*-nitrophenyl-β-2-acetamido-2-deoxy-D-glucopyranoside) [73]
0.16 <35> ((N-acetyl-D-glucosamine)$_4$, <35>, Ex 1 [55]) [55]
0.17 <32> (*p*-nitrophenyl-acetylchitotrioside) [47]
0.17 <4> (*p*-nitrophenyl-β-2-acetamido-2-deoxy-D-galactopyranoside) [72]
0.19 <17> (*p*-nitrophenyl-β-2-acetamido-2-deoxy-D-galactopyranoside) [22]
0.19 <53> (*p*-nitrophenyl-β-2-acetamido-2-deoxy-D-galactopyranoside) [82]
0.2 <12, 20, 27> (*p*-nitrophenyl-β-2-acetamido-2-deoxy-D-galactopyranoside, <12>, colonic mucosa, form I [21]) [21, 26, 37]
0.23 <25> ((N-acetyl-D-glucosamine)$_3$) [79]
0.24 <24, 30> (*p*-nitrophenyl-β-2-acetamido-2-deoxy-D-glucopyranoside) [34, 46]
0.25 <25> (4-methylumbelliferyl-β-N-acetylglucosaminide, <25>, D246N [35]) [35]
0.25 <20> (*p*-nitrophenyl-2-acetamido-2-deoxy-β-D-glucopyranoside) [26]
0.25 <25, 51> (*p*-nitrophenyl-β-2-acetamido-2-deoxy-D-glucopyranoside) [54, 78]
0.26 <12> (*p*-nitrophenyl-β-2-acetamido-2-deoxy-D-glucopyranoside, <12>, colonic tumor, form B [21]) [21]
0.27 <44> (2,4-dinitrophenyl-N-acetyl-β-D-glucosaminide) [64]
0.27 <18> (*p*-nitrophenyl-β-2-acetamido-2-deoxy-D-galactopyranoside) [66]
0.27 <1, 7> (*p*-nitrophenyl-β-2-acetamido-2-deoxy-D-galactopyranoside, <7>, β-NAHA-IA [7]) [1, 7]
0.28 <54> (*p*-nitrophenyl-β-2-acetamido-2-deoxy-D-galactopyranoside) [83]
0.28 <8, 32> (*p*-nitrophenyl-β-2-acetamido-2-deoxy-D-glucopyranoside, <8>, low molecular weight form EII [8]) [8, 47]
0.29 <1> (*p*-nitrophenyl-β-2-acetamido-2-deoxy-D-glucopyranoside) [1]
0.3 <36> (*p*-nitrophenyl-β-2-acetamido-2-deoxy-D-glucopyranoside) [57]
0.3-0.4 <4> (4-methylumbelliferyl-β-N-acetylgalactosaminide, <4>, liver and brain enzymes type S [19]) [19]
0.31 <22> (N,N-diacetylchitobiose) [73]
0.31 <21, 35> (*p*-nitrophenyl-β-2-acetamido-2-deoxy-D-galactopyranoside, <35>, Ex 2 [55]) [29, 55]
0.31 <12, 46, 53> (*p*-nitrophenyl-β-2-acetamido-2-deoxy-D-glucopyranoside, <12>, colonic mucosa, form B [21]; <46>, passing through enzyme [67]) [21, 67, 82]
0.32 <18> (*p*-nitrophenyl-β-2-acetamido-2-deoxy-D-glucopyranoside, <18>, form B, pH 5.5 [76]) [76]
0.322 <16> (*p*-nitrophenyl-β-2-acetamido-2-deoxy-D-galactopyranoside) [20]
0.33 <36> (N,N-diacetylchitobiose) [57]
0.33 <12, 26> (*p*-nitrophenyl-β-2-acetamido-2-deoxy-D-galactopyranoside, <12>, colonic mucosa: form A; colonic tumor: form A [21]) [21, 36]
0.34 <7> (*p*-nitrophenyl-β-2-acetamido-2-deoxy-D-galactopyranoside, <7>, β-NAHA-IB [7]) [7]

0.34 <44> (*p*-nitrophenyl-*β*-2-acetamido-2-deoxy-D-glucopyranoside) [64]
0.35 <22> (N,N,N,N,N-diacetylchitopentaose) [73]
0.35 <12, 18> (*p*-nitrophenyl-*β*-2-acetamido-2-deoxy-D-galactopyranoside, <12,18>, N-acetylgalactosaminidase [30]) [30]
0.35 <45> (*p*-nitrophenyl-*β*-2-acetamido-2-deoxy-D-glucopyranoside) [65]
0.37 <19> (*p*-nitrophenyl-*β*-2-acetamido-2-deoxy-D-glucopyranoside, <19>, form IV [24,25]) [24, 25]
0.38 <22> (N,N,N,N-diacetylchitotetraose) [73]
0.38 <18> (*p*-nitrophenyl-*β*-2-acetamido-2-deoxy-D-glucopyranoside, <18>, form A, pH 5.5 [76]) [76]
0.39 <22> (N,N,N-diacetylchitotriose) [73]
0.397 <35> (*p*-nitrophenyl-*β*-2-acetamido-2-deoxy-D-glucopyranoside) [56]
0.42 <12> (4-methylumbelliferyl-*β*-N-acetylgalactosaminide, <12>, N-acetylgalactosaminidase [13]) [13]
0.42 <35> (*p*-nitrophenyl-*β*-2-acetamido-2-deoxy-D-glucopyranoside, <35>, Ex 2 [55]) [55]
0.43 <17, 47> (*p*-nitrophenyl-*β*-2-acetamido-2-deoxy-D-glucopyranoside) [22, 69]
0.433 <35> (*p*-nitrophenyl-*β*-2-acetamido-2-deoxy-D-galactopyranoside) [56]
0.45 <12> (4-methylumbelliferyl-*β*-N-acetylgalactosaminide, <12>, N-acetylglucosaminidase [13]) [13]
0.45 <2, 46> (*p*-nitrophenyl-*β*-2-acetamido-2-deoxy-D-glucopyranoside, <46>, adherent enzyme [67]) [2, 17, 67]
0.46 <8> (*p*-nitrophenyl-*β*-2-acetamido-2-deoxy-D-galactopyranoside, <8>, low molecular weight form EII [8]) [8]
0.47 <3> (*p*-nitrophenyl-*β*-2-acetamido-2-deoxy-D-galactopyranoside) [3]
0.48 <3, 10> (*p*-nitrophenyl-*β*-2-acetamido-2-deoxy-D-glucopyranoside) [3, 11]
0.49 <7, 9, 35> (*p*-nitrophenyl-*β*-2-acetamido-2-deoxy-D-glucopyranoside, <7>, *β*-NAHA-IA [7]; <35>, Ex 1 [55]) [7, 10, 55]
0.51 <34> (*p*-nitrophenyl-2-acetamido-2-deoxy-*β*-D-galactopyranoside) [49]
0.51 <7> (*p*-nitrophenyl-*β*-2-acetamido-2-deoxy-D-glucopyranoside, <7>, *β*-NAHA-IB [7]) [7]
0.52 <50> (N,N-diacetylchitobiose) [75]
0.52 <18> (*p*-nitrophenyl-*β*-2-acetamido-2-deoxy-D-glucopyranoside) [66]
0.527 <38> (phenyl-N-acetyl-*β*-D-glucosaminide) [61]
0.53 <12> (*p*-nitrophenyl-*β*-2-acetamido-2-deoxy-D-glucopyranoside, <12>, colonic mucosa, form I [21]) [21]
0.54 <2, 12, 18> (*p*-nitrophenyl-*β*-2-acetamido-2-deoxy-D-galactopyranoside, <12, 18>, N-acetylhexosaminidase [30]) [2, 17, 30]
0.55 <4> (4-methylumbelliferyl-2-acetamido-2-deoxy-*β*-D-glucopyranoside, <4>, form P [77]) [77]
0.55 <19> (*p*-nitrophenyl-*β*-2-acetamido-2-deoxy-D-galactopyranoside, <19>, form II [24, 25]) [24, 25]
0.555 <4> (4-methylumbelliferyl-2-acetamido-2-deoxy-*β*-D-glucopyranoside, <4>, with 0.01% bovine or human serum albumin [72]) [72]
0.57 <32> (*p*-nitrophenyl-*β*-2-acetamido-2-deoxy-D-galactopyranoside) [47]

0.58 <4> (4-methylumbelliferyl-2-acetamido-2-deoxy-β-D-glucopyranoside) [72]
0.58 <7, 34> (*p*-nitrophenyl-β-2-acetamido-2-deoxy-D-galactopyranoside, <7>, β-NAHA-II [7]) [7, 50]
0.59 <32> (*p*-nitrophenyl-thio-β-N-acetyl-D-glucosaminide) [47]
0.6 <12> (*p*-nitrophenyl-β-2-acetamido-2-deoxy-D-glucopyranoside, <12>, caput, Hex A [39]) [39]
0.61 <12> (*p*-nitrophenyl-β-2-acetamido-2-deoxy-D-galactopyranoside, <12>, caput, Hex A [39]) [39]
0.61 <7, 12> (*p*-nitrophenyl-β-2-acetamido-2-deoxy-D-glucopyranoside, <12>, caput, Hex B [39]; <7>, β-NAHA-II [7]) [7, 39]
0.617 <38> (*p*-nitrophenyl-β-2-acetamido-2-deoxy-D-galactopyranoside) [61]
0.62 <19> (*p*-nitrophenyl-β-2-acetamido-2-deoxy-D-glucopyranoside, <19>, form II [24, 25]) [24, 25]
0.625 <16> (*p*-nitrophenyl-β-2-acetamido-2-deoxy-D-glucopyranoside) [20]
0.63 <9, 12> (*p*-nitrophenyl-N-acetyl-β-D-galctosaminide, <12>, caput, Hex B [39]) [10, 39]
0.64 <35> ((N-acetyl-D-glucosamine)$_2$, <35>, Ex 2 [55]) [55]
0.64 <21> (*p*-nitrophenyl-2-acetamido-2-deoxy-β-D-glucopyranoside) [29]
0.64 <35> (*p*-nitrophenyl-β-2-acetamido-2-deoxy-D-galactopyranoside, <35>, Ex 1 [55]) [55]
0.65 <4> (4-methylumbelliferyl-2-acetamido-2-deoxy-β-D-glucopyranoside, <4>, form A [77]) [77]
0.65 <6> (*p*-nitrophenyl-β-2-acetamido-2-deoxy-D-glucopyranoside) [6]
0.66 <12> (*p*-nitrophenyl-β-2-acetamido-2-deoxy-D-galactopyranoside, <12>, caudal, Hex A [39]) [39]
0.67 <35> ((N-acetyl-D-glucosamine)$_2$, <35>, Ex 1 [55]) [55]
0.68 <12> (*p*-nitrophenyl-β-2-acetamido-2-deoxy-D-galactopyranoside, <12>, caudal, Hex B [39]) [39]
0.68 <12> (*p*-nitrophenyl-β-2-acetamido-2-deoxy-D-glucopyranoside, <12>, caudal, Hex A and Hex B [39]) [39]
0.7 <25> (dansyl-Asn(GlcNAc)$_2$Man) [79]
0.72 <12, 18> (*p*-nitrophenyl-β-2-acetamido-2-deoxy-D-glucopyranoside, <12, 18>, N-acetylglucosaminidase [30]) [30]
0.73 <12> (*p*-nitrophenyl-β-2-acetamido-2-deoxy-D-glucopyranoside, <12>, colonic tumor, form A [21]) [21]
0.74 <4> (4-methylumbelliferyl-2-acetamido-2-deoxy-β-D-glucopyranoside, <4>, hexosaminidase B [9]) [9]
0.76 <7, 12> (*p*-nitrophenyl-β-2-acetamido-2-deoxy-D-glucopyranoside, <7>, β-NAHA-IC [7]; <12>, colonic mucosa, form A [21]) [7, 21]
0.77 <45> (4-methylumbelliferyl-2-acetamido-2-deoxy-β-D-glucopyranoside) [65]
0.78 <26> ((N-acetyl-D-glucosamine)$_4$) [36]
0.787 <35> ((N-acetyl-D-glucosamine)$_2$) [56]
0.8 <4> (4-methylumbelliferyl-2-acetamido-2-deoxy-β-D-glucopyranoside, <4>, hexosaminidase A [9]) [9]

0.8 <12, 18> (*p*-nitrophenyl-β-2-acetamido-2-deoxy-D-glucopyranoside, <12, 18>, N-acetylhexosaminidase [30]) [30]
0.8-0.9 <4> (4-methylumbelliferyl-β-N-acetylglucosaminide, <4>, liver and brain enzymes type S [19]) [19]
0.83 <26> ((N-acetyl-D-glucosamine)$_3$) [36]
0.83 <4, 34> (*p*-nitrophenyl-β-2-acetamido-2-deoxy-D-glucopyranoside) [50, 72]
0.85 <18> (*p*-nitrophenyl-β-2-acetamido-2-deoxy-D-glucopyranoside, <18>, form B, pH 4.5 [76]) [76]
0.86 <44> (*p*-nitrophenyl-β-2-acetamido-2-deoxy-D-galactopyranoside) [64]
0.87 <35> ((N-acetyl-D-glucosamine)$_3$, <35>, Ex 1 [55]) [55]
0.87 <13> (*p*-nitrophenyl-β-2-acetamido-2-deoxy-D-galactopyranoside) [14]
0.87 <18> (*p*-nitrophenyl-β-2-acetamido-2-deoxy-D-glucopyranoside, <18>, form A, pH 4.55 [76]) [76]
0.9 <27> (4-methylumbelliferyl β-D-glucosaminide, <27>, PI [60]) [60]
0.91 <27> (4-methylumbelliferyl β-D-glucosaminide, <27>, PII [60]) [60]
0.93 <45> (N,N-diacetylchitobiose) [65]
0.94 <28, 34> (*p*-nitrophenyl-β-2-acetamido-2-deoxy-D-glucopyranoside) [38, 49]
1 <4> (4-methylumbelliferyl-2-acetamido-2-deoxy-β-D-glucopyranoside, <4>, hexosaminidase I2 [9]) [9]
1 <10> (*p*-nitrophenyl-β-2-acetamido-2-deoxy-D-galactopyranoside) [11]
1 <54> (phenyl-N-acetyl-β-D-galactosaminide) [83]
1 <54> (phenyl-N-acetyl-β-D-glucosaminide) [83]
1.05 <45> (*p*-nitrophenyl-β-2-acetamido-2-deoxy-D-galactopyranoside) [65]
1.18 <54> (*p*-nitrophenyl-β-2-acetamido-2-deoxy-D-glucopyranoside) [83]
1.23 <45> (4-methylumbelliferyl-2-acetamido-2-deoxy-β-D-galactopyranoside) [65]
1.24 <29> (*p*-nitrophenyl-β-2-acetamido-2-deoxy-D-galactopyranoside) [43]
1.3 <44> (*p*-nitrophenyl-thio-β-N-acetyl-D-glucosaminide) [64]
1.4 <15> (*p*-nitrophenyl-β-2-acetamido-2-deoxy-D-glucopyranoside) [18]
1.5 <48> (*p*-nitrophenyl-β-2-acetamido-2-deoxy-D-galactopyranoside, <48>, extract of wild type strain, purified major and minor isozyme [70]) [70]
1.5 <48> (*p*-nitrophenyl-β-2-acetamido-2-deoxy-D-glucopyranoside) [70]
1.53 <12> (*p*-nitrophenyl-β-2-acetamido-2-deoxy-D-galactopyranoside, <12>, corpus, Hex A and Hex B [39]) [39]
1.55 <12> (*p*-nitrophenyl-β-2-acetamido-2-deoxy-D-glucopyranoside, <12>, corpus, Hex A and Hex B [39]) [39]
1.7 <26> ((N-acetyl-D-glucosamine)$_2$) [36]
1.71 <7> (*p*-nitrophenyl-β-2-acetamido-2-deoxy-D-galactopyranoside, <7>, β-NAHA-IC [7]) [7]
1.8 <4> (4-methylumbelliferyl-2-acetamido-2-deoxy-β-D-glucopyranoside-6-sulfate) [51]
1.8 <27> (*p*-nitrophenyl-β-2-acetamido-2-deoxy-D-glucopyranoside) [37]
1.92 <13> (*p*-nitrophenyl-β-2-acetamido-2-deoxy-D-glucopyranoside) [14]
1.96 <18> (*p*-nitrophenyl-β-2-acetamido-2-deoxy-D-glucopyranoside, <18>, form A, pH 3.55 [76]) [76]

2 <50> (*p*-nitrophenyl-β-2-acetamido-2-deoxy-D-glucopyranoside) [75]
2.3 <36> (*p*-nitrophenyl-β-2-acetamido-2-deoxy-D-galactopyranoside) [57]
2.38 <18> (*p*-nitrophenyl-β-2-acetamido-2-deoxy-D-glucopyranoside, <18>, form B, pH 3.55 [76]) [76]
2.58 <47> (glycopeptide monomer) [69]
3.5 <26> ((N-acetyl-D-glucosamine)$_6$) [36]
3.7 <26> ((N-acetyl-D-glucosamine)$_5$) [36]
4.1 <32> ((N-acetyl-D-glucosamine)$_2$) [47]
5.67 <25> (4-methylumbelliferyl-β-N-acetylglucosaminide, <25>, R162H [35]) [35]
7.8 <48> (*p*-nitrophenyl-β-2-acetamido-2-deoxy-D-galactopyranoside, <48>, extract of mutant strains [70]) [70]
8 <32> ((N-acetyl-D-glucosamine)$_3$) [47]
27 <24> (pyridylamino-glycoside B) [34]
30 <24> (pyridylamino-glycoside C) [34]
440 <24> (pyridylamino-glycoside A) [34]
Additional information <19> (<19>, K_M 0.056, new form I, after 4 months of storage at 4°C [24]) [24]

pH-Optimum

2.2 <2> (<2>, substrate: *p*-nitrophenyl-β-2-acetamido-2-deoxy-D-glucopyranoside [17]) [17]
2.5 <2> (<2>, substrate: N,N-diacetylchitobiose and *p*-nitrophenyl-β-2-acetamido-2-deoxy-D-glucopyranoside [2]) [2]
3 <25> (<25>, SpHex wild type [35]) [35]
3-4.5 <10> [11]
3.4 <18, 30> (<18>, N-acetylgalactosamnidase activity [66]; <30>, substrate: *p*-nitrophenyl-β-2-acetamido-2-deoxy-D-galactopyranoside [46]) [46, 66]
3.5 <38, 48> (<38>, forms I-IV, substrate: phenyl-N-acetyl-β-D-glucosaminide [61]; <48>, extract of mutant strains [70]) [61, 70]
3.5-4 <4, 21> (<4>, substrate: 4-methylumbelliferyl-2-acetamido-2-deoxy-β-D-glucopyranoside-6-sulfate [51]; <21>, substrate: *p*-nitrophenyl-β-2-acetamido-2-deoxy-D-galactopyranoside [29]) [29, 51]
3.7 <2, 30> (<2>, substrate: *p*-nitrophenyl-β-2-acetamido-2-deoxy-D-galactopyranoside [2, 17]; <30>, substrate: *p*-nitrophenyl-β-2-acetamido-2-deoxy-D-glucopyranoside [46]) [2, 17, 46]
3.8 <12, 18> (<12, 18>, substrate: *p*-nitrophenyl-β-2-acetamido-2-deoxy-D-galactopyranoside, N-acetylhexosaminidase [30]) [30]
3.8 <25> (<25>, D246N [35]) [35]
4 <5, 9, 13, 14, 15, 28> (<5, 14, 15, 28>, substrate: *p*-nitrophenyl-β-2-acetamido-2-deoxy-D-glucopyranoside [5, 16, 18, 38]; <9, 13>, substrate: *p*-nitrophenyl-β-2-acetamido-2-deoxy-D-galactopyranoside [10, 14]) [5, 10, 14, 16, 18, 38]
4-4.5 <25> (<25>, substrate: dansyl-Asn(GlcNAc)$_2$Man [79]) [79]
4-5 <8> (<8>, substrate: *p*-nitrophenyl-β-2-acetamido-2-deoxy-D-glucopyranoside, low molecular weight form EII [8]) [8]
4-5.5 <24> [34]

4.1-4.3 <27> (<27>, substrate: *p*-nitrophenyl-*β*-2-acetamido-2-deoxy-D-galactopyranoside or -glucosaminide [37]) [37]
4.2 <12, 13, 18> (<12, 13>, substrate: *p*-nitrophenyl-*β*-2-acetamido-2-deoxy-D-glucopyranoside [14, 21]; <12>, substrate: *p*-nitrophenyl-*β*-2-acetamido-2-deoxy-D-glucopyranoside [21]; <12, 18>, substrate: *p*-nitrophenyl-*β*-2-acetamido-2-deoxy-D-glucopyranoside, N-acetylhexosaminidase [30]; <12>, both isozymes [39]) [14, 21, 30, 39]
4.3 <54> (<54>, N-acetylgalactosaminidase activity [83]) [83]
4.4 <1, 19, 50> (<1>, substrate: *p*-nitrophenyl-*β*-2-acetamido-2-deoxy-D-galactopyranoside and (N-acetyl-D-glucosamine)$_{3-6}$ [1]; <19>, substrate: *p*-nitrophenyl-*β*-2-acetamido-2-deoxy-D-galactopyranoside, form III [24]; <50>, N-acetylgalactosaminidase activity [75]) [1, 24, 75]
4.5 <54> (<54>, N-acetylglucosamnidase activity, first optimum [83]) [83]
4.5 <9, 17, 18> (<9>, substrate: *p*-nitrophenyl-*β*-2-acetamido-2-deoxy-D-glucopyranoside [10]; <17>, substrate: *p*-nitrophenyl-*β*-2-acetamido-2-deoxy-D-galactopyranoside and -glucopyranoside [22]; <18>, N-acetylglucosamnidase activity [66]) [10, 22, 66]
4.5-5 <4, 7, 29> (<7>, all 4 isozymes, substrates: *p*-nitrophenyl-*β*-2-acetamido-2-deoxy-D-glucopyranoside and *p*-nitrophenyl-*β*-2-acetamido-2-deoxy-D-galactopyranoside [7]) [7, 43, 72]
4.6 <3, 16, 19> (<3>, substrate: *p*-nitrophenyl-*β*-2-acetamido-2-deoxy-D-galactopyranoside [3]; <16>, substrate: *p*-nitrophenyl-*β*-2-acetamido-2-deoxy-D-glucopyranoside [20]; <19>, substrate: *p*-nitrophenyl-*β*-2-acetamido-2-deoxy-D-galactopyranoside, form IV [24]) [3, 20, 24]
4.6-5 <51> [78]
4.7 <23, 54> (<54>, N-acetylglucosamnidase activity, second optimum [83]) [33, 83]
4.8 <19, 37, 46> (<19>, substrate: *p*-nitrophenyl-*β*-2-acetamido-2-deoxy-D-glucopyranoside, form IV [24]; <37>, N-acetylglucosamnidase activity [59]) [24, 59, 67]
5 <1, 3, 4, 19, 36, 48, 53> (<1>, substrate: (N-acetyl-D-glucosamine)$_2$ [1]; <3, 53>, substrate: *p*-nitrophenyl-*β*-2-acetamido-2-deoxy-D-glucopyranoside [3, 82]; <4>, substrate: 4-methylumbelliferyl-*β*-N-acetylglucosaminide or -acetylgalactosaminide, liver and brain enzymes type S [19]; <19>, substrate: *p*-nitrophenyl-*β*-2-acetamido-2-deoxy-D-glucopyranoside, forms II and III [24]; <19>, substrate: *p*-nitrophenyl-*β*-2-acetamido-2-deoxy-D-galactopyranoside, form II [24]; <53>, substrate: N,N-diacetylchitobiose [82]; <48>, extract of wild type strain [70]) [1, 3, 19, 24, 57, 70, 82]
5-5.5 <6, 33> (<6>, substrate: *p*-nitrophenyl-*β*-2-acetamido-2-deoxy-D-glucopyranoside [6]) [6, 48]
5-6 <21, 35> (<21>, substrate: *p*-nitrophenyl-*β*-2-acetamido-2-deoxy-D-glucopyranoside [29]; <35>, Ex 1, Ex 2 and alimentary canal enzyme [55]) [29, 55]
5.2 <1, 12, 18> (<1>, substrate: *p*-nitrophenyl-*β*-2-acetamido-2-deoxy-D-glucopyranoside [1]; <12>, substrate: 4-methylumbelliferyl-*β*-N-acetylgalactosaminide, N-acetylglucosaminidase [13]; <12, 18>, N-acetylglucosaminidase,

substrate: p-nitrophenyl-β-2-acetamido-2-deoxy-D-glucopyranoside [30]) [1, 13, 30]
5.4 <37, 45, 50> (<37>, N-acetylgalactosaminidase activity [59]; <45>, chromogenic substrates [65]; <50>, N-acetylglucosaminidase activity [75]) [59, 65, 75]
5.5 <12, 18, 19> (<12, 18>, N-acetylgalactosaminidase, substrate: p-nitrophenyl-β-2-acetamido-2-deoxy-D-galactopyranoside [30]; <19>, substrate: p-nitrophenyl-β-2-acetamido-2-deoxy-D-galactopyranoside, form I [24]) [24, 30]
5.5-6 <22, 35> (<22>, Endo-BI and -BII [32]) [32, 56]
5.5-6.5 <8> (<8>, substrate: p-nitrophenyl-β-2-acetamido-2-deoxy-D-galactopyranoside, low molecular weight form EII [8]) [8]
5.7 <27> (<27>, PI and PII [60]) [60]
5.8 <16, 26> (<16>, substrate: p-nitrophenyl-β-2-acetamido-2-deoxy-D-galactopyranoside [20]; <26>, substrate: (N-acetyl-D-glucosamine)2 [36]) [20, 36]
5.9 <22> [73]
6 <12, 19, 25> (<12>, substrate: 4-methylumbelliferyl-β-N-acetylgalactosaminide, N-acetylgalactosaminidase [13]; <19>, substrate: p-nitrophenyl-β-2-acetamido-2-deoxy-D-glucopyranoside, form I [24]) [13, 24, 54]
6.2 <20> (<20>, substrate: p-nitrophenyl-β-2-acetamido-2-deoxy-D-galactopyranoside or -glucosaminide [26]) [26]
6.5 <25, 32, 45> (<25>, R162H [35]; <45>, substrate: N,N'-diacetylchitobiose [65]) [35, 47, 65]
6.8 <47> (<47>, crude enzyme preparation [69]) [69]
7 <26> (<26>, substrate: (N-acetyl-D-glucosamine)$_{3-6}$ [36]) [36]
7-8 <26> (<26>, substrate: p-nitrophenyl-β-2-acetamido-2-deoxy-D-glucopyranoside [36]) [36]
7.7 <47> (<47>, purified enzyme [69]) [69]
Additional information <4> [77]

pH-Range

2-10 <16> [20]
2.5-5 <2, 10> (<2>, substrate: p-nitrophenyl-β-2-acetamido-2-deoxy-D-galactopyranoside [17]; <10>, active within this range [11]) [11, 17]
3-6.5 <51> [78]
4.5 <13, 46> (<13>, active [14]; <46>, precipitous drop at values below [67]) [14, 67]
4.5-8.5 <8> (<8>, low molecular weight form EII [8]) [8]
4.8 <37> (<37>, sharp decline of N-acetylglucosaminidase activity on either side of this value [59]) [59]
5.4 <37> (<37>, sharp decline of N-acetylgalactosaminidase activity on either side of this value [59]) [59]
5.5-9.5 <26> (<26>, full activity at pH 7.5 after preincubation for 15 min in that range, substrate: p-nitrophenyl-β-2-acetamido-2-deoxy-D-glucopyranoside [36]) [36]
6 <7> (<7>, β-NAHA-II, no hydrolysis at pH 6.0 and above, β-NAHA-IA, -IB and IC show 20-25% of maximum activity at pH 6.0 [7]) [7]
Additional information <44> [64]

Temperature optimum (°C)

35-37 <16> [20]
35-40 <26> (<26>, at pH 7.5, substrate: *p*-nitrophenyl-β-2-acetamido-2-deoxy-D-glucopyranoside [36]) [36]
40-45 <19> (<19>, all 4 forms [24]) [24]
45 <35> (<35>, alimentary canal enzyme [55,56]) [55, 56]
45-50 <30> (<30>, substrate: *p*-nitrophenyl-β-2-acetamido-2-deoxy-D-galactopyranoside or -glucosaminide [46]) [46]
50 <24, 33> (<24>, at pH 5.2 [34]) [34, 48]
55 <10, 29, 35, 38> (<35>, Ex 1 [55]; <38>, forms II and IV [61]) [11, 43, 55, 61]
55-65 <38> (<38>, form I [61]) [61]
60 <28, 38> (<38>, form III [61]) [38, 61]
65 <23> [33]
75 <32> [47]

Temperature range (°C)

0-40 <30> (<30>, at pH 4.5 [46]) [46]
10-70 <16> [20]

4 Enzyme Structure

Molecular weight

15300 <34> (<34>, gel filtration, Sephadex G-75 [49]) [49]
24000 <19> (<19>, form II, SDS-PAGE [24]) [24]
28000 <19> (<19>, form I, SDS-PAGE [24]) [24]
29000 <4> (<4>, gel electrophoresis [4]; <4>, hexosaminidase B [9]) [3, 9]
30000 <19> (<19>, form IV, SDS-PAGE [24]) [24]
30500 <19> (<19>, form III, SDS-PAGE [24]) [24]
33500 <47> (<47>, gel filtration [69]) [69]
38000 <47> (<47>, SDS-PAGE [69]) [69]
46000 <25> (<25>, gel filtration [79]) [79]
49500 <25> (<25>, sedimentation equilibrium [79]) [79]
52000 <25> (<25>, SDS-PAGE [79]) [79]
53000 <11> (<11>, lysosomal enzyme, SDS-PAGE [12]) [12]
55000 <12, 25, 29> (<25>, recombinant SpHex, western blot analysis [35]; <12>, SDS-PAGE, Hex B [39]; <29>, SDS-PAGE [43]; <12>, neutral N-acetylgalactosaminidase, SDS-PAGE [13]) [13, 35, 39, 43]
56000 <30> (<30>, SDS-PAGE [46]) [46]
60000 <23> (<23>, SDS-PAGE [33]) [33]
61000 <18, 35> (<18,35>, SDS-PAGE [23,68]) [23, 68]
62000 <19> (<19>, new form I, after 4 months of storage at 4°C [24]) [24]
62500 <18, 34> (<34>, gel filtration, Bio-Gel-60 [49]; <18>, gel filtration [23]) [23, 49]
63000 <12> (<12>, SDS-PAGE, Hex A [39]) [39]
64000 <33> (<33>, SDS-PAGE [48]) [48]

65000 <2> (<2>, SDS-PAGE [2,17]) [2, 17]
66000 <14> (<14>, SDS-PAGE [16]) [16]
68000 <11> (<11>, egg white-enzyme, SDS-PAGE [12]) [12]
69000 <24, 34> (<24, 34>, SDS-PAGE [34, 49]) [34, 49]
69380 <26> (<26>, amino acid calculation [36]) [36]
70000 <8, 26, 39> (<26>, SDS-PAGE [36]; <39>, gel filtration [62]; <8>, low molecular weight form EII, SDS-PAGE [8]) [8, 36, 62]
72000 <19, 36> (<19>, form II, gel filtration [24]; <36>, gel filtration [57]) [24, 57]
74000 <13, 32> (<13, 32>, SDS-PAGE [14, 47]) [14, 47]
75000 <22, 32> (<22, 32>, gel filtration [47, 73]; <22>, sucrose density gradient centrifugation, SDS-PAGE [73]) [47, 73]
82000 <12> (<12>, neutral N-acetylgalactosaminidase, sucrose density gradient centrifugation [13]) [13]
84000 <19> (<19>, form I, gel filtration [24]) [24]
88000 <12> (<12>, neutral N-acetylglucosaminidase, sucrose density gradient centrifugation [13]) [13]
92000 <51> (<51>, SDS-PAGE [78]) [78]
95000 <16> (<16>, SDS-PAGE [20]) [20]
100000 <4, 16, 21> (<4>, hexosaminidase A [9]; <16, 21>, gel filtration [20, 29]) [9, 20, 29]
100000-110000 <4> (<4>, hexosaminidase I2, gradient gel electrophoresis [9]) [9]
105000 <4> (<4>, sedimentation equilibrium using untreated enzyme, carboxymethylated enzyme in 6 M guanidine-HCl, meniscus depletion experiments with fresh enzyme prepatation at pH 11.5 [72]) [72]
107000 <4> (<4>, gel filtration [72]) [72]
108000 <45> (<45>, native PAGE [65]) [65]
110000 <33> (<33>, gel filtration [48]) [48]
112000 <4, 37> (<37>, native PAGE [59]) [44, 59]
118000 <18> (<18>, A form, gel filtration [58]) [58]
118000-147000 <17> (<17>, gel filtration Sephadex G-200 [22]) [22]
119000 <35> (<35>, gel filtration [56]) [56]
120000 <2, 4, 29, 38> (<2, 29, 38>, gel filtration [2, 17, 43, 61]; <4>, enzyme after dialysis and freeze-drying, 6 M guanidine-HCl, pH 7.2 [72]) [2, 17, 43, 61, 72]
120000-145000 <17> (<17>, gel filtration Sephadex G-100 [22]) [22]
124000 <37> (<37>, gel filtration [59]) [59]
125000 <8, 35, 53> (<35>, Ex 2 and alimentary canal enzyme, gel filtration [55]; <35>, SDS-PAGE [56]; <8>, low molecular weight form EII, gel filtration [8]; <35, 53>, gel filtration [68, 82]) [8, 55, 56, 68, 82]
127000 <23> (<23>, gel filtration [33]) [33]
130000 <4, 12> (<4>, Hex I2, gel filtration [53]; <12>, neutral N-acetylglucosaminidase, SDS-PAGE [13]) [13, 53]
132000 <14> (<14>, gel filtration [16]) [16]
133000 <3> (<3>, gel filtration [3]) [3]
135000 <35> (<35>, Ex 1, gel filtration [55]) [55]

136000 <49> (<49>, sedimentation equilibrium, gel filtration [71]) [71]
140000 <4, 18> (<18>, sedimentation equilibrium [76]; <4>, gel filtration [77]) [76, 77]
141000 <10, 50> (<10>, sedimentation equilibrium [11]; <50>, gel filtration [75]) [11, 75]
143000 <10> (<10>, gel filtration [11]) [11]
149000 <44> (<44>, gel filtration and amino acid composition [64]) [64]
150000 <6, 19, 20, 22, 24, 28> (<19>, form IV, gel filtration [24]; <6, 20, 24, 28>, gel filtration [6, 26, 34, 38]; <22>, native PAGE, Endo-BII [32]) [6, 24, 26, 32, 34, 38]
152000 <4> (<4>, brain enzyme type S, gel filtration [19]) [19]
154000 <4> (<4>, liver enzyme type S, gel filtration [19]) [19]
160000 <27> (<27>, gel filtration [37]) [37]
170000 <5, 13> (<5>, β-hexosaminidase II + III [5]; <13>, at pH 7.0, gel filtration [14]) [5, 14]
180000 <19> (<19>, form III, gel filtration [24]) [24]
190000 <18> (<18>, gel filtration [66]) [66]
210000 <7, 12, 22> (<7>, gel filtration, all 4 isozymes [7]; <12>, neutral N-acetylgalactosaminidase, gel filtration [13]; <22>, native PAGE, Endo-BI [32]) [7, 13, 32]
220000 <18> (<18>, B form, sedimentation equilibrium analysis [58]) [58]
223000 <12> (<12>, gel filtration, Hex B [39]) [39]
234000 <18> (<18>, B form, gel filtration [58]) [58]
250000 <12> (<12>, gel filtration, Hex A [39]) [39]
260000 <4> (<4>, pH 2.5, fresh enzyme prepatation, meniscus depletion experiments [72]) [72]
330000 <13> (<13>, at pH 4.5, gel filtration [14]) [14]
525000 <12> (<12>, neutral N-acetylglucosaminidase, gel filtration [13]) [13]
Additional information <18> [76]

Subunits

dimer <2, 4, 7, 8, 10, 11, 14, 18, 23, 24, 33, 35> (<7>, β-NAHA-II, 1 * 62000 + 1 * 64000 [7]; <8>, 2 * 70000, low molecular weight form EII [8]; <4>, hexosaminidase A, major components: 1 * 56000 + 1 * 29000, minor components: 26000 - 22000, SDS-PAGE [9]; <4>, hexosaminidase I2, major components: 1* 56000 + 1 + 29000, minor components: 65000 - 57000, SDS-PAGE [9]; <10>, 2 * 68000, SDS-PAGE [11]; <11>, 1 * 68000 + 1 * 53000, SDS-PAGE, microsomal enzyme [12]; <4>, 1 * 54000 + 1 * 26000, placental enzyme, SDS-PAGE [15]; <14>, 2 * 66000, SDS-PAGE [16]; <2>, 2 * 65000, SDS-PAGE [17]; <4, 23>, Hex A: α β, Hex B: β β [31, 33, 41, 42, 44, 52, 53]; <4>, Hex S: α α [31]; <23>, 2 * 60000, SDS-PAGE [33]; <24>, 2 * 69000 [34]; <33>, 2 * 64000, SDS-PAGE [48]; <4>, major Hex I2 form: α, β, 1 * 66000 + 1 * 66000, β chain 2 residues short of being identical to α chain, SDS-PAGE [53]; <35>, Ex 1, homodimer: 2 * 67500; Ex 2: 1 * 67500 + 1 * 57500, SDS-PAGE [55]; <35>, alimentary canal enzyme: 1 * 67500 + 1 * 57500, SDS-PAGE [55, 56]; <18>, A and B form: 1 * 55000 + 1 * 25000, SDS.PAGE [58]; <18>, 1 * 53000 + 1 * 13400,

SDS-PAGE [66]; <35>, 2 * 61000, SDS-PAGE [68]) [7, 8, 9, 11, 12, 15, 16, 17, 31, 33, 34, 41, 42, 44, 48, 52, 53, 55, 56, 58, 66, 68]
hexamer <19> (<19>, 6 * 30500, SDS-PAGE, form III [24]) [24]
monomer <32, 34, 47> (<32>, 1 * 74000 [47]; <34>, 1 * 62500-69000, SDS-PAGE [49]; <47>, 1 * 36000, SDS-PAGE [69]) [47, 49, 69]
pentamer <19> (<19>, 5 * 30000, SDS-PAGE, form IV [24]) [24]
tetramer <4, 38> (<4>, 1 * 68000 + 1 * 59600 + 1 * 55600 + 1 * 29500, SDS-PAGE, mucolipidosis enzyme [15]; <38>, 2 * 27000 + 2 * 34000, SDS-PAGE [61]) [15, 61]
trimer <4, 19, 28> (<4>, 1 * 58000 + 1 * 55000 + 1 * 29000, SDS-PAGE, normal enzyme [15]; <19>, 3 * 28000, SDS-PAGE, form I [24]; <19>, 3 * 24000, SDS-PAGE, form II [24]; <28>, 1 * 64000 + 1 * 57000 + 1 * 51000, SDS-PAGE [38]) [15, 24, 38]
Additional information <4, 18, 37> (<37>, major band: 68900, minor band: 65900, SDS-PAGE [59]) [53, 59, 76]

Posttranslational modification

glycoprotein <4, 7, 8, 10, 18, 19, 23, 35, 37, 44, 49> (<8>, low molecular weight form EII [8]; <18>, neutral hexose [58]; <18, 35>, neutral sugar [68, 76]; <4, 18, 49>, sialylation [4, 71, 72, 76]; <4>, modification at the carbohydrate moiety of hexosaminidase B causes appearance of acidic variant form [4]; <4, 37, 44, 49>, galactosylation [15, 59, 64, 71, 72]; <4>, increased amount of galactose present on mucolipidosis enzyme compared to normal enzyme [15]; <19, 37, 44>, fucosylation [24, 59, 64]; <19, 23, 37, 44, 49>, mannosylation [24, 33, 59, 64, 71]; <19, 37, 44, 49>, N-acetyl-glucosaminylation [24, 59, 64, 71]; <37>, N-acetyl-galactosaminylation [59]; <4, 18>, glucosaminylation [72, 76]; <4>, carbohydrate modification [15, 45]; <4>, mucolipidosis enzyme shows decreased carbohydrate composition compared to normal enzyme [15]; <4>, carbohydrate units on enzymes isolated from I-cell or Pseudo-Hurler patients show modification to hybrid- or complex-type oligosaccharide [45]) [4, 7, 8, 11, 15, 19, 24, 31, 33, 44, 45, 51, 53, 58, 59, 64, 66, 68, 71, 72, 76]

5 Isolation/Preparation/Mutation/Application

Source/tissue

alimentary canal <35> [55, 56]
amniotic fluid <4> [41, 77]
brain <4, 12, 18, 27> [13, 19, 30, 37]
cell extract <20, 40, 41, 42, 44, 47, 52, 53> [26, 63, 64, 69, 81, 82]
chorionic villus sample cell <4> [41]
colonic mucosa <12> [21]
colonic tumor <12> [21]
cotyledon <7, 19, 34> (<19>, forms I and II [24, 25]) [7, 24, 25, 49]
culture filtrate <2, 8, 10, 16, 22, 24, 25, 29, 32, 37, 50, 52> (<8>, low molecular weight form [8]) [2, 8, 11, 17, 20, 34, 43, 47, 59, 74, 75, 79, 81]

digestive fluid <45> [65]
digestive gland <17, 49> [22, 71, 80]
egg <11> (<11>, egg white [12]) [12]
embryo <19> (<19>, forms III and IV [24, 25]) [24, 25]
endosperm <19> (<19>, forms I and II [24]; <19>, form II [25]) [24, 25]
epididymis <12, 18> (<12>, from caput, corpus and caudal region of epididymis [39]) [27, 39]
fibroblast <4> (<4>, skin fibroblast [40]) [40, 41]
foetal membrane <4> [77]
fruit <6> [6]
gastric mucosa <27> [60]
granulocyte <4> [51]
haemolymph <23, 35> [33, 68]
hepatopancreas <38> [61]
integument <35> [55]
internal organs <14> [16]
larva <23> [33]
leaf <1, 28> (<1>, primary leaves [1]) [1, 38]
leukocyte <4> (<4>, lysate [31]; <4>, from peripheral blood [40]) [31, 40, 41, 51, 52]
liver <4, 33, 43> [9, 19, 28, 48, 63]
lymphoblast <4> [40, 45]
mammary gland <18> [58]
mid-gut gland <30> [46]
mononuclear cell <4> [51]
mycelium <25, 36> [54, 57]
oviduct <11> [12]
placenta <4> [4, 15, 44, 53, 77]
plasma <4> (<4>, desalinated [31]) [31]
roe <5> [5]
seed <3, 7, 19, 21, 34> (<19, 34>, germinating seeds [24, 25, 49, 50]; <21> meal [29]) [3, 7, 24, 25, 29, 49, 50]
semen <15, 18> [18, 66]
serum <4, 46> (<4>, cord serum and maternal serum [77]) [40, 52, 67, 72, 77]
skin fibroblast <4> [28]
spleen <18> [76]
sporangiophore <36> [57]
testis <18, 54> [23, 83]
urine <4> [15, 42]
viscera <13> [14]
Additional information <23> (<23>, no activity in saliva [33]) [33]

Localization

acrosome <18> [66]
cytoplasm <15, 47> [18, 69]

extracellular <2, 4, 8, 10, 16, 18, 22, 24, 25, 29, 32, 37, 45, 50, 52> (<4>, in mucolipidosis patients [15]; <4>, in newborns with hypoxia excretion is higher than in other groups [42]; <18>, in milk of cows with mastitis of the mammary gland high concentrations of enzymes are found [58]) [2, 8, 11, 15, 17, 20, 34, 42, 43, 47, 58, 59, 65, 74, 75, 79, 81]
lysosome <4, 11, 27> (<4>, in healthy humans [15]) [12, 15, 31, 40, 41, 42, 44, 45, 51, 53, 60]
membrane <40, 41, 42> [63]
microsome <11, 23> [12, 33]
periplasm <26, 52> [36, 81]
protein body <7> [7]

Purification

<1> (2500fold, 5 major isoenzymes [1]) [1]
<2> (1400fold [2]; 1396fold to homogeneity [17]) [2, 17]
<3> (113fold [3]) [3]
<4> (1500fold [4]; hexosaminidase A, B and I2 purified 2699, 10320 and 13077fold [9]; to homogeneity, normal enzyme 64863fold, I-cell enzyme 20140fold [15]; liver enzyme type S 13500fold, brain enzyme type S 5240fold, liver enzymes types A and B [19]; partial [28]; 195fold to homogeneity [72]) [4, 9, 15, 19, 28, 44, 72]
<5> (β-hexosaminidase II: 202fold , β-hexosaminidase III: 426fold, still very heterogenous [5]) [5]
<6> (to homogeneity [6]) [6]
<7> (4 isozymes, β-NAHA-II was purified 3200fold to apparent homogeneity, β-NAHA-IA 770fold, β-NAHA-IB 1750fold, β-NAHA-IC 500fold [7]) [7]
<8> (low molecular weight form EII, 3500fold to homogeneity [8]) [8]
<9> (partial purification, 30fold, more extensive purification requires bovine serum albumin for retention of activity [10]) [10]
<10> (to homogeneity [11]) [11]
<11> (lysosomal enzyme 550fold, microsomal enzyme 756fold, egg white-enzyme 2206fold [12]) [12]
<12> (neutral N-acetylgalactosaminidase, 2000-2500fold to homogeneity [13]; colonic mucosa: form A 4800fold, form B 5110fold, form I 4100fold, colonic tumors: form A 5090fold, form B 5620fold [21]; neutral N-acetylglucosaminidase, 300-330fold, 85-90% pure [13]; N-acetylhexosaminidase: 63fold; N-acetylglucosaminidase and -galactosaminidase [30]; Hex A: 161fold, Hex B: 189fold [39]) [13, 21, 30, 39]
<13> (820fold [14]) [14]
<14> (1560fold, fairly homogenous [16]) [16]
<15> (226fold to homogeneity [18]) [18]
<16> (68fold [20]) [20]
<17> (330fold [22]) [22]
<18> (310fold [27]; N-acetylhexosaminidase: 63fold, N-acetylglucosaminidase and -galactosaminidase [30]; A form, still heterogenous: 63fold, B form, close to homogeneity: 92fold [58]; 135fold to homogeneity [66]; A form: 310fold, B form: 260fold, to homogeneity [76]) [23, 27, 30, 58, 66, 76]

<19> (form I: 2262fold, form II: 2102fold, form III: 20380fold, form IV: 5061fold [24]; form I: 185fold, form II: 670fold, form III: 20880fold, form IV: 4740fold [25]) [24, 25]
<20> (784fold [26]) [26]
<21> (to homogeneity [29]) [29]
<22> (to homogeneity, Endo-BI: 18fold [32]; 7000fold to homogeneity [74]) [32, 73, 74]
<23> (to homogeneity, 1564fold [33]) [33]
<24> (13.2fold to homogeneity [34]) [34]
<25> (SpHex-MBP fusion protein [35]; to homogeneity [54, 79]) [35, 54, 79]
<26> (40fold to homogeneity [36]) [36]
<27> (form A: 1300-1500fold to homogeneity [37]; forms PI and PII, each 200fold [60]) [37, 60]
<28> (256fold to homogeneity [38]) [38]
<29> (<29>, 70fold [43]) [43]
<30> (7.6fold [46]) [46]
<32> (387fold to homogeneity [47]) [47]
<33> (699fold [48]) [48]
<34> (partial, 132.9-176.6fold [49]) [49]
<35> (Ex 1: 3000fold, Ex 2: 3200fold [55]; 64fold to homogeneity [56]; 3350fold to homogeneity [68]) [55, 56, 68]
<36> (partial, 5-10fold [57]) [57]
<37> (to homogeneity, 1200fold [59]) [59]
<38> (types I-IV 150fold, type I to homogeneity [61]) [61]
<39> [62]
<40> (30fold [63]) [63]
<43> (30fold [63]) [63]
<44> (2 isozymes, 153fold to homogeneity [64]) [64]
<45> (160fold to homogeneity [65]) [65]
<46> (100fold [67]) [67]
<47> (1094fold to near homogeneity [69]) [69]
<48> (stage-specific enzyme: 2200fold, vegetative enzyme: 1400fold [70]) [70]
<49> (to homogeneity [71]) [71]
<50> (46.7 fold to homogeneity [75]) [75]
<51> (241fold, partial [78]) [78]
<53> (200fold to homogeneity [82]) [82]

Crystallization

<4> (hexosaminidase B, space group P6122 i.e. enantiomorph, 3.2 A resolution [44]) [44]
<21> [29]

Cloning

<25> (into E. coli [35]) [35]
<26> (into E. coli [36]) [36]

Engineering

D246N <25> [35]
E314Q <25> [35]
R162H <25> [35]
R178H <4> (<4>, from patient with the genetic disease GM2-gangliosidosis B1 variant, mutated Hex A, altered active site of α subunit [31]) [31]

Application

analysis <5, 17, 21, 22, 25, 30, 34, 37, 49, 53> [5, 22, 29, 32, 35, 46, 49, 54, 59, 80, 82]
medicine <4, 18> (<4>, diagosis of Tay-Sachs disease [41, 51, 52]; <18>, veterinary medicine: increased level of enzyme in secreted milk can be detected by diagnosis test for mastitis of the mammary gland of cows [58]) [41, 51, 52, 58]

6 Stability

pH-Stability

1.5 <2> (<2>, activity is lost below [17]) [17]
2 <50> (<50>, complete inactivation after 30 min at 25°C [75]) [75]
2-4 <2> (<2>, relatively stable, retains 85% of activity after 45 hours at 5°C [17]) [17]
2.5-10 <24> (<24>, not less than 80% of activity remains, at 5°C for 1 h [34]) [34]
3 <50> (<50>, 40% inactivation after 30 min at 25°C [75]) [75]
3-9 <24, 28> (<24>, not less than 80% of activity remains, at 37°C for 1 h [34]; <28>, stable for 4 hr at room temperature [38]) [34, 38]
3.3 <17> (<17>, 50% of activity are lost [22]) [22]
3.5 <25> (<25>, half-maximal activity [79]) [79]
3.5-5.5 <30> (<30>, stable [46]) [46]
3.8-4.6 <17> (<17>, at 37°C stable for 2 hours [22]) [22]
4 <51> (<51>, half-maximal activity [78]) [78]
4-7.5 <13, 27> (<13>, stable [14]; <27>, stable for 2 hr at 37°C [37]) [14, 37]
4-8 <25> (<25>, stable at 30°C for 30 min [54]) [54]
4-10 <29> (<29>, stable, at 37°C [43]) [43]
4-11 <33, 50> (<33,50>, stable [48,75]) [48, 75]
4.5 <18> (<18>, at 37°C, stable for 15 min [76]) [76]
4.5-7 <23> (<23>, at 4°C for 24 hours and at 37°C for 60 min [33]) [33]
5-6 <14> (<14>, stable [16]) [16]
5-8.5 <35> (<35>, alimentary canal enzyme, stable [55]) [55]
5-9 <35> (<35>, Ex 2, stable [55]) [55]
5.5 <25> (<25>, half-maximal activity [79]) [79]
5.5-8.5 <35> (<35>, stable [56]) [56]
6 <7, 26, 51> (<7>, β-NAHA-II, no hydrolysis above at pH 6.0 and above, β-NAHA-IA, -IB and IC show 20-25% of maximum activity at pH 6.0 [7]; <26>,

50% of activity, substrate: *p*-nitrophenyl-β-2-acetamido-2-deoxy-D-glucopyranoside [36]; <51>, half-maximal activity [78]) [7, 36, 78]
6-7 <22, 46> (<22>, stable overnight when maintained at 4°C [32]; <46>, activity is reduced by 23% [67]) [32, 67]
6-8 <35> (<35>, Ex 1, stable [55]) [55]
6.2 <36> (<36>, 80% of activity [57]) [57]
6.3 <17> (<17>, 50% of activity are lost [22]) [22]
6.5 <32> (<32>, stable [47]) [47]
6.5-8 <6> (<6>, fairly stable [6]) [6]
7 <7, 13, 37, 53> (<7>, loss of 25% of acivity if pH is raised to 8.0 [7]; <13>, inactive [14]; <37, 53>, stable around pH 7.0 [59, 82]) [7, 14, 59, 82]
7-9 <10> (<10>, stable within this range [11]) [11]
7.5 <26> (<26>, no activity, substrate: (N-acetyl-D-glucosamine)$_2$ [36]) [36]
8-8.3 <26> (<26>, 50-60% of activity with (N-acetyl-D-glucosamine)$_{3-6}$ as substrate, no activity with (N-acetyl-D-glucosamine)$_2$ as substrate [36]) [36]
8.5 <22> (<22>, most stable at this pH, stability rapidly decreasing above or below [73]) [73]
8.5-9 <43> (<43>, inactivation at high pH like 8.5-9.0 [63]) [63]
9 <26> (<26>, 50% of activity, substrate: *p*-nitrophenyl-β-2-acetamido-2-deoxy-D-glucopyranoside [36]) [36]
10 <28> (<28>, rapidly inactivated above, at room temperature [38]) [38]
Additional information <21, 44> [29, 64]

Temperature stability

0 <25> (<25>, pH from 3.0-8.0, stable for 24 h [79]) [79]
0-37 <4> (<4>, stable up to 24 h [51]) [51]
4-42 <26> (<26>, stable for 15 min, at pH 7.5 [36]) [36]
5 <2> (<2>, stable at this temperature between pH 2.0-4.0 [2]) [2]
20 <25, 26> (<25>, stable for at least 48 h at room temperature, wild type and mutant enzymes [35]; <26>, stable for at least 24 h at room temperature [36]) [35, 36]
25 <25> (<25>, pH from 3.0-8.0, stable for 24 h [79]) [79]
28 <29> (<29>, stable [43]) [43]
30 <24, 28, 35, 45> (<24>, below: completely stable for 30 min [34]; <28>, stable below for 15 min at pH 7.0 [38]; <35>, stable below, for 10 min [56]; <45>, rather stable at pH 5.0-7.0 [65]) [34, 38, 56, 65]
35-60 <28> (<28>, for 15 min at pH 7.0, gradual inactivation [38]) [38]
37 <4, 18, 19, 26, 29, 30, 50> (<4>, at pH 3.0, form P: loss of 3% of activity after 30 min, form A: loss of 22% of activity after 30 min [77]; <18>, in presence of 1% bovine serum albumin and 0.1% sodium azide at pH 4.5 stable for at least 72 h [23]; <18>, stable for 60 min at pH 4.5 [66]; <18>, at pH 4.5, stable for 15 min [76]; <19>, at its optimum pH both hexosaminidase activities of each of the 4 forms remain stable over 5 hours [24]; <26>, stable for at least 24 h [36]; <29>, 25% inactivation after 30 min [43]; <30>, stable for 60 min at pH 3.5-5.5 [46]; <50>, crude enzyme, relatively stable [75]) [23, 24, 36, 43, 46, 66, 75, 76, 77]

40 <13, 22, 35> (<13>, stable for 30 min [14]; <22>, for 10 min, stable [32]; <35>, alimentary canal enzyme, stable below [55]) [14, 32, 55]
45 <2, 8, 10> (<8>, low molecular weight form EII, stable for 10 min [8]; <10>, stable for 10 min [11]; <2>, below 45°C stable for 10 min at pH 2.8 [17]) [8, 11, 17]
48 <35> (<35>, Ex 1, stable below [55]) [55]
50 <2, 4, 12, 13, 17, 19, 50> (<2>, unstable above [2]; <4>, purified brain enzyme, for 50 min, loses about 70% of activity [19]; <4>, 95% of activity is destroyed after 2 hours [28]; <4>, rapid inactivation, retarded by acidification [51]; <4>, at pH 5.0, form P: almost stable for 40 min, form A: loss of 70% of activity during 40 min [77]; <12>, mucosal enzymes, relatively slow inactivation, tumor B-form shows greater stability than mucosal B-form [21]; <12>, N-glucosaminidase: complete loss of activity after 30 min with or without addition of bovine serum albumin, N-galactosaminidase: loss of 20% of activity after 30 min with bovine serum albumin [13]; <13>, loses 30% of activity after 30 min [14]; <17>, stable for 100 min [22]; <19>, both hexosaminidase activities of form I decrease rapidly, forms II, III and IV show biphasic denaturation curves [24]; <50>, 50% inactivation after 10 min [75]) [2, 13, 14, 19, 21, 22, 24, 28, 51, 75, 77]
50-52 <4> (<4>, Hex A is labile, Hex B is stable, at lower sample concentrations Hex B is less stable at 52°C [41]) [41]
52 <4> (<4>, for 120 min, 90-95% of activity remains [4]; <4>, only Hex A and not Hex B is deactivated [52]) [4, 52]
55 <12, 15, 27, 29, 33, 48> (<12>, mucosal enzymes, rapid and almost complete inactivation after 5 min [21]; <15>, purified enzyme loses 85% of activity after 20 min [18]; <27>, PI: complete inactivation after 20 min, PII: after 1 h 30% of activity remain [60]; <29>, 60% inactivation after 30 min, complete inactivation after 120 min [43]; <33>, stable below [48]; <48>, major peak: half-life of 17 min at pH 5.0, minor peak: half-life of more than 100 min at pH 5.0 [70]) [18, 21, 43, 48, 60, 70]
56 <54> (<54>, at pH 4.5, loss of 30% of activity after 6 min [83]) [83]
57.5 <8> (<8>, low molecular weight form EII, 50% of activity remains after 10 min [8]) [8]
58 <35> (<35>, alimentary canal enzyme, 50% of activity remains [55]) [55]
60 <12, 13, 16, 19, 23, 32, 36, 50> (<12>, mucosal enzymes, rapid and almost complete inactivation after 5 min [21]; <13>, loses almost all activity after 5 min [14]; <16>, temperatures above quickly inactivate the enzyme [20]; <19>, complete inactivation of all 4 forms after 20 min [24]; <23>, stable for 60 min at pH 4.7 [33]; <32>, stable for 10 min [47]; <36>, complete loss of activity after 40 min, biphasic denaturation curve [57]; <50>, complete inactivation after 10 min [75]) [14, 20, 21, 24, 33, 47, 57, 75]
62 <35> (<35>, Ex 1, 50% of activity remains [55]) [55]
65 <10, 24, 26, 28> (<10>, complete inactivation within 10 min [11]; <24>, over 50% of activity remains after 30 min [34]; <26>, complete loss of activity after 15 min at pH 7.5 [36]; <28>, for 15 min at pH 7.0, rapid inactivation [38]) [11, 34, 36, 38]
70 <36> (<36>, 95% of activity are lost after 2 min [57]) [57]

80 <23, 32> (<23>, completely inactivated at 80°C and above [33]; <32>, retains 28% of activity after 10 min [47]) [33, 47]
90 <32> (<32>, complete inactivation after 10 min [47]) [47]
Additional information <12, 21, 45> [29, 39, 65]

Organic solvent stability

Triton X-100 <1, 25, 54> (<1>, stabilizes activity of purified enzyme [1]; <25>, loss of 70% of activity with 0.2% after 5 h at 37°C [79]; <54>, prevents loss of activity caused by dilution [83]) [1, 79, 83]
acetonitrile <32> [47]
β-mercaptoethanol <16, 18> (<18>, denaturation in subunits [66]; <16>, with 2 mM hydrolysis of *p*-nitrophenyl-β-2-acetamido-2-deoxy-D-glucopyranoside is reduced to 22%, hydrolysis of *p*-nitrophenyl-β-2-acetamido-2-deoxy-D-galactopyranoside to 20% [20]) [20, 66]
dimethyl sulfoxide <4, 32> (<32>, 80%, 18.0% of activity remains [47]; <4>, with 7.5% 14% of activity remains [51]) [47, 51]
dimethylformamide <4, 32> (<32>, 80%, 13.9% of activity remains [47]; <4>, with 7.5% 9% of activity remains [51]) [47, 51]
dioxane <32> [47]
dithiothreitol <4, 18> (<4>, reduction, produces fragments of 63000, 56000, 36000 and 28000 Da [53]; <18>, 2 mM cause total inhibition [76]) [53, 76]
ethanol <32> [47]
guanidine-HCl <18, 32> (<32>, 4 M at 37°C for 30 min, 54% of activity remains [47]; <18>, denaturation in subunits [66]) [47, 66]
n-butanol <32> [47]
n-propanol <32> [47]
pyridine <32> [47]
sodium dodecyl sulfate <25, 32> (<32>, 0.1% and 1% SDS decreases activity to 66% and 26% [47]; <25>, complete inactivation with 0.2% in less than 2 h at 37°C [79]) [47, 79]
urea <18, 32> (<32>, 8 M at 37°C for 30 min, 91% of activity remains [47]; <18>, denaturation in subunits [66]) [47, 66]

General stability information

<2>, relatively stable to freezing and thawing in 0.05 M citrate buffer pH 4.5 [2]
<4>, resistant to mechanical inactivation by homogenization or freezing and thawing [51]
<5>, high purification results in great loss of activity [5]
<9>, purification by N-acetylglucosamine-agarose columns results in extreme unstability of enzyme, extensive purification requires bovine serum albumin for retention of activity [10]
<14>, unstable at pH 4.0, i.e. the enzyme's pH optimum [16]
<16>, lyophilization results in marked decrease in activity [20]
<18>, no loss of activity on repeated freezing and thawing [23]
<18>, unstable during chromatography and dialysis [76]

<25>, stable to repeated freeze-thaw cycles and to lyophilization, 20-30% of activity are lost if enzyme concentration is decreased to 0.02 mg/ml, stable with 0.1 mg bovine serum albumin per ml [79]
<32>, at 70°C for 10 min: most stable in potassium phosphate buffer, more stable under alkaline than acidic conditions [47]
<35>, lyophilization reduces activity to 34.5% [56]
<37>, unstable to freezing and thawing [59]
<47>, enzyme unstable, extensive purification only in presence of 20% glycerol possible [69]
<51>, stable after lyophilization [78]

Storage stability

<1>, -20°C, after 6 weeks no significant loss of activity [1]
<4>, -70°C, hexosaminidase A, B and I2 separated from each other [9]
<4>, 4°C, liver enzyme type S, concentrated by ultrafiltration to 0.18 mg/ml, in 0.04 citrate buffer and in presence of 0.2 mg/ml bovine serum albumin stable for at least 3 months [19]
<5>, -20°C, in 10 mM sodium acetate buffer pH 4.0, very unstable [5]
<5>, -20°C, in 5-50 mM sodium citrate buffer pH 4.0, very stable, even after repeated freeze-thawing [5]
<7>, -20°C, lyophilized [7]
<8>, -20°C, low molecular weight form EII, concentrated by ultrafiltration [8]
<9>, 4°C, enzyme purified in presence of bovine serum albumin, storage in presence of 0.04% w/v sodium azide, enzyme retains 76% of activity after 5 months [10]
<10>, -20°C, enzyme preparation concentrated by ultrafiltration [11]
<11>, lyophilization [12]
<14>, 0°C, in 10% glycerol, pH 5.0-6.0, no loss of activity in 4 months [16]
<16>, -20°C, purified enzyme, stable [20]
<16>, refrigerated temperatures, stable for several days [20]
<18>, -20°C, crude extract [76]
<18>, -20°C, stable for at least 1 year [23]
<18>, 2°C, in $(NH_4)_2SO_4$, stable [76]
<19>, 4°C, after 2 months of storage in phosphate buffer pH 5.0 only form I shows decrease of activity by 50-60%, after 4 months a new form of enzyme with a molecular weight of 62000 Da and a K_m of 0.056 mM appears [24]
<20>, 4°C, purified enzyme stable for few weeks [26]
<22>, -20°C, crude enzyme, stable [74]
<24>, -20°C [34]
<25>, 4°C, crude supernatant fraction stable for at least a year [79]
<25>, 4°C, purified enzyme after sterilization stable for at least 6 months [79]
<25>, 4°C, wild type and mutant enzymes, stable for at least 2 weeks [35]
<26>, -70°C, purified enzyme, stable for at least 6 months [36]
<34>, -20°C, stable for several weeks [49]
<35>, -20°C, after 27 days 87.1% of activity remains [56]
<35>, 4°C, not stable, rapid decrease of activity after the first 12 days [56]

<36>, 4°C, stable for 6 weeks [57]
<37>, -10°C, crude culture supernatant [59]
<37>, 4°C, purified enzyme, stable for several months in presence of 30% $(NH_4)_2SO_4$ in 50 mM phosphate buffer, pH 7.0 [59]
<46>, 4°C, stable for several months [67]
<47>, -20°C, in presence of 20% glycerol stable for at least 2 months, even more stable in 50% glycerol [69]
<47>, -70°C, in presence of 20% glycerol stable for at least 3 months, even more stable in 50% glycerol [69]
<50>, 0°C, half-life of the purified enzyme is 6 days [75]
<51>, 4°C, enzyme solution unstable [78]
<51>, frozen [78]
<12, 18>, -20°C, N-acetyl-glucosaminidase and -galactosaminidase completely lose activity after several days [30]
<12, 18>, -20°C, N-acetyl-hexosaminidase, in presence of ammonium sulfate, after several weeks: precipitate devoid of enzymatic activity forms but can be removed by centrifugation, further purification: loss of activity after several days at -20°C [30]
<40, 41, 42>, -80°C, crude cell-free extract, 0.25-2.0 mg protein/ml [63]

References

[1] Barber, M.S.; Ride, J.P.: Purification and properties of a wheat leaf N-acetyl-β-D-hexosaminidase. Plant Sci., **60**, 163-172 (1989)
[2] Ohtakara, A.: Chitinase and β-N-acetylhexosaminidase from Pycnoporus cinnabarinus. Methods Enzymol., **161**, 462-470 (1988)
[3] Gers-Barlag, H.; Bartz, I.; Ruediger, H.: β-N-Acetylhexosaminidase from soybean. Phytochemistry, **27**, 3739-3741 (1988)
[4] Kinoshita, K.; Taniguchi, N.; Makita, A.; Narita, M.; Oikawa, K.: Purification and characterization of β-N-acetylhexosaminidase I from human placenta. J. Biochem., **104**, 827-831 (1988)
[5] DeGaspari, R.; Li, S.C.; Li, Y.T.: A GM2-specific β-hexosaminidase from the roe of striped mullet (Mugil cephalus). J. Biol. Chem., **263**, 1325-1328 (1988)
[6] Nakagawa, H.; Enomoto, N.; Asakawa, M.; Uda, Y.: Purification and characterization of α-mannosidase and β-N-acatylhexosaminidase from watermelon fruit. Agric. Biol. Chem., **52**, 2223-2230 (1988)
[7] Harley, S.M.; Beevers, L.: Isozymes of β-N-acetylhexosaminidase from pea seeds (Pisum sativum L.). Plant Physiol., **85**, 1118-1122 (1987)
[8] Yamamoto, K.; Tsuji, Y.; Matsushita, S.; Kumagai, H.; Tochikura, T.: Purification and properties of β-N-acetylhexosaminidase from Mucor fragilis grown in bovine blood. Appl. Environ. Microbiol., **51**, 1019-1023 (1986)
[9] Dewji, N.N.; De-Keyser, D.R.; Stirling, J.L.: Purification and characterization of β-N-acetylhexosaminidase I2 from human liver. Biochem. J., **234**, 157-162 (1986)

[10] Poulton, J.E.; Thomas, M.A.; Ottwell, K.K.; McCormick, S.J.: Partial purification and characterization of a β-N-acetylhexosaminidase from black cherry (Prunus serotina Ehrh.) seeds. Plant Sci., **42**, 107-114 (1985)

[11] Yamamoto, K.; Lee, K.M.; Kumagai, H.; Tochikura, T.: Purification and characterization of β-N-acetylhexosaminidase from Penicillium oxalicum. Agric. Biol. Chem., **49**, 611-619 (1985)

[12] Ogawa, Y.; Nakamura, R.; Sato, Y.: Purification of β-N-acetylhexosaminidase from egg white abd the microsomal and lysosomal fractions of hen oviduct. Agric. Biol. Chem., **47**, 2085-2089 (1983)

[13] Izumi, T.; Suzuki, K.: Neutral β-N-acetylhexosaminidases of rat brain. Purification and enzymatic and immunological characterization. J. Biol. Chem., **258**, 6991-6999 (1983)

[14] Uda, Y.; Itoh, T.: Purification and characterization of β-N-acetylhexosaminidase from the ascidian, Halocynthia roretzi. J. Biochem., **93**, 847-855 (1983)

[15] Kress, B.C.; Hirani, S.; Freeze, H.H.; Little, L.; Miller, A.L.: Mucolipidosis III β-N-acetyl-D-hexosaminidase A. Purification and properties. Biochem. J., **207**, 421-428 (1982)

[16] Shigeta, S.; Matsuda, A.; Oka, S.: Purification and characterization of a β-N-acetylhexosaminidase of sea-squirt. J. Biochem., **92**, 163-172 (1982)

[17] Ohtakara, A.; Yoshida, M.; Murakami, M.; Izumi, T.: Purification and characterization of β-N-acetylhexosaminidase from Pycnoporus cinnabarinus. Agric. Biol. Chem., **45**, 239-247 (1981)

[18] Farooqui, A.A.; Srivastava, P.N.: Isolation of β-N-acetylhexosaminidase from rabbit semen and its role in fertilization. Biochem. J., **191**, 827-834 (1980)

[19] Potier, M.; Teitelbaum, J.; Melancon, S.B.; Dallaire, L.: Purification and some properties of liver and brain β-N-acetyl-hexosaminidase S. Biochim. Biophys. Acta, **566**, 80-87 (1979)

[20] Eriquez, L.A.; Pisano, M.A.: Purification and characterization of an extracellular β-N-acetylhexosaminidase from Paecilomyces persicinus. J. Bacteriol., **137**, 620-626 (1979)

[21] Mian, N.; Herries, D.G.; Cowen, D.M.; Batte, E.A.: The multiple forms and kinetic properties of the N-acetyl-β-D-hexosaminidases from colonic tumours and mucosa of rats treated with 1,2-dimethylhydrazine. Biochem. J., **177**, 319-330 (1979)

[22] Calvo, P.; Reglero, A.; Cabezas, J.A.: Purification and properties of β-N-acetylhexosaminidase from the mollusc Helicella ericetorum Müller. Biochem. J., **175**, 743-750 (1978)

[23] Sarber, R.L.; Distler, J.J.; Jourdian, G.W.: β-N-Acetylhexosaminidase from bovine testes. Methods Enzymol., **50**, 520-523 (1978)

[24] Bouquelet, S.; Spik, G.: Properties of four molecular forms of N-acetyl-β-D-hexosaminidase isolated from germinating seeds of fenugreek (Trigonella foenum graecum). Eur. J. Biochem., **84**, 551-559 (1978)

[25] Bouquelet, S.; Spik, G.: Characterization and localization of four forms of N-acetyl-β-D-hexosaminidase from fenugreek (Trigonella foenum graecum) germinated seeds. FEBS Lett., **63**, 95-101 (1976)

[26] Edwards, R.G.; Thomas, P.; Westwood, J.H.: The purification and properties of a β-N-acetylhexosaminidase from Trichomonas foetus. Biochem. J., **151**, 145-148 (1975)
[27] Pokorny, M.; Glaudemans, P.J.: Purification of N-acetyl β-D-hexosaminidase from bull epididymis by affinity chromatography. FEBS Lett., **50**, 66-69 (1975)
[28] Dawson, G.; Propper, R.L.; Dorfman, A.: Partial purification of β-N-acetylhexosaminidase A by affinity chromatography. Biochem. Biophys. Res. Commun., **54**, 1102-1110 (1973)
[29] Li, S.-C.; Li, Y.-T.: Studies on the glycosidases of jack bean meal, III. Crystallisation and properties of β-N-acetylhexosaminidase. J. Biol. Chem., **245**, 5153-5160 (1970)
[30] Frohwein, Y.Z.; Gatt, S.: Isolation of β-N-acetylhexosaminidase, β-N-acetylglucosaminidase, and β-N-acetylgalactosaminidase from calf brain. Biochemistry, **6**, 2775-2782 (1967)
[31] Pérez, L.F.; Ribeiro, H.M.; Casal, J.A.; Pinto, R.A.; Sá Miranda, M.C.; Tutor, J.C.: Thermodynamic characterization of the mutated isoenzyme A of β-N-acetylhexosaminidase in GM2-gangliosidosis B1 variant. Clin. Chim. Acta, **285**, 45-51 (1999)
[32] Yamamoto, K.; Tanaky, T.; Fujimori, K.; Kang, C.-S.; Ebihara, H.; Kanamori, J.; Kadowaki, S.; Tochikura, T.; Kumagai, H.: Characterization of Bacillus sp. endo-β-N-acetylglucosaminidase and its application to deglycosylation of hen ovomucoid. Biotechnol. Appl. Biochem., **28**, 235-242 (1998)
[33] Del Pino, F.B.; Brandelli, A.; Termignoni, C.; Gonzales, J.C.; Henriques, J.A.P.; Dewes, H.: Purification and characterization of β-N-acetylhexosaminidase from bovine tick Boophilus microplus (Ixodide) larvae. Comp. Biochem. Physiol. B, **123**, 193-200 (1999)
[34] Koga, K.; Iwamoto, Y.; Sakamoto, H.; Hatano, K.; Sano, M.; Kato, I.: Purification and characterization of β-N-acetylhexosaminidase from Trichoderma harzianum. Agric. Biol. Chem., **55**, 2817-2823 (1991)
[35] Mark, B.L.; Wasney, G.A.; Salo, T.J.S.; Khan, A.M.; Cao, Z.; Robbins, P.W.; James, M.N.G.; Triggs-Raine, B.L.: Structural and functional characterization of Streptomyces plicatus β-N-acetylhexosaminidase by comparative molecular modeling and site-directed mutagenesis. J. Biol. Chem., **273**, 19618-19624 (1998)
[36] Keyhani, N.O.; Roseman, S.: The chitin catabolic cascade in the marine bacterium Vibrio furnissii: Molecular cloning, isolation, and characterization of a periplasmic β-N-acetylhexosaminidase. J. Biol. Chem., **271**, 33425-33432 (1996)
[37] Garcia-Alonso, J.; Reglero, A.; Cabezas, J.A.: Purification and properties of β-N-acetylhexosaminidase A from pig brain. Int. J. Biochem., **22**, 645-651 (1990)
[38] Chang, C.-T.; Young, F.-P.; Chang, M.-H.; Sung, H.-Y.: Purification and properties of β-N-acetylhexosaminidase from cabbage. Biochem. Mol. Biol. Int., **45**, 371-380 (1998)

[39] Hall, J.C.; Kochins, J.G.; Perez, F.M.: Purification and characterization of N-acetyl-β-D-hexosaminidase in different anatomical regions of the adult rat epididymis. Biochem. Int., **28**, 613-620 (1992)

[40] Ribeiro, M.G.; Pinto, R.A.; Dos Santos, M.R.; Maia, M.; Sá Miranda, M.C.: Biochemical characterization of β-acetylhexosaminidase in different biological specimens from eleven patients with GM2-gangliosidosis B1 variant. J. Inher. Metab. Dis., **14**, 715-720 (1991)

[41] Prence, E.M.; Zalewski, I.; Natowicz, M.R.: Unusual thermolability properties of β-hexosaminidase: Studies of enzyme from cultured cells and clinical implications. Am. J. Med. Genet., **65**, 320-324 (1996)

[42] Pérez, L.F.; Martínez, M.I.; Ribeiro, H.M.; Pinto, R.A.; Sá Miranda, M.C.; Tutor, J.C.: Characterization of the isoenzyme profile of β-N-acetylhexosaminidase in the urine of newborns. Clin. Chem. Lab. Med., **37**, 765-769 (1999)

[43] Geimba, M.P.; Riffel, A.; Brandelli, A.: Purification and characterization of β-N-acetylhexosaminidase from the phytopathogenic fungus Bipolaris sorokiniana. J. Appl. Microbiol., **85**, 708-714 (1998)

[44] Church, W.B.; Swenson, L.; James, M.N.G.: Crystallization of human β-N-acetylhexosaminidase B. J. Mol. Biol., **227**, 577-580 (1992)

[45] Miller, A.L.; Norton, V.; Robertson, R.; Jenks, M.; Yeh, R.Y.; Wright, D.: Light and heavy lysosomes: characterization of N-acetyl-β-D-hexosaminidase isolatad from normal and I-cell disease lymphoblasts. Glycobiology, **3**, 313-318 (1993)

[46] Sakai, T.; Nakanishi, Y.; Kato, I.: Purification and characterization of β-N-acetyl-D-hexosaminidase from the mid-gut gland of scallop (Patinopecten yessoensis). Biosci. Biotechnol. Biochem., **57**, 965-968 (1993)

[47] Sakai, K.; Narihara, M.; Kasama, Y.; Wakayama, M.; Moriguchi, M.: Purification and characterization of thermostable β-N-acetylhexosaminidase of Bacillus stearothermophilus CH-4 isolated from chitin-containing compost. Appl. Environ. Microbiol., **60**, 2911-2915 (1994)

[48] Koga, D.; Hoshika, H.; Matsushita, M.; Tanaka, A.; Ide, A.; Kono, M.: Purification and characterization of β-N-acetylhexosaminidase from the liver of a prawn, Penaeus japonicus. Biosci. Biotechnol. Biochem., **60**, 194-199 (1996)

[49] Pócsi, I.; Kiss, L.; Nánási, P.: Studies on the N-acetyl-β-D-hexosaminidase B from germinating Lupinus luteus L. seeds: I. Purification and characterization. Biochim. Biophys. Acta, **1039**, 110-118 (1990)

[50] Pócsi, I.; Kiss, L.; Zsoldos-Mádi, V.; Pintér, I.: Studies on the N-acetyl-β-D-hexosaminidase B from germinating Lupinus luteus L. seeds: II. Mechanism and inhibition with some 2-acetamido-2-deoxyaldono(1-4)lactones. Biochim. Biophys. Acta, **1039**, 119-122 (1990)

[51] Den Tandt, W.R.; Scharpe, S.: Characteristics of hexosaminidase A in homogenates of white blood cells using methylumbelliferyl-N-acetyl-β-D-glucosaminide-6-sulphate as substrate. Clin. Chim. Acta, **199**, 231-236 (1991)

[52] Prence, E.M.; Natowicz, M.R.; Zalewski, I.: Unusual thermolability properties of leukocyte β-hexosaminidase: Implications in screening for carriers of Tay-Sachs disease. Clin. Chem., **39**, 1811-1814 (1993)
[53] Mahuran, D.J.: Characterization of human placental β-hexosaminidase I2: Proteolytic processing intermediates of hexosaminidase A. J. Biol. Chem., **265**, 6794-6799 (1990)
[54] Iwamoto, T.; Okiura, T.; Sasaki, T.; Inaoka, M.: Kinds of glycosidase and properties of β-N-acetylglucosaminidase bound to mycelia of Streptomyces sp.. J. Ferment. Technol., **65**, 593-596 (1987)
[55] Koga, D.; Shimazaki, C.; Yamamoto, K.; Inoue, K.; Kimura, S.; Ide, A.: β-N-acetyl-D-glucosaminidases from integument of the silkworm, Bombyx mori: Comparative biochemistry with the pupal alimentary canal enzyme. Agric. Biol. Chem., **51**, 1679-1681 (1987)
[56] Koga, D.; Nakashima, M.; Matsukara, T.; Kimura, S.; Ide, A.: Purification and properties of β-N-acetyl-D-glucosaminidase from alimentary canal of the silkworm, Bombyx mori. Agric. Biol. Chem., **50**, 2357-2368 (1986)
[57] Cohen, R.J.: β-N-Acetylglucosaminidase from Phycomyces blakesleeanus. Plant Sci., **43**, 93-101 (1986)
[58] Kitchen, B.J.; Masters, C.J.: Purification and properties of bovine mammary gland N-acetyl-β-D-glucosaminidase. Biochim. Biophys. Acta, **831**, 125-132 (1985)
[59] Bedi, G.S.; Shah, R.H.; Bahl, O.P.: Studies on Turbatrix aceti β-N-acetylglucosaminidase; 1. Purification and physicochemical characterization. Arch. Biochem. Biophys., **233**, 237-250 (1984)
[60] Devery, R.; Collins, P.; Johnson, A.; Watson, G.: Purification of pig gastric-mucosal N-acetyl-β-D-glucosaminidase. Biochem. Soc. Trans., **12**, 460 (1984)
[61] Ceccarini, C.; D'Aniello, A.; Cacace, M.G.; Atkinson, P.H.: Purification and characterization of a β-N-acetylglucosaminidase from Octopus vulgaris. Determination of specificity by using 360-MHz ^{1}H-NMR spectroscopy. Eur. J. Biochem., **132**, 469-476 (1983)
[62] Ishikawa, F.; Oishi, K.; Aida, K.: Purification and immunological properties of chitin-binding hemagglutinin and β-N-acetylglucosaminidase produced by Conidiobolus lamprauges. Agric. Biol. Chem., **47**, 149-151 (1983)
[63] Wheeler, P.R.; Bharadwaj, V.P.; Gregory, D.: N-Acetyl-β-glucosaminidase, β-glucuronidase and acid phosphatase in Mycobacterium leprae. J. Gen. Microbiol., **128**, 1063-1071 (1982)
[64] Jones, C.S.; Kosman, D.J.: Purification, properties, kinetics, and mechanism of β-N-acetylglucosaminidase from Aspergillus niger. J. Biol. Chem., **255**, 11861-11869 (1980)
[65] Mommsen, T.P.: Chitinase and β-N-acetylglucosaminidase from the digestive fluid of the spider, Cupiennius salei. Biochim. Biophys. Acta, **612**, 361-372 (1980)
[66] Khar, A.; Anand, S.R.: Studies on the glycosidases of semen. Purification and properties of β-N-acetylglucosaminidase from bull sperm. Biochim. Biophys. Acta, **483**, 141-151 (1977)

[67] Jain, R.S.; Binder, R.L.; Walz, C.; Buck, C.A.; Warren, L.: Purification of β-N-acetyl-D-glucosaminidase of the horseshoe crab by affinity chromatography. J. Chromatogr., **136**, 141-146 (1977)
[68] Kimura, S.: Insect haemolymph exo-β-N-acetylglucosaminidase from Bombyx mori. Biochim. Biophys. Acta, **446**, 399-406 (1976)
[69] Yem, D.W.; Wu, H.C.: Purification and properties of β-N-acetylglucosaminidase from Escherichia coli. J. Bacteriol., **125**, 324-331 (1976)
[70] Dimond, R.L.; Loomis, W.F.: Vegetative isozyme of N-acetylglucosaminidase in Dictyostelium discoideum. J. Biol. Chem., **249**, 5628-5632 (1974)
[71] Phizackerley, P.J.R.; Bannister, J.V.: The characterization of N-acetylglucosaminidases from the limpet Patella vulgata L.. Biochim. Biophys. Acta, **362**, 129-135 (1974)
[72] Verpoorte, J.A.: Isolation and characterization of the major β-N-acetyl-D-glucosaminidase from human plasma. Biochemistry, **13**, 793-799 (1974)
[73] Berkeley, R.C.W.; Brewer, S.J.; Ortiz, J.M.; Gillespie, J.B.: An exo-β-N-acetylglucosaminidase from Bacillus subtilis B. Characterization. Biochim. Biophys. Acta, **309**, 157-168 (1973)
[74] Ortiz, J.M.; Gillespie, J.B.; Berkeley, R.C.W.: An exo-β-N-acetylglucosaminidase from Bacillus subtilis B. Extraction and purification. Biochim. Biophys. Acta, **289**, 174-186 (1972)
[75] Reyes, F.; Byrde, R.J.: Partial purification and properties of a β-N-acetylglucosaminidase from the fungus Sclerotina fructigena. Biochem. J., **131**, 381-388 (1973)
[76] Verpoorte, J.A.: Purification of two β-N-acetyl-D-glucosaminidases from beef spleen. J. Biol. Chem., **247**, 4787-4793 (1972)
[77] Stirling, J.L.: Separation and characterization of N-acetyl-β-glucosaminidases A and P from maternal serum. Biochim. Biophys. Acta, **271**, 154-162 (1972)
[78] Mitchell, E.D.; Houston, C.W.; Latimer, S.B.: Purification and properties of a β-N-acteylaminoglucohydrolase from malted barley. Phytochemistry, **15**, 1869-1871 (1976)
[79] Trimble, R.B.; Tarentino, A.L.; Evans Aumick, G.; Maley, F.: Endo-β-N-acetylglucosaminidase L from Streptomyces plicatus. Methods Enzymol., **83**, 603-610 (1982)
[80] Bannister, J.V.; Phizackerley, P.J.R.: Purification of an N-acetylglucosaminidase from the limpet Patella vulgata L.. FEBS Lett., **29**, 313-317 (1973)
[81] Jenkinson, H.F.; Shepherd, M.G.: A mutant of Candida albicans deficient in β-N-acetylglucosaminidase (chitobiase). J. Gen. Microbiol., **133**, 2097-2106 (1987)
[82] Sone, Y.; Misaki, A.: Purification and characterization of β-D-mannosidase and β-N-acetyl-D-hexosaminidase of Tremella fuciformis. J. Biochem., **83**, 1135-1144 (1978)
[83] Woollen, J.W.; Heymorth, R.; Walker, P.G.: Studies on glucosaminidases. 3. Testicular N-acetyl-β-D-glucosaminidase and N-acetyl-β-D-galactosaminidase. Biochem. J., **78**, 111-116 (1961)

β-N-Acetylgalactosaminidase 3.2.1.53

1 Nomenclature

EC number
3.2.1.53

Systematic name
β-N-acetyl-D-galactosaminide N-acetylgalactosaminohydrolase

Recommended name
β-N-acetylgalactosaminidase

Synonyms
Bacillus β-GlcNAcase <3> [3]
N-acetyl-β-D-galactosaminidase
N-acetyl-β-galactosaminidase
N-acetylgalactosaminidase
β-D-N-acetylgalactosaminidase
β-N-acetylgalactosaminidase
β-acetylgalactosaminidase

CAS registry number
9054-43-7

2 Source Organism

<1> *Rattus norvegicus* (rat [1]) [1]
<2> *Bos taurus* (calf [2]) [2]
<3> *Bacillus sp.* (AT173-1 [3]) [3]

3 Reaction and Specificity

Catalyzed reaction
hydrolysis of terminal non-reducing N-acetyl-D-galactosamine residues in N-acetyl-β-D-galactosaminides

Reaction type
hydrolysis of O-glycosyl bond

Natural substrates and products
S N-acetyl-β-D-galactosaminides + H_2O <1, 2> [1, 2]
P ?

S β-N-acetylgalactosaminylated oligosaccarides + H_2O <3> [3]
P ?

Substrates and products

S 4-methylumbelliferyl β-N-acetylgalactosaminide + H_2O <1> [1]
P ?
S N-acetyl-β-D-galactosaminides + H_2O <1-3> (<1, 2> e.g. 4-methylumbelliferyl-β-N-acetylgalactosaminide [1, 2]; <3> substrate specificity towards β-GalNAc residues is markedly higher than those of any other β-GalNacase [3]) [1-3]
P N-acetyl-D-galactosamine + an alcohol <1, 2> (<1,2> e.g. 4-methylumbelliferone [1, 2]) [1, 2]
S β-phenyl-N-acetylgalactosamines + H_2O <2> (<2> substrate specificity 85% [2]) [2]
P ?
S *p*-nitrophenyl β-N-acetylgalactosaminide + H_2O <2, 3> [2, 3]
P ?

Inhibitors

$AgNO_3$ <2> [2]
Cu^{2+} <3> (<3> enzyme activity completely inhibited in the presence of 10 mM $CuSO_4$ [3]) [3]
Fe^{2+} <3> (<3> enzyme activity completely inhibited in the presence of 10 mM $FeSO_4$ [3]) [3]
N-acetylgalactosamine <1-3> (<2> competitive inhibition [2]; <3> relative activity in presence of 10 mM 50%, in presence of 50 mM 12% [3]) [1-3]
N-acetylglucosamine <2> (<2> no inhibition at N-acetylglucosamine concentration of 40 mM [2]) [2]
Zn^{2+} <3> (<3> enzyme activity completely inhibited in the presence of 10 mM $ZnSO_4$ [3]) [3]
p-hydroxymercuribenzoate <2> [2]
urea <3> (<3> relative activity in presence of 2 M 32% [3]) [3]

Activating compounds

N-ethylmaleimide <2> (<2> 1mM solution increase the reaction rates 2-3fold [2]) [2]
dithiothreitol <3> [3]

Metals, ions

Mg^{2+} <3> (<3> relative activity in presence of $MgSO_4$ 10 mM 104% [3]) [3]

Specific activity (U/mg)

9.8 <1> [1]
15.4 <3> [3]

K_m-Value (mM)

0.35 <2> (*p*-nitrophenyl β-acetylgalactosaminide) [2]
0.42 <1> (4-methylumbelliferyl β-N-acetylgalactosaminide) [1]
1 <3> (*p*-nitrophenyl β-acetylgalactosaminide) [3]

pH-Optimum

5.5 <2> (<2> *p*-nitrophenyl β-N-acetylgalactosaminide [2]) [2]
6 <1, 3> (<1, 3> 4-methylumbelliferyl β-N-acetylgalactosaminide [1, 3]) [1, 3]

pH-Range

4 <2> (<2> not active below, *p*-nitrophenyl β-N-acetylgalactosaminide [2]) [2]
5.5-9.5 <3> (<3> enzyme retains more than 80% activity [3]) [3]

Temperature optimum (°C)

45 <3> [3]

Temperature range (°C)

4-45 <3> (<3> enzyme gradually inactivated in solution at 4°C, reactivated by incubation with DTT [3]) [3]

4 Enzyme Structure

Molecular weight

48000 <3> (<3> SDS-PAGE, enzyme denatured with SDS and 2-mercaptoethanol [3]) [3]
54000 <3> (<3> native enzyme, gel filtration [3]) [3]
82000 <1> (<1> sucrose density gradient centrifugation [1]) [1]
210000 <1> (<1> gel filtration [1]) [1]

5 Isolation/Preparation/Mutation/Application

Source/tissue

brain <1, 2> [1, 2]
culture medium <3> (<3> AT 173-1, culture broth centrifuged to remove cells [3]) [3]
liver <2> [2]
spleen <2> [2]

Localization

lysosome <1> [1]

Purification

<1> [1]
<2> [2]
<3> [3]

6 Stability

Organic solvent stability

acetone <2> (<2> inactivation [2]) [2]

General stability information

<2>, cysteine, stabilization [2]

<2>, dithiothreitol, stabilization [2]

Storage stability

<2>, -20 C, inactivation after several weeks [2]

References

[1] Izumi, T.; Suzuki, K.: Neutral β-N-acetylhexosaminidases of rat brain. Purification and enzymatic and immunological characterization. J. Biol. Chem., **258**, 6991-6999 (1983)

[2] Frohwein, Y.Z.; Gatt, S.: Isolation of β-N-acetylhexominidase, β-N-acetylglucosaminidase, and β-N-acetylgalactosaminidase from calf brain. Biochemistry, **6**, 2775-2782 (1967)

[3] Tanaka, A.; Ozaki, S.: Purification and characterization of β-N-acetylgalactosaminidase from Bacillus sp. AT173-1. J. Biochem., **122**, 330-336 (1997)

Cyclomaltodextrinase 3.2.1.54

1 Nomenclature

EC number
3.2.1.54

Systematic name
cyclomaltodextrin dextrin-hydrolase (decyclizing)

Recommended name
cyclomaltodextrinase

Synonyms
CD-ase
H-17 CDase <10> [9]
H-17 thermostable CDase <10> [9]
Lsp26X-Mdase <12> [13]
cyclodextrinase
cycloheptaglucanase
cyclohexaglucanase
cyclomaltodextrin dextrin-hydrolase
cytoplasmic decycling maltodextrinase <12> [13]
neopullulanase <14> [18]
thermophilic CDase <5> [10]

CAS registry number
37288-41-8

2 Source Organism

<1> *Bacillus coagulans* (gram positive bacterium, isolated from soil [1]) [1, 5]
<2> *Bacillus macerans* (ATCC8514 [2, 3]) [2, 3]
<3> *Bacteroides distasonis* (C18-7 [4]) [4]
<4> *Bacteroides ovatus* (3524 [4]) [4]
<5> *Clostridium thermohydrosulfuricum* (39E [6, 10]) [6, 10]
<6> *Bacillus circulans* [7]
<7> *Bacillus sp.* (I-5 [17, 19]) [17, 19]
<8> *Bacillus sphaericus* (E-244 [8]; ATCC 7055 [12]) [8, 12]
<9> *Bacillus stearothermophilus* (HY-1 [14, 20]) [14, 20]
<10> *Bacillus subtilis* (H-17, transformed with Bacillus caldolyticus C2 DNA [9]) [9]
<11> *Flavobacterium sp.* [11]

<12> *Lactobacillus sp.* [13]
<13> *Klebsiella oxytoca* (M5a1 [15, 16]) [15, 16]
<14> *Alicyclobacillus acidocaldarius* (ATCC27009 [18]) [18]

3 Reaction and Specificity

Catalyzed reaction

cyclomaltodextrin + H_2O = linear maltodextrin

Reaction type

hydrolysis of O-glycosyl bond

Natural substrates and products

S cyclomaltodextrins + H_2O <1, 2> [1, 2]
P linear maltodextrins <1, 2> [1, 2]

Substrates and products

S 2-O-α-maltosylglucose + H_2O <1> [5]
P ?
S 4-O-α-D-glucosylsucrose + H_2O <1> [5]
P ?
S acarbose + H_2O <7> (<7> pseudotetrasaccharide inhibitor of glucosidases and α-amylases, binding the active sites of the enzyme [17, 19]) [17, 19]
P ?
S α-cyclodextrin <9> [14]
P maltotriose + maltose + glucose <9> [14]
S α-cyclodextrin + H_2O <3> [4]
P linear maltodextrins <3> [4]
S α-cyclodextrin + H_2O <3, 10, 11> [4, 9, 11]
P glucose + maltose + maltotriose <3, 11> [4, 11]
S α-cyclodextrin + H_2O <10> [9]
P maltose + glucose <10> [9]
S amylopectin + H_2O <5, 9, 10> (<5> rate of hydrolysis less than 5% of the rate of α-cyclodextrin hydrolysis [10]) [9, 10, 14, 20]
P maltose <10> (<10> degraded slowly by cleavage of maltose units from the nonreducing ends [9]) [9]
S amylose + H_2O <5, 9, 10> (<5> 53% hydrolysis [6]; <5> less than 5% of the rate of α-cyclodextrin hydrolysis [10]; <10> 12% [9]) [6, 9, 10, 14, 20]
P ?
S β-cyclodextrin + H_2O <3, 8> [4, 12]
P glucose + maltose + maltotriose + maltotetraose <3, 8> [4, 12]
S β-cyclodextrin + H_2O <5> [6]
P glucose + maltose + long chain linear oligomers <5> [6]
S β-cyclodextrin + H_2O <6, 10> [7, 9]
P maltose + maltotriose + maltotetraose + maltopentaose + maltoheptaose + glucose <6> [7]

S β-cyclodextrin + H_2O <8, 10> [9, 12]
P maltose + glucose <10> [9]
S branched cyclodextrins + H_2O <8> [12]
P glucose + maltose <8> [12]
S cyclodextrin + H_2O <4> (<4> α and β [4]) [4]
P glucose <4> [4]
S cyclomaltodextrin + H_2O <1-7, 8, 9, 11-14> (<1> α-, β- and γ [1]; <2> opening the ring of the cyclic dextrins forming a linear molecule with corresponding number of glucose units [3]; <5> α-cyclodextrin 100% hydrolysis, β-cyclodextrin 67% hydrolysis, γ-cyclodextrin 8% [6]) [1-5, 7, 10-18, 20]
P maltodextrin <3, 4, 6> (<4> linear, glucose only [4]; <3> series of maltooligomers [4]; <6> maltooligosaccharides [7]) [4, 7]
S cyclomaltoheptaose + H_2O <1> [5]
P ?
S cyclomaltohexaose + H_2O <1> [5]
P ?
S γ-cyclodextrin + H_2O <8, 11> [11, 12]
P glucose + maltose + maltotriose + maltotetraose + maltopentaose + maltohexaose + maltoheptaose + maltooctaose <11> [11]
S glucosyl-α-cyclodextrin + H_2O <8> (<8> branched α-cyclodextrin, synonym 6-O-α-D-glucosylmaltohexaose [8]) [8]
P 6,3-O-α-D-glucosylmaltohexaose <8> [8]
S glucosyl-β-cyclodextrin + H_2O <8> (<8> branched β-cyclodextrin, synonym 6-O-α D-glucosylmaltoheptaose [8]) [8, 12]
P 6,4-O-α-D-glucosylmaltoheptaose <8> [8]
S glycogen + H_2O <5> (<5> relative rate of hydrolysis 0,1% [10]) [10]
P ?
S hydroxyethyl-β-cyclodextrin + H_2O <12> [13]
P ?
S hydroxypropyl-β-cyclodextrin + H_2O <12> [13]
P ?
S isopanose + H_2O <1> [5]
P ?
S maltodextrins + H_2O <1, 2, 5, 6, 9-11, 13> (<2, 5, 9, 10> linear, enzyme degrades the linear molecules by splitting off maltose units from the nonreducing end of the chain [3, 9, 10, 14]; <11> 6 to 8 membered maltooligosaccharides [11]) [1, 3, 7, 9-11, 14, 15]
P maltose + glucose <10> [9]
S maltoheptaose + H_2O <2, 5, 6, 8, 10, 11> [3, 7, 9, 10, 11, 12]
P glucose + maltose <8> [12]
S maltohexaose + H_2O <1, 2, 5, 6, 10, 11> [1, 3, 5, 7, 9, 10, 11]
P maltotetraose + maltose <1, 2, 5, 6, 10> [3, 5, 7, 9, 10]
S maltopentaose + H_2O <1, 2, 5, 6, 8, 10> [1, 3, 5, 7, 9, 10, 12]
P maltose + maltotriose <1, 2> [3, 5]
S maltosyl-α-cyclodextrin + H_2O <8> (<8> branched α-cyclodextrin, synonym 6-O-α-D-maltosylcyclomaltohexaose [8]) [8, 12]

P 6,3-O-α-D-maltosylmaltohexaose <8> [8]
S maltosyl-β-cyclodextrin <8> (<8> branched β-cyclodextrin, synonym 6-O-α-D-maltosylcyclomaltoheptaose [8]) [8, 12]
P 6,4-O-α-D-maltosylmaltoheptaose <8> [8]
S maltotetraose + H_2O <1, 2, 5, 6, 10> [1, 3, 5, 7, 9, 10]
P maltose <1, 2> [3, 5]
S maltotriitol + H_2O <1> [5]
P ?
S maltotriose + H_2O <1, 5, 6, 10, 11> (<1, 5> slowly [1, 10]) [1, 5, 7, 9, 10, 11]
P ?
S methyl-β-cyclodextrin + H_2O <12> [13]
P ?
S phenyl α-maltoside + H_2O <1> [5]
P ?
S pullulan + H_2O <9, 14> [14, 18, 20]
P panose + maltose + glucose <9, 14> [14, 18]
S pullulan + H_2O <5, 7, 8, 10-12> (<5> rate of hydrolysis less than 5% of the rate of α-cyclodextrin hydrolysis [10]) [9-13, 17]
P glucose + panose + isopanose + large saccharides containing α 1,6-glucosidic bonds <11> [11]
S starch + H_2O <5, 7, 8-12, 14> (<5> 50% hydrolysis [6]; <10> 10% [9]) [6, 9-14, 17, 18, 20]
P glucose + maltose <8, 10, 11> (<10> degraded slowly by cleavage of maltose units from the nonreducing ends [9]; <11> glucose minor early product, + maltotriose + maltotetraose [11]) [9, 11, 12]
S Additional information <1> (<1, 5> not: maltose [1, 10]; <1> slowly: short chain amylose [1]; <1> hydrolyses by detaching one maltose unit, recognizes maltose residues at non-reducing end of various maltosides and hydrolyzes (1$\rightarrow$ 4)-, (1$\rightarrow$ 2)-, (1$\rightarrow$ 3)-and (1$\rightarrow$ 6)-α-D-glucosidic linkages and the glucosidic linkages between D-glucose and phenol, D-glucose and D-glucitol, and D-glucose and D-fructose adjacent to the maltose molecule [5]) [1, 5]
P ?

Inhibitors

Ag^+ <5-7> [7, 10, 17]
Cr^{3+}<7> (<7> relative activity 58.4% [17]) [17]
Cu^{2+} <5-7, 9> (<9> relative activity 1.6% [20]) [7, 10, 14, 17, 20]
Fe^{2+} <1> (<1> reduces activity to79% [1]) [1]
Fe^{3+} <1, 7, 9> (<1> reduce activity to 50% [1]; <9> slightly inhibition [14]; <9> relative activity 33% [20]; <7> relative activity 37.7% [17]) [1, 14, 17, 20]
Hg^{2+} <1, 5, 6, 9> (<1> reduces activity to 24% [1]; <9> relative activity 1.6% [20]) [1, 7, 10, 14, 20]
Li^+ <7> (<7> relative activity 75.5% [17]) [17]
Mg^{2+} <7> (<7> relative activity 43% [17]) [17]
Mn^{2+} <7> (<7> relative activity 64.2% [17]) [17]

N-bromosuccinimide <5, 9> (<9> 0.1 mM, relative activity 8%, 1.0 mM, relative activity 4.6% [20]) [10, 20]
Ni^{2+} <7> (<7> relative activity 72.2% [17]) [17]
Pb^{2+} <9> [14]
SDS <6> (<6> 0.1% [7]) [7]
Zn^{2+} <5, 7, 9> (<7> relative activity 5.6% [17]; <9> relative activity 1.5% [20]) [10, 14, 17, 20]
acarbose <8> [12]
diethyldicarbonate <9> (<9> DEPC 10 mM, relative activity 2.3% [20]) [20]
dodecylbenzenesulfonate <6> (<6> 0.1% [7]) [7]
p-chloromercuribenzoate <5, 6> [6, 7, 10]

Activating compounds

2-deoxy-D-glucose <8> (<8> 0.01 M, relative activity 140% [12]) [12]
EDTA <7> (<7> 5 mM EDTA, activity increased more than 2fold, relative activity 237.2% [17]) [17]
EGTA <7> (<7> increases activity about 2.3fold, relative activity 236.3% [17]) [17]
glucosamine <8> (<8> 0.01 M, relative activity 133% [12]) [12]
glucose <8> (<8> 0.01 M, relative activity 143% [12]) [12]
glucose 6-phosphate <8> (<8> 0.005 M, relative activity 111% [12]) [12]
maltose <8> (<8> 0.01 M, relative activity 140% [12]) [12]
miglitol <8> (<8> 1 mg/ml, relative activity 107% [12]) [12]

Metals, ions

Ba^{2+} <7> (<7> relative activity 101.5% [17]) [17]
Ca^{2+} <7> (<7> activated and stabilized by Ca^{2+}, relative activity 128.7% [17]) [17]
Mg^{2+} <1, 9> (<1, 9> increase activity to 112%/102% [1, 14]) [1, 14]

Specific activity (U/mg)

0.12 <1> (<1> crude enzyme [1]) [1]
5 <2> (<2> crude enzyme, purified 1fold [2]) [2]
14.5 <5> [6]
14.8 <6> (<6> gel filtration [7]) [7]
33.4 <9> (<9> affinity chromatography [14]) [14]
60 <11> (<11> CM-Servacel, pH 6.7 [11]) [11]
62 <5> (<5> gel filtration [10]) [10]
100 <11> (<11> β-cyclodextrin as substrate [11]) [11]
125 <12> (<12> β-cyclodextrin as substrate [13]) [13]
148 <7> (<7> crude enzyme, purification 1fold [17]) [17]
175 <2> [2]
366.3 <8> (<8> homogenous by SDS-PAGE [12]) [12]
Additional information <1, 2> [1, 2]

K_m-Value (mM)

0.244 <11> (β-cyclodextrin) [11]
0.47 <1> (γ-cyclodextrin) [1]
0.8 <6> (γ-cyclodextrin) [7]

0.8 <6> (maltoheptaose) [7]
0.83 <7> (β-cyclodextrin) [19]
0.91 <10> (maltoheptaose) [9]
0.92 <7> (γ-cyclodextrin) [19]
0.93 <10> (α-cyclodextrin) [9]
1.2 <13> (γ-cyclodextrin) [15]
1.2 <6> (maltopentaose) [7]
1.23 <7> (α-cyclodextrin) [19]
1.3 <5> (γ-cyclodextrin) [10]
1.3 <6> (maltohexaose) [7]
1.4 <1> (maltohexaose) [1]
1.67 <9> (β-cyclodextrin) [14]
1.79 <9> (α-cyclodextrin) [14]
1.8 <6> (β-cyclodextrin) [7]
2.1 <5> (β-cyclodextrin) [10]
2.11 <7> (pullulan) [19]
2.3 <1> (maltopentaose) [1]
2.5 <5> (α-cyclodextrin) [10]
2.5 <9> (γ-cyclodextrin) [14]
2.62 <2> (α-cyclodextrin) [2]
2.65 <2> (β-cyclodextrin) [2]
2.8 <1> (β-cyclodextrin) [1]
3.01 <7> (soluble starch) [19]
3.2 <6> (α-cyclodextrin) [7]
3.9 <8, 13> (β-cyclodextrin) [12, 15]
4 <8> (α-cyclodextrin) [12]
4 <1> (maltotetraose) [1]
4.2 <8> (γ-cyclodextrin) [12]
4.5 <1> (maltotriose) [1]
5.3 <1> (maltotriose) [5]
5.5 <1> (phenyl α-maltoside) [5]
5.8 <8> (maltoheptaose) [12]
6.1 <8> (maltohexaose) [12]
7.4 <6> (maltotetraose) [7]
7.5 <8> (maltotetraose) [12]
8.3 <9> (pullulan) [14]
10 <1> (α-cyclodextrin) [1]
12.5 <6> (maltotriose) [7]
14 <8> (maltotriose) [12]
17 <1> (4-O-α-D-glucosylsucrose) [5]
21 <1> (2-O-α-maltosylglucose) [5]
28 <1> (isopanose) [5]
50 <13> (α-cyclodextrin) [15]
110 <1> (maltotriitol) [5]

pH-Optimum

5-7.2 <12> [13]

5.5 <14> [18]
5.9 <5> [10]
6 <5, 6> [6, 7]
6 <7> (<7> soluble starch, the optimum pH depending on the substrate [17]) [17]
6-6.5 <8> (<8> 97% activity at pH 7 [12]) [12]
6-7 <7> (<7> pullulan as substrate [17]) [17]
6-7.5 <11> [11]
6.2 <1, 9> [1, 14, 20]
6.2-6.4 <2> [2]
7 <3, 4, 13> (<3, 4, 13> crude enzyme [4, 15]) [4, 15]
7 <7> (<7> β-cyclodextrin as substrate [17]) [17]

pH-Range
4-7 <14> [18]
4-10 <5> [10]
5-7.5 <9> [14]
5-8 <1, 5, 6> [1, 6, 7]
5.8-7.8 <11> [11]
6-7.6 <2> [2]
Additional information <4> (<4> crude enzyme [4]) [4]

Temperature optimum (°C)
23 <13> [15]
37 <3, 4> (<3,4> pH 7, expected conditions of the human colon environment [4]) [4]
40 <8> [12]
40-50 <7> (<7> in the presence of Ca^{2+} [17]) [17]
42 <4> (<4> crude enzyme, incubation at 37°C and higher for longer than 24 h, significant loss of activity [4]) [4]
50 <1, 6> [1, 7]
50 <7> (<7> pullulan, optimum temperature variable depending on the substrate [17]) [17]
55 <3, 9, 14> (<3> crude enzyme, incubation at 4-55°C for 48 h, little loss of activity [4]) [4, 14, 18, 20]
65 <5> (<5> β-cyclodextrin as substrate [6, 10]) [6, 10]

Temperature range (°C)
10-34 <13> [15]
10-70 <4> [4]
20-70 <14> [18]
30-55 <2> [2]
30-55 <6> (<6> no activity at 60°C [7]) [7]
30-70 <3> [4]
30-80 <7> [17]
40-70 <9> [14]
50-75 <5> [10]
50-80 <5> [6]

4 Enzyme Structure

Molecular weight

61000 <9> (<9> SDS-PAGE [20]) [14, 20]
62000 <1, 11, 12> (<1> SDS-PAGE [1]) [1, 11, 13]
63000 <7> (<7> MW deduced from the nucleotide sequence 64884 Da) [17, 19]
66000 <5, 14> (<5, 14> monomer SDS-PAGE [10, 18]) [10, 18]
68030 <5> (<5> denatured protein estimated by SDS-PAGE, native enzyme determined by gel filtration [10]) [10]
69000 <13> (<13> gel filtration [15]) [15]
91200-95000 <8> (<8> gel filtration [12]) [12]
126000 <6> (<6> MW estimation, gel filtration [7]) [7]

Subunits

dimer <6> (<6> 2 * 67000, molecular weight estimated to be 126000 by gel filtration, 67000 by SDS-PAGE [7]) [7]

5 Isolation/Preparation/Mutation/Application

Source/tissue

colon <3, 4> (<3, 4> cell-bound in bacteroids [4]) [4]

Localization

cell-bound <11> [11]
cytoplasm <13, 14> [15, 18]
intracellular <1, 3, 4> [1, 4]

Purification

<1> (not completely purified, minor contamination in ultracentrifugation [1]; purified to homogeneity [5]) [1, 5]
<2> (purified 40fold, highly purified, not completely free from contaminations [2, 3]) [2, 3]
<5> (partially purified 205fold [6]; gene encoding enzyme cloned in E. coli, expressed and purified 12fold from E. coli [10]) [6, 10]
<6> (purified 252fold [7]) [7]
<7> (gene encoding enzyme cloned in E. coli, expressed and purified from E. coli [17]) [17, 19]
<8> [8, 12]
<9> (purified 230fold [14]) [14]
<10> [9]
<11> (purified to 94% homogeneity [11]) [11]
<12> (purified to 93% homogeneity [13]) [13]
<13> [15, 16]

Cloning

<5> (plasmid pPG22, expression in E. coli [10]) [10]
<7> (CDase I-5 gene, pBR322, expression in E. coli MC1061 [17, 19]) [17, 19]

<13> (gene cymH, pCYMH, expression in E. coli JM109 [15]) [15]
<14> (cdaA gene, pJM30, expression in E. coli DK6F) [18]

Application
analysis <13> (<13> synthesis of radiolabeled linear and cyclic maltodextrins as tools for metabolism investigation [16]) [16]

6 Stability

pH-Stability
4.8-7.2 <12> (<12> 30°C, 6 h stable [13]) [13]
5.5 <5> [6]
5.8-7.6 <9> [14]
6-7 <8> (<8> 100% stable [12]) [12]
6-7.3 <1> (<1> 40°C, 2 h stable [1]) [1]
6-10 <6> (<6> 40°C, 30 min stable [7]) [7]
6.8 <3, 4> (<3, 4> 37°C [4]) [4]

Temperature stability
4-55 <3> (<3> 48 h, little loss of activity [4]) [4]
37 <4> (<4> 24 h, significant loss of activity [4]) [4]
40 <1, 8, 13> (<1> stable up at pH 6.0-7.3 on 2 h incubation [1]; <8> 100% stable for 30 min up to 40°C [12]; <13> irreversibel inactivated at 40°C [15]) [1, 12, 15]
45 <1> (<1> stable at temperatures below , pH 7 [1]) [1]
50 <9> (<9> stable at temperatures below [14]) [14]
60 <7, 9, 11> (<11> enzyme is not thermostable, above 60°C more than 80% activity loss [11]; <7> thermal activity more than 80% in presence of Ca^{2+} [17]) [11, 14, 17]
60-70 <5> (<5> enzyme half-live 402 min at 60°C, 180 min at 65°C, 75 min at 70°C [6]) [6]
65 <5, 10, 12> (<5> incubation for 1 h led to a 15% loss of activity [10]; <12> 10 min at 65°C [13]) [9, 10, 13]

Storage stability
<6>, 4°C, one month without any loss of activity [7]

References

[1] Kitahata, S.; Taniguchi, M.; Beltran, S.D.; Sugimoto, T.; Okada, S.: Purification and some properties of cyclodextrinase from Bacillus coagulans. Agric. Biol. Chem., **47**, 1441-1447 (1983)
[2] DePinto, J. A.; Campbell, L. L.: Purification and properties of the cyclodextrinase of Bacillus macerans. Biochemistry, **7**, 121-125 (1968)

[3] DePinto, J. A. ; Campbell, L. L.: Pattern of action of the amylase and the cyclodextrinase of Bacillus macerans. Arch. Biochem. Biophys., **125**, 253-258 (1968)
[4] Antenucci, R.N.; Palmer, J.K.: Enzymatic degradation of α- and β-cyclodextrins by bacteroids of the human colon. J. Agric. Food Chem., **32**, 1316-1321 (1984)
[5] Kitahata, S.; Okada, S.: Hydrolytic action on various maltosides by an enzyme from Bacillus coagulans. Carbohydr. Res., **137**, 217-225 (1985)
[6] Badal, C.S.; Zeikus, J.G.: Characterization of thermostable cyclodextrinase from Clostridium thermohydrosulfuricum 39E. Appl. Environ. Microbiol., **9**, 2941-2943 (1990)
[7] Yoshida, A.; Iwasaki, Y.; Akiba, T.; Horikoshi, K.: Purification and properties of cyclomaltodextrinase from alkalophilic Bacillus sp.. J. Ferment. Bioeng., **71**, 226-229 (1991)
[8] Oguma, T.; Kikuchi, M.; Mizusawa, K.: Hydrolysis of branched cyclodextrins by a cyclodextrin-hydrolyzing enzyme from Bacillus sphaericus E-244. FEBS Lett., **290**, 13-16 (1991)
[9] Bradley, M.K.; Lindsay, J.A.: Reclassification of a thermostable α-glucosidase from Bacillus subtilis H-17 as a cyclomaltodextrinase. Enzyme Microb. Technol., **14**, 194-196 (1992)
[10] Podkovyrov, S.M.; Zeikus, J.G.: Structure of the gene encoding cyclomaltodextrinase from Clostridium thermohydrosulfuricum 39E and characterization of the enzyme purified from Escherichia coli. J. Bacteriol., **8**, 5400-5405 (1992)
[11] Bender, H.: Purification and characterization of a cyclodextrin-degrading enzyme from Flavobacterium sp.. Appl. Microbiol. Biotechnol., **39**, 714-719 (1993)
[12] Galvin, N.M.; Kelly, C.T.; Fogarty, W.M.: Purification and properties of the cyclodextrinase of Bacillus sphaericus ATCC 7055. Appl. Microbiol. Biotechnol., **42**, 46-50 (1994)
[13] Bender, H.: Purification and characterization of a soluble, cytoplasmic decycling maltodextrinase from a Lactobacillus species, strain 26X, isolated from kitchen waste water. Appl. Microbiol. Biotechnol., **43**, 838-843 (1995)
[14] Zhong, W.; Zheng, Z.S.; Jun, Y.S.: Purification and properties of cyclodextrinase from Bacillus stearothermophilus HY-1. Appl. Biochem. Biotechnol., **59**, 63-75 (1996)
[15] Feederle, R.; Pajatsch, M.; Kremmer, E.; Böck, A.: Metabolism of cyclodextrins by Klebsiella oxytoca M5a1: purification and characterisation of a cytoplasmically located cyclodextrinase. Arch. Microbiol., **165**, 206-212 (1996)
[16] Pajatsch,M.; Böck, A.; Boos, W.: Enzymatic preparation of radiolabeled linear maltodextrins and cyclodextrins of high specific activity from [^{14}C]maltose using amylomaltase, cyclodextrin glucosyltransferase and cyclodextrinase. Carbohydr. Res., **307**, 375-379 (1998)
[17] Kim, T.-J.; Shin, J.-H.; Oh, J.-H.; Kim, M.-J.; Lee, S.-B.; Ryu, S.; Kwon, K.; Kim, J.-W.; Choi, E.-H.; Robyt, J.F.; Park, K.-H.: Analysis of the gene encoding cyclomaltodextrinase from alkalophilic Bacillus sp. I-5 and cgaracteri-

zation of enzymatic properties. Arch. Biochem. Biophys., **353**, 221-227 (1998)

[18] Matzke, J.; Herrmann, A.; Schneider, E.; Bakker, E.P.: Gene cloning, nucleotide sequence and biochemical properties of a cytoplasmic cyclomaltodextrinase (neopullunase) from Alicyclobacillus acidocaldarius, reclassification of a group of enzymes. Fems Microbiol. Lett., **183**, 55-61 (2000)

[19] Kim, M.-J.; Park, W.-S.; Lee, H.-S.; Kim, T.-J.; Shin, J.-H.; Yoo, S.-H.; Cheong, T. K.; Ryu, S.; Kim, J.-C.; Kim, K.-W.; Moon, T.-W.; Robyt, J.F.; Park, K.-H.: Kinetics and inhibition of cyclomaltodextrinase from alkalophilic Bacillus sp. I-5. Arch. Biochem. Biophys., **373**, 110-115 (2000)

[20] Yang, S.J.; Wang, Z.; Zhang, S. Z.: Purification and properties of cyclodextrinase from Bacillus stearothermophilus HY-1. Ann. N. Y. Acad. Sci., **799**, 425-428 (1996)

α-N-Arabinofuranosidase 3.2.1.55

1 Nomenclature

EC number
3.2.1.55

Systematic name
α-L-arabinofuranoside arabinofuranohydrolase

Recommended name
α-N-arabinofuranosidase

Synonyms
EC 3.2.1.79 (formerly, created 1972, incorporated 1976)
L-arabinosidase
acid arabinosidase <3> [3]
α-L-arabinanase
α-L-arabinofuranoside hydrolase
α-L-arabinosidase
α-L-arafase I <73> [21]
α-L-arafase II <73> [21]
α-arabinofuranosidase
α-arabinosidase
arabinosidase
arabinoxylan arabinofuranohydrolase <41, 47, 51> [34, 38, 40]
polysaccharide α-L-arabinofuranosidase

CAS registry number
9067-74-7

2 Source Organism

<1> *Aspergillus niger* (strain KI [1, 20, 24]; strain 5195 [9]; strain N400(CBS 120-49) [15, 29]; strain 5-16 [20, 25]; 3M43 [30]; strain B [35]) [1-10, 12, 14-20, 22-32, 34-37, 39, 40, 43]
<2> *Corticium rolfsii* (plant-pathogenic oxalate forming fungus, IFO 4878 [2]) [2, 5, 6, 8, 9, 22, 29]
<3> *Rhodotorula flava* (yeast, strain IFO 0407, also called Cryptococcus flavus, Torula flava, Chromotorula flava [3]) [3, 5, 6, 25, 37]
<4> *Botrytis cinerea* [3, 6]
<5> *Botrytis alli* [3]
<6> *Sclerotinia libertiana* [3, 6]

<7> *Sclerotinia sclerotiorum* [3, 26, 31]
<8> *Sclerotina fructigena* [6, 7, 31]
<9> *Monilinia fructigena* [11, 16]
<10> *Clostridium acetobutylicum* (ATCC 824 [10, 16, 23, 25, 26, 37]) [10, 16, 22, 23, 25, 26, 37]
<11> *Clostridium felsineum* (var. sikokianum [3, 8]) [3, 5, 6, 8]
<12> *Bacillus subtilis* (F-11 [5, 9, 17, 25]; strain 3-6 [8, 25, 34]) [3, 6, 7, 9, 10, 15, 16, 17, 22, 23, 25, 29, 31, 34, 36, 39, 40, 44]
<13> *Ruminococcus albus* (8 [7, 17, 23-26, 28, 29]) [7, 10, 16, 17, 23-26, 28, 29, 36]
<14> *Streptomyces sp.* (no. 17-1 [5, 7, 10, 17, 22, 23, 25]; no. 003 [5]; no. 065 [7]) [5, 7, 10, 16, 17, 22, 23, 24, 25, 29]
<15> *Streptomyces purpurascens* (IFO 3389 [5, 6, 8, 15-17, 23-26, 34]) [5, 6, 8, 10, 15, 16, 17, 22-26, 28, 31, 34, 36]
<16> *Scopolia japonica* [7, 13, 14, 17, 25, 37]
<17> *Coniothyrium diplodiella* [3]
<18> *Gloeosporium kaki* [3]
<19> *Fusarium oxysporum* (f. lupini [3]; F3 [36]) [3, 36]
<20> *Butyrivibrio fibrisolvens* (GS113 [16, 23, 25, 26, 28, 32, 37]; NCFB 2249 [23]) [16, 22, 23, 25, 26, 28, 32, 37]
<21> *Bacillus sp.* (strain 430 [25]; D3 [39]) [16, 17, 25, 39, 40]
<22> *Pseudomonas sp.* [16]
<23> *Bacteroides ruminicola* [16]
<24> *Bacteroides succinogenes* (synonym *Fibrobacter succinogenes* [16]) [16]
<25> *Bacteroides ovatus* [16, 39, 40]
<26> *Ruminococcus flavefaciens* [16]
<27> *Streptomyces diastaticus* (strain ET/BW200 [17]) [17, 25, 26, 37]
<28> *Streptomyces diastatochromogenes* (strain 065 [25]) [17, 25, 39]
<29> *Trichoderma reesei* (filamentous fungus, VTT-D-79125 [24]; PC-3-7 [33]) [17, 22, 24-26, 29, 31, 33, 35, 36, 43]
<30> *Bacillus stearothermophilus* (L1 [18,26]; T-6 [26, 37]) [18, 26, 28, 34, 37, 40, 41]
<31> *Streptomyces lividans* (66strain 1326; mutant IAF10-164 [22]; 66 [26]) [22, 26, 33, 34, 39, 43, 44]
<32> *Dichomitus squalens* [22, 24, 25, 36, 37]
<33> *Pseudomonas fluorescens* [22, 30, 39, 42]
<34> *Clostridium stercoarium* [22, 28, 34, 39, 40, 42, 43]
<35> *Bacteroides xylanolyticus* (X5-1, DSM 3808 [23]) [23]
<36> *Aspergillus nidulans* [24, 31, 35, 43]
<37> *Aspergillus terreus* (VTT-D-82209 [27, 31]; CECT 2663 [31, 37]) [29, 31, 32, 34, 35, 37]
<38> *Thermobacillus xylanilyticus* (D3 [34]) [34]
<39> *Penicillium purpurogenum* [35]
<40> *Pichia capsualata* (X91 [37]) [37]
<41> *Hordeum vulgare* (barley, cv. Maud [38]; L.cv. Clipper [40]) [38, 40, 43]
<42> *Streptomyces chartreusis* (GS901 [39, 42, 43]) [39-43]
<43> *Pseudomonas cellulosa* [41, 42]
<44> *Streptomyces thermoviolaceus* (OPC-520 [43]) [43]

<45> *Geobacillus stearothermophilus* (T-6 [44]; mutant E175A [44]) [44]
<46> *Talaromyces emersonii* [29]
<47> *Aspergillus awamori* (CMI 142717 [29]; IFO4033 [34, 35, 37, 39]) [29, 32, 34-39, 43]
<48> *Streptomyces massasporeus* (IFO 3841 [6, 25]) [5, 6, 25]
<49> *Erwinia carotovora* (IAM 1024 [8]) [5, 6, 8, 31]
<50> *Raphanus sativus* (radish [21, 34]) [21, 34]
<51> *Bifidobacterium adolescentis* [34, 38, 42]
<52> *Cytophaga xylanolytica* [34, 39, 40]
<53> *Penicillium capsulatum* [35]
<54> *Thermotoga maritima* [39, 40]
<55> *Streptomyces coelicolor* [39, 40, 43, 44]
<56> *Aureobasidium pullulans* [34, 35, 37]
<57> *Aspergillus aculeatus* [25]
<58> *Aspergillus sojae* [27, 43]
<59> *Glycine max* (soybean [14, 25]) [14, 25]
<60> *Bacillus halodurans* [44]
<61> *Agrobacterium tumefaciens* [44]
<62> *Mesorhizobium loti* [44]
<63> *Streptomyces chattanoogensis* [43]
<64> *Sinorhizobium meliloti* [44]
<65> *Arabidopsis thaliana* [39, 40]
<66> *Streptomyces aureofaciens* (strain KA 204 [6]) [5, 6]
<67> *Streptomyces flaveolus* (strain KA1143 [6]) [5, 6]
<68> *Streptomyces lavendulae* (strain KA140 [6]) [5, 6]
<69> *Streptomyces olivochromogenes* (IFO3178 [6]) [5, 6]
<70> *Botrytis fabae* [6]
<71> *Myrothecium verrucaria* [6]
<72> *Daucus carota* (carrot [14]) [14]
<73> *Spinacia oleracea* [21]

3 Reaction and Specificity

Catalyzed reaction

hydrolysis of terminal non reducing α-L-arabinofuranoside residues in α-L-arabinosides

Reaction type

hydrolysis of O-glycosyl bond

Natural substrates and products

S arabinooligosaccharide + H_2O <16> [13]
P L-arabinose + ?
S beet arabinan + H_2O <49> [8]
P arabinotriose + ?
S beet arabinan + H_2O <1-3, 10, 12-16, 19, 20, 30, 31, 36, 37, 39-43, 49, 59> (<1, 16> extracted from beet pulp [1, 3, 13, 24]; <1> prepared from sugar

beet chips [4, 15]; <3> α-1,5-, 1,3-arabinan [3]) [1-6, 8-10, 13-16, 22, 26, 29, 32, 35-37, 40-42]
P L-arabinose + ?
S birchwood xylan + H_2O <36, 39> [24, 35]
P L-arabinose + ?
S bran extract + H_2O <3> [3]
P L-arabinose + ?
S citrus galactan + H_2O <1> [29]
P ?
S corn endosperm xylan + H_2O <20> [16]
P L-arabinose + ?
S grape glycoside + H_2O <40> [37]
P ?
S grape monoterpenyl arabinofuranosylglucoside + H_2O <1> [29, 37]
P ?
S gum arabic + H_2O <1, 36> [9, 12, 24]
P L-arabinose + ?
S larchwood arabinogalactan + H_2O <39-41> [35, 37, 40]
P L-arabinose + ?
S larchwood xylan + H_2O <38> [34]
P L-arabinose + ?
S oat spelt xylan + H_2O <1, 10, 20-25, 27-30, 34, 37-39, 44, 47, 58> [16, 17, 26-30, 33-35, 43]
P L-arabinose + ?
S rye flour arabinoxylan + H_2O <31, 37> [22, 29]
P L-arabinofuranose + ?
S rye flour arabinoxylan+ H_2O <37> [29]
P L-arabinose + ?
S softwood Kraft pulp + H_2O <30> [18]
P lignin + ?
S softwood Kraft pulp + H_2O <37> (<37> arabinoglucuronoxlan I and II [29]) [29]
P L-arabinose + ?
S sugar beet arabinan + H_2O <41, 73> (<73> α-L-arafase II [21]) [21, 40]
P L-arabinose + ?
S sugar beet pectin + H_2O <1, 3> [3, 30]
P L-arabinose + ?
S wheat L-arabino-D-xylan + H_2O <1, 38, 43, 47> [9, 29, 30, 32, 34, 41, 42]
P L-arabinose + ?
S wheat bran + H_2O <37, 39, 58> [27, 32, 35]
P L-arabinose + ?
S wheat straw xylan + H_2O <19, 39> [35, 36]
P L-arabinose + ?
S wheat-flour arabinoxylan + H_2O <31, 41> [22, 40]
P L-arabinofuranose + ?

Substrates and products

S 1,3-arabinan + H_2O <14> [5]
P L-arabinose
S 1,5-L-arabinan + H_2O <1-3, 14, 27, 43, 59> [1-5, 9, 14, 17, 29, 41]
P L-arabinose + ?
S 1,5-α-L-arabinobiose + H_2O <15, 40, 43> [10, 37, 42]
P L-arabinose
S 1,5-α-L-arabinofuranohexaose + H_2O <1, 40, 41, 43> [37, 40, 42]
P L-arabinose + L-arabinopentaose
S 1,5-α-L-arabinotetraose + H_2O <40, 43> [37, 42]
P L-arabinose
S 1,5-α-L-arabinotriose + H_2O <1, 15, 40, 43> [10, 37, 41, 42]
P L-arabinose
S 1,5-α-L-arabinotriose + H_2O <1> [19]
P L-arabinose + arabinobiose
S 1,5-methyl α-L-arabinofuranobioside + H_2O <42> [39]
P methyl α-L-arabinofuranose
S 2,5-dinitrophenyl β-D-xylopyranoside + H_2O <45> [44]
P 2,5-dinitrophenol + β-D-xylopyranose
S 2-O-α-L-arabinofuranosyl-α-L-arabinofuranoside + H_2O <42> [42]
P α-L-arabinofuranose
S 2-nitrophenyl β-D-xylopyranoside + H_2O <45> [44]
P 2-nitrophenol + β-D-xylopyranose
S 3,4-dinitrophenyl β-D-xylopyranoside + H_2O <45> [44]
P 3,4-dinitrophenol + β-D-xylopyranose
S 3-O-α-L-arabinofuranosyl-α-L-arabinofuranoside + H_2O <42> [42]
P α-L-arabinofuranose
S 5-O-α-L-arabinofuranosyl-α-L-arabinofuranoside + H_2O <42> [42]
P α-L-arabinofuranose
S L-arabinan + H_2O <1, 10, 12, 15, 19, 30, 31, 37, 40-43, 58, 66-72> (<43> preferentially removes α 1,2 and α 1,3-linked arabinofuranose side chains [42]) [5, 6, 9, 10, 12, 14, 19-20, 22, 25-27, 29, 36, 37, 39, 40, 42]
P $(1,5\text{-}\alpha\text{-L-arabinose})_{25}$ + ?
S L-arabinofuranosyl-D-glucoside + H_2O <1> [29]
P L-arabinose + D-glucose
S L-arabinofuranoxylobiose + H_2O <46> [29]
P L-arabinose + xylobiose
S L-arabinogalactan + H_2O <1, 14, 19, 30, 32, 34, 36, 37, 40-42> (<30> low activity on the substrate [18]) [5, 10, 18, 22, 24, 28, 30, 32, 36, 37, 39, 40]
P L-arabinose + ?
S L-arabinoglucuronoxylan + H_2O <37> [29, 32]
P L-arabinose + ?
S L-arabinopentaose + H_2O <40, 43> [37, 42]
P L-arabinotetraose + L-arabinose
S L-arabinoxylan + H_2O <1, 3, 14, 19, 27, 29-31, 34, 38, 39, 41-44, 51, 58> (<3> arabinofuranosyl-1,3-xyloglycan [3]; <30> low activity on the substrate [18]; <43> preferentially removes α-1,2- and α-1,3-linked arabino-

furanose side chains [41, 42]) [1, 3, 5, 10, 12, 17, 18, 22, 27, 28, 30, 33-35, 37-43]
P L-arabinose + ?
S L-arabinoxylohexaose + H_2O <31> [22]
P L-arabinose + ?
S L-arabinoxylopentaose + H_2O <31> [22]
P L-arabinose + ?
S L-arabinoxylotetraose + H_2O <31> [22]
P L-arabinose + ?
S L-arabinoxylotriose + H_2O <31> [22]
P L-arabinose + ?
S O-α-L-arabinofuranosyl-(1-3-)-O-β-D-xylopyranosyl-(1-4)-D-xylopyranose + H_2O <1, 12> [20, 25]
P L-arabinose + O-β-D-xylopyranosyl-(1-4)-D-xylopyranose
S O-β-D-xylopyranosyl-(1-4)-[O-α-L-arabinofuranosyl-(1-3)]-O-β-D-xylopyranosyl-(1-4)-D-xylopyranose + H_2O <12> [25]
P L-arabinose + O-β-D-xylopyranosyl-(1-4)-O-β-D-xylopyranosyl-(1-4)-D-xylopyranose
S alkyl-α-arabinofuranoside + H_2O <34> [28]
P L-arabinose + alkyl alcohol
S α-L-arabinofuranosyl 1-naphthol + H_2O <9> [11]
P α-L-arabinofuranose + 1-naphthol
S α-L-arabinofuranosyl 3,4-dinitrophenol + H_2O <9> [11]
P α-L-arabinofuranose + 3,4-dinitrophenol
S α-L-arabinofuranosyl 3,5-dinitrophenol + H_2O <9> [11]
P α-L-arabinofuranose + 3,5-dinitrophenol
S α-L-arabinofuranosyl 4-bromoisoquinoline + H_2O <9> [11]
P α-L-arabinofuranose + 4-bromoisoquinoline
S α-L-arabinofuranosyl 4-methylumbelliferone + H_2O <9, 36, 43> [11, 24, 42]
P methylumbelliferone + α-L-arabinofuranose
S α-L-arabinofuranosyl 4-phenol + H_2O <9> [11]
P α-L-arabinofuranose + phenol
S α-L-arabinofuranosyl 4-phenylphenol + H_2O <9> [11]
P α-L-arabinofuranose + 4-phenylphenol
S α-L-arabinofuranosyl-1,3-arabinose + H_2O <16> [13]
P L-arabinose
S α-L-arabinofuranosyl-1,5-arabinose + H_2O <16> [13]
P L-arabinose
S α-L-arabinofuranotriose + H_2O <1> [19]
P L-arabinose
S arabino-4-O-methylglucuronoxylan + H_2O <46> [29]
P ?
S arabino-4-O-methylglucuronoxylan + H_2O <46> [29]
P ?
S arabinoxylobiose + H_2O <20, 31> [16, 22]
P L-arabinose + xylobiose
S aryl-α-arabinofuranoside + H_2O <34, 39, 43> [28, 35, 41]

P L-arabinose + aryl alcohol
S base-soluble carbohydrates + H_2O <13> (<13> BSC of alfalfa cell walls [10]) [10]
P ?
S carboxymethyl cellulose + H_2O <30> [26]
P ?
S carboxymethyl-arabinan + H_2O <27> (<27> debranched substrate from sugar beets [17]) [17]
P L-arabinose + ?
S methylumbelliferyl-α-L-arabinofuranoside + H_2O <20, 34, 39> [16, 28, 35]
P methylumbelliferone
S *p*-nitrophenyl L-arabinoside + H_2O <9> [11]
P L-arabinose + 4-nitrophenol
S *p*-nitrophenyl-α-D-glucoside + H_2O <10> (<10> 9% of the rate exhibited on *p*-nitrophenyl-α-L-arabinofuranoside [10]) [10]
P *p*-nitrophenol + α-D-glucose
S *p*-nitrophenyl-α-L-arabinofuranoside + H_2O <1, 9, 10, 12-16, 19, 27, 29-31, 34, 35, 37-45, 58, 73> (<12> arabinofuranosidase I [9]) [5-7, 9-13, 16-21, 22, 23, 27-29, 32-44]
P *p*-nitrophenol + L-arabinose
S *p*-nitrophenyl-β-D-galactopyranoside + H_2O <34> (<34> 0.33% of activity observed with *p*-nitrophenyl-α-L-arabinofuranoside [28]) [28]
P 4-nitrophenol + β-D-galactopyranose
S *p*-nitrophenyl-β-D-glucopyranoside + H_2O <34> (<34> 0.26% of activity observed with *p*-nitrophenyl-α-L-arabinofuranoside [28]) [28]
P 4-nitrophenol + β-D-glucopyranose
S *p*-nitrophenyl-β-D-xylopyranoside + H_2O <34, 73> (<34> 0.44% of activity observed with *p*-nitrophenyl-α-L-arabinofuranoside [28]; <73> α-L-arafase I [21]) [21, 28]
P 4-nitrophenol + β-D-xylopyranose
S phenyl α-L-arabinofuranoside + H_2O <1-65> [1-4, 7, 9-13, 15-18, 20, 22-26, 28-44]
P L-arabinose + phenol
S phenyl-α-L-arabinofuranoside + H_2O <1, 49> [8, 12, 29]
P L-arabinose + phenol
S polygalacturonase-soluble carbohydrates + H_2O <13> (<13> PGSC of alfalfa cell walls [10]) [10]
P ?
S Additional information <1, 3, 10, 12, 14, 15, 19, 20, 30, 31, 35, 37-42, 44, 49, 58, 59> (<1> no release of arabinose from linear arabinan [19]; <1> O-β-D-xylopyranosyl-(1-4)-O-[α-L-arabinofuranosyl-(1-3)]-O-β-D-xylopyranosyl-(1-4)-D-xylopyranose, O-β-D-xylopyranosyl-(1-2)-O-α-L-arabinofuranosyl-(1-3)-O-β-D-xylopyranosyl(1-4)-O-β-D-xylopyranosyl-(1-4)-D-xylopyranose, gum arabic and arabinoxylan are no substrates [20]; <3> no enzyme activity on L-arabinose and other mono-and oligosaccharides, inactive on *p*-nitrophenyl α-D-galactopyranoside and *p*-nitro-

phenyl β-D-galactopyranoside [3]; <10> no activity against *p*-nitrophenylglycosides, except *p*-nitrophenyl-α-D-glucoside, no activity on carboxymethylcellulose, arabinogalactan, arabinoxylan or oat spelt xylan, larchwood xylan, CMC and rye arabinoxylan are not degraded [10]; <12> arabinofuranosidase I does not hydrolyze L-arabinan, arabinofuranosidase II has no activity on polygalacturonic acid, carboxymethylcellulose, larch galactan, citrus pectic galactan or the synthetic substrates *p*-nitrophenyl derivatives of α- and β-D-xylopyranosides, α- and β-D-glucopyranosides, β-D-fucopyranoside and β-D-galactopyranoside [9]; <12> O-β-D-xylopyranosyl-(1-2)-O-α-L-arabinofuranosyl-(1-3)-O-β-D-xylopyranosyl-(1-4)-O-β-D-xylopyranosyl-(1-4)-D-xylopyranose, arabinoxylan, gum arabic and arabinogalactan are no substrates [25]; <14> inactive on gum arabic and *p*-nitrophenyl-α- and β-galactopyranosides, purified enzyme is inactive on xylan, galactan, soluble starch, carboxymethyl cellulose, xyloside, glycoside, mannoside, fucoside, N-acetyl-glucosaminides and N-acetylgalactosaminides [5]; <15> inactive on *p*-nitrophenyl β-D-galactopyranoside, phenyl β-L-arabinopyranoside and *p*-nitrophenyl glycosides like α-D-galactopyranoside, α-and β-D-glucopyranosides, α- and β-D-xylopyranosides, α-L-fucopyranoside, α-D-mannopyranoside and N-acetyl β-D-glucosaminide [6]; <15> inactive towards arabinan, arabinoxylan and arabinogalactan [6, 22]; <19> does not hydrolyze *p*-nitrophenyl-α-L-arabinopyranoside [36]; <20> no release of sugars from cellulose, carboxymethyl cellulose or ara-arabinogalactan, no activity on arabinoxylobiose linked to coumaric or ferulic acid [16]; <30> no activity detected with β-L-arabinopyranoside, β-D-cellobiose, α-L-fucopyranoside, β-L-fucopyranoside, α-L-rhamnopyranoside, α-L-arabinopyranoside, β-D-galactopyranoside, β-D-glucopyranoside, α-D-mannopyranoside, β-D-xylopyranoside and arabinogalactan [26]; <31> no activity on *p*-nitrophenyl glycosides like α-L-arabinopyranose, β-D-glucose, β-D-galactopyranose, β-D-cellobiose, α-L-mannopyranose, oat spelts xylan, xylan or arabinogalactan [22]; <35> no activity with oat spelt xylan or arabinogalactan, not able to hydrolyse *p*-nitrophenyl-α-L-arabinopyranoside, xylobiose and xylotriose are not hydrolysed [23]; <37> lacks the ability to hydrolyze efficiently α-1,5-linked arabinan, able to release α-1,3-linked arabinose subunits as only product from polymeric arabinoglucuronoxylan [29]; <38> does not hydrolyze *p*-nitrophenyl-α-L-arabinopyranoside or gum arabic [34]; <39> less than 2% activity with *p*-nitrophenyl-β-D-xylopyranoside, *p*-nitrophenyl α-L-arabinopyranoside, *o*-nitrophenyl-β-D-xylopyranoside, *p*-nitrophenyl-β-D-glucopyranoside, *p*-nitrophenyl-β-D-mannopyranoside, *p*-nitrophenyl-β-D-fucopyranoside, *p*-nitrophenyl-α-D-galactopyranoside [35]; <40> shows very little or no activity towards *p*-nitrophenyl α-L-rhamnopyranoside, *p*-nitrophenyl-β-D-glucopyranoside, *p*-nitrophenyl-β-D-xylopyranoside and *p*-nitrophenyl-β-D-cellobioside [37]; <41> no activity against *p*-nitrophenyl α-L-arabinopyranoside or *p*-nitrophenyl β-D-xylopyranoside, inactive against linear or branched-chain arabinan [38]; <41> enzyme is unable to hydrolyse 4-nitrophenyl arabinopyranosides, xylopyranosides, mannopyranosides, glucopyranosides or galactopyranosides, enzyme does not hydrolyse laminarin, β-glucan, gum arabic or carboxymethylxylan [40]; <42> does not hydrolyse gum arabic or arabinoxylo-oligosaccharides, p-nitrophenyl α-arabinopyranoside, p-nitrophenyl-β-D-xylopyranoside or p-nitrophenyl-β-D-galactopyranoside [39]; <42> does not attack arabinoxylans [41]; <44> no hydrolysis observed with birchwood xylan, linear 1,5-arabinan, arabinogalactan, gum arabic, p-nitrophenyl α-L-arabinopyranoside, p-nitrophenyl β-D-xylopyranoside, p-nitrophenyl β-D-galactopyranoside or p-nitrophenyl β-D-glucopyranoside [43]; <49> inactive on 1,5-arabinan, arabinoxylan, arabinogalactan and p-nitrophenyl α-L-arabinofuranoside [8]; <58> enzyme cannot hydrolyze p-nitrophenyl β-D-galactopyranoside [27]; <59> enzyme does not degrade arabinogalactan or arabinoxylan, arabinose can not be hydrolyzed by the purified enzyme, β-1,3 bonding arabinoside are not good substrates [14]) [3, 5, 6, 8-10, 14, 16, 20, 22, 23, 26, 27, 34-41, 43]

P ?

Inhibitors

1-α-L-arabinofuranosylmethyl-3-p-nitrophenyltriazene <9> [11]
Ag^+ <1, 3, 14, 15, 73> (<1> inhibited by $AgNO_3$ [12]; <73> α-L-arafase I [21]) [3, 5, 6, 12, 21]
Ba^{2+} <59> [14]
Ca^{2+} <14> [5]
Cd^{2+} <44> [43]
Co^{2+} <38, 44> [34, 43]
Cu^{2+} <3, 16, 34, 37, 38, 40, 44, 59> [3, 13, 14, 28, 31, 34, 37, 43]
D-glucono-1,5-lactone <73> (<73> α-L-arafase I [21]) [21]
D-xylose <73> (<73> α-arafase I [21]) [21]
EDTA <44, 58, 59> [14, 27, 43]
Fe^{2+} <30> (<30> 1 mM $FeSO_4$ inhibits the activity by 25% [26]) [26]
Hg^{2+} <3, 14, 15, 30, 34, 36-38, 40, 44, 49, 59, 73> (<14> 1 mM $Hg(Ch_3COO)_2$ [5]; <30,38> 1 mM $HgCl_2$ inhibits the activity by 98% [26,34]; <73> α-L-arafase I [21]) [3, 5, 6, 8, 14, 21, 24, 26, 28, 31, 34, 37, 43]
L-arabino-γ-lactone <15, 59> [6, 14]
L-arabinose <1> (<1> competitive inhibitor [12]) [12]
Mn^{2+} <58> [27]
Ni^{2+} <44> [43]
SDS <34, 73> (<73> α-L-arafase I [21]) [21, 28]
Sn^{2+} <3> [3]
Zn^{2+} <16, 34> [13, 28]
ethanol <34> (<34> slight inhibition [28]) [28]
glucose <37> (<37> reduces original activity of ABF2 and ABF3 by 20% [31]) [31]
mercuric acetate <1> [9]
mercurous acetate <1> [12]
p-chloromercuribenzoate <14, 40, 44> [5, 37, 43]
p-hydroxymercuribenzoate <34, 6> [3, 24, 28]
sodium tetrathionate <1> [9]

Activating compounds

2-mercaptoethanol <40> (<40> 1 mM, relative activity 102% [37]) [37]
$CdCl_2$ <1> (<1> 1 mM, relative activity 103.1% [20]) [20]
D-galactono-1,4-lactone <59> (<59> 0.5 mM, relative enzyme activity 102% [14]) [14]
EDTA <1> (<1> 1 mM, relative activity 105.8% [20]) [20]
L-arabinose <20, 29> (<20, 29> inducer of AF activity [26, 33]) [26, 33]
SDS <1> (<1> 1 mM, relative activity 101.7% [20]) [20]
Triton X-100 <38> [34]
arabitol <37, 39> (<37, 39> best inducer [31, 35]) [31, 35]
lactose <36> (<36> induces enzyme activity [24]) [24]
monoiodoacetic acid <1> (<1> 1 mM, relative activity 102.7% [20]) [20]
sorbose <29> (<29> inducer of AF activity [33]) [33]
wheat bran <27> (<27> best inducer for enzyme production [17]) [17]
xylan <20> (<20> induces enzyme activity [26]) [26]
xylose <36> (<36> induces enzyme activity [24]) [24]

Metals, ions

Ca^{2+} <30, 40, 44, 59> (<40> 1 mM $CaCl_2$, relative activity 105% [37]; <30, 44> weak stimulation of the enzyme activity [26, 43]; <59> 0.5 mM $CaCl_2$, relative enzyme activity 139% [14]) [14, 26, 37, 43]
Co^{2+} <59> (<59> 0.5 mM $CoCl_2$, relative enzyme activity 111% [14]) [14]
Fe^{2+} <40> (<40> 1 mM $FeCl_2$, relative activity 105% [37]) [37]
K^+ <49> (<49> 2 mM KCl, relative activity 102% [8]) [8]
Mg^{2+} <30, 49> (<30> small stimulating effect [26]; <49> $MgSO_4$ 2 mM, relative activity 102% [8]) [8, 26]
Mn^{2+} <30> (<30> small stimulating effect [26]) [26]
Zn^{2+} <59> (<59> activated by 0.5 mM $ZnSO_4$, relative enzyme activity 180% [14]) [14]

Specific activity (U/mg)

0.061 <29> (<29> isoenzyme AF35, oat spelt xylan as substrate [33]) [33]
0.61 <73> (<73> α-L-arafase I [21]) [21]
1.8 <10> (<10> *p*-nitrophenyl-α-D-glucoside as substrate [10]) [10]
2.5 <73> (<73> α-L-arafase II [21]) [21]
2.6 <29> (<29> isoenzyme AF53, oat spelt xylan as substrate [33]) [33]
3.16 <42> (<42> isoenzyme II [39]) [39]
3.4 <14> [5, 17]
4.184 <41> [38]
6.3 <41> [40]
6.42 <19> [36]
7.1 <20> [16]
7.6 <9> [11]
9.4 <9> (<9> isoenzyme AFIII [11]) [11]
10.4 <42> (<42> isoenzyme I [39]) [39]
10.5 <37> (<37> isoenzyme ABF3 [31]) [31]
19.2 <10> [10]
19.4 <10> (<10> *p*-nitrophenyl-α-L-arabinofuranoside as substrate [10]) [10]

32.5 <3> [3]
36.3 <40> [37]
40 <49> [8]
44 <2> [2]
48.8 <12> [17]
50.4 <37> (<37> isoenzyme ABF2 [31]) [31]
58.4 <39> [35]
69.5 <58> [27]
76.6 <37> (<37> isoenzyme ABF1 [31]) [31]
82.9 <15> [6]
95 <20, 30> (<20,30> extracellular enzyme activity [26]) [26]
95.2 <59> [14]
101 <30> [18]
105 <27> (<27> isoenzyme C1 [17]) [17]
108 <35> [23]
120 <29> (<29> isoenzyme AF53 [33]) [33]
126 <30> (<30> purified enzyme [18]) [18]
126.7 <1> [20]
131.2 <27> (<27> isoenzyme C1 [17]) [17]
153 <31> [22]
196 <29> (<29> isoenzyme AF35 [33]) [33]
320 <12> [25]
350 <38> (<38> 60°C [34]) [34]
366.6 <1> [17]
388.6 <1> [12]
390 <1> [12]
396.6 <1> [1]
490 <38> (<38> 75°C [34]) [34]
611 <16> [13]
655.5 <13> [7, 17]
749.1 <30> [26]
883 <34> (<34> recombinant arfB gene product [28]) [28]

K_m-Value (mM)

0.017 <1> (L-arabinan) [9]
0.042 <43> (*p*-nitrophenyl-α-L-arabinofuranoside, <43> mutant E194A [42]) [42]
0.08 <15> (*p*-nitrophenyl-α-L-arabinofuranoside) [16]
0.082 <15> (*p*-nitrophenyl-α-L-arabinofuranoside) [6, 10]
0.09 <9> (α-L-arabinofuranosyl 4-phenylphenol) [11]
0.1 <45> (2,5-dinitrophenyl β-D-xylopyranoside, <45> mutant E175A [44]) [44]
0.18 <43> (*p*-nitrophenyl-α-L-arabinofuranoside) [42]
0.22 <9> (α-L-arabinofuranosyl 1-naphthol) [11]
0.22 <30> (*p*-nitrophenyl-α-L-arabinofuranoside) [18]
0.26 <43> (4-methylumbelliferyl-α-L-arabinofuranoside) [42]
0.26 <1> (beet arabinan) [10]

0.27 <43> (wheat arabinoxylan) [42]
0.31 <37> (p-nitrophenyl-α-L-arabinofuranoside, <37> isoenzyme ABF3 [31]) [31]
0.33 <43> (sugar beet arabinan) [42]
0.35 <37> (p-nitrophenyl-α-L-arabinofuranoside, <37> isoenzyme ABF1 [31]) [31]
0.42 <30> (p-nitrophenyl-α-L-arabinofuranoside) [26]
0.47 <37> (p-nitrophenyl-α-L-arabinofuranoside, <37> isoenzyme ABF2 [31]) [31]
0.48 <9> (α-L-arabinofuranosyl 4-methylumbelliferone) [11]
0.48 <1> (p-nitrophenyl-α-L-arabinofuranoside, <1> strain B [35]) [24, 35]
0.5 <35> (p-nitrophenyl-α-L-arabinofuranoside) [23]
0.5 <41> (p-nitrophenyl-α-L-arabinofuranoside) [38]
0.5 <38> (p-nitrophenyl-α-L-arabinose, <38> 75°C [34]) [34]
0.52 <1> (p-nitrophenyl-α-L-arabinofuranoside, <1> α-L-arabinofuranosidase B [15]) [15]
0.53 <59> (p-nitrophenyl-α-L-arabinofuranoside) [14]
0.6 <31> (p-nitrophenyl-α-L-arabinofuranoside) [22]
0.6 <1> (p-nitrophenyl-α-L-arabinofuranoside) [24]
0.66 <20> (p-nitrophenyl-α-L-arabinofuranoside) [16]
0.68 <1> (p-nitrophenyl-α-L-arabinofuranoside, <1> α-L-arabinofuranosidase A [15]) [15]
0.7 <20> (p-nitrophenyl-α-L-arabinofuranoside) [16]
0.71 <9> (α-L-arabinofuranosyl 3,4-dinitrophenol) [11]
0.72 <38> (p-nitrophenyl-α-L-arabinose, <38> 60°C [34]) [34]
0.77 <41> (p-nitrophenyl-α-L-arabinofuranoside) [40]
0.8 <9> (p-nitrophenyl-α-L-arabinofuranoside) [16]
0.82 <9> (p-nitrophenyl-α-L-arabinofuranoside) [11]
0.83 <9> (α-L-arabinofuranosyl 3,5-dinitrophenol) [11]
0.9 <45> (3,4-dinitrophenyl β-D-xylopyranoside, <45> mutant E175A [44]) [44]
0.9 <29> (p-nitrophenyl-α-L-arabinofuranoside, <29> isoenzyme AF53 [33]) [33]
1.1 <29> (p-nitrophenyl-α-L-arabinofuranoside, <29> isoenzyme AF35 [33]) [33]
1.2 <29> (p-nitrophenyl-α-L-arabinofuranoside) [24, 35]
1.2 <73> (p-nitrophenyl-α-L-arabinofuranoside, <73> α-L-arafase I [21]) [21]
1.23 <39> (p-nitrophenyl-α-L-arabinofuranoside) [35]
1.3 <13> (p-nitrophenyl-α-L-arabinofuranoside) [7, 16]
1.58 <43> (arabinobiose) [42]
1.6 <13> (p-nitrophenyl-α-L-arabinofuranoside) [10]
1.64 <32> (p-nitrophenyl-α-L-arabinofuranoside) [24]
1.7 <43> (arabinobiose, <43> mutant E194A [41]) [42]
2 <1> ((1-5)-L-arabinan) [9]
2 <10> (beet arabinan) [10]

2.2 <73> (*p*-nitrophenyl-α-L-arabinofuranoside, <73> α-L-arafase II [21]) [21]
2.7 <36> (*p*-nitrophenyl-α-L-arabinofuranoside) [24]
2.83 <2> (phenyl α-L-arabinofuranoside) [9]
2.86 <2> (phenyl α-L-arabinofuranoside) [2]
3.6 <14> (*p*-nitrophenyl-α-L-arabinofuranoside) [5, 10, 16]
4 <13> (*p*-nitrophenyl-α-L-arabinofuranoside) [10]
4 <10> (*p*-nitrophenyl-α-L-arabinofuranoside) [10, 16]
4.1 <9> (α-L-arabinofuranosyl 4-bromoisoquinoline) [11]
4.1 <1> (*p*-nitrophenyl-α-L-arabinofuranoside) [24]
4.4 <45> (3,4-dinitrophenyl β-D-xylopyranoside, <45> wild-type Abfa T-6 [44]) [44]
4.6 <9> (α-L-arabinofuranosyl 4-phenol) [11]
4.7 <1> (phenyl α-L-arabinofuranoside) [9]
4.9 <1> (*p*-nitrophenyl-α-L-arabinofuranoside) [10, 16]
5 <1> (*p*-nitrophenyl-α-L-arabinofuranoside) [12]
5 <1> (phenyl-α-L-arabinofuranoside) [12]
5.3 <45> (2-nitrophenyl β-D-xylopyranoside, <45> mutant E175A [44]) [44]
8 <45> (2,5-dinitrophenyl β-D-xylopyranoside, <45> wild-type Abfa T-6 [44]) [44]
9.1 <3> (phenyl α-L-arabinofuranoside) [3]
10 <27> (phenyl-α-L-arabinofuranoside, <27> isoenzyme C1 [17]) [17]
12.5 <27> (phenyl-α-L-arabinofuranoside, <27> isoenzyme C2 [17]) [17]
15.3 <45> (2-nitrophenyl β-D-xylopyranoside, <45> wild-type Abfa T-6 [44]) [44]
Additional information <1, 2, 10, 41> (<1> arabinan 0.26 g/L, 1,5-L-arabinan 20.4 g/l [12]; <1> sugar beet arabinan 0.24 g/l, isoenzyme B [15]; <2> arabinan 8.47 g/l, 1,5-arabinan 28.6 g/l [2,9]; <10> arabinan 2.0 mg/ml [10]; <41> arabinoxylan, 9.0 mg/ml [40]) [2, 9, 10, 12, 15, 40]

pH-Optimum

2 <3> [3, 36]
2.5 <2> (<2> towards phenyl α-L-arabinofuranoside or beet arabinan [2]) [2, 9]
3.4 <1> (<1> isoenzyme A [15]) [15, 29]
3.5 <32> [24]
3.5-4.5 <37> [29]
3.7 <1> (<1> strain B [34]) [24, 35]
3.8 <1> (<1> isoenzyme B [15]) [12, 15, 29]
3.8-4 <1> (<1> towards phenyl α-L-arabinofuranoside or beet arabinan [1, 24]) [1, 9, 10, 16, 24, 29]
3.9 <1> [29]
4 <1, 9, 29, 39> [4, 16, 20, 24, 33, 35]
4.1 <1, 9> (<9> isozyme high molecular weight form [11]) [11, 24]
4.2 <41> [38]
4.3 <41> [40]
4.5-5.5 <1> [24]

4.8 <16, 59, 73> [13, 14, 21]
5 <1, 9, 34, 37, 44, 47, 58> (<1> commercial enzyme preparation [15]; <9> AFIII, isozyme low-Mw form [11]; <37> isoenzyme ABF1 and ABF2 [31]) [11, 15, 27-29, 31, 43]
5-5.5 <10> [10, 16]
5-6.5 <27> [17]
5.5 <36, 37, 43> (<36> crude enzyme preparation [24]; <37> isoenzyme ABF3 [31]) [24, 31, 42]
5.5-6 <30, 35> [23, 26]
5.6-6.2 <38> [34]
6 <14, 19, 20, 31, 40, 49> [5, 8, 10, 16, 22, 36, 37]
6-6.5 <1, 20> [16, 24]
6.5 <12, 15> [6, 9, 10, 16]
6.9 <13> (<13> with *p*-nitrophenyl-α-L-arabinofuranoside as substrate [7]) [7, 10, 16]
7 <12, 30, 36, 42> (<42> isoenzyme II [39]) [18, 24, 25, 39]

pH-Range
1.5-3 <2> [2]
2-5.5 <1> [12]
2.5-8 <1, 37> [15, 20, 29]
3-10 <40> [37]
3.5-5 <41> [38]
3.5-6.3 <16> (<16> 50% activity detected [13]) [13]
3.5-7.5 <27> [17]
4-10 <14, 30> [5, 26]
4-11 <38> [34]

Temperature optimum (°C)
30 <36> [24]
45 <20, 59> [14, 16]
46 <1> (<1> isoenzyme A [15]) [15]
50 <1, 20, 35, 37, 39, 40, 42, 58> (<1> commercial enzyme preparation [15]; <37> isoenzyme ABF1 [31]; <42> isoenzyme II [39]) [15, 23, 24, 27, 31, 35, 37, 39]
55 <1, 20, 36, 42> (<36> crude enzyme preparation [24]) [16, 24, 39]
55-60 <19> [36]
56 <1> (<1> isoenzyme B [15]) [15]
60 <1, 12, 19, 29, 31, 32, 37, 41, 44, 47, 58> (<37> isoenzyme ABF1 [31]) [20, 22, 24, 25, 27, 31, 33, 35, 36, 38, 43]
65 <36, 37> (<37> isoenzyme ABF3 [31]) [31]
70 <30, 34> (<30> at pH 7 and pH 8 [18]) [18, 26, 28]
75 <38> [34]

Temperature range (°C)
20-60 <1> [20]
20-65 <37> [29]
24-75 <37> [31]

25-50 <27> [17]
25-75 <1> [15]
30-80 <41> [38]
30-90 <34> [28]
45-80 <30> [26]
50-70 <41> [38]

4 Enzyme Structure

Molecular weight

32000 <1, 47> (<1> strain 3M43, laser desorption spectrometry, gel filtration [30]) [29, 30]
34000 <58> (<58> gel filtration [27]) [27]
34000 <9> (<9> SDS-PAGE [11]) [11]
34300 <58> (<58> SDS-PAGE [27]) [27]
35000 <9, 29> (<9> SDS-PAGE [11]; <29> AF35, SDS-PAGE [33]) [11, 33]
35990 <44> (<44> calculated from amino acid sequence [43]) [43]
36000 <36> (<36> SDS-PAGE [24]) [24]
37000 <42, 44> (<42> isoenzyme II [39]; <44> SDS-PAGE [43]) [39, 43]
38000 <27> (<27> isoenzyme C_1 [17]) [17]
39000 <37> (<37> isoenzyme A, SDS-PAGE [29]) [29]
40000 <9> (<9> gel filtration [11]) [11, 16]
51000 <39> (<39> gel filtration [35]) [35]
52000 <34> (<34> zymogram staining [28]) [28]
53000 <1, 29> (<1> gel filtration [1]; <1> gel filtration [12]; <1> strain KI [24]; <29> AF53, SDS-PAGE [33]) [1, 9, 10, 12, 16, 24, 29, 33, 35, 36]
55000 <43> (<43> Abf51A, antibody cross reaction, Western blot [41]) [41]
56000 <38> (<38> SDS-PAGE [34]) [34]
56070 <38> (<38> amino acid sequence analysis [34]) [34]
56090 <38> (<38> mass spectrometry [34]) [34]
57000 <43> (<43> SDS-PAGE [42]) [42]
58000 <39> (<39> SDS-PAGE [35]) [35]
59000 <37> (<37> isoenzyme B1+B2, SDS-PAGE [29]) [29]
60000 <1, 27, 32, 41> (<27> isoenzyme C2 [17]; <41> isoenzyme Ara1, SDS-PAGE [38]) [17, 24, 29, 35, 37, 38]
61000 <12> [25, 29]
62000 <16, 47> (<16> gel filtration [13]) [13, 35]
65000 <12, 41, 47> (<41> SDS-PAGE [40]) [9, 10, 16, 36, 40]
66000 <19, 41> (<19> gel filtration, SDS-PAGE [36]; <41> isoenzyme Ara1, SDS-PAGE [38]) [36, 38]
67000 <1> (<1> α-L-arabinofuranosidase B [15]; <1> SDS-PAGE [20]) [15, 20, 29]
68000 <73> (<73> α-L-arafase II [21]) [21]
68280 <41> (<41> isoenzyme AXAH, encoded by cDNA [40]) [40]
68760 <41> (<41> isoenzyme AXAH-II, encoded by cDNA [40]) [40]
69000 <41> (<41> isoenzyme Ara3, SDS-PAGE [38]) [38]

72490 <31> (<31> calculated from DNA and amino acid sequence analysis [22]) [22]
75000 <13> [36]
78500 <37> (<37> isoenzyme ABF3 [31]) [31]
80000 <42> (<42> isoenzyme I [39]) [39]
82000 <37> (isoenzyme ABF2 [31]) [31]
83000 <1> (<1> α-L-arabinofuranosidase A [15]; <1> strain N400 [29]) [13, 29]
87000 <59> (<59> SDS-PAGE [14]) [14]
90000 <10, 37> (<10> SDS-PAGE [10]; <37> isoenzyme ABF21 [31]) [10, 31]
92000 <14> (<14> strain 17-1, gel filtration [5, 16]) [5, 10, 16]
94000 <10> (<10> gel filtration [10]) [10, 16]
108000 <30> (<30> gel filtration, Sepharose Cl 6B [18]) [18]
110000 <30> (<30> gel filtration [18]) [18]
114800 <30> (<30> gel filtration, Superose 12 [18]) [18]
118000 <73> (<73> α-L-arafase I [21]) [21]
128000 <1> [24, 29]
195000 <34> (<34> gel filtration [28]) [28]
240000 <20> [16]
250000 <40> (<40> native-PAGE [37]) [37]
256000 <30> [26]
284000 <30> (<30> gel filtration [26]) [26, 37]
305000 <13> (<13> disc gel electrophoresis [7]) [7]
310000 <13> (<13> gel filtration [7]) [7, 10, 16, 28]
364000 <35> (<35> gel filtration [23]) [23]
380000 <31> (<31> gel filtration [22]) [22]
495000 <15> (<15> gel filtration [6]) [6, 10, 16]

Subunits

dimer <30> (<30> 1 * 52500 + 1 * 57500, SDS-PAGE [18]; <30> strain L1 [26]) [18, 26]
hexamer <31, 35> (<31> 6 * 69000, SDS-PAGE [22]; <35> 6 * 61000, SDS-PAGE [23]) [22, 23]
monomer <1, 9, 10, 12, 14, 29, 39, 58> (<1> 1 * 53000 [16]; <1> 1 * 33000, strain 3M43, SDS-PAGE [30]; <9> 1 * 35000 [16]; <10> 1 * 90000, SDS-PAGE [10]; <10> 1 * 94000 [16]; <12> 1 * 65000 [16]; <14> 1 * 92000 [16]; <39> 1 * 58000, SDS-PAGE [35]; <58> 1 * 34300, SDS-PAGE [27]) [10, 11, 16, 27, 35]
octamer <15, 20> (<15> 8 * 62000 [6,10,16]; <20> 8 * 31000 [16]) [6, 10, 16, 26]
tetramer <13, 30, 34, 40> (<13> 4 * 75000, SDS-PAGE [7, 10, 16, 28]; <30> 4 * 64000, SDS-PAGE [26, 37]; <34> 4 * 52000, SDS-PAGE [28]; <40> 4 * 72000, SDS-PAGE [37]) [7, 10, 16, 26, 28, 37]

5 Isolation/Preparation/Mutation/Application

Source/tissue

callus <16> [7, 13, 14, 17, 25, 37]
conidium <29> [33]
cotyledon <59> [14, 25]
culture fluid <1> [1]
grain <41> [40]
leaf <73> (<73> mesophyll tissue [21]) [21]
malt <41> [38]
mycelium <29> [33]
seed <50, 59> [14, 21, 34]
seedling <41, 59> [14, 25, 40]

Localization

cell membrane <73> [21]
cytoplasm <20, 30, 31> [16, 23, 26]
extracellular <1, 30, 13, 19, 30, 34-36, 39> [7, 10, 15, 18, 20, 23, 24, 28, 30, 31, 35, 36]
intercellular layer <73> [21]
intracellular <1, 40> [20, 29, 37]
outer membrane <43> [41]

Purification

<1> [1, 2, 3, 4, 6, 7, 9, 12, 14-16, 19, 20, 22, 24, 27, 29-31, 34, 37]
<2> [2, 9, 22]
<3> [3]
<7> [31]
<9> (isoenzyme designated AFIII [11]) [11]
<10> [10, 16, 37]
<12> [9, 10, 25]
<13> [10, 16, 17, 23, 24, 26, 28, 29]
<14> [5]
<15> [6]
<16> [13, 16, 37]
<19> [36]
<20> [16, 23, 25, 26, 37]
<21> [25]
<27> (2 isoenzymes, C1 and C2 [17]) [17, 25, 26, 37]
<29> (isoenzymes AF35 and AF53 [33]) [22, 33]
<30> [18, 26, 34, 37, 41]
<31> [22, 26, 34]
<32> [22, 24]
<34> (recombinant protein, arfB gene product [28]) [28]
<35> [23]
<37> (3 isoenzymes [29, 31]) [29, 31, 35, 37]
<38> (recombinant protein, expressed in E. coli [34]) [34]
<39> [35]

<40> [37]
<41> (3 isoenzymes, Ara1, Ara2 and Ara3 [38]) [38, 40, 43]
<42> (2 isoenzymes [39]) [39-41, 43]
<44> (recombinant enzyme, expressed in E. coli [43]) [43]
<45> (recombinant enzyme, expressed in E. coli BL21(DE3) [44]) [44]
<47> [32, 35, 37]
<49> [8]
<51> [34]
<53> (2 isoenzymes [35]) [35]
<56> [35]
<58> [27]
<59> [14]
<73> (2 isoenzymes, α-L-arafase I and α-L-arafase II [21]) [21]

Crystallization

<1> [1, 2, 3, 5, 6, 12, 16, 24, 25, 27, 30, 34, 37]

Cloning

<1> (gene, encoding AF, heterologous cloned in E. coli [22]; ABF2 gene cloned and expressed in Saccharomyces cerevisiae [35]) [22, 35, 42]
<20> (XS gene cloned and expressed in E. coli, the recombinant 60 kDa protein shows XS and AF activities [16]; gene xylB encoding AF, heterologous cloned in E. coli [22]) [16, 22]
<29> (α-AF cDNA cloned from Trichoderma reesei RUT C-30 chromosomal DNA using PCR technique, trying to express this gene in the yeast Schizosaccharomyces pombe, expression in Saccharomyces cerevisiae [33, 35, 43]) [33, 35, 43]
<31> (gene abfA encoding α-L-arabinofuranosidase homologously expressed in Streptomyces lividans by complementation of mutant IAF10-164 and DNA sequence determined, multicopy vector pIJ702 permits overproduction of the enzyme [22]) [22, 26, 33, 34]
<33> (xynC gene, encoding AF, heterologous expressed in E. coli [22]) [22]
<34> (gene, encoding AF, heterologous expressed in E. coli JM83 [22, 28]) [22, 28, 34]
<36> (abfB gene, nucleotide sequence determined [35, 43]) [35, 43]
<38> (gene AbfD3 isolated, inserted into pET24 vector and expressed in E. coli [34]) [34]
<41> (cDNA clones isolated [40, 43]) [40, 43]
<42> (gene cloning of 2 isoenzymes, PCR products subcloned into pCR2.1 plasmids [39, 43]) [39, 43]
<43> (gene abf51A amplified by PCR [42]; 2 genes, abf51A and abf62A, transposon mutagenesis [41]) [41, 42]
<44> (cloning and sequencing of the gene stxIV encoding α-L-arabinofuranosidase, expressed in E. coli JM109 [43]) [43]
<45> (cloning of the abfA T-6 gene, PCR, T7 polymerase expression vector pET9d, expressed in E. coli BL21(DE3) [44]) [44]
<51> (arfI and arfII, genes encoding α-L-arabinofuranosidase [34]) [34]
<55> (cDNA isolation [40]) [40]

Application

agriculture <1, 19, 45> (<1> processing of agricultural products like fruit, vegetables and cereals [15]; <1> enzymes capable of acting on arabinoglucuronoxylan have been reported to improve the digestibility of animal feed [29]; <19> arabinoxylans are important animal feed components [36]; <45> bioconversion of lignocellulose material for the improvement of animal feedstock digestibility [44]) [15, 29, 36, 44]
analysis <1, 34> (<1> commercially available enzyme, potential for use in the structural analysis of complex carbohydrates [19]; <34> enzymatic tool for structural analysis of oligo- and polysaccharides [28]) [28]
degradation <34, 36> (<34> application potential for industrial purposes, biomass degradation and refining, extremely low end-product inhibition by arabinose further increases its applicability [28]; <36> enzyme of industrial interest because of its ability to improve hydrolytic action of other glycanases on hemicellulolytic polysaccharides [24]) [19, 24, 28]
medicine <1> (<1> enzyme has been shown to be of interest in connection with a new concept of cancer chemotherapy [4]) [4]
nutrition <1, 2, 12, 37, 40, 45, 58> (<1, 2, 12> food industry, preparation of wines and fruit juices [9]; <1> quality of fruit juices can be improved by enzymes hydrolyzing haze-forming arabinans and arabinogalactans [29]; <37> utilization of the enzyme in food and feed, potential application in enology [29]; <40> application of the enzyme in vinification process, fermentation of Muscat grape juice [37]; <45> aromatic potential in wine, exploitation in flavor improvement and wine aromatization [44]; <58> most commonly used in soy sauce manufacture for producing koji mold [27]) [9, 27, 29, 37, 44]
paper production <1, 12, 30, 37, 39, 45> (<1, 12> enzyme becomes very important in the effective use of hemicellulose [20, 25]; <30> extraction of lignin from wood chips, essential step in the preparation of pulp fibres [18]; <37> utilization in pulp and paper industry [29]; <39> enzyme may take part in the cellulose pulp biobleaching process [35]; <45> potential industrial use in biobleaching of paper pulp, bioconversion of lignocellulose material to fermentative products and for the improvement of animal feedstock digestibility [44]) [18, 20, 25, 29, 35, 44]

6 Stability

pH-Stability

1.5 <3> (<3> highly acid stable, retaining 82% of its activity after being maintained for 24 h [37]) [37]
1.5-10 <2, 3> (<2> acid stable, even after storage at pH 2 and 5°C for 72 h, 83% enzymic activity remains [2, 9]; <3> acid stable, retains 82% of its activity after being maintained for 24 h at pH 1.5 at 30°C [3]) [2, 3, 9]
2-5 <3> (<3> quite stable for 72 h when maintained at 2°C [3]) [3]
2.5-8.5 <1> (<1> considerable stability [12]) [12]
3-4 <10> (<10> poor stability at low pH [10]) [10]

3-4.4 <37> [31]
3.5-5.5 <36> (<36> unstable above [24]) [24]
4-7 <1, 29> (<29> isoenzyme AF35 [33]) [20, 33]
4-8 <20> (<20> fairly stable [16]) [16]
4-9 <14> [10, 16]
4-10 <14> (<14> stable for 72 h at 2°C, loses 38% of its original activity at pH 9.0 [5]) [5]
4-12 <38> [34]
4.5-7 <29> (<29> isoenzyme AF53 [33]) [33]
5-9 <58> [27]
5-11 <49> [8]
5.5-8 <10> [10, 16]
5.5-9 <35> [23]
6 <31> (<31> stable for 12 h at 4°C [22]) [22]
6-8 <40> [37]
7 <1> (<1> extremely stable at [12]) [9, 12]
7-10 <12> (<12> enzyme becomes inactive rapidly above pH 10 and slowly below pH 7 [25]) [25]

Temperature stability
10-60 <40> [37]
30 <1> (<1> stable up to [20]) [20]
37 <16> [13]
40 <29> (<29> stable up to [33]) [33]
45 <20, 59> (<20> relatively stable up to [16]) [16, 25]
50 <27, 35, 58> (<27> after 3 h incubation isoenzyme C1 loses 90% of its original activity, C2 loses only 40% [17]; <35> stable up to [23]; <58> stable up to [27]; <59> stable up to [14]) [14, 17, 23, 27]
50-55 <39> (<39> half-life of 2 h at 50°C and of 25 min at 55°C [35]) [35]
50-60 <1> (<1> enzyme is stable for 10 min at 50°C from pH 3-pH 8, at 60°C stable at pH 4 for 10 min, less stable at pH 5.5 and stable again at pH 7 [9]) [9]
50-80 <44> (<44> stable up to 50°C, retains 50% original activity at 80°C [43]) [43]
55 <20> (<20> stable to [16]) [16]
60 <31, 36, 37, 41> (<31> loses activity completely after 10 min [43]; <36> half-life time 1.5 h, inactivation after 2 h t 70°C [24]; <37> isoenzymes ABF1 and ABF3 seemed to be quite stable [31]; <41> half-life 8 min [38]) [24, 31, 38, 43]
60-80 <30> (<30> at pH 7 enzyme is completely stable at 60°C for at least 80 min, retains 50% of its maximum activity after 75 min at 70°C and loses all activity after 15 min at 80°C [18]) [18]
65 <37> (<37> isoenzyme ABF2 retains 7% activity after 4 h [31]) [31]
68 <31> (<31> half-life 12 min [22]) [22]
70 <37> (<37> isoenzymes ABF1 and ABF3, half-life 2 h and 5 h [31]) [31]
70-90 <34> (<34> remarkably thermostable, half of its activity remains after 0.7 h at 90°C [28]) [28]

85 <34> (<34> recombinant arfB gene product, stable up to [28]) [28]
90 <38> (<38> highly stable up to [34]) [34]

Oxidation stability
<35>, O2-stable [23]

Organic solvent stability
ethanol <34> (<34> resistance to ethanol, neutral detergents and NaCl [28]) [28]

General stability information
<1>, completely denatured in 6 M urea [12]
<16>, stable enzyme [13]

Storage stability
<1>, 5°C, if aseptically stored no significant loss of activity over a 3-months period [12]
<2>, 30°C, purified enzyme at low pH values stable for 5 h to 20 days [2]
<16>, 4°C, after storage at various pH of 2.5-11 for 20 h, activity completely retained above pH 3.5 [13]
<16>, 4°C, storage at pH 5.5 for 1 month, loss of activity less than 5% [13]
<20>, -20°C, purified enzyme loses less than 20% of its activity in 30 days [16]
<39>, 4°C, at pH 2.5-7.5 retains more than 60% activity after 3 days, at pH 9 loses 90% of its activity in 20 h [35]

References

[1] Kaji, A., Tagawa, K.: Purification, crystallization and amino acid composition of α-L-arabinofuranosidase from Aspergillus niger. Biochim. Biophys. Acta, **207**, 456-464 (1970)
[2] Kaji, A., Yoshihara, O.: Properties of purified α-L-arabinofuranosidase from Corticium rolfsii. Biochim. Biophys. Acta, **250**, 367-371 (1971)
[3] Uesaka, E., Sato, M., Raiju, M., Kaji, A.: α-L-Arabinofuranosidase from Rhodotorula flava. J. Bacteriol., **133**, 1073-1077 (1978)
[4] Waibel, R., Amadò, R., Neukom, H.: Purification of an α-L-arabinofuranosidase from Aspergillus niger by substrate affinity chromatography. J. Chromatogr., **197**, 86-91 (1980)
[5] Kaji, A.; Sato, M.; Tsutsui, Y.: An α-L-arabinofuranosidase produced by wild-type Streptomyces sp. No. 17-1. Agric. Biol. Chem., **45**, 925-931 (1981)
[6] Komae, K.; Kaji, A.; Sato, M.: An α-L-arabinofuranosidase from Streptomyces purpurascens IFO 3389. Agric. Biol. Chem., **46**, 1899-1905 (1982)
[7] Greve, L.C., Labavitch, J.M., Hungate, R.E.: α-L-Arabinofuranosidase from Ruminococcus albus 8: purification and possible role in hydrolysis of alfalfa cell wall. Appl. Environ. Microbiol., **47**, 1135-1140 (1984)
[8] Kaji, A.; Shimokawa, K.: New exo-type arabinase from Erwinia carotovora IAM 1024. Agric. Biol. Chem., **48**, 67-72 (1984)

[9] Whitaker, J.R.: Pectic substances, pectic enzymes and haze formation in fruit juices. Enzyme Microb. Technol., **6**, 341-349 (1984)
[10] Lee, S.F., Forsberg, C.W.: Purification and characterization of an α-L-arabinofuranosidase from Clostridium acetobutylicum ATCC 824. Can. J. Microbiol., **33**, 1011-1016 (1987)
[11] Kelly, M.A., Sinnott, M.L., Herrchen, M.: Purification and mechanistic properties of an extracellular α-L-arabinofuranosidase from Monilinia fructigena. Biochem. J., **245**, 843-849 (1987)
[12] Tagawa, K., Kaji, A.: α-L-Arabinofuranosidase from Aspergillus niger. Methods Enzymol., **160**, 707-712 (1988)
[13] Uchida, T., Tanaka, M.: α-L-Arabinofuranosidase from Scopolia japonica. Methods Enzymol., **160**, 712-719 (1988)
[14] Hatanaka, H.; Imaoka, H.; Tajima, S.; Kasai, T.: Purification and properties of an α-L-arabinofuranosidase from cotyledons of soybean seedlings. Agric. Biol. Chem., **55**, 2599-2605 (1991)
[15] v.d. Veen, P.; Flipphi, M.J.A.; Voragen, A.G.J.; Visser, J.: Induction, purification and characterisation of arabinases produced by Aspergillus niger. Arch. Microbiol., **157**, 23-28 (1991)
[16] Hespell, R.B.; O'Bryan, P.: Purification and characterization of an α-L-arabinofuranosidase from Butyrivibrio fibrisolvens GS113. Appl. Environ.Microbiol., **58**, 1082-1088 (1992)
[17] Tajana, E.; Fiechter, A.; Zimmermann, W.: Purification and characterization of two α-L-arabinofuranosidases from Streptomyces diastaticus. Appl. Environ. Microbiol., **58**, 1447-1450 (1992)
[18] Bezalel, L.; Shoham, Y.; Rosenberg, E.: Characterization and delignification activity of a thermostable α-L-arabinofuranosidase from Bacillus stearothermophilus. Appl. Microbiol. Biotechnol., **40**, 57-62 (1993)
[19] Lerouge, P.; O'Neill, M.A.; Darvill, A.G.; Albersheim, P.: The purification of commercially available endo-α-L-arabinases and α-L-arabinosidase for use in the structural analysis of pectic polysaccharides. Carbohydr. Res., **243**, 373-378 (1993)
[20] Kaneko, S.; Shimasaki, T.; Kusakabe, I.: Purification and properties of intracellular α-L-arabinofuranosidase from Aspergillus niger 5-16. Biosci. Biotechnol. Biochem., **57**, 1161-1165 (1993)
[21] Hirano, Y.; Tsumuraya, Y.; Hashimoto, Y.: Characterization of spinach leaf α-L-arabinofuranosidases and β-galactosidases and their synergistic action on an endogenous arabinogalactan-protein. Physiol. Plant., **92**, 286-296 (1994)
[22] Manin, C.; Shareek, F.; Morosoli, R.; Klueppel, D.: Purification and characterization of an α-L-arabinofuranosidase from Streptomyces lividans 66 and DNA sequence of the gene (abfA). Biochem. J., **302**, 443-449 (1994)
[23] Schyns, P.J.Y.M.J.; de Frankrijker, J.; Zehnder, A.J.B.; Stams, A.J.M.: Production, purification and characterization of an α-L-arabinofuranosidase from Bacteroides xylanolyticus X5-1. Appl. Microbiol. Biotechnol., **42**, 548-554 (1994)

[24] Fernandez-Espinar, M.T.; Pena, J.L.; Pinaga, F.; Valles, S.: α-L-Arabinofuranosidase production by Aspergillus nidulans. FEMS Microbiol. Lett., **115**, 107-112 (1994)

[25] Kaneko, S.; Sano, M.; Kusakabe, I.: Purification and some properties of α-L-arabinofuranosidase from Bacillus subtilis 3-6. Appl. Environ. Microbiol., **60**, 3425-3428 (1994)

[26] Gilead, S.; Shoham, Y.: Purification and characterization of α-L-arabinofuranosidase from Bacillus stearothermophilus T-6. Appl. Environ. Microbiol., **61**, 170-174 (1995)

[27] Kimura, I.; Sasahara, H.; Tajima, S.: Purification and characterization of two xylanases and an arabinofuranosidase from Aspergillus sojae. J. Ferment. Bioeng., **80**, 334-339 (1995)

[28] Schwarz, W.H.; Bronnenmeier, K.; Krause, B.; Lottspeich, F.; Staudenbauer, W.L.: Debranching of arabinoxylan: Properties of the thermoactive recombinant α-L-arabinofuranosidase from Clostridium stercoarium (ArfB). Appl. Microbiol. Biotechnol., **43**, 856-860 (1995)

[29] Luonteri, E.; Siika-aho, M.; Tenkanen, M.; Viikari, L.: Purification and characterization of three α-arabinosidases from Aspergillus terreus. J. Biotechnol., **38**, 279-291 (1995)

[30] Scott, M.; Connerton, I.F.; Harris, G.W.; Gravesen, T.N.; Madrid, S.M.; Mikkelsen, J.D.: Crystallization and preliminary X-ray analysis of arabinofuranosidase C from Aspergillus niger strain 3M43. Acta Crystallogr. Sect. D, **53**, 222-223 (1997)

[31] Le Clinche, F.; Pinaga, F.; Ramon, D.; Valles, S.: α-L-Arabinofuranosidases from Aspergillus terreus with potential application in enology: Induction, purification, and characterization. J. Agric. Food Chem., **45**, 2379-2383 (1997)

[32] Luonteri, E.; Kroon, P.A.; Tenkanen, M.; Teleman, A.; Williamson, G.: Activity of an Aspergillus terreus α-arabinofuranosidase on phenolic-substituted oligosaccharides. J. Biotechnol., **67**, 41-48 (1999)

[33] Nogawa, M.; Yatsui, K.; Tomioka, A.; Okada, H.; Morikawa, Y.: An α-L-arabinofuranosidase from Trichoderma reesei containing a noncatalytic xylan-binding domain. Appl. Environ. Microbiol., **65**, 3964-3968 (1999)

[34] Debeche, T.; Cummings, N.; Connerton, I.; Debeire, P.; O'Donohue, M.J.: Genetic and biochemical characterization of a highly thermostable α-L-arabinofuranosidase from Thermobacillus xylanilyticus. Appl. Environ. Microbiol., **66**, 1734-1736 (2000)

[35] De Ioannes, P.; Peirano, A.; Steiner, J.; Eyzaguirre, J.: An α-L-arabinofuranosidase from Penicillium purpurogenum: Production, purification and properties. J. Biotechnol., **76**, 253-258 (2000)

[36] Christakopoulos, P.; Katapodis, P.; Hatzinikolaou, D.G.; Kekos, D.; Macris, B.J.: Purification and characterization of an extracellular α-L-arabinofuranosidase from Fusarium oxysporum. Appl. Biochem. Biotechnol., **87**, 127-133 (2000)

[37] Yanai, T.; Sato, M.: Purification and characterization of a novel α-L-arabinofuranosidase from Pichia capsulata X91. Biosci. Biotechnol. Biochem., **64**, 1181-1188 (2000)

[38] Ferre, H.; Borberg, A.; Duus, J.OE.; Thomsen, K.K.: A novel type of arabinoxylan arabinofuranohydrolase isolated from germinated barley. Analysis of substrate preference and specificity by nano-probe NMR. Eur. J. Biochem., **267**, 6633-6641 (2000)
[39] Matsuo, N.; Kaneko, S.; Kuno, A.; Kobayashi, H.; Kusakabe, I.: Purification, characterization and gene cloning of two α-L-arabinofuranosidases from Streptomyces chartreusis GS901. Biochem. J., **346**, 9-15 (2000)
[40] Lee, R.C.; Burton, R.A.; Hrmova, M.; Fincher, G.B.: Barley arabinoxylan arabinofuranohydrolases: Purification, characterization and determination of primary structures from cDNA clones. Biochem. J., **356**, 181-189 (2001)
[41] Beylot, M.-H.; Emani, K.; McKie, V.A.; Gilbert, H.J.; Pell, G.: Pseudomonas cellulosa expresses a single membrane-bound glycoside hydrolase family 51 arabinofuranosidase. Biochem. J., **385**, 599-605 (2001)
[42] Beylot, M.-H.; McKie, V.A.; Voragen, A.G.J.; Doeswijk-Voragen, C.H.L.; Gilbert, H.J.: The Pseudomonas cellulosa glycoside hydrolase family 51 arabinofuranosidase exhibits wide substrate specificity. Biochem. J., **358**, 607-614 (2001)
[43] Tsujibo, H.; Takada, C.; Wakamatsu, Y.; Kosaka, M.; Tsuji, A.; Miyamoto, K.; Inamori, Y.: Cloning and expression of an α-L-arabinofuranosidase gene (stxIV) from Streptomyces thermoviolaceus OPC-520. Biosci. Biotechnol. Biochem., **66**, 434-438 (2002)
[44] Shallom, D.; Belakhov, V.; Solomon, D.; Gilead-Gropper, S.; Baasov, T.; Shoham, G.; Shoham, Y.: The identification of the acid-base catalyst of α-arabinofuranosidase from Geobacillus stearothermophilus T-6, a family 51 glycoside hydrolase. FEBS Lett., **514**, 163-167 (2002)

Glucuronosyl-disulfoglucosamine glucuronidase 3.2.1.56

1 Nomenclature

EC number
3.2.1.56

Systematic name
3-D-glucuronsyl-N2,6-disulfo-β-D-glucosamine glucuronohydrolase

Recommended name
glucuronosyl-disulfoglucosamine glucuronidase

Synonyms
chondro-1-3-glycuronidase <1> (<1> hydrolyses 1,3-linked unsaturated disaccharides [4]) [4]
chondroglycuronidase <1> [4]
glucuronidase, glucuronosyldisulfoglucosamine
glycuronidase
heparan sulphate glycuronidase <1> [3]
heparo-glycuronidase <1> (<1> hydrolyses 1,4-linked disaccharides, derived from heparin and heparan sulfate [4]) [4]

CAS registry number
37288-42-9

2 Source Organism

<1> *Flavobacterium heparinum* (gram-negative soil bacterium, also known as Cythophaga heparinia [4]) [1-4]

3 Reaction and Specificity

Catalyzed reaction
3-D-glucuronosyl-N2,6-disulfo-β-D-glucosamine + H_2O = D-glucuronate + $N^{2,6}$-disulfo-D-glucosamine

Reaction type
hydrolysis of O-glycosyl bond

Substrates and products
S 1,3-(4-deoxy-α-L-threo-4-enopyranosyluronicacid)-(2-N-acetylgalactosamine-6-O-sulfate) + ? <1> [4]

P ?
S 3-D-glucuronosyl-$N^{2,6}$-disulfo-β-D-glucosamine + H_2O <1> [1, 4]
P $N^{2,6}$-disulfo-D-glucosamine + D-glucuronate <1> [1, 4]
S disaccharide + SO_4^{2-} <1> (<1> overview disulfated disaccharides [1, 2, 3]) [1, 2, 3]
P glucosamine 2,6-disulfate + uronic acid <1> (<1> α, β-keto acid [1, 2, 4]) [1, 2, 4]
S glucosamine 2,6-disulfate <1> [2]
P glucosamine-N-SO_4 + SO_4^{2-} <1> [2]
S heparan sulfate + H_2O <1> (<1> unusual glycuronidase, hydrolyses all disaccharides obtained from heparan sulfate [3]) [3]
P ?
S heparin + H_2O <1> [2, 3]
P trisulfated disaccharide + disulfated disaccharide + glucosamine -2,6-disulfate + glucosamine-N-SO_4 + SO_4^{2-} <1> [2, 3, 4]

Specific activity (U/mg)
21 <1> [3]
29.4 <1> (<1> gel filtration [4]) [4]
50 <1> (<1> enzyme used in solution [3]) [3]
259 <1> (<1> crude, enzymatic activity measured by the disappearance of substrate [2]) [2]
Additional information <1> (<1> 6.470 mg/h [2]) [2]

K_m-Value (mM)
163 <1> (1,3-(4-deoxy-α-L-threo-4-enopyranosyluronicacid)-(2-N-acetylgalactosamine-6-O-sulfate)) [4]

pH-Optimum
6.5 <1> [1, 4]

pH-Range
4.5-9 <1> [4]
5.5-8 <1> [1]

Temperature optimum (°C)
30 <1> (<1> relatively temperature insensitive [1]) [1]
37 <1> (<1> activity decrease of 50% after 3.2 h [4]) [4]

Temperature range (°C)
5-45 <1> [1]
10-60 <1> [4]

4 Enzyme Structure

Molecular weight
41800 <1> (<1> SDS-PAGE [4]) [4]

5 Isolation/Preparation/Mutation/Application

Purification

<1> (not highly purified, but 25fold over the preparation described before [3]; purified by 639fold to apparent homogeneity [4]) [3, 4]

6 Stability

Storage stability

<1>, -70°C [4]

References

[1] Dietrich, C.P.: Enzymic degradation of heparin. A glucosaminidase and a glycuronidase from Flavobacterium heparinum. Biochemistry, **8**, 2089-2094 (1969)

[2] Dietrich, C.P.; Silva, M.E.; Michelacci, Y.M.: Sequential degradation of heparin in Flavobacterium heparinum. Purification and properties of five enzymes involved in heparin degradation. J. Biol. Chem., **248**, 6408-6415 (1973)

[3] Hovingh, P.; Linker, A.: Specificity of flavobacterial glycuronidases acting on disaccharides derived from glycosaminoglycans. Biochem. J., **165**, 287-293 (1977)

[4] Kenan, G.U.; Linhardt, R.J.; Laliberte, M.; Kangfu, G.U.; Zimmermann, J.: Purification, characterization and specificity of chondroitin lyases and glycuronidase from Flavobacterium heparinum. Biochem. J., **312**, 569-577 (1995)

Isopullulanase 3.2.1.57

1 Nomenclature

EC number
3.2.1.57

Systematic name
pullulan 4-glucanohydrolase (isopanose-forming)

Recommended name
isopullulanase

CAS registry number
37288-43-0

2 Source Organism

<1> *Aspergillus niger* (ATCC 9642, IPU F1 and IPU F2 [6,7]) [1, 4-7]
<2> *Arthrobacter globiformis* (T6, single enzyme with G2-dextranase EC 3.2.1.94 and isopullanase activity [3]) [2, 3]

3 Reaction and Specificity

Catalyzed reaction
hydrolysis of pullulan to isopanose (6-α-maltosylglucose)

Reaction type
hydrolysis of O-glycosyl bond

Natural substrates and products
S pullulan + H_2O <1> [1, 4, 6, 7]
P isopanose + $(Glc)_4$ <1> (<1> small amount [1]) [1, 4, 6, 7]
S pullulan + H_2O <2> [2, 3]
P isopanose + $(Glc)_6$ <2> [2, 3]

Substrates and products
S α-glucosylmaltotriose + H_2O <1> [1]
P isomaltose + maltose
S α-maltosylmaltose + H_2O <1> [1, 6]
P isopanose + glucose
S azidopullulan + H_2O <1> [5]
P ?

S panose + H_2O <1> [1, 6, 7]
P isomaltose + glucose
S pullulan + H_2O <1> [1, 4, 6, 7]
P isopanose + $(Glc)_4$ <1> (<1> small amount [1]) [1, 4, 6, 7]
S pullulan + H_2O <2> [2, 3]
P isopanose + $(Glc)_6$ <2> [2, 3]

Inhibitors

Ag^+ <2> (<2> 5 mM, partial inhibition [3]) [3]
Fe^{3+} <1, 2> (<2> 5 mM complet inactivation [3]) [3, 6]
Hg^{2+} <2> (<2> 5 mM complet inactivation [3]) [3]
$KMnO_4$ <2> (<2> 5 mM complet inactivation [3]) [3]
N-bromosuccinimide <1, 2> (<2> 5 mM complete inactivation [3]) [3, 6]

Specific activity (U/mg)

27 <1> (<1> F2 [6]) [6]
27.3 <1> (<1> F1 [6]) [6]
480 <1> [1]

pH-Optimum

3-3.5 <1> [1]
3-4 <1> [6]
3.5 <1> (<1> recombinant enzyme [7]) [7]
3.5-4 <1> [4]
4.5 <2> [2]
4.8 <2> [3]

pH-Range

2.8-8 <2> (<2> 4°C [3]) [3]
4.1-8 <2> (<2> 45°C [3]) [3]

Temperature optimum (°C)

30-40 <1> [4]
40 <1> (<1> at pH 3.5 [1]; <1> F2 at pH 3.5 [6]) [1, 6]
40-45 <1> (<1> recombinant enzyme [7]) [7]
45 <1> (<1> F1 at pH 3.5 [6]) [6]
60 <2> [2, 3]

4 Enzyme Structure

Molecular weight

59000 <1> (<1> F1 and F2 deglycosylated [6]) [6, 7]
69000 <1> (<1> F2, SDS-PAGE [6]) [6]
71000 <1> (<1> F1, SDS-PAGE [6]) [6]
74000 <1> (<1> gel filtration [1]) [1]
91000 <1> (<1> recombinant enzyme, SDS-PAGE [7]) [7]

Posttranslational modification

glycoprotein <1> (<1> n-glycosylated [7]) [6, 7]

5 Isolation/Preparation/Mutation/Application

Localization

cell-bound <1> [6, 7]
extracellular <1> [6, 7]

Purification

<1> [1, 6, 7]
<2> [3]

6 Stability

pH-Stability

2-7 <1> (<1> at 4°C 12 h [6]) [6]
2-8 <1> (<1> 5°C, 24 h [1]) [1]
3-7 <1> (<1> recombinant enzyme, at 4°C 12 h [7]) [7]
4-7 <1> [4]

Temperature stability

35 <1> (<1> F2, below, 30 min at pH 3.5 [6]) [6]
45 <1> (<1> F1, below, 30 min at pH 3.5 [6]) [6]
50 <1> (<1> 30 min at pH 3.7-4.5 [1]) [1]
50 <2> (<2> below completely stable [3]) [3]
55 <1> (<1> F2, loss of activity [6]) [6]
60 <1, 2> (<1> F1, loss of activity [6]; <2> 30 min, stable up to [2]) [2, 6]
70 <2> (<2> 10 min, 80% loss of activity [3]) [3]
80 <2> (<2> inactivated [3]) [3]

Oxidation stability

<2>, photooxidation in presence of rose bengal [3]

References

[1] Sakano, Y.; Higuchi, M.; Kobayashi, T.: Pullulan 4-glucanohydrolase from Aspergillus niger. Arch. Biochem. Biophys., **153**, 180-187 (1972)
[2] Tago, M.; Aoji, M.; Sakano, Y.; Kobayashi, T.; Sawai, T.: Isopullanase from Arthrobacter globiformis T6. Agric. Biol. Chem., **35**, 971-973 (1971)
[3] Okada, G.; Takayanagi, T.; Miyahara, S.; Sawai, T.: An isomalto-dextranase accompanied by isopullulanase activity from Arthrobacter globiformis T6. Agric. Biol. Chem., **52**, 829-836 (1988)
[4] Sakano. Y.; Masuda, N.; Kobayashi, T.: Hydrolysis of pullulan by a novel enzyme from Aspergillus niger. Agric. Biol. Chem., **35**, 971-973 (1972)
[5] Ball, D.H.; Wiley, B.J.; Reese E.T.: Effect of substitution at C-6 on the susceptibility of pullulan to pullulanases. Enzymatic degradation of modified pullulans. Can. J. Microbiol., **38**, 324-327 (1992)

[6] Aoki, H.; Yopi; Padmajanti, A.; Sakano, Y.: Two components of cell-bound isopullulanase from Aspergillus niger ATCC 9642 - Their purification and enzymatic properties. Biosci. Biotechnol. Biochem., **60**, 1795-1798 (1996)
[7] Aoki, H.; Yopi; Sakano, Y.: Molecular cloning and heterologous expression of the isopullulanase gene from Aspergillus niger A.T.C.C. 9642. Biochem. J., **323**, 757-764 (1997)

Glucan 1,3-β-glucosidase 3.2.1.58

1 Nomenclature

EC number
3.2.1.58

Systematic name
1,3-β-glucan glucohydrolase

Recommended name
glucan 1,3-β-glucosidase

Synonyms
β-1,3-glucan exo-hydrolase
exo (13)-β-glucanase
exo-1,3-β-D-glucanase
exo-1,3-β-glucanase
exo-1,3-β-glucosidase
exo-1,3-β-glucosidase
exo-1,3-β-glucosidase
exo-β-(13)-D-glucanase
exo-β-(13)-glucanohydrolase
exo-β-1,3-D-glucanase
exo-β-1,3-glucanase

CAS registry number
9073-49-8

2 Source Organism

<1> *Euglena gracilis* [1, 2]
<2> *Saccharomyces cerevisiae* (strains X14, ts-136Me10, S288C [27]) [3, 9, 10, 17, 25, 27]
<3> *Coniothyrium minitans* [4]
<4> *Trichoderma viride* [4]
<5> *Cryptococcus diffluens* [5]
<6> *Helix pomatia* [6]
<7> *Basidiomycete sp.* (QM806 [7, 13, 18, 24]) [7, 13, 18, 24]
<8> *Schizosaccharomyces japonicus* (var. versatilis [8]) [8]
<9> *Candida albicans* [11, 16, 38, 39, 40, 41, 45]
<10> *Trichoderma harzianum* [12, 37]
<11> *Atriplex littoralis* [14]

<12> *Pinus pinaster* [15]
<13> *Neurospora crassa* [19]
<14> *Streptomyces sp. 1228* [20]
<15> *Gossypium hirsutum* [21]
<16> *Porodisculus pendulus* [22]
<17> *Valerianella olitoria* [23]
<18> *Zea mays* [26, 29]
<19> *Candida utilis* [27, 34]
<20> *Geotrichum lactis* [27]
<21> *Kluyveromyces phaseolosporus* [27]
<22> *Strongylocentrotus purpuratus* (sea urchin [28]) [28]
<23> *Trichoderma reesei* [30]
<24> *Streptomyces murinus* [31]
<25> *Kluyveromyces aestuarii* [32]
<26> *Fungi imperfecti* [33]
<27> *Achlya bisexualis* [35]
<28> *Coccidioides immitis* [36]
<29> *Acremonium persicinum* [42]
<30> *Aspergillus saitoi* [43]
<31> *Hordeum vulgare* [44, 47]
<32> *Cochliobolus carbonum* [46]

3 Reaction and Specificity

Catalyzed reaction

successive hydrolysis of 1,3-β-D-glucose units from the non-reducing ends of 1,3-β-D-glucans, releasing α-glucose

Natural substrates and products

S laminarin + H_2O <2> [3]
P α-D-glucose

Substrates and products

S 4-methyl-umbelliferyl-β-D-glucoside <2, 9> [9, 11, 25]
P 4-methylumbelliferone + D-glucose <2, 9> [9, 11, 25]
S CM-cellulose + H_2O <6> (<6> very weak hydrolysis [6]) [6]
P D-glucose <6> [6]
S CM-pachyman + H_2O <1, 29> (<1> very weak hydrolysis [1]) [1, 42]
P D-glucose <1, 29> [1, 42]
S N-N'-diacetyl-chitobioside + H_2O <28> [36]
P ?
S N-acetyl-β-D-galactosaminide + H_2O <28> [36]
P ?
S N-acetyl-β-D-glucosaminide + H_2O <28> [36]
P ?
S carboxymethylpachyman + H_2O <23> [30]
P D-glucose <23> [30]

S cellobiose + H_2O <2, 15, 28, 31> [17, 21, 36, 47]
P ?
S cellodextrin + H_2O <6> [6]
P D-glucose <6> [6]
S cellotetraose + H_2O <18, 31> [26, 47]
P ?
S cellotriose + H_2O <31> [47]
P ?
S curdlan + H_2O <10, 16, 29> [12, 22, 42]
P ?
S gentiobiose + H_2O <6, 8, 15, 16, 31> [6, 8, 21, 22, 47]
P D-glucose <6> [6]
S glucan + H_2O <1, 6, 7, 18, 26, 27, 29, 31> (<1, 7> Claviceps glucan [1, 13]; <1, 6, 7, 26, 29> yeast glucan [1, 6, 13, 33, 42]; <1, 6, 31> barley β-glucan [1, 6, 47]; <31> extracted and commerical barley β-glucan [47]; <1,6> oat β-glucan [1, 6]; <7> Ganoderma applanatum glucan [24]; <7> Sclerotium rolfsii glucan [24]; <18> Avena caryopsis glucan [26]; <27> Achlya neutral glucan [35]; <29> Acremonium persicinum extracellular glucan [42]) [1, 6, 13, 24, 26, 33, 35, 42, 47]
P D-glucose + gentiobiose <1, 6, 7> (<7> and laminaribiose [13]; <6> D-glucose only [6]) [1, 6, 13, 24]
S laminaran + H_2O <7, 13, 15, 16, 28> (<7> from Eisenia bicyclis [24]) [19, 21, 22, 24, 36]
P D-glucose + gentiobiose + gentiotriose <7, 16> (<16> almost all D-glucose [22]) [24, 22]
S laminaribiose + H_2O <6, 10, 15, 16, 29, 31> [6, 12, 21, 22, 42, 47]
P D-glucose <6> [6]
S laminariheptaose + H_2O <29, 31> [42, 47]
P ?
S laminarihexaose + H_2O <10, 16, 26, 29, 31> [12, 22, 33, 42, 47]
P D-glucose + laminarioligosaccharides <10, 31> [12, 47]
S laminarin + H_2O <1-3, 9, 6-11, 14, 17, 18, 19, 22, 24-27, 29, 30, 31> (<29> oxidized laminarin, reduced laminarin [42]) [1-4, 6-14, 16-18, 20, 23, 25, 26, 28, 29, 31-35, 37, 42, 43, 44, 47]
P D-glucose + laminaribiose + gentiobiose + oligosaccharides + laminaritriose <1, 2, 6-9, 18, 19, 25, 26, 29> (<1, 2, 6-9, 18, 19, 25, 26, 29> different combinations and amounts of following products: D-glucose (main product), laminaribiose, gentiobiose, oligosaccharides, laminaritriose [1, 3, 6, 8, 11, 13, 17, 25, 26, 29, 32, 33, 34, 42]) [1, 3, 6, 8, 11, 13, 17, 25, 26, 29, 32, 33, 34, 42]
S laminarin + H_2O <7, 29> (<7, 29> periodate-oxidized and reduced [7, 42]) [7, 42]
P glycerol + glycoaldehyde + D-glucose <7> [7]
S laminaripentaose + H_2O <1, 5, 6, 10, 16, 26, 29, 31> [1, 5, 6, 12, 22, 33, 42, 47]

P D-glucose + laminaribiose + laminaritriose + laminaritetraose <1, 5, 6> (<1, 5, 6> different combinations and amount of this products, depending on organism [1, 5, 6]) [1, 5, 6]
S laminaritetraose + H_2O <1, 6, 10, 16, 26, 29, 30, 31> [1, 6, 12, 22, 33, 42, 43, 47]
P D-glucose + laminaribiose + laminaritriose <10, 1, 6> (<10, 1, 6> different combinations and amount of these products, depending on organism [1, 6, 12]) [1, 6, 12]
S laminaritriose + H_2O <1, 6, 10, 16, 29, 31> [1, 6, 12, 22, 42, 47]
P D-glucose + laminaribiose <10, 1, 6> (<6> D-glucose only [6]) [1, 6, 12]
S lichenan + H_2O <13, 15, 18, 31> [19, 21, 26, 47]
P ?
S lichenin + H_2O <1, 6> [1, 6]
P D-glucose <1, 6> [1, 6]
S luteose + H_2O <6> [6]
P D-glucose <6> [6]
S maltose + H_2O <15> [21]
P ?
S neutral glucan + H_2O <27> (<27> from Achlya [35]) [35]
P D-glucose + laminaribose <27> [35]
S nigeran + H_2O <15> [21]
P ?
S *o*-nitrophenyl-β-D-glucopyranoside + H_2O <15> [21]
P *o*-nitrophenol + D-glucose <15> [21]
S oligoglucosides from pachyman + H_2O <1> [1]
P D-glucose <1> [1]
S *p*-nitrophenyl-β-D-glucopyranoside <2, 27, 28, 31> [3, 25, 27, 35, 36, 44]
P *p*-nitrophenol + D-glucose <2, 27, 28, 31> [3, 25, 27, 35, 36, 44]
S *p*-nitrophenyl-β-D-glucoside <2, 5, 8, 9, 15, 18, 25, 31> [5, 8-10, 16, 17, 21, 26, 32, 47]
P D-glucose + nitrophenol + glucan <2> (incompletely degraded) [10]
S pachyman + H_2O <1, 6, 7, 8, 13, 15, 26, 29, 30> [1, 6, 8, 19, 21, 24, 33, 42, 43]
P D-glucose <1> [1]
S paramylon + H_2O <1> [1]
P D-glucose <1> [1]
S pendulan + H_2O <16> [22]
P D-glucose <16> [22]
S phosphoglucan + H_2O <27> [35]
P ?
S pullulan + H_2O <15> [21]
P ?
S pustulan + H_2O <2, 8, 13, 15, 23, 25, 31> [8, 9, 17, 19, 21, 25, 30, 32, 47]
P D-glucose <8, 23> [8, 30]
S salicin + H_2O <2, 13> [17, 19]
P ?
S schizophyllan + H_2O <16, 29> [22, 42]

P D-glucose + gentiobiose <29> [42]
S scleroglucan + H_2O <29> [42]
P ?
S sclerotan + H20 <3, 4> (<3, 4> glucan from Sclerotinia sclerotiorum cell wall [4]) [4]
P D-glucose + gentiobiose <4> [4]
S sophorose + H_2O <15> [21]
P ?

Inhibitors

$(NH_4)_2MoO_4$ <29> [42]
1-deoxynojirimycin <28> [36]
2-deoxy-D-glucose <8> (<8> slight [8]) [8]
2-fluoroglucosylpyranoside <9> (<9> mechanism-based [41]) [41]
5-amino-1H-tetrazole <7> (<7> diazotized [18]) [18]
Ag^+ <9, 10> [11, 12]
$AgNO_3$ <6> [6]
$BaCl_2$ <29> [42]
$CoCl_2$ <13> [19]
$CoSO_4$ <26> [33]
Cu^{2+} <27> [35]
$CuCl_2$ <13, 29> [19, 42]
$CuSO_4$ <6, 24, 26> [6, 31, 33]
D-erythritol <22> [28]
D-glucono-δ-lactone <6, 8, 9, 11, 15, 19, 22, 26, 31> (<8> K_i: 0,9 mM [8]) [6, 8, 11, 14, 21, 28, 33, 34, 44]
DL-threitol <22> [28]
$FeCl_2$ <6> [6]
$FeCl_3$ <13, 29> [19, 42]
Hg^{2+} <8-10, 18, 19, 26, 27, 28, 30> [8, 11, 12, 29, 33-36, 43]
$HgCl_2$ <6, 13, 24, 26, 29> [6, 19, 31, 33, 42]
KCN <6> [6]
$KMnO_4$ <29> [42]
Li_2SO_4 <26> [33]
$MgSO_4$ <29> [42]
$MnCl_2$ <24, 26, 29> [31, 33, 42]
N-Bromosuccinimide <6, 26, 29> [6, 33, 42]
Na_2HAsO_4 <26> [33]
$NiCl_2$ <24> [31]
Pb^{2+} <16> [22]
$SnCl_2$ <29> [42]
Zn^{2+} <8, 28> [8, 36]
$ZnSO_4$ <24, 26, 29> [31, 33, 42]
acetamide <29> [42]
ammonium molybdate <6> [6]
ethanol <26> (<26> inhibition reversible when up to 80% ethanol concentration used [33]) [33]

indole-3-acetic acid <8> (<8> slight [8]) [8]
iodoacetic acid <29> [42]
nojirimycin <15> [21]
p-chloromercuribenzoate <26> [33]
p-hydroxymercuribenzoate <29> [42]
phenyl mercuric nitrate <6> [6]
tetranitromethane <7> [18]

Metals, ions

$AgNO_3$ <26> (<26> activity enhanced by 15% [33]) [33]
Ca^{2+} <24, 28> (<24,28> enhanced activity [31,36]) [31, 36]
Co^{2+} <24> (<24> enhanced activity [31]) [31]
Mg^{2+} <24> (<24> enhanced activity [31]) [31]
Zn^{2+} <24> (<24> enhanced activity [31]) [31]

Specific activity (U/mg)

0.696 <31> [47]
0.7 <2> (<2> isoenzyme I [10]) [10]
0.845 <13> [19]
2.5 <2> (<2> isoenzyme II [10]) [10]
4.7 <31> (<31> isoenzyme I [44]) [44]
5.8 <10> [37]
6.7 <24> [31]
12.62 <9> [11]
12.8 <31> (<31> isoenzyme II [44]) [44]
13.2 <2> [25]
20.7 <19> [34]
24.2 <6> [6]
27.7 <16> [22]
52.1 <28> [36]
65 <9> [39]
169.4 <26> [33]
180 <8> [8]
215 <9> [40]
250 <25> [32]
261 <29> (<29> enzyme I [42]) [42]
292 <29> (<29> enzyme II [42]) [42]
335 <29> (<29> enzyme III [42]) [42]
400 <7> [24]
593 <10> [12]
612 <23> [30]
7840 <27> [35]

K_m-Value (mM)

0.0018 <29> (laminarin, <29> enzyme I, based on molecular mass for laminarin of 5300 [42]) [42]
0.0041 <29> (laminarin, <29> enzyme IIII, based on molecular mass for laminarin of 5300 [42]) [42]

0.0043 <29> (laminarin, <29> enzyme II, based on molecular mass for laminarin of 5300 [42]) [42]
0.06 <10> (laminaritetraose) [12]
0.066 <26> (laminarin) [33]
0.095 <31> (cellobiose) [47]
0.111 <1> (laminaripentaose) [1]
0.18 <19> (laminarin, <19> molecular weight of 4000 assumed [34]) [34]
0.23 <25> (laminarin, <25> or 1.7 mg/ml [32]) [32]
0.255 <31> (laminaribiose) [47]
0.28 <10> (laminaripentaose, <10> or 2.08 mg/ml [12]) [12]
0.36 <10> (laminarihexaose) [12]
0.37 <25> (pustulan, <25> or 4.5 mg/ml [32]) [32]
0.41 <28> (*p*-nitrophenyl-β-D-glucopyranoside) [36]
0.83 <26> (laminarihexaose) [33]
0.85 <10> (laminarin, <10> molecular weight postulated to be 2448 [12]) [12]
1.03 <28> (laminaran) [36]
1.4 <31> (4-nitrophenyl β-D-glucoside) [44]
1.49 <10> (laminaritriose) [12]
1.9 <9> (methylumbelliferyl-β-glucoside, <9> Q330E [45]) [45]
2 <9> (methylumbelliferyl-β-glucoside) [45]
2.3 <26> (laminaripentaose) [33]
4.07 <2> (*p*-nitrophenyl-β-D-glucopyranoside, <2> under glycosylated form [25]) [25]
5 <25> (*p*-nitrophenyl-β-D-glucoside) [32]
6.6 <19> (*p*-nitrophenyl-β-D-glucopyranoside) [34]
18.6 <26> (laminaritetraose) [33]
Additional information <1, 6, 8, 9, 11, 16, 19, 22, 23, 27, 30> (<1> laminarin 0.008% (w/v) [1]; <6> laminarin 1.22 mg/ml [6]; <8> laminarin 6.25 mg/ml [8]; <11> laminarin 0.09 g/l [14]; <16> laminaran 0.080 mg/ml, laminaribiose 4.06 mg/ml, laminarihexaose 0.086 mg/ml, laminaripentaose 0.107 mg/ml, laminaritetraose 0.125 mg/ml, laminaritriose 0.172 mg/ml [22]; <19> yeast cell wall 1,3-β-glucan 2.5 mg/ml [34]; <22> laminarin sigmoidal kinetics, Hill coefficient 1.6, half-maximum velocity at 58 *g/ml [28]; <23> laminarin 0.28 mg/ml [30]; <27> laminarin 0.67 mg/ml, neutral glucan from Achlya 1.11 mg/ml, phosphoglucan 2.1 mg/ml [35]; <30> laminarin 2.5 mg/ml [43]; <6> lichenin 2.22 mg/ml [6]; <8> pustulan 166 mg/ml [8]; <9> laminarin 1.7 mg/ml [11]; <9> laminarin 3.8 mg/ml, E230Q 2.6, E330Q 2.9, Q330E 2.9 [45]) [1, 6, 8, 11, 14, 22, 28, 30, 34, 35, 43, 45]

pH-Optimum

3.8 <30> [43]
4.4-4.5 <6> [6]
4.5 <13> [19]
4.5-5 <16, 18> [22, 26]
4.6 <10> [12]
4.7-5.2 <1> [1]

5 <8, 19, 25, 29, 31> [8, 32, 34, 42, 47]
5-5.5 <23> [30]
5.4 <22> [28]
5.5-6 <9> [11]
5.7 <11, 14> [14, 20]
5.8 <15> [21]
6 <24, 26, 27> [31, 33, 35]
6.5 <29> (<29> enzyme I [42]) [42]
8 <28> [36]

pH-Range

2-8 <25> [32]
3-6.5 <6> [6]
3-7 <1> [1]
4-7 <10> [37]
4.5-5.5 <10> [37]
4.5-7 <29> (<29> enzyme II and III [42]) [42]
5-7 <9> [11]
5-7.5 <29> (<29> enzyme I [42]) [42]
6-7 <27> [35]

Temperature optimum (°C)

30-40 <27> [35]
37 <19> [34]
37-40 <28> [36]
40 <10> [37]
45 <13, 16> [19, 22]
46 <10> [12]
50 <9, 24, 26, 30> (<9> for short incubation times {15 min} [11]) [11, 31, 33, 43]
53 <6> [6]
55 <29> (<29> enzyme II [42]) [42]
60 <22, 29> (<29> enzyme III [42]) [28, 42]
65 <14, 29> (<29> enzyme I [42]) [20, 42]

4 Enzyme Structure

Molecular weight

20000 <19> (<19> gel filtration [34]) [34]
21000 <19> (<19> PAGE [34]) [34]
31500 <10> (<10> gel filtration [12]) [12]
43000 <2> (<2> under glycosylated form [25]) [25]
43000 <25> (<25> gel filtraion [32]) [32]
43000 <26> (<26> gel filtration [33]) [33]
45600 <9> [38]
57000 <30> [43]
58800 <2> [25]

59500 <16> (<16> TSK G3000SW, gel filtration [22]) [22]
60000 <18> (<18> gel filtration [29]) [29]
68000 <22> (<22> subunit weight [28]) [28]
70000 <23> (<23> gel filtration [30]) [30]
78000 <10> (<10> SDS-PAGE and isoelectric focussing [37]) [37]
82000 <13> (<13> disc gel electrophoresis, varying polyacrylamide gel concentrations [19]) [19]

Subunits

? <2, 6, 9, 16, 19-22, 27, 28, 29, 31, 32> (<22> x * 68000 [28]; <6> x * 82000, SDS-PAGE [6]; <2> x * 85000, deglycosylated form, SDS-PAGE [10]; <9> x * 107000, SDS-PAGE [11]; <16> x * 55500, SDS-PAGE [22]; <19-21, 2> x * 61000, SDS-PAGE, different yeasts [27]; <27> x * 30000, SDS-PAGE [35]; <28> x * 120000, SDS-PAGE [36]; <9> x * 42000, SDS-PAGE [39]; <29> x * 81000-89000, SDS-PAGE [42]; <32> x * 73000, SDS-PAGE [46]; <31> x * 66000, SDS-PAGE [47]; <31> x * 69000, isoenzyme I, SDS-PAGE [44]; <31> x * 71000, isoenzyme II, SDS-PAGE [44]) [6, 10, 11, 22, 27, 28, 35, 36, 39, 42, 44, 46, 47]
dimer <2, 9> (<2> 1 * 47000 + 1 * 48000, SDS-PAGE, deglycosylated form [10]; <9> 1 * 63000 + 1 * 44000, SDS-PAGE [11]) [10, 11]
Additional information <19-21, 2> (<19-21, 2> exo-β-glucanases from yeast are aggregates of many functional units [27]) [27]

5 Isolation/Preparation/Mutation/Application

Source/tissue

commercial preparation <30> (<30> from Aspergillus saitoi [43]) [43]
culture medium <1, 3, 4, 14, 9, 10, 29> [2, 4, 20, 37-42, 45]
egg <22> [28]
fiber <15> [21]
leaf <17> (<17> young [23]) [23]
seed <31> [44]
seedling <18, 31> [26, 29, 47]

Localization

cell wall <11, 12, 13, 15, 17, 18, 28, 31> [14, 15, 19, 21, 23, 26, 29, 36, 47]
extracellular <2, 14, 9, 10, 29> [3, 20, 37-42, 45, 46]
intracellular <2> [3]
periplasm <2, 9> [11, 16, 17]
protoplast <2> [17]

Purification

<1> (53fold [1]) [1]
<2> (4661fold [25]) [25]
<6> [6]
<8> (418.6fold [8]) [8]
<9> (1333fold [39]) [39]

<9> (2 mg pure protein per 10 liters culture [38]) [38]
<9> (3fold [40]) [40]
<9> (432.7fold [11]) [11]
<10> [12]
<10> (22.5fold [37]) [37]
<13> (224.4fold [19]) [19]
<16> (620fold [22]) [22]
<18> (partial purification [29]) [29]
<19> (16.9fold [34]) [34]
<22> (800-1500fold [28]) [28]
<25> (8333fold [32]) [32]
<26> [33]
<27> (109fold [35]) [35]
<28> [36]
<29> (enzyme I 22fold, enzyme II 25fold, enzyme III 28fold [42]) [42]
<30> (23fold [43]) [43]
<31> (58fold [47]) [47]
<31> (isoenzyme I: 47fold, isoenzyme II: 128fold [44]) [44]

Crystallization
<9> (resolution 1.9 A [38]; resolution 1.85 A [41]) [38, 41]

Cloning
<9> (in Saccharomyces cervisiae [40, 41, 45]) [40, 41, 45]

Engineering
E230Q <9> [45]
E330Q <9> [45]

6 Stability

pH-Stability
3-8 <25> [32]
3.1 <2> (<2> 44% loss of activity after 75 min, underglycosylated form 93% loss of activity after 75 min [25]) [25]
4-6 <24> [31]
4-7 <13> [19]
4.5-10 <29> (<29> enzyme I, II and III [42]) [42]
5.5-10.5 <16> [22]
7 <7> (<7> unstable above [18]) [18]

Temperature stability
30 <24> (<24> looses activity at higher temperatures [31]) [31]
35 <26> (<26> stable below 35°C [33]) [33]
45 <2, 27> (<2> stable for 45 min, underglycosylated form after 45 min 35% loss of activity [25]; <27> stable below 45°C [35]) [25, 35]
50 <6, 8, 16, 29> (<16> stable below 50°C [22]; <29> enzyme II and III, stable below 50°C [42]; <6> up to 50°C [6]; <8> up to 50°C [8]) [6, 8, 22, 42]

52 <25> (<25> stable at 52°C for 20 min [32]) [32]
55 <29> (<29> enzyme I, stable below 55°C [42]) [42]
60 <23> (<23> after 15 min 100% loss of activity [30]) [30]

Storage stability

<1>, -20°C, 5 months, 25% loss of activity [1]
<2>, -20°C, several months, no loss of activity [3]
<2>, 4°C, 1 week, no loss of activity [3]
<6>, room temperature, several hours [6]
<8>, 1-4°C, sodium succinate buffer pH 5.0-5.5, 3 months, no loss of activity [8]
<8>, freeze drying, 50% activity loss [8]
<8>, frozen, 8 months, 50% activity loss [8]
<23>, -15°C, 10 days, 21% activity loss, 4°C, 10 days, 15% activity loss [30]
<29>, -20°C, loss of activity in some samples [42]
<29>, crude and purified enzyme preparations, -80°C, 12 months, no loss of activity [42]
<29>, crude enzyme preparations, 4°C, 5 weeks, 5% activity loss [42]
<29>, purified enzyme 4°C, 4 months, no loss of activity [42]

References

[1] Barras, D.R.; Stone, B.A.: II. Purification and properties of the β-1,3-glucan exo-hydrolase. Biochim. Biophys. Acta, **191**, 342-353 (1969)
[2] Barras, D.R.; Stone, B.A.: I. The Nature of the hydrolases. Biochim. Biophys. Acta, **191**, 329-341 (1969)
[3] Cortat, M.; Matile, P.; Wiemken, A.: Isolation of glucanase-containing vesicles from budding Yeast. Arch. Mikrobiol., **82**, 189-205 (1972)
[4] Jones, D.; Gordon, A.H.; Bacon, J.S.D.: Co-operative action by endo- and exo-β-(1 leads to 3)-glucanases from parasitic fungi in the degradation of cell-wall glucans of Sclerotinia sclerotiorum (Lib.) de Bary. Biochem. J., **140**, 47-55 (1974)
[5] Dabbagh, R.; Conant, N.F.; Burns, R.O.: Effect of temperature on saprophytic cryptococci: observations relating to wall biosynthesis at non-permissive growth temperatures. J. Gen. Microbiol., **85**, 190-202 (1974)
[6] Marshall, J.J.; Grand, R.J.A.: Comparative studies on β-glucan hydrolases. Isolation and characterization of an exo(1 yields 3)-β-glucanase from the snail, Helix pomatia. Arch. Biochem. Biophys., **167**, 165-175 (1975)
[7] Nelson, T.E.: The attack mechanism of an exo-1,3-β-glucosidase from Basidiomycete sp. QM 806. Biochim. Biophys. Acta, **377**, 139-145 (1975)
[8] Fleet, G.H.; Phaff, H.J.: Glucanases in Schizosaccharomyces. Isolation and properties of an exo-β-glucanase from the cell extracts and culture fluid of Schizosaccharomyces japonicus var. versatilis. Biochim. Biophys. Acta, **401**, 318-332 (1975)

[9] Nebreda, A.R.; Villa, T.G.; Villanueva, J.R.; del Rey, F.: Cloning of genes related to exo-β-glucanase production in Saccharomyces cerevisiae: characterization of an exo-β-glucanase structural gene. Gene, **47**, 245-259 (1986)

[10] Ramirez, M.; Hernandez, L.M.; Larriba, G.: A similar protein portion for two exoglucanases secreted by Saccharomyces cerevisiae. Arch. Microbiol., **151**, 391-389 (1989)

[11] Molina, M.; Cenamor, R.; Sanchez, M.; Nobela, C.: Purification and some properties of Candida albicans exo-1,3-β-glucanase. J. Gen. Microbiol., **135**, 309-314 (1989)

[12] Kitamoto, Y.; Kono, R.; Shimotori, A.; Mori, N.; Ichikawa, Y.: Purification and some properties of an exo-β-1,3-glucanase from Trichoderma harzianum. Agric. Biol. Chem., **51**, 3385-3386 (1987)

[13] Ryan, E.M.; Ward, O.P.: Studies on the production and properties of the yeast lytic enzyme of Basidiomycete sp. QM806. Biotechnol. Lett., **9**, 405-410 (1987)

[14] Boucaud, J.; Bigot, J.; Devaux, J.: Partial purification and characterization of exo-and-endo-1,3-β-D glucanases from Atriplex leaf tissues, their involvement in cell extension growth and the effects of salinity. J. Plant Physiol., **128**, 337-349 (1987)

[15] Llamazares, J.; Acebes, J.L.; Zarra, I.: Cell wall autolysis in Pinus pinaster aiton hypocotyls. Enzymatic activities involved. J. Plant Physiol., **127**, 11-22 (1987)

[16] Molina, M.; Cenamor, R.; Nobela, C.: Exo-1,3-β-glucanase activity in Candida albicans: effect of the yeast-to-mycelium transition. J. Gen. Microbiol., **133**, 609-617 (1987)

[17] Cenamor, R.; Molina, M.; Galdona, J.; Sanchez, M.; Nombela, C.: Production and secretion of Saccharomyces cervisiae β-glucanases: Differences between protoplast and periplasmic enzymes. J. Gen. Microbiol., **133**, 619-628 (1987)

[18] Jeffcoat, R.; Kirkwood, S.: Implication of histidine at the active site of exo-β-(1-3)-D-glucanase from Basidiomycete sp. QM 806. J. Biol. Chem., **262**, 1088-1091 (1987)

[19] Hiura, N.; Kobayashi, M.; Nakajima, T.; Matsuda, K.: Purification and some properties of two wall-associated (β1,3)glucanases from Neurospora crassa cells. Agric. Biol. Chem., **50**, 2461-2467 (1986)

[20] Bielecki, S.; Antczak, T.; Galas, E.: β-1,3-Glucanase from Streptomyces sp. 1228. Eur. Congr. Biotechnol., 3rd Ed., **2**, 489-496 (1984)

[21] Bucheli, P.; Dürr, M.; Buchala, A.J.; Meier, H.: β-Glucanases in developing cotton (Gossypium hirsutum L.) fibres. Planta, **166**, 530-536 (1985)

[22] Iwamuro, Y.; Aoki, M.; Mikami, Y.: Purification and some properties of an exo-β-1,3-glucanase from Porodisculus pendulus. J. Ferment. Technol., **63**, 405-409 (1985)

[23] Lienart, Y.; Barnoud, F.: β-D-Glucan-hydrolase actvities in pure cell-wall-enriched fractions from Valerianella olitoria cells. Planta, **165**, 68-75 (1985)

[24] Nanjo, F.; Usui, T.; Suzuki, T.: Mode of action of an exo-β-(1,3)-D-glucanase on the laminaran from Eisenia bicyclis. Agric. Biol. Chem., **48**, 1523-1532 (1984)

[25] Sanchez, A.; Villanueva, J.R.; Villa, T.G.: Effect of tunicamycin on exo-1,3-β-D-glucanase synthesis and secretion by cells and protoplasts of Saccharomyces cerevisiae. J. Gen. Microbiol., **128**, 3051-3060 (1982)
[26] Huber, D.J.; Nevins, D.J.: Exoglucanases from Zea mays L. seedlings: their role in β-D-glucan hydrolysis and their potential role in extension growth. Planta, **155**, 467-472 (1982)
[27] Sanchez, A.; Nebreda, A.R.; Villa, T.G.: Self-associating properties of yeast exo-β-glucanases are dependent on the ionic strength. FEBS Lett., **145**, 213-216 (1982)
[28] Talbot, C.F.; Vacquier, V.D.: The purification and characterization of an exo-β (1,3)-glucanohydrolase from sea urchin eggs. J. Biol. Chem., **257**, 742-746 (1982)
[29] Huber, D.J.; Nevins, D.J.: Partial purification of endo- and exo-β-D-glucanase enzymes from Zea mays L. seedlings and their involvement in cell-wall autohydrolysis. Planta, **151**, 206-214 (1981)
[30] Bamforth, C.W.: The adaptability, purification and properties of exo-β 1,3-glucanase from the fungus Trichoderma reesei. Biochem. J., **191**, 863-866 (1980)
[31] Yokotsuka, K.; Kakehashi, Y.; Oyama, S.; Kato, A.; Akatsuka, S.; Yoshitake, S.; Kushida, T.: Production and properties of lytic enzyme and exo-β-1,3-glucanase from Streptomyces murinus. J. Ferment. Technol., **56**, 599-605 (1978)
[32] Lachance, M.-A.; Villa, T.G.; Phaff, H.J.: Purification and partial characterization of an exo-β-glucanase from the yeast Kluyveromyces aestaurii. Can. J. Biochem., **55**, 1001-1006 (1977)
[33] Nagasaki, S.; Saito, K.; Yamamoto, S.: Purification and characterization of an exo β-1,3-glucanase from a Fungi imperfecti. Agric. Biol. Chem., **41**, 493-502 (1977)
[34] Villa, T.G.; Notario, V.; Benitez, T.; Villanueva, J. R.: Purification of an exo-1,3-β-glucanase from Candida utilis. Can. J. Biochem., **54**, 927-934 (1976)
[35] Xin Du; Mullins, J.T.: Glucan 1,3-β-glucosidase activities of Achlya bisexualis: synthesis and properties. Mycologia, **91**, 353-358 (1999)
[36] Kruse, D.; Cole, G.T.: A seroreactive 120-kilodalton β-1,3-glucanase of Coccidioides immitis which may participate in spherule morphogenesis. Infect. Immun., **60**, 4350-4363 (1992)
[37] Lorito, M.; hayes, C.K.; Di Pietro, A.; Woo, S.L.; Harman, G.E.: Purification, characterization, and synergistic activity of a glucan 1,3-β-glucosidase and an N-acetyl-β-glucosaminidase from Trichoderma harzianum. Phytopathology, **84**, 398-405 (1994)
[38] Cutfield, S.; Brooke, G.; Sullivan, P.; Cutfield, J.: Crystallization of the exo(1,3)-βglucanase from Candida albicans. J. Mol. Biol., **225**, 217-218 (1992)
[39] Chambers, R.S.; Broughton, M.J.; Cannon, R.D.; Carne, A.; Emerson, G.W.; Sullivan, P.A.: An exo-β-(1,3)-glucanase of Candida albicans: purification of the enzyme and molecular cloning of the gene. J. Gen. Microbiol., **139**, 325-334 (1993)

[40] Chambers, R.S.; Waldern, A.R.; Brooke, G.S.; Cutfield, J.F.; Sullivan, P.A.: Identification of a putative active site residue in the exo-β-(1,3)-glucanase of Candida albicans. FEBS Lett., **327**, 366-369 (1993)

[41] Cutfield, S.M.; Davies, G.J.; Murshudov, G.; Anderson, B.F.; Moody, P.C.E.; Sullivan, P.A.; Cutfield, J.F.: The structure of the exo-β-(1,3)-glucanase from Candida albicans in native and bound forms: Relationship between a pocte an groove in family 5 glycosyl hydrolases. J. Mol. Biol., **294**, 771-783 (1999)

[42] Pitson, S.M.; Seviour, R.J.; McDougall, B.M.; Woodward, J.R.; Stone, B.A.: Purification and characterization of three extracellular (1,3)-β-D-glucan glucohydrolases from the filamentous fungus Acremonium persicinum. Biochem. J., **308**, 733-741 (1995)

[43] Kasahara, S.; Nakajima, T.; Miyamoto, C.; Wada, K.; Furuichi, Y.; Ichishima, E.: Characterization and mode of action of exo-1,3-β-D-glucanase from Aspergillus saitoi. J. Ferment. Bioeng., **74**, 238-240 (1992)

[44] Hrmova, M.; Harvey, A.J.; Wang, J.; Shirley, N.J.; Jones, G.P.; Stone, B.A.; Hoj, P.B.; Fincher, G.B.: Barley β-D-glucan exohydrolases with β-D-glucosidase activity. J. Biol. Chem., **271**, 5277-5286 (1996)

[45] Mackenzie, L.F.; Brooke, G.S.; Cutfield, J.F.; Sullivan, P.A.; Withers, S.G.: Identification of Glu-330 as the catalytic nucleophile of Candida albicans exo-β-(1,3)-glucanase. J. Biol. Chem., **272**, 3161-3167 (1997)

[46] Nikolskaya, A.N: Pitkin, J.W.; Schaeffer, H.J.; Ahn, J.-H.; Walton, J.D.: EXG1p, a novel exo-β1,3-glucanase from the fungus Cochliobolus carbonum, contains a repeated motiv present in other proteins that interact with polysaccharides. Biochim. Biophys. Acta, **142**, 632-636 (1998)

[47] Kotake, T.; Nakagawa, N.; Takeda, K.; Sakurai, N.: Purification an characterization of wall-bound exo-1,3-β-D-glucanase from barley (Hordeum vulgare L.) seedlings. Plant Cell Physiol., **38**, 194-200 (1997)

Glucan endo-1,3-α-glucosidase 3.2.1.59

1 Nomenclature

EC number
3.2.1.59

Systematic name
1,3(1,3;1,4)-α-D-glucan 3-glucanohydrolase

Recommended name
glucan endo-1,3-α-glucosidase

Synonyms
cariogenanase
cariogenase
endo-(1→3)-α-glucanase
endo-1,3-α-D-glucanase
endo-1,3-α-glucanase
endo-1,3-α-glucanase
glucanase, endo-1,3-α-mutanase

CAS registry number
9075-84-7

2 Source Organism

<1> *Bacteroides oralis* [1]
<2> *Flavobacterium sp.* [8]
<3> *Trichoderma viride* [9]
<4> *Trichoderma harzianum* [3, 10, 12]
<5> *Pseudomonas sp.* [2]
<6> *Streptomyces chartreusis* [4]
<7> *Bacillus circulans* (HU-M1 [11]) [5, 11]
<8> *Streptomyces sp.* (ki-8 [6]) [6, 7]
<9> *Penicillium purpurogenum* [12]

3 Reaction and Specificity

Catalyzed reaction
endohydrolysis of 1,3-α-D-glucosidic linkages in isolichenin, pseudonigeran and nigeran

Reaction type

hydrolysis of O-glycosyl bond

Natural substrates and products

S glucan <7> [5]
S isolichenin <7> [5]
S nigeran <7> [5]
S pseudonigeran <7> [5]

Substrates and products

S glucan + H_2O <2> (<2> insoluble, sticky glucan of Streptococcus mutans [8]) [8]
P isomaltose + nigerose + nigerotriose + oligosaccharides <2> [8]
S glucan + H_2O <3, 6-8> (<6> glucan produced from sucrose by Streptococcus mutans, [4]; <3, 7, 8> α-1,3-glucan, [5, 7, 9]) [4, 5, 7, 9]
P nigerose + glucose <7, 8> [5, 7]
S lentinus α-1,3-glucan + H_2O <8> [7]
P ?
S limit glucan + H_2O <5> [2]
P ?
S mutan + H_2O <8> [7]
P ?
S nigeran + H_2O <7> [5]
P ?
S nigeropentaose + H_2O <8> [7]
P glucose + nigerose <8> [7]
S nigerotetraose + H_2O <8> [7]
P glucose + nigerose <8> [7]
S nigerotriose + H_2O <8> [7]
P glucose + nigerose <8> [7]
S pseudonigeran + H_2O <8> (<7> not [5]) [7]
P nigerose + α-D-glucose <8> [7]
S Additional information <2, 3, 7, 8> (not: nigerose, <8> [7]) [5, 7, 8, 9]
P ?

Inhibitors

Ag^+ <6> [4]
Ag^+ <2, 3, 5, 7> (<3> irreversible inhibition [9]) [2, 5, 8, 9]
Cu^{2+} <6, 7> [4, 5]
EDTA <5> (<6> no effect [4]) [2]
Fe^{2+} <3, 4> (<6> no effect [4]; <3> reversible inhibition [9]) [3, 9]
Fe^{3+} <5> [2]
Hg^{2+} <2, 5-8> (<3> irreversible inhibition [9]) [2, 4, 5, 7, 8]
Mn^{2+} <6-8> [4, 5, 7]
Pb^{2+} <2, 5, 7> [2, 5, 8]
Zn^{2+} <3, 5, 7> (<3> reversible inhibition [9]) [2, 5, 9]
iodoacetamide <7> [5]
merthiolate <6> [4]
p-chloromercuribenzoic acid <5> [2]

Activating compounds

EDTA <7> [5]

Metals, ions

Ca^{2+} <5> (<5> increases activity [2]; <6, 7> no effect [4, 5]) [2]
Co^{2+} <5> (<5> increases activity [2]; <7> no effect [5]) [2]
Additional information <7> (<7> no requirement of divalent metal [5]) [5]

Specific activity (U/mg)

0.831 <2> [8]
0.968 <6> [4]
7.52 <7> [5]
11.2 <7> [5]
78.1 <5> [2]
Additional information <3> [9]

K_m-Value (mM)

0.46 <3> (pseudonigeran) [9]
0.89 <6> (glucan) [4]
80 <5> (limit glucan) [2]
Additional information <7, 8> [5, 7]

pH-Optimum

3.5-4.5 <4, 9> [12]
4.5 <3> [9]
5 <5> [2]
5.4 <8> [7]
5.5-6 <6> [4]
6.3 <2> [8]
6.9 <7> [11]
7.5-8.5 <7> [5]

pH-Range

3-7 <8> (<8> pH 3: 30% of optimum activity, pH 7: 20% of optimum activity [7]) [7]
3-8 <3> [9]
4.5-8 <5> (<5> sharp decrease below pH 4.5, gradual decrease up to pH 8.0 [2]) [2]
4.5-8.5 <2> (<2> pH 4.5: 40% of optimal activity, pH 8.5: 70% of optimal activity [8]) [8]

Temperature optimum (°C)

42 <2> [8]
50 <3, 7> [9, 11]
50-55 <4, 9> [12]
55 <6> [4]
56 <5> [2]
60 <8> [7]

Temperature range (°C)

30-80 <8> (<8> 30°C: 18% of optimum activity, 80°C: 25% of optimum activity [7]) [7]

4 Enzyme Structure

Molecular weight

47000 <3> (<3> gel filtration [9]) [9]
65000 <2> (<2> SDS-PAGE [8]) [8]
68000 <6> (<6> SDS-PAGE [4]) [4]
75000 <4> (<4> SDS-PAGE and MALDI-MS [12]) [12]
78000 <8> (<8> SDS-PAGE [6]) [6]
134000 <7> (<7> SDS-PAGE [5]) [5]
160000 <7> (<7> SDS-PAGE [11]) [11]
180000 <7> (<7> HPLC [11]) [11]
279000 <5> (<5> aggregated form, gel filtration [2]) [2]

Subunits

oligomer <6> (<6> possibly oligomeric or aggregated form, molecular weight larger than 300000, x * 68000, SDS-PAGE, Streptomyces chartreusis) [4]

Posttranslational modification

no glycoprotein <7> [5]

5 Isolation/Preparation/Mutation/Application

Source/tissue

culture filtrate <3, 7> [5, 9]
culture supernatant <2, 5, 7, 8> [2, 6, 8, 11]

Localization

cytoplasm <1> (<1> only [1]) [1]
extracellular <7> [5]

Purification

<2> [8]
<3> [9]
<4> [12]
<5> (NRRL-B-12324 [2]) [2]
<6> [4]
<7> [5, 11]
<8> [6, 7]
<9> [12]

Cloning

<4, 9> (expression in Aspergillus oryzae [12]) [12]

6 Stability

pH-Stability

3.8 <7> (<7> 50% loss of activity after 10 min [5]) [5]
4.2 <7> (<7> 50% loss of activity after 30 min [5]) [5]
5.25-10.5 <7> (<7> 30°C, stable for at least 1.5 h [5]) [5]

Temperature stability

37 <5> (<5> pH 5.5, 3 days, 3% loss of activity [2]) [2]
40 <7> (<7> 5% loss of activity after 90 min [5]) [5]
45 <7> (<7> 25% loss of activity after 90 min [5]) [5]
50 <3, 7, 8> (<7> complete loss of activity after 20 min [5]; <8> stable up to 10 min [7]; <3> rapid inactivation [9]) [5, 7, 9]
60 <2, 5> (<2> complete loss of activity after 30 min [8]; <5> loss of activity at or above [2]) [2, 8]
65 <6> (<6> pH 6.0, 10 min, complete inactivation [4]) [4]

General stability information

<7>, freezing, -15°C, appreciable loss in activity [5]

Storage stability

<5>, 4°C, 18 months, 17% loss of activity [2]
<7>, 4°C, 2 years, pH 6, stable [5]

References

[1] Takahshi, N.; Satoh, Y.; Takamori, K.: Subcellular localization of D-glucanases in Bacteroides oralis Ig4a. J. Gen. Microbiol., **131**, 1077-1082 (1985)
[2] Simonson, L.G.; Gaugler, R.W.; Lamberts, B.L.; Reiher, D.A.: Purification and properties of endo-1,3-α-D-glucanase from Pseudomonas. Biochim. Biophys. Acta, **715**, 189-195 (1982)
[3] Stephen, E.R.; Nasim, A.: Production of protoplasts in different yeasts by mutanase. Can. J. Microbiol., **27**, 550-553 (1981)
[4] Takehara, T.; Inoue, M.; Morioka, T.; Yokogawa, K.: Purification and properties of endo-α-1,3-glucanase from a Streptomyces chartreusis strain. J. Bacteriol., **145**, 729-735 (1981)
[5] Meyer, M.T.; Phaff, H.J.: Purification and properties of (1→3)-α-glucanasea from Bacillus circulans WL-12. J. Gen. Microbiol., **118**, 197-208 (1980)
[6] Imai, K.; Kikuta, T.; Kobayashi, M.; Matsuda, K.: An α-1,3-glucanase from Sreptomyces sp. KI-8 Production and purification. Agric. Biol. Chem., **41**, 1339-1346 (1977)
[7] Imai, K.; Kobayashi, M.; Matsuda, K.: Properties of an α-1,3-glucanase from Streptomyces sp. KI-8. Agric. Biol. Chem., **41**, 1889-1895 (1977)
[8] Ebisu, S.; Kato, K.; Kotani, S.; Misaki, A.: Isolation and purification of Flavobacterium α-1,3-glucanase-hydrolyzing, insoluble, sticky glucan of Streptococcus mutans. J. Bacteriol., **124**, 1489-1501 (1975)

[9] Hasegawa, S.; Nordin, J.H.: Enzymes that hydrolyze fungal cell wall polysaccharides. I. Purification and properties of an endo-α-D-(1-3)-glucanase from Trichoderma. J. Biol. Chem., **244**, 5460-5470 (1969)
[10] Quivey, R.G.: Raffinose-induced mutase production from Trichoderma harzianum. FEMS Microbiol. Lett., **112**, 307-312 (1993)
[11] Matsuda, S.; Kawanami, Y.; Takeda, H.; Ooi, T.; Kinoshita, S.: Purification of mutanase from Bacillus circulans. J. Ferment. Bioeng., **83**, 593-595 (1997)
[12] Fuglsang, C.C.; Berka, R.M.; Wahleithner, J.A.; Kauppinen, S.; Shuster, J.R.; Rasmussen, G.; Halkier, T.; Dalboge, H.; Henrissat, B.: Biochemical analysis of recombinant fungal mutanases. J. Biol. Chem., **275**, 2009-2018 (2000)

Glucan 1,4-α-maltotetraohydrolase 3.2.1.60

1 Nomenclature

EC number
3.2.1.60

Systematic name
1,4-α-D-glucan maltotetraohydrolase

Recommended name
glucan 1,4-α-maltotetraohydrolase

Synonyms
G4-amylase
G4-exo-α-amylase
exo-malto tetraohydrolase
exo-maltotetraohydrolase
exomaltotetraohydrolase
maltotetraohydrolase, exo-

CAS registry number
37288-44-1

2 Source Organism

<1> *Pseudomonas stutzeri* (NRRL B-3389 [2, 3, 11]; MO-19 [7]; mutant E219Q [14]) [1-5, 7, 11, 14, 15]
<2> *Pseudomonas saccharophila* (IAM 1504 [6]) [6, 9]
<3> *Bacillus circulans* (MG-4 [8]) [8]
<4> *Pseudomonas sp.* (IMD 353 [10]; MS300 [13]) [10, 13]
<5> *Bacillus sp.* (GM8901 [12]) [12]

3 Reaction and Specificity

Catalyzed reaction
hydrolysis of 1,4-α-D-glucosidic linkages in amylaceous polysaccharides, to remove successive maltotetraose residues from the non-reducing chain ends (<4> endo-acting mechanism producing maltotetraose in the α-anomeric form [10]; <5> exo-cleavage mechanism [12])

Reaction type
hydrolysis of O-glycosyl bond

Natural substrates and products

S Additional information <4, 5> (<5> five alkaline amylases are produced in culture broth [12]; <4> more amylase production under high hydrostatic pressure than under atmospheric pressure [13]) [12, 13]
P ?

Substrates and products

S amylopectin + H_2O <1-3, 5> (<2> recombinant and native enzyme [9]) [1, 3, 7-9, 12]
P maltotetraose <1-3> [1, 8, 9]
S amylose + H_2O <1-3, 5> (<2> recombinant and native enzyme [9]) [3, 8, 9, 12]
P maltotetraose <1, 3> [3, 8]
S glycogen + H_2O <1-3> (<1> oyster and shellfish glycogen [3]; <2> recombinant and native enzyme, oyster and rabbit liver glycogen [9]) [3, 7-9]
P maltotetraose <1> [3]
S maltoheptaose + H_2O <1, 5> [1, 12]
P maltotetraose + maltotriose <1> [1]
S maltohexaose + H_2O <1, 5> [1, 12]
P maltotetraose + maltose <1> [1]
S maltopentaose + H_2O <5> [12]
P ?
S starch + H_2O <1-5> (<2> recombinant and native enzyme [9]; <4> amylase A and B potato starch [13]) [2-13, 15]
P maltotetraose <1-5> [2, 3, 8-13, 15]
S Additional information <4> (<4> constitutive maltotetraose-producing amylase using glycose as carbon source [10]) [10]
P ?

Inhibitors

Ag^+ <2> (<2> $AgNO_3$, recombinant and native enzyme [9]) [9]
Al^{3+} <1> (<1> F1 and F2 form 30% inhibition [3]) [3]
Co^{2+} <1> (<1> F1 and F2 form 47% inhibition [3]) [3]
Cu^{2+} <1-4> (<1> F1 and F2 form 92% inhibition [3]; <2> $CuSO_4$, recombinant and native enzyme [9]; <4> 1 mM $CuCl_2$, amylase A, 98% inhibition, amylase B, complete inhibition [13]) [3, 8, 9, 13]
EDTA <5> (<5> 1 mM, 82% inhibition [12]) [12]
Fe^{2+} <1> (<1> F1 form, 96% F2 form, 98% inhibition [3]) [3]
Fe^{3+} <1, 2, 4> (<1> F1 form, 75% F2 form, 76% inhibition [3]; <2> $FeCl_3$, recombinant and native enzyme [9]; <4> 1 mM $FeCl_3$, amylase A, 73% inhibition, amylase B, 87% inhibition [13]) [3, 9, 13]
Hg^{2+} <1, 3-5> (<1> F1 and F2 form, complete inhibition [3]; <5> 1 mM $HgCl_2$, complete inhibition [12]; <4> 1 mM $HgCl_2$, amylase A, 99% inhibition, amylase B, 97% inhibition [13]) [3, 8, 12, 13]
N-bromosuccinimide <1> (<1> F1 form, pH 6.5 complete inhibition, pH 8.0, 93% inhibition, F2 form, pH 6.5, 95% inhibition, pH 8.0, 28% inhibition [3]) [3]
Ni^{2+} <1> (<1> F1 form 62%, F2 form 67% inhibition [3]) [3]

Pb^{2+} <2> (<2> $Pb(OAc)_2$, recombinant and native enzyme [9]) [9]
Zn^{2+} <1-4> (<1> F1 form 94%, F2 form 96% inhibition [3]; <4> 1 mM $ZnCl_2$, amylase A 99% inhibition, amylase B, 97% inhibition [13]) [3, 8, 9, 13]
phenylmethylsulfonyl fluoride <5> (<5> 1 mM, 4°C, 10 h incubation, 76% inhibition [12]) [12]

Metals, ions

Ag^+ <5> (<5> 1 mM $AgNO_3$ activates [12]) [12]
Ca^{2+} <5> (<5> 1 mM $CaCl_2$ activates [12]) [12]
Co^{2+} <5> (<5> 1 mM $CoCl_2$ activates [12]) [12]
Cu^{2+} <5> (<5> 1 mM $CuCl_2$ activates [12]) [12]
Mg^{2+} <5> (<5> 1 mM $MgCl_2$ activates [12]) [12]

Specific activity (U/mg)

102 <1> (<1> F2 form [3]) [3]
104 <1> (<1> F1 form [3]) [3]
158 <5> [12]
313 <1> (<1> G4-1 form [7]) [7]
364 <1> (<1> G4-2 form [7]) [7]
1544 <4> (<4> amylase A [13]) [13]
1775 <4> (<4> amylase B [13]) [13]
2590 <1> [1]
Additional information <1, 3, 4> [2, 8, 10, 12]

K_m-Value (mM)

0.32 <1> (amylose, <1> F-1 and F-2 form [3]) [3]
5.7 <1> (starch, <1> F-2 form [3]) [3]
5.9 <1> (starch, <1> F-1 form [3]; <1> shellfish glycogen, F-1 and F-2 form [3]) [3]
6.7 <1> (oyster glycogen, <1> F-2 form [3]) [3]
7.1 <1> (oyster glycogen, <1> F-1 form [3]) [3]
Additional information <1> [7]

pH-Optimum

6.3 <4> (<4> amylase A, 40°C [13]) [13]
6.5 <2> (<2> recombinant and native enzyme [9]) [9]
7 <1> (<1> G4-1 and G4-2 form [7]) [7]
7.5 <3> [8]
7.8 <4> (<4> amylase B, 40°C [13]) [13]
7.8-8.2 <1> (<1> F1 and F2 form [3]) [3]
8 <1> [1]
10.5-12 <5> [12]

pH-Range

4-9 <2> (<2> recombinant and native enzyme, pH 4.0, 30%, pH 9.0, 50% of maximal activity [9]) [9]
4.5-10.5 <1> (<1> 95% of maximum activity at pH 7 and pH 9.5, pH 4.5, 40% of maximal activity, pH 10.5, 82% of maximal activity [1]) [1]
6-13 <5> (<5> pH 6.0, 30%, pH 13.0, 55% of maximal activity [12]) [12]

Temperature optimum (°C)

45 <1, 2> (<1> pH 8.0, F1 and F2 form [3]; <2> recombinant and native enzyme [9]) [3, 9]
47 <1> [1]
50 <1, 3, 4> (<1> G4-1 and G4-2 form [7]; <4> amylase A and B [13]) [7, 8, 13]
55 <1> (<1> immobilized enzyme [4]) [4]
60 <5> [12]

Temperature range (°C)

15-60 <1> (<1> 15°C, 20% of maximal activity, 60°C, 30% of maximal activity [1]) [1]
25-65 <1> (<1> F1 and F2 form 25°C, 60% of maximal activity, 65°C, 20% of maximal activity [3]) [3]
30-80 <5> (<5> 30°C, 40% of maximal activity, 80°C, 20% of maximal activity [12]) [12]
40-55 <3> (<3> 40°C and 55°C, 80% of maximal activity [8]) [8]

4 Enzyme Structure

Molecular weight

10000 <3> (<3> gel filtration [8]) [8]
46000 <1> (<1> G4-2 form, SDS-PAGE [7]) [7]
47200 <1> (<1> calculation from sequence of amino acid [15]) [15]
55000 <1> (<1> F1 and F2 form, SDS-PAGE [3]) [3]
56000 <2> (<2> recombinant and native enzyme, SDS-PAGE [9]) [9]
57000 <1> (<1> G4-1 form, SDS-PAGE [7]) [7]
97000 <5> (<5> SDS-PAGE [12]) [12]
125000 <1> (<1> gel filtration [1]) [1]

Subunits

oligomer <1> (<1> x * 12500, dimers, tetramers, hexamers, octamers or decamers, concentration dependent [1]) [1]

5 Isolation/Preparation/Mutation/Application

Localization

extracellular <1, 5> [1, 2, 12]

Purification

<1> (two active forms, F1 and F2 [3]; two active forms, G4-1 and G4-2 [7]) [1-3, 7]
<3> (partial [8]) [8]
<4> (amylase A and B [13]) [10, 13]
<5> [12]

Crystallization
<1> (<1> mutant E219Q [14]; <1> X-ray structure analysis [15]) [14, 15]

Cloning
<2> (<2> expression in E. coli [6, 9]) [6, 9]

Application
analysis <1> (<1> bioconversion of starch into maltotetraose in a membrane recycle bioreactor [11]) [11]

6 Stability

pH-Stability
5.5-11 <1> (<1> 30°C, 60 min, presence of $CaCl_2$, stable between [3]) [3]
6-13 <5> (<5> stable between, 85-100% of maximal activity [12]) [12]
6.5-8.5 <1> (<1> absence of Ca^{2+}, 30°C, 60 min stable between [3]) [3]
6.5-9.5 <1> (<1> G4-1 and G4-2 form, stable between [7]) [7]
7-8 <4> (<4> amylase A, 40°C, 30 min, stable between [13]) [13]
7-9 <3, 4> (<3> stable between [8]; <4> amylase B, 40°C, 30 min, stable between [13]) [8, 13]

Temperature stability
40 <1-4> (<1> F1 and F2 form, stable below on the treatment at pH 8.0, 60 min in the presence or absence of 5 mM $CaCl_2$ [3]; <3> unstable below [8]; <2> recombinant and native enzyme stable on heating in 10 mM acetate buffer, pH 6.0, 30 min, 40°C and below [9]; <4> amylase A and B, pH 7, full activity below [13]) [3, 8, 9, 13]
50 <1> (<1> G4-1-form heat-sensitive than G4-2 form [7]) [7]
55 <1> (<1> F1 and F2 form, complete loss of activity in the absence of Ca^{2+} [3]) [3]
60 <1, 2, 4, 5> (<1> F1 and F2 form, complete loss of activity in the presence of Ca^{2+} [3]; <5> 1 mM $CaCl_2$ and 1% soluble starch enhance thermostability [12]; <2> recombinant and native enzyme, complete loss of activity [9]; <4> 10 min, amylase A and B, inactivation [13]) [3, 9, 12, 13]
Additional information <2, 3> (<2> 2 mM $CaCl_2$, thermal stability of the recombinant and native enzymes increase [9]; <3> Ca^{2+} protects enzyme from heat [8]) [8, 9]

General stability information
<1>, Ca^{2+} stabilizes [3]
<1>, Ca^{2+} stabilizes immobilized enzyme [5]
<1>, dialysis, 4°C, prolonged periods, complete loss of activity [1]

References

[1] Robyt, J.F.; Ackerman, R.J.: Isolation, purification, and characterization of a maltotetraose-producing amylase from Pseudomonas stutzeri. Arch. Biochem. Biophys., **145**, 105-114 (1971)

[2] Dellweg, H.; John, M.; Schmidt, J.: Amylases from Pseudomonas stutzeri. Affinity chromatography of exo-maltotetraohydrolase using Sephadex G-100 as an adsorbent. Eur. J. Appl. Microbiol. Biotechnol., **1**, 191-198 (1975)

[3] Sakano, Y.; Kashiyama, E.; Kobayashi, T.: Purification of α-maltotetraose-forming exo-amylase of Pseudomonas stutzeri-two forms of the amylase and their enzymatic properties. Agric. Biol. Chem., **47**, 1761-1768 (1983)

[4] Kimura, T.; Ogata, M.; Yoshida, M.; Nakakuki, T.: Continuous production of maltotetraose using immobilized Pseudomonas stutzeri. Biotechnol. Bioeng., **32**, 669-676 (1988)

[5] Kimura, T.; Ogata, M.; Yoshida, M.; Nakakuki, T.: Stability of immobilized maltotetraose-forming amylase from Pseudomonas stutzeri. Biotechnol. Bioeng., **33**, 845-855 (1989)

[6] Zhou, J.; Takano, T.; Kobayashi, S.: Cloning of exo-maltotetrahydrolase gene from Pseudomonas saccharophila in Escherichia coli. Agric. Biol. Chem., **53**, 301-302 (1989)

[7] Nakada, T.; Kubota, M.; Sakai, S.; Tsujisaka, Y.: Purification and characterization of two forms of maltotetraose-forming amylase from Pseudomonas stutzeri. Agric. Biol. Chem., **54**, 737-743 (1990)

[8] Takasaki, Y.; Shinohara, H.; Tsuruhisa, M.; Hayashi, S.; Imada, K.: Maltotetraose-producing amylase from Bacillus sp. MG-4. Agric. Biol. Chem., **55**, 1715-1720 (1991)

[9] Zhou, J.; Baba, T.; Takano, T.; Kobayashi, S.; Arai, Y.: Properties of the enzyme expressed by the Pseudomonas saccharophila maltotetrahydrolase gene (mta) in Escherichia coli. Carbohydr. Res., **223**, 255-261 (1992)

[10] Fogarty, W.M.; Bourke, A.C.; Kelly, C.T.; Doyle, E.M.: A constitutive maltotetraose-producing amylase from Pseudomonas sp. IMD 353. Appl. Microbiol. Biotechnol., **42**, 198-203 (1994)

[11] Woo, G.-J.; McCord, J. D.: Bioconversion of starches into maltotetraose using Pseudomonas stutzeri maltotetraohydrolase in a membrane recycle bioreactor: effect of multiple enzyme systems and mass balance study. Enzyme Microb. Technol., **16**, 1016-1020 (1994)

[12] Kim, T.U.; Gu, B.G.; Jeong, J.Y.; Byun, S.M.; Shin, Y.C.: Purification and characterization of a maltotetraose-forming alkaline α-amylase from an alkalophilic Bacillus strain, GM8901. Appl. Environ. Microbiol., **61**, 3105-3112 (1995)

[13] Kobayashi, S.; Takaki, Y.; Kobata, K.; Takami, H.; Inoue, A.: Characterization of maltotetraohydrolase produced Pseudomonas sp. MS300 isolated from the deepest site of the Mariana Trench. Extremophiles, **2**, 401-407 (1998)

[14] Yoshioka, Y.; Hasegawa, K.; Matsuura, Y.; Katsube, Y.; Kubota, M.: Crystal structures of a mutant maltotetraose-forming exo-amylase cocrystallized with maltopentaose. J. Mol. Biol., **271**, 619-628 (1997)
[15] Morishita, Y.; Hasegawa, K.; Matsuura, Y.; Katsube, Y.; Kubota, M.; Sakai, S.: Crystal structure of a maltotetraose-forming exo-amylase from Pseudomonas stutzeri. J. Mol. Biol., **267**, 661-672 (1997)

Mycodextranase 3.2.1.61

1 Nomenclature

EC number
3.2.1.61

Systematic name
1,3-1,4-α-D-glucan 4-glucanohydrolase

Recommended name
mycodextranase

Synonyms
dextranase, myco-

CAS registry number
9047-04-5

2 Source Organism

<1> *Penicillium melinii* (QM 1931 [2, 3]) [1-7]
<2> *Streptomyces sp.* (J-13-3 [6]) [6]
<3> *Streptomyces griseus* (IFO 3355 [6]) [6]
<4> *Chloridium sp.* [6]
<5> *Trichoderma sp.* [6]
<6> *Cladosporium sp.* [6]
<7> *Bacillus circulans* (NHB-1 [7]) [7]

3 Reaction and Specificity

Catalyzed reaction
endohydrolysis of 1,4-α-D-glucosidic linkages in α-D-glucans containing both 1,3- and 1,4-bonds (<1, 4> endo-multichain mechanism [2, 6]; <2, 7> enzyme hydrolyzes nigeran tetrasaccharide by an endo-type of action [6, 7])

Reaction type
hydrolysis of O-glycosyl bond

Natural substrates and products
S nigeran + H_2O <1> [2]
P nigerose + 4-α-D-nigerosylglucose <1> [2]

Substrates and products

S nigeran + H_2O <1-7> (<1> both fast and slow enzyme [2]) [1-7]

P nigerose + α-D-glucopyranosyl-1,3-α-D-glucopyranosyl-1,4-α-D-glucopyranosyl-1,3-D-glucose <1, 2> (<1> no production of glucose [2]) [1, 2, 6]

Inhibitors

Ag^+ <1> (<1> fast enzyme [2]) [2]
Hg+ <1, 2> (<1> fast enzyme [2]) [2, 6]
Hg^{2+} <2, 7> [6, 7]
Mn^{2+} <7> [7]
N-bromosuccinimide <1, 7> (<1> fast enzyme [2]) [2, 7]
N-chlorosuccinimide <1, 7> (<1> fast enzyme [2]) [2, 7]
maltose <1> [2]
maltotriose <1> [2]
nigeran tetrasaccharide <1> [2]
nigeran trisaccharide <1> [2]
nigerose <1> [2]
p-chloromercuribenzoic acid <2> [6]
urea <1> [2]

Specific activity (U/mg)

2.53 <2> [6]
7.88 <7> [7]
27 <1> [2]

K_m-Value (mM)

Additional information <2, 7> [6, 7]

pH-Optimum

6 <2, 7> [6, 7]

pH-Range

2-9 <1> (<1> fast and slow enzyme pH 2 15%, pH 9 75% of maximal activity [2]) [2]
6-7 <7> (<7> stable between [7]) [7]
6-8 <2> (<2> stable between [6]) [6]

Temperature optimum (°C)

50 <2, 7> [6, 7]
70 <1> (<1> fast and slow enzyme [2]) [2]

Temperature range (°C)

23-80 <1> (<1> 23°C: fast enzyme 7%, slow enzyme 16% of maximal activity, 80°C: fast enzyme 50%, slow enzyme 30% of maximal activity [2]) [2]

4 Enzyme Structure

Molecular weight

40000 <1> (<1> sedimentation equilibrium fast and slow enzyme [2]) [2]
120000 <7> (<7> gel filtration [7]) [7]

Subunits

monomer <1, 2, 7> (<1> 1 * 40000, guanidine hydrochloride treatment [2]; <2> 1 * 68000, SDS-PAGE [6]; <7> 1 * 120000, SDS-PAGE [7]) [2, 6, 7]
Additional information <1> (<1> fast and slow enzyme have similar but not identical amino acid compositions [2]) [2]

Posttranslational modification

glycoprotein <1> (<1> fast form: 16% carbohydrate, slow form: 15% carbohydrate, only mannose and glucose [2]; <1> chemical constitution of carbohydrate portion [4]) [2-4]

5 Isolation/Preparation/Mutation/Application

Localization

extracellular <1> [3]

Purification

<1> (fast and slow enzyme, different electrophoretic mobilities [2]) [2]
<2> [6]
<7> [7]

6 Stability

pH-Stability

4-5 <1> (<1> stable between [2]) [2]
5-10 <7> (<7> 80% of maximal activity after incubation at 4°C for 48 h [7]) [7]

Temperature stability

50 <7> (<7> enzyme stable up to 50°C [7]) [7]
60 <2, 7> (<2> 10 min, 60% of maximal activity [6]; <7> 10 min, 67% of maximal activity [7]) [6, 7]
70 <2, 7> (<2> 10 min, 22% of maximal activity [6]; <7> 10 min, 17% of maximal activity [7]) [6, 7]

Storage stability

<1>, 4°C, 2 months, no loss of activity, fast and slow enzyme [2]

References

[1] Tung, K.K.; Nordin, J.H.: Structure of the tetrasaccharide produced by the hydrolysis of nigeran by the enzyme mycodextranase. Biochim. Biophys. Acta, **158**, 154-156 (1968)
[2] Tung, K.K.; Rosenthal, A.; Nordin, J.H.: Enzymes that hydrolyze fungal cell wall polysaccharides. II. Purification and properties of mycodextranase, an endo-D-(1→4) glucanase from Penicillium melinii. J. Biol. Chem., **246**, 6722-6736 (1971)
[3] Samuel, O.; Nordin, J.H.: A unique system for the study of the biosynthesis and secretion of a specific glycoprotein enzyme. Biochem. Biophys. Res. Commun., **45**, 1376-1383 (1971)
[4] Rosenthal, A.L.; Nordin, J.H.: Enzymes that hydrolyze fungal cell wall polysaccharides.The carbohydrate constitution of mycodextranase, an endo-D-(1→4) glucanase from Penicillium melinii. J. Biol. Chem., **256**, 5295-5303 (1975)
[5] Marchessault, R.H.; Revol, J.-F.; Bobbitt, F.; Nordin, J.H.: Enzymatic depolymerization of lamellar single crystals of nigeran. Biopolymers, **19**, 1069-1080 (1980)
[6] Okazaki, K.; Abe, T.; Saruwatari, K.; Kato, F.; Maruyama, K.; Tagawa, K.: Purification and properties of mycodextranase from Streptomyces sp. J-13-3. Biosci. Biotechnol. Biochem., **56**, 1401-1405 (1992)
[7] Okazaki, K.; Jin, S.Y.; Sawamura, Y.; Hayakawa, S.: Purification and properties of mycodextranase from Bacillus circulans NHB-1. J. Ferment. Bioeng., **79**, 370-372 (1995)

Glycosylceramidase 3.2.1.62

1 Nomenclature

EC number
3.2.1.62

Systematic name
glycosyl-N-acylsphingosine glycohydrolase

Recommended name
glycosylceramidase

Synonyms
LPH <2> [6, 7, 11]
cerebrosidase
glycosyl ceramide glycosylhydrolase
lactase-phlorizin hydrolase
phloretin-glucosidase
phloridzin glucosidase
phlorizin β-glucosidase
phlorizin hydrolase

CAS registry number
9033-10-7

2 Source Organism

<1> *Ovis aries aries* (sheep) [1]
<2> *Homo sapiens* [2, 3, 5-7, 9, 11, 13-15]
<3> *Rattus sp.* [4, 10, 17, 19, 20, 22]
<4> *Oryctolagus cuniculus* (rabbit) [8, 9]
<5> *Sus scrofa* [12, 15]
<6> *Mus musculus* [16]
<7> *Macaca radiata* (monkey) [18, 21]
<8> *hamster* [23]

3 Reaction and Specificity

Catalyzed reaction
glycosyl-N-acylsphingosine + H_2O = N-acylsphingosine + a sugar (broad specificity, cf. EC 3.2.1.45, glucosylceramidase, and EC 3.2.1.46, galactosylcer-

amidase. Also hydrolyses phlorizin to phloretin and glucose. The intestinal enzyme is a complex that also catalyses the reaction of EC 3.2.1.108, lactase)

Reaction type

hydrolysis of O-glycosyl bond

Natural substrates and products

S (2-hydroxymethylphenyl)-β-D-glucopyranoside + H_2O <7> (<7> salicin [18]; <7> low activity [18]) [18]
P 2-hydroxymethylphenol + β-D-glucose <7> [18]
S 3-fucosyllactose + H_2O <3> (<3> only oligosaccharides containing galactose as the glycone in β-linkage hydrolysed [19]) [19]
P β-D-glucose + 3-fucosylgalactose <3> [19]
S 4-O-β-D-galactopyranosyl-D-N-acetylglucosamine + H_2O <6> (<6> poor activity [16]) [16]
P β-D-galactose + N-acetylglucosamine <6> [16]
S 4-O-β-D-galactopyranosyl-D-fructoside + H_2O <6> (<6> very low activity [16]) [16]
P β-D-galactose + D-fructose <6> [16]
S 4-O-β-D-galactopyranosyl-D-glucopyranoside + H_2O <1-7> (<1, 3> acts on dietary glycosides before absorption [1, 4]; <1, 4> hydrolysis located on a specific active site for lactose hydrolysis [1, 8]; <2> 4fold higher activity for the N- and O-glycosylated than for the N-glycosylated molecule [7]; <3-5, 7> two enzymatic activities, one for lactose- and one for phlorizin-and lactose-hydrolysis [8, 9, 15, 18-20]; <2, 3, 6, 7> best substrate [14, 16, 18, 20, 22]) [1-9, 14-22]
P D-galactose + β-D-glucose <1-7> [1-9, 14-22]
S 4-O-β-D-glucopyranosyl-D-glucose + H_2O <2, 3, 5-7> (<2, 3, 5-7> cellobiose [14-16, 18-20, 22]) [14-16, 18-20, 22]
P β-D-glucose + D-glucose <2, 3, 5-7> [14-16, 18-20, 22]
S 6-O-β-D-glucopyranosyl-D-glucose + H_2O <2, 5, 6> (<2, 5, 6> gentiobiose [14-16]; <5, 6> low activity [15, 16]) [14-16]
P β-D-glucose + D-glucose <2, 5, 6> [14-16]
S N-palmitoyldihydroglucosylceramide + H_2O <3> [17]
P β-D-glucose + ? <3> [17]
S cellotetraose + H_2O <2, 5> (<2, 5> considerable activity [15]) [15]
P ? + β-D-glucose <2, 5> [15]
S cellotriose + H_2O <2, 5> (<2, 5> considerable activity [15]) [15]
P ? + β-D-glucose <2, 5> [15]
S cellulose + H_2O <2, 5> (<2, 5> very low activity [15]) [15]
P ? + β-D-glucose <2, 5> [15]
S galactosylceramide + H_2O <6> [16]
P β-D-galactose + ceramide <6> [16]
S galactosylsphingosine + H_2O <6> (<6> very low activity [16]) [16]
P β-D-galactose + N-acylsphingosine <6> [16]
S glucosylceramide + H_2O <6> [16]
P β-D-glucose + ceramide <6> [16]
S glucosylsphingosine + H_2O <6> (<6> very low activity [16]) [16]

P β-D-glucose + N-acylsphingosine <6> [16]
S lacto-N-tetraose + H_2O <3> [19]
P β-D-glucose + ? <3> [19]
S lactosylceramide + H_2O <6> (<6> low activity [16]) [16]
P β-D-glucose + galactosylceramide <6> [16]
S phlorizin + H_2O <1-8> (<1, 4, 5> hydrolysis located on a specific active site for phlorizin hydrolysis [1, 8, 15]; <5> 5% of the activity of lactose hydrolysis [15]; <6, 7> low activity [16, 18]; <3> only substrate that is hydrolysed at the active site for phlorizin hydrolysis [19]) [1, 8, 14-16, 18-23]
P phloretin + β-D-glucose <1-8> [1, 8, 14-16, 18-23]

Substrates and products
S (2-hydroxymethylphenyl)-β-D-glucopyranoside + H_2O <7> (<7> salicin [18]; <7> low activity [18]) [18]
P 2-hydroxymethylphenol + β-D-glucose <7> [18]
S 2-naphthyl-β-D-galactopyranoside + H_2O <5> (<5> very low activity [15]) [15]
P naphthol + β-D-galactopyranose <5> [15]
S 2-nitrophenyl-β-D-galactopyranoside + H_2O <2, 3, 5> (<5> lower activity than for 4-nitrophenyl-β-D-galactopyranoside [15]) [14, 15, 22]
P 2-nitrophenol + β-D-galactose <2, 3, 5> [14, 15, 22]
S 2-nitrophenyl-β-D-glucopyranoside + H_2O <7> [18]
P 2-nitrophenol + β-D-glucose <7> [18]
S 3-fucosyllactose + H_2O <3> (<3> only oligosaccharides containing galactose as the glycone in β-linkage hydrolysed [19]) [19]
P β-D-glucose + 3-fucosylgalactose <3> [19]
S 4'-deoxyphlorizin + H_2O <3> (<3> low activity [20]) [20]
P 4'-deoxyphloritin + β-D-glucose <3> [20]
S 4-O-β-D-galactopyranosyl-D-N-acetylglucosamine + H_2O <6> (<6> poor activity [16]) [16]
P β-D-galactose + N-acetylglucosamine <6> [16]
S 4-O-β-D-galactopyranosyl-D-fructoside + H_2O <6> (<6> very low activity [16]) [16]
P β-D-galactose + D-fructose <6> [16]
S 4-O-β-D-galactopyranosyl-D-glucopyranoside + H_2O <1-7> (<1, 3> acts on dietary glycosides before absorption [1, 4]; <1,4> hydrolysis located on a specific active site for lactose hydrolysis [1, 8]; <2> 4fold higher activity for the N- and O-glycosylated than for the N-glycosylated molecule [7]; <3-5, 7> two enzymatic activites, one for lactose- and one for phlorizin- and lactose-hydrolysis [8, 9, 15, 18-20]; <2, 3, 6, 7> best substrate [14, 16, 18, 20, 22]) [1-9, 14-22]
P D-galactose + β-D-glucose <1-7> [1-9, 14-22]
S 4-O-β-D-glucopyranosyl-D-glucose + H_2O <2, 3, 5-7> (<2, 3, 5-7> cellobiose [14-16, 18-20, 22]) [14-16, 18-20, 22]
P β-D-glucose + D-glucose <2, 3, 5-7> [14-16, 18-20, 22]
S 4-methylumbelliferyl-α-L-arabinoside + H_2O <6> [16]
P 4-methylumbelliferone + α-L-arabinose <6> [16]

S 4-methylumbelliferyl-β-D-galactoside + H_2O <3, 6> [16, 19]
P 4-methylumbelliferone + β-D-galactose <3, 6> [16, 19]
S 4-methylumbelliferyl-β-D-glucoside + H_2O <6> [16]
P 4-methylumbelliferone + β-D-glucose <6> [16]
S 4-methylumbelliferyl-β-D-xyloside + H_2O <6> (<6> low activity [16]) [16]
P 4-methylumbelliferone + β-D-xylose <6> [16]
S 4-nitrophenyl-β-D-fucoside + H_2O <6> [16]
P 4-nitrophenol + β-D-fucose <6> [16]
S 4-nitrophenyl-β-D-galactopyranoside + H_2O <2, 3, 5, 7> (<2> low activity [14]) [14, 15, 18, 22]
P 4-nitrophenol + β-D-galactose <2, 3, 5, 7> [14, 15, 18, 22]
S 4-nitrophenyl-β-D-glucopyranoside + H_2O <2, 3, 5, 7> [14, 15, 18, 19, 22]
P 4-nitrophenol + β-D-glucose <2, 3, 5, 7> [14, 15, 18, 19, 22]
S 6-O-β-D-glucopyranosyl-D-glucose + H_2O <2, 5, 6> (<2, 5, 6> gentiobiose [14-16]; <5, 6> low activity [15, 16]) [14-16]
P β-D-glucose + D-glucose <2, 5, 6> [14-16]
S 6-bromonaphthyl-β-D-galactopyranoside <5> (<5> very low activity [15]) [15]
P 6-bromonaphthol + β-D-galactopyranose <5> [15]
S N-palmitoyldihydroglucosylceramide + H_2O <3> [17]
P β-D-glucose + ? <3> [17]
S N-palmitoyldihydroglucosylcerebroside + H_2O <7> (<7> 1.5% of the activity towards lactose [18]) [18]
P β-D-glucose + N-palmitoyldihydrocerebroside <7> [18]
S β-methylglucoside + H_2O <7> [18]
P methanol + β-D-glucose <7> [18]
S cellotetraose + H_2O <2, 5> (<2, 5> considerable activity [15]) [15]
P ? + β-D-glucose <2, 5> [15]
S cellotriose + H_2O <2, 5> (<2, 5> considerable activity [15]) [15]
P ? + β-D-glucose <2, 5> [15]
S cellulose + H_2O <2, 5> (<2, 5> very low activity [15]) [15]
P ? + β-D-glucose <2, 5> [15]
S daidzein-7-glucopyranoside + H_2O <1> [1]
P daidzein + β-D-glucose <1> [1]
S galactosyl-N-palmitoylceramide + H_2O <3> [20]
P β-D-galactose + N-palmitoylceramide <3> [20]
S galactosyl-N-palmitoyldihydroceramide + H_2O <3> [20]
P β-D-galactose + N-palmitoyldihydroceramide <3> [20]
S galactosylceramide + H_2O <6> [16]
P β-D-galactose + ceramide <6> [16]
S galactosylsphingosine + H_2O <6> (<6> very low activity [16]) [16]
P β-D-galactose + N-acylsphingosine <6> [16]
S genistein-7-glucopyranoside + H_2O <1> [1]
P genistein + β-D-glucose <1> [1]
S glucosyl-N-lignoceroyldihydroceramide + H_2O <3> [20]
P β-D-glucose + N-lignoceroyldihydroceramide <3> [20]

S glucosyl-N-palmitoylceramide + H_2O <3> [20]
P β-D-glucose + N-palmitoylceramide <3> [20]
S glucosyl-N-palmitoyldihydroceramide + H_2O <3> [20]
P β-D-glucose + N-palmitoyldihydroceramide <3> [20]
S glucosyl-N-stearoyldihydroceramide + H_2O <3> [20]
P β-D-glucose + N-stearoylhihydroceramide <3> [20]
S glucosylceramide + H_2O <6> [16]
P β-D-glucose + ceramide <6> [16]
S glucosylsphingosine + H_2O <6> (<6> very low activity [16]) [16]
P β-D-glucose + N-acylsphingosine <6> [16]
S lacto-N-tetraose + H_2O <3> [19]
P β-D-glucose + ? <3> [19]
S lactosyl-N-lignoceroyldihydroceramide + H_2O <3> [20]
P β-D-glucose + D-galactose + N-lignoceroyldihydroceramide <3> [20]
S lactosyl-N-palmitoyldihydroceramide + H_2O <3> [20]
P β-D-glucose + D-galactose + N-palmitoyldihydroceramide <3> [20]
S lactosyl-N-stearoyldihydroceramide + H_2O <3> [20]
P β-glucose + D-galactose + N–stearoyldihydroceramide <3> [20]
S lactosylceramide + H_2O <6> (<6> low activity [16]) [16]
P β-D-glucose + galactosylceramide <6> [16]
S phenyl-β-D-galactopyranoside + H_2O <5> (<5> low activity [15]) [15]
P phenol + β-D-galactopyranose <5> [15]
S phloretin-β-D-galactoside + H_2O <3> (<3> very low activity [22]) [22]
P phloretin + β-D-galactose <3> [22]
S phlorizin + H_2O <1-8> (<1, 4, 5> hydrolysis located on a specific active site for phlorizin hydrolysis [1, 8, 15]; <5> 5% of the activity of lactose hydrolysis [15]; <6, 7> low activity [16, 18]; <3> only substrate that is hydrolysed at the active site for phlorizin hydrolysis [19]) [1, 8, 14-16, 18-23]
P phloretin + β-D-glucose <1-8> [1, 8, 14-16, 18-23]
S quercetin-3-glucopyranoside + H_2O <1> [1]
P quercetin + β-D-glucose <1> [1]
S quercetin-4-glucopyranoside + H_2O <1> (<1> flavone glucosides more efficiently hydrolysed than isoflavone glucosides, major contribution to the hydrolysis is from the lactase site [1]) [1]
P quercetin + β-D-glucose <1> [1]

Inhibitors

2',4'-dinitrophenyl-2-fluoro-2-deoxy-β-D-glucopyranoside <1> (<1> rapidly inhibits phlorizin hydrolase activity by covalently binding to the glutamic acid residue of the active site, lactase activity more slowly inhibited, pre-incubation with glucal protects the lactase active site from inactivation [1]) [1]
2-oxoglutaric acid <7> (<7> 50% inhibition of the tartaric acid-activated phlorizin hydrolase at 20 mM [21]) [21]
D-galactono-1,4-lactone <3, 8> (<3> competitive inhibition of lactase activity, K_i: 0.35 mM [22]; <8> competitive inhibition of lactase activity, not phlorizin activity, K_i: 2 mM [23]) [22, 23]

D-glucono-1,5-lactone <3, 8> (<3> competitive inhibition of lactase activity, K_i: 0.027 mM [22]; <8> competitive inhibition of phlorizin hydrolysis, K_i: 0.18 mM, competitive inhibition of lactase activity, K_i: 0.275 mM [23]) [22, 23]
NaCl <7> (<7> 30% inhibition of the tartaric acid-stimulated phlorizin hydrolase at 20 mM [21]) [21]
SO_4^{2-} <7> (<7> strong inhibition of the tartaric acid-activated phlorizin hydrolase at 20 mM [21]) [21]
Tris(hydroxymethyl)aminomethane <2, 3, 5> (<2> competitive inhibition of lactase acitiviy, phlorizin activity not affected, K_i: 12 mM [14]; <5> competitive inhibition of lactase activity, K_i: 5.1 mM [15]; <3> competitive inhibition of lactase activity, K_i: 4.5 mM [22]) [14, 15, 22]
cellobiose <3> (<3> inhibition of galactosylceramide hydrolysis [20]) [20]
cerebroside <7> (<7> from brain, competitive inhibition of lactase activity, K_i: 1.11 mM [18]) [18]
conduritol-B-epoxide <4> (<4> 90% inactivation of lactase hydrolysis, 55% inactivation of phlorizin hydrolysis, inactivation by covalent binding to Glu1747 and Glu1271 [8]) [8]
deoxygalactonojirimycin <4> (<4> competitive inhibition, K_i: 0.076 mM [8]) [8]
deoxynojirimycin <4> (<4> competitive inhibition, K_i: 0.142 mM [8]) [8]
galactosylceramide <3> (<3> competitive inhibition of glucosyldihydroceramide hydrolysis [20]) [20]
glucosyldihydroceramide <3> (<3> competitive inhibition of galactosylceramide hydrolysis [20]) [20]
maleic acid <7> (<7> 55% inhibition for the tartaric acid-stimulated phlorizin hydrolase at 20 mM [21]) [21]
oxalacetic acid <7> (<7> 50% inhibition of the tartaric acid-stimualted phlorizin hydrolase at 20 mM [21]) [21]
oxalic acid <7> (<7> 47% inhibition of the tartaric acid-activated phlorizin hydrolase at 20 mM [21]) [21]
phlorizin <2, 3, 5, 7> (<2> competitive inhibition of lactase activity; K_i: 0.44 mM [14]; <5> competitive inhibition of lactase activity, K_i: 0.37 mM [15]; <7> competitive inhibition of lactase activity, K_i: 0.18 mM [18]; <3> competitive inhibition of galactosylceramide hydrolysis, K_i: 0.35 mM [20]; <3> competitive inhibition of lactase activity, K_i: 0.047 mM [22]) [14, 15, 18, 20, 22]
pyruvic acid <7> (<7> strongly inhibitory for the tartaric acid-stimulated phlorizin hydrolase [21]) [21]
sodium cholate <7> (<7> competitive inhibition of lactase activity, in the presence of bovine serum albumin [18]) [18]

Activating compounds

L-ascorbic acid <7> (<7> 1.6fold stimulation only for phlorizin activity [21]) [21]
L-tartaric acid <7> (<7> most stimulating effect only for phlorizin activity [21]) [21]

cis-aconitic acid <7> (<7> more than 2fold stimulation only for phlorizin activity [21]) [21]
citric acid <7> (<7> highest stimulation at 50 mM and 30°C only for phlorizin activity [21]) [21]
isocitric acid <7> (<7> 1.7fold stimulation only for phlorizin activity [21]) [21]
pyruvic acid <7> (<7> 0.5fold stimulation only for phlorizin activity [21]) [21]
taurodeoxycholate <6> (<6> activation [16]) [16]

Turnover number (min^{-1})
30 <1> (phlorizin) [1]
126 <1> (daidzein-7-β-D-glucopyranoside) [1]
378 <1> (quercetin-3-β-D-glucopyranoside) [1]
390 <1> (genistein-7-β-D-glucopyranoside) [1]
450 <1> (quercetin-4-β-D-glucopyranoside) [1]
2028 <1> (4-O-β-D-galactopyranosyl-D-glucopyranoside) [1]

Specific activity (U/mg)
0.45 <7> (<7> phlorizin activity [18]) [18]
1.78 <6> (<6> galactosylceramide hydrolysis [16]) [16]
2.67 <6> (<6> glucosylceramide hydrolysis [16]) [16]
4.8 <4> [8]
5-6 <7> (<7> lactase activity [21]) [21]
15.33 <6> (<6> lactose hydrolysis [16]) [16]
16.1 <7> (<7> lactase activity [18]) [18]
20 <3> [20]
22 <3> [17]
32 <2> [14]

K_m-Value (mM)
0.002 <1> (phlorizin) [1]
0.01 <6> (glucosylceramide) [16]
0.044 <1> (quercetin-4-glucopyranoside) [1]
0.046 <1> (quercetin-3-glucopyranoside) [1]
0.065 <6> (galactosylceramide) [16]
0.085 <1> (genistein-7-glucopyranoside) [1]
0.1 <8> (phlorizin) [23]
0.149 <1> (daidzein-7-glucopyranoside) [1]
0.18 <6> (phlorizin) [16]
0.24 <6> (4-methylumbelliferyl-β-D-galactoside) [16]
0.24 <6> (galactosylsphingosine) [16]
0.27 <6> (lactosylceramide) [16]
0.36 <6> (4-nitrophenyl-β-D-fucoside) [16]
0.37 <6> (glucosylsphingosine) [16]
0.4 <3> (phloretin-β-D-galactoside) [22]
0.4 <3, 7> (phlorizin) [18, 22]
0.44 <2> (phlorizin) [14]

0.45 <6> (4-methylumbelliferyl-α-L-arabinoside) [16]
0.51 <3> (galactosylceramide) [20]
0.53 <6> (4-methylumbelliferyl-β-D-glucoside) [16]
0.62 <3> (glucosyldihydroceramide) [20]
0.73 <3> (N-palmitoyldihydrocerebroside, <3> from proximal intestine isolated phlorizin hydrolase activity [17]) [17]
1.1 <3> (4-O-β-D-glucopyranosyl-D-glucopyranoside) [22]
1.12 <3> (N-palmitoyldihydrocerebroside, <3> from distal intestine isolated phlorizin hydrolase activity [17]) [17]
1.3 <6> (4-O-β-D-glucopyranosyl-D-glucopyranoside) [16]
1.4 <6> (4-methylumbelliferyl-β-D-xyloside) [16]
2.78 <6> (4-O-β-D-glucopyranosyl-D-glucopyranoside) [16]
4.1 <3> (4-nitrophenyl-β-D-glucopyranoside) [22]
4.4 <2> (4-O-β-D-glucopyranosyl-D-glucospyranoside) [14]
4.8 <6> (4-O-β-D-galactopyranosyl-D-glucosamine) [16]
5.1 <5> (4-O-β-D-galactopyranosyl-D-glucopyranoside, <5> slightly higher catalytic efficiency than for the human enzyme [15]) [15]
8.8 <1> (4-O-β-D-galactopyranosyl-D-glucopyranoside) [1]
13.1 <3> (2-nitrophenyl-β-D-galactopyranoside) [22]
14 <2> (4-O-β-D-galactopyranosyl-glucopyranoside, <2> both mannose-rich precursor form and mature form [7]) [7]
14.8 <3> (4-O-β-D-galactopyranosyl-D-glucopyranoside, <3> from proximal intestine isolated lactase [17]) [17]
19.1 <3> (4-O-β-D-galactopyranosyl-D-glucopyranoside, <3> from distal intestine isolated lactase [17]) [17]
21 <2> (4-O-β-D-galactopyranosyl-D-glucopyranoside) [14]
21.8 <7> (4-O-β-D-galactopyranosyl-D-glucopyranoside) [18]
22 <6> (6-O-β-D-glucopyranosyl-D-glucopyranoside) [16]
22.3 <3> (4-nitrophenyl-β-D-galactopyranoside) [22]
28 <6> (4-O-β-D-galactopyranosyl-D-glucopyranoside) [16]
40 <3> (4-O-β-D-galactopyranosyl-D-glucopyranoside) [22]
41.7 <3> (4-O-β-D-galactopyranosyl-D-glucopyranoside) [19]
67 <6> (4-O-β-D-galactopyranosyl-D-fructoside) [16]

pH-Optimum

5.2 <6> (<6> hydrolysis of galactosyl- or glucosylceramide in the presence of taurodeoxycholate [16]) [16]
5.2-5.6 <7> (<7> phlorizin activity in acetate buffer [21]) [21]
5.5 <8> [23]
5.5-5.7 <6> (<6> hydrolysis of disaccharides or artificial substrates [16]) [16]
5.6 <7> (<7> lactase activity, phlorizin activity in the acidic pH range [18]; <7> phlorizin activity at pH 3.3-3.5 in citrate buffer [21]) [18, 21]
5.6-6.2 <5> [15]
5.8-6 <2> [14]
6 <3> (<3> both isoforms [17]) [17, 20]

4 Enzyme Structure

Molecular weight

130000 <3> (<3> mature form, proteolytically cleaved from precursor form, SDS-PAGE [10]) [10]
135000 <4> (<4> SDS-PAGE [8]) [8]
145000 <2> (<2> SDS-PAGE, mature form [3]) [3]
160000 <2, 5> (<2, 5> cleaved mature form, SDS-PAGE [6, 7, 11-13]) [6, 7, 11-13]
215000 <2> (<2> mannose-rich form, SDS-PAGE [5]; <2> mannose-rich precursor form, SDS-PAGE [7, 11]) [5, 7, 11]
220000 <2, 3> (<2, 3> single-chain precursor, subsequently proteolytically processed to its active form, SDS-PAGE [3, 10]) [3, 10]
230000 <2> (<2> mannose-rich form, SDS-PAGE [5]) [5]
245000 <2, 5> (<2, 5> high-mannose precursor form, SDS-PAGE [12, 13]) [12, 13]
280000 <2> (<2> gel filtration, hydrophilic form [14]) [14]
290000-300000 <6> (<6> gel filtration [16]) [16]
320000 <2, 5> (<2, 5> gel filtration, amphiphilic form [14, 15]) [14, 15]

Subunits

dimer <2, 5> (<2> 2 * 160000, SDS-PAGE [14] ; <5> 2 * 160000, SDS-PAGE [15]) [14, 15]
monomer <3, 4> (<4> 1 * 135000, SDS-PAGE [8] ; <3> 1 * 130000, SDS-PAGE [10] ; <2, 5> 1 * 160000, SDS-PAGE, mature brush-border form after proteolytic cleavage of the precursor form [11-13]) [8-13]

Posttranslational modification

glycoprotein <2, 3, 5> (<2> various modified glycoforms do not affect the transport from the Golgi to the cell surface [6]; <2> mature LPH O- and N-glycosylated, no sialic acids [7]; <5> endo H treatment [12]; <3> 17% carbohydrate content, consisting of galactose, mannose, fucose, N-acetylglucosamine [19]) [2, 3, 5-7, 11, 12, 19]

5 Isolation/Preparation/Mutation/Application

Source/tissue

enterocyte <3> [4]
epithelium <2> [11]
intestine <1-8> (<1, 2, 4-8> small intestine [1, 8, 11-23]; <3> highest activity in proximal jejenum, lowest in proximal and distal ends of the small intestine [4]; <5> proximal jejenum [15]) [1, 2, 4, 6-23]

Localization

brush border membrane <1-4, 8> (<1> luminal site [1]; <4> anchored via the C-terminal hydrophobic stretch in C(in)-N(out) orientation [9]) [1, 8, 9, 17, 19, 22, 23]

endoplasmic reticulum-Golgi <3> [10]
microvillar membrane <2, 3, 5, 8> [10, 12, 14, 15, 23]
plasma membrane <2> (<2> transfected COS-1 cells) [2, 3]

Purification

<1> (to homogeneity [1]) [1]
<2> (Helix pomatia lectin chromatography [7]; to homogeneity, immunoadsorbent chromatography [14]) [7, 14]
<3> (to homogeneity, immunoaffinity chromatography [10]; two isoenzymes from proximal and distal intestine, to homogeneity, chromatography techniques [17]; partial [19, 22]) [10, 17, 19, 20, 22]
<4> (to homogeneity, chromatography techniques [8]; antibody affinity chromatography [9]) [8, 9]
<5> (to homogeneity, immunoadsorbent chromatography, amphiphilic form [15]) [15]
<6> (to homogeneity, chromatography steps [16]) [16]
<7> (to homogeneity, chromatography techniques [18]; highly purified [21]) [18, 21]

Cloning

<2> (expression of mutants in COS-7 cells [2]; expression in COS-1 cells [3, 5]) [2, 3, 5]

Engineering

D1711N <2> (<2> introduction of a glycosylation site, reduced transport rate to the plasma membrane [2]) [2]
I1697N <2> (<2> introduction of a glycosylation site, reduced transport rate to the plasma membrane [2]) [2]
P1743S <2> (<2> introduction of a glycosylation site, reduced transport rate to the plasma membrane [2]) [2]

6 Stability

pH-Stability

3.2 <5> (<5> little activity below [15]) [15]
4.2-9 <2> (<2> little activity outside this range [14]) [14]

Temperature stability

4 <2> (<2> no detectable loss of activity for several weeks [14]) [14]
45 <3, 6> (<6> more than 90% of activity lost within 20 min [16]; <3> stable for 60 min [17]) [16, 17]
46 <3> (<3> 34-70% remaining activity after 30 min, depending on substrates [20]) [20]
49 <5> (<5> lactase activity faster destroyed than phlorizin activity [15]) [15]
55 <7> (<7> complete inactivation [18]) [18]

General stability information

<2>, sensitive to repeated freezing and thawing [14]

<3>, at neutral pH and 37°C, lactase acitivity rapidly lost, partially prevented by *p*-chloromercuribenzoate, more stable if attached to its border-membrane [19]

Storage stability

<2, 3>, -20°C, several months [14, 19]

References

[1] Day, A.J.; Canada, F.J.; Diaz, J.C.; Kroon, P.A.; Mclauchlan, R.; Faulds, C.B.; Plumb, G.W.; Morgan, M.R.A.; Williamson, G.: Dietary flavonoids and isoflavone glycosides are hydrolysed by the lactase site of lactase phlorizin hydrolase. FEBS Lett., **468**, 166-170 (2000)

[2] Jacob, R.; Weiner, J.R.; Stadge, S.; Naim, H.Y.: Additional N-glycosylation and its impact on the folding of intestinal lactase-phlorizin hydrolase. J. Biol. Chem., **275**, 10630-10637 (2000)

[3] Wüthrich, M.; Sterchi, E.E.: Human lactase-phloizin hydrolase expressed in COS-1 cells is proteolytically processed by the lysosomal pathway. FEBS Lett., **405**, 321-327 (1997)

[4] Estrada, G.; Krasinski, S.D.; Montgomery, R.K.; Grand, R.J.; Garcia-Valero, J.; Lopez-Tejero, M.D.: Quantitative analysis of lactase-phlorizin hydrolase expression in the absorptive enterocytes of newborn rat small intestine. J. Cell. Physiol., **167**, 341-348 (1996)

[5] Naim, H.Y.: The pro-region of human intestinale lactase-phlorizin hydrolase is enzymatically inactive towards lactose. Biol. Chem. Hoppe-Seyler, **376**, 255-258 (1995)

[6] Naim, H.Y.: Processing and transport of human small intestinal lactase-phlorizin hydrolase (LPH). Role of N-linked oligosaccharide modification. FEBS Lett., **342**, 302-307 (1994)

[7] Naim, H.Y.; Lentze, M.J.: Impact of O-glycosylation on the function of human intestinal lactase-phlorizin hydrolase. Characterization of glycoforms varying in enzyme activity and localization of O-glycoside addition. J. Biol. Chem., **267**, 25494-25504 (1992)

[8] Wacker, H.; Keller, P.; Falchetto, R.; Legler, G.; Semenza, G.: Location of the two catalytic sites in intestinal lactase-phlorizin hydrolase. J. Biol. Chem., **267**, 18744-18752 (1992)

[9] Mantei, N.; Villa, M.; Enzler, T.; Wacker, H.; Boll, W.; James, P.; Hunziker, W.; Semenza, G.: Complete primary structure of human and rabbit lactase-phlorizin hydrolase: implications for biosynthesis, membrane anchoring and evolution of the enzyme. EMBO J., **7**, 2705-2713 (1988)

[10] Büller, H. A.; Montgomery, R.K.; Sasak, W.V.; Grand, R.J.: Biosynthesis, glycosylation, and intracellular transport of intestinal lactase-phlorizin hydrolase in rat. J. Biol. Chem., **262**, 17206-17211 (1987)

[11] Naim, H.Y.; Sterchi, E.E.; Lentze, M.J.: Biosynthesis and maturation of lactase-phlorizin hydrolase in the human small intestinal epithelial. Biochem. J., **241**, 427-434 (1987)
[12] Danielsen, E.M.; Skovbjerg, H.; Noren, O.; Sjöström, H.: Biosynthesis of intestinal microvillar proteins. Intracellular processing of lactase-phlorizin hydrolase. Biochem. Biophys. Res. Commun., **122**, 82-90 (1984)
[13] Skovbjerg, H. ; Danielsen, E.M. ; Noren, O. ; Sjöström, H.: Evidence for biosynthesis of lactase-phlorizin hydrolase as a single-chain high-molecular weight precursor. Biochim. Biophys. Acta, **798**, 247-251 (1984)
[14] Skovbjerg, H.; Sjöström, H.; Noren, O.: Purification and characterization of amphiphilic lactase/phlorizin hydrolase from human small intestine. Eur. J. Biochem., **114**, 653-661 (1981)
[15] Skovbjerg, H.; Noren, O.; Sjöström, H.; Danielsen, E.M.; Enevoldsen, B.S.: Further characterization of intestinal lactase/phlorizin hydrolase. Biochim. Biophys. Acta, **707**, 89-97 (1982)
[16] Kobayashi, T.; Suzuki, K.: The glycosylceramidase in the murine intestine. Purification and substrate specificity. J. Biol. Chem., **256**, 7768-7773 (1981)
[17] Cousineau, J.; Green, J.R.: Isolation and characterization of the proximal and distal forms of lactase-phlorizin hydrolase from the small intestine of the suckling rat. Biochim. Biophys. Acta, **615**, 147-157 (1980)
[18] Ramaswamy, S.; Radhakrishnan, A.N.: Lactase-phlorizin hydrolase complex from monkey small intestine. Purification, properties and evidence for two catalytic sites. Biochim. Biophys. Acta, **403**, 446-455 (1975)
[19] Birkenmeier, E.; Alpers, D.H.: Enzymatic properties of rat lactase-phlorizin hydrolase. Biochim. Biophys. Acta, **350**, 100-112 (1974)
[20] Leese, H.J.; Semenza, G.: On the identity between the small intestine enzymes phlorizin hydrolase and glycosylceramidase. J. Biol. Chem., **248**, 8170-8173 (1973)
[21] Hamaswamy, S.; Radhakrishnan, A.N.: Lactase-phlorizin hydrolase complex from monkey small intestine: Stimulation of phlorizin hydrolase activity by organic acids. Biochem. Biophys. Res. Commun., **54**, 197-204 (1973)
[22] Kraml, J.; Kolinska, J.; Ellederova, D.; Hirsova, D.: β-Glucosidase (phlorizin hydrolase) activity of the lactase fraction isolated from the small intestinal mucosa of infant rats, and the relationship between β-glucosidases and β-galactosidases. Biochim. Biophys. Acta, **258**, 520-530 (1972)
[23] Malathi, P.; Crane, R.K.: Phlorizin hydrolase: A β-glucosidase of hamster intestinal brush border membrane. Biochim. Biophys. Acta, **173**, 245-256 (1969)

1,2-α-L-Fucosidase 3.2.1.63

1 Nomenclature

EC number
3.2.1.63

Systematic name
2-α-L-fucopyranosyl-β-D-galactoside fucohydrolase

Recommended name
1,2-α-L-fucosidase

Synonyms
1,2-α-L-fucosidase
α-(1→ 2)-L-fucosidase
α-L-fucosidase II
fucosidase, 1,2-α-L-

CAS registry number
37288-45-2

2 Source Organism

<1> *Aspergillus niger* [1]
<2> *Prunus dulcis* (almond [2]) [2]
<3> *Bifidobacterium bifidum* (strain: VIII-210 [3]) [3]
<4> *Bifidobacterium infantis* (strain: VIII-240 [3]) [3]
<5> *Ruminococcus torquens* (strains: VIII-239 and IX-70 [3]) [3]
<6> *Ruminococcus sp. AB* (strain: VI-268 [3]) [3]
<7> *Clostridium perfringens* [4]
<8> *Bacillus fulminans* [5]

3 Reaction and Specificity

Catalyzed reaction
methyl-2-α-L-fucopyranosyl-β-D-galactoside + H_2O = L-fucose + methyl β-D-galactoside (high specificity for non-reducing terminal L-fucose residues α 1,2-linked to D-galactose residues)

Reaction type
hydrolysis of O-glycosyl bond

Natural substrates and products

S submaxillary glycoprotein + H_2O <7> (<7> degradation of blood group substances [4]) [4]
P L-fucose + ? <7> [4]

Substrates and products

S 2'-fucosyllactose <8> [5]
P ?
S 2'-fucosyllactose + H_2O <2> [2]
P fucose + lactose <2> [2]
S 2-O-α-L-fucopyranosyl-D-galactose + H_2O <1, 2> [1, 2]
P L-fucose + galactose <1, 2> [1, 2]
S 2-fucosyllactitol + H_2O <2> [2]
P fucose + lactitol
S O-α-L-fucose-(1-2)-O-β-D-galactose-(1-4)-D-glucose + H_2O <1, 7, 8> (<1, 7, 8> i.e. 2-O-α-L-fucosyllactose [1, 4, 5]) [1, 4, 5]
P L-fucose + lactose <1, 7, 8> [1, 4, 5]
S α-Fuc-(1-2)-β-Gal-(1-3)-β-GlcNAc-(1-3)-β-Gal-(1-4)-β-Glu + H_2O <1> (<1> i.e. lacto-N-fucopentaose I [1]) [1]
P α-L-fucose + β-Gal-(1-3)-β-GlcNAc-(1-3)-β-Gal-(1-4)-β-Glu <1> [1]
S lacto-N-fucopentaose I <8> [5]
P ?
S methyl-2-α-L-fucopyranosyl-β-D-galactoside + H_2O <1> [1]
P L-fucose + methyl β-D-galactoside <1> [1]
S mucin + H_2O <1, 8> (<1> porcine and canine, submaxillary [1]; <8> porcine submaxillary and glycopeptide of desialyzed mucin, [5]) [1, 5]
P L-fucose + ? <1> [1]
S Additional information <1, 2, 7> (<7> not: *p*-nitrophenyl-α-L-fucoside, 2-O-, 3-O- and 4-O-α-L-fucopyranosylfucoses, 3-O- and 4-O-α-L-fucopyranosyl-D-galactoses, action on oligosaccharides and glycoproteins, not: simple methyl nitrophenyl fucosides [4]; <1,2> highly specific for non-reducing terminal L-fucose residues linked to D-galactose residues by 1,2-α linkage [1, 2]) [1, 2, 4, 5]
P ?

Inhibitors

Cu^{2+} <7> [4]
EDTA <7, 8> [4, 5]
Fe^{2+} <7> [4]
Fe^{3+} <7> [4]
Hg^{2+} <7, 8> [4, 5]
$MgCl_2$ <2> [2]
Mn^{2+} <8> [5]
SDS <8> [5]
Zn^{2+} <7> [4]
deoxycholate <8> [5]
iodoacetamide <7> [4]

p-chloromercuribenzoate <7> [4]
p-chloromercuriphenylsulfonate <2> (<2> 1 mM, 12% inhibition [2]) [2]

Metals, ions
$CaCl_2$ <2> (<2> slight activation [2]) [2]

Specific activity (U/mg)
0.00186 <2> [2]
0.12 <8> [5]
6.7 <1> [1]
70 <7> [4]

K_m-Value (mM)
0.083 <1> (methyl 2-O-α-L-fucopyranosyl-β-D-galactoside) [1]
0.175 <8> (glycoprotein, hog, submaxillary) [5]
0.33 <8> (porcine submaxillary mucin) [5]
0.675 <2> (2-fucosyllactitol) [2]
1.6 <8> (glycopeptide desialyzed porcine submaxillary mucin) [5]
2 <8> (α-Fuc-(1-2)-β-Gal-(1-3)-β-GlcNAc-(1-3)-β-Gal-(1-4)-β-Glu) [5]
2.5 <8> (2'-fucosyllactose) [5]

pH-Optimum
3.8 <1> [1]
5-6.5 <2> [2]
6 <7> [4]

pH-Range
2.4-7.5 <1> [1]
4.5-8 <7> [4]

Temperature optimum (°C)
37 <1> (<1> assay at [1]) [1]

4 Enzyme Structure

Molecular weight
70000-80000 <8> (<8> gel filtration [5]) [5]
200000 <7> (<7> value greater than, multiple isozymic forms [4]) [4]

5 Isolation/Preparation/Mutation/Application

Source/tissue
emulsin <2> [2]
Additional information <1> (<1> commercial preparation [1]) [1]

Purification
<1> [1]
<2> (partial [2]) [2]
<7> [4]
<8> [5]

6 Stability

Temperature stability

50 <8> (<8> 5 min, no loss of activity [5]) [5]
60 <8> (<8> 5 min, 7% loss of activity [5]) [5]
70 <8> (<8> 5 min, 77% loss of activity [5]) [5]
80 <8> (<8> 5 min, complete loss of activity [5]) [5]

General stability information

<7>, dialysis, water: rapid loss of activity, buffers: precipitation with considerable inactivation below pH 5.5 [4]
<7>, lyophilization causes inactivation [4]
<7>, repeated freezing and thawing causes inactivation [4]

References

[1] Bahl, O.P.: Glycosidases of Aspergillus niger. II. Purification and general properties of 1,2-α-L-fucosidase. J. Biol. Chem., **245**, 299-304 (1970)
[2] Ogata-Arakawa, M.M.; Muramatsu, T.; Kobata, A.: α-L-Fucosidases from almond emulsin: characterization of the two enzymes with different specificities. Arch. Biochem. Biophys., **181**, 353-358 (1977)
[3] Larson, G.; Falk, P.; Hoskin, L.C.: Degradation of human intestinal glycosphingolipids by extracellular glycosidases from mucin-degrading bacteria of the human fecal flora. J. Biol. Chem., **263**, 10790-10798 (1988)
[4] Aminoff, D.; Frurukawa, K.: Enzymes that destroy blood group specificity. I. Purification and properties of α-L-fucosidase from Clostridium perfringens. J. Biol. Chem., **245**, 1659-1669 (1970)
[5] Kochibe, N.: Purification and properties of α-L-fucosidase from Bacillus fulminans. J. Biochem., **74**, 1141-1149 (1973)

2,6-β-Fructan 6-levanbiohydrolase 3.2.1.64

1 Nomenclature

EC number
3.2.1.64

Systematic name
2,6-β-D-fructan 6-β-D-fructofuranosyl fructohydrolase

Recommended name
2,6-β-fructan 6-levanbiohydrolase

Synonyms
2,6-β-D-fructan 6-levanbiohydrolase
2,6-β-fructan 6-levanbiohydrolase
LDE
β2,6-fructan-6-levanbiohydrolase
levanbiohydrolase, 2,6-β-fructan 6-

CAS registry number
37288-46-3

2 Source Organism

<1> *Pseudomonas sp.* (strain 11 [1]) [1]
<2> *Streptomyces exfoliatus* (strain F3-2 [2]) [2]

3 Reaction and Specificity

Catalyzed reaction
hydrolysis of 2,6-β-D-fructan so as to remove successive levanbiose residues from the end of the chain

Reaction type
hydrolysis of O-glycosyl bond

Natural substrates and products
S levan + H_2O <1, 2> (levan i.e. 2,6-β-D-fructan) [1, 2]
P levanbiose <1, 2> (6-β-fructosylfructose) [1, 2]

Substrates and products
S 1-fructosylnystose + H_2O <2> (<2> at 81.4% the rate of levan [2]) [2]
P β-D-fructose + α-D-glucose + sucrose + 1-kestose + nystose <2> [2]

S 1-kestose + H_2O <2> (<2> at 33.3% the rate of levan [2]) [2]
P sucrose + β-D-fructose + α-D-glucose <2> [2]
S levan + H_2O <1, 2> (<1, 2> levan i.e. 2,6-β-D-fructan [1, 2]; <2> smallest substrate to be hydrolyzed is levantriose [2]) [1, 2]
P levanbiose <1, 2> (6-β-fructosylfructose) [1, 2]
S nystose + H_2O <2> (<2> at 73.4% the rate of levan [2]) [2]
P β-D-fructose + α-D-glucose + sucrose + 1-kestose <2> [2]
S sucrose + H_2O <2> (<2> at 6.5% the rate of levan [2]) [2]
P β-D-fructose + α-D-glucose <2> [2]

Specific activity (U/mg)
105 <2> [2]

pH-Optimum
5.5 <2> [2]

4 Enzyme Structure

Molecular weight
60000 <2> (<2> gel filtration [2]) [2]

Subunits
monomer <2> (<2> 1 * 54000, SDS-PAGE [2]) [2]

5 Isolation/Preparation/Mutation/Application

Source/tissue
culture supernatant <1, 2> [1, 2]

Localization
extracellular <1, 2> [1, 2]

6 Stability

pH-Stability
3.5-8 <2> [2]

Temperature stability
50 <2> (<2> stable up to [2]) [2]

References

[1] Avigad, G.; Zelikson, R.: Cleavage of fructans to levanbiose by a specific hydrolase. Bull. Res. Counc. Isr., **11 A4**, 253-257 (1963)
[2] Saito, K.; Kondo, K.; Kojima, I.; Yokota, A.; Tomita, F.: Purification and characterization of 2,6-β-D-fructan 6-levanbiohydrolase from Streptomyces exfoliatus F3-2. Appl. Environ. Microbiol., **66**, 252-256 (2000)

Levanase 3.2.1.65

1 Nomenclature

EC number
3.2.1.65

Systematic name
2,6-β-D-fructan fructanohydrolase

Recommended name
levanase

Synonyms
levan hydrolase

CAS registry number
9041-11-6

2 Source Organism

<1> *Bacillus subtilis* [1, 2, 3, 4, 5, 7, 9, 10, 11]
<2> *Actinomyces viscosus* [3, 5]
<3> *Escherichia coli* [1, 2, 4, 5, 11]
<4> *Enterobacter cloacae* (var. Laevanivorans [6]) [6]
<5> *Pseudomonas caryophylli* [6]
<6> *Streptococcus mutans* [6]
<7> *Streptococcus salivarius* [6, 7]
<8> *Streptococcus sanguis* [6]
<9> *Arthrobacter sp.* [8]
<10> *yeasts* (Saccharomyces anamensis, Sacch. fermentati, Sacch. florenzani, Sacch. fragilis, Sacch. rosei, Schizosaccharomyces pombe, Schwanniomyces alluvius, Schw. castellii, Torulopsis colliculosa [6]; Sacch. cerevisiae [6, 10]) [6, 10]
<11> *Lactobacillus plantarum* [9, 10]
<12> *Lactobacillus casei* [10]
<13> *Pseudomonas strain K-52* [12]
<14> *Rhodotorula sp.* [13]
<15> *Streptomyces sp. 366L* [14]
<16> *Pseudomonas sp. No. 43* [15]
<17> *Bacillus sp. L7* [16]

3 Reaction and Specificity

Catalyzed reaction

random hydrolysis of 2,6-β-D-fructofuranosidic linkages in 2,6-β-D-fructans containing more than 3 fructose units (levans)

Reaction type

hydrolysis of O-glycosyl bond

Natural substrates and products

S levans + H_2O <1, 2> [1, 3, 4, 6, 10, 11]
P oligofructosides <1, 2> [1, 3, 4, 6, 10, 11]

Substrates and products

S ? <7> [7]
P fructose <7> [7]
S D-(+)-xylose + H_2O <13> [12]
P ?
S cellulose + H_2O <13> [12]
P ?
S dextran + H_2O <13> [12]
P ?
S fructose + H_2O <13> [12]
P ?
S galactose + H_2O <13> [12]
P ?
S glucose + H_2O <13> [12]
P ?
S inulin + H_2O <1, 2, 14> [1, 3, 4, 5, 6, 10, 11, 13]
P ?
S lactose + H_2O <13> [12]
P ?
S levan + H_2O <1, 2, 9, 10> [1, 3, 4, 6, 8, 10, 11]
P oligofructosides <1, 2, 9, 10> [1, 3, 4, 6, 8, 10, 11]
S levan + H_2O <14, 17> [13, 16]
P fructose <14, 17> [13, 16]
S levan + H_2O <15> [14]
P levanheptaose <15> [14]
S levan + H_2O <16> [15]
P levanbiose <16> [15]
S maltose + H_2O <13> [12]
P ?
S raffinose + H_2O <13> [12]
P ?
S soluble starch + H_2O <13> [12]
P ?
S sucrose + H_2O <1, 2, 14> [1, 2, 3, 4, 6, 10, 11, 13]
P ?

S xylan + H_2O <13> [12]
P ?

Inhibitors

2-mercaptoethanol <16> [15]
Ag^+ <1, 7, 9, 13, 15, 16, 17> [7, 8, 11, 12, 14, 15, 16]
Al^{3+} <16> [15]
Ba^{2+} <2, 16> [3, 15]
Ca^{2+} <14, 16, 17> [13, 15, 16]
Co^{2+} <14, 15, 16> [13, 14, 15]
Cu^{2+} <7, 9, 13, 14, 17> [7, 8, 12, 13, 16]
EDTA <2, 13, 14, 16> [3, 12, 13, 15]
Fe^{2+} <13> [12]
Fe^{3+} <2, 17> [3, 16]
Hg^{2+} <1, 7, 9, 13, 14, 17> [3, 7, 8, 11, 12, 13, 16]
$KMnO_4$ <14, 16> [13, 15]
Mg^{2+} <14, 16, 17> [13, 15, 16]
Mn^{2+} <2, 15, 16, 17> [3, 14, 15, 16]
Na^+ <17> [16]
Pb^{2+} <9> [8]
SDS <15> [14]
Tris buffer <9> [8, 6]
Zn^{2+} <2, 15, 17> [3, 14, 16]
ammonium oxalate <14, 16> [13, 15]
ammonium persulfate <14> [13]
ascorbic acid <14, 16> [13, 15]
diethyl dicarbonate <13> [12]
glutathione <14> [13]
hydrazine sulfate <14> [13]
iodoacetate <15> [14]
p-chloromercuribenzoate <7, 15> [7, 14]
sodium citrate <14> [13]
sodium thioglycollate <14, 16> [13, 15]
sucrose <17> (weak inhibitory effect at high concentrations) [16]

Activating compounds

2-mercaptoethanol <14> [13]

Specific activity (U/mg)

0.09 <11> (<11> U/ml, inulin [10]) [10]
0.3 <1> (<1> BG4103-strain [4]) [4]
0.4 <1> (<1> QB167-strain [4]) [4]
0.5 <12> (<12> U/ml, inulin [10]) [10]
30.5 <16> (<16> U/mg [15]) [15]
34.4 <14> [13]
39.2 <13> [12]
97.4 <1> [11]
128 <17> (<17> U/mg [16]) [16]

500 <9> [8]
810 <7> [7]
851 <2> [3]
4309 <15> (<15> U/mg [14]) [14]

K_m-Value (mM)
0.0012 <1> (levan) [11]
0.5 <9> (levan) [8]
0.97 <14> (levan) [13]
6.8 <1> (inulin) [11]
10 <9> (tetraose) [8]
65 <1> (sucrose) [11]

pH-Optimum
4 <2> (<2> not active below [3]) [3]
5.5 <1, 17> (<17> no activity loss at pH 6.0 and 60°C for 20 min [16]) [5, 11, 16]
6 <14> [13]
6 <2> [3, 5]
6.5 <7> (<7> no activity below pH 5.5 [7]) [7]
7 <13, 14, 15, 16> (<15> rapidly decreased activity below and above pH 7.0 [14]) [12, 13, 14, 15]

pH-Range
5.4-6.6 <2> [5]
5.5-6.5 <14> [13]
5.5-6.5 <9> (<9> in presence of levan extended to pH 4.6-7.5 [8]) [8]
6-9 <13> (<13> 4°C [12]) [12]
6-10 <15> (<15> 20°C [14]) [14]

Temperature optimum (°C)
35 <13> [12]
40 <14, 15, 16> (<15> pH 7.0; rapidly inactivated above 45°C [14]) [13, 14, 15]
45 <2> (<2> not active above 55°C [3]) [3]
50 <17> [16]
55 <1> (<1> rapidly inactivated at 60°C [11]) [11]

Temperature range (°C)
30-40 <13> [12]
40-50 <7> (<7> not active above 55°C [7]) [7]
45-55 <9> [8]

4 Enzyme Structure

Molecular weight
36000 <16> (<16> gel filtration [15]) [15]
38000 <13> (<13> SDS-PAGE and gel filtration [12]) [12]

39000 <14> (<14> SDS-PAGE [13]) [13]
73000 <1> (<1> SDS-PAGE [4]) [4]
74000 <1> (<1> in vitro transcription and translation experiments [2]) [2]
75000 <1> [11]
75000 <1> (<1> SDS-PAGE [1]) [1]
75870 <1> (<1> DNA sequence analysis [4, 7]) [4, 7]
78000 <15> (<15> SDS-PAGE [14]) [14]
80000 <15> (<15> gel filtration [14]) [14]
86000 <17> (<17> SDS-PAGE and DNA sequence analysis [16]) [16]
89000 <2> (<2> SDS-PAGE [3]) [3]
100000 <7> (<7> SDS-PAGE [7]) [7]

Subunits

monomer <2, 7, 15> (<2, 7, 15> SDS-PAGE [3, 7, 14]; <15> gel filtration [14]) [3, 7, 14]

5 Isolation/Preparation/Mutation/Application

Source/tissue

dental plaque <2> [3]
soil <16, 17> [15, 16]

Localization

extracellular <1, 2, 16, 17> [1, 2, 3-5, 10, 11, 15, 16]
intracellular <3> (<3> small extent transported into periplasmic space [5]) [5]

Purification

<1> (<1> expression in E. coli [4]) [4, 11]
<2> [3]
<7> [7]
<9> [8]
<13> [12]
<14> [13]
<15> [14]
<16> [15]
<17> [16]

Cloning

<1> (<1> expression in E. coli [1, 2, 4, 5, 11]; <1> expression in Lactobacillus plantarum [9]; <1> expression in Lactobacillus-E. coli [10]) [1, 2, 4, 5, 9, 10, 11]
<17> (<17> expression in B. subtilis [16]) [16]

Application

agriculture <11, 12, 14, 15, 16> (<14> slime dissolving for more effective penetration of biocides inside the cell wall [13]; <15, 16> plant differentiation, growth promoting activities for some enteric bacteria [14, 15]) [10, 13, 14, 15]

medicine <1, 2, 13, 15, 16> (<1> mitogenic for human B lymphocytes and activates the alternative complement pathway [3]; <2> pathogenicity of periodontal disease [3]; <13, 15, 16> enhancing effect on the immune system and other physiological functions [12, 14, 15]) [3, 12, 14, 15]
nutrition <1, 15, 16> (<1> production of pure fructose syrups [11, 12]; <15, 16> energy source [14, 15]; <15, 16> production of noncaloric sweeteners [14, 15]) [11, 12, 14, 15]
synthesis <1, 14> (<1> synthesis of various products such as ethanol or aceton-butanol [11]; <14> paper industry [13]; <15, 16> protection from aging; lowering of water activity [14, 15]) [11, 13-15]

6 Stability

pH-Stability

4.6-7.5 <9> [8]
5.5-6.5 <14> [13]
6-8 <16> (<16> 20°C; stable up to 50°C at pH 7.0; activity decreases rapidly below pH 6.0 and above pH 8.0 [15]) [15]

Temperature stability

-25 <1, 6, 8> (<1, 6, 8> stable for several weeks [5, 6]) [5, 6]
-21 <4, 5> (<4> rapidly inactivated above [6]) [6]
4 <14> [13]
40 <16> (<16> stable up to 50°C but rapidly inactivated above [15]) [15]

Storage stability

<2>, 4°C, phosphate buffer, pH 6.0 [3]
<9>, -15°C, several months [8]

References

[1] Schoergendoerfer, K.; Schwab, H.; Lafferty, R.M.: Molecular characterization of Bacillus subtilis levanase and a C-terminal deleted derivative. J. Biotechnol., **7**, 247-258 (1988)
[2] Schoergendoerfer, K.; Schwab, H.; Lafferty, R.M.: Nucleotide sequence of a cloned 2.5 kb PstI-EcoRI Bacillus subtilis DNA fragment coding for levanase. Nucleic Acids Res., **15**, 9606 (1987)
[3] Igarashi, T.; Takahashi, M.; Yamamoto, A.; Etoh, Y.; Takamori, K.: ATCC19246. Infect. Immun., **55**, 3001-3005 (1987)
[4] Martin, I.; De Barbouille, M.; Ferrari, E.; Klier, A.; Rapoport, G.: Characterization of the levanase gene of Bacillus subtilis which shows homology to yeast invertase. Mol. Gen. Genet., **208**, 177-184 (1987)
[5] Friehs, K.; Schoergendoerfer, K.; Schwab, H.; Lafferty, R.M.: Cloning and phenotypic expression in Escherichia coli of a Bacillus subtilis gene fragment coding for sucrose hydrolysis. J. Biotechnol., **3**, 333-341 (1986)

[6] Fuchs, A.; de Bruijn, J.M.; Niedeveld, C.J.: Bacteria and yeasts as possible candidates for the production of inulinases and levanases. Antonie Leeuwenhoek, **51**, 333-351 (1985)
[7] Takahashi, N.; Mizuno, F.; Takamori, K.: Isolation and properties of levanase from Streptococcus salivarius KTA-19. Infect. Immun., **42**, 231-236 (1983)
[8] Avigad, G.; Bauer, S.: Fructan hydrolases. Methods Enzymol., **8**, 621-628 (1966)
[9] Holts, P.; Ferain, T.; Garmyn, D.; Bernhard, N.; Delcour, J.: Use of homologous expression-secretion signals and vector-free stable chromosomal integration in engineering of Lactobacillus plantarum for α-amylase and levanase expression. Appl. Environ. Microbiol., **60**, 1401-1413 (1994)
[10] Wanker, E.; Leer, R.J.; Pouwels, P.H.; Schwab, H.: Expression of Bacillus subtilis levanase gene in Lactobacillus plantarum and Lactobacillus casei. Appl. Microbiol. Biotechnol., **43**, 297-303 (1995)
[11] Wanker, E.; Huber, A.; Schwab, H.: Purification and characterization of the Bacillus subtilis levanase produced in Escherichia coli. Appl. Environ. Microbiol., **61**, 1953-1958 (1995)
[12] Kang, S.K.; Lee, S.O.; Lim, Y.S.; Jang, K.L.; Lee, T.H.: Purification and characterization of a novel levanoctaoase-producing levanase from Pseudomonas strain K-52. Biotechnol. Appl. Biochem., **27**, 159-166 (1998)
[13] Chaudhary, A.; Gupta, L.K.; Gupta, J.K.; Banerjee, U.C.: Purification and properties of levanase from Rhodotorula sp.. J. Biotechnol., **46**, 55-61 (1996)
[14] Lim, Y.S.; Kang, S.K.; Lee, S.O.; Lee, J.D.; Lee, T.H.: Purification and charakterization of a levanase from Streptomyces sp. 366L. J. Biotechnol., **61**, 33-41 (1998)
[15] Kang, E.J.; Lee, S.O.; Lee, J.D.; Lee, T.H.: Purification and characterization of a levanbiose-producing levanase from Pseudomonas sp. No. 43. Biotechnol. Appl. Biochem., **29**, 263-268 (1999)
[16] Miasnikov, A.N.: Characterization of a novel endo-levanase and its gene from Bacillus sp. L7. FEMS Microbiol. Lett., **154**, 23-28 (1997)

Quercitrinase 3.2.1.66

1 Nomenclature

EC number
3.2.1.66

Systematic name
quercitrin 3-L-rhamnohydrolase

Recommended name
quercitrinase

CAS registry number
37288-47-4

2 Source Organism

<1> *Aspergillus flavus* (synthesized by only a few members of the Aspergillus flavus group [1]) [1]

3 Reaction and Specificity

Catalyzed reaction
quercitrin + H_2O= L-rhamnose + quercetin

Reaction type
hydrolysis of O-glycosyl bond

Natural substrates and products
S quercitrin + H_2O <1> (<1> quercitrin is identical with quercetin 3-L-rhamnoside, growth on quercitrin as carbon source [1]) [1]
P L-rhamnose + quercetin <1> [1]

Substrates and products
S myricitrin + H_2O <1> (<1> myricitrin is identical with myricetin-3-L-rhamnoside [1]) [1]
P L-rhamnose + myricetin <1> [1]
S quercitrin + H_2O <1> (<1> quercitrin is identical with quercetin-3-L-rhamnoside [1]) [1]
P L-rhamnose + quercetin <1> [1]
S robinin + H_2O <1> (<1> robinin is identical with kaempferol-3-robinoside-7-rhamnoside [1]) [1]
P rhamnose + kaempferol-3-robinoside <1> [1]

pH-Optimum
6.4 <1> [1]

pH-Range
4.5-8 <1> [1]

5 Isolation/Preparation/Mutation/Application

Source/tissue
culture medium <1> [1]
mycelium <1> [1]

Localization
extracellular <1> (<1> 78% of the total activity [1]) [1]

6 Stability

Temperature stability
35 <1> (<1> stable above [1]) [1]

Storage stability
<1>, -17°C, several weeks, little loss of activity [1]

References

[1] Westlake, D.W.S.: Microbial degradation of quercitrin. Can. J. Microbiol., **9**, 211-220 (1963)

Galacturan 1,4-α-galacturonidase 3.2.1.67

1 Nomenclature

EC number
3.2.1.67

Systematic name
poly(1,4-α-D-galacturonide)galacturonohydrolase

Recommended name
galacturan 1,4-α-galacturonidase

Synonyms
PGase
carrot exopolygalacturonase <10> [14]
corn pollen polygalacturonase <1> [1]
exo-D-galacturonanase
exo-D-galacturonase
exo-polygalacturonase
exoPG
exoPGase
exolytic PGase
exopoly-D-galacturonase
exopolygalacturonase
macerating enzyme <6> [8]
oligogalacturonate hydrolase <15> [17]
peach exopolygalacturonase <10> [14]
poly 1,4-α-D-galacturonide-galacturonohydrolase
poly(galacturonate)hydrolase

CAS registry number
9045-35-6

2 Source Organism

<1> *Zea mays* (corn [1]) [1]
<2> *Daucus carota* (carrot [2, 5, 12, 14, 23]) [2, 5, 12, 14, 23]
<3> *Lycopersicon esculentum* (tomato [4]) [4]
<4> *Butyrivibrio fibrisolvens* (strain 781, isolated from bovine rumen [6]) [6]
<5> *Trichoderma reesei* (strain QM9414 [7]) [7]
<6> *Penicillium oxalicum* [8]
<7> *Musa sapientum* (banana, var. GROS MICHEL [10]) [10]

<8> *Malus domestica* (apple, Borkh. cv Cox Orange Pippin [11]) [11]
<9> *Aspergillus niger* [13, 21, 21]
<10> *Prunus persica* (peach [14, 15]) [14, 15]
<11> *Colletotrichum lindemuthianum* (fungal plant pathogen, bean [16]) [16]
<12> *Colletotrichum truncatum* [3]
<13> *Colletotrichum capsici* [3]
<14> *Curvularia pallescens* [3]
<15> *Bacillus sp.* (aerobic Bacillus, isolate B [17]) [17]
<16> *Marchantia polymorpha* (liverwort, cell line HYH-2F [2, 9]) [2, 9]
<17> *Fusarium oxysporum* (plant pathogen filamentous fungus, f. sp. melonis [18]; f.sp. f. sp. radicis lycopersici [22]) [18, 22]
<18> *Geotrichum lactis* (yeast-like fungus, ATCC48590 [19, 20]) [19, 20]
<19> *Aspergillus tubingensis* (filamentous fungus, wild-type strain [24]) [24, 25]
<20> *Yersinia enterolitica* (enteropathogen, ATCC 49397 [26]) [26]

3 Reaction and Specificity

Catalyzed reaction

(1,4-α-D-galacturonide)$_n$ + H_2O = (1,4-α-D-galacturonide)$_{n-1}$ + D-galacturonate

Reaction type

hydrolysis of O-glycosyl bond

Natural substrates and products

S pectic acid + H_2O <1-16> [1-17]
P D-galacturonide + ? <12, 13, 14> [3]

Substrates and products

S (1,4-α-D-galacturonide)$_2$ + H_2O <2-5, 7, 9, 10, 15, 19> (<2, 4, 7> specific substrate for exo-D-galactoronanase [5, 6, 10]; <3> slowly, reaction rate increases with increasing galacturonan chain length [4]) [4-7, 10, 12, 13, 14, 17, 24]
P D-galacturonic acid <9, 15> [13, 17]
S (1,4-α-D-galacturonide)$_3$ + H_2O <3, 4, 7, 15, 19> (<15> rate of activity maximal with this substrate [17]) [4, 6, 10, 17, 24]
P galacturonic acid <15> (<15> primary product [17]) [17]
S (1,4-α-D-galacturonide)$_4$ + H_2O <2, 4, 7, 10, 19> [6, 10, 14, 24]
P D-galacturonate <2> [14]
S (1,4-α-D-galacturonide)$_4$ + H_2O <2, 3, 7, 9, 15> (<9> substrate of preference [13]) [4, 10, 12, 13, 17]
P trigalacturonic acid + D-galacturonic acid <2, 15> (<15> primary product galacturonic acid [17]) [12, 17]
S (1,4-α-D-galacturonide)$_5$ + H_2O <2> [23]
P D-galactopyranuronic acid <2> [23]
S (1,4-α-D-galacturonide)$_5$ + H_2O <3, 4, 7, 19> [4, 6, 10, 24]

P D-galacturonic acid + digalacturonic acid + trigalacturonic acid + tetragalacturonic acid <3> [4]
S (1,4-α-D-galacturonide)$_6$ + H_2O <1, 2, 4, 7, 10, 19> [1, 10, 6, 14, 24]
P D-galacturonic acid + (1,4-α-D-galacturonide)5 <1> (<1> products increase with time, gradually tetragalacturonic acid and trigalacturonic acid appears, exoenzyme removes monomeric units from the substrate [1]) [1]
S (1,4-α-D-galacturonide)$_7$ + H_2O <2, 7, 10, 19> [10, 14, 24]
P ?
S (1,4-α-D-galacturonide)$_8$ + H_2O <2, 7, 10> [10, 14]
P ?
S (1,4-α-D-galacturonide)$_n$ + H_2O <1-20> (<1> large galacturonans, release of monomeric units from the non-reducing ends of the substrate chains [1]; <3> polygalacturonic acid I, II, III [4]; <7> 51% degraded [10]; <10> partial hydrolysis [15]) [1-18, 20, 23-26]
P D-galacturonate + digalacturonic acid + trigalacturonic acid <1-17> [1-18, 23]
S (1,4-α-D-galacturonide)$_n$ + H_2O <2, 10-13, 14, 18> (<12, 13, 14> carbon source, induce the production of pectolytic enzymes [3]) [3, 14, 16, 19, 20]
P D-galacturonic acid <11> [16]
S 4,5-dehydrogalacturonosyl-α-D-trigalacturonic acid + H_2O <9> (<9> 4,5-unsaturated tetragalacturonic acid [13]) [13]
P (1,4-α-D-galacturonide)$_3$ <9> [13]
S arabinogalactan + H_2O <11> (<11> exopolygalacturonase still contains arabinase activity, limits the usefulness on any pectic polysaccharide with arabinose residues [16]) [16]
P ?
S oligo-D-galacturonate + H_2O <2, 7, 9, 10, 15, 19> (<15> highly specific for saturated oligogalacturonides, inactive towards unsaturated [17]) [10, 12-14, 17, 24, 25]
P D-galacturonic acid <2, 10, 15> [14, 17]
S *p*-nitrophenyl-α-D-galactopyranuronide + H_2O <2, 9, 11> (<2, 9, 11> cleaves all substrates with 4-deoxy-L-threo-hexo-4-eno-pyranosyluronate unit at the nonreducing end [12, 13, 16]) [12, 13, 16]
P D-galacturonic acid + 4-nitrophenol <11> [16]
S pectic acid + H_2O <1-3, 5-7, 9, 10, 12-18> (<2> degraded not fully [12]; <15> with ASPA, acid soluble pectid acid only galacturonic acid as product [17]; <6, 9, 12-14> lytic cleavage of NaPP, degradation in vitro [3, 8, 21]; <9> 2 exoforms of enzyme, PGI and PGII, prefer NaPPas substrate over pectin [21]) [1-10, 12-15, 17, 19-23]
P D-galacturonic acid + ? <1, 3, 5, 9, 17> (<5> oligo-D-galacosiduronic acids in the first h [7]; <9> traces of oligogalacturonic acids found [13]) [1, 4, 7, 13, 22]
S pectin + H_2O <2, 4, 10, 16-18> (<10> PGII solubilizes pectin very effective [15]; <17> 5% activity on substrate [18, 22]; <19> apple pectin [25]) [2, 6, 9, 14, 15, 18-20, 22]
P digalacturonate + D-galacturonic acid <4, 18> (<4> unsaturated, digalacturonate main end product [6]) [6, 20]

S xylogalacturonan + H_2O <19> (<19> terminal galacturonic acid residue carries a single xylose substitution [25]) [25]
P xylogalacturonate <19> [25]

Inhibitors

Ba^{2+} <2, 10, 16> (<2> 1 mM reduce activity 15-28% [2]) [2, 9, 15]
Ca^{2+} <17> [22]
Cd^{2+} <3> [4]
Cu^{2+} <2, 16> (<2> 1 mM reduces activity 15-28% [2]; <16> reduces activity 20-53% [2, 9]) [2, 9]
D-galactopyranuronic acid <2, 19> (<19> catabolite repression, completely inhibited, inhibitory constant 0.3 mM [24]) [23, 24, 25]
EDTA <1, 3, 8, 10> (<1> chelating agent [1]) [1, 4, 11, 15]
Hg^{2+} <2, 3, 16> (<2> 1 mM reduces activity 15-28% [2]) [2, 4, 9]
Mg^{2+} <8> [11]
Mn^{2+} <16> [2, 9]
Na^{+} <3> [4]
NaCl <2, 10> (<10> above 0.15 M [15]) [12, 14, 15]
β-glycosylamine <2> (<2> of ammonium D-galactopyranuronate, competitive inhibition [23]) [23]
citrate <1, 3, 8, 10> (<1> chelating agent [1]) [1, 4, 11, 15]
glucose <18> (<18> catabolite repression, even in presence of inducers [19, 20]) [19, 20]

Specific activity (U/mg)

0.08 <16> [2]
0.35 <2> [2]
0.44 <2> (<2> 2 several forms of exo-D-galacturonanase [12]) [12]
0.56 <2> (<2> 2 several forms of exo-D-galacturonanase [12]) [12]
0.69 <2> (<2> pentagalacturonic acid [23]) [23]
1.44 <9> (<9> digalacturonic acid [13]) [13]
1.7 <4> (<4> sodium pectate [6]) [6]
1.9 <9> (<9> sodium pectate [13]) [13]
2.86 <15> [17]
3.5 <11> [16]
3.8 <9> (<9> enzyme form PGII [21]) [21]
4.74 <16> [9]
5.45 <4> (<4> digalacturonic acid [6]) [6]
7.5 <9> (<9> enzyme form PGI [21]) [21]
8.8 <17> [18]
16 <1> [1]
31.49 <17> [22]
35.3 <3> [4]
183 <11> [16]

K_m-Value (mM)

0.012 <8> ((1,4-α-D-galacturonide)$_n$) [11]
0.012-0.015 <2> (pectic acid) [14]

0.032 <16> ((1,4-α-D-galacturonide)$_n$) [2]
0.074-0.159 <2> ((1,4-α-D-galacturonide)$_8$, <2> heptagalacturonate [14]) [14]
0.12 <9> (pectic acid, <9> enzyme form PGII [21]) [21]
0.24 <9> (pectic acid, <9> enzyme form PGI [21]) [21]
0.32 <19> ((1,4-α-D-galacturonide)$_7$) [24]
0.423 <2> (pectic acid) [12]
0.43 <2> ((1,4-α-D-galacturonide)$_n$) [2]
0.44 <19> ((1,4-α-D-galacturonide)$_6$) [24]
0.49 <18> (pectin) [19, 20]
0.53 <19> ((1,4-α-D-galacturonide)$_5$) [24]
0.63-0.97 <2> ((1,4-α-D-galacturonide)$_3$) [14]
0.64 <17> (pectic acid) [22]
0.68 <19> ((1,4-α-D-galacturonide)$_3$) [24]
0.9 <18> (pectic acid) [19, 20]
0.94 <19> ((1,4-α-D-galacturonide)$_3$) [24]
1.44 <19> ((1,4-α-D-galacturonide)$_2$) [24]
2.2-2.8 <2> ((1,4-α-D-galacturonide)$_2$) [14]
2.24 <4> ((1,4-α-D-galacturonide)$_2$) [6]
3.1 <19> ((1,4-α-D-galacturonide)$_n$, <19> recombinant enzyme [24]) [24]
3.2 <19> ((1,4-α-D-galacturonide)$_n$, <19> wild type enzyme [24]) [24]
6 <17> ((1,4-α-D-galacturonide)$_n$) [18]

pH-Optimum

3 <12, 13> [3]
3.6-3.8 <16> [2, 9]
4.2 <19> [24]
4.4 <7> (<7> depending upon chain length, digalacturonate, tetragalacturonate [10]) [10]
4.5-5 <8> (<8> polygalacturonic acid [11]) [11]
4.5-5.5 <2, 3> (<2, 3> polygalacturonate [4, 14]) [4, 14]
4.6 <7> (<7> depending upon chain length, hexagalacturonate, heptagalacturonate [10]) [10]
4.7 <2> [14]
4.8 <2> [2]
4.9 <7> (<7> polygalacturonate [10]) [10]
5 <2, 3, 9, 14, 17, 18> (<9> all 3 forms of enzyme [21]) [3, 4, 18, 19-21, 23]
5.1 <2> (<2> pectate, native enzyme [5, 12]) [5, 12]
5.2 <9> (<9> digalacturonate [13]) [13]
5.3 <1, 2> (<1> polygalacturonate, pH optimum independend of Ca^{2+} concentration [1]; <2> immobilized enzyme [5]) [1, 5]
5.5 <10> (<10> polygalacturonate [15]) [14, 15]
5.6 <4, 17> (<4> pectate [6]) [6, 22]
6.3-6.6 <15> (<15> digalacturonate [17]) [17]

pH-Range

3-5 <13, 19> [3, 24]
3.5-6.5 <17> [18]

3.5-7 <18> [19]
4-7.5 <1> (<1> polygalacturonate [1]) [1]

Temperature optimum (°C)
40 <17, 18> [18-20]
45 <9> (<9> pectate [13]) [13]
50 <2, 7> (<2> pectate [5]; <7> digalacturonate [10]) [5, 10]
60 <17> [22]
60-70 <2> (<2> pectate, optimum temperature for enzyme action [12]) [12]
70 <2> [23]

Temperature range (°C)
30-50 <19> (<19> 50°C 50% activity retained [24]) [24]
30-60 <9> [13]
60 <7> (<7> not active above, digalacturonate [10]) [10]

4 Enzyme Structure

Molecular weight
41000 <10> (<10> PGII, gel filtration [15]) [15]
47000 <3> (<3> gel filtration [4]) [4]
48000 <2> (<2> gel filtration [2, 23]) [2, 23]
51000 <1> (<1> gel filtration [1]) [1]
53000 <18> (<18> gel filtration [19, 20]) [19, 20]
55000 <2> (<2> gel filtration [14]) [14]
58000 <8, 17> (<8, 17> gel filtration [11, 18]) [11, 18]
68000 <10, 17> (<10> PGI, gel filtration [14, 15]; <17> SDS-PAGE [22]) [14, 15, 22]
76000 <16> (<16> gel filtration [2, 9]) [2, 9]
78000 <19> (<19> SDS-PAGE [24]) [24]

5 Isolation/Preparation/Mutation/Application

Source/tissue
cortical tissue <8> [11]
fruit <3> [4]
peel <7> [10]
pollen <1> [1]
root <2> [2, 12, 14]

Localization
extracellular <9, 11, 17-19> [13, 16, 18, 19, 20, 22, 24]
intracellular <4> [6]
periplasm <18> [19]

Purification
<1> (partially [1]) [1]

<2> (partially [14]) [2, 5, 12, 14, 23]
<3> [4]
<4> [6]
<5> [7]
<6> [8]
<7> [10]
<8> (partially [11]) [11]
<9> (partially [13]) [13, 21]
<10> (partially [14]) [14, 15]
<11> (partially, preparation contains contaminating activities [16]) [16]
<15> [17]
<16> [2, 9]
<17> [18, 21]
<18> [19]
<19> (wild-type and recombinant enzyme [24]) [24]

Cloning

<19> (gene pgaX cloned by reverse genetics, pIM361, overexpression of the enzyme by transformants [24]) [24, 25]
<20> (gene pehY, pYEII-4-H192, cloned and expressed in E. coli [26]) [26]

Application

analysis <19> (<19> studying fine structure of xylogalacturonan in pectin [25]) [25]
nutrition <9, 19> (<9> enzymes degrading pectic substances to improve cloud stability of fruit and vegetable nectars, clarification of fruit juice and wine, production of baby food, commercial preparations of pectolytic enzymes for food processing industry [21]; <19> potent enzyme for production of D-galacturonate from pectin, starting compound in synthesis of ascorbic acid [25]) [21, 25]
synthesis <9, 19> (<9> pectic enzymes used to generate protoplasts in cell cultures [21]; <9> as adjunct to cellulases and hemicellulases in cellulosic biomass treatment [21]; <19> pectic enzymes industrially important in processing of agricultural products, production of pectinolytic enzymes, specifical enzyme for modification of pectins [24]; <19> preparing methylated pectins [25]) [21, 24, 25]

6 Stability

pH-Stability

3-6.5 <16> [2, 9]
3.5-6.5 <17> [18]
3.5-7 <2> [2]
5.2 <9> [13]

Temperature stability

30 <2> (<2> not stable above [5]) [5]

40 <9> (<9> not stable above [13]) [13]
50 <7> (<7> rapid inactivation at temperatures above 60°C [10]) [10]
55 <2, 16> (<2, 16> 30 min barely inactivates enzyme, one-half of activity destroyed when heated to 65°C, completely destroyed at 80°C [2, 9]) [2, 9]
60 <17> (<17> stable below 60°C, rapidly inactivated above 60°C, 10 min at 65°C, 40% activity lost [22]) [22]
60-70 <2> (<2> highly stable [12]) [12]

General stability information

<2>, stable, even at high temperatures, relatively stable in acidic and neutral, unstable in alkaline media [12]
<6>, rather unstable [8]
<7>, stable in acidic medium [10]
<17>, stable at room temperature for 24 h [18]

Storage stability

<2>, -20°C, 2 months [2]
<9>, 0-4°C, 3 months 90% activity retained, PGII 28°C 24 h 85% activity retained, PGI is unstable, cannot be stored [21]
<16>, -20°C, 1 month [2, 9]

References

[1] Pressey,R.; Reger, B.J.: Polygalcturonase in pollen from corn and other grasses. Plant Sci., **59**, 57-62 (1989)
[2] Konno, H.: Galacturan 1,4-α-galacturonidase from carrot Daucus carota and liverwort Marchantia polymorpha. Methods Enzymol., **161**, 373-380 (1988)
[3] Ikotun, T.; Balogun, O.: In vitro and in vivo production of pectolytic enzymes by some phytopathogenic fungi. J. Basic Microbiol., **27**, 347-354 (1987)
[4] Pressey, R.: Exopolygalacturonase in tomato fruit. Phytochemistry, **26**, 1867-1870 (1987)
[5] Heinrichova, K.; Lehoczki, M.; Zliechovcova, D.: Immobilization of exo-D-galacturonanase by coupling to a polyacrylamide type support. J. Gen. Microbiol., **131**, 2291-2298 (1986)
[6] Heinrichova, K.; Wojciechowicz, M.; Ziolecki, A.: An exo-D-galacturonanase of Butyrivibrio fibrisolvens from the bovine rumen. J. Gen. Microbiol., **131**, 2053-2058 (1985)
[7] Markovic, O.; Slezarik, A.; Labudova, I.: Purification and characterization of pectinesterase and polygalacturonase from Trichoderma reesei. FEMS Microbiol. Lett., **27**, 267-271 (1985)
[8] Ikotun, T.: Isolation, purification and assay of the macerating enzyme produced by Penicillium oxalicum Curie and Thom. Z. Allg. Mikrobiol., **24**, 247-252 (1984)
[9] Konno, H.; Yamasaki, Y.; Katoh, K.: Exopolygalacturonase from suspension cultures of Marchantia polymorpha. Plant Physiol., **73**, 216-222 (1983)

[10] Heinrichova, K.: Purification and a partial characterization of exo-D-galacturonanase of banana. Collect. Czech. Chem. Commun., **47**, 2433-2439 (1982)
[11] Bartley, I. M.: Exopolygalacturonase of apple. Phytochemistry, **17**, 213-216 (1978)
[12] Heinrichova, K.: Isolation, characterization and mode of action of exo-D-galacturonanase from carrot. Collect. Czech. Chem. Commun., **42**, 3214-3221 (1977)
[13] Heinrichova, K.; Rexova-Benkova, L.: Purification and characterization of an extracellular exo-D-galacturonanase of Aspergillus niger. Biochim. Biophys. Acta, **422**, 349-356 (1976)
[14] Pressey, R.; Avants, J.K.: Modes of action of carrot and peach exopolygalacturonases. Phytochemistry, **14**, 957-961 (1975)
[15] Pressey, R.; Avants, J.K.: Separation and characterization of endopolygalacturonase and exopolygalacturonase from peaches. Plant Physiol., **52**, 252-256 (1973)
[16] Keegstra, K.; English, P.D.; Albersheim, P.: Four glycosidases secreted by Colletotrichum lindemuthianum. Phytochemistry, **11**, 1873-1880 (1972)
[17] Hasegawa, S.; Nagel, C.W.: Isolation of an oligogalacturonate hydrolase from a Bacillus species. Arch. Biochem. Biophys., **124**, 513-520 (1968)
[18] Martinez, M.J.; Alconada, M.T.; Gillen, F.; Vasquez, C.; Reyes, F.: Pectic activities from Fusarium oxysporum f. sp. melonis: Purification and characterization of an exopolygalacturonase. FEMS Microbiol. Lett., **81**, 145-150 (1991)
[19] Pardo, C.; Lapena, M.A.; Gacto, M.: Purification and characterization of an extracellular exopolygalacturonase from Geotrichum lactis. Can. J. Microbiol., **37**, 974-977 (1991)
[20] Pardo, C.; Gacto, M.: Production and properties of a nexopolygalacturonic acid hydrolase from Geotrichum lactis. Microbios, **71**, 193-202 (1992)
[21] Behere, A.; Satyanarayan, V.; Pawal-Desai, S.R.: Separation and limited characterization of three polygalacturonases from Aspergillus niger. Enzyme Microb. Technol., **15**, 158-161 (1993)
[22] Vasquez, C.; Patino, B.; Martinez, M.J.: Purification and characterization of an exopolygalacturonase produced by Fusarium oxysporum f. sp. radicis lycopersici. FEMS Microbiol. Lett., **110**, 193-196 (1993)
[23] Dzurova, M.; Linek, K.; Stratilova, E.: Inhibition of exopolxgalacturonase from carrot by reaction products and by their basic analogues. Biologia (Bratisl.), **50**, 595-599 (1995)
[24] Kester, H.C.M.; Kusters-Van Someren, M.A.; Mueller, Y.; Visser, J.: Primary structure and characterization of an exopolxgalacturonase from Aspergillus tubingensis. Eur. J. Biochem., **240**, 738-746 (1996)
[25] Kester, H.C.M.; Benen, J.A.E.; Visser, J.: The exopolygalacturonase from Aspergillus tubingensis is also active on xylogalacturonan. Biotechnol. Appl. Biochem., **30**, 53-57 (1999)
[26] Liao, C.-H.; Revear, L.; Hotchkiss, A.; Savary, B.: Genetic and biochemical characterization of an exopolygalacturonase and a pectate lyase from Yersinia enterolitica. Can. J. Microbiol., **45**, 396-403 (1999)

Isoamylase 3.2.1.68

1 Nomenclature

EC number
3.2.1.68

Systematic name
glycogen α-1,6-glucanohydrolase

Recommended name
isoamylase

Synonyms
Cythophaga isoamylase <9> [7]
Pseudomonas isoamylase <3> [8]
SU1 isoamylase <7> [16]
alkaline isoamylase <4> (<4> alkaliphilic Bacillus sp KSM-3309, isolated from soil [11, 14]) [11, 14]
α 1,6-glucan hydrolase [16]
bacterial isoamylase <3, 4> [14, 15]
debranching enzyme
glycogen 6-glucanohydrolase <2, 3, 4> [1, 3, 11]
isoamylase II <7> [16]
maize Sugary-1 isoamylase <7> [16]
pancreatic isoamylase isoA <5> [13]
plant isoamylase (enzyme from plant sources attacks amylopectin and its β-dextrin but no oligosaccharide α-dextrin) [7]
potato isoamylase <1> [2]
rice isoamylase <8> [17]
yeast isoamylase [7]

CAS registry number
9067-73-6

2 Source Organism

<1> *Solanum tuberosum* (potato, L. cul. var. "Danshaku" [2]
<2> *Lipomyces kononenkoae* (yeast, strain IGC 4051) [3]
<3> *Pseudomonas amyloderamosa* (gram-negative soil bacterium, strain SB-15 [7, 8]; SMP1 [11, 15]; strain MI-414, isomylase-hyperproducing mutant of SB-15 [9]; strain JD210, isoamylase-hyperproducing mutant [15]; strain

WU-5315, isoamylase hyperproducing mutant derived from JD-210 [12]; strain KlC, revertant strain of Kl [4, 5]) [1, 4-12, 15]
<4> *Bacillus sp.* (KSM.3309, gram-positive, isolated from soil [11, 14]; KSM-K16 [14]) [11, 14]
<5> *Homo sapiens* [13]
<6> *Flavobacterium sp.* [15]
<7> *Zea mays* [16]
<8> *Oryza sativa* (L. cv. Fujikari) [17]
<9> *Cytophaga sp.* [7]

3 Reaction and Specificity

Catalyzed reaction
hydrolysis of 1,6-α-glucosidic linkages in glycogen, amylopectin and their β-limit dextrins

Reaction type
hydrolysis of O-glycosyl bond

Natural substrates and products
S starch + H_2O <1-3, 7> [1-8, 16]
P ?

Substrates and products
S D-glucose + H_2O <2> (<2> 59% of the activity with dextrin as substrate [3]) [3]
P ?
S α-D-glucan + H_2O <3> (<3> cleavage of α-D-glucosidic linkages of certain branched α-D-glucans [1]) [1]
P ?
S amylopectin + H_2O <1-4, 7, 8> (<1-4, 7> or the β-limit dextrins, cleaves the branching points completely in vitro [1-8, 10, 11, 16]; <3> waxy maize amylopectin, potato amylopectin, amylopectin β-dextrin [7]; <4> specific cleavage of α 1,6-glucosidic inter-chain linkages producing essentially linear chains [11]) [1-8, 10, 11, 16, 17]
P maltose + maltooligosaccharides <1-3> (<1, 3> hydrolysis of β-limit dextrins forming maltotriose as main reaction product [2, 8, 10]) [1-8, 10]
S amylopectin phi-dextrin + H_2O <3> (<3> substrate on which all 3 types of debranching enzymes act [7]) [7]
P maltotetraose <3> [7]
S amylose + H_2O <1> (<1> short chain [2]) [2]
P ?
S branched chain oligosaccharides + H_2O <3> [8]
P maltotriose + higher maltosaccharides <3> (<3> no detectable maltose released, characteristic action is hydrolysis of α 1,6-glucosidic linkages, but α 1,6 glucosidic linkages attaching maltose residues to chains of α 1,4-linked glucose residues are not readily hydrolyzed [8]) [8]

S dextrin + H_2O <2> (<2> enzyme activity 100% [3]) [3]
P ?
S glycogen + H_2O <1-4, 7, 8> (<1-4, 7> or the β-limit dextrins; cleaves the branching points completely in vitro [1-8, 11, 16]; <3> oyster glycogen, glycogen β-dextrin, rabbit liver glycogen [7]) [1-8, 11, 16, 17]
P maltose + maltooligosaccharides <1-3> [1-8]
S maltopentaose + H_2O <1> (<1> sugar immobilized on Sepharose 4B [2]) [2]
P ?
S maltose + H_2O <3, 6> (<3, 6> iam gene induced by maltose [9, 15]) [8, 9, 15]
P ?
S maltotriose + H_2O <2> (<2> activity 28% [3]) [3]
P ?
S panose + H_2O <2> (<2> acts very slowly, small amounts of product formed after 24 h [3]) [3]
P D-glucose + maltose <2> [3]
S phythoglycogen + H_2O <1, 7, 8> (<1> highly branched polysaccharide, complete hydrolysis of the α-1,6-glucosidic bonds [2]) [2, 16, 17]
P ?
S saccharose + H_2O <2> (<2> activity 67% [3]) [3]
P ?
S starch + H_2O <1, 2> (<1> soluble starch, immobilized on Sepharose 4B [2]) [2, 3, 16]
P ?
S Additional information <1, 3> (<1, 3> distinguished from α-dextrin endo-1,6-α-glucosidase, EC 3.2.1.41, by the inability of isoamylase to attack pullulan, and by limited action on α-limit dextrins, action on glycogen however is complete in contrast to limited action by α-dextrin glucanohydrolase, 1,6-linkage hydrolysed only at branch point [1, 2]; direct debranching enzyme, attacks unmodified glycogen and/or amylopectin with hydrolysis of the 1,6-bond, smallest substrate is not known [1, 2, 7]) [1, 2, 7]
P ?

Inhibitors

$(NH_4)_2Mo_2O_7$ <8> (<8> 1 mM 22% inhibition, 10 mM 100% inhibition [17]) [17]
2,4-dinitrofluorobenzene <3> [6]
2-hydroxy-5-nitrobenzyl bromide <3> [6]
Ag^{2+} <3> [6]
Cu^{2+} <3> (<3> relative activity 78% [6]) [6]
$CuCl_2$ <3, 8> (<3> 1 mM, 30% inhibition [8]; <8> 0.1 mM reduce activity to 3% [17]) [8, 17]
D-glucono-δ-lactone <3> (<3> relative activity 87% [6]) [6]
D-glucose <3> (<3> 10 mM, 0.4% amylose, relative activity 95% [6]) [6]
Hg^{2+} <3, 4> (<4> 52% inhibition [11]) [6, 11]

$HgCl_2$ <3, 8> (<3> 1mM, 41% inhibition; 0.1 mM 34% inhibition [8]; <8> 1 mM, complete inhibition [17]) [8, 17]
N-bromosuccinimide <3, 4> (<4> 0.01 mg abolishes enzyme activity almost completely [11]) [6, 11]
NaF <3> (<3> 1 mM, 19% inhibition [8]) [8]
SDS <3> (<3> 0,1%, activity completely lost [6]) [6]
amylose <3> [6]
β-cyclodextrin <8> (<8> 29%, slightly inhibitory at 10 mM [17]) [17]
cellobiose <3> (<3> 10 mM, 0.4% amylose, relative activity 88% [6]) [6]
corn amylose <3> [6]
cyclomaltoheptaose <3> (<3> 10mM, 0.4% amylose, relative activity 93% [6]) [6]
guanidine hydrochloride <3> (<3> 3 M, activity completely lost [6]) [6]
iodine <3> (<3> relative activity 93% [6]) [6]
iodoacetate <3, 8> (<3> 10 mM, 13% inhibition [8]; <8> 21%, slightly inhibition [17]) [6, 8, 17]
isomaltose <3> (<3> 10 mM, 0.4% amylose, relative activity 82% [6]) [6]
maltose <3> (<3> 10 mM, inhibition slightly, with 0.4% amylose relative activity 90% [6]) [6]
maltotetraose <3> (<3> 10 mM, 0.4% amylose, relative activity 50% [6]) [6]
maltotriose <3> (<3> competitive inhibitor [5]; 10 mM, 0.4% amylose, relative activity 53% [6]) [5, 6]
oligosaccharides with *α*-1,4-glucosidic linkages <3> [6]
p-chloromercuribenzoate <1, 3, 8> (<1> reversible inactivation [2]; <3> 1 mM 34% inhibition [8]; <3> 1 mM, complete inhibition [8, 17]) [2, 8, 17]
phenylmercuriacetate <3> [6]
succinic anhydride <3> (<3> relative activity 94% [6]) [6]
urea <3> (<3> 20% activity retained [6]) [6]
xylose <3> (<3> 10 mM, 0.4% amylose, relative activity 97% [6]) [6]

Activating compounds

Triton X-100 <1> (<1> reagent in the preincubation mixture at concentrations 0.05-1% activates the enzyme by 45-72% [2]) [2]

Specific activity (U/mg)

3 <6> (<6> pullulan, recombinant isoamylase [15]) [15]
8 <1> [2]
50 <6> [15]
51.2 <8> (<8> rabbit liver glycogen [17]) [17]
52 <2> [3]
54.6 <6> (<6> amylose, recombinant isoamylase [15]) [15]
59 <3> [9]
59.1 <3> [8]
120 <6> (<6> potato starch, recombinant isoamylase [15]) [15]
154 <6> (<6> rice starch, recombinant isoamylase [15]) [15]
172 <6> (<6> corn amylopectin, recombinant isoamylase [15]) [15]
174 <6> (<6> rabbit liver glycogen, recombinant isoamylase [15]) [15]
182 <6> (<6> oyster glycogen, recombinant isoamylase [15]) [15]

190 <3> [5]
469 <4> [11]

K_m-Value (mM)

Additional information <4, 8>(<4> potatoamylopectin 0.59 mg/ml-1, oyster glycogen 0.45 mg/ml-1 [11]; <8> amylopectin 7.2 mg/ml-1, phytoglycogen 11.9 mg/ml-1, β-limit dextrin 13.1 mg/ml-1, glycogen 21.8 mg/ml-1 [17]) [11, 17]

pH-Optimum

3-4 <3> (<3> starch, 0.1 M acetate-HCl buffer, pH 3.5, 0.1 M phosphate-citric acid buffer, pH 4 [8]) [8, 15]
5.5-6 <1> (<1> amylopectin [2]) [2]
5.6 <2> (<2> starch [3]) [3]
6-7 <7> [16]
6.3 <6> (<6> only isoamylase with neutral pH optimum [15]) [15]
6.5-7 <8> [17]
9 <4> [11, 14]

pH-Range

1-5 <3> (<3> starch, 0.1 M acetate-HCl buffer [8]) [8]
2.2-7.8 <3> (<3> 0.1 M phosphate-citric acid buffer [8]) [8]
4-8 <1> (<1> not active below, amylopectin [2]) [2]
6-7 <6> [15]
7.5-9.5 <4> [11]

Temperature optimum (°C)

25-37 <7> [16]
30 <2, 8> (<2> starch [3]) [3, 17]
52 <3> (<3> starch, 10 min reaction at pH 3.5 [8]) [8]
55 <4> [11]

Temperature range (°C)

30-60 <3> (<3> not active above, starch [8]) [8]
60 <1> (<1> not active above, amylopectin [2]) [2]

4 Enzyme Structure

Molecular weight

50000 <3> (<3> 2 subunits, not linked covalently [6]) [6]
65000 <2, 4> (<2> gel filtration, 0.5-m Bio-Gel A [3]; <4> SDS-PAGE, gel filtration [11]) [3, 11]
75000 <7> (<7> SU1 [16]) [16]
79000 <7> (<7> maize gene su1 recombinant polypeptide, expressed in E. coli [16]) [16]
80000 <3> [10]
80810 <3> (<3> SB-15, DNA sequence analysis [1]) [1]

83000 <6, 8> (<6> SDS-PAGE, co-purified with a second minor band of 75000, probably C-terminal truncation of the native enzyme in vivo or during purification [15]) [15, 17]
86000-94000 <3> (<3> SB-15, gel filtration, SDS-PAGE [1]; 86000, undegraded polypeptide chain, SDS-PAGE [4]) [1, 4, 15]
88000 <3> (<3> calculated by comparison with standard proteins [4]) [4]
90000 <3> (<3> undegraded polypeptide chain [4]) [4]
94000 <3> (<3> sedimentation equilibrium [6]; <3> native enzyme, centrifugation analysis [4, 6]) [4, 6-8]
95000 <3> (<3> gel electrophoresis, low speed sedimentation [8]) [8]
120000 <9> [7]
141000 <7> (<7> isoamylase II, gel filtration [16]) [16]
178000 <1> (<1> native enzyme [2]) [2]
340000 <8> (<8> native enzyme, gel filtration, TSKgel G3000SWxl [17]) [17]
490000 <8> (<8> native enzyme, gel filtration, TSKgel G4000SWxl [17]) [17]
520000 <1> (<1> gel filtration [2]) [2]

Subunits

dimer <1, 3> (<1> 1 * 83000 + 1 * 95000 [2]; <3> 2 * 46000, sedimentation equilibrium; 2 * 52000, gel filtration [4]; dimers probably obtained by partial proteolytic degradation [4]; 2 * 50000, not linked covalently [6]) [2, 4, 6]
monomer <3, 4> (<3> gel electrophoresis [4]; <4> purified enzyme has monomeric structure [11]) [4, 11]
oligomer <8> (<8> probably homo-tetramer and homohexamer [17]) [17]

5 Isolation/Preparation/Mutation/Application

Source/tissue

endosperm <7> [16]
kernel <7> [16]
tuber <1> [2]

Localization

extracellular <2, 3, 6> [1, 3, 5, 8, 15]

Purification

<1> [2]
<2> [3]
<3> [5, 8, 9, 10, 12]
<4> [11]
<6> [15]
<7> [16]
<8> [17]

Crystallization

<3> (crystallized from ammonium sulfate solution, crystals belong to orthorhombic space group P212121 [10]) [1, 4-6, 10]

Cloning

<3> (<3> plasmid pIAM275, pUC9 carrying iam gene, E. coli, recominant plasmid does not direct the synthesis of isoamylase in E. coli, sequenced [9]) [1, 9]
<6> (<6> iam gene cloned and expressed in E. coli JM101, pMON17481, production of a nonsecreted signal peptide-less isoamylase [15]) [15]
<7> (<7> gene su1, expressed in E. coli, pAR4) [16]
<8> (<8> isoamylase gene is a single copy in rice genome, located on chromsome 8 [17]) [17]

Application

analysis <3, 9> (<3> elucidating of the finestructures of various polysaccharides [1, 5, 9]; <9> structural determination of glycogen and starch components [7]) [1, 5, 7, 9, 15]
detergent <4> (<4> effective additive in dishwashing and laundry detergents [11, 14]) [11, 14]
medicine <5> (<5> diagnosis of acute pancreatitis in humans [13]) [13]
nutrition <2, 3, 4> (<3> industrial production of amylose, maltose and D-glucose from starch, alone or in combination with β-amylase and glucoamylase production [1, 5, 9]; <2> production of high-maltose syrups and highly purified maltose by a combination of debranching enzymes [3]; <6> industrial production of various syrups from starch [15]; <3> strain MI-414 for industrial use, produces 10fold more isoamylase than SB-15 [9]) [1, 3, 5, 9, 11]

6 Stability

pH-Stability

2.5-7.5 <3> [8]
3.5-5.5 <3> [5]
5-11.5 <4> (<4> more than 50% of the original activity detectable between pH 6.6-10.5 [11]) [11]
6.5-7 <6> (<6> assayed at 22°C enzyme displays broader activity and stability optima of pH 5.0 -7.5 [15]) [15]

Temperature stability

10-40 <8> (<8> temperature treatment does not influence the enzyme activity [17]) [17]
40 <4> (<4> stable up to 40°C, at temperature below 30°C and above 65°C enzyme is almost inactive [11]) [11]
45 <1, 3> (<3> not stable above [8]) [2, 8]
50 <1> (<1> amylopectin, incubated in the absence of substrate enzyme stable only up to 45°C [2]) [2]
60 <3> (<3> activity decreased about 95% [8]) [8]
65 <3> (<3> heating 10 min at 65°C results in complete loss of activity [8]) [8]

General stability information

<2>, very unstable, even at low temperatures [3]
<3>, quite stable in maltose-containing buffer [12]

Storage stability

<3>, 4°C for 5 months 85% enzyme activity retained, preservation with 0.1% potassium sorbate isoamylase activity was not reduced until 2 months of incubation [12]
<3>, 4°C several months [8]

References

[1] Amemura, A.; Chakraborty, R.; Fujita, M.; Noumi, T.; Masamitsu, F.: Cloning and nucleotide sequence of the isoamylase gene from Pseudomonas amyloderamosa SB-15. J. Biol. Chem., **263**, 9271-9275 (1988)
[2] Ishizaki, Y.; Taniguchi, H.; Maruyama, Y.; Nakamura, M.: Debranching enzymes of potato tubers (Solanum tuberosum L.). I. Purification and some properties of potato isoamylase. Agric. Biol. Chem., **47**, 771-779 (1983)
[3] Spencer-Martins, I.: Extracellular isoamylase produced by the yeast Lipomyces kononenkoae. Appl. Environ. Microbiol., **44**, 1253-1257 (1982)
[4] Amemura, A.; Konishi, Y.; Harada, T.: Molecular weight of the undegraded polypeptide chain of Pseudomonas amyloderamosa isoamylase. Biochim. Biophys. Acta, **611**, 390-393 (1980)
[5] Kato, K.; Konishi, Y.; Amemura, A.; Harada, T.: Affinity chromatography of Pseudomonas isoamylase on cross-linked amylose gel. Agric. Biol. Chem., **41**, 2077-2080 (1977)
[6] Kitagawa, H.; Amemura, A.; Harada, T.: Studies on the inhibition and molecular properties of crystalline Pseudomonas isoamylase. Agric. Biol. Chem., **39**, 989-994 (1975)
[7] Lee, E.Y.C.; Whelan, W.J.: Glycogen and starch debranching enzymes. The Enzymes, 3RD ED. (Boyer, P.D., ED.), **5**, 191-234 (1972)
[8] Yokobayashi, K.; Misaki, A.; Harada, T.: Purification and properties of Pseudomonas isoamylase. Biochim. Biophys. Acta, **212**, 458-469 (1970)
[9] Fujita, F.; Sakai, S.; Futai, M.; Amemura, A.: Characterization of an isoamylase-hyperproducing mutant of Pseudomonas amyloderamosa. Agric. Biol. Chem., **54**, 2315-2321 (1990)
[10] Katsuya, Y.; Mezaki, Y.; Kubota, M.; Matsuura, Y.: Three-dimensional structure of Pseudomonas isoamylase at 2.2 A resolution. J. Mol. Biol., **281**, 885-897 (1998)
[11] Ara, K.; Saeki, K.; Ito, S.: Purification and characterization of an alkaline isomylase from an alkalophilic strain of Bacillus. J. Gen. Microbiol., **139**, 781-786 (1993)
[12] Lin, L.-L.; Fang, T.-Y.; Chu, W.-S.; Hsu, W.-H.: Improved elution of isoamylase adsorbed on raw starch and the preservation of purified enzyme. Lett. Appl. Microbiol., **19**, 383-385 (1994)

[13] Clave, P.; Guillaumes, S.; Blanco, I.; Nabau, N.; Merce, J.; Farre, A.; Marruecos, L.; Lluis, F.: Amylase, lipase, pancreatic isoamylase, and phospholipase A in diagnosis of acute pancreatitis. Clin. Chem., **41**, 1129-1134 (1995)
[14] Ito, S.; Kobayashi, T.; Ara, K.; Ozaki, K.; Kawai, S.; Hatada, Y.: Alkaline detergent enzymes from alkaliphiles: enzymatic properties, genetics, and structures. Extremophiles, **2**, 185-190 (1998)
[15] Krohn, M.; Barry, G.F.; Kishore, G.M.: An isoamylase with neutral pH optimum from a Flavobacterium species: cloning, characterization and expression of the iam gene. Mol. Gen. Genet., **254**, 469-478 (1997)
[16] Rahman, A.; Wong, K.-s.; Jane, J.-l.; Myers, A. M.; James, M. G.: Characterization of SU1 isoamylase, a determinant of storage starch structure in maize. Plant Physiol., **117**, 425-435 (1998)
[17] Fujita, N.; Kubo, A.; Francisco, P. B. Jr.; Nakakita, M.; Harada, K.; Minaka, N.; Nakamura, Y.: Purification, characterization, and cDNA structure of isoamylase from developing endosperm of rice. Planta, **208**, 283-293 (1999)

Amylopectin 6-glucanohydrolase 3.2.1.69

1 Nomenclature

EC number

3.2.1.69 (deleted, included in EC 3.2.1.41)

Recommended name

amylopectin 6-glucanohydrolase

Glucan 1,6-α-glucosidase 3.2.1.70

1 Nomenclature

EC number
3.2.1.70

Systematic name
1,6-α-D-glucan glucohydrolase

Recommended name
glucan 1,6-α-glucosidase

Synonyms
α-1,6-glucosidase
dextran glucosidase
exo-1,6-α-glucosidase
exodextranase
glucodextranase

CAS registry number
37288-48-5

2 Source Organism

<1> *Arthrobacter globiformis* [1, 2, 6, 7, 9]
<2> *Lipomyces lipofer* [4]
<3> *Streptococcus mitis* [5, 8]
<4> *Brevibacterium fuscum var. dextranlyticum* [10]
<5> *Achromobacter sp.* [11]
<6> *Bacteroides oralis* [3]
<7> *Streptococcus mutans* [12, 13]

3 Reaction and Specificity

Catalyzed reaction
hydrolysis of successive glucose residues from 1,6-α-D-glucans and derived oligosaccharides

Reaction type
hydrolysis of O-glycosyl bond

Natural substrates and products

S dextrans + H_2O <1-6> [1-11]
P β-D-glucose <1-6> [1-11]

Substrates and products

S dextrans + H_2O <1-7> [1-13]
P β-D-glucose <1-7> [1-13]
S isomaltodextrins + H_2O <1-6> [1-11]
P β-D-glucose <1-6> [1-11]
S isomaltopentaose + H_2O <7> [12]
P β-D-glucose <7> [12]
S isomaltose + H_2O <7> [12, 13]
P β-D-glucose <7> [12, 13]
S isomaltotetraose + H_2O <7> [12]
P β-D-glucose <7> [12]
S isomaltotriose + H_2O <7> [12, 13]
P β-D-glucose <7> [12, 13]
S panose + H_2O <7> [13]
P β-D-glucose <7> [13]

Inhibitors

Cu^{2+} <1, 4> [1, 2, 7, 10]
D-glucose <2> [4]
EDTA <1> [7]
Fe^{3+} <1> [1, 2]
Hg^{2+} <1, 4> [1, 2, 10]
$KMnO_4$ <1> [1, 2]
N-bromosuccinimide <1, 4> [1, 2, 10]
Pb^{2+} <1> [1, 2, 7]
Zn^{2+} <1> [1, 2, 7]
iodine <4> [10]
methyl-α-D-glucoside <1> [2]
p-chloromercuribenzoate <4> [10]
phenyl α-D-glucoside <1> [2]

Turnover number (min^{-1})

5714 <1> (dextran) [2]

Specific activity (U/mg)

25.6 <1> [1, 2]
34.3-38.4 <3, 4> [8, 10]
50 <2> [4]

K_m-Value (mM)

0.012 <2> (dextran) [4]

pH-Optimum

4-4.4 <5> [11]
4.5-5 <2> [4]
6 <1> [1, 2, 9]

pH-Range
3-8 <2, 5> [4, 11]

Temperature optimum (°C)
45 <1, 2> [1, 2, 4]

Temperature range (°C)
60 <2> (<2> not active above [4]) [4]

4 Enzyme Structure

Molecular weight
29000 <2> (<2> gel filtration [4]) [4]
70000 <1> (<1> gel filtration [6]) [6]
120000 <1> (<1> PAGE [2]) [2]

Subunits
monomer <1> (<1> SDS-PAGE [1, 6]) [1, 6]

5 Isolation/Preparation/Mutation/Application

Localization
cytoplasm <3, 6> [3, 5]
extracellular <1, 2, 5> [1, 2, 4, 6, 11]

Purification
<1> [1, 2, 6, 7, 9]
<2> [4]
<3> [8]
<4> [10]
<5> (<5> partially [11]) [11]

Crystallization
<1> [6]

Cloning
<7> [12]

6 Stability

pH-Stability
4.5-6 <2> [4]
5-11 <4> [10]
5.5-7.5 <1> [2]

Temperature stability
55 <1, 2, 4> (<1, 2, 4> not stable above [2, 4, 10]) [2, 4, 10]

References

[1] Okada, G.; Unno, T.: A Glucodextranase accompanied by glucoamylase activity from Arthrobacter globiformis I42. Agric. Biol. Chem., **53**, 223-228 (1989)
[2] Okada, G.; Unno, T.; Sawai, T.: A simple purification procedure for and further characterization of a glucodextranase from Arthrobacter globiformis I42. Agric. Biol. Chem., **52**, 2169-2176 (1988)
[3] Takahashi, N.; Satoh, Y.; Takamori, K.: Subcellular localization of D-glucanases in Bacteroides oralis Ig4a. J. Gen. Microbiol., **131**, 1077-1082 (1985)
[4] Ramos, A.; Spencer-Martins, I.: Extracellular glucose-producing exodextranase of the yeast Lipomyces lipofer. Antonie Leeuwenhoek, **49**, 183-190 (1983)
[5] Linder, L.; Andersson, C.; Sund, M.L.; Shockman, G.D.: Protoplast formation and localization of enzymes in Streptococcus mitis. Infect. Immun., **40**, 1146-1154 (1983)
[6] Ohya, T.; Suzuki, H.; Sawai, T.: Crystallization of exo-1,6-α-glucosidase (glucodextranase) from Arthrobacter globiformis I42. Agric. Biol. Chem., **44**, 203-204 (1980)
[7] Ohya, T.; Sawai, T.; Uemura, S.; Abe, K.: Some catalytic properties of an exo-1,6-α-glucosidase (glucodextranase) from Arthrobacter globiformis I42. Agric. Biol. Chem., **42**, 571-577 (1978)
[8] Wheatley, M.A.; Moo-Young, M.: Degradation of polysaccharides by endo- and exoenzymes: dextran-dextranase model systems. Biotechnol. Bioeng., **19**, 219-233 (1977)
[9] Sawai, T.; Yamaki, T.; Ohya, T.: Purification and some properties of Arthrobacter globiformis exo-1,6-α-glucosidase. Agric. Biol. Chem., **40**, 1293-1299 (1976)
[10] Sugiura, M.; Ito, A.; Yamaguchi, T.: Studies on dextranase. II. New exo-dextranase from Brevibacterium fuscum var. dextranlyticum. Biochim. Biophys. Acta, **350**, 61-70 (1974)
[11] Sawai, T.; Toriyama, K.; Yano, K.: A bacterial dextranase releasing only isomaltose from dextrans. J. Biochem., **75**, 105-112 (1974)
[12] Russel, R.R.B.; Ferretti, J.J.: Nucleotide sequence of the dextran glucosidase (dexB) gene of Streptococcus mutans. J. Gen. Microbiol., **136**, 803-810 (1990)
[13] Whiting, G.C.; Sutcliffe, I.C.; Russel, R.R.B.: Metabolism of polysaccharides by the Streptococcus mutans dexB gene product. J. Gen. Microbiol., **139**, 2019-2026 (1993)

Glucan endo-1,2-β-glucosidase 3.2.1.71

1 Nomenclature

EC number
3.2.1.71

Systematic name
1,2-β-D-glucan glucanohydrolase

Recommended name
glucan endo-1,2-β-glucosidase

CAS registry number
37288-49-6

2 Source Organism

<1> *Aspergillus auratus* [1]
<2> *Aspergillus aureolus* [1]
<3> *Aspergillus quadricinctus* [1]
<4> *Aspergillus sydowii* [1]
<5> *Aspergillus unguis* [1]
<6> *Beauveria bassiana* [1]
<7> *Fusarium oxysporum* [1]
<8> *Penicillium brefeldianum* [1]
<9> *Penicillium funicolosum* [1]
<10> *Penicillium melinii* [1]
<11> *Penicillium parvum* [1]
<12> *Penicillium quadrilineatum* [1]
<13> *Penicillium verruculosum* [1]
<14> *Acremonium sp.* [2]
<15> *Cytophaga arvensicola* [3]

3 Reaction and Specificity

Catalyzed reaction
random hydrolysis of 1,2-glucosidic linkages in 1,2-β-D-glucans

Reaction type
hydrolysis of O-glycosyl bond

Natural substrates and products

S β-1,2-glucan +H_2O <1-13> [1]

P oligosaccharides <1-13> (<1-13> β-1,2-linked, including sophorose [1]) [1]

Substrates and products

S β-1,2-glucan + H_2O <1-13> [1]

P oligosaccharides <1-13> (<1-13> β-1,2-linked, including sophorose [1]) [1]

S cyclic 1,2-β-D-glucan + H_2O <15> [3]

P oligosaccharides <15> (<15> β-1,2-linked with small amount of glucose [3]) [3]

S cyclic 1,2-β-D-glucan + H_2O <14> (<14> not laminaran, curdlan, carboxymethyl-cellulose [2]) [2]

P sophorose <14> (<14> 56% yield [2]) [2]

pH-Optimum

4 <1-13> [1]
4-4.5 <14> [2]
6.5 <15> [3]

pH-Range

2.8-7 <1-13> (<1-13> at pH 7.0 activity very low, at pH 2.8 25% of optimum [1]) [1]

Temperature optimum (°C)

35 <15> [3]
60 <1-13> [1]

Temperature range (°C)

40-60 <1-13> (<1-13> activity rises over range [1]) [1]

4 Enzyme Structure

Molecular weight

36000 <14> (<14> SDS-PAGE [2]) [2]

5 Isolation/Preparation/Mutation/Application

Localization

extracellular <14> [2]
intracellular <15> (<15> highly bound [3]) [3]

Purification

<14> [2]
<15> (partially [3]) [3]

6 Stability

pH-Stability

5.5-8 <14> (<14> 40°C, 2 h [2]) [2]

Temperature stability

40 <14> (<14> stable below, 2 h, 20 mM acetate buffer [2]) [2]

References

[1] Reese, E.T.; Parrish, F.W.; Mandels, M.: β-D-1,2-Glucanases in fungi. Can. J. Microbiol., **7**, 309-317 (1961)

[2] Kitahata, S.; Edagawa, S.: Cyclic (1,2)-β-D-glucan-hydrolyzing enzymes from Acremonium sp. 15. Purification and some properties of endo-(1,2)-β-D-glucanase and β-D-glucosidase. Agric. Biol. Chem., **51**, 2701-2708 (1987)

[3] Mendoza, N.S.; Amemura, A.: (1,2)-β-D-Glucan-hydrolyzing enzymes in Cytophaga arvensicola: partial purification an some properties of endo-(1,2)-β-glucanase and β-D-glucosidase specific for (1,2)-and (1,3)-linkages. J. Ferment. Technol., **61**, 473-481 (1983)

Xylan 1,3-β-xylosidase 3.2.1.72

1 Nomenclature

EC number
3.2.1.72

Systematic name
1,3-β-D-xylosidase

Recommended name
xylan 1,3-β-xylosidase

Synonyms
1,3-β-D-xylan xylohydrolase
β-1,3'-xylanase
exo-1,3-β-xylosidase
exo-β-1,3'-xylanase
xylosidase, exo-1,3-β-

CAS registry number
37288-50-9

2 Source Organism

<1> *Chaetomium globosum* (strain A2 [1]) [1]

3 Reaction and Specificity

Catalyzed reaction
hydrolysis of successive xylose residues from the non-reducing termini of 1,3-β-D-xylans

Reaction type
hydrolysis of O-glycosyl bond

Natural substrates and products
S β-1,3-xylan + H_2O <1> [1]
P Additional information <1> (<1>, xylose + and small amounts of glucose [1]) [1]

Substrates and products
S β-1,3-xylan + H_2O <1> [1]
P Additional information <1> (<1>, xylose + and small amounts of glucose [1]) [1]

Specific activity (U/mg)
Additional information <1> [1]

5 Isolation/Preparation/Mutation/Application

Source/tissue
culture medium <1> [1]

Localization
extracellular <1> [1]

Purification
<1> [1]

References

[1] Fukui, S.; Suzuki, T.; Kitahara, K.; Miwa, T.: Studies on β-1,3'-xylanase. I. Existence of a new enzyme, β-1,3'-xylanase. J. Gen. Appl. Microbiol., **6**, 270-282 (1969)

Licheninase 3.2.1.73

1 Nomenclature

EC number
3.2.1.73

Systematic name
1,3-1,4-β-D-glucan glucanohydrolase

Recommended name
licheninase

Synonyms
1,3-1,4-β-D-glucan 4-glucanohydrolase
1,3;1,4-β-glucan 4-glucanohydrolase
1,3;1,4-β-glucan endohydrolase
β-(1→ 3), (1→ 4)-D-glucan 4-glucanohydrolase

CAS registry number
37288-51-0

2 Source Organism

<1> *Bacteroides succinogenes* [1, 3]
<2> *Bacillus subtilis* [2, 4, 10, 19, 20]
<3> *Hordeum vulgare* (isoenzyme I and II [5]) [5, 6, 7, 16, 20, 22]
<4> *Bacillus sp.* (IMET B 376 [8]) [8, 25]
<5> *Bacillus laterosporus* [9, 10]
<6> *Bacillus pumilus* [9, 10, 11]
<7> *Bacillus amyloliquefaciens* (parental enzymes, hybrid β-glucanase enzyme H1 which contains the 107 amino-terminal residues of mature Bacillus amyloliquefaciens β-glucanase and the 107 carboxyl-terminal amino acid residues of Bacillus β-glucanase and hybrid enzyme H2 which consists of the 105 amino-terminal residues from the Bacillus macerans enzyme and the carboxyl-terminal 107 amino acids from Bacillus amyloliquefaciens [12]; parental enzymes and hybrid enzymes containing 16, 36, 78, or 152 amino acid N-terminal sequence derived from Bacillus amyloliquefaciens 1,3-1,4-β-D-glucan glucanohydrolase followed by a C-terminal segment derived from Bacillus macerans 1,3-1,4-β-D-glucan glucanohydrolase [19]) [9, 10, 12, 19, 21]
<8> *Bacillus circulans* [9, 10]
<9> *Bacillus polymyxa* [9, 10]

<10> *Bacillus macerans* (wild-type and mutant enzymes W101F, W101Y, E103D, E103Q, D105N, D105K, E107D, E107Q, and E107H [15]; parental enzymes, hybrid β-glucanase enzyme H1 which contains the 107 amino-terminal residues of mature Bacillus amyloliquefaciens β-glucanase and the 107 carboxyl-terminal amino acid residues of Bacillus β-glucanase and hybrid enzyme H2 which consists of the 105 amino-terminal residues from the Bacillus macerans enzyme and the carboxyl-terminal 107 amino acids from Bacillus amyloliquefaciens [12]; parental enzymes and hybrid enzymes containing 16, 36, 78, or 152 amino acid N-terminal sequence derived from Bacillus amyloliquefaciens 1,3-1,4-β-D-glucan glucanohydrolase followed by a C-terminal segment derived from Bacillus macerans 1,3-1,4-β-D-glucan glucanohydrolase [19]) [10, 12, 15, 19]
<11> *Orpinomyces sp.* (strain PC-2 [13]) [13]
<12> *Oryza sativa* (L. cv. Kirara 397 [14]) [14]
<13> *Triticum aestivum* [17]
<14> *Bacillus brevis* [18]
<15> *Clostridium thermocellum* [23]
<16> *Bacillus licheniformis* [24]

3 Reaction and Specificity

Catalyzed reaction

hydrolysis of 1,4-β-D-glucosidic linkages in β-D-glucans containing 1,3- and 1,4-bonds

Reaction type

hydrolysis of O-glycosyl bond

Substrates and products

S β-D-glucan + H_2O <1, 2, 4-13, 15> (<1, 12>, β-D-glucan from oat [3, 14]; <2, 4-12, 13, 15>, β-D-glucan from barley [8, 9, 11, 12, 13, 14, 17, 19, 23]) [3, 4, 8, 9, 11-14, 17, 19, 23]
P Additional information <11, 12> (<11>, main products are triose and tetraose [13]; <12>, oligosaccharides are formed as the major products, no glucose [14]) [13, 14]
S carboxymethyl β-glucan + H_2O <2> (<2>, carboxymethyl β-glucan from barley [4]) [4]
P Additional information <2> (<2>, trisaccharide, 53%, tetrasaccharide, 25% [4]) [4]
S lichenan + H_2O <1, 4, 6, 7, 10, 14, 15> [1, 8, 11, 12, 15, 18, 23]
P ?
S lichenin + H_2O <1, 2, 5-12> (<12>, lichenin from Cetralia islandica and from Usnea barbata [14]) [3, 9, 13, 14, 19]
P Additional information <11> (<11>, main products are triose and tetraose [13]) [13]
S Additional information <1, 4, 6, 11-13> (<1>, no hydrolysis of starch [3]; <1, 4, 6, 11, 12>, no hydrolysis of carboxymethylcellulose [3, 8, 11, 13, 14];

<1, 4>, no hydrolysis of carboxymethylpachyman [3, 8]; <6>, no hydrolysis of laminaran [11]; <1, 11, 12, 13>, no hydrolysis of laminarin [3, 13, 14, 17]; <1, 11>, no hydrolysis of xylan [3, 13]; <11, 13>, no hydrolysis of pustulan [13, 17]) [3, 8, 11, 13, 14, 17]

P ?

Inhibitors

(3,4)-epoxybutyl β-D-cellobioside <7> (<7>, identification of Glu105 at the active site [21]) [21]
2-mercaptoethanol <15> (<15>, 10 mM, 28% inhibition [23]) [23]
$CaCl_2$ <15> (<15>, 10 mM, 33% inhibition [23]) [23]
EDTA <4, 15> (<6>, no effect [11]; <15>, 10 mM, 18% inhibition [23]) [8, 23]
Hg^{2+} <6> [11]
PCMB <4> (<6>, no effect [11]) [8]
SDS <15> (<15>, 100 mM, 28% inhibition [23]) [23]
dithiothreitol <15> (<15>, 1 mM, 21% inhibition [23]) [23]
epoxyalkyl cellobiosides <2, 3> (<2, 3>, irreversible active-site directed inactivation, stereospecific requirements for inhibition [20]) [20]
glucono-δ-lactone <4> [8]
Additional information <14> (<14>, no inhibition up to a concentration of 10% SDS-PAGE [18]) [18]

Turnover number (min^{-1})

12 <10> (lichenan, mutant enzyme E107D [15]) [15]
33 <10> (lichenan, mutant enzyme E103D [15]) [15]
42 <10> (lichenan, mutant enzyme E107H [15]) [15]
102 <10> (lichenan, mutant enzyme W101Y [15]) [15]
174 <10> (lichenan, mutant enzyme W101F [15]) [15]
186 <10> (lichenan, mutant enzyme D105N [15]) [15]
61380 <10> (lichenan, <10>, wild-type enzyme [15]) [15]

Specific activity (U/mg)

10.4 <7, 10> (<7, 10>, hybrid enzyme H2 which consists of the 105 amino-terminal residues from the Bacillus macerans enzyme and the carboxyl-terminal 107 amino acids from Bacillus amyloliquefaciens [12]) [12]
387 <12> [14]
1329 <4> [8]
1330 <7> [12]
1372 <1> [3]
3659 <11> [13]
3722 <7, 10> (<7, 10>, hybrid β-glucanase enzyme H1 which contains the 107 amino-terminal residues of mature Bacillus amyloliquefaciens β-glucanase and the 107 carboxyl-terminal amino acid residues of Bacillus β-glucanase [12]) [12, 19]
5030 <10> [12]
Additional information <2, 3, 6> [4, 7, 11]

K_m-Value (mM)

Additional information <1, 2, 4, 7, 10, 11> (<10>, wild-type and mutant enzymes [15]) [3, 4, 8, 12, 13, 15]

pH-Optimum

4.5 <12> [14]
5 <3> [6]
5.5-6 <8> [10]
5.6-6.6 <7, 10> (<10>, hybrid β-glucanase enzyme H1 which contains the 107 amino-terminal residues of mature Bacillus amyloliquefaciens β-glucanase and the 107 carboxyl-terminal amino acid residues of Bacillus β-glucanase [12]) [12]
5.8-6.2 <11> [13]
5.8-6.3 <1> [3]
6-7 <4, 7, 9> [8, 10, 12]
6-7.5 <10> [12]
6-8 <10> [10]
6.5 <2, 7> [10, 19]
6.5-7 <2> [4]
6.5-7.5 <2> [10]
7 <6> [11]
7-8 <5, 7, 10> (<7, 10>, hybrid enzyme H2 which consists of the 105 amino-terminal residues from the Bacillus macerans enzyme and the carboxyl-terminal 107 amino acids from Bacillus amyloliquefaciens [10]) [10]
7.6 <2, 7> [19]
8-9 <15> [23]
8-10 <14> [18]
8.5 <6> [10]

pH-Range

3-6 <12> (<12>, pH 3.0: about 50% of maximal activity, pH 6.0: about 40% of maximal activity [14]) [14]
5-12.5 <15> (<15>, half-maximal activity at pH 5.0 and at pH 12.5 [23]) [23]
5.2-8 <2> (<2>, 50% of maximal activity at pH 5.2 and pH 8.0 [4]) [4]
5.4-7 <11> (<11>, about 80% of maximal activity at pH 5.4 and at pH 7.0 [13]) [13]
5.5-6 <10> (<10>, mutant enzyme E107H [15]) [15]
5.5-7 <7> (<7>, more than 80% of maximal activity at pH 5.5 and pH 7.0 [19]) [19]
5.9-6.5 <2> (<2>, more than 80% of maximal activity at pH 5.9 and pH 6.5 [19]) [19]
5.9-8.5 <10> (<10>, more than 80% of maximal activity at pH 5.9 and pH 8.5 [19]) [19]
6-7 <10> (<10>, wild-type enzyme and mutant enzymes W101F, W101Y, E103D, D105N and E107D [15]) [15]
6-11 <14> (<14>, pH 6.0: about 55% of maximal activity, pH 11.0: about 45% of maximal activity [18]) [18]

Temperature optimum (°C)

45 <9, 11> [10, 13]
50 <1> (<1>, lichenin or oat β-D-glucan as substrate [3]) [3]
50-55 <5, 6, 8> [10]
50-60 <6> [11]
55 <2, 7> [10, 19]
60 <2, 10> [4, 10]
65 <10> [19]
65-70 <14> [18]
80 <15> [23]

Temperature range (°C)

30-60 <1> (<1>, 30°C, about 30% of maximal activity with oat β-D-glucan, about 45% of maximal activity with lichenin, 60°C: about 20% of maximal activity with lichenin and oat β-D-glucan as substrate [3]) [3]
37-65 <2, 7> (<2 7>, more than 80% of maximal activity at 37°C and at 65°C [19]) [19]
40-55 <9> (<9>, about 60% of maximal activity at 40°C and at 55°C [10]) [10]
40-60 <7> (<7>, 40°C: about 60% of maximal activity, 60°C: about 20% of maximal activity [10]) [10]
40-65 <2, 5, 6> (<2>, about 40% of maximal activity at 40°C and at 65°C [10]; <5>, 40°C: about 80% of maximal activity, 65°C: about 60% of maximal activity [10]; <6>, 40°C: about 60% of maximal activity, 65°C: about 50% of maximal activity [10]) [10]
40-80 <14> (<14>, 40°C: about 40% of maximal activity, 80°C: about 30% of maximal activity [18]) [18]
45-65 <8> (<8>, 45°C: about 80% of maximal activity, 65°C: about 90% of maximal activity [10]) [10]
55-75 <10> (<10>, more than 80% of maximal activity at 55°C and at 75°C [19]) [19]
55-90 <15> (<15>, 55°C: about 50% of maximal activity, 90°C: about 30% of maximal activity [23]) [23]

4 Enzyme Structure

Molecular weight

12000 <6> (<6>, discrepancy between molecular weight determined by gel filtration and low speed equilibrium sedimentation is probably due to a moderate interaction between the enzyme and the Sephadex polmer, gel filtration [11]) [11]
27000 <6> (<6>, discrepancy between molecular weight determined by gel filtration and low speed equilibrium sedimentation is probably due to a moderate interaction between the enzyme and the Sephadex polmer, low speed sedimentation without reaching equilibrium [11]) [11]
27700 <3> (<3>, enzyme I, ultracentrifugation [7]) [7]

32000 <12> (<12>, gel filtration [14]) [14]
33800 <3> (<3>, enzyme II, ultracentrifugation [7]) [7]
37200 <1> (<1>, gel filtration [3]) [3]

Subunits

? <2-4, 11, 13, 14, 15> (<2>, x * 24000, SDS-PAGE [4]; <4>, x * 26000, SDS-PAGE [8]; <11>, x * 27000, recombinant enzyme, SDS-PAGE [13]; <2>, x * 27500, enzyme inactivated by epoxyalkyl β-D-oligoglucosides [2]; <11>, x * 27929, calculation from nucleotide sequence [13]; <14>, x * 29000, SDS-PAGE [18]; <3, 13>, x * 30000, SDS-PAGE [6, 17]; <3>, x * 32085, calculation from nucleotide sequence [16]; <15>, x * 35000, recombinant enzyme, SDS-PAGE [23]) [2, 4, 6, 8, 13, 16, 17, 18, 23]
monomer <1, 3, 12> (<3>, 1 * 28000, enzyme I, SDS-PAGE [7]; <3>, 1 * 28000, enzyme II, SDS-PAGE [7]; <12>, 1 * 31000, SDS-PAGE [14]; <1>, 1 * 35200, SDS-PAGE [3]) [3, 7, 14]

Posttranslational modification

glycoprotein <3> (<3>, enzyme I contains 0.7% carbohydrate of which three residues are probably N-acetylglucosamine, enzyme II contains 3.6% carbohydrate [7]) [7]
proteolytic modification <11> (<11>, a N-terminal signal peptide of 29 amino residues is cleaved from the enzyme secreted from Orpinomyces whereas 21 amino acid residues of the signal peptide are removed when the enzyme is produced by E. coli [13]) [13]

5 Isolation/Preparation/Mutation/Application

Source/tissue

aleurone layer <3> (<3>, isoenzyme II is synthesized exclusively in the aleurone layer, isoenzyme III is detected in scutellum and aleurone layer [5]) [5]
culture medium <4, 11> (<11>, recombinant enzyme expressed in E. coli [13]) [8, 13]
scutellum <3> (<3>, isoenzyme I is synthesized predominantly in the scutellum, isoenzyme III is detected in scutellum and aleurone layer [5]) [5]
seed <13> (<13>, germinated [17]) [17]

Localization

cytoplasm <14> [18]
extracellular <4, 11, 14> (<11>, recombinant enzyme expressed in E. coli [13]) [8, 13, 18]
periplasmic space <11, 14> [13, 18]

Purification

<1> [3]
<2> [4]
<3> [6, 7]
<4> [8]
<6> [11]

<7> (parental enzymes and hybrid enzymes containing 16, 36, 78, or 152 amino acid N-terminal sequence derived from Bacillus amyloliquefaciens 1,3-1,4-β-D-glucan glucanohydrolase followed by a C-terminal segment derived from Bacillus macerans 1,3-1,4-β-D-glucan glucanohydrolase [19]; parental enzymes, hybrid β-glucanase enzyme H1 which contains the 107 amino-terminal residues of mature Bacillus amyloliquefaciens β-glucanase and the 107 carboxyl-terminal amino acid residues of Bacillus β-glucanase and hybrid enzyme H2 which consists of the 105 amino-terminal residues from the Bacillus macerans enzyme and the carboxyl-terminal 107 amino acids from Bacillus amyloliquefaciens [12]) [12, 19]
<10> (parental enzymes and hybrid enzymes containing 16, 36, 78, or 152 amino acid N-terminal sequence derived from Bacillus amyloliquefaciens 1,3-1,4-β-D-glucan glucanohydrolase followed by a C-terminal segment derived from Bacillus macerans 1,3-1,4-β-D-glucan glucanohydrolase [19]; parental enzymes, hybrid β-glucanase enzyme H1 which contains the 107 amino-terminal residues of mature Bacillus amyloliquefaciens β-glucanase and the 107 carboxyl-terminal amino acid residues of Bacillus β-glucanase and hybrid enzyme H2 which consists of the 105 amino-terminal residues from the Bacillus macerans enzyme and the carboxyl-terminal 107 amino acids from Bacillus amyloliquefaciens [12]; wild type and mutant enzymes W101F, W101Y, E103D, E103Q, D105N, D105K, E107D, E107Q, and E107H [15]) [12, 15]
<11> (recombinant enzyme [13]) [13]
<12> [14]
<13> [17]

Crystallization

<4> (hybrid enzyme [25]) [25]
<10> [15]
<16> [24]

Cloning

<1> (expression in E. coli [1,3]) [1, 3]
<3> (transgenic barley expressing a protein-engineered, thermostable (1,3-1,4)-β-glucanase during germination [16]; expression in E. coli, isolation and characterization of the gene [22]) [16, 22]
<7> (hybrid β-glucanase enzyme H1 which contains the 107 amino-terminal residues of mature Bacillus amyloliquefaciens β-glucanase and the 107 carboxyl-terminal amino acid residues of Bacillus β-glucanase and hybrid enzyme H2 which consists of the 105 amino-terminal residues from the Bacillus macerans enzyme and the carboxyl-terminal 107 amino acids from Bacillus amyloliquefaciens, expression in E. coli [12]; hybrid enzymes containing 16, 36, 78, or 152 amino acid N-terminal sequence derived from Bacillus amyloliquefaciens 1,3-1,4-β-D-glucan glucanohydrolase followed by a C-terminal segment derived from Bacillus macerans 1,3-1,4-β-D-glucan glucanohydrolase, expression in E. coli [19]) [12, 19]
<10> (hybrid β-glucanase enzyme H1 which contains the 107 amino-terminal residues of mature Bacillus amyloliquefaciens β-glucanase and the 107

carboxyl-terminal amino acid residues of Bacillus β-glucanase and hybrid enzyme H2 which consists of the 105 amino-terminal residues from the Bacillus macerans enzyme and the carboxyl-terminal 107 amino acids from Bacillus amyloliquefaciens, expression in E. coli [12]; hybrid enzymes containing 16, 36, 78, or 152 amino acid N-terminal sequence derived from Bacillus amyloliquefaciens 1,3-1,4-β-D-glucan glucanohydrolase followed by a C-terminal segment derived from Bacillus macerans 1,3-1,4-β-D-glucan glucanohydrolase, expression in E. coli [19]) [12, 19]
<11> (expression in E. coli [13]) [13]
<14> (expression in E. coli [18]) [18]
<15> (expression in E. coli [22]) [23]

Engineering

D105K <10> (<10>, activity is reduced to less than 1% [15]) [15]
D105N <10> (<10>, activity is reduced to less than 1% [15]) [15]
E103D <10> (<10>, activity is reduced to less than 1% [15]) [15]
E103Q <10> (<10>, activity is reduced to less than 1% [15]) [15]
E107D <10> (<10>, activity is reduced to less than 1% [15]) [15]
E107H <10> (<10>, activity is reduced to less than 1% [15]) [15]
E107Q <10> (<10>, activity is reduced to less than 1% [15]) [15]
W101F <10> (<10>, activity is reduced to less than 1% [15]) [15]
W101Y <10> (<10>, activity is reduced to less than 1% [15]) [15]
Additional information <3, 7, 10> (<7, 10>, construction of hybrid β-glucanase enzyme H1 which contains the 107 amino-terminal residues of mature Bacillus amyloliquefaciens β-glucanase and the 107 carboxyl-terminal amino acid residues of Bacillus β-glucanase. Hybrid enzyme H2 consists of the 105 amino-terminal residues from the Bacillus macerans enzyme and the carboxyl-terminal 107 amino acids from Bacillus amyloliquefaciens. Hybrid enzyme H1 exhibits increased thermostability especially in an acidic environment compared to both parental enzymes. Hybrid enzyme H2 is more thermolabile than the naturally occuring β-glucanases [12]; <3>, protein-engineered, thermostable (1,3-1,4)-β-glucanase: the codons for hybrid H(A12-M)ΔY13 are modified to match thoses of the gene encoding barley (1,3-1,4)-β-glucanase isoenzyme EII [16]; <7,10>, hybrid enzymes containing 16, 36, 78, or 152 amino acid N-terminal sequence derived from Bacillus amyloliquefaciens 1,3-1,4-β-D-glucan glucanohydrolase followed by a C-terminal segment derived from Bacillus macerans 1,3-1,4-β-D-glucan glucanohydrolase, expression in E. coli [19]) [12, 16, 19]

Application

synthesis <15> (<15>, the unusually resistance against inactivation by heat, ethanol or ionic detergents makes the enzyme highly suitable for industrial application in the mashing process of beer brewing [23]) [23]

6 Stability

pH-Stability

2.5-9 <2> (<2>, 4°C or 20°C, stable for 20 h [4]) [4]

3.4-9.8 <11> (<11>, 4°C, stable for at least 24 h [13]) [13]

3.5-7 <7, 10> (<7,10>, hybrid β-glucanase enzyme H1 which contains the 107 amino-terminal residues of mature Bacillus amyloliquefaciens β-glucanase, stable [12]) [12]

4-5 <3> (<3>, 25°C, 72-96 h, stable [6]) [6]

4.1 <7, 10> (<7>, 65°C, 1 h, 95% loss of activity [19]; <10>, 65°C, 1 h, complete inactivation [19]; <7, 10>, 65°C, 1 h, hybrid enzymes containing 16, 36, 78, or 152 amino acid N-terminal sequence derived from Bacillus amyloliquefaciens 1,3-1,4-β-D-glucan glucanohydrolase followed by a C-terminal segment derived from Bacillus macerans 1,3-1,4-β-D-glucan glucanohydrolase, 85%-95% of the initial activity is retained [19]) [19]

4.8-6 <7> (<7>, stable [12]) [12]

4.8-6.4 <10> (<10>, stable [12]) [12]

5-6 <4> (<4>, 60°C, 10 min, stable [8]) [8]

5-7 <6> (<6>, 10 min, stable [11]) [11]

5.6-6 <7, 10> (<7, 10>, hybrid enzyme H2 which consists of the 105 amino-terminal residues from the Bacillus macerans enzyme and the carboxyl-terminal 107 amino acids from Bacillus amyloliquefaciens, stable [12]) [12]

Temperature stability

40 <3, 11, 12> (<3>, pH 5.5, 10 min, stable [6]; <11>, recombinant enzyme, pH 6.0, 24 h, stable [13]; <12>, 10 min, stable [14]) [6, 13, 14]

45 <11> (<11>, recombinant enzyme, pH 6.0, 24 h, 28% loss of activity [13]) [13]

50 <11, 12> (<11>, recombinant enzyme, pH 6.0, 24 h, 41% loss of activity [13]; <12>, 10 min, about 25% loss of activity [14]) [13, 14]

55 <11> (<11>, recombinant enzyme, pH 6.0, 24 h, 70% loss of activity [13]) [13]

60 <1, 2, 4, 6, 12> (<1>, 10 min, 50% loss of activity, crude enzyme extract [3]; <2>, 15 min, pH 6.5, stable [4]; <3>, pH 5.5, 10 min, complete loss of activity [6]; <4>, pH 5.0-6.0, 10 min, stable. pH 6, in presence of 10 mM $CaCl_2$ and serum albumin, stable for 120 min [8]; <6>, 10 min, about 35% loss of activity [11]; <12>, 10 min, complete inactivation [14]) [3, 4, 6, 8, 11, 14]

65 <7, 10> (<7>, half-life: 25 min [12]; <7,10>, hybrid β-glucanase enzyme H1 which contains the 107 amino-terminal residues of mature Bacillus amyloliquefaciens β-glucanase is stable for more than 1 h at pH 5.5. The 107 carboxyl-terminal amino acid residues of Bacillus β-glucanase. Hybrid enzyme H2 which consists of the 105 amino-terminal residues from the Bacillus macerans enzyme and the carboxyl-terminal 107 amino acids from Bacillus amyloliquefaciens irreversible thermoinactivated within 20-25 min [12]) [12]

70 <7, 10, 14> (<14>, 1 h, 15% loss of activity [18]; <7, 10>, pH 6.0, 1 h, hybrid enzymes containing 16, 36, 78, or 152 amino acid N-terminal se-

quence derived from Bacillus amyloliquefaciens 1,3-1,4-β-D-glucan glucanohydrolase followed by a C-terminal segment derived from Bacillus macerans 1,3-1,4-β-D-glucan glucanohydrolase exhibit 90% of the initial activity, the parental enzymes retain 5% or less of the initial activity [19]) [18, 19]
75 <14> (<14>, 1 h, 25% loss of activity [18]) [18]
Additional information <2, 4-10> (<2, 4-10>, particularly thermostable in presence of Ca^{2+} [8,10]; <7, 10>, Ca^{2+} stabilizes hybrid enzymes and parental enzymes against thermal inactivation [12]) [8, 10, 12]

Organic solvent stability
ethanol <15> (<15>, stable up to concentrations up to 10% w/v [23]) [23]

General stability information
<2>, stable to repeated freeze-thaw cycles [4]
<4>, particularly thermostable in presence of Ca^{2+} [8]
<7, 10>, Ca^{2+} stabilizes hybrid enzymes and parental enzymes against thermal inactivation [12]

Storage stability
<3>, -15°C, stable for 12 months [7]

References

[1] Irvin, J.E.; Teather, R.M.: Cloning and expression of a Bacteroides succinogenes mixed-linkage β-glucanase (1,3-1,4-β-D-glucan 4-glucanohydrolase) gene in Escherichia coli. Appl. Environ. Microbiol., **54**, 2672-2676 (1988)

[2] Hoj, P.H.; Rodriguez, E.B.; Stick, R.V.; Stone, B.A.: Differences in active site structure in a family of β-glucan endohydrolases deduced from the kinetics of inactivation by epoxyalkyl β-oligoglucosides. J. Biol. Chem., **264**, 4939-4947 (1989)

[3] Erfle, J.D.; Teather, R.M.; Wood, P.J.; Irvin, J.E.: Purification and properties of a 1,3-1,4-β-D-glucanase (lichenase, 1,3-1,4-β-D-glucan 4-glucanohydrolase, EC 3.2.1.73) from Bacteroides succinogenes cloned in Escherichia coli. Biochem. J., **255**, 833-841 (1988)

[4] McCleary, B.V.: Lichenase from Bacillus subtilis. Methods Enzymol., **160**, 572-575 (1988)

[5] Stuart, I.M.; Loi, L.; Fincher, G.B.: Development of (1→3,1→4)-β-D-glucan endohydrolase isoenzymes in isolated scutella and aleurone layers of barley (Hordeum vulgare). Plant Physiol., **80**, 310-314 (1986)

[6] Yamashita, H.; Hayase, F.; Kato, H.: β-Glucanases in malt that form insoluble material. Agric. Biol. Chem., **49**, 1313-1320 (1985)

[7] Woodward, J.R.; Fincher, G.B.: Purification and chemical properties of two 1,3;1,4-β-glucan endohydrolases from germinating barley. Eur. J. Biochem., **121**, 663-669 (1982)

[8] Borriss, R.: Purification and characterization of an extracellular β-glucanase from Bacillus IMET B 376. Z. Allg. Mikrobiol., **21**, 7-17 (1981)

[9] Borriss, R.; Zemek, J.: β-1,3-1,4-Glucanase in sporenbildenden Mikroorganismen. III. Substratspezifität und Wirkungsweise einiger Bacillus-β-Glucan Hydrolasen. Zentralbl. Bakteriol. Parasitenkd. Infektionskr. Hyg., Abt. II, **135**, 696-703 (1980)
[10] Borriss, R.; Zemek, J.: β-1,3-1,4-Glucanase in sporenbildenden Mikroorganismen. IV. Eigenschaften einiger Bacillus-β-Glucan-Hydrolasen. Zentralbl. Bakteriol. Parasitenkd. Infektionskr. Hyg., Abt. II, **136**, 63-69 (1981)
[11] Suzuki, H.; Kaneko, T.: Degradation of barley glucan and lichenan by a Bacillus pumilus enzyme. Agric. Biol. Chem., **40**, 577-586 (1976)
[12] Borriss, R.; Olsen, O.; Thomsen, K.K.; von Wettstein, D.: Hybrid Bacillus endo-(1-3,1-4)-β-glucanases: construction of recombinant genes and molecular properties of the gene products. Carlsberg Res. Commun., **54**, 41-54 (1989)
[13] Chen, H.; Li, X.-L.; Ljungdahl, L.G.: Sequencing of a 1,3-1,4-β-D-glucanase (lichenase) from the anaerobic fungus Orpinomyces strain PC-2: properties of the enzyme expressed in Escherichia coli and evidence that the gene has a bacterial origin. J. Bacteriol., **179**, 6028-6034 (1997)
[14] Akiyama, T.; Kaku, H.; Shibuya, N.: Purification, characterization and NH_2-terminal sequencing of an endo-(1→3,1→4)-β-glucanase from rice (Oryza sativa L.) bran. Plant Sci., **134**, 3-10 (1998)
[15] Hahn, M.; Olsen, O.; Politz, O.; Borriss, R.; Heinemann, U.: Crystal structure and site-directed mutagenesis of Bacillus macerans endo-1,3-1,4-β-glucanase. J. Biol. Chem., **270**, 3081-3088 (1995)
[16] Jensen, L.G.; Olsen, O.; Kops, O.; Wolf, N.; Thomsen, K.K.; von Wettstein, D.: Transgenic barley expressing a protein-engineered, thermostable (1,3-1,4)-β-glucanase during germination. Proc. Natl. Acad. Sci. USA, **93**, 3487-3491 (1996)
[17] Lai, D.M.L.; Hoj, P.B.; Fincher, G.B.: Purification and characterization of (1→3,1→4)-β-glucan endohydrolases from germinated wheat (Triticum aestivum). Plant Mol. Biol., **22**, 847-859 (1993)
[18] Louw, M.E.; Reid, S.J.: Characterization, cloning and sequencing of a thermostable endo-(1,3-1,4)β-glucanase-encoding gene from an alkalophilic Bacillus brevis. Appl. Microbiol. Biotechnol., **38**, 507-513 (1993)
[19] Olsen, O.; Borrisss, R.; Simon, O.; Thomsen, K.K.: Hybrid Bacillus (1-3,1-4)-β-glucanases: engineering thermostable enzymes by construction of hybrid genes. Mol. Gen. Genet., **225**, 177-185 (1991)
[20] Hoj, P.B.; Rodriguez, E.B.; Iser, J.R.; Stick, R.V.; Stone, B.A.: Active site-directed inhibition by optically pure epoxyalkyl cellobiosides reveals differences in active site geometry of two 1,3-1,4-β-D-glucan 4-glucanohydrolase. J. Biol. Chem., **266**, 11628-11631 (1991)
[21] Hoj, P.B.; Condron, R.; Traeger, J.C.; McAuliffe, J.C.; Stone, B.A.: Identification of glutamic acid 105 at the active site of Bacillus amyloliquefaciens 1,3-1,4-β-D-glucan 4-glucanohydrolase using epoxide-based inhibitors. J. Biol. Chem., **267**, 25059-25066 (1992)
[22] Litts, J.C.; Simmons, C.R.; Karrer, E.E.; Huang, N.; Rodriguez, R.L.: The isolation and characterization of a barley 1,3-1,4-β-glucanase gene. Eur. J. Biochem., **194**, 831-838 (1990)

[23] Schimming, S.; Schwarz, W.H.; Staudenbauer, W.L.: Properties of a thermoactive β-1,3-1,4-glucanase (lichenase) from Clostridium thermocellum expressed in Escherichia coli. Biochem. Biophys. Res. Commun., **177**, 447-452 (1991)
[24] Hahn, M.; Pons, J.; Planas, A.; Querol, E.; Heinemann, U.: Crystal structure of Bacillus licheniformis 1,3-1,4-β-D-glucan 4-glucanohydrolase at 1.8 A resolution. FEBS Lett., **374**, 221-224 (1995)
[25] Hahn, M.; Keitel, T.; Heinemann, U.: Crystal and molecular structure at 0.16-nm resolution of the hybrid Bacillus endo-1,3-1,4-β-D-glucan 4-glucanohydrolase H(A16-M). Eur. J. Biochem., **232**, 849-858 (1995)

Glucan 1,4-β-glucosidase 3.2.1.74

1 Nomenclature

EC number
3.2.1.74

Systematic name
1,4-β-D-glucan glucohydrolase

Recommended name
glucan 1,4-β-glucosidase

Synonyms
β-1,4-β-glucanase
β-glucosidase
exo-1,4-β-glucanase
exo-1,4-β-glucosidase
exo-β-1,4-glucanase
exo-β-1,4-glucosidase
exocellulase

CAS registry number
37288-52-1

2 Source Organism

<1> *Torulopsis wickerhamii* [1]
<2> *Ruminococcus albus* (AR67 [20]) [2, 20, 21]
<3> *Penicillium funiculosum* [3, 7]
<4> *Trichoderma reesei* (QM 9414 [4]) [4]
<5> *Botrytis cinerea* [5]
<6> *Chrysosporium pannorum* [5]
<7> *Cladosporium cladosporioides* [5]
<8> *Cladosporium resinae* [5]
<9> *Fusarium culmorum* [5]
<10> *Fusarium equiseti* [5]
<11> *Fusarium oxysporum* [5]
<12> *Fusarium oxysporum f. sp. dianthi* [5]
<13> *Fusarium pose* [5]
<14> *Fusarium solani* [5]
<15> *Gilmaniella humicola* [5]
<16> *Myrothecium striatisporum* [5]

<17> *Phoma sp.* [5]
<18> *Rhizopus arrhizus* [5]
<19> *Scopulariopsis brevicaulis* [5]
<20> *Stachybotrys chartarum* [5]
<21> *Stemphylium botryosum* [5]
<22> *Thamnidium elegans* [5]
<23> *Thielavia basicola* [5]
<24> *Trichoderma koningii* [5]
<25> *Trichoderma polysporum* [5]
<26> *Trichoderma viride* [5]
<27> *Verticillium nigrescens* [5]
<28> *Sporotrichum pulverulentum* (also called *Chrysosporium lignorum* [6]) [6]
<29> *Acetobacter xylinum* (BPR2001 [8]) [8]
<30> *Prevotella ruminicola* [9]
<31> *Irpex lacteus* [10]
<32> *Cellulomonas fimi* [11, 22]
<33> *Ruminococcus flavefaciens* (FD-1 [12]) [12]
<34> *Anaerocellum thermophilum* [13]
<35> *Thermobifida fusca* [14]
<36> *Hordeum vulgare* [15]
<37> *Pseudomonas fluorescens* (subsp. cellulosa [16]) [16]
<38> *Clostridium stercorarium* [17]
<39> *Clostridium papyrosolvens* [18]
<40> *Thermotoga maritima* [19]

3 Reaction and Specificity

Catalyzed reaction

hydrolysis of 1,4-linkages in 1,4-β-D-glucans, to remove successive glucose units

Reaction type

hydrolysis of O-glycosyl bond

Substrates and products

S 4-methylumbelliferyl 4-O-(β-D-glucopyranosyl)-β-D-xylopyranoside + H_2O <35> [14]
P ?
S 4-methylumbelliferyl-β-D-glucopyranoside + H_2O <35> [14]
P ?
S 4-methylumbelliferyl-β-D-glucoside + H_2O <29, 37> [8, 16]
P ?
S 4-methylumbelliferyl-β-cellobioside + H_2O <35, 37> [14, 16]
P ?
S 4-methylumbelliferyl-β-cellotrioside + H_2O <35> [14]
P ?

S Avicel + H_2O <2, 3, 28, 31, 34, 38> (<38> H_3PO_4 or HCl swollen [17]) [3, 6, 10, 13, 17, 21]
P D-glucose + cellotetraose + cellotriose + cellobiose <3, 34, 38> (<3> D-glucose only [3]; <38> cellobiose and cellotriose only [17]) [3, 13, 17]
S amygdalin + H_2O <30> [9]
P ?
S arbutin + H_2O <30> [9]
P ?
S β-glucan + H_2O <1> (<1> from yeast and from barley [1]) [1]
P ?
S carboxymethyl cellulose + H_2O <2, 31, 33, 35, 40> [10, 12, 14, 19, 21]
P ?
S cellobiose + H_2O <1, 2, 3, 4, 29, 30, 36, 37> (<31> only isoenzyme I [10]) [1, 2, 3, 4, 7, 8, 9, 15, 16, 20, 21]
P D-glucose + cellotetraose + cellotriose <3, 37> (<37> D-glucose only [16]) [3, 16]
S cellodextrins + H_2O <2, 4, 34, 40> [4, 13, 19, 20]
P D-glucose + cellotriose + cellobiose <2, 4, 34, 40> (<4> D-glucose only [4]; <40> at high substrate concentrations also transglycosylation reactions [19]) [4, 13, 19, 20]
S celloheptaose + H_2O <1, 38> [1, 17]
P ?
S cellohexaose + H_2O <1, 4, 36, 37, 38> [1, 4, 15, 16, 17]
P D-glucose <1, 37, 38> (<37> in the early stages also cellopentaose and cellotetraose [16]; <38> cellotetraose, cellotriose and cellobiose only [17]) [1, 16, 17]
S cellooligomers + H_2O <2> [2]
P D-glucose + cellooligomers <2> [2]
S cellooligomers + H_2O <2> [21]
P cellobiose + cellooligomers <2> [21]
S cellopentaose + H_2O <1, 2, 3, 4, 30, 35, 36, 37, 38> [1, 2, 3, 4, 7, 9, 14, 15, 16, 17, 21]
P D-glucose <37, 38> (<38> cellobiose and cellotriose only [17]) [16, 17]
S cellotetraose + H_2O <2, 3, 4, 30, 31, 35, 36, 37, 38> [2, 3, 4, 7, 9, 10, 14, 15, 16, 17, 20, 21]
P D-glucose + cellotriose + cellobiose <31, 37, 38> (<38> cellobiose only [17]) [10, 16, 17]
S cellotriose + H_2O <1, 2, 3, 4, 30, 35, 36, 37, 40> [1, 2, 3, 4, 7, 9, 14, 15, 16, 19, 20, 21]
P D-glucose + cellobiose <37, 40> (<37> D-glucose only [16]) [16, 19]
S cellulose + H_2O <2, 3, 29, 31, 35, 38> (<29> H_3PO_4 swollen and regenerated [8]; <2> ball-milled [21]; <3> Walseth cellulose [3]; <3, 35> H_3PO_4 swollen [7, 14]; <31> bacterial cellulose powder, not dried [10]; <38> cellulose HBS and cellulose MN300 [17]; <35> bacterial microcrystalline cellulose [14]) [3, 7, 8, 10, 14, 17, 21]
P D-glucose <3> (<3> D-glucose + cellobiose [7]) [3, 7]
S esculin + H_2O <2> [20]

P ?
S gentiobiose + H_2O <1, 2, 3, 30> [1, 7, 9, 20]
P ?
S hydroquinone-β-D-glucopyranoside + H_2O <1> [1]
P hydroquinone + D-glucose <1> [1]
S laminaribiose + H_2O <1, 2, 3> [1, 7, 20]
P ?
S laminarin + H_2O <2> [20]
P ?
S laminaritriose + H_2O <1> (<1> from barley [1]) [1]
P ?
S lichenan + H_2O <2, 29> [8, 20]
P ?
S n-heptyl-β-D-glucopyranoside + H_2O <1> [1]
P n-heptanol + D-glucose <1> [1]
S n-hexyl-β-D-glucopyranoside + H_2O <1> [1]
P n-hexanol + D-glucose <1> [1]
S n-octyl-β-D-glucopyranoside + H_2O <1> [1]
P n-octanol + D-glucose <1> [1]
S n-pentyl-β-D-glucopyranoside + H_2O <1> [1]
P n-pentanol + D-glucose <1> [1]
S *o*-nitrophenyl β-D-glucopyranoside + H_2O <3> [7]
P ?
S *p*-nitrophenyl β-D-lactoside + H_2O <31> [10]
P ?
S *p*-nitrophenyl-β-D-cellobioside + H_2O <2, 31, 33, 37, 32> [2, 10, 12, 16, 20, 22]
P D-glucose + *p*-nitrophenyl-β-D-glucoside <2> [2]
S *p*-nitrophenyl-β-D-cellobioside + H_2O <2> [21]
P cellobiose + *p*-nitrophenol <2> [21]
S *p*-nitrophenyl-β-D-cellotrioside + H_2O <29> [8]
P D-glucose + *p*-nitrophenyl-β-D-cellobioside + *p*-nitrophenyl-β-D-cellotetraoside + *p*-nitrophenyl-β-D-glucoside <29> [8]
S *p*-nitrophenyl-β-D-glucopyranoside + H_2O <1, 2, 29, 30, 36> [1, 2, 8, 9, 15, 20]
P 4-nitrophenol + D-glucose <1, 2, 29, 30, 36> (<36> when incubated with 100 mM substrate, products are also 4-nitrophenyl-β-laminaribioside, 4-nitrophenyl-β-cellobioside and 4-nitrophenyl-β-gentiobioside [15]) [1, 2, 8, 9, 15, 20]
S phenyl-β-D-glucopyranoside + H_2O <1> [1]
P phenol + D-glucose <1> [1]
S prunasin + H_2O <30> [9]
P ?
S salicin + H_2O <2> [20]
P ?
S sophorose + H_2O <1, 3> [1, 7]
P ?

S trehalose + H_2O <2> (<2> slight hydrolysis [20]) [20]
P ?
S xylan + H_2O <38> [17]
P ?
S xylobiose + H_2O <1> [1]
P ?

Inhibitors
1,2-anhydro-myo-inositol <36> [15]
$CaCl_2$ <39> [18]
$CoCl_2$ <39> [18]
Cu^{2+} <2> [2, 21]
$CuCl_2$ <39> [18]
D-cellobiose <1, 35> (<1, 35> competitive inhibiton [1, 14]) [1, 14]
D-glucono-1,5-lactone <3> [3, 7]
D-glucose <1, 3> (<1> competitive inhibiton [1]) [1, 7]
EDTA <39> [18]
$FeSO_4$ <39> [18]
Hg^{2+} <2, 29> (<29> slight inhibition [8]) [2, 8]
$MgCl_2$ <39> [18]
N-bromosuccinimide <29, 39> [8, 18]
NH_4Cl <39> [18]
NaCl <39> [18]
Nonidet P40 <39> [18]
SDS <29> [8]
Zn^{2+} <2> [2, 21]
$ZnSO_4$ <39> [18]
iodoacetoamide <2, 39> [2, 18, 21]
p-chloromercuribenzoate <2> [2, 21]

Metals, ions
$MnCl_2$ <39> (<39> activity increased by 58% [18]) [18]

Turnover number (min^{-1})
1920 <4> (cellobiose) [4]

Specific activity (U/mg)
0.0037 <38> [17]
0.025 <38> (<38> cellulose MN300 [17]) [17]
0.03 <38> (<38> cellulose HBS [17]) [17]
0.041 <38> (<38> avicel [17]) [17]
0.102 <38> (<38> H_3PO_4 swollen avicel [17]) [17]
0.108 <38> (<38> HCl swollen avicel [17]) [17]
0.128 <38> (<38> celloheptaose [17]) [17]
0.236 <38> (<38> cellohexaose [17]) [17]
0.285 <34> [13]
0.34 <38> (<38> xylan [17]) [17]
0.42 <38> (<38> cellopentaose [17]) [17]
0.684 <38> (<38> cellotetraose [17]) [17]

6 <2> [20]
7 <3> [3]
13 <2> [2]
14.5 <31> (<31> bacterial cellulose powder, not dried, isoenzyme II [10]) [10]
15.1 <31> (<31> bacterial cellulose powder, not dried, isoenzyme I [10]) [10]
20.9 <31> (<31> avicel, isoenzyme II [10]) [10]
21.1 <31> (<31> avicel, isoenzyme I [10]) [10]
21.1 <31> (<31> isoenzyme I [10]) [10]
25.5 <31> (<31> carboxymethyl cellulose, isoenzyme I [10]) [10]
26 <31> (<31> carboxymethyl cellulose, isoenzyme II [10]) [10]
27.1 <31> (<31> *p*-nitrophenyl β-D-lactoside, isoenzyme II [10]) [10]
27.3 <31> (<31> *p*-nitrophenyl β-D-lactoside, isoenzyme I [10]) [10]
30.94 <39> [18]
33.2 <31> (<31> *p*-nitrophenyl β-D-celloside, isoenzyme II [10]) [10]
34 <31> (<31> *p*-nitrophenyl β-D-celloside, isoenzyme I [10]) [10]
276.7 <1> [1]
Additional information <2, 28, 29> (<2> 20.8 OD405/mg [21]; <29> 840 U/mg [8]) [6, 8, 21]

K_m-Value (mM)

0.028 <37> (cellohexaose) [16]
0.29 <36> (cellohexaose) [15]
0.37 <3> (cellopentaose) [7]
0.41 <36> (cellopentaose) [15]
0.42 <30> (*p*-nitrophenyl-β-D-glucoside) [9]
0.44 <3> (cellotriose) [7]
0.7 <2, 32> (*p*-nitrophenyl-β-D-cellobioside) [20, 22]
0.77 <3> (cellotetraose) [7]
0.77 <3> (*o*-nitrophenyl β-D-glucopyranoside) [7]
0.89 <36> (cellotetraose) [15]
0.97 <36> (cellotriose) [15]
1.2 <37> (cellobiose) [16]
2.2 <2> (*p*-nitrophenyl-β-D-glucoside) [2]
2.67 <36> (cellobiose) [15]
2.8 <1> (*p*-nitrophenyl-β-D-glucopyranoside) [1]
3.7 <29> (cellotriose) [8]
4 <2> (*p*-nitrophenyl-β-D-glucopyranoside) [20]
5 <30> (arbutin) [9]
6 <2> (cellopentaose) [20]
6.7 <2> (cellotetraose) [20]
8.7 <2> (cellotriose) [20]
10 <3> (cellobiose) [7]
10.7 <30> (gentiobiose) [9]
12.5 <2> (cellobiose) [20]
13.5 <30> (cellobiose) [9]
26 <2> (cellobiose) [2]

127 <3> (cellulose, <3> H3OP4 swollen [7]) [7]
Additional information <3, 36, 39>(<3> Walseth cellulose 10 mg/ml, cellopentaose 0.83 mg/ml, cellotriose 0.55 mg/ml, cellotriose 0.55 mg/ml [3]; <36> laminarioligosaccharide degree of polymerization 2, 5.37, laminarioligosaccharide degree of polymerization 3, 2.77, laminarioligosaccharide degree of polymerization 4, 0.52 [15]; <39> cellobiose 15 mg/ml [18]) [3, 15, 18]

pH-Optimum
4.3 <1> [1]
4.5 <4> [4]
5 <3, 31, 39> (<31> both isoenzymes [10]) [3, 7, 10, 18]
5-6 <34, 38> [13, 17]
5.5 <29> [8]
6-7.5 <40> [19]
6.5 <2> [2, 20]
6.8 <2> [21]

Temperature optimum (°C)
30-35 <2> [2]
37 <2> [21]
40 <29> [8]
45 <2> [20]
50 <1, 3, 31, 39> (<31> both isoenzymes [10]) [1, 3, 10, 18]
70 <4, 38> [4, 17]
85 <34> (<34> avicelase activity [13]) [13]
95 <40> [19]
95-100 <34> (<34> carboxymethyl cellulase activity [13]) [13]

4 Enzyme Structure

Molecular weight
48600 <28> (<28> ultracentrifugation [6]) [6]
53000 <31> (<31> isoenzyme II [10]) [10]
56000 <31> (<31> isoenzyme I [10]) [10]
64000 <2> (<2> gel filtration [20]) [20]
65000 <3> (<3> gel filtration [7]) [7]
151000 <1> (<1> gel filtration [1]) [1]
200000 <2> (<2> gel filtration [21]) [21]
Additional information <4> (<4> 70000 g/mol [4]) [4]

Subunits
? <2> (<2> x * 58000, SDS-PAGE [20]) [20]
? <29> (<29> x * 81200, SDS-PAGE [8]) [8]
? <3> (<3> x * 20000, SDS-PAGE [3]) [3]
? <30> (<30> x * 87000, SDS-PAGE [9]) [9]
? <32> (<32> x * 47300, nonglycosylated form, SDS-PAGE [22]) [22]
? <32> (<32> x * 49300, glycosylated form, SDS-PAGE [22]) [22]

? <33> (<33> x * 112000, SDS-PAGE [12]) [12]
? <34> (<34> x * 23000, SDS-PAGE [13]) [13]
? <37> (<37> x * 80000, SDS-PAGE [16]) [16]
? <38> (<38> x * 87000, SDS-PAGE [17]) [17]
? <39> (<39> x * 85000, SDS-PAGE [18]) [18]
? <40> (<40> x * 29000, SDS-PAGE [19]) [19]
dimer <1> (<1> 2 * 83500 [1]) [1]
dimer <2> (<2> α,β, 2 * 100000, SDS-PAGE [21]) [21]

5 Isolation/Preparation/Mutation/Application

Source/tissue

commercial preparation <31> (<31> Driselase, from Irpex lacteus [10]) [10]
culture medium <1, 2, 3, 4, 28, 29, 34, 35, 38, 39> [1, 3, 4, 6, 7, 8, 13, 14, 17, 18, 21]
seedling <36> [15]

Localization

cell envelope <37> (<37> expressed in E. coli [16]) [16]
extracellular <1, 2, 3, 4, 28, 29, 34, 35, 38, 39> [1, 3, 4, 6, 7, 8, 13, 14, 17, 18, 21]
membrane <30> [9]

Purification

<1> [1]
<2> (39fold [21]) [21]
<2> (520fold [2]) [2]
<2> (66.7fold [20]) [20]
<3> [3]
<3> [7]
<4> [4]
<28> (15fold [6]) [6]
<29> (<29> 3368fold [8]) [8]
<32> [11]
<32> [22]
<33> [12]
<34> (4.8fold [13]) [13]
<38> (12fold [17]) [17]
<39> (154.84fold [18]) [18]

Crystallization

<32> (resolution 2.0 A [11]) [11]

Cloning

<35> (E. coli and Streptomyces lividans [14]) [14]
<2, 30, 31, 32, 34, 37> (E. coli [9, 10, 11, 13, 16, 20, 22]) [9, 10, 11, 13, 16, 20, 22]

Engineering

A229V <35> (<35> loop mutant, weaker binding affinities to all tested ligands [14]) [14]
E495D <35> (<35> active site mutant, weaker binding affinities to all tested ligands [14]) [14]
E495N <35> (<35> active site mutant, weaker binding affinities to all tested ligands [14]) [14]
G234A <35> (<35> loop mutant, weaker binding affinities to all tested ligands [14]) [14]
G234D <35> (<35> loop mutant, weaker binding affinities to all tested ligands [14]) [14]
G234N <35> (<35> loop mutant, weaker binding affinities to all tested ligands [14]) [14]
G234S <35> (<35> loop mutant, higher binding affinities for 4-methylumbelliferyl β-cellotrioside and cellotriose [14]) [14]
G234S/G284P <35> (<35> loop mutant, higher binding affinity for 4-methylumbelliferyl β-cellotrioside [14]) [14]
G284A <35> (<35> loop mutant, weaker binding affinities to all tested ligands [14]) [14]
G284I <35> (<35> loop mutant, weaker binding affinities to all tested ligands [14]) [14]
G284P <35> (<35> loop mutant, weaker binding affinities to all tested ligands [14]) [14]
G284P/A229V <35> (<35> loop mutant, weaker binding affinities to all tested ligands [14]) [14]
G331A <35> (<35> active site mutant, weaker binding affinities to all tested ligands [14]) [14]
G496P <35> (<35> loop mutant, weaker binding affinities to all tested ligands [14]) [14]
H326A <35> (<35> active site mutant, weaker binding affinities to all tested ligands [14]) [14]
H326S <35> (<35> active site mutant, higher binding affinities for 4-methylumbelliferyl β-cellobioside and cellotriose [14]) [14]
L230C <35> (<35> loop mutant, weaker binding affinities to all tested ligands [14]) [14]
L230C/D512C <35> (<35> loop mutant, weaker binding affinities to all tested ligands [14]) [14]
N233C <35> (<35> loop mutant, weaker binding affinities to all tested ligands [14]) [14]
N233C/D506C <35> (<35> loop mutant, weaker binding affinities to all tested ligands [14]) [14]
W329C <35> (<35> active site mutant, weaker binding affinities to all tested ligands [14]) [14]
W332A <35> (<35> active site mutant, weaker binding affinities to all tested ligands [14]) [14]

W332Y <35> (<35> active site mutant, weaker binding affinities to all tested ligands [14]) [14]

6 Stability

pH-Stability

3-5 <31> (<31> isoenzyme II [10]) [10]
3-9 <31> (<31> isoenzyme I [10]) [10]
3.6-7 <29> [8]
5.5-8 <2> [21]
6-7.5 <2> (<2> most stable at pH 6-7.5 [2]) [2]

Temperature stability

30 <2, 29> (<2, 29> stable below 30°C [2, 8, 21]; <29> inactivated above 50°C [8]) [2, 8, 21]
37 <2> (<2> 10% activity loss after 3 h at 37°C [20]; <2> 20% activity loss after 10 min, 75% activity loss after 15 h at 37°C [21]) [20, 21]
50 <3, 4, 31> (<3> stable at 50°C, activity loss above 70°C [7]; <31> both isoenzymes stable below 50°C [10]; <4> loses less than 3% activity after 7 hours at 50°C [4]) [4, 7, 10]
55 <1> (<1> stable below 55°C [1]) [1]
95 <40> (<40> 50% activity loss after 30 min at 95°C [19]) [19]

References

[1] Himmel, M.E.; Tucker, M.P.; Lastick, S.M.; Oh, K.K.; Fox, J.W.; Spindler, D.D.; Grohmann, K.: Isolation and characterization of a 1,4-β-D-glucan glucohydrolase from the yeast, Torulopsis wickerhamii. J. Biol. Chem., **261**, 12948-12955 (1986)

[2] Ohmiya, K.; Shimizu, S.: β-Glucosidase from Ruminococcus albus. Methods Enzymol., **160**, 408-414 (1988)

[3] Rao, M.; Mishra, C.: Glucohydrolase from Penicillium funiculosum. Appl. Microbiol. Biotechnol., **30**, 130-134 (1989)

[4] Schmid, G.; Wandrey, C.: Characterization of a cellodextrin glucohydrolase with soluble oligomeric substrates: Experimental results and modeling of concentration-time-course data. Biotechnol. Bioeng., **33**, 1445-1460 (1989)

[5] Augustin, J.; Zemek, J.; Kuniak, B.; Fassatiova, O.: Production of β-glucosidase, exo-β-1,4-glucanase and endo-β-1,4-glucanase by selected microscopic fungi. Folia Microbiol., **26**, 14-18 (1981)

[6] Eriksson, K.-E.; Petterson, B.: Extracellular enzyme system utilized by the fungus Sporotrichum pulverulentum (Chrysosporium lignorum) for the breakdown of cellulose. 3. Purification and physico-chemical characterization of an exo-1,4-β-glucanase. Eur. J. Biochem., **51**, 213-218 (1975)

[7] Wood, T.M.; McCrae, S.I.: Purification and some properties of a (1-4)-β-D-glucan glucohydrolase associated with the cellulase from the fungus Penicillium funiculosum. Carbohydr. Res., **110**, 291-303 (1982)
[8] Tahara, N.; Tonouchi, N.; Yano, H.; Yoshinaga, F.: Purification an characterization of exo-1,4-β-glucosidase from Acetobacter xylinum BPR2001. J. Ferment. Bioeng., **85**, 589-594 (1998)
[9] Wulff-Strobel, C.; Wilson, D.B.: Cloning, sequencing, and characterization of a membrane-associated Prevotella ruminicola B14 β-glucosidase with cellodextrinase and cyanoglycosidase activities. J. Bacteriol., **177**, 5884-5890 (1995)
[10] Hamada, N.; Ishikawa, K.; Fuse, N.; Kodaira, R.; Shimosaka, M.; Amano, Y.; Kanda, T.; Okazaki, M.: Purification, characterization and gene analysis of exo-cellulase II (Ex-2) from the white rot basidiomycete Irpex lacteus. J. Biosci. Bioeng., **87**, 442-451 (1999)
[11] Bedarkar, S.; Gilkes, N.R.; Kilburn, D.G.; Kwan, E.; Rose, D.R.; Miller Jr.; R.C.; Warren, R.A.J.; Withers, S.G.: Crystallization and preliminary X-ray diffraction analysis of the catalytic domain fo Cex, an exo-β-1,4-glucanase and β-1,4-xylanase from the bacterium Cellulomonas fimi. J. Mol. Biol., **228**, 693-695 (1992)
[12] Doerner, K.C.; Gardner, R.M.; Schook, L.B.; Mackie, R.I.; White, B.A.: Inhibition of the exo-β-1,4-glucanase from Ruminococcus flavefaciens FD-1 by a specific monoclonal antibody. Enzyme Microb. Technol., **16**, 2-9 (1994)
[13] Zverlov, V.; Mahr, S.; Riedel, K.; Bronnenmeier, K.: Properties and gene structure of a bifunctional celluloytic enzyme (CelA) from the extreme thermophile Ánaerocellum thermophilum' with separate glycosyl hydrolase family 9 and 48 catalytic domains. Microbiology, **144**, 457-465 (1998)
[14] Zhang, S.; Irwin, D.C.; Wilson, D.B.: Site-directed mutation of noncatalytic residues of Thermobifida fusca exocellulase Cel6B. Eur. J. Biochem., **267**, 3101-3115 (2000)
[15] Hrmova, M.; MacGregor, E.A.; Biely, P.; Stewart, R.J.; Fincher, G.B.: Substrate binding and catalytic mechanism of a barley β-D-glucosidase/(1,4)-β-D-glucan exohydrolase. J. Biol. Chem., **273**, 11134-11143 (1998)
[16] Rixon, J.E.; Ferreira, L.M.A.; Durrant, A.J.; Laurie, J.I.; Hazlewood, G.O.; Gilbert, H.J.: Characterization of the gene celD and its encoded product 1,4-β-D-glucan glucohydrolase D from Pseudomonas fluorescens subsp. cellulosa. Biochem. J., **285**, 947-955 (1992)
[17] Bronnemeier, K.; Rücknagel, K.P.; Staudenbauer, W.L.: Purification and properties of a novel type of exo-1,4-β-glucanase (Avicelase II) from the cellulolytic thermophile Clostridium stercorarium. Eur. J. Biochem., **200**, 379-385 (1991)
[18] Sharmila, T.; Sreeramulu, G.; Nand, K.: Purification and characterization of β-1,4-glucosidase from Clostridium papyrosolvens. Biotechnol. Appl. Biochem., **27**, 175-179 (1998)
[19] Bronnenmeier, K.; Kern, A.; Liebl, W.; Staudenbauer, W.L.: Purification of Thermotoga maritima enzymes for the degradation of cellulosic materials. Appl. Environ. Microbiol., **61**, 1399-1407 (1995)

[20] Ware, C.E.; Lachke, A.H.; Gregg, K.: Mode of action and substrate specificity of a purified exo-1,4-β-D-glucosidase cloned from the cellulolytic bacterium Ruminococcus albus AR67. Biochem. Biophys. Res. Commun., **171**, 777-786 (1990)
[21] Ohmiya, K.; Shimizu, S.: Cellobiosidase from Ruminococcus albus. Methods Enzymol., **160**, 391-398 (1988)
[22] Gilkes, N.R.; Warren, A.J.; Miller Jr.; R.C.; Kilburn, D.G.: Precise excision of the cellulose binding domains from two Cellulomonas fimi cellulases by a homologous protease an the effect on catalysis. J. Biol. Chem., **263**, 10401-10407 (1988)

Glucan endo-1,6-β-glucosidase 3.2.1.75

1 Nomenclature

EC number
3.2.1.75

Systematic name
1,6-β-D-glucan glucanohydrolase

Recommended name
glucan endo-1,6-β-glucosidase

Synonyms
β-1,6-glucan 6-glucanohydrolase
β-1,6-glucan hydrolase
β-1,6-glucanase-pustulanase
β-1-6-glucan hydrolase
endo-(1-6)-β-D-glucanase
endo-1,6-β-glucanase
endo-β-1,6-glucanase

CAS registry number
37278-39-0

2 Source Organism

<1> *Neurospora crassa* (strain IFO-6068 [1]) [1, 7]
<2> *Trichoderma harzianum* (CECT 2313 [20]) [2, 17, 19, 20, 22]
<3> *Chlamys albidus* [3, 18]
<4> *Penicillium brefeldianum* (QM 1873 [4]) [4]
<5> *Flavobacterium M64* [5]
<6> *Acinetobacter sp.* [6]
<7> *Gibberella fujikuroi* [12, 13, 15]
<8> *Saccharomyces cerevisiae* [7]
<9> *Trichoderma viride* [7, 14]
<10> *Mucor hiemalis* [8]
<11> *Bacillus circulans* (WL-12 [9, 10]) [9, 10]
<12> *Rhizopus chinensis* (R-69 [11]) [11]
<13> *Penicillium sp.* (10 different species [14]) [14]
<14> *Aspergillus sp.* (4 different species [14]) [14]
<15> *Beauveria bassiana* [14]
<16> *Fusarium roseum* [14]

<17> *Poria cocos* [14]
<18> *Trichoderma lignorum* [14]
<19> *Sporotrichum pruinosum* [14]
<20> *Aspergillus niger* [15]
<21> *Aspergillus saitoi* [15]
<22> *Aspergillus usamii* [15]
<23> *Fusarium lini* [15]
<24> *Fusarium oxysporum* [15]
<25> *Saccharomycopsis fibuligera* (NCYC 451 [21]) [21]
<26> *Helminthosporium sesanum* [15]
<27> *Macrosporium bataticola* [15]
<28> *Acremonium persicinum* [16]
<29> *Trichoderma sp.* (several strains: A.11.178, A.9.3., B.1.2, T.16.C, T.2.C [17]) [17]

3 Reaction and Specificity

Catalyzed reaction
random hydrolysis of 1,6-linkages in 1,6-β-D-glucans

Reaction type
hydrolysis of O-glycosyl bond

Substrates and products
S amylopectin + H_2O <12> [11]
P ?
S amylose + H_2O <1, 12> [1, 11]
P ?
S β-glucan + H_2O <1, 2, 4, 6, 10, 11, 12, 28> (<1> Neurospora crassa cell wall β-glucan [1]; <4> Ecklonia radiata glucan, poor substrate [4]; <6, 10, 11, 12, 28> yeast glucans [6, 8, 9, 11, 16]; <2> Saccharomyces cerevisiae glucan [20]) [1, 4, 6, 8, 9, 11, 16, 20]
P glucose + laminari-oligosaccharides + gentio-oligosaccharides <6> [6]
S carboxymethyl cellulose + H_2O <5, 10> [5, 8]
P ?
S carboxymethyl-pachyman + H_2O <4> (<4> poor substrate [4]) [4]
P ?
S cellulose + H_2O <1, 12> [1, 11]
P ?
S dextran + H_2O <10, 12> (<12> low molecular weight [11]) [8, 11]
P ?
S gentiobiose + H_2O <4, 5> (<4> poor substrate [4]) [4, 5]
P D-glucose <4, 5> [4, 5]
S gentiohexaose + H_2O <7, 10> [8, 13]
P gentiobiose + gentiotriose + gentiotetraose <10> [8]
S gentiopentaose + H_2O <6, 7, 10, 12> (<6> reduced [6]) [6, 8, 11, 13]

P D-glucose + gentiotetraose + gentiotriose + gentiobiose <6, 10, 12> (<10, 12> gentiobiose and gentiotriose only [8, 11]) [6, 8, 11]
S gentiopentaosylsorbitol + H_2O <6> [6]
P gentiotetraose + gentiobiose + D-glucose <6> [6]
S gentiotetraose + H_2O <6, 7, 10, 11, 12> (<6> reduced [6]) [6, 8, 9, 11, 13]
P D-glucose + gentiotriose + gentiobiose <6, 10, 12> (<10, 12> gentiobiose only [8, 11]) [6, 8, 11]
S gentiotriose + H_2O <6> [6]
P D-glucose + gentiobiose <6> [6]
S gentiotriosylsorbitol + H_2O <6> [6]
P ?
S glycogen + H_2O <1, 12> [1, 11]
P ?
S gyrophoran + H_2O <7> [13]
P ?
S islandican + H_2O <10> [8]
P ?
S lactose + H_2O <12> [11]
P D-glucose <12> [11]
S laminaran + H_2O <1, 7, 10> (<7> from Eisenia bicyclis [12]) [1, 8, 12]
P ?
S laminarin + H_2O <2, 6, 11, 12, 28, 25> [6, 9, 11, 16, 20, 21]
P gentiobiose + D-glucose <11> [9]
S lichenan + H_2O <1> [1]
P ?
S lutean + H_2O <6, 7, 20-24, 26-28> [6, 13, 15, 16]
P D-glucose + gentiobiose + gentiotriose <7, 20-24, 26, 27> [13, 15]
S luteose + H_2O <5, 12> [5, 11]
P gentiobiose + gentiotriose + gentiotetraose <12> [11]
S *p*-nitrophenyl-β-D-glucoside + H_2O <5, 11> [5, 9]
P 4-nitrophenol + β-D-glucose <5, 11> [5, 9]
S pachyman + H_2O <4, 11, 12> (<4,11> poor substrate [4, 9]) [4, 9, 11]
P gentiobiose <11> (<11> trace [9]) [9]
S pustulan + H_2O <1, 2, 3, 6, 11, 13, 28, 29> [1, 3, 6, 9, 10, 14, 16-20, 22]
P glucose + gentiobiose <1, 2, 3, 6, 11, 13> (<3> glucose, gentiobiose and gentiotriose [3]; <6> glucose and oligosaccharides [6]; <11> also gentiopentaose, gentiotetraose and gentiotriose [9]; <13> gentiotetraose and longer chains, gentiotriose, gentiotriose and glucose [14]; <2> glucose, gentiotetraose, gentiobiose [20]) [1, 3, 6, 9, 14, 20]
S starch + H_2O <12> (<12> soluble, from corn, from potato, from wheat [11]) [11]
P ?
S succinoglycan + H_2O <5> [5]
P ?
S Additional information <7> [12]
P ?

Inhibitors

2-hydroxy-5-nitrobenzyl bromide <28> [16]
Cd^{2+} <10> [8]
Co^{2+} <1> [1]
Cu^{2+} <1, 3, 10, 12, 28> [1, 3, 8, 11, 16]
Fe^{2+} <1, 3, 12> [1, 3, 11]
Fe^{3+} <1, 10> [1, 8]
Hg^{2+} <1, 10, 11, 12, 28> [1, 8, 10, 11, 16]
$KMnO_4$ <28> [16]
$MgSO_4$ <28> [16]
Mn^{2+} <3, 28> [3, 16]
N-acetylimidazole <28> [16]
N-bromosuccinimide <28> [16]
N-ethyl-5-phenylisoxazolium 3'-sulfonate <28> [16]
Sn^{2+} <28> [16]
Zn^{2+} <3, 10, 12, 28> [3, 8, 11, 16]
dicyclohexylcarbodiimide <28> [16]
glucono-δ-lactone <11> [10]
p-hydroxymercuribenzoate <28> [16]
Additional information <3> (<3> ethanol extracts of various marine invertebrates: Geodinella spherastrosa, Spirastrella digitata, Spirastrella vagabunda, Chondrilla australiensis, Phakelia cardius, Desmoxya lunata, Mycale monanchorata, Clathria frondifera, Clathria lendenfeldi, Iotrochota baculifera, Phorbas sp., Callyspongia diffusa, Callyspongia sp., Callyspongia subarmigera, Arenosclera sp., Amphimedon sp., Oceanapia imperfecta, Sigmadocia symbiotica, Dysidera herbacea, Dysidea fragilis, Thorecta cacos, Ircinia sp. [18]) [18]

Specific activity (U/mg)

0.22 <5> [5]
0.7 <12> [11]
1.654 <3> [3]
5 <25> (<25> intracellular enzyme [21]) [21]
5.18 <10> [8]
13.4 <1> [1]
75.4 <25> (<25> extracellular enzyme [21]) [21]
131.2 <11> [10]
146 <28> [16]
150 <2> [20]
204 <2> [22]
405 <4> [4]
Additional information <6, 7> (<6> 224.7 U/mg, one U being 1 mg of liberated reducing sugar from pustulan after 2 h incubation [6]; <7> 34.3 reducing sugars corresponding to 4.43 micro mol of glucose/min/mg [13]) [6, 13]

K_m-Value (mM)

0.00012 <7> (lutean, <7> assumed molecular weight for lutean 650000 Da [13]) [13]

0.014 <7> (gentiotetraose) [13]
0.0753 <28> (pustulan, <28> assumed molecular weight for pustulan 17000 Da, or K_M of 1.28 mg/ml [16]) [16]
Additional information <1, 2, 3, 4, 10, 11, 13, 19, 28>(<1> pustulan 0.8 mg/ml [1]; <3> pustulan 0.38 mg/ml [3]; <4> pustulan 2.78 mg/ml [4]; <10> islandican 0.173% [8]; <11> pustulan 0.29 mg/ml [9]; <13, 19> pustulan 0.4-1.3 mg/ml [14]; <28> lutean 1.38 mg/ml, laminarin 1.67 mg/ml [16]; <2> pustulan 0.8 mg/ml [20]) [1, 3, 4, 8, 9, 14, 16, 20]

pH-Optimum
4.2 <4> [4]
4.2-4.6 <3> [3]
4.5 <10> [8]
5 <1, 2, 7, 28> [1, 13, 16, 22]
5-5.5 <7, 20-24, 26, 27> [15]
5.5 <6, 11, 25> [6, 10, 21]
5.5-6 <12> [11]
6 <11> (<11> in Tris/succinate buffer [9]) [9]
6.5 <11> (<11> in sodium phosphate buffer [9]) [9]
Additional information <9, 13-19> (<9, 13-19> pH 4 +/- 0.4 for six species [14]) [14]

pH-Range
2-7 <9, 13-19> [14]

Temperature optimum (°C)
50 <1, 2, 4, 25> [1, 4, 20, 21]
50-60 <9, 13-19> [14]
55 <10> [8]
60 <12> [11]

4 Enzyme Structure

Molecular weight
20000 <3> (<3> gel filtration [3]) [3]
20000-25000 <2> (<2> gel filtration [20]) [20]
30000 <12> (<12> gel filtration [11]) [11]
30500 <12> (<12> sedimentation equilibrium [11]) [11]
37000 <7> (<7> gel filtration [13]) [13]
43000 <25> (<25> extracellular enzyme, gel filtration [21]) [21]
44300 <28> (<28> gel filtration [16]) [16]
50000 <6> (<6> gel filtration [6]) [6]
51000 <25> (<25> intracellular enzyme, gel filtration [21]) [21]
54000 <11> (<11> SDS-PAGE, gel filtration [9]) [9]
55000 <6> (<6> PAGE [6]) [6]

Subunits
? <2> (<2> x * 43000, SDS-PAGE [22]) [22]

? <2> (<2> x * 51000, SDS-PAGE [20]) [20]
? <28> (<28> x * 42700, SDS-PAGE [16]) [16]
? <4> (<4> x * 44000, SDS-PAGE [4]) [4]

5 Isolation/Preparation/Mutation/Application

Source/tissue

commercial preparation <2> (<2> NOVOZYM SP-116 [2]) [2]
crystalline style <3> [3, 18]
culture filtrate <1, 2, 4, 6, 7, 10, 11, 12, 28, 29, 25> [1, 4, 6, 8-13, 16, 17, 20, 21, 22]

Localization

extracellular <1, 2, 4, 6, 7, 9, 10, 11, 12, 28, 29, 25> [1, 4, 6-13, 16, 17, 20, 21, 22]
intracellular <25> [21]

Purification

<1> (618fold [1]) [1]
<2> (<2> 154fold, 68fold expressed in Pichia pastoris [22]) [22]
<3> (486.5fold [3]) [3]
<4> (7080fold [4]) [4]
<5> (140fold [5]) [5]
<6> (102.1fold [6]) [6]
<7> [13]
<11> (39.5fold [10]) [10]
<12> [11]
<25> (<25> intracellular enzyme 100fold, extracellular enzyme 75fold [21]) [21]
<28> (<28> 89fold [16]) [16]

Cloning

<2> (expression in Saccharomyces cerevisiae [19]; expression in Pichia pastoris [22]) [19, 22]

6 Stability

pH-Stability

3-7 <10> [8]
4-6 <12> [11]
4.5-9.5 <28> [16]
4.9-5.5 <3> [3]
5-6 <4> (<4> at -20°C [4]) [4]
5-8 <1> [1]
5-8 <4> [4]

Temperature stability

30 <7> (<7> stable for a long time at 30°C, activity decrease at 40°C [13]) [13]
30-60 <10> [8]
50 <1, 28> (<1> stable blow 50°C [1]; <28> short-term stability up to 50°C [16]) [1, 16]
65 <12> (<12> stable below 65°C [11]) [11]
70 <9, 13-19> (<9, 13-19> inactivated in less than 10 min at 70°C [14]) [14]

Storage stability

<3>, 4°C, 6 months, no loss of activity [3]
<4>, -20°C, pH 6, 2 months, no loss of activity [4]
<11>, -25°C, 1 month, less than 10% loss of activity [9]
<11>, 2°C, 1 week, 20% loss of activity [9]
<28>, 4°C, 4 months, no loss of activty, -70°C, 12 months, no loss of activity [16]

References

[1] Hiura, N.; Nakajima, T.; Matsuda, K.: Purification and some properties of an endo-β-1,6-glucanase from Neurospora crassa. Agric. Biol. Chem., **51**, 3315-3321 (1987)
[2] Dubourdieu, D.; Desplanques, C.; Villettaz, J.-C.; Ribereau-Gayon, P.: Investigations of an industrial β-D-glucanase from Trichoderma harzianum. Carbohydr. Res., **144**, 277-287 (1985)
[3] Rudakova, V.Y.; Shevchenko, N.M.; Elyakova, L.A.: Isolation and some properties of endo-β-1,6-glucanase from marine bivalve Chlamys abbidus. Comp. Biochem. Physiol. B, **81**, 677-682 (1985)
[4] Schep, G.P.; Shepherd, M.G.; Sullivan, P.: Purification and properties of a β-1,6-glucanase from Penicillium brefeldianum. Biochem. J., **233**, 707-714 (1984)
[5] Abe, J.; Amemura, A.; Harada, T.: An endo-(1-6)-β-D-glucanase of Flavobacterium M64 hydrolyzing the octasaccharide repeating unit of succinoglycan to two tetrasaccharides. Agric. Biol. Chem., **44**, 1877-1884 (1980)
[6] Katohda, S.; Suzuki, F.; Katsuki, S.; Sato, T.: Purification and some properties of endo-β-1,6-glucanase from Acinetobacter sp.. Agric. Biol. Chem., **43**, 2029-2034 (1979)
[7] Del Rey, F.; Garcia-Acha, I.; Nombela, C.: The regulation of β-glucanase synthesis in fungi and yeast. J. Gen. Microbiol., **110**, 83-89 (1979)
[8] Miyazaki, T.; Oikawa, N.: An endo-(1-6)-β-D-glucanase from Mucor hiemalis. Carbohydr. Res., **48**, 209-216 (1976)
[9] Rombouts, F.M.; Phaff, H.J.: Lysis of yeast cell walls. Lytic β-(1-6)-glucanase from Bacillus circulans WL-12. Eur. J. Biochem., **63**, 109-120 (1976)
[10] Fleet, G.H.; Phaff, H.J.: Lysis of yeast cell walls: glucanases from Bacillus circulans WL-12. J. Bacteriol., **119**, 207-219 (1974)

[11] Yamamoto, S.; Kobayashi, R.; Nagasaki, S.: Purification and properties of an endo β-1,6-glucanase from Rhizopus chinensis R-69. Agric. Biol. Chem., **38**, 1493-1500 (1974)
[12] Shibata, Y.: Enzymatic hydrolysis of glucans containing β-1,3- and β-1,6-linkages. III. Gibberella β-1,6-glucan 6-glucanohydrolase operative in the selective hydrolysis of β-1,3-glucosidic linkages in Eisenia laminaran. J. Biochem., **75**, 85-92 (1974)
[13] Shibata, Y.; Fukimbara, T.: A β-1,6-glucan 6-glucanohydrolase from Gibberella fujikuroi. J. Ferment. Technol., **51**, 216-226 (1973)
[14] Reese, E.T.; Parrish, F.W.; Mandels, M.: β-D-1,6-Glucanases in fungi. Can. J. Microbiol., **8**, 327-329 (1962)
[15] Shibata, Y.; Fukimbara, T.: Enzymic hydrolysis of glucans containing β-1,3- and β-1,6-linkages: (I) β-1,6-glucan hydrolase of fungi. J. Ferment. Technol., **50**, 388-396 (1972)
[16] Pitson, S.M.; Seviour, R.J.; McDougall, B.M.; Stone, B.A.; Sadek, M.: Purification and characterization of an extracellular (1-6)-β-glucanase from the filamentous fungus Acremonium persicinum. Biochem. J., **316**, 841-846 (1996)
[17] Soler, A.; de la Cruz, J.; Llobell, A.: Detection of β-1,6-glucanase isozymes from Trichoderma strains in sodium dodecyl sulphate-polyacrylamide gel electrophoresis and isoelectrofocusing gels. J. Microbiol. Methods, **35**, 245-251 (1999)
[18] Zvyagintseva, T.N.; Elyakova, L.A.; Krasohin, V.B.: The search for effectors of β-D-glucanases among marine invertebrates. Comp. Biochem. Physiol., **102B**, 187-191 (1992)
[19] Lora, J.M.; de la Cruz, J.; Llobell, A.; Benítez, Pintor-Toro, J.A.: Molecular characterization and heterologous expression of an endo-β-1,6-glucanase gene from the mycoparasitic fungus Trichoderma harzianum. Mol. Gen. Genet., **247**, 639-645 (1995)
[20] de la Cruz, J.; Llobell, A.: Purification and properties of a basic endo-β-1,6-glucanase (BGN16.1) from the antagonistic fungus Trichoderma harzianum. Eur. J. Biochem., **265**, 245-151 (1999)
[21] Mulenga, D.K.; Berry, D.R.: Isolation and characterization of a uniqe endo-β-1,6-glucanase from the yeast Saccharomycopsis fibuligera NCYC 451. Microbios, **80**, 143-154 (1994)
[22] Bom, I.J.; Dielbandhoesing, S.K.; Harvey, K.N.; Oomes, S.J.C.M.; Klis, F.M.; Brul, S.: A new tool for studying the molecular architecture of the fungal cell wall: one-step purification of recombinant Trichoderma β-(1-6)-glucanase expressed in Pichia pastoris. Biochim. Biophys. Acta, **1425**, 419-424 (1998)

L-Iduronidase 3.2.1.76

1 Nomenclature

EC number
3.2.1.76

Systematic name
glycosaminoglycan α-L-iduronohydrolase

Recommended name
L-iduronidase

Synonyms
α-L-iduronidase

CAS registry number
9073-56-7

2 Source Organism

<1> *Homo sapiens* [1, 3-6, 8-14, 16, 18-21]
<2> *Rattus norvegicus* [2, 15]
<3> *Sus scrofa* [7, 23]
<4> *Bos taurus* [15]
<5> *canine* [17]
<6> *Mus musculus* [22]

3 Reaction and Specificity

Catalyzed reaction
hydrolysis of unsulfated α-L-iduronosidic linkages in dermatan sulfate

Natural substrates and products
S α-L-iduronisyl-(α-1,3)-anhydrotalitol 4-sulfate + H_2O <1> [6, 11]
S α-L-iduronisyl-(α-1,4)-anhydrotalitol 4-sulfate + H_2O <1> [6]
S α-L-iduronisyl-(α-1,4)2,5-anhydro-D-mannitol + H_2O <1> [11]
S α-L-iduronisyl-(α-1,4)2,5-anhydro-D-mannitol 6-sulfate + H_2O <1> [11]
S anhydromannitol iduronide + H_2O <1> [5]

Substrates and products
S 4-methylumbelliferyl-α-L-iduronide + H_2O <1-4> [7, 11, 15]
P ?

S IdoA-GlcNAc$_6$S-IdoA$_2$S-anM$_6$S + H_2O <1> (O-(α-L-iduronic acid)-(1,4)-D-O-(α-N-acetylglucosamine 6-sulphate)-L-O-(α-iduronic acid 2-sulphate)-O-D-2,5-anhydro-D-mannitol 6-sulphate [16]) [16]
P ?
S α-L-iduronisyl-(α-1,3)-anhydrotalitol 4-sulfate + H_2O <1> [6, 11]
P ?
S α-L-iduronisyl-(α-1,4)-anhydrotalitol 4-sulfate + H_2O <1> [6]
P ?
S α-L-iduronisyl-(α-1,4)2,5-anhydro-D-mannitol + H_2O <1> [11]
P ?
S α-L-iduronisyl-(α-1,4)2,5-anhydro-D-mannitol 6-sulfate + H_2O <1> [11]
P ?
S anhydromannitol iduronide + H_2O <1> [5]
P ?
S phenyl α-L-iduronide + H_2O <1, 2> [2, 3, 5]
P ?

Inhibitors

2(S)-carboxy-3(R),4(R),5(S)-trihydroxypiperidine <1> [8]
2,5-anhydro-D-mannitol <1> [11]
2,5-anhydro-D-mannitol 6-sulfate <1> [11]
Cu^{2+} <1, 3> [5, 7, 18]
$CuCl_2$ <1> [11]
D-saccharolactone <2> [2]
Fe^{3+} <1> [5]
Na_2SO_4 <1> [11, 16]
NaCl <1> [11]
SO_4^{2-} <3> [7]
α-L-idosyl-(α-1,4)2,5-anhydro-D-mannitol 6-sulfate <1> [10]
p-chloromercuribenzoate <1> [5, 14]

Activating compounds

2-mercaptoethanol <1> [5, 14]
dithiothreitol <1> [5, 14]
glutathione <1> [5, 14]

Specific activity (U/mg)

2.08 <1> [10]
16 <1> (<1> phenyl α-L-iduronide as substrate [5]) [5]
34.9 <3> [7]
50 <1> [6]
87.5 <1> (<1> soluble form [13]) [13]
109 <1> (<1> membrane associated form [13]) [13]
250 <1> [14]
270 <1> (<1> anhydromannitol iduronide as substrate [5]) [5]
5000-10000 <1> [20]
126000 <5> [17]

K_m-Value (mM)

0.001 <1> (IdoA-GlcNAc$_6$S-IdoA$_2$S-anM$_6$S, <1> pH 3.3 [16]) [16]
0.001 <1> (IdoA-GlcNH$_6$S-IdoA$_2$S-anM$_6$S, <1> pH 3 [16]) [16]
0.001 <1> (IdoA-GlcNS$_6$S-IdoA$_2$S-anM$_6$S, <1> pH 3.6 [16]) [16]
0.002 <1> (IdoA-GlcNS$_6$S-IdoA$_2$S-anM$_6$S, <1> pH 4.8 [16]) [16]
0.002 <1> (α-L-iduronisyl-(α-1,4)2,5-anhydro-D-mannitol 6-sulfte, <1> pH 2.7 [16]) [16]
0.003 <1> (IdoA-GlcNAc$_6$S-IdoA$_2$S-anM$_6$S, <1> pH 4.8 [16]) [16]
0.01 <1> (α-L-iduronisyl-(α-1,4)2,5-anhydro-D-mannitol 6-sulfate, <1> pH 4.8 [16]) [16]
0.016 <1> (α-L-iduronisyl-(α-1,4)2,5-anhydro-D-mannitol, <1> pH 4.8 [16]) [16]
0.02 <1> (α-L-iduronisyl-(α-1,4)2,5-anhydro-D-mannitol, <1> pH 3 [16]) [16]
0.04 <1> (4-methylumbelliferyl-α-L-iduronide, <1> pH 4.8 [16]) [16]
0.07 <3> (4-methylumbelliferyl-α-L-iduronide) [7]
0.08 <1> (α-L-iduronisyl-(α-1,3)-anhydrotalitol 4-sulfate) [11]
0.09 <1> (4-methylumbelliferyl-α-L-iduronide) [14]
0.1 <1> (4-methylumbelliferyl-α-L-iduronide, <1> pH 2.8 [16]) [16]
0.12 <1> (α-L-iduronisyl-(α-1,4)2,5-anhydro-D-mannitol 6-sulfate) [11]
0.14 <1> (α-L-iduronisyl-(α-1,4)2,5-anhydro-D-mannitol) [11]
0.18 <1> (4-methylumbelliferyl-α-L-iduronide) [20]
0.2 <2> (phenyl α-L-iduronide) [2]
0.31 <1> (phenyl α-L-iduronide, <1> L-iduronidase II [4]) [4]
0.45 <4> (4-methylumbelliferyl-α-L-iduronide) [15]
0.45 <1> (phenyl α-L-iduronide, <1> L-iduronidase I [4]) [4]
0.82 <1> (4-methylumbelliferyl-α-L-iduronide) [11]
1.5 <1> (phenyl α-L-iduronide) [5, 14]
1.92 <1> (4-methylumbelliferyl-α-L-iduronide) [16]
9 <1> (anhydromannitol iduronide) [5, 14]

pH-Optimum

3-3.5 <1, 3> (<1> both forms [4]) [4, 7]
3.5 <2> [2]
4 <1> (<1> phenyl α-L-iduronide as substrate [5]) [5, 14]
4.4 <1> (<1> anhydromannitol iduronide as substrate [5]) [5]
4.5 <1> (<1> anhydromannitol as substrate [14]) [10, 14]

pH-Range

3.5-7.5 <1> [10]

4 Enzyme Structure

Molecular weight

60000 <1> (<1> gel filtration [5]; <1> sedimentation equilibrium [14]) [5, 14]
65000 <1> (<1> gel filtration, form I and II from liver [6]) [6]

67000 <1> (<1> gel filtration, L-iduronidase II [4]) [4]
68000 <1> (<1> gel filtration, soluble form [13]) [13]
70000 <3> (<3> gel filtration [7]) [7]
85000 <1> (<1> gel filtration, membrane associated form [13]) [13]
87000 <1> (<1> gel filtration, L-iduronidase I [4]) [4]

Subunits

dimer <1> (<1> 2 * 31000, SDS-PAGE [5]; <1> 2 * 30000, SDS-PAGE [1]) [5, 14, 20]

Posttranslational modification

glycoprotein <1, 3> (<1> contains glucosamine [4]) [4, 5, 7, 14]
proteolytic modification <1> (<1> 84000Da to 69000Da [20]) [9, 18-20]

5 Isolation/Preparation/Mutation/Application

Source/tissue

amniotic fluid <1> [3]
brain <1, 3> (<3> low activity [23]) [21, 23]
colon <3> (<3> moderate activity [23]) [23]
diaphragm <3> (<3> low activity [23]) [23]
fibroblast <1> [1, 3, 4, 9, 12, 19]
heart <3> (<3> low activity [23]) [23]
ileum <3> (<3> moderate activity [23]) [23]
iliopsas muscle <3> (<3> low activity [23]) [23]
kidney <1, 3> (<3> high activity [23]) [5, 6, 14, 23]
liver <1-4> (<3> high activity [23]) [1, 2, 6, 7, 10, 11, 15, 16, 23]
lung <1, 3> (<3> moderate activity [23]) [6, 13, 23]
lymph node <3> (<3> moderate activity [23]) [23]
pancreas <3> (<3> moderate activity [23]) [23]
preputial gland <2> [15]
skin <1-3> (<3> low activity [23]) [1-3, 23]
spleen <3> (<3> high activity [23]) [23]
stomach <3> (<3> moderate activity [23]) [23]
testis <5> [17]
urine <1> [14]

Localization

lysosome <2> [2]
membrane-bound <1, 3> [7, 13]
soluble <1> [13]

Purification

<1> [5, 6, 10, 13, 14, 16, 20]
<4> [15]
<5> [17]

Cloning

<1> (<1> expression in CHO-K1 cells [18]) [18]
<1> (<1> expression in retroviral transduced fibroblasts [21]) [21]
<1> (<1> human-mouse somatic cell hybrids [12]) [12]
<5> (<5> expression in COS-1 cells [17]) [17]
<6> (<6> expression in COS-1 cells [22]) [22]

6 Stability

pH-Stability

3 <1> (<1> rapidly inactivated [10]) [10]
8 <1> (<1> rapidly inactivated [10]) [10]

Temperature stability

55 <1> (<1> both forms are stable over 30 min [4]) [4]
65 <1> (<1> L-iduronidase II is more labile [4]) [4]

Storage stability

<1>, 4°C, 0.03M sodium dimethylglutarate buffer, 10% glycerol, 0.5 M NaCl, pH 6 or 3.7, 18 months, 80% of activity retained [10]

References

[1] Matalon, R.; Cifonelli, J.A.; Dorfmann, A.: L-Iduronidase in cultured human fibroblasts and liver. Biochem. Biophys. Res. Commun., **42**, 340-345 (1971)

[2] Weissmann, B.; Santiago, R.: α-L-Iduronidase in lysosomal extracts. Biochem. Biophys. Res. Commun., **46**, 1430-1433 (1972)

[3] Hall, C.W.; Neufeld, E.F.: α-L-Iduronidase activity in cultured skin fibroblasts and amniotic fluid cells. Arch. Biochem. Biophys., **158**, 817-821 (1993)

[4] Shapiro, L.J.; Hall, C.W.; Leder, I.G.; Neufeld, E.F.: The relationship of α-L-iduronidase and hurler corrective faktor. Arch. Biochem. Biophys., **172**, 156-161 (1976)

[5] Rome, L.H.; Garvin, A.J.; Neufeld, E.F.: Human kidney α-L-iduronidase: purification and characterization. Arch. Biochem. Biophys., **189**, 344-353 (1978)

[6] Clements, P.R.; Brooks, D.A.; McCourt, P.A.G.; Hopwood, J.J.: Immunopurification and characterization of human α-L-iduronidase with the use of monoclonal antibodies. Biochem. J., **259**, 199-208 (1989)

[7] Ohshita, T.; Sakuda, H.; Nakasone, S.; Iwamasa, T.: Purification, characterization and localization of pig liver α-L-iduronidase. Eur. J. Biochem., **179**, 201-207 (1989)

[8] di Bello, I.C.; Dorling, P.; Fellows, L.; Winchester, B.: Specifc inhibition of human β-D-glucuronidase and α-L-iduronidase by a trihydroxy pipecolic acid of plant origin. FEBS Lett., **176**, 61-64 (1984)

[9] Myerowitz R.; Neufeld, E.F.: Maturation of α-L-iduronidase in cultured human fibroblasts. J. Biol. Chem., **256**, 3044-3048 (1981)
[10] Clements, P.R.; Brooks, D.A.; Saccone, G.T.P; Hopwood, J.J.: Human α-L-iduronidase 1. purification, monoclonal antibody production, native and subunit molecular mass. Eur. J. Biochem., **152**, 21-28 (1985)
[11] Clements, P.R.; Muller, V.; Hopwood, J.J.: Human α-L-iduronidase 2. catalytic properties. Eur. J. Biochem., **152**, 29-34 (1985)
[12] Schuchmann, E.H.; Astrin, K.H.; Aula, P.; Desnick, R.J.: Regional assignment of the structural gene for human α-L-iduronidase. Proc. Natl. Acad. Sci. USA, **81**, 1169-1173 (1984)
[13] Schuchmann, E.H.; Guzmann, N.A.; Desnick, R.J.: Human α-L-iduronidase. J. Biol. Chem., **259**, 3132-3140 (1984)
[14] Rome, L.H.: α-L-Iduronidase from human kidney. Methods Enzymol., **83**, 578-583 (1982)
[15] Kosaka, H.; Isemura, M.; Ono, T.; Nishimura, Y.; Kato, K.: Studies on the α-L-iduronidase activity of β-glucuronidase preparations from bovine liver, rat liver, and rat preputial gland. J. Biochem., **88**, 69-75 (1980)
[16] Freeman, C.; Hopwood, J.J.: Human α-L-iduronidase. Biochem. J., **282**, 899-908 (1992)
[17] Stoltzfus, L.J.; Sosa-Pineda, B.; Moskowitz, S.M.; Menon, K.P.; Dlott, B.; Hooper, L.; Teplow, D.B.; Shull, R.M.; Neufeld, E.F.: Cloning and characterization of cDNA encoding canine α-L-iduronidase. J. Biol. Chem., **267**, 6570-6575 (1992)
[18] Unger, E.G.; Durrant, J.; Anson, D.S.; Hopwood, J.J.: Recombinant α-L-iduronidase: characterization of the purified enzyme and correction of mucopolysaccharidosis type I fibroblasts. Biochem. J., **304**, 43-49 (1994)
[19] Brooks, D.A.; Harper, G.S.; Gibson, G.J.; Ashton, L.J.; Taylor, J.A.; McCourt, P.A.D.; Freeman, C.; Clements, P.R.; Hoffmann, J.W.; Hopwood, J.J.: Hurler Syndrome: a patient with abnormally high levels of α-L-iduronidase. Biochem. Med. Metab. Biol., **47**, 211-220 (1992)
[20] Kakkis, E.D.; Matynia, A.; Jonas, A.J.; Neufeld, E.F.: Overexpression of the human lysosomal enzyme α-L-iduronidase in chinese hamster ovary cells. Protein Expression Purif., **5**, 225-232 (1994)
[21] Stewart, K.; Brown, O.A.; Morelli, A.E.; Fairbairn, L.J.; Lashford, L.S.; Cooper, A.; Hatton, C.E.; Dexter, T.M.; Castro, M.G.; Lowenstein, P.R.: Uptake of α-(L)-iduronidase produced by retrovirally transduced fibroblasts into neuronal and glial cells in vitro. Gene Therapy, **4**, 63-75 (1997)
[22] Clarke, L.A.; Nasir, J.; Zhang, H.; McDonald, H.; Applegarth, D.A.; Hayden, M.R.; Toone, J.: Murine α-L-iduronidase: cDNA isolation and expression. Genomics, **24**, 311-316 (1994)
[23] Sakuda, H.; Kusaba, A.; Ohshita, T.; Iwamasw, T.: Tissue and cellular distribution of α-L-iduronidase in the pig. J. Histochem. Cytochem., **38**, 785-792 (1990)

Mannan 1,2-(1,3)-α-mannosidase 3.2.1.77

1 Nomenclature

EC number
3.2.1.77

Systematic name
1,2-1,3-α-D-mannan mannohydrolase

Recommended name
mannan 1,2-(1,3)-α-mannosidase

Synonyms
α-D-mannan mannohydrolase
exo-1,2-1,3-α-mannosidase

CAS registry number
37288-53-2

2 Source Organism

<1> *Penicillium charlesii* [1]
<2> *Arthrobacter sp.* [2-4, 6]
<3> *Xanthomonas campestris pv. manihotis* [5]
<4> *Streptomyces sp.* (M-03) [6]
<5> *Pseudomonas sp.* (M-15) [6]
<6> *Bacillus sp.* (M-25) [6]
<7> *Acinetobacter sp.* (M-30) [6]

3 Reaction and Specificity

Catalyzed reaction
hydrolysis of 1,2- and 1,3-linkages in yeast mannan, releasing mannose

Reaction type
hydrolysis of O-glycosyl bond

Natural substrates and products
S mannan + H_2O <1-7> [1-6]

Substrates and products
S mannan + H_2O <1-7> [1-6]
P mannose + 1,6-α-D-mannan-backbone <1-7> (<1-7> further slow degradation [1-6]) [1-6]

Inhibitors

Cu^{2+} <2> [3]
EDTA <2> [3]
Mg^{2+} <2> [3]
Zn^{2+} <2> [3]

Metals, ions

Ca^{2+} <2> (<2> restores activity after EDTA treatment [3]) [3]

K_m-Value (mM)

0.01 <2> (mannan, <2> based on molecular weight of 22000 from intact S. cerevisiae mannan [3]) [3]
0.5-2 <2> (oligosaccharides from mannan, <2> molecular weight of mannan multiplied by factor 50 describes molar concentration of side chains in the mannan molecule [3]) [3]

pH-Optimum

6 <4, 7> [6]
6.5 <5, 6> [6]
6.5-7 <2> (<2> mannan [3]) [3]
7 <2> [6]

pH-Range

4.5-8 <7> [6]
5-7.5 <6> [6]
5-8 <4, 5> [6]
5-8.5 <2> [6]
5-9 <2> (<2> mannan [3]) [3]

5 Isolation/Preparation/Mutation/Application

Localization

extracellular <1, 2> [1-3, 6]

Purification

<2> (partially [3]) [3]
<3> [5]

6 Stability

Temperature stability

30 <2, 4-7> (<2, 4-7> stable for 3 h [6]) [6]
60 <2> (<2> completely inactivated [6]) [6]
70 <2> (<2> not stable above [3]) [3]

Storage stability

<2>, -10°C, indefinitely, [3]
<2>, 0°C, unstable on storage [6]

<2>, 20°C, one drop of toluene, 1 day, 15% decline in activity [3]
<2>, 4°C, 2 days, [3]

References

[1] Preston, J.F.; Lapis, E.; Gander, J.E.: The appearence of exo-α-mannosidase in cultures of Penicillium charlesii. Arch. Microbiol., **88**, 71-76 (1973)
[2] Jones, G.H.; Ballou, C.E.: Studies on the structure of yeast mannan. II. Mode of action of the Arthrobacter α-mannosidase on yeast mannan. J. Biol. Chem., **244**, 1052-1059 (1969)
[3] Jones, G.H.; Ballou, C.E.: Studies on the structure of yeast mannan. I. Purification and some properties of an α-mannosidase from an Arthrobacter species. J. Biol. Chem., **244**, 1043-1051 (1969)
[4] John, E.; Hampel W.A.: Enzyme formation by a yeast cell wall lytic Arthrobacter species: formation of α-mannanase. Appl. Microbiol. Biotechnol., **35**, 60-64 (1991)
[5] Wong-Madden S.T.; Landry D.: Purificaion and characterization of novel glycosidases from the bacterial genus Xanthomonas. Glycobiology, **5**, 19-28 (1995)
[6] Malek M.A.; Berry D.R.: Isolation and partial characterization of yeast mannan hydrolyzing enzymes from bacterial isolates. Microbios, **83**, 229-241 (1995)

Mannan endo-1,4-β-mannosidase 3.2.1.78

1 Nomenclature

EC number
3.2.1.78

Systematic name
1,4-β-D-mannan mannanohydrolase

Recommended name
mannan endo-1,4-β-mannosidase

Synonyms
β-1,4-mannan 4-mannanohydrolase
β-D-mannanase
β-mannanase B
endo-1,4-β-mannanase
endo-β-1,4-mannase
endo-β-mannanase
mannanase, endo-1,4-β-

CAS registry number
37288-54-3

2 Source Organism

<1> *Caldibacillus cellulovorans* [31]
<2> *Lactuca sativa* [3, 6, 18, 19]
<3> *Aspergillus niger* (commercial preparations [4, 17]) [4, 17, 23]
<4> *Medicago sativa* (lucerne [5, 7, 17]) [5, 7, 17]
<5> *Streptomyces sp.* [8]
<6> *Trigonella foenum-graecum* (fenugreek [8, 16]) [8, 16]
<7> *Aeromonas hydrophila* (anearogenes [9]) [9]
<8> *Tyromyces palustris* [10]
<9> *Chlorella fusca* [11, 13]
<10> *Bacillus stearothermophilus* [35]
<11> *Penicillium purpurogenum* [14]
<12> *Bacillus sp.* (M-I, M-II, M-III [15]) [15]
<13> *Helix pomatia* (snail [7, 17]) [7, 17]
<14> *Irpex lacteus* [17]
<15> *Lycopersicon esculentum* [2, 24]
<16> *Basidiomycetes sp.* (commercial preparation [7]) [7]

<17> *Bacillus subtilis* (commercial preparation [7, 17]; KU-1 [28]; KK01 [30]) [7, 17, 28, 30]
<18> *Cyamopsis tetragonoloba* (guar [17]) [17]
<19> *Leucaena leucocephala* [7]
<20> *Gleditsia triacanthos* [7]
<21> *Thermotoga neapolitana* (5068 [20, 31]) [20, 32, 34]
<22> *Sclerotium rolfsii* [21]
<23> *Lilium testaceum* [22]
<24> *Vibrio sp.* (strain MA-138 [25, 33]) [25, 33]
<25> *Aspergillus aculeatus* [26]
<26> *Dictyoglomus thermophilum* (Rt46B.1 [27]) [27]
<27> *Caldicellulosiruptor sp.* (Rt8B.4 [29]) [29]
<28> *Lemonniera terrestris* [12]
<29> *Alatospora acuminata* [12]
<30> *Articulospora tetracladia* [12]
<31> *Gyoerffyella rotula* [12]
<32> *Gyoerffyella speciosa* [12]
<33> *Lemonniera aquatica* [12]
<34> *Penicillium wortmanni* [1]
<35> *Aspergillus giganteus* [1]
<36> *Paecilomyces varioti* [1]
<37> *Penicillium verruculosum* [1]
<38> *Penicillium funiculosum* [1]
<39> *Penicillium ochro-chloron* [1]
<40> *Aspergillus luchuensis* [1]
<41> *Fusicoccum sp.* [1]

3 Reaction and Specificity

Catalyzed reaction
random hydrolysis of 1,4-β-D-mannosidic linkages in mannans, galactomannans and glucomannans

Reaction type
hydrolysis of O-glycosyl bond

Natural substrates and products
S guar gum + H_2O <34, 35, 38> [1]
P mannose + mannobiose + mannotriose <34, 35, 38> (<35> mannobiose + mannotriose + mannotetraose [1]) [1]
S ivory nut mannan + H_2O <24> [33]
P mannobiose + mannotriose <24> [33]
S locust bean gum + H_2O <17, 24, 10> [30, 33, 35]
P mannose + galactose <24> [33]
S mannan + H_2O <9> [11, 13]
P ?

Substrates and products

S galactoglucomannan + H_2O <2-4, 12-14, 17, 18, 22> [3, 4, 6, 8, 15, 17, 18, 19, 21]
P ?
S galactomannans + H_2O <3, 4, 7, 11, 13, 16, 17, 19, 20, 22, 26> [4, 5, 7, 9, 14, 17, 21, 27]
P hydrolyzed galactomannans <3, 4, 13, 16, 17, 19, 20> [4, 7, 8, 17]
S glucomannans + H_2O <2-4, 7, 8, 12-14, 17, 18, 22, 23, 24> [3, 4, 6, 7, 9, 10, 15, 17, 18, 19, 21, 22, 25, 32]
P oligosaccharides <7, 8, 24> (<7,8> of various size [9,10]) [9, 10, 25, 32]
S mannans + H_2O <2-4, 7, 12-14, 17, 18, 22, 24, 34-41> (<7> codium, coffee [9]) [1, 3, 6, 9, 15, 17, 21, 25, 30, 33]
P oligosaccharides <34-41, 12, 17, 24> (<34-41> dimers, trimers and other [1]) [1, 15, 25, 30, 32]
S mannopentaose + H_2O <2-4, 7, 12-14, 17, 18, 24> [3, 4, 5, 6, 9, 15, 17, 18, 19, 33]
P mannose + mannobiose + mannotriose + mannotetraose <7, 24> [9, 33]
S mannotetraose + H_2O <3, 4, 5, 7, 11, 12, 13, 16, 17, 19, 20> [4, 5, 7, 8, 9, 14, 15, 17]
P mannobiose + mannose + mannotriose <3, 7, 12> (<3, 12> small amounts of mannose and mannotriose [4, 15, 17]) [4, 9, 15, 17]
S mannotriose + H_2O <3, 11, 12> (<7> not [9]) [4, 9, 14, 15, 17]
P mannobiose + mannose <3, 12> (<11> not [14]) [4, 14, 15, 17]
S Additional information <34-41, 3, 7, 12, 22, 26> (<6> transglycosylation reaction, Streptomyces: transfer of one mannose unit from oligosaccharides, fenugreek: transfer of oligomannose residues [8, 16]; <3, 7, 12> not: mannobiose [4, 8, 9, 15, 16, 17]) [1, 4, 8, 9, 15, 16, 17, 21, 27]
P ?

Inhibitors

Ag^+ <7, 12, 17> [9, 15, 28]
Cr^{3+}<17> [28]
Cu^{2+} <7, 17> [9, 28]
Fe^{3+} <7> [9]
Hg^{2+} <7, 11, 17> [9, 14, 28]
Mn^{2+} <17> [28]
N-bromosuccinimide <12> [15]
Pb^{2+} <7> [9]
Zn^{2+} <7> [9]
cycloheximide <2> [3, 6, 18, 19]
p-chloromercuribenzoate <7> [9]

Activating compounds

cysteine <12> (<12> slight activation [15]) [15]

Metals, ions

Ca^{2+} <7> (<7> stimulates [9]) [9]
Mg^{2+} <11> (<11> slight activation [14]) [14]

Specific activity (U/mg)

21 <11> [14]
39 <3> [17]
60 <25> [26]
125 <18> [17]
201 <7> [9]
380 <22> (<22> 41.9 kDa protein [1, 21]) [1, 21]
470 <12> [15]
475 <22> (<22> 61.2 kDa protein [1, 21]) [1, 21]
514 <17> [17]
Additional information <2-4, 9, 12-14, 17, 18> [3, 4, 6, 7, 11, 13, 15, 17, 18, 19]

K_m-Value (mM)

1.9 <4> (soluble mannan) [5, 17]
Additional information <2-4, 12-14, 16-20> (<12> copra β-mannan: 2.0 mg/ml, locust bean β-mannan: 3.8 mg/ml, konjak β-mannan: 7.7 mg/ml, carob galactomannan: 0,9 mg/ml [15, 17]) [3, 6, 7, 15, 17, 18, 19]

pH-Optimum

2.9 <22> (<22> 61.2 kDa protein [21]) [21]
3 <3, 16> [4, 7, 17]
3.3 <22> (<22> 41.9 kDa protein [21]) [21]
3.6 <3> [4, 17]
4 <18> [17]
4.5 <4, 23> [7, 22]
4.5-5.5 <13> [17]
5 <11, 25, 26> (<25> recombinant protein [26]) [14, 26, 27]
5-6 <17> [17]
5.5 <7> [9]
5.5-7.5 <10> [35]
6 <34> [1]
6.5 <24> [33]
6.9 <21> [34]
7 <17> [28]
7.1 <17, 21> (<21> at 91°C [32]) [30, 32]
7.5 <24> [25]
8.5 <12> (<12> M-III, at 65°C [15]) [15]
9 <12> (<12> M-I, M-II, at 60°C [15]) [15]
Additional information <2-4, 11-14, 16-20> [3, 4, 6, 7, 14, 15, 17, 18, 19]

pH-Range

3-8 <34, 36-41> [1]
3-8 <3, 11> (<11> pH 3: 15% of maximal activity [14]) [14, 17]
4.5-8 <35> [1]
6-10 <12> [15]
Additional information <2-4, 12-14, 17, 18, 34> [1, 3, 4, 6, 7, 11, 13, 15, 17, 18, 19]

Temperature optimum (°C)

40 <23, 24> [21, 22, 33]
45 <24> [25]
50-55 <17> [28]
50-60 <17> (<17> for four enzyme forms [30]) [30]
60 <12, 25> (<12> M-I, M-II [15]; <25> recombinant protein from Aspergillus oryzae, SDS-PAGE [15, 26]) [15, 26]
60-70 <34-41> [1]
65 <12> (<12> M-III [15]) [15]
70 <11> [14]
72 <22> (<22> 41.9 kDa protein [21]) [21]
74 <22> (<22> 61.2 kDa protein [21]) [21]
80 <3, 26> [4, 17, 27]
90-92 <21> [32, 34]
Additional information <2-4, 11-14, 17, 18> [3, 4, 5, 6, 7, 14, 15, 17, 18, 19]

Temperature range (°C)

30-70 <34> [1]
30-80 <11> (<11> 30°C: about 25% of maximal activity, 80°C: about 30% of maximal activity [14]) [14]
50-70 <12> [15]
Additional information <2-4, 12-14, 15, 17, 18> [2, 3, 4, 6, 7, 11, 13, 15, 17, 18, 19]

4 Enzyme Structure

Molecular weight

24000 <7> (<7> gel filtration [9]) [9]
25000 <9> (<9> gel filtration [11, 13]) [11, 13]
27000 <9> (<9> gel filtration [11, 13]) [11, 13]
33000 <23> (<23> gel filtration [22]) [22]
37000 <13, 17> (<13> SDS-PAGE [7, 17]; <17> SDS-PAGE [7, 17]) [7, 17]
38950 <15> (<15> calculation from sequence of DNA [24]) [24]
39000 <17> (<17> SDS-PAGE, 40000 by gel filtration [28]) [28]
40000 <3, 26> [23, 27]
41000 <24> (<24> recombinant protein expressed in E. coli, SDS-PAGE [25]) [25]
41000 <4> (<4> SDS-PAGE [7, 17]) [7, 17]
41900 <22> [21]
42000 <12> (<12> M-III, SDS-PAGE [15]) [15]
43100 <24> (<24> calculation from sequence of DNA [25]) [25]
44000 <9> (<9> SDS-PAGE [11, 13]) [11, 13]
45000 <3> [7, 17]
45000 <25> (<25> recombinant protein expressed in Aspergillus oryzae, SDS-PAGE [26]) [26]
46000 <2> (<2> SDS-PAGE [3, 6, 18, 19]) [3, 6, 18, 19]

49000 <24> (<24> SDS-PAGE [33]) [33]
53000 <16> (<16> SDS-PAGE [7]) [7]
55000 <21> [32]
57000 <11> (<11> SDS-PAGE [14]) [14]
58000 <12> (<12> M-I, SDS-PAGE [15]) [15]
59000 <12> (<12> M-II, SDS-PAGE [15]) [15]
61200 <22> [21]
65000 <21> (<21> SDS-PAGE [34]) [20, 34]
70000 <9> (<9> gel filtration [11, 13]) [11, 13]
162000 <10> (<10> dimeric protein, SDS-PAGE [35]) [35]

Subunits
dimer <10> (<10> 2 * 73000, SDS-PAGE [35]) [35]

Posttranslational modification
glycoprotein <24> [33]

5 Isolation/Preparation/Mutation/Application

Source/tissue
bulb <23> [22]
commercial preparation <3, 4, 13, 16, 17, 19, 20> [4, 5, 7, 17]
culture filtrate <34-41, 4, 11, 14, 17, 21> [1, 5, 7, 14, 17, 20, 23, 25, 28, 30, 32]
culture medium <7> [9]
endosperm <2-4, 8, 12-14, 17, 18> [3, 4, 5, 6, 7, 10, 15, 17, 18, 19]
gut <3, 4, 13, 16, 17, 19, 20> [4, 5, 7, 17]
seed <2, 4, 15> [2, 3, 5, 6, 7, 17, 18, 19, 24]

Localization
extracellular <12> [15]

Purification
<1> (recombinant protein containing active protein domain [31]) [31]
<2> [6, 18]
<4> [17]
<7> [9]
<9> [11, 13]
<10> [35]
<11> [14]
<12> (M-I, M-II, M-III [15]) [15]
<13> [17]
<14> [17]
<17> (four enzyme forms [30]) [17, 28, 30]
<18> [17]
<21> [34]
<24> (recombinant protein expressed in E. coli [25]) [25, 33]
<25> (recombinant protein expressed in Aspergillus oryzae [26]) [26]

Cloning

<1> [31]
<15> [24]
<24> (expression in E. coli [25]) [25]
<25> (expression in Aspergillus oryzae [26]) [26]
<26> [27]
<27> [29]

6 Stability

pH-Stability

2.5-10 <25> (<25> recombinant protein expressed in Aspergillus oryzae [26]) [26]
4-5 <22> [21]
4-8 <4> (40°C, 1 h <4>) [5, 17]
4.5-8 <11> [14]
4.5-9 <27> [29]
5-8 <13> [17]
5-9 <7> [9]
Additional information <2-4, 11-14, 16-20> [7, 14, 15, 17, 19]

Temperature stability

39 <3> (<3> 4 h, stable between pH 5 and 7 [17]) [7, 17]
40 <7> (<7> stable up to) [9]
50 <3> [23]
50 <4, 13, 17> (<4, 13, 17> stable below [7, 17]) [7, 17]
60 <16, 17> (<16> stable up to [7]; <17> for 1 h [7, 28]) [7, 28]
65 <11, 10> (<11> stable up to [14]; <10> completely stable at [35]) [14, 35]
70 <3, 25> (<3> stable up to [4]; <25> recombinant protein from Aspergillus oryzae [4, 26]) [4, 17, 26]
80 <26> (<26> no loss of activity over 16 h [27]) [27]
91 <21> (<21> half-life 13 h [32, 34]) [32, 34]
Additional information <2-4, 11-14, 17, 18> [3, 6, 7, 14, 15, 17, 18, 19]

Organic solvent stability

acetone <34> (<34> may be used in preparations [1]) [1]

Storage stability

<3>, 4°C, stable for 18 h, pH 4-10 [17]
<7>, dialyzed [9]
<34-41>, cold [1]

References

[1] Reese, E.T.; Shibata, Y.: β-Mannanases of fungi. Can. J. Microbiol., **11**, 167-18 (1965)

[2] Groot, S.P.C.; Kieliszewska-Rokicka, B.; Vermeer, E.; Karsson, C.M.: Gibberellin-induced hydrolysis of endosperm cell walls in gibberellin-deficient tomato seeds prior to radicle protrusion. Planta, **174**, 500-504 (1988)
[3] Halmer, P.; Bewley, J.D.; Thorpe, T.A.: Enzyme to break down lettuce endosperm cell wall during gibberellin- and light-induced germination. Nature, **258**, 716-718 (1975)
[4] Yamazaki, N.; Sinner, M.; Dietrichs, H.H.: Isolierung und Eigenschaften einer β-1,4-Mannanase aus Aspergillus niger. Holzforschung, **30**, 101-109 (1976)
[5] McCleary, B.V.: Purification of a β-mannanase enzyme from lucerne seed by substrate affinity chromatography. Phytochemistry, **17**, 651-653 (1978)
[6] Halmer, P.; Bewley, J.D.: Mannanase production by the lettuce endosperm. Planta, **144**, 333-340 (1979)
[7] McCeary, B.V.: Modes of action of β-mannanase enzymes of diverse origin on legume seed galactomannans. Phytochemistry, **18**, 757-763 (1979)
[8] Coulombel, C.; Clermont, S.; Foglietti, M.J.; Percheron, F.: Transglycosylation reactions catalysed by two β-mannanases. Biochem. J., **195**, 333-335 (1981)
[9] Araki, T.: Purification and characterization of an endo-β-mannanase from Aeromonas sp. F-25. J. Fac. Agric. Kyushu Univ., **27**, 89-98 (1983)
[10] Shimizu, K.; Ishihara, M.: Isolation and characterization of oligosaccharides from the hydrolyzate of larch wood glucomannan with endo-β-D-mannanase. Agric. Biol. Chem., **47**, 949-955 (1983)
[11] Loos, E.; Meindl, D.: Cell wall-lytic activity in Chlorella fusca. Planta, **160**, 357-362 (1984)
[12] Zemek, J.; Marvanova, L.; Kuniak, L.; Kadlecikova, B.: Hydrolytic enzymes in aquatic Hyphomycetes. Folia Microbiol., **30**, 363-372 (1985)
[13] Loos, E.; Meindl, D.: Cell-wall-bound lytic activity in Chlorella fusca: function and characterization of an endo-mannanase. Planta, **166**, 557-562 (1985)
[14] Park, G.G.; Kusakabe, I.; Komatsu, Y.; Kobayashi, H.; Yasui, T.; Murakami, K.: Purification and some properties of β-mannanase from Penicillium purpurogenum. Agric. Biol. Chem., **51**, 2709-2716 (1987)
[15] Akino, T.; Nakamura, N.; Horikoshi, K.: Characterization of three β-mannanase of an alkalophilic Bacillus sp.. Agric. Biol. Chem., **52**, 773-779 (1988)
[16] Spyropoulos, C.G.; Reid, J.S.G.: Water stress and galactomannan breakdown in germinated fenugreek seeds. Stress affects the production and the activities in vivo of galactomannan-hydrolysing enzymes. Planta, **174**, 473-478 (1988)
[17] McCleary, B.: β-D-Mannanase. Methods Enzymol., **160**, 596-610 (1988)
[18] Dulson, J., Bewley, J.D.: Mannanse from Lactuca sativa: Metabolic requirements for production and partial purification. Phytochemistry, **28**, 363-369 (1989)
[19] Halmer, P.: De novo synthesis of mannanase by the endosperm of Lactuca sativa. Phytochemistry, **28**, 371-379 (1989)
[20] Duffaud, G.D.; McCutchen, C.M.; Leduc, P.; Parker, K.N.; Kelly, R.M.: Purification and characterisation of extremely thermostable β-mannanase, β-

mannosidase, and α-galactosidase from the hyperthermophilic eubacterium Thermotoga neapolitana 5068. Appl. Environ. Microbiol., **63**, 169-177 (1997)
[21] Gübitz, G.M.; Hayn, M.; Sommerhauer, M.; Steiner, W.: Mannan-degrading enzymes from Sclerotium rolfsii: Characterisation and synergism of two endo β-mannanases and β-mannosidase. Bioresour. Technol., **58**, 127-135 (1996)
[22] Wozniewski, T.; Blaschek, W.; Fransz, G.: Isolation and characterization of an endo-β-mannanase of Lilium testaceum bulbs. Phytochemistry, **31**, 3365-3370 (1992)
[23] Ademark, P.; Carga, A.; Medve, J.; Harjunpää, V.; Drakenberg, T.; Tjerneld, F.; Stalbrand, H.: Softwood hemicellulose-degrading enzymes from Aspergillus niger: Purification and properties of a β-mannase. J. Biotechnol., **63**, 199-210 (1998)
[24] Bewley, J.D.; Burton, R.A.; Morohashi, Y.; Fincher, G.B.: Molecular cloning of a cDNA encoding a (1→4)-β-mannan endohydrolase from the seeds of germinated tomato (Lycopersicon esculentum). Planta, **203**, 454-459 (1997)
[25] Tamaru, Y.; Araki, T.; Morishita, T.; Kimura, T.; Sakka, K.; Ohmiya, K.: Cloning, DNA sequencing, and expression of the β-1,4-mannanase gene from a marine bacterium, Vibrio sp. strain MA-138. J. Ferment. Bioeng., **83**, 201-205 (1997)
[26] Christgau, S.;, Kauppinen, S.; Vind, J.; Kofod, L.V.; Dalboge, H.: Expression cloning, purification and charakterisation of a β-1,4-mannanase from Aspergillus aculeatus. Biochem. Mol. Biol. Int., **33**, 917-925 (1994)
[27] Gibbs, M.D.; Reeves, R.A.; Sunna, A.; Bergquist, P.L.: Sequencing and expression of a β-mannanase gene from the extreme thermophile Dictyoglomus thermophilum Rt46B.1, and characteristics of the recombinant enzyme. Curr. Microbiol., **39**, 351-357 (1999)
[28] Zakaria, M.M.;, Yamamoto, S.; Yagi, T.: Purification and characterization of an endo-1,4-β-mannanase from Bacillus subtilis KU-1. FEMS Microbiol. Lett., **158**, 25-31 (1998)
[29] Gibbs, M.D.; Elinder, A.U.; Reeves, R.A.; Bergquist, P.L.: Sequencing, cloning and expression of a β-1,4-mannanase gene, manA, from the extremely thermophilic anaerobic bacterium, Caldicellulosiruptor Rt8B.4. FEMS Microbiol. Lett., **141**, 37-43 (1996)
[30] Hossain, M.Z.; Abe, J.; Hizukuri, S.: Multiple forms of β-mannanase from Bacillus sp. KK01. Enzyme Microb. Technol., **18**, 95.98 (1996)
[31] Sunna, A.; Gibbs, M.D.; Chin, C.W.J.; Nelson, P.J.; Bergquist, P.L.: A gene encoding a novel multidomain β-1,4-mannanase from Caldibacillus cellulovorans and action of the recombinant enzyme on kraft pulp. Appl. Environ. Microbiol., **66**, 664-670 (2000)
[32] McCutchen, C.M.; Duffaud, G.D.; Leduc, P.; Petersen, A.R.H.; Tayal, A.; Khan, S.A., Kelly, R.M.: Characterization of extremely thermostable enzymatic breakers (α-1,6-galactosidase and β-1,4-mannanase) from the hyperthermophilic bacterium Thermotoga neapolitana 5068 for hydrolysis of guar gum. Biotechnol. Bioeng., **52**, 332-339 (1996)

[33] Tamaru, Y.; Araki, T.; Amagoi, H.; Mori, H.; Morishita, T.: Purification and characterization of an extracellular mannanase from a marine bacterium, Vibrio sp. strain MA-138. Appl. Environ. Microbiol., **61**, 4454-4458 (1995)
[34] Duffaud, G.D.; McChutchen, C.M.; Leduc, P.; Parker, K.; Kelly, R.M.: Purification and characterisation of extremely thermostable β-mannanase, β-mannosidase, and α-galactosidase from the hyperthermophilic eubacterium Thermotoga neapolitana 5068. Appl. Environ. Microbiol., **63**, 169-177 (1997)
[35] Talbot, G.; Sygusch, J.: Purification and characterization of themostable β-mannanase and α-galactosidase from Bacillus stearothermophilus. Appl. Environ. Microbiol., **56**, 3505-3510 (1990)

α-L-Arabinofuranoside hydrolase 3.2.1.79

1 Nomenclature

EC number

3.2.1.79 (deleted, included in EC 3.2.1.55)

Recommended name

α-L-arabinofuranoside hydrolase

Fructan β-fructosidase 3.2.1.80

1 Nomenclature

EC number
3.2.1.80

Systematic name
β-D-fructan fructohydrolase

Recommended name
fructan β-fructosidase

Synonyms
exo-β-D-fructosidase
exo-β-fructosidase
fructan exohydrolase
fructofuranosidase, polysaccharide β
polysaccharide β-fructofuranosidase

CAS registry number
37288-56-5

2 Source Organism

<1> *Streptococcus mutans* (GS-5 [1, 5]; ingbritt [4]) [1, 4, 5]
<2> *Streptococcus salivarius* [2]
<3> *Helianthus tuberosus* (Jerusalem artichoke) [3]
<4> *Actinomyces naeslundii* (T14V) [6]
<5> *Avena sativa* (oat) [7]
<6> *Hordeum vulgare* (barley) [8]
<7> *Chicorium intybus* [9]

3 Reaction and Specificity

Catalyzed reaction
hydrolysis of terminal, non-reducing 2,1- and 2,6-linked β-D-fructofuranose residues in fructans

Reaction type
hydrolysis of O-glycosyl bond

Natural substrates and products

S fructans + H_2O <1, 3> (<1, 3> degradation [1, 3]) [1, 3]
P D-fructose

Substrates and products

S 6G,6-kestotetraose + H_2O <5> (<5> 100%, chicorium inulin 11.9%, Aerobacter levan 8.5%, sucrose 0% rate of hydrolysis) [7]
P D-fructose + neokestose <5> [7]
S inulin + H_2O <1, 2, 4, 7> [1, 2, 4, 6, 9]
P D-fructose <1, 2> [1, 2, 4]
S levan + H_2O <1, 2, 4> [1, 2, 6]
P D-fructose <1, 2> [1, 2]
S neokestin+ H_2O <5> [7]
P D-fructose <5> [7]
S neokestose + H_2O <5> [7]
P D-fructose + sucrose <5> [7]
S raffinose + H_2O <1, 2, 4> [1, 2, 4, 6]
P D-fructose + α-D-galactopyranosyl-1,6-α-D-glucopyranoside <2> [2]
S sucrose + H_2O <1, 2, 4> [1, 2, 6]
P D-fructose + D-glucose <2> [2]
S Additional information <2> (not: melizitose, dextran, pseudonigeran) [2]
P ?

Inhibitors

Ag^+ <1, 2> [1, 2]
$CaCl_2$ <7> (<7> partially [9]) [9]
Cu^{2+} <1, 2> (<1> partially [1]) [1, 2]
EDTA <7> (<7> partially [9]) [9]
Fe^{3+} <1, 2> [2, 4]
Hg^{2+} <1, 2> [1, 2, 4]
Tris <2> [2]
Zn^{2+} <1> [4]
p-chloromercuribenzoate <2> (<2> presence of cysteine protects against inactivation but no recovery [2]) [2]
sucrose <7> [9]

Activating compounds

EDTA <2> (<2> slight stimulatuion after 30 min of incubation [2]) [2]
KCN <2> (<2> slight stimulatuion after 30 min of incubation [2]) [2]

Metals, ions

Co^{2+} <1> (<1> slight stimulation [1]) [1]
Mn^{2+} <1, 2> (<1, 2> slight stimulation [1, 2]) [1, 2]

pH-Optimum

4.5 <1> (<1> sucrose [1]) [1]
4.5-5.5 <5> (<5> neokestin [7]) [7]
5 <1> (<1> sucrose [4]) [4]
5-5.5 <7> (<7> inulin [9]) [9]

5.5 <1> (<1> fructan polymers [1]) [1]
5.5 <1> (<1> levan [4]) [4]
6 <1> (<1> linulin [4]) [4]
7 <2> [2]

pH-Range
3.5-7 <1> (<1> little fructanase activity at or above pH 7.0 [1]) [1]
4.5-8 <2> [2]

Temperature optimum (°C)
35 <7> (<7> inulin [9]) [9]
50 <2> [2]
55 <1> (<1> levan [4]) [4]

Temperature range (°C)
20-55 <2> (<2> rapid inactivation at 60°C [2]) [2]
40-60 <1> [4]

4 Enzyme Structure

Molecular weight
60000 <7> (<7> gel-filtration [9]) [9]
64000 <7> (<7> SDS-PAGE [9]) [9]
83000-85000 <2> (<2> gel filtration, SDS-PAGE [2]) [2]

Subunits
? <1, 5, 6> (<6> x * 33000, SDS-PAGE [8]; <5> x * 43000, SDS-PAGE [7]; <1> x * 127000, SDS-PAGE [4]; <1> x * 140000, SDS-PAGE [1]) [1, 4, 7, 8]

5 Isolation/Preparation/Mutation/Application

Source/tissue
tuber <3> [3]

Localization
vacuole <3, 6> [3, 8]

Purification
<1> [1, 4]
<2> [2]
<3> [3]
<5> [7]
<6> [8]
<7> [9]

Cloning
<1, 4> (expression in E. coli [1, 5, 6]) [1, 5, 6]

6 Stability

pH-Stability

4.5 <2> (<2> almost complete loss of activity at [2]) [2]
7 <2> (<2> 37°C, 30 min [2]) [2]

Temperature stability

37 <2> (<2> pH 7.0, 30 min [2]) [2]
60 <2> (<2> rapid inactivation [2]) [2]

General stability information

<2>, freezing and thawing accelerates inactivation [2]

Storage stability

<2>, 4°C, 0.02 mg/ml, 20 mM phosphate buffer, pH 7.0, stable at least for several days, 90% loss of activity after 2 weeks [2]
<2>, higher stability at -20°C than at 4°C [2]

References

[1] Burne, R.A.; Schilling, K.; Bowen, W.H., Yasbin, R.E.: Expression, purification, and characterization of an exo-β-D-fructosidase of Streptococcus mutans. J. Bacteriol., **169**, 4507-4517 (1987)

[2] Takahashi, N.; Mizuno, F.; Takamori, K.: Purification and preliminary characterization of exo-β-D-fructosidase in Streptococcus salivarius KTA-19. Infect. Immun., **47**, 271-276 (1985)

[3] Frehner, M.; Keller, F.; Wiemken, A.: Localisation of fructan metabolism in the vacuoles isolated from protoplasts of jerusalem artichoke tubers (Helianthus tuberosus L.). J. Plant Physiol., **116**, 197-208 (1984)

[4] Igarashi, T.; Yamamoto, A.; Goto, N.: Characterization of an exo-β-D-fructosidase from Streptococcus mutants ingbritt. Microbiol. Immunol., **36**, 643-647 (1992)

[5] Burne, R.A.; Penders, J.E.C.: Characterization of the Streptococcus mutans GS-5 fruA gene encoding exo-β-D-fructosidase. Infect. Immun., **60**, 4621-4632 (1992)

[6] Norman, J.M.; Bunny K.L.; Giffard, P.M.: Characterization of levJ, a sucrase/fructanase encoding gene from Actinomyces naeslundii T14V, and comparison of its product with other sucrose-cleaving enzymes. Gene, **152**, 93-98 (1994)

[7] Henson, C.A.; Livingston III, D.P.: Purification and characterization of an oat fructan exohydrolase that preferentially hydrolyzes β-2,6-fructans. Plant Physiol., **110**, 639-644 (1996)

[8] Henson, C.A.; Livingston III, D.P.: Characterization of a fructan exohydrolase purified from barley stems that hydrolyzes multiple fructofuranosidic linkages. Plant Physiol. Biochem., **36**, 715-720 (1998)

[9] De Roover, J.; Van Laere, A.; De Winter, M.; Timmermanns, J.W.; Van den Ende, W.: Purification and properties of a second fructan exohydrolase from the roots of Chicorium intybus. Physiol. Plant., **106**, 28-34 (1999)

Agarase 3.2.1.81

1 Nomenclature

EC number
3.2.1.81

Systematic name
agarose 3-glycanohydrolase

Recommended name
agarase

Synonyms
β-agarase

CAS registry number
37288-57-6

2 Source Organism

<1> *Pseudomonas atlantica* [1, 4, 6, 7, 14, 15]
<2> *Streptomyces coelicolor* [2, 3, 5, 15, 17]
<3> *Vibrio harveyi* [4]
<4> *Cytophaga flevensis* [4, 9]
<5> *Pseudomonas sp.* (W7 [4]) [4, 14]
<6> *Pseudomonas-like bacteria* [8]
<7> *Cytophaga sp.* [8, 10, 11, 12, 13]
<8> *Vibrio sp.* [14, 15, 19]
<9> *Pseudoalteromonas antarctica* [16]
<10> *Alteromonas sp.* (E-1 [20]) [18, 20]

3 Reaction and Specificity

Catalyzed reaction
hydrolysis of 1,3-β-D-galctosidic linkages in agarose, giving the tetramer as the predominant product

Reaction type
hydrolysis of O-glycosyl bond

Natural substrates and products
S agarose + H_2O <1-10> [1-16, 18, 19]

P neoagarotetraose
S agarose + H_2O <10> [20]
P neoagarobiose

Substrates and products

S agaropectin + H_2O <4> [9]
P ?
S agarose + H_2O <1-10> [1-16, 18, 19]
P neoagarotetraose
S agarose + H_2O <10> [20]
P neoagarobiose
S alkali treated porphyran + H_2O <7> [11, 12]
P ?
S neoagarohexaitol + H_2O <7> [10]
P ?
S neoagarohexaose + H_2O <7> [10]
P neoagarobiose + neoagarotetraose
S neoagarooctaose + H_2O <1, 6> [1, 8]
P neoagarotetraose <6> (<6> β-agarose IIb [8]) [8]
S neoagarooctaose + H_2O <6, 7> [8]
P neoagarohexaose + neoagarobiose <6, 7> (<6> β-agarase I [8]) [8]
S polysaccharides with neoagarobiose units + H_2O <4> [9]
P neoagarobiose + neoagarotetraose
S porphyran + H_2O <1, 7> [6, 10-13]
P neoagaro-oligosaccharides <1, 7> (<1, 7> methylated, sulfated or unsubstituted [6, 10-13]) [6, 10-13]

Inhibitors

Ca^{2+} <2> (<2> in higher concentrations, 10-20 mM [3]) [3]
Cu^{2+} <10> [20]
Fe^{2+} <10> [20]
Hg^{2+} <10> [20]
KCl <10> (<10> 50% inhibition at 0.1 M [18]) [18]
L-galactose 6-sulfate <7> [12]
Mn^{2+} <2, 10> (<2> in higher concentrations, 10-20 mM [3]) [3, 20]
NH_4Cl <10> (<10> 50% inhibition at 0.1 M [18]) [18]
NaCl <10> (<10> 50% inhibition at 0.1 M [18]) [18]
NaCl <3> (<3> at low concentrations, 1%, no effect between 3-8% [4]) [4]
Zn^{2+} <10> [20]

Activating compounds

EDTA <10> (<10> reverses inhibition by bivalent metal cations [20]) [20]

Metals, ions

K^+ <10> (<10> slightly activating [20]) [20]
Na^+ <10> (<10> slightly activating [20]) [20]
NaCl <5> (<5> 0.9 M [14]) [14]

Specific activity (U/mg)

9.3 <8> [19]
22 <3> [4]
34 <10> [20]
43.3 <5> [14]
234 <10> [18]
290 <9> [16]
616 <2> [3]
7142 <2> [17]

K_m-Value (mM)

1.7 <10> (neoagarohexaose) [18]

pH-Optimum

5.2 <7> (<7> porphyran as substrate [13]) [13]
5.5 <8> [15]
6 <2, 3> (<2> in imidazol or acetate buffers [3]) [3, 4]
6-7 <1> [7]
6.3 <4> [9]
6.5 <10> [18]
6.7 <6> (<6> β-agarase I and IIb [8]) [8]
6.8 <7> (<7> porphyran as substrate [13]) [13]
7 <1, 9> [15, 16]
7.2 <7> (<7> porphyran [12]) [12]
7.4 <7> (<7> porphyran as substrate [13]) [13]
7.5 <10> [20]
7.5 <2> (<2> in Tris buffer, activity significant lower [3]) [3]
7.8 <5> [14]
8 <8> [19]

pH-Range

5-8.5 <2> (<2> in imidazol or acetate buffers [3]) [3, 17]
5-9 <3> (<3> about 60% of activity [4]) [4]
5.5-8 <7> [12]
6-9 <4, 10> [9, 20]

Temperature optimum (°C)

20-40 <5> [14]
30 <1, 4, 8> (<4> neoagarohexaose [9]) [7, 9, 15, 19]
34 <7> (<7> porphyran as substrate [13]) [13]
35 <4> (<4> in assays with agar as substrate, to avoid gelling [9]) [9]
37 <2> [17]
38 <6> (<6> β-agarase I [8]) [8]
40 <1, 10> (<1> a higher rate of reaction for 1 h [7]) [7, 20]
40-41 <7> [12]
43 <6> (<6> β-agarase IIb [8]) [8]

Temperature range (°C)

25-50 <2> [17]

4 Enzyme Structure

Molecular weight

20000 <8> (<8> SDS-PAGE [15]) [15]
21000 <5> (SDS-PAGE [4]) [4]
26000 <4> [4]
26500 <4> (<4> gel filtration [9]) [9]
28000 <2> (<2> SDS-PAGE [5]) [5]
29000 <2> (<2> calculation from sequence of DNA [2]) [2]
32000 <1, 2> (<1> SDS-PAGE [7]) [4, 7, 15, 17]
33000 <9> (<9> SDS-PAGE [16]) [16]
35130 <2> (<2> calculation from sequence of an open reading frame [2]) [2]
49000 <3> (<3> gel filtration [4]) [4]
50000 <3> (<3> SDS-PAGE [4]) [4]
52000 <10> (<10> gel filtration [18]) [18]
55000 <1> [14]
59000 <5> (<5> SDS-PAGE [14]) [14]
61000 <6> (<6> gel filtration, β-agarase IIb [8]) [8]
63000 <5> (SDS-PAGE [4]) [4]
63600 <6> (<6> analytical ultracentrifugation, β-agarase IIb [8]) [8]
64000 <6> (<6> SDS-PAGE, β-agarase IIb [8]) [8]
72000 <8> (<8> SDS-PAGE [19]) [19]
82000 <10> (<10> SDS-PAGE [20]) [20]
105000 <8> [14]
127000 <6> (<6> gel filtration, β-agarase I [8]) [8]
140000 <6> (<6> SDS-PAGE, β-agarase I [8]) [8]
180000 <10> (gel filtration [20]) [20]
210000 <6> (<6> analytical ultracentrifugation, β-agarase I [8]) [8]

Subunits

dimer <6> (<6> α_2, 2 * 14000, agarase I [8]) [8]

Posttranslational modification

glycoprotein <6> (<6> galactose, N-acetylglucosamine and glucosamine [8]) [8]
proteolytic modification <2> [2, 3]

5 Isolation/Preparation/Mutation/Application

Localization

extracellular <1, 2, 9, 10> [1, 2, 5, 8, 16, 18]
intracellular <2> [3]

Purification

<1> [7]
<2> [3, 17]
<3> [4]

<5> [14]
<6> [8]
<8> [15]
<9> [16]
<10> [18]

Cloning
<2> (expression in Streptomyces lividans, natural Dag- [3]) [3, 5, 17]
<1, 8> (expression in E. coli [1, 14]) [1, 14]

6 Stability

pH-Stability
6-9 <4> (<4> beyond this values the enzyme is rather unstable [9]) [9]

Temperature stability
30 <10> (<10> inhibition above 30°C, after 20 min 35°C 50% of activity [18]) [18]
40 <3> (<3> not stable above [4]) [4]
40 <4> (<4> loss in activity 5% [9]) [9]
45 <3> (<3> about 50% of activity [4]) [4]
47.5 <4> (<4> loss in activity 53% [9]) [9]
55 <3> (<3> about 25% of activity [4]) [4]
60 <8> (<8> inactive after 15 min [15]) [15, 19]

Storage stability
<1>, 20°C, two months [7]
<4>, -20°C, more than 2 years [9]
<6>, -20°C, at least 6 months [8]
<8>, -20°C, 20 mM Tris-HCl, more than 12 months [19]
<8>, 4°C, 20 mM Tris-HCl buffer, more than 12 months [15]
<9>, -20°C, a year [16]
<10>, 4°C, Tris-HCl with 0.4 M NaCl, pH 7.5, more than 6 months [18]

References

[1] Belas, R.; Bartlett, D.; Silverman, M.: Cloning and gene replacement mutagenesis of a Pseudomonas atlantica agarase gene. Appl. Environ. Microbiol., **54**, 30-37 (1988)
[2] Buttner, M.J.; Fearnley, I.M.; Bibb, M.J.: The agarase gene (dagA) od Streptomyces coelicolor A3(2): nucleotide sequence and transcriptional analysis. Mol. Gen. Genet., **209**, 101-109 (1987)
[3] Bibb, M.J.; Jone, G.H.; Joseph, R.; Buttner, M.J.; Ward, J.M.: The agarase gene (dagA) of Streptomyces coelicolor A3(2): affinity purification and characterization of the cloned gene product. J. Gen. Microbiol., **133**, 2089-2096 (1987)

[4] Fukasawa, S.; Kobayashi, H.: Properties of the agarase from a luminous bacterium, Vibrio harveyi. Agric. Biol. Chem., **51**, 269-270 (1987)
[5] Kendall, K.; Cullum, J.: Cloning and expression of an extracellular-agarase gene from Streptomyces coelicolor A3(2) in Streptomyces lividans 66. Gene, **29**, 315-321 (1984)
[6] Morrice, L.M.; McLean, M.W.; Long, W.F.; Williamson, F.B.: β-Agarases I and II from Pseudomonas atlantica, substrate specificities. Eur. J. Biochem., **137**, 149-154 (1983)
[7] Morrice, L.M.; McLean, M.W.; Williamson, F.B.; Long, W.F.: β-agarases I and II from Pseudomonas atlantica, purifications and some properties. Eur. J. Biochem., **135**, 553-558 (1983)
[8] Malmqvist, M.: Purification and characterization of two different agarose-degrading enzymes. Biochim. Biophys. Acta, **537**, 31-43 (1978)
[9] van der Meulen, H.J.; Harder, W.: Production and characterization of the agarase of Cytophaga flevensis. Antonie Leeuwenhoek, **41**, 431-447 (1975)
[10] Duckworth, M.; Turvey, J.R.: The specifity of an agarase from a Cytophaga species. Biochem. J., **113**, 693-696 (1969)
[11] Duckworth, M.; Turvey, J.R.: The action of a bacterial agarase on agarose, porphyran and alkali-treated porphyran. Biochem. J., **113**, 687-692 (1969)
[12] Duckworth, M.; Turvey, J.R.: An extracellular agarase from a Cytophaga species. Biochem. J., **113**, 139-142 (1969)
[13] Turvey, J.R.; Christison, J.: The hydrolysis of algal galactans by enzymes from a Cytophaga species. Biochem. J., **105**, 311-316 (1967)
[14] Ha, J.-C.; Kim, G.-T.; Kim, S.-K.; Oh, J.-H.; Kong, I.-S.: β-Agarase from Pseudomonas sp. W7: purification and recombinant enzyme from Escherichia coli and the effects of salt on its activity. Biotechnol. Appl. Biochem., **26**, 1-6 (1997)
[15] Sugano, Y.; Terada, I.; Arita, M.; Noma, M.; Matsumoto, T.: Purification and characterization of a new agarase from a marine bacterium, Vibrio sp. strain JT0107. Appl. Environ. Microbiol., **59**, 1549-1554 (1993)
[16] Vera, J.; Alvarez, R.; Murano, E.; Slebe, J.C.; Leon, O.: Identification of a marine agrolytic Pseudoalteromonas isolate and characterization of its extracellular agarase. Appl. Environ. Microbiol., **64**, 4378-4383 (1998)
[17] Parro, V.; Vives, C.; Godia, F.; Mellado, R.P.: Overproduction and purification of an agarase of bacterial origin. J. Biotechnol., **58**, 59-66 (1997)
[18] Leon, O.; Quintana, L.; Peruzzo, P.; Slebe, J.C.: Purification and properties of an extracellular agarase from Alteromonas sp. strain C-1. Appl. Environ. Microbiol., **58**, 4060-4063 (1992)
[19] Sugano, Y.; Nagae, H.; Inagaki, K.; Yamamoto, T.; Terada, I.; Yamazaki, Y.: Production and characteristics of some new β-agarases from a marine bacterium Vibrio sp. strain JT0107. J. Ferment. Bioeng., **6**, 549-554 (1995)
[20] Kirimura, K.; Masuda, N.; Iwasaki,Y.; Nakagawa, H.; Kobayashi, R.; Usami, S.: Purification and characterization of a novel β-agarase from an alkalophilic bacterium, Alteromonas sp. E-1. J. Biosci., **87**, 436-441 (1999)

Exo-poly-α-galacturonosidase 3.2.1.82

1 Nomenclature

EC number
3.2.1.82

Recommended name
exo-poly-α-galacturonosidase

Synonyms
exopolygalacturanosidase
exopolygalacturonosidase
polygalacturonate hydrolase

CAS registry number
37288-58-7

2 Source Organism

<1> *Selenomonas ruminantium* [1]
<2> *Erwinia chrysanthemi* [2, 4, 6]
<3> *Erwinia carotovora* [3]
<4> *Pseudomonas sp.* [5]
<5> *Clostridium thermosaccharolyticum* (protein complex with polygalacturonate hydrolase activity [7]) [7]
<6> *Ralstonia solanacearum* [8]

3 Reaction and Specificity

Catalyzed reaction
hydrolysis of pectic acid from the non-reducing end, releasing digalacturonate

Reaction type
hydrolysis of O-glycosyl bond

Natural substrates and products
S pectin + H_2O <1-6> [1-8]
P digalacturonate

Substrates and products
S D-galacturonan + H_2O <2> [2]

P digalacturonate
S pectate + H_2O <1, 5> [1, 7]
P digalacturonate
S polygalacturonates + H_2O <1-6> (<1> tri-, tetra-, nonagalacturonate [1]) [1-8]
P digalacturonate <1-6> [1-8]

K_m-Value (mM)
0.02 <5> (pectin) [7]
0.05 <2> (D-galacturonan) [4]
0.08 <5> (pectate) [7]
3.1 <5> (tetra-pentagalacturonate) [7]
4.1 <5> (hexagalacturonate) [7]
1002 <1> (nonagalacturonate) [1]
1159 <1> (trigalacturonate) [1]

pH-Optimum
5 <5> (<5> pectate [7]) [7]
6 <2> (<2> D-galacturonan [4]) [4]
6.6 <1> (<1> tetragalacturonate [1]) [1]
7 <1> (<1> pectate [1]) [1]

pH-Range
2.5-7 <5> (<5> pectate [7]) [7]
3-9 <2> (<2> D-galacturonan [4]) [4]

Temperature optimum (°C)
40 <1> (<1> pectate [1]) [1]
70 <5> (<5> pectate [7]) [7]

Temperature range (°C)
40-70 <5> (<5> activity decreases rapidly at higher temperatures [7]) [7]

4 Enzyme Structure

Molecular weight
43000 <2> (<2>, gel filtration [4]) [4]
58000-65000 <2> (<2>, sedimentation equilibrium [4]) [4]
67000 <2> (<2>, SDS-PAGE of purified protein [4]) [4]
68000 <2> (<2>, activity-stained SDS-PAGE [2]) [2]
74200 <6> (<6>, predicted by DNA sequencing [8]) [8]
79000 <6> (<6>, SDS-PAGE of recombinant protein [8]) [8]

5 Isolation/Preparation/Mutation/Application

Localization
extracellular <2-6> [2-8]

Purification

<1> [1]
<2> [4]
<4> [5]
<5> (partial [7]) [7]

Cloning

<2> (expression in E. coli [6]) [6]
<6> (expression in E. coli [8]) [8]

6 Stability

Temperature stability

70 <5> (<5> stable for more than 1 h [7]) [7]

References

[1] Heinrichova, K.; Wojciechowicz, M.; Ziolecki, A.: The pectinolytic enzyme of Selenomonas ruminantium. J. Appl. Bacteriol., **66**, 169-174 (1989)
[2] Ried, J.L.; Collmer, A.: Activity stain for rapid characterization of pectic enzymes in isoelectric focusing and sodium dodecyl sulfate-polyacrylamide gels. Appl. Environ. Microbiol., **50**, 615-622 (1985)
[3] Kegoya, Y.; Masuda, H.; Hatanaka, C.: Secretion of an exo-poly-α-D-galacturonosidase by Erwinia carotovora during the early logarithmic growth phase. Agric. Biol. Chem., **48**, 1911-1912 (1984)
[4] Collmer, A.; Whalen, C.H.; Beer, S.V.; Bateman, D.F.: An exo-poly-α-D-galacturonosidase implicated in the regulation of extracellular pectate lyase production in Erwinia chrysanthemi. J. Bacteriol., **149**, 626-634 (1982)
[5] Hatanaka, C.; Imamura, T.: Production of an exopolygalacturonase liberating digalacturonic acid by a Pseudomonas. Agric. Biol. Chem., **38**, 2267-226 (1974)
[6] Hang He, S.; Collmer, A.: Molecular cloning, nucleotide sequence, and marker exchange mutagenesis of the exo-poly-α-D-galacturonosidase-encoding pehX gene of Erwinia chrysanthemi EC16. J.Bacteriol., **172**, 4988-4995 (1990)
[7] van Rijssel, M.; Gerwig, G.J.; Hansen, T.A.: Isolation and characterisation of an extracellular glycosylated protein complex from Clostridium thermosaccharolyticum with pectin methylesterase and polygalacturonate hydrolase activity. Appl. Environ. Microbiol., **59**, 828-836 (1993)
[8] Huang, Q.; Allen, C.: An exo-poly-α-D-galacturonosidase, pehB, is required for wild-type virulence of Ralstonia solanacearum. J. Bacteriol., **179**, 7369-7378 (1997)

κ-Carrageenase 3.2.1.83

1 Nomenclature

EC number
3.2.1.83

Systematic name
κ-carrageenan 4-β-D-glycanohydrolase

Recommended name
κ-carrageenase

Synonyms
carrageenase, κ-
κ-carrageenanase

CAS registry number
37288-59-8

2 Source Organism

<-1> no activity in *Tricladium angulatum* [9]
<1> *Pseudomonas carrageenovora* (strain NCIMB 302 and strain 9 [6]) [1, 2, 5, 6]
<2> *Diadema setosum* (Leske [3, 4]) [3, 4]
<3> *Cytophaga sp.* (1k-C783 [7, 8, 10]) [7, 8, 10]
<4> *Alatospora acuminata* [9]
<5> *Alatospora flagellata* [9]
<6> *Anguillospora crassa* [9]
<7> *Anguillospora longissima* [9]
<8> *Articulospora tetracladia* [9]
<9> *Calcalispora hiemalis* [9]
<10> *Clavariopsis aquatica* [9]
<11> *Culicidospora gravida* [9]
<12> *Dactylella aquatica* [9]
<13> *Dendrospora erecta* [9]
<14> *Flagellospora penicillioides* [9]
<15> *Gyoerffyella gemallipara* [9]
<16> *Gyoerffyella rotula* [9]
<17> *Gyoerffyella speciosa* [9]
<18> *Heliscella stellata* [9]
<19> *Heliscus lugdunensis* [9]
<20> *Lateriramulosa uni-inflata* [9]

<21> *Lemonniera aquatica* [9]
<22> *Lemonniera centrosphaera* [9]
<23> *Lemonniera cornuta* [9]
<24> *Lemonniera terrestris* [9]
<25> *Margaritispora aquatica* [9]
<26> *Pleuropedium tricladioides* [9]
<27> *Pyramidospora densa* [9]
<28> *Sigmoidea prolifera* [9]
<29> *Taeniospora gracilis* [9]
<30> *Tetrachaetum elegans* [9]
<31> *Tetracladium marchalianum* [9]
<32> *Tetracladium maxilliforme* [9]
<33> *Tetracladium setigerum* [9]
<34> *Tricellula aquatica* [9]
<35> *Tricladium splendens* [9]
<36> *Varicosporium elodeae* [9]
<37> *Volucrispora graminea* [9]
<38> *Pseudoalteromonas carrageenovora* [11]
<39> *Alteromonas carrageenovora* [12]
<40> *bacterium* (strain Dsij, isolated from Delesseria sanguinea [13]) [13]

3 Reaction and Specificity

Catalyzed reaction

hydrolysis of 1,4-β-D-linkages between D-galactose 4-sulfate and 3,6-anhydro-D-galactose in various carrageenans (<1>, endo or random mechanism [1])

Reaction type

hydrolysis of O-glycosyl bond

Natural substrates and products

S Additional information <1, 3> (<1>, induction by kappa-carrageenan [6]; <3>, glucose, cellobiose and galactose inhibit enzyme production [7]) [6, 7]
P ?

Substrates and products

S κ-carrageenan + H_2O <1-40> [1-13]
P Additional information <1, 40> (<1>, 3-O-(3,6-anhydro-α-D-galactopyranosyl)-D-galactopyranose-4-O-sulfate, i.e. neocarrabiose-4-O-sufate, as the major degradation product [1, 2]; <40>, kappa-neocarratetraose [13]) [1, 2, 13]
S neocarrahexaitol + H_2O <39> [12]
P Additional information <39> (<39>, neocarrabiitol and β-neocarratetraose are initially formed [12]) [12]
S *p*-nitrophenyl-β-D-galactopyranoside + H_2O <2> [4]
P ?

S Additional information <2, 39> (<2>, raffinose and *p*-nitrophenyl-α-D-galactopyranoside are not hydrolyzed [4]; <39>, the enzyme also catalyzes transglycosylation [12]) [4, 12]
P ?

Inhibitors
$AgNO_3$ <3> (<3>, 1 mM, complete inhibition [7]) [7]
$CuSO_4$ <3> (<3>, 1 mM, 31.3% inhibition [7]) [7]
$HgCl_2$ <3> (<3>, 1 mM, complete inhibition [7]) [7]
KCl <3> (<3>, 1 mM, 27.4% inhibition [7]) [7]
$MgCl_2$ <3> (<3>, 1 mM, 15.4% inhibition [7]) [7]
$MnCl_2$ <3> (<3>, 1 mM, 15.4% inhibition [7]) [7]
NaCl <3, 40> (<3>, 1.5% or above, complete inhibition [7]; <40>, above 150 mM [13]) [7, 13]
$ZnCl_2$ <3> (<3>, 1 mM, 50% inhibition [7]) [7]

Metals, ions
NaCl <40> (<40>, optimal activity at 150 mM [13]) [13]

Specific activity (U/mg)
6 <3> (<3> measurement of production of reducing sugar [10]) [10]
Additional information <1, 40> [1, 13]

K_m-Value (mM)
0.244 <40> (kappa-carrageenan) [13]

pH-Optimum
7 <3, 40> (<40>, in MOPS buffer [13]) [7, 13]
7.2 <40> (<40>, in Tris buffer [13]) [13]
7.5 <1> [1]
7.6 <2, 3> [3, 4, 10]
8 <1> (<1>, 15 min assay [2]) [2]

pH-Range
5-8 <40> (<40>, pH 5.0: about 50% of maximal activity, pH 8.0: about 75% of maximal activity [13]) [13]
5-9 <2> (<2>, broad range of activity [3, 4]) [3, 4]
6-10 <1> (<1>, about 45% of maximal activity [2]) [2]
7.4-8.4 <3> (<3>, about 60% of maximal activity at pH 7.4 and at pH 8.4 [10]) [10]

Temperature optimum (°C)
25 <3> [10]
25-50 <3> [7]
40 <1> (<1>, 15 min assay [1]) [1]

Temperature range (°C)
15-45 <3> (<3>, 15°C: about 40% of maximal activity, 45°C: about 50% of maximal activity [10]) [10]
20-60 <1> (<1>, 20°C: about 60% of maximal activity, 60°C: about 25% of maximal activity, 15 min assay [2]) [2]

4 Enzyme Structure

Molecular weight

32940 <3> (<3>, electrospray mass spectrometry [10]) [10]

Subunits

? <1, 3, 39, 40> (<1,39>, x * 35000, SDS-PAGE [2,12]; <40>, x * 40000, SDS-PAGE [13]; <3>, x * 100000, SDS-PAGE [10]) [2, 10, 12, 13]

Posttranslational modification

proteolytic modification <39> (<39>, processed upon secretion from a 44000 Da preproprotein to a 32000 Da secretion product, first by N-terminal removal of the 25-amino-acid signal peptide, then by cleavage of a 96-amino-acid C-terminal region from amino acids 302-397 [12]) [12]

5 Isolation/Preparation/Mutation/Application

Source/tissue

culture medium <1, 3> (<3>, synthesized and released from resting cells and growing cells [8]) [1, 7, 8, 10]
digestive tract <2> [3, 4]

Localization

extracellular <1, 3, 39> [1, 7, 8, 10, 12]

Purification

<1> [1, 2]
<2> (partial [3]) [3]
<3> [10]
<38> (recombinant enzyme expressed in E. coli [11]) [11]
<39> [12]
<40> [13]

Crystallization

<38> (recombinant enzyme expressed in E. coli [11]) [11]

Cloning

<38> (expression in E. coli [11]) [11]

Application

analysis <1> (<1>, studies of carrageenan structure [5]) [5]
molecular biology <1> (<1>, potential use in the field of seaweed biotechnology for obtaining protoplasts, and in molecular biology to prevent severe separation problems occuring in the presence of phycocolloid oligomers [6]) [6]
pharmacology <1> (<1>, potential use: production of defined phycocolloid oligomers for pharmacy and immunology [6]) [6]

6 Stability

pH-Stability

4 <40> (<40>, 1 h, about 65% loss of activity [13]) [13]
4.5 <40> (<40>, 1 h, about 40% loss of activity [13]) [13]
6 <40> (<40>, 1 h, about 30% loss of activity [13]) [13]
7 <40> (<40>, 1 h, stable [13]) [13]
8.5 <40> (<40>, 1 h, about 50% loss of activity [13]) [13]
9 <40> (<40>, 1 h, about 75% loss of activity [13]) [13]

Temperature stability

5 <1> (<1>, stable for 2 weeks [1]) [1]
30 <40> (<40>, 2 h, stable [13]) [13]
40 <1, 40> (<1>, stable for 3.5 h [1]; <40>, stable for up to 1 h [13]) [1, 13]
50 <1, 3> (<1>, 10 min, inactivated [1]; <3>, 15 min, more than 70% loss of activity [7]) [1, 7]
55 <3> (<3>, 10 min, complete inactivation [7]) [7]

Storage stability

<2>, -20°C, stable for several months [3]
<40>, -20°C, stable for at least 16 months [13]

References

[1] Weigl, J.; Yaphe, W.: The enzymic hydrolysis of carrageenan by Pseudomonas carrageenovora: purification of a kappa-carrageenase. Can. J. Microbiol., **12**, 939-947 (1966)

[2] McLean, M.M.; Williamson, F.B.: κ-Carrageenase from Pseudomonas carrageenovora. Eur. J. Biochem., **93**, 553-558 (1979)

[3] Benites, L.V.; Macaranas, J.M.: Partial purification of a carragenase from the tropical sea urchin Diadema setosum. Proc. Int. Seaweed Symp., **9**, 353-359 (1977)

[4] Benites, L.V.; Macaranas, J.M.: Purification and characterization of a kappa-carragenase from the tropical sea urchin, Diadema setosum Leske. Inf. Ser.-N. Z. Dep. Sci. Ind. Res., 137 (Proc. Int. Symposium Mar. Biogeogr. Evol. South. Hemishere), **2**, 371-376 (1978)

[5] McLean, M.W.; Williamson, F.B.: Enzymes from Pseudomonas carrageenovora. Application to studies of carrageenan structure. Proc. Int. Seaweed Symp., **10**, 479-484 (1981)

[6] Dyrset, N.; Lystad, K.Q.; Levine, D.W.: Development of a fermentation process for production of a κ-carrageenase from Pseudomonas carrageenovora. Enzyme Microb. Technol., **20**, 418-423 (1997)

[7] Sarwar, G.; Sakata, T.; Kakimoto, D.: The production and characteristics of carrageenase from marine Cytophaga. Nippon Suisan Gakkaishi, **49**, 1689-1694 (1983)

[8] Sarwar, G.; Oda, H.; Sakawa, T.; Kakimoto, D.: Potentiality of artificial sea water salts for the production of carrageenase by a marine Cytophaga sp. Microbiol. Immunol., **29**, 405-411 (1985)
[9] Zemek, J.; Marvanova, L.; Kuniak, L.; Kadlecikova, B.: Hydrolytic enzymes in aquatic hyphomycetes. Folia Microbiol., **30**, 363-372 (1985)
[10] Sarwar, G.; Matayoshi, S.; Oda, H.: Purification of a kappa-carrageenase from marine Cytophaga species. Microbiol. Immunol., **31**, 869-877 (1987)
[11] Michel, G.; Barbeyron, T.; Flament, D.; Vernet, T.; Kloareg, B.; Dideberg, O.: Expression, purification, crystallization and preliminary X-ray analysis of the kappa-carrageenase from Pseudomonas carrageenovora. Acta Crystallogr. Sect. D, **55**, 918-920 (1999)
[12] Potin, P.; Richard, C.; Barbeyron, T.; Henrissat, B.; Gey, C.; Petillot, Y.; Forest, E.; Dideberg, O.; Rochas, C.; Kloareg, B.: Processing and hydrolytic mechanism of the cgkA-encoded kappa-carrageenase of Alteromonas carrageenovora. Eur. J. Biochem., **228**, 971-975 (1995)
[13] Potin, P.; Sanseau, A.; le Gall, Y.; Rochas, C.; Kloareg, B.: Purification and characterization of a new kappa-carrageenase from a marine Cytophaga-like bacterium. Eur. J. Biochem., **201**, 241-247 (1991)

Glucan 1,3-α-glucosidase 3.2.1.84

1 Nomenclature

EC number
3.2.1.84

Systematic name
1,3-α-D-glucan 3-glucohydrolase

Recommended name
glucan 1,3-α-glucosidase

Synonyms
exo-1,3-α-glucanase
glucosidase II

CAS registry number
9073-99-8

2 Source Organism

<1> *Vigna radiata* (mung bean [1, 13, 15, 25]) [1, 13, 15, 25]
<2> *Bos taurus* (cow [2]; calf [7]) [2, 7]
<3> *Rattus norvegicus* [3, 5, 8, 9, 14, 16, 18, 19, 21, 23]
<4> *Sus scrofa* [4, 6, 20, 22]
<5> *Cladosporium resinae* [10]
<6> *Trichoderma sp.* [11]
<7> *Aspergillus nidulans* [12]
<8> *Glycine max* (soybean [15]) [15]
<9> *Schizosaccharomyces pombe* [17]
<10> *Homo sapiens* [24]

3 Reaction and Specificity

Catalyzed reaction
hydrolysis of terminal 1,3-α-D-glucosidic linkages in 1,3-α-D-glucans ($Glc_2Man_9GlcNAc_2 + H_2O = Man_9GlcNAc_2$ + glucose <2, 3> [2, 7, 8, 9])

Reaction type
hydrolysis of O-glycosyl bond

Natural substrates and products

S $Glc_2Man_9GlcNAc_2$ + H_2O <2, 3, 4> [2, 5-9]
P $Man_9GlcNAc_2$ + α-D-glucose

Substrates and products

S 4-methylumbelliferyl-α-D-glucopyranoside + H_2O <1, 4> [1, 6]
P 4-methylumbelliferone + α-D-glucose
S $Glc_2Man_9GlcNAc_2$ + H_2O <2, 3, 4> [2, 5-9]
P $Man_9GlcNAc_2$ + glucose
S α-1,3-glucans + H_2O <5> (<5> exo-wise hydrolysis from non-reducing ends [10]) [10]
P α-D-glucose
S maltose + H_2O <3> [8]
P α-D-glucose
S *p*-nitrophenyl-α-D-glucopyranoside + H_2O <3> [8]
P *p*-nitrophenol + α-D-glucose
S pseudonigeran + H_2O <5, 6> [10, 11]
P α-D-glucose

Inhibitors

1-deoxynojirimycin <1, 3, 4> [20, 21, 25]
2,6-dideoxy-2,6-imino-7-O-(β-D-glucopyranosyl)-D-glycero-L-glucoheptitol <1> (<1> MDL [1]) [1, 25]
2-propanol <3> [18]
6-epicastanospermine <1> (<1> poor inhibitor [13]) [13]
Ag^+ <2, 5> [2, 10]
Ca^{2+} <3> [18]
D-glucono-1,5-lactone <3> [8, 22]
Fe^{2+} <5> [10]
Fe^{3+} <5> [10]
Hg^{2+} <2, 5> [2, 10]
Mg^{2+} <3> [18]
N-5-carboxypentyl-1-deoxynojirimycin <4> [20]
NH_4^+ <3> [18]
Tris <3, 4> [6, 8]
bromoconduritol <3> (<3> binds to the low- but not high-affinity site for maltose and *p*-nitrophenyl-α-D-glucoside [14]) [14]
castanospermine <1> [1]
ethanol <3> [18]
glucose <3> [8]
glycerol <3> [8]
isobutanol <3> [18]
maltose <2-4> [2, 6, 8]
methanol <3> [18]
n-butanol <3> [18]
n-propanol <3> [18]
nigerose <1, 3> [21, 25]
p-chloromercuribenzenesulfonate <3> [8]

p-nitrophenyl-α-D-glucoside <2> [7]
p-nitrophenyl-α-D-mannoside <2> [7]
p-nitrophenyl-β-D-glucoside <2> [7]
p-nitrophenyl-β-D-mannoside <2> [7]
turanose <2, 4> [2, 6]
validamine <3> [21]
valienamine <3> [21]
valiolamine <3> [21]

Activating compounds

2-deoxy-D-glucose <3> (<3> activation substrate p-nitrophenyl-α-D-glucoside [8]) [8]
isomaltose <4> (<4> lowering K_m, 142% activation at 50 mM [6]) [6]
mannose <4> (<4> increasing V_{max}, 172% activation at 10 mM [6]) [6]
starch <4> (<4> increasing V_{max}, 160% activation at 7 mg/ml [6]) [6]

Specific activity (U/mg)

0.096 <2> (<2> 4-methylumbelliferyl α-D-glucopyranoside as substrate [2]) [2]
0.151 <4> (<4> 4-methylumbelliferyl α-D-glucopyranoside as substrate [6]) [6]
0.442 <3> (<3> p-nitrophenyl α-D-glucopyranoside as substrate [5]) [5]
0.501 <4> [20]
0.617 <3> (<3> p-nitrophenyl-α-D-glucopyranoside as substrate [18]) [18]
0.782 <3> (<3> active peptide fragment after trypsinization [5]) [5]
11 <6> [11]

K_m-Value (mM)

0.013 <2> (4-methylumbelliferyl α-D-glucopyranoside) [2]
0.019 <4> (4-methylumbelliferyl α-D-glucopyranoside) [6]
0.43 <3> (maltose, <3> high-affinity site [19]) [19]
0.78 <3> (p-nitrophenyl α-D-glucopyranoside, <3> high-affinity site [23]) [23]
0.85 <3> (p-nitrophenyl α-D-glucopyranoside) [8]
1.2 <3> (p-nitrophenyl α-D-glucopyranoside) [18]
4.8 <3> (maltose) [8]
7.1 <6> (pseudonigeran) [11]
13 <5> (pseudonigeran) [10]
57.7 <3> (maltose, <3> low-affinity site [19]) [19]
481 <3> (p-nitrophenyl α-D-glucopyranoside, <3> high-affinity site [23]) [23]

pH-Optimum

4.5 <5, 6> [10, 11]
5.5 <7> [12]
6-7.5 <4> [6]
6.5 <4> [20]
6.5-7 <2, 3> (<3> 0.75 mM maltose [18]) [2, 18]
6.6-7 <3> (<3> depending on substrate [8]) [8]

6.7 <2> [7]
6.8 <3> [18]
6.8-7.5 <1> [25]
7-8 <3> (<3> 30 mM maltose [19]) [19]

pH-Range
3-7.5 <3> (<3> citrate/phosphate buffer [18]) [18]
3-8 <7> [12]
6.8-8.2 <3> (<3> HEPES buffer [18]) [18]

Temperature optimum (°C)
47 <5> [10]
50 <7> [12]
55 <6> [11]

Temperature range (°C)
30-60 <7> [12]

4 Enzyme Structure

Molecular weight
36000 <7> (<7> gel filtration [12]) [12]
47000 <6> (<6> gel filtration [11]) [11]
88000 <5> (<5> gel filtration, SDS-PAGE [10]) [10]
100000-123000 <3> (<3> immunoprecipitation [3]; <3> SDS-PAGE with mercaptoethanol [5]) [3, 5]
107000 <3> (<3> may be a proteolytic degradation product of the 112000 protein [18]) [18]
110000 <8> [15]
112000 <4> [20]
220000 <1> (<1> gel filtration [25]) [25]
260000-290000 <2, 3> (<2> gel filtration [2]; <3> gel electrophoresis, gel filtration [8]) [2, 8]
400000 <4> (<4> non denaturing PAGE [6]) [6]
425000 <3> (<3> gel filtration [18]) [18]

Subunits
dimer <13, 9, 10> (<1> α_2, 2 * 110000, SDS-PAGE [15]; <3> $\alpha\beta$ 1 * 104000 + 1 * 58000 [16]; <10> $\alpha\beta$ 1 * 110000 + 1 * 80000 [24]) [15-17, 24, 25]
tetramer <2-4> (<2> 4 * 63000-65000, SDS-PAGE, comparison with other values [2]; <3> SDS-PAGE [8]; <3> α_4, 4 * 106000 [18]; <4> 4 * 100000, SDS-PAGE [6]) [2, 6, 8, 18, 20]

Posttranslational modification
glycoprotein <2, 3, 4, 10> (<2> N-linked oligosaccharides, high mannose type [2]; <4> also sialic acid [22]) [2, 3, 5, 6, 22, 24]
proteolytic modification <3> [3]

5 Isolation/Preparation/Mutation/Application

Source/tissue

culture homogenate <5> [10]
culture medium <5> [10]
hepatoma cell line <3> [3]
kidney <4> [4, 6, 22]
liver <2, 3> [5, 7-9, 14, 16, 18, 20, 21, 24]
mammary gland <2> [2]

Localization

Golgi apparatus <4> [22]
endocytic elements <4> [22]
endoplasmic reticulum <2, 3, 4> (<2, 4> rough and smooth [4, 7]; <3, 4> lumen [6, 16]) [4, 6, 7, 16]
extracellular <5> [10]
microsome <1-4> [2, 5, 6, 8, 15, 18, 20]
plasma membrane <4> [22]
post-Golgi apparatus <4> (<4> predominantly [22]) [22]
soluble <3> [3]
Additional information <4> (<4> subcellular localisation [4]) [4]

Purification

<1> [25]
<2> [2, 7]
<3> [5, 8, 16, 18, 23]
<4> [6, 20]
<5> [10]
<6> [11]

Cloning

<9> (expression in E. coli [17]) [17]
<10> (expression in COS-1 cells [24]) [24]

6 Stability

pH-Stability

3.5-6 <6> (<6> at 30°C [11]) [11]
3.5-7 <5, 6> (<5, 6> at 4°C [10, 11]) [10, 11]
4-7 <5> (<5> at 30°C [10]) [10]
6 <1, 4> (<1, > inactivation below [6, 25]) [6, 25]
6-8 <3> (<3> with 5 mM mercaptoethanol [8]) [8]

Temperature stability

36 <3> [8]
38 <5> (<5> at pH 7.0 [10]) [10]

40 <3> (<3> inactivation [8]; <3> 100% of activity after 30 min, 10 mM potassium phosphate buffer, pH 7, 0.1% Triton X-100, 5 mM 2-mercaptoethanol [18]) [8, 18]
43 <5> (<5> up to at pH 3.5 [10]) [10]
47 <3, 5> (<3> 47% of activity [18]; <5> up to at pH 4.5 [10]) [10, 18]
50 <6> (<6> up to [11]) [11]

General stability information

<2, 4>, glycerol, stabilization [6, 7]

Storage stability

<3>, -70°C, 0.9% NaCl, several months [5]
<3>, -70°C, 100 mM phosphate buffer, pH 7, 100 mM maltose, 5 mM mercaptoethanol, several weeks [8]
<3>, 4-6°C, 100 mM phosphate buffer, pH 7, 10% glycerol, 5 mM mercaptoethanol, several days [8]
<3>, 4°C, 10 mM potassium phosphate buffer, pH 7, 0.1% Triton X-100, 5 mM 2-mercaptoethanol, 14 days [18]
<3>, 4°C, 10 mM potassium phosphate buffer, pH 7, 0.1% Triton X-100, 5 mM 2-mercaptoethanol, 8 days ethanol treated rats [18]
<3>, freezing inactivates the enzyme [18]

References

[1] Kaushal, G.P.; Pan, Y.T.; Tropea, J.E.; Mitchell, M.; Liu, P.; Elbein, A.D.: Selective inhibition of glycoprotein-processing enzymes. Differential inhibition of glucosidases I and II in cell culture. J. Biol. Chem., **263**, 17278-17283 (1988)

[2] Saxena, S.; Shailubhai, K.; Dong-Yu, B.; Vijay, I.K.: Purification and characterization of glucosidase II involved in N-linked glycoprotein processing in bovine mammary gland. Biochem. J., **247**, 563-570 (1987)

[3] Strous, G.J.; Van Kerkhof, P.; Brok, R.; Roth, J.; Brada, D.: Glucosidase II, a protein of the endoplasmic reticulum with high mannose oligosaccharide chains and a rapid turnover. J. Biol. Chem., **262**, 3620-3625 (1987)

[4] Lucocq, J.M.; Brada, D.; Roth J.: Immunolocalization of the oligosaccharide trimming enzyme glucosidase II. J. Cell Biol., **102**, 2137-2146 (1986)

[5] Hino, Y.; Rothman, J.E.: Glucosidase II, a glycoprotein of the endoplasmic reticulum membrane. Proteolytic cleavage into enzymatically active fragments. Biochemistry, **24**, 800-805 (1985)

[6] Brada, D.; Dubach, U.C.: Isolation of a homogeneous glucosidase II from pig kidney microsomes. Eur. J. Biochem., **141**, 149-156 (1984)

[7] Tabas, I.; Kornfeld, S.: N-Asparagine-linked oligosaccharides: processing. Methods Enzymol., **83**, 416-429 (1982)

[8] Burns, D.M.; Touster, O.: Purification and characterization of glucosidase II, an endoplasmic reticulum hydrolase involved in glycoprotein biosynthesis. J. Biol. Chem., **257**, 9991-10000 (1982)

[9] Grinna, L.S.; Robbins, P.W.: Substrate specificities of rat liver microsomal glucosidases which process glycoproteins. J. Biol. Chem., **255**, 2255-2258 (1980)
[10] Tsunoda, A.; Sakano, Y.; Kobayashi, T.: Purification and properties of an exo-α-1,3-glucanase from Cladosporium resinae. Agric. Biol. Chem., **42**, 1045-1053 (1978)
[11] Tsunoda, A.; Nagaki, T.; Sakano, Y.; Kobayashi, T.: Purification and properties of an exo-α-1,3-glucanase from Trichoderma viride. Agric. Biol. Chem., **41**, 939-943 (1977)
[12] Zonneveld, B.J.M.: A new type of enzyme, and exo-splitting α-1,3 glucanase from non-induced cultures of Aspergillus nidulans. Biochim. Biophys. Acta, **258**, 541-547 (1972)
[13] Molyneux, R.J.; Roitman, J.N.; Dunnheim, G.; Szumilo, T.; Elbein, A.D.: 6-Epicastanospermin, a novel indolizidine alkaloid that inhibits α-glucosidase. Arch. Biochem. Biophys., **251**, 450-457 (1986)
[14] Alonso, J.F.; Santa-Cecilia, A.; Calvo, P.: Effect of bromoconduritol on glucosidase II from rat liver, a new kinetic model for the binding and hydrolysis of the substrate. Eur. J. Biochem., **215**, 37-42 (1993)
[15] Kaushal, G.P.; Zeng, Y.; Elbein, A.D.: Biosynthesis of glucosidase II in suspension-cultured soybean cells. J. Biol. Chem., **268**, 14536-14542 (1993)
[16] Trombetta, E.S.; Simons, J.F.; Helenius, A.: Endoplasmic reticulum glucosidase II is composed of a catalytic subunit, conserved from yeast to mammals, and a tightly bound noncatalytic HDEL-containing subunit. J. Biol. Chem., **271**, 27509-27516 (1996)
[17] D'Alessio, C.; Fernàndez, F.; Trombetta, E.S.; Parodi, A.J.: Genetic evidence for the heterodimeric structure of glucosidase II: the effect of disrupting the subunit-encoding genes on glycoprotein folding. J. Biol. Chem., **274**, 25899-25905 (1996)
[18] Santa-Cecilia, A.; Alonso, J.F.; Calvo, P.: Glucosidase II from control and ethanol-treated rats: purification and properties. Biol. Chem., **372**, 373-380 (1991)
[19] Alonso, J.F.; Santa-Cecilia, A.; Chinchetru, M.A.; Calvo, P.: Characterization of the maltase activity of glucosidase II from rat liver: kinetic model. Biol. Chem., **374**, 977-982 (1993)
[20] Hentges, A.; Bause, E.: Affinity purification and characterization of glucosidase II from pig liver. Biol. Chem., **378**, 1031-1038 (1997)
[21] Takeuchi, M.; Kamata, K.; Yoshida, M.; Kameda, Y.; Matsui, K.: Inhibitory effect of pseudo-aminosugars on oligosaccharide glucosidases I and II and on lysosomal α-glucosidase from rat liver. J. Biochem., **108**, 42-46 (1990)
[22] Brada, D.; Kerjaschki, D.; Roth, J.: Call type-specific post-Golgi apparatus localization of a "resident" endoplasmic reticulum glycoprotein, glucosidase II. J. Cell Biol., **110**, 309-318 (1990)
[23] Alonso, J.F.; Santa-Cecilia, A.; Calvo, P.: Glucosidase II from rat liver microsomes: kinetic model for binding and hydrolysis. Biochem. J., **278**, 721-727 (1991)

[24] Treml, K.; Meimaroglou, D.; Hentges, A.; Bause, E.: The α- and β-subunits are required for expression of catalytic activity in the hetero-dimeric glucosidase II complex from human liver. Glycobiology, **10**, 493-502 (2000)
[25] Kaushal, G.P.; Pastuszak, I.; Hatanka, K.-i.; Elbein, A.D.: Purification to homogeneity and properties of glucosidase II from mung bean seedlings and suspension-cultured soybean cells. J. Biol. Chem., **265**, 16271-16279 (1990)

6-Phospho-β-galactosidase 3.2.1.85

1 Nomenclature

EC number
3.2.1.85

Systematic name
6-phospho-β-D-galactoside 6-phosphogalactohydrolase

Recommended name
6-phospho-β-galactosidase

Synonyms
6-phospho-β-D-galactosidase <2> [9, 12]
BGase-III <5> [7]
P-β-gal <7> (<7> I and II [17]) [17]
PGALase <8> [18]
β-D-phosphogalactosidase <6> [6]
β-D-phosphogalactoside galactohydrolase <3, 6> [1, 10, 11]
β-Pgal <6> [11]
β-phosphogalactosidase <6> [5]
phospho-β-galactosidase <1, 2, 4, 5> [2, 3, 7, 8]

CAS registry number
37237-42-6

2 Source Organism

<1> *Streptococcus mutans* [8]
<2> *Staphylococcus aureus* [3, 9, 12-15]
<3> *Streptococcus cremoris* (HP [10]) [10]
<4> *Streptococcus lactis* [2, 4]
<5> *Klebsiella sp.* (strain RE 1755A, Lac^+Gal^+ mutant CT-1 [7]) [7]
<6> *Lactobacillus casei* (subsp. casei 64H [1, 6]; L. casei 9595 [11]) [1, 5, 6, 11, 14, 18]
<7> *Lactobacillus gasseri* (JCM 1031 [16, 17]) [16, 17]
<8> *Lactococcus lactis* [14, 18]

3 Reaction and Specificity

Catalyzed reaction

a 6-phospho-β-D-galactoside + H_2O = an alcohol + 6-phospho-D-galactose

Reaction type

hydrolysis of O-glycosyl bond

Natural substrates and products

S lactose-6-phosphate + H_2O <1-8> [1-8, 10-18]
P D-glucose + β-D-galactose-6-phosphate <1-8> [1-8, 10-18]

Substrates and products

S *o*-nitrophenyl-β-D-galactopyranoside 6-phosphate + H_2O <1-8> [3-17]
P *o*-nitrophenol + β-D-galactose-6-phosphate <1-8> [3-17]
S *o*-nitrophenyl-β-D-glucopyranoside 6-phosphate <2, 6, 8> (<2, 6, 8> only little activity [14]) [14]
P *o*-nitrophenol + β-D-glucose-6-phosphate <2, 6, 8> [14]
S *p*-nitrophenyl-β-D-galactopyranoside 6-phosphate + H_2O <2, 4, 6, 8> [2, 14]
P *p*-nitrophenol + β-D-galactose-6-phosphate <2, 4, 6, 8> [2, 14]
S *p*-nitrophenyl-β-D-glucopyranoside 6-phosphate <2, 6, 8> (<2, 6, 8> only little activity [14]) [14]
P *p*-nitrophenol + β-D-glucose-6-phosphate <2, 6, 8> [14]

Inhibitors

2,4-dinitrophenyl-2-deoxy-2-bromo-β-D-galactopyranosyl-6-phosphate <2> (<2> fluoro-, chloro- and bromo-derivatives, complete inactivation, most rapid with fluoro-, followed by chloro- and bromo-derivatives [15]) [15]
2,4-dinitrophenyl-2-deoxy-2-chloro-β-D-galactopyranosyl-6-phosphate <2> (<2> fluoro-, chloro- and bromo-derivatives, complete inactivation, most rapid with fluoro-, followed by chloro- and bromo-derivatives [15]) [15]
2,4-dinitrophenyl-2-deoxy-2-fluoro-β-D-galactopyranosyl-6-phosphate <2> (<2> fluoro-, chloro- and bromo-derivatives, complete inactivation, most rapid with fluoro-, followed by chloro- and bromo-derivatives [15]) [15]
2-phosphoglycerate <1> (<1> 26% inhibition at 4.2 mM [8]) [8]
ATP <1> (<1> 100% inhibition at 2.1 mM [8]) [8]
D-fructose-1,6-diphosphate <1> (<1> 19% inhibition at 4.2 mM [8]) [8]
D-fructose-6-phosphate <1> (<1> 20% inhibition at 4.2 mM [8]) [8]
D-galactose-6-phosphate <1, 2, 3, 5> (<1> 55% inhibition at 4.2 mM [8]; <2, 3> competitive inhibition [10, 14]; <2> weak inhibition [14]) [7, 8, 10, 13, 14]
D-glucose-6-phosphate <1, 7> (<1> 38% inhibition at 4.2 mM [8]; <7> 63% of activity [17]) [8, 17]
N-ethylmaleimide <7> [16, 17]
iodoacetamide <7> [16, 17]
p-chloromercuribenzoic acid <7> [16, 17]

p-hydroxymercuribenzoate <3> (<3> 83.4% of control activity at 1 mM, 12.2% at 2 mM [10]) [10]
p-nitrophenyl-β-thiogalactoside 6-phosphate <2> (<2> competitive inhibition [14]) [14]
p-nitrophenyl-β-thioglucoside 6-phosphate <2> (<2> competitive inhibition [14]) [14]
phosphoenolpyruvate <1> (<1> 97% inhibition at 1 mM [8]) [8]

Activating compounds

2-mercaptoethanol <7> (<7> lower effect on P-β-gal II than P-β-gal I [17]) [16, 17]
EDTA <3, 7> (<3> slight stimulation to 118.2% by 2 mM [10]; <7> 30% increase of activity [16]) [10, 16]
Fe^{2+} <7> (<7> 60% increase of activity [16, 17]) [16, 17]
L-cysteine <7> [16]
dithiothreitol <3, 7> (<3> enhanced activiy after dialysis by addition up to 0.75 mM [10]) [10, 16]
glutathione <7> [16]

Metals, ions

Ca^{2+} <7> (<7> 78% of activity at 1 mM $CaCl_2$ [17]) [17]
Co^{2+} <1> (<1> 14% inhibition at 5 mM [8]) [8]
Cu^{2+} <1, 2> (<1> 100% inhibition at 5 mM [8]) [8, 9]
Hg^{2+} <2, 7> (<7> complete inhibition at 1 mM $HgCl_2$ [17]) [9, 12, 16, 17]
Li^{+} <3> (<3> 82% of control activity at 2 mM [10]) [10]
Mg^{2+} <3, 7> (<3> 74.5% of control activity at 2 mM [10]; <7> slight inhibition at 1 mM $MgCl_2$ [17]) [10, 17]
Mn^{2+} <2> (<2> 50% inactivation to 10 mM [9];<7> slight inhibition at 1 mM $MnCl_2$ [17]) [9, 12]
Zn^{2+} <1, 7> (<1> 57% inhibition at 5 mM [8]; <7> 98.4% inhibition at 1 mM $ZnSO_4$ [17]) [8, 16, 17]

Turnover number (min^{-1})

120 <6> (p-nitrophenyl-β-D-glucopyranoside 6-phosphate) [14]
240 <6> (o-nitrophenyl-β-D-glucopyranoside 6-phosphate) [14]
240 <2> (p-nitrophenyl-β-D-glucopyranoside 6-phosphate) [14]
300 <8> (p-nitrophenyl-β-D-glucopyranoside 6-phosphate) [14]
420 <8> (o-nitrophenyl-β-D-glucopyranoside 6-phosphate) [14]
420 <6> (p-nitrophenyl-β-D-galactopyranoside 6-phosphate) [14]
540 <2> (o-nitrophenyl-β-D-glucopyranoside 6-phosphate) [14]
1080 <2> (p-nitrophenyl-β-D-galactopyranoside 6-phosphate) [14]
1320 <8> (p-nitrophenyl-β-D-galactopyranoside 6-phosphate) [14]
3420 <6> (o-nitrophenyl-β-D-galactopyranoside 6-phosphate) [14]
3780 <2> (o-nitrophenyl-β-D-galactopyranoside 6-phosphate) [14]
4980 <8> (o-nitrophenyl-β-D-galactopyranoside 6-phosphate) [14]

Specific activity (U/mg)

0.004 <5> (<5> with citrate-grown cell-extracts of CT-1 [7]) [7]
0.032 <5> (<5> with lactose-grown cell-extracts of CT-1 [7]) [7]

0.28 <4> (<4> *o*-nitrophenyl-β-D-galactopyranoside 6-phosphate as substrate, maximum of activity with Streptococcus lactis strain MG1299 after growth on glucose [4]) [4]
0.28-0.37 <6> (<6> *o*-nitrophenyl-β-D-galactopyranoside 6-phosphate as substrate [5]) [5]
1.45 <4> (<4> *p*-nitrophenyl-β-D-galactopyranoside 6-phosphate as substrate, 35 mM phosphate buffer, pH 6.8 [2]) [2]
2.44 <4> (<4> *o*-nitrophenyl-β-D-galactopyranoside 6-phosphate as substrate, maximum of activity with Streptococcus lactis strain MG1299 after growth on lactose [4]) [4]
20.9 <1> (<1> purified enzyme [8]) [8]
23.8 <3> (<3> *o*-nitrophenyl-β-D-galactopyranoside 6-phosphate as substrate [10]) [10]
28.9 <6> (<6> *o*-nitrophenyl-β-D-galactopyranoside 6-phosphate as substrate, maximum of activity [11]) [11]
38-47 <2> (<2> at 30°C [9, 12]) [9, 12]
100 <2> (<2> maximal activity of hybrid with protein A [13]) [13]

K_m-Value (mM)
0.04 <8> (*o*-nitrophenyl-β-D-glucopyranoside 6-phosphate) [14]
0.09 <8> (*p*-nitrophenyl-β-D-galactopyranoside 6-phosphate) [14]
0.1 <8> (*p*-nitrophenyl-β-D-glucopyranoside 6-phosphate) [14]
0.15 <6> (*o*-nitrophenyl-β-D-glucopyranoside 6-phosphate) [14]
0.16 <2> (*o*-nitrophenyl-β-D-glucopyranoside 6-phosphate) [14]
0.19 <1> (*o*-nitrophenyl-β-D-galactopyranoside 6-phosphate) [8]
0.2 <6> (*p*-nitrophenyl-β-D-galactopyranoside 6-phosphate) [14]
0.3 <2> (*p*-nitrophenyl-β-D-galactopyranoside 6-phosphate) [14]
0.4 <8> (*o*-nitrophenyl-β-D-galactopyranoside 6-phosphate) [14]
0.4 <2> (*p*-nitrophenyl-β-D-glucopyranoside 6 phosphate) [14]
0.59 <3> (*o*-nitrophenyl-β-D-galactopyranoside 6-phosphate) [10]
0.6 <6> (*p*-nitrophenyl-β-D-glucopyranoside 6-phosphate) [14]
0.6-3 <2> (*o*-nitrophenyl-β-D-galactopyranoside 6-phosphate, <2> 2.0 hybrid with protein A [13]) [9, 12, 13, 14]
0.76-0.8 <7> (*o*-nitrophenyl-β-D-galactopyranoside 6-phosphate) [16, 17]
1.3-1.59 <6> (*o*-nitrophenyl-β-D-galactopyranoside 6-phosphate) [11, 14]
3 <5> (*o*-nitrophenyl-β-D-galactopyranoside 6-phosphate, <5> with strain CT-1 [7]) [7]
7.3 <5> (*o*-nitrophenyl-β-D-galactopyranoside 6-phosphate, <5> with strain RE1755A [7]) [7]

pH-Optimum
4.3 <5> [7]
5 <6> [11]
5-5.5 <7> [16, 17]
6.5 <1> (<1> most active in Tris-HCl buffer [8]) [8]
7 <2> [9, 12]

pH-Range

4-7 <6> [11]
5-7 <7> (<7> at 37°C for 15 min [16]) [16]
5-8 <3> (<3> nearly constant activity [10]) [10]

Temperature optimum (°C)

30 <7> (<7> P-β-gal II [17]) [17]
37 <6> (<6> o-nitrophenyl-β-D-galactopyranoside 6-phosphate [11]) [11]
45-52 <3> (<3> o-nitrophenyl-β-D-galactopyranoside 6-phosphate as substrate [10]) [10]
50 <7> (<7> P-β-gal I [17]) [17]

4 Enzyme Structure

Molecular weight

35000-56000 <6> (<6> gel filtration, SDS-PAGE [5, 6]) [5, 6]
40000 <1> (<1> strain SL1, gel filtration [8]) [8]
46000-53000 <2> (<2> gel filtration, SDS-PAGE [9, 12]) [3, 9, 12]
53990 <6> (<6> DNA sequence analysis [1]) [1]
55100-56500 <4> (<4> DNA sequence analysis, SDS-PAGE [2]) [2]
58000 <4, 7> (<7> SDS-PAGE [16]) [4, 16]
67600 <3> (<3> gel filtration [10]) [10]
130000 <6> (<6> gel filtration [11]) [11]

Subunits

monomer <2> (<2> 1 * 50000, SDS-PAGE [9, 12]) [9, 12]

5 Isolation/Preparation/Mutation/Application

Localization

cytoplasm <1> [8]

Purification

<1> [8]
<2> (hybrid with protein A [13]) [9, 12-14]
<3> [10]
<6> [11, 14]
<7> [16, 17]
<8> [14]

Crystallization

<8> (hanging drop method [18]) [18]

Cloning

<2, 4, 6, 8> (expressed in E.coli [1-4, 6, 13, 14, 18]) [1-4, 6, 13, 14, 18]

Engineering

E160Q <2> (<2> 1000fold reduction in activity compared to wild type [14]) [14]
E75Q <2> (<2> complete loss of activity [14]) [14]
G332R <2> (<2> complete loss of catalytic activity, but same behavior during purification and in affinity to antisera as wild type [14]) [14]
Additional information <2> (<2> hybrid protein with protein A [13]) [13]

6 Stability

pH-Stability

4-9 <7> (<7> at 37°C for 15 min [16]) [16, 17]

Temperature stability

40-60 <7> (<7> P-β-gal II, stable below 40°C, inactivated at 60°C [17]) [17]
50 <5> (<5> rapid thermal inactivation, about 50% of activity after 3 min with extracts of strain CT-1, after 5 min with RE1755A [7]) [7]
55 <2> (<2> 50% of activity after 5 min in Tris-buffer [9]) [9]
70 <3> (<3> stable up to for 5 min [10]) [10]

Storage stability

<1>, -70°C, purified enzyme, at least 6 months [8]

References

[1] Porter, E.V.; Chassy, B.M.: Nucleotide sequence of the β-D-phosphogalactoside galactohydrolase gene of Lactobacillus casei: comparison to analogous pbg genes of other gram-positive organisms. Gene, **62**, 263-276 (1988)
[2] Boizet, B.; Villeval, D.; Slos, P.; Novel, M.; Novel, G.; Mercenier, A.: Isolation and structural analysis of the phospho-β-galactosidase gene from Streptococcus lactis Z268. Gene, **62**, 249-261 (1988)
[3] Breidt jr., F.; Stewart, G.C.: Cloning and expression of the phospho-β-galactosidase gene of Staphylococcus aureus in Escherichia coli. J. Bacteriol., **166**, 1061-1066 (1986)
[4] Maeda, S.; Gasson, M.J.: Cloning, expression and location of the Streptococcus lactis gene for phospho-β-D-galactosidase. J. Gen. Microbiol., **132**, 331-340 (1986)
[5] Jimeno, J.; Casey, M.; Hofer, F.: The occurence of β-galactosidase and β-phosphogalactosidase in Lactobacillus casei strains. FEMS Microbiol. Lett., **25**, 275-278 (1984)
[6] Lee, L.-J.; Hansen, J.B.; Jagusztyn-Krynicka, E.K., Chassy, B.M.: Cloning and expression of the β-D-phosphogalactoside galactohydrolase gene of Lactobacillus casei in Escherichia coli K-12. J. Bacteriol., **152**, 1138-1146 (1982)
[7] Hall, B.G.: Lactose metabolism involving phospho-β-galactosidase in Klebsiella. J. Bacteriol., **138**, 691-698 (1979)

[8] Calmes, R.; Brown, A.T.: Regulation of lactose catabolism in Streptococcus mutans: purification and regulatory properties of phospho-β-galactosidase. Infect. Immun., **23**, 68-79 (1979)
[9] Hengstenberg, W.; Morse, M.L.: 6-Phospho-β-D-galactosidase. Methods Enzymol., **42**, 491-494 (1975)
[10] Johnson, K.G.; McDonald, I.J.: β-D-Phosphogalactoside galactohydrolase from Streptococcus cremoris HP: purification and enzyme properties. J. Bacteriol., **117**, 667-674 (1974)
[11] Premi, L.; Sandine, W.E.; Elliker, P.R.: Lactose-hydrolyzing enzymes of Lactobacillus species. Appl. Microbiol., **24**, 51-57 (1972)
[12] Hengstenberg, W.; Penberthy, W.K.; Morse, M.L.: Purification of the staphylococcal 6-phospho-β-D-galactosidase. Eur. J. Biochem., **14**, 27-32 (1970)
[13] Witt, E.; Stewart, G.C.; Hengstenberg, W.: A new protein that replaces the use of secondary antibodies engineered from the two staphylococcal enzymes protein A and 6-phospho-β-galactosidase. Protein Eng., **5**, 267-271 (1992)
[14] Witt, E.; Frank, R.; Hengstenberg, W.: 6-Phospho-β-galactosidases of gram-positive and 6-phospho-β-glucosidase B of gram-negative bacteria: comparison of structure and function by kinetic and immunological methods and mutagenesis of the lacG gene of Staphylococcus aureus. Protein Eng., **6**, 913-920 (1993)
[15] Staedtler, P.; Hoenig, S.; Frank, R.; Withers, S.G.; Hengstenberg, W.: Identification of the active site nucleophile in 6-phospho-β-galactosidase from Staphylococcus aureus by labeling with synthetic inhibitors. Eur. J. Biochem., **232**, 658-663 (1995)
[16] Suzuki, M.; Saito, T.; Itoh, T.: Purification and characterization of 6-phospho-β-galactosidase from Lactobacillus gasseri JCM 1031. Biosci. Biotechnol. Biochem., **60**, 139-141 (1996)
[17] Suzuki, M.; Saito, T.; Itoh, T.: Coexistence of two kinds of 6-phospho-β galactosidase in the cytosol of Lactobacillus gasseri JCM 1031 - purification and characterization of 6-phospho-β-galactosidase II. Biosci. Biotechnol. Biochem., **60**, 708-710 (1996)
[18] Wiesmann, C.; Hengstenberg, W.; Schulz, G.E.: Crystal structures and mechanism of 6-phospho-β-galactosidase from Lactococcus lactis. J. Mol. Biol., **269**, 851-860 (1997)

6-Phospho-β-glucosidase 3.2.1.86

1 Nomenclature

EC number
3.2.1.86

Systematic name
6-phospho-β-D-glucosyl-(1,4)-D-glucose glucohydrolase

Recommended name
6-phospho-β-glucosidase

Synonyms
phospho-β-glucosidase
phospho-β-glucosidase A
phosphocellobiase

CAS registry number
37205-51-9

2 Source Organism

<1> *Escherichia coli* (K-12 [1, 3, 5]) [1, 3, 5, 7]
<2> *Aerobacter aerogenes* [2, 4]
<3> *Fusobacterium mortiferum* [6]
<4> *Lactococcus lactis* [8]

3 Reaction and Specificity

Catalyzed reaction
6-phospho-β-D-glucosyl-(1,4)-D-glucose + H_2O = D-glucose 6-phosphate + D-glucose

Reaction type
hydrolysis of O-glycosyl bond

Natural substrates and products
S cellobiose 6-phosphate + H_2O <2> [2, 4]
P D-glucose 6-phosphate + D-glucose <2> [2, 4]

Substrates and products
S 4-methylumbelliferyl-β-D-glucopyranside 6-phosphate + H_2O <3> [6, 7]
P 4-methylumbelliferone + D-glucopyransoide 6-phosphate

S arbutin 6-phosphate + H_2O <1-3> [1, 2, 4-7]
P hydroquinone + D-glucose 6-phosphate
S arbutin 6-phosphate + H_2O <3> [6]
P hydroquinone + D-glucose 6-phosphate
S cellobiitol monophosphate + H_2O <2> [2, 4]
P ?
S cellobiose 6-phosphate + H_2O <1-3> [2, 4, 6, 7]
P D-glucose 6-phosphate + D-glucose
S esculin 6-phosphate + H_2O <3> [6]
P ?
S gentiobiose monophosphate + H_2O <1-3> [2, 4, 6, 7]
P D-glucose 6-phosphate + D-glucose
S laminaribiose 6-phosphate + H_2O <3> [6]
P D-glucose 6-phosphate + D-glucose
S methyl β-D-glucoside monophosphate + H_2O <2, 3> [2, 4, 6]
P ?
S methyl β-glucoside 6-phosphate + H_2O <1> [3, 7]
P ?
S *o*-nitrophenyl-β-D-galactopyranoside 6-phosphate + H_2O <1, 4> [3, 8]
P *o*-nitrophenol + D-galactose 6-phosphate
S *p*-nitrophenyl-β-D-glucopyranoside 6-phosphate + H_2O <1, 3> (<1> isoenzyme B [3]) [3, 5, 6, 7]
P *p*-nitrophenyl + D-glucopyranside 6-phosphate
S *p*-nitrothiophenyl β-glucoside 6-phosphate + H_2O <1> [3]
P ?
S phenyl β-D-glucoside 6-phosphate + H_2O <1, 2> [2-4]
P phenol + glucose 6-phosphate
S salicin 6-phosphate + H_2O <1, 2> (<1> isoenzyme B [1, 3, 5]) [1-7]
P 2-hydroxymethylphenol + D-glucose 6-phosphate
S sophorose 6-phosphate + H_2O <3> [6]
P D-glucose 6-phosphate + D-glucose

Inhibitors

2'-AMP <1> [3]
2'-AMP + 3'-AMP <1> [3]
3',5'-AMP <1> [3]
3'-AMP <1> [3]
AMP <1> [3]
AMP <1> [3]
ATP <1> [3]
CMP <1> [3]
D-glucono-δ-lactone <3> [6]
DL-α-glycerol-phosphate <1> [3]
NADH <1> [3]
fructose-1,6-diphosphate <1> [3]
phosphoenolpyruvate <1> [3]

Specific activity (U/mg)

0.54 <2> [2, 4]
85 <1> [3]

K_m-Value (mM)

0.0092 <3> (4-methylumbelliferyl-β-D-glucopyranoside 6-phosphate) [6]
0.0291 <3> (*p*-nitrophenyl-β-D-glucopyranoside 6-phosphate) [6]
0.0875 <3> (laminaribiose 6-phosphate) [6]
0.0886 <3> (salicin 6-phosphate) [6]
0.0977 <3> (arbutin 6-phosphate) [6]
0.11 <1> (*p*-nitrothiophenyl β-glucoside 6-phosphate, <1> isoenzyme B [3]) [3]
0.1212 <3> (sophorose 6-phosphate) [6]
0.1384 <3> (methyl-β-glucoside 6-phosphate) [6]
0.18 <1> (*p*-nitrophenyl-β-D-glucopyranoside 6-phosphate, <1> isoenzyme B [3]) [3]
0.1823 <3> (cellobiose 6-phosphate) [6]
0.19 <1> (methyl β-glucoside 6-phosphate, <1> isoenzyme B [3]) [3]
0.23 <2> (cellobiose 6-phosphate) [4]
0.23-0.5 <2> (arbutin 6-phosphate) [2]
0.23-0.5 <2> (cellobiitol monophosphate) [2]
0.23-0.5 <2> (cellobiose 6-phosphate) [2]
0.23-0.5 <2> (gentiobiose monophosphate) [2]
0.23-0.5 <2> (methyl β-D-glucoside monophosphate) [2]
0.23-0.5 <2> (phenyl β-D-glucoside 6-phosphate) [2]
0.23-0.5 <2> (salicin 6-phosphate) [2]
0.24 <2> (gentiobiose phosphate) [4]
0.25 <2> (cellobiitol monophosphate) [4]
0.25 <1> (*o*-nitrophenyl β galactoside 6-phosphate, <1> isoenzyme B [3]) [3]
0.27 <2> (methyl β-D-glucoside phosphate) [4]
0.28 <1> (methyl β-glucoside 6-phosphate, <1> isoenzyme A [3]) [3]
0.3 <1> (4-methylumbelliferyl-β-D-glucopyransoide 6-phosphate) [7]
0.31 <2> (arbutin 6-phosphate) [4]
0.33 <1> (*p*-nitrophenyl-β-D-glucopyranoside 6-phosphate, <1> isoenzyme A [3]) [3]
0.3355 <3> (gentiobiose 6-phosphate) [6]
0.35 <1> (arbutin 6-phosphate) [7]
0.44 <1> (*p*-nitrophenyl-β-D-glucopyranoside 6-P) [7]
0.44 <1> (salicin 6-phosphate) [7]
0.46 <1> (phenyl β-glucoside 6-phosphate, <1> isoenzyme B [3]) [3]
0.5 <2> (phenyl β-D-glucoside 6-phosphate) [4]
0.5 <2> (salicin 6-phosphate) [4]
0.51 <1> (salicin 6-phosphate, <1> isoenzyme B [3]) [3]
0.6 <1> (phenyl β-glucoside 6-phosphate, <1> isoenzyme A [3]) [3]
1.3 <1> (cellobiose 6-phosphate) [7]
2.22 <1> (methyl β-glucoside 6-phosphate) [7]
12.5 <1> (gentiobiose monophosphate) [7]

pH-Optimum

6.3 <1> (<1> isoenzyme A, *p*-nitrophenyl β-D-glucospyranoside [3]) [3]
6.5-7 <3> [6]
7-8 <2> (<2> cellobiose 6-phosphate [2]) [2, 4]
7-9 <1> (<1> isoenzyme B, *p*-nitrophenyl β-D-glucospyranoside [3]) [3]

pH-Range

4-10 <1> [3]

Temperature optimum (°C)

34-38 <1> [7]
35-40 <3> [6]

4 Enzyme Structure

Molecular weight

21000 <1> (<1> gel filtration [7]) [7]
50000 <1> (<1> SDS-PAGE [7]) [7]
52000 <2> (<2> sucrose density gradient centrifugation [2, 4]) [2, 4]
53000 <3> (<3> SDS-PAGE [6]) [6]
53560 <3> (<3> MALDI-MS [6]) [6]
54000 <3> (<3> gel filtration chromatography [6]) [6]
65000 <1> (<1> isoenzyme B, SDS-PAGE [3]) [3]
68000 <1> (<1> isoenzyme A, SDS-PAGE [3]) [3]
132000 <1> (<1> isoenzyme B, molecular sieve chromatography together with sedimentation velocity measurement [3]) [3]
142000 <1> (<1> isoenzyme A, molecular sieve chromatography together with sedimentation velocity measurement [3]) [3]

Subunits

dimer <1> (<1> homodimer, 2 * 68000 isoenzyme A, 2 * 65000 isoenzyme B, SDS-PAGE [3]) [3]
tetramer <1> (<1> 4 * 50000 [7]) [7]

5 Isolation/Preparation/Mutation/Application

Purification

<1> (63 fold [3]) [3]
<2> (14 fold [2,4]) [2, 4]
<3> (23 fold [6]) [6]

6 Stability

pH-Stability

4-10 <1> (<1> for more than 1 h [3]) [3]

Temperature stability

10-28 <1> (<1> isoenzyme B [3]) [3]
25-45 <1> (<1> isoenzyme A [3]) [3]

Storage stability

<1>, isoenzyme A, 4°C, 1 year, sterile starting buffer, 0% loss in activity [3]
<1>, isoenzyme B, -20°C, 2 months, sterile starting buffer, 50% loss of activity [3]
<2>, 4°C, 2 months, purified enzyme, 60% loss of activity [4]
<3>, -20°C, 6 months, no significant decline in activity [6]

References

[1] Schnetz, K.; Toloczyki, C.; Rak, B.: β-Glucoside (bgl) operon of Escherichia coli K-12: nucleotide sequence, genetic organization, and possible evolutionary relationship to regulatory components of two Bacillus subtilis genes. J. Bacteriol., **169**, 2579-2590 (1987)
[2] Anderson, R.L.; Palmer, R.E.: Phosphocellobiase. Methods Enzymol., **42 C**, 494-497 (1975)
[3] Wilson, G.; Fox, C.F.: The β-glucoside system of Escherichia coli. IV. Purification and properties of phospho-β-glucosidases A and B. J. Biol. Chem., **249**, 5586-5598 (1974)
[4] Palmer, R.E.; Anderson, R.L.: Cellobiose metabolism in Aerobacter aerogenes. 3. Cleavage of cellobiose monophosphate by a phospho-glucosidase. J. Biol. Chem., **247**, 3420-3423 (1972)
[5] Prasad, I.; Young, B.; Schaefler, S.: Genetic determination of the constitutive biosynthesis of phospho-β-glucosidase A in Escherichia coli K-12. J. Bacteriol., **114**, 909-915 (1973)
[6] Thompson, J.; Robrish, S.A.; Bouma, C.L.; Freedberg, D.I.; Folk, J.E.: Phospho-β-glucosidase from Fusobacterium mortiferum: Purification, clonig, and inactivation by 6-phosphoglucono-sigma-lactone. J. Bacteriol., **179**, 1636-1645 (1997)
[7] Thompson, J.; Ruvinov, S.B.; Freedberg, D.I.; Hall, B.G.: Cellobiose-6-phosphate hydrolase (CelF) of Escherichia coli: Characterization and assignment to the unsual family 4 of glycosylhydrolases. J. Bacteriol., **181**, 7339-7345 (1999)
[8] Simons, G.; Nijhuis, M.; de Vos, W.: Integration and gene replacement in the Lactococcus lactis lac operon: Induction of a cryptic phospho-β-glucosidase in LacG-deficient strains. J. Bacteriol., **175**, 5168-5175 (1993)

Capsular-polysaccharide endo-1,3-α-galactosidase 3.2.1.87

1 Nomenclature

EC number
3.2.1.87

Systematic name
Aerobacter-capsular-polysaccharide galactohydrolase

Recommended name
capsular-polysaccharide endo-1,3-α-galactosidase

Synonyms
galactohydrolase, capsular polysaccharide
polysaccharide depolymerase

CAS registry number
62213-16-5

2 Source Organism

<1> *Aerobacter aerogenes* (phage induced [1, 2]) [1, 2]

3 Reaction and Specificity

Catalyzed reaction
random hydrolysis of 1,3-α-D-galactosidic linkages in Aerobacter aerogenes capsular polysaccharide (hydrolyzes the galactosyl-α-1,3-D-galactose linkages only in the complex substrate, bringing about depolymerization)

Reaction type
hydrolysis of O-glycosyl bond

Natural substrates and products
S [D-Gal-(1-3)(D-GlcUA-(1-2))-α-D-Man-(1-3)-α-D-Gal]$_n$ + H_2O <1> (<1> capsular, polysaccharide produced by Aerobacter aerogenes, hydrolyses the galactosyl-α-1,3-D-galactose linkages only in the complex substrate, bringing about depolymerization, the enzyme is phage-induced and serves to degrade the capsular polysaccharide of the host [1, 2]) [1, 2]
P D-Gal-(1-3-)(D-GlcUA-(1-2))-α-D-Man-(1-3)-α-D-Gal <1> [1, 2]

Substrates and products

S [D-Gal-(1-3)(D-GlcUA-(1-2))-α-D-Man-(1-3)-α-D-Gal]$_n$ + H_2O <1> (<1> capsular, polysaccharide produced by Aerobacter aerogenes, hydrolyzes the galactosyl-α-1,3-D-galactose linkages only in the complex substrate, bringing about depolymerization [1, 2]) [1, 2]

P D-Gal-(1-3-)(D-GlcUA-(1-2))-α-D-Man-(1-3)-α-D-Gal <1> [1, 2]

Inhibitors

$FeCl_3$ <1> [2]

Metals, ions

Additional information <1> (<1> metal ion not required [2]) [2]

Specific activity (U/mg)

235 <1> [2]

pH-Optimum

5.2 <1> [2]

pH-Range

3.2-8 <1> [2]

Temperature optimum (°C)

55 <1> [2]

4 Enzyme Structure

Molecular weight

379000 <1> (<1> sedimentation coefficient, diffusion coefficient [2]) [2]

Subunits

heptamer <1> (<1> heptamer: 4 * 63200, 3 * 36400 or octamer: 4 * 63200, 4 * 36400, SDS-PAGE [2]) [2]

octamer <1> (<1> octamer: 4 * 63200, 4 * 36400 or heptamer: 4 * 63200, 3 * 36400, SDS-PAGE [2]) [2]

5 Isolation/Preparation/Mutation/Application

Localization

cytosol <1> [2]

Purification

<1> [2]

6 Stability

Temperature stability

68 <1> (<1> 52% inactivation after 10 min [2]) [2]

References

[1] Yurewicz, E.C.; Ghalambor, M.A.; Heath, E.C.: The structure of Aerobacter aerogenes capsular polysaccharide. J. Biol. Chem., **246**, 5596-5606 (1971)
[2] Yurewicz, E.C.; Ghalambor, M.A.; Duckworth, D.H.; Heath, E.C.: Catalytic and molecular properties of a phage-induced capsular polysaccharide depolymerase. J. Biol. Chem., **246**, 5607-5616 (1971)

β-L-Arabinosidase 3.2.1.88

1 Nomenclature

EC number
3.2.1.88

Systematic name
β-L-arabinoside arabinohydrolase

Recommended name
β-L-arabinosidase

Synonyms
arabinosidase,β-L-
β-L-arabinosidase
vicianosidase

CAS registry number
39361-63-2

2 Source Organism

<1> *Cajanus indicus* [1, 2]

3 Reaction and Specificity

Catalyzed reaction
H_2O + a β-L-arabinoside = L-arabinose + an alcohol

Reaction type
hydrolysis of O-glycosyl bond

Substrates and products
S *p*-nitrophenyl β-L-arabinoside + H_2O <1> [1, 2]
P *p*-nitrophenol + β-L-arabinose <1> [1]
S Additional information <1> (<1> no hydrolysis of: *p*-nitrophenyl α-D-galactoside, *p*-nitrophenyl α-D-fucoside, *p*-nitrophenyl β-D-galactoside [2]) [2]
P ?

K_m-Value (mM)
0.8 <1> (*p*-nitrophenyl β-L-arabinoside) [1]
0.83 <1> (*p*-nitrophenyl-β-L-arabinoside) [2]

Additional information <1>(<1> very small effect of pH in the range 2.5-7.2, no influence of alcohols on K_m [1]) [1]

pH-Optimum
3-4.6 <1> [2]

pH-Range
2-6.5 <1> [2]

Temperature optimum (°C)
30 <1> (<1> assay at [1, 2]) [1, 2]

4 Enzyme Structure

Molecular weight
25900 <1> (<1> gel filtration [2]) [2]

5 Isolation/Preparation/Mutation/Application

Source/tissue
seed <1> [1, 2]

Purification
<1> [2]

6 Stability

pH-Stability
2-6.5 <1> (<1> 30°C, 30 min [2]) [2]

Temperature stability
4-40 <1> [1]

Oxidation stability
<1>, photooxidation in presence of methylene blue [1]

References

[1] Dey, P.M.: Further characterization of β-L-arabinosidase from Cajanus indicus. Biochim. Biophys. Acta, **746**, 8-13 (1983)
[2] Dey, P.M.: β-L-Arabinosidase from Cajanus indicus: a new enzyme. Biochim. Biophys. Acta, **302**, 393-398 (1973)

Arabinogalactan endo-1,4-β-galactosidase 3.2.1.89

1 Nomenclature

EC number
3.2.1.89

Systematic name
arabinogalactan 4-β-D-galactanohydrolase

Recommended name
arabinogalactan endo-1,4-β-galactosidase

Synonyms
arabinogalactanase
endo-1,4-β-galactanase
galactanase

CAS registry number
58182-40-4

2 Source Organism

<1> *Penicillium citrinum* [1-3]
<2> *Bacillus subtilis* [4, 7, 9-11]
<3> *Botrytis cinerea* [5]
<4> *Sclerotinia sclerotiorum* [6]
<5> *Fusarium roseum* [8]
<6> *Aspergillus aculeatus* [12-14]
<7> *Aspergillus niger* [15, 16]
<8> *Bacillus spec.* (strains S-2 and S-39) [17]
<9> *Aspergillus sojae* [18]

3 Reaction and Specificity

Catalyzed reaction
endohydrolysis of 1,4-β-galactosidic linkages in arabinogalactans

Reaction type
hydrolysis of O-glycosyl bond

Natural substrates and products
S arabinogalactan + H_2O <1> [1-3]
P galactose

Substrates and products

S arabinogalactan + H_2O <1-5> [1-11]
P galactose
S galactobiose + H_2O <1> (<1> by galactanase I [3]) [3]
P galactose
S galactotetraose + H_2O <1> (<1> by galactanase I [3]) [3]
P galactose
S galactotriose + H_2O <1> (<1> by galactanase I [3]) [3]
P galactose
S *o*-nitrophenyl-β-D-galactopyranoside + H_2O <1> [1]
P *o*-nitrophenol + galactose + pyranose
S *p*-nitrophenyl-β-D-galactoside + H_2O <1> [3]
P *p*-nitrophenol + galactose

Inhibitors

Ag^+ <2, 9> [10, 18]
EDTA <1, 2> (<1> no effect [3]; <2> complete inactivation [9]) [3, 9]
Fe^{3+} <1, 2, 9> [3, 10, 18]
Hg^{2+} <1, 2, 9> [3, 10, 18]
Mn^{2+} <9> [18]
arabinose <1> (<1> *o*-nitrophenyl-β-D-galactoside hydrolyzing activity [3]) [3]
galactose <1> (<1> *o*-nitrophenyl-β-D-galactoside hydrolyzing activity [3]) [3]
glucose <1> (<1> *o*-nitrophenyl-β-D-galactoside hydrolyzing activity [3]) [3]
lactose <1> (<1> *o*-nitrophenyl-β-D-galactoside hydrolyzing activity [3]) [3]
xylose <1> (<1> *o*-nitrophenyl-β-D-galactoside hydrolyzing activity [3]) [3]

Specific activity (U/mg)

46.4 <2> [4]
166 <2> (<2> galactanase I [9]) [9]
180 <2> (<2> galactanase III [9]) [9]
185 <2> [10]
199 <2> (<2> galactanase II [9]) [9]
224 <1> (<1> galactanase II [3]) [3]
238 <1> (<1> galactanase I [3]) [3]
660 <9> [18]
1790 <2> [7]
6593 <7> [15]

pH-Optimum

3.5 <7> [15, 16]
4 <3, 6, 8> (<3> galactanase I [5]; <8> S-39 [17]) [5, 12, 14, 17]
4.5 <1, 3, 4, 9> (<3> galactanase II [5]) [3, 5, 6, 18]
5 <5> [8]
6 <2> [4, 7]
10 <8> (<8> S-2 [17]) [17]

pH-Range
4-8 <2> (<2> inactive below and above [7]) [7]

Temperature optimum (°C)
40 <8> (<8> S-39 [17]) [17]
45-65 <6> [14]
48 <2> [7]
50 <8, 9> (<8> S-2 [17]) [17, 18]
50-55 <6, 7> [12, 16]
55 <1, 7> [3, 15]

Temperature range (°C)
6-48 <2> (<2> loses activity above, inactive at 60°C [7]) [7]

4 Enzyme Structure

Molecular weight
22000-24000 <4> (<4> gel filtration [6]) [6]
32000 <2, 7> (<2> galactanase III, gel filtration [9]) [9, 15]
35000 <1, 2> (<2> galactanase I, gel filtration [9]) [9, 3]
36000 <2, 8> (<2> galactanase II, gel filtration [9]; <8> S-2 [17]) [9, 17]
37000 <2> (<2> sedimentation equilibrium [10]) [10]
38000 <6, 7> (<6> western blotting [12]; <7> SDS-PAGE [16]) [12, 16]
43000 <6> (<6> calculation from sequence of DNA [13]) [13]
44000 <6> (<6> recombinant enzyme [13]) [13, 14]
66000-67000 <5> (<5> gel filtration [8]) [8]

Subunits
monomer <1, 9> (<1> 1 * 35000 SDS-PAGE [3]; <9> 1 * 39700 SDS-PAGE [18]) [3, 18]

Posttranslational modification
glycoprotein <6> (<6> N-glycosylation, mannose residues [13, 14]) [13, 14]
proteolytic modification <6> [12]

5 Isolation/Preparation/Mutation/Application

Localization
extracellular <1-9> [1-18]

Purification
<1> [3]
<2> [7, 9]
<3> (<3> partial [5]) [5]
<4> (<4> partial [6]) [6]
<6> [14]
<7> [15, 16]

<8> [17]
<9> [18]

Crystallization

<1> (<1> galactanase I forms a hexagonal crystal [3]) [3]
<2> (<2> concentrated enzyme solution crystallized in the cold [10]) [10]
<6> (<6> hanging drops, vapour-diffusion, 30% polyethylene glycol 400, 0.2 M $CaCl_2$, 0.1 M Na HEPES, pH 7.5, orthorhombic space group [13]) [13]

Cloning

<6> (expression in Saccharomyces cerevisiae [14]) [14]
<6> (expression in potato [12]) [12]
<6> (overexpression in Aspergillus oryzae [13]) [13]

6 Stability

pH-Stability

2-8 <6> [14]
4-7 <7> [16]
4-10 <1> [3]
5-9.5 <2> (<2> at 30°C [10]) [10]
5-12 <8> (<8> S-39 [17]) [17]
5.5-10.5 <2> (<2> in presence of Ca^{2+} [9]) [9]
6-10 <9> [18]
7-12 <8> (<8> S-2 [17]) [17]

Temperature stability

30 <7> (<7> above 30°C the activity is reduced after 30 min [16]) [16]
45 <8> (<8> S-2, at pH 10 for 15 min stable [17]) [17]
50 <2, 6> (<2, 6> stable below [9, 14]) [9, 14]
55 <1> (<1> stable for 10 min at pH 5.5 [3]) [3]
60 <6, 8, 9> (<6> stable below [14]; <8> S-39 at pH 4 for 15 min stable [17]; <9> complete loss of activity [18]) [14, 17, 18]
70 <6> (<6> stable below [14]) [14]

Storage stability

<4>, -20°C loss of activity [6]
<4>, 4°C, 50 mM sodium acetate, pH 5.0, several weeks [6]

References

[1] Nakano, H.; Takenishi, S.; Watanabe, Y.: Formation of transfer products from soybean arabinogalactan and glycerol by galactanase from Penicillium citrinum. Agric. Biol. Chem., **52**, 1913-1921 (1988)
[2] Nakano, H.; Takenishi, S.; Watanabe, Y.: Action of galactanase from Penicillium citrinum on *o*-nitrophenyl-β-D-galacopyranoside. Agric. Biol. Chem., **50**, 3005-3012 (1986)

[3] Nakano, H.; Takenishi, S.; Watanabe, Y.: Purification and properties of two galactanases from Penicillium citrinum. Agric. Biol. Chem., **49**, 3445-3454 (1985)
[4] Voragen, A.-G.-J.; Geerst, F.; Pilnik, W.: Hemi-cellulases in enzymic fruit procession. Util. Enzymes Technol. Aliment., Symp. Int., 497-502 (1982)
[5] Urbanek, H.; Zalewska-Sobczak, J.: 1.4-β-Galactanases and 1.3-β-galactanases of Botrytis cinerea isolate infecting apple. Biochem. Physiol. Pflanz., **181**, 321-329 (1986)
[6] Bauer, W.D.; Bateman, D.F.; Whalen, C.H.: Purification of an endo-β-1.4 galactanase produced by Sclerotinia sclerotiorum: effects on isolated plant cell walls and potato tissue. Phytopathology, **67**, 862-868 (1977)
[7] Labavitch, J.M.; Freeman, L.E.; Albersheim, P.: Structure of plant cell walls: purification and characterization of a β-1.4-galactanase which degrades a structural component of the primary cell walls of dicots. J. Biol. Chem., **251**, 5904-5910 (1976)
[8] Mullen, J.M.; Bateman, D.F.: Polysaccharide degrading enzymes produced by Fusarium roseum "Avanaceum" in culture and during pathogenesis. Physiol. Plant Pathol., **6**, 233-246 (1975)
[9] Emi, S.; Yamamoto, T.: Purification and properties of several galactanases of Bacillus subtilis var. amylosacchariticus. Agric. Biol. Chem., **36**, 1945-1954 (1972)
[10] Emi, S.; Fukumoto, J.; Yamamoto, T.: Studies of hemicellulolytic enzymes of Bacillus subtilis: part I purification, crystallization and some properties of arabinogalactanase. Agric. Biol. Chem., **35**, 1891-1898 (1971)
[11] Dekker, R.F.H.: Biodegradation of the hemicelluloses. Biosynthesis and Biodegradation of Wood Components, (Higuchi,T., ed.), 505-532 (1985)
[12] Oxenboll Sorensen, S.; Pauly, M.; Bush, M.; Skjot, M.; McCann, M.C.; Borkhardt, B.; Ulvkov, P.: Pectin engineering: modification of potato pectin by in vivo expression of an endo-1.4-β-D-galactanase. Proc. Natl. Acad. Sci. USA, **97**, 7639-7644 (2000)
[13] Ryttersgaard, C.; Poulsen, J.-C.N.; Christgau, S.; Sandal, T.; Dalboge, H.; Larsen, S.: Crystallization and preliminary x-ray studies of β-1,4-galactanase from Aspergillus aculeatus. Acta Crystallogr. Sect. D, **55**, 929-930 (1999)
[14] Christgau, S.; Sandal, T.; Dalboge, H.: Expression cloning, purification and characterization of a β-1,4-galactanase from Aspergillus aculeatus. Curr. Genet., **27**, 135-141 (1995)
[15] Yamaguchi, F.; Inoue, S.-i.; Hatanaka, C.: Purification and properties of endo-β-1,4-D-galactanase from Aspergillus niger. Biosci. Biotechnol. Biochem., **59**, 1742-1744 (1995)
[16] Lahaye, M.; Thibault, J.V. & J.-F.: Endo-β-1,4-D-galactanase from Aspergillus niger var. aculeatus: purification and some properties. Carbohydr. Polym., **15**, 431-444 (1991)
[17] Tsumura, K.; Hashimoto, Y.; Akiba, T.; Horikoshi, K.: Purifications and properties of galactanases from alkalophilic Bacillus sp. S-2 and S-39. Agric. Biol. Chem., **55**, 1265-1271 (1991)
[18] Kimura, I.; Yoshioka, N.; Tajima, S.: Purification and characterization of an endo-1,4-β-D-galactanase from Aspergillus sojae. J. Ferment. Bioeng., **85**, 48-52 (1998)

Arabinogalactan endo-1,3-β-galactosidase 3.2.1.90

1 Nomenclature

EC number
3.2.1.90 (deleted, not sufficiently characterized)

Recommended name
arabinogalactan endo-1,3-β-galactosidase

Cellulose 1,4-β-cellobiosidase 3.2.1.91

1 Nomenclature

EC number
3.2.1.91

Systematic name
1,4-β-D-glucan cellobiohydrolase

Recommended name
cellulose 1,4-β-cellobiosidase

Synonyms
1,4-β-D-glucan cellobiohydrolase
1,4-β-glucan cellobiohydrolase
1,4-β-glucan cellobiosidase
C1 cellulase
CBH 1
CBH II
Cel6A
avicelase
avicelase II
β-1,4-glucan cellobiohydrolase
β-1,4-glucan cellobiosylhydrolase
cellobiohydrolase
cellobiohydrolase I
cellobiohydrolase II
cellobiohydrolase, exo-
cellobiosidase
cellobiosidase, 1,4-β-glucan
cellulase, C1
exo-β-1,4-glucan cellobiohydrolase
exo-cellobiohydrolase
exocellobiohydrolase
exoglucanase

CAS registry number
37329-65-0

2 Source Organism

<1> *Chrysosporium thermophile* [1]

<2> *Trichoderma viride* (Exo I /II/III [2]) [2, 16, 18, 29, 30, 31]
<3> *Trichoderma reesei* (CBH I, CBH II [15, 43, 45]; CBH I is exo-glucanase, CHB II is endo-glucanase [43]; engineered enzymes [41]; Cel6A [49, 50]; CBH I [55]) [3, 9, 11, 15, 16, 19, 24, 26, 40-43, 45, 47, 49, 50, 55]
<4> *Trichoderma koningii* (CBH 1, CBH 1a [4, 28, 32]) [4, 28, 32]
<5> *Cellulomonas uda* [5]
<6> *Irpex lacteus* [6, 33]
<7> *Trichoderma harzianum* [12]
<8> *Cellulomonas fimi* (gene cloned in E. coli [17]; gene cloned in Saccharomyces cerevisiae [34]) [17, 34]
<9> *Sclerotium rolfsii* [7]
<10> *Bacteroides succinogenes* [8, 37, 38]
<11> *Aspergillus ficuum* (CBH I, CBH II [10]) [10]
<12> *Penicillium purpurogenum* (2 forms: I/II [13]) [13]
<13> *Penicillium sp.* [14]
<14> *Clostridium stercorarium* [20, 51]
<15> *Fusarium lini* (2 forms: I/II [21]) [21]
<16> *Humicola insolens* [22, 39]
<17> *Humicola grisea* [22]
<18> *Penicillium pinophilum* (2 forms: I/II [23]) [23]
<19> *Ruminococcus albus* [25]
<20> *Penicillium funicolosum* [27]
<21> *Ruminococcus flavefaciens* [35]
<22> *Mucor pusillus* [36]
<23> *Thermoascus aurantiacus* [36]
<24> *Chaetomium thermophile* [36]
<25> *Sporotrichum thermophile* [36]
<26> *Torula thermophila* [36]
<27> *Humicola lanuginosa* [36]
<28> *Allescheria terrestris* [36]
<29> *Aspergillus fumigatus* [36]
<30> *Planerochaete chrysosporium* (CBH I [44]) [44]
<31> *Clostridium thermocellum* (CBH III [48]) [46, 48, 57]
<32> *Bacillus circulans* [52]
<33> *Aspergillus niger* [53]
<34> *Thermotoga sp.* (strain FjSS3-B.1 [54]) [54]
<35> *Bacillus subtilis* (168 [56]) [56]

3 Reaction and Specificity

Catalyzed reaction

hydrolysis of 1,4-β-D-glucosidic linkages in cellulose and cellotetraose, releasing cellobiose from the non-reducing ends of the chains

Reaction type

hydrolysis of O-glycosyl bond

Substrates and products

S 2-chloro-4-nitrophenyl-β-lactoside <3> [55]
P ?
S avicel + H_2O <2, 4-6, 12-15, 31> (<10>, not [37]) [2, 4-6, 13, 14, 21, 46, 48, 51]
P cellobiose <4, 6, 13, 31> (<6> plus traces of cellotriose, [6]) [4, 6, 14, 46]
S carboxymethylcellulose + H_2O <3, 31, 32> [43, 48, 52, 57]
P glucose + cellobiose + cellutetraose <3> (<3> CBH II, endo-glucanase [43]) [43]
S carboxymethylcellulose + H_2O <2, 3, 13, 14, 34> (<4>, not significantly [4]; <2, 10>, not [31, 37]; <4>, little activity [32]) [2, 14, 43, 51, 54]
P cellobiose <3> (<3> CBH I, exo-glucanase [43]) [43]
S cellodextrins + H_2O <1> [1]
P cellobiose + glucose + cellotriose <1> [1]
S cellohexaose + H_2O <4, 6, 32> [6, 32, 52]
P cellotriose + cellobiose + cellotetraose <32> [52]
S celloligosaccharides + H_2O <10> [37]
P cellobiose + H_2O <10> [37]
S cellopentaose + H_2O <6, 12, 32> [6, 13, 52]
P cellobiose + glucose + cellotriose <6, 32> [6, 52]
S cellotetraose + H_2O <1, 4, 5, 12, 14, 32> [1, 4, 5, 13, 32, 51, 52]
P cellobiose + glucose + cellotriose <1, 5> (<1, 5> different combinations and proportions of products [1, 5]; <32> preferably cellobiose [52]) [1, 5, 52]
S cellotriitol + H_2O <1> [1]
P cellobiose + sorbitol <1> [1]
S cellotriose + H_2O <1, 6, 10, 32> [1, 6, 37, 52]
P cellobiose + glucose <1, 32> [1, 52]
S cellulose + H_2O <1-34> [1-37, 40-44, 46-48, 50-54, 57]
P cellobiose + glucose + cellotriose <1-29> (<1-29> different combinations and proportions of products depending on organism [1-37]; <32> only cellobiose [52]) [1-37, 52]
S lichenan + H_2O <31> [48]
P ?
S methylumbelliferyl β-D-cellobioside + H_2O <34> [54]
P ?
S methylumbelliferyl-lactoside + H_2O <3> [43]
P methylumberlliferone + lactose
S *p*-nitrophenyl-β-D-cellobioside + H_2O <3, 8, 10, 19, 21, 30-32, 34> [8, 17, 25, 35, 37, 38, 44, 45, 48, 52, 54]
P *p*-nitrophenol + cellobiose <10, 30> [8, 44]
S *p*-nitrophenyl-lactoside + H_2O <1, 10, 31> [1, 8, 37, 48]
P *p*-nitrophenol + lactose
S xylan + H_2O <14, 31, 32> [48, 51, 52, 57]
P xylobiose + xylotetraose + xylohexaose <32> [52]

S Additional information <1, 2, 4, 6, 10, 11, 21> (<1, 6, 10> not: cellobiose [1,6,37]; <3> cellobiose slightly [24]; <6> hydrolyses internal glucosidic linkages of β-1,3, 1,4-D-glucan [6]) [1, 6, 10, 28, 30, 33, 35, 37]

P ?

Inhibitors

1,10-phenanthroline <21> [35]
Ag^{+} <1> (<1> slight [1]) [1]
Cu^{2+} <1, 31> (<1> slight [1]) [1, 48]
D-xylose <15> [21]
EDTA <21, 33> (<21,33> slight [35, 53]) [35, 53]
EGTA <21> [35]
Fe^{3+} <1, 31> [1, 48, 57]
Hg^{2+} <1, 31, 33> [1, 48, 53]
N-bromosuccinimide <1, 9, 11> [1, 7, 10]
PCMB <33> [53]
Pb^{2+} <1, 33> (<1> slight [1]) [1, 53]
SDS <31> [48]
Zn^{2+} <31> [48, 57]
cellobiose <1, 3, 15, 21, 30, 31, 34> (<3> CBH I [43]; <30> presence of cellobiose dehydrogenase relieves inhibition [44]) [1, 3, 21, 35, 43, 44, 46, 54, 57]
ethanol <31> [48]
glucose <35> [56]
kanamycin <33> [53]
lactose <1, 31> [1, 46]
neomycin <33> [53]
suramin <33> (<33> slight [53]) [53]
urea <15> [21]
viomycin <33> [53]

Metals, ions

Ca^{2+} <21> (<21> required for maximal activity [35]) [35]
Co^{2+} <5, 33> (<5, 33> activates [5, 53]) [5, 53]
Mg^{2+} <16, 32> (<16> hexa-co-ordinate [39]; <32> activates [52]) [39, 52]
Mn^{2+} <5, 33> (<5, 33> activates [5, 53]) [5, 53]

Specific activity (U/mg)

0.037 <14> [51]
0.0571 <5> [5]
0.114 <11> (<11> CBH II [10]) [10]
0.151 <11> (<11> CBH I [10]) [10]
0.32 <1> [1]
1.4 <31> (<31> cellulose [48]) [48]
1.8 <31> (<31> xylan [48]) [48]
2.1 <31> (<31> avicel [48]) [48]
2.9 <33> [53]
4.9 <34> [54]
6.67 <7> [12]

11.93 <10> [8]
14.26 <32> [52]
18 <31> (<31> *p*-nitrophenyl lactose [48]) [48]
18.7 <31> (<31> *p*-nitrophenyl-β-D-cellobioside [48]) [48]
18.8 <31> (<31> carboxymethyl cellulose [48]) [48]
24.6 <31> (<31> lichenan [48]) [48]
Additional information <6, 10, 13, 15> [6, 14, 21, 38]

K_m-Value (mM)

0.1 <10> (*p*-nitrophenyl-β-D-cellobioside) [8]
0.21 <10> (*p*-nitrophenyl-β-D-cellobioside, <10> extracellular [38]) [38]
0.27 <10> (*p*-nitrophenyl-β-D-cellobioside, <10> periplasmic [38]) [38]
0.384 <30> (*p*-nitrophenyl-β-D-cellobioside) [44]
0.4 <10> (*p*-nitrophenyl-β-D-cellobioside) [37]
0.46 <3> (2-chloro-4-nitrophenyl-β-lactoside) [55]
0.57 <3> (2-chloro-4-nitrophenyl-β-lactoside, <3> D214N [55]) [55]
0.68 <3> (2-chloro-4-nitrophenyl-β-lactoside, <3> E212Q [55]) [55]
0.78 <3> (2-chloro-4-nitrophenyl-β-lactoside, <3> E217Q [55]) [55]
3 <1> (*p*-nitrophenyl-lactoside) [1]
3.08 <21> (*p*-nitrophenyl-β-D-cellobioside) [35]
12 <2> (avicel, <2> exo III [2]) [2]
44.1 <2> (avicel, <2> exo II [2]) [2]
Additional information <7, 15, 20>(<15> cellulose: 1.0 mg/ml [21]; <20> H_3PO_4-swollen cellulose: 0.0067 mM [27]) [12, 21, 27]

pH-Optimum

2.5 <18> (<18> isozyme I [23]) [23]
3.5-4 <1> (<1> amorphous cellulose [1]) [1]
4 <13> [14]
4-4.5 <12> (<12> isozyme II [13]) [13]
4-5 <1, 12> (<1> crystalline cellulose [1]; <12> isozyme I [13]) [1, 13]
4-6 <22-29> (<22-29> thermophilic fungi [36]) [36]
4.2 <9> (<9> avicel [7]) [7]
4.5 <7, 9, 18, 32, 35> (<9> cellulose [7]; <18> isozyme II [23]) [7, 12, 23, 52, 56]
4.5-5.5 <15> [21]
4.5-7 <2> (<2> exo III [2]) [2]
4.8 <2> [30]
5 <1, 6> (<1> *p*-nitrophenyl-lactoside [1]) [1, 6]
5-6 <14> [51]
5.5 <33> [53]
5.5-5.7 <2> (<2> exo I [2]) [2]
5.5-6.5 <4> (<4> avicel [4]) [4]
6.2 <10> [8]
6.5 <31> [48]
6.8 <19> [25]
6.8-7.8 <34> [54]
Additional information <20> [27]

pH-Range

4-6 <33> [53]
5-7.5 <10> (<10> pH 5: 50% of maximal activity, pH 7.5: 25% of maximal activity at [37]) [37]
5.1-9 <34> (<34> 50% of maximal activity [54]) [54]
5.9-6.2 <10> [37]
Additional information <10> [8]

Temperature optimum (°C)

37 <9, 19, 35> (<9> avicel [7]) [7, 25, 56]
39-45 <21> [35]
40 <3, 32> (<3> assay at [43]) [43, 52]
40-50 <4> [4]
45-50 <10> [37, 38]
50 <5, 7, 9, 33> (<9> cellulose [7]) [5, 7, 12, 53]
55 <12, 15> (<12> isozyme I [13]) [13, 21]
55-70 <2> (<2> thermophilic fungi [30]) [30]
60 <12, 13, 20, 31> (<12> isozyme II [13]) [13, 14, 27, 46, 48]
70 <14> [51]
105 <34> [54]

Temperature range (°C)

23-60 <10> (<10> 10% of maximal activity [37]) [37]
60-70 <20> (<20> 60°C: optimum, 70°C: appreciable amount of activity [27]) [27]

4 Enzyme Structure

Molecular weight

22000 <34> (<34> gel filtration [54]) [54]
40000 <10> (<10> gel filtration [37]) [37]
41800 <2> (<2> sedimentation equilibrium analysis [30]) [30]
42000 <9> (<9> gel filtration [7]) [7]
46000 <2> (<2> chromatography of reduced and alkylated enzyme [31]) [31]
46300 <18, 20> (<18> isozyme I, gel filtration [23]; <20>, gel filtration [27]) [23, 27]
50000 <15> (<15> isozyme II, gel filtration [21]) [21]
50300 <3> (<3> light scattering [16]) [16]
50700 <18> (<18> isozyme II, gel filtration) [23]
52500 <33> (<33> gel filtration [53]) [53]
57000 <15> (<15> isozyme I, gel filtration [21]) [21]
60000 <32> (<32> gel filtration [52]) [52]
62000-68000 <3, 4> (<3, 4> gel filtration [4, 11]) [4, 11]
64000 <12, 13> (<13> gel filtration [14]; <12>, II, gel filtration [13]) [13, 14]
65000 <6> (<6> gel filtration [6]) [6]
84000 <12> (<12> isozyme I, gel filtration [13]) [13]
200000 <19> (<19> gel filtration [25]) [25]

230000 <21> (<21> gel filtration [35]) [35]
Additional information <3> [19]

Subunits

? <3, 5, 8, 10, 11, 14, 16, 31> (<5> x * 81000, SDS-PAGE [5]; <10> x * 75000, SDS-PAGE [8]; <11> CBH I, x * 128000, SDS-PAGE [10]; <3> CHH II, 53000, SDS-PAGE [15]; <8> x * 58000, SDS-PAGE [17]; <16> x * 72000, SDS-PAGE [22]; <14> x * 87000, SDS-PAGE [51]; <31> x * 75000, SDS-PAGE [57]) [5, 8, 10, 15, 17, 22, 51, 57]
dimer <19, 21> (<19> 2 * 100000, SDS-PAGE [25]; <21> 2 * 118000, SDS-PAGE [35]) [25, 35]
monomer <9> (<9> carboxymethylated derivative [7]) [7]

Posttranslational modification

glycoprotein <1-4, 10-13, 16> (<1> 8% reducing sugar [1]; <4>, CBH 1, 9% reducing sugar, CBH 1a 33% reducing sugar [4]; <10> 8-16% reducing sugar [8]; <11> CBH I: 10.7% reducing sugar, CBH II 8.2% reducing sugar [10]; <2> attached carbohydrate in addition to glycoprotein constituents [18]; <16> α-linked mannose and N-linked glycolysation [39]) [1, 2, 4, 8, 10, 13-15, 18, 21-13, 29, 30, 31, 39]

5 Isolation/Preparation/Mutation/Application

Source/tissue

commercial preparation <3, 8> [17, 43, 45]
culture filtrate <2, 3, 6, 7, 10, 13, 15, 19, 21, 34> [6, 8, 11, 12, 14, 16, 21, 24, 25, 29, 30, 35, 54]

Localization

cytoplasm <10> [37]
extracellular <1, 10> [1, 38, 37]
periplasm <10> [37, 38]

Purification

<1> [1]
<2> [2, 18, 29, 30, 31]
<3> (CBH I and CBH II [45]; cloned in E. coli [47,48]) [3, 11, 15, 16, 19, 24, 26, 43, 45, 47, 48]
<5> [5]
<6> [6]
<7> [12]
<8> (gene cloned in E. coli [17]) [17]
<9> [7]
<10> [8, 38]
<11> [10]
<12> [13]
<13> [14]
<14> (avicellase II [51]) [51]

<15> [21]
<16> [22]
<17> [22]
<18> [23]
<19> [25]
<20> [27]
<21> [35]
<30> [44]
<31> [46, 57]
<32> [52]
<33> [53]
<34> [54]
<35> (overexpressed in E. coli [56]) [56]

Crystallization
<3> (engineered enzymes [41, 49, 55]) [40-42, 49, 55]
<16> [39]

Cloning
<3> (cloned in E. coli [9, 47]) [9, 47]
<8> (cloned in Saccharomyces cerevisiae; cloned in E. coli [17] [17, 34]) [17, 34]
<31> (cloned in E. coli [46, 48]) [46, 48]

Engineering
D214N <3> (<3> impaired catalytic activity [55]) [55]
E212Q <3> (<3> impaired catalytic activity [55]) [55]
E217Q <3> (<3> impaired catalytic activity [55]) [55]
W272A <3> [50]
W272D <3> [50]
Y169F <3> [49]
Y31A <3> (<3> structural role minor, but does not bind strongly to cellulose [41]) [41]
Y32A <3> (<3> structural role minor, but does not bind strongly to cellulose [41]) [41]
Y5A <3> (<3> complete loss of binding affinity to cellulose [41]) [41]

6 Stability

pH-Stability
2-7 <11> (<11> 4°C, 24 h, CBH II [10]) [10]
2.5-6 <11> (<11> 4°C, 24 h, CBH I [10]) [10]
3-10 <13> (<13> 6°C, 48 h [14]) [14]
3.5-9.5 <16, 17> [22]
4.5-5 <9> (<9> highest stability [7]) [7]
4.5-7 <32> [52]
5.5-8 <5, 19> (<19> highest stability [25]; <5> 25°C, 24 h [5]) [5, 25]

Temperature stability

40 <1> (<1> 3 hours [1]) [1]
50 <32> [52]
60 <6, 13> (<6> pH 5.5, 10 min, complete loss of activity [6]; <13> stable for 10 min [14]) [6, 14]
65 <11, 16, 17> (<11> 60% loss of activity after 10 min, CBH I [10]; <16,17> stable for 5 min [22]) [10, 22]
70 <11> (<11> CHB II: 40% loss of activity after 10 min [10]) [10]
78 <2> (<2> complete denaturation after 3 min [30]) [30]
80 <1> (<1> 50% loss of activity after 3 hours [1]) [1]
100 <6> (<6> 95% loss of activity after 10 min [6]) [6]

General stability information

<10>, bovine serum albumin stabilizes [8, 38]
<10>, ribonuclease stabilizes [38]
<15>, dilute solutions, unstable [21]
<31>, Ca^{2+} and cysteine stabilizes [57]
<33>, glycerol stabilizes [53]
<9, 20>, freezing and thawing, stable [7, 27]

Storage stability

<9>, -15°C, pH 4.5, 50 mM citrate buffer [7]
<15>, -20°C, concentrated solutions, several months [21]
<16, 17>, 0-4°C, 0.05 M acetate buffer, pH 5.0, at least 2 weeks [22]

References

[1] Fracheboud, D.; Canevascini, G.: Isolation, purification, and properties of the exocellulase from Sporotrichum (Chrysosporium) thermophile. Enzyme Microb. Technol., **11**, 220-229 (1989)
[2] Voragen, A.G.J.; Beldman, G.; Rombouts, F.M.: Cellulase of a mutant strain of Trichoderma viride QM 9414. Methods Enzymol., **160**, 243-251 (1988)
[3] Schuelein, M.: Cellulases of Trichoderma reesei. Methods Enzymol., **160**, 234-242 (1988)
[4] Wood, T.M.: Cellulase of Trichoderma koningii. Methods Enzymol., **160**, 221-233 (1988)
[5] Nakamura, K.; Kitamura, K.: Cellulases of Cellulomonas uda. Methods Enzymol., **160**, 211-216 (1988)
[6] Kanda, T.; Nisizawa, K.: Exocellulase of Irpex lacteus (Polyporus tulipiferae). Methods Enzymol., **160**, 403-408 (1988)
[7] Sadana, J.C.; Patil, R.V.: 1,4-β-D-Glucan cellobiohydrolase from Sclerotium rolfsii. Methods Enzymol., **160**, 307-314 (1988)
[8] Huang, L.; Forsberg, C.W.; Thomas, D.Y.: Purification and characterization of a chloride-stimulated cellobiosidase from Bacteroides succinogenes S85. J. Bacteriol., **170**, 2923-2932 (1988)

[9] Shoemaker, S.; Schweickart, V.; Ladner, M.; Gelfand, M.; Kwok, S.; Myambo, K.; Innis, M.: Molecular cloning of exo-cellobiohydrolase I derived from Trichoderma resei strain L27. Bio/Technology, **1**, 691-696 (1983)
[10] Hayashida, S.; Mo, K.; Hosoda, A.: Production and characteristics of avicel-digesting and non-avicel-digesting cellobiohydrolases from Aspergillus ficum. Appl. Environ. Microbiol., **54**, 1523-1529 (1988)
[11] Riske, F.J.; Eveleigh, D.E.; Macmillan, J.D.: Purification of cellobiohydrolase I from Trichoderma reesei using monoclonal antibodies in an immunomatrix. J. Ind. Microbiol., **1**, 259-264 (1986)
[12] Sidhu, M.S.; Kalra, M.K.; Sandhu, D.K.: Purification and characterization of cellulolytic enzymes from Trichoderma harzianum. Folia Microbiol., **31**, 293-302 (1986)
[13] Kamagata, Y.; Sasaki, H.; Takao, S.: Purification and some properties of two cellobiohydrolases from Penicillium purpurogenum. J. Ferment. Technol., **64**, 211-217 (1986)
[14] Funaguma, T.; Tsuji, H.; Hara, A.: Purification and some properties of cellulose 1,4-β-cellobiosidase from a strain of Penicillium sp.. J. Ferment. Technol., **64**, 77-80 (1986)
[15] Bhikhabhai, R.; Johansson, G.; Pettersson, G.: Isolation of cellulolytic enzymes from Trichoderma reesei QM 9414. J. Appl. Biochem., **6**, 336-345 (1984)
[16] Odegaard, B.H.; Anderson, P.C.; Lovrien, R.E.: Resolution of the multienzyme cellulase complex of Trichoderma reesei QM9414. J. Appl. Biochem., **6**, 156-183 (1984)
[17] Gilkes, N.R.; Langsford, M.L.; Kilburn, D.G.; Miller, R.C.; Warren, R.A.J.: Mode of action and substrate specificities of cellulases from cloned bacterial genes. J. Biol. Chem., **259**, 10455-10459 (1984)
[18] Alurralde, J.L.; Ellenrieder, G.: Effect of attached carbohydrates on the activity of Trichoderma viride cellulases. Enzyme Microb. Technol., **6**, 467-690 (1984)
[19] Shoemaker, S.; Watt, K.; Tsitovsky, G.; Cox, R.: Characterization and properties of cellulases purified from Trichoderma reesei strain L27. Bio/Technology, **1**, 687-690 (1983)
[20] Creuzet, N.; Berenger, J.-F.; Frixon, C.: Characterization of exoglucanase and synergistic hydrolysis of cellulose in Clostridium stercorarium. FEMS Microbiol. Lett., **20**, 347-350 (1983)
[21] Mishra, C.; Vaidya, M.; Rao, M.; Deshpande, V.: Purification and properties of two exocellobiohydrolases from a cellulolytic culture of Fusarium lini. Enzyme Microb. Technol., **5**, 430-434 (1983)
[22] Hayashida, S.; Ohta, K.; Mo, K.: Cellulases of Humicola insolens and Humicola grisea. Methods Enzymol., **160**, 323-332 (1988)
[23] Wood, T.M.: Cellobiohydrolases of Penicillium pinophilum. Methods Enzymol., **160**, 398-403 (1988)
[24] Nummi, M.; Niku-Paavola, M.-L.; Lappalainen, A.; Enari, T.-M.; Raunio, V.: Cellobiohydrolase from Trichoderma reesei. Biochem. J., **215**, 677-683 (1983)

[25] Ohmiya, K.; Shimizu, M.; Taya, M.; Shimizu, S.: Purification and properties of cellobiosidase from Ruminococcus albus. J. Bacteriol., **150**, 407-409 (1982)
[26] Faegerstam, L.G.; Pettersson, L.G.: The 1,4-β-glucan cellobiohydrolase of Trichoderma reesei QM 9414. FEBS Lett., **119**, 97-100 (1980)
[27] Wood, T.M.; McCrae, S.I.; MacFarlane, C.C.: The isolation, purification and properties of the cellobiohydrolase component of Penicillium funiculosum cellulase. Biochem. J., **189**, 51-56 (1980)
[28] Wood, T.M.; McCrae, S.I.: The cellulase of Trichoderma koningii. Purification and properties of some endoglucanase components with special reference to their action on cellulose when acting alone and in synergism with the cellobiohydrolase. Biochem. J., **171**, 61-72 (1978)
[29] Gum, E.K.; Brown, R.D.: Structural characterization of a glycoprotein cellulase, 1,4-β-D-glucan cellobiohydrolase C from Trichoderma viride. Biochim. Biophys. Acta, **446**, 371-386 (1976)
[30] Berghem, L.E.R.; Pettersson, L.G.; Axioe-Fredriksson, U.-B.: The mechanism of enzymatic cellulose degradation. Eur. J. Biochem., **53**, 55-62 (1975)
[31] Berghem, L.E.R.; Pettersson, L.G.: The mechanism of enzymatic cellulose degradation. Purification of a cellulolytic enzyme from Trichoderma viride active on highly ordered cellulose. Eur. J. Biochem., **37**, 21-30 (1973)
[32] Wood, T.M.; McCrae, S.I.: The purification and properties of the C 1 component of Trichoderma koningii cellulase. Biochem. J., **128**, 1183-1192 (1972)
[33] Kanda, T.; Yatomi, H.; Makishima, S.; Amano, Y.; Nisizawa, K.: Substrate specificities of exo- and endo-type cellulases in the hydrolysis of β-(1-3)- and β-(1-4)-mixed D-glucans. J. Biochem., **105**, 127-132 (1989)
[34] Wong, W.K.R.; Curry, C.; Parekh, R.S.; Parekh, S.R.; Wayman, M.; Davies, R.W.; Kilburn, D.G.; Skipper, N.: Wood hydrolysis by Cellulomonas fimi endoglucanase and exoglucanase coexpressed as secreted enzymes in Saccharomyces cerevisiae. Bio/Technology, **6**, 713-719 (1988)
[35] Gardner, R.M.; Doerner, K.C.; White, B.A.: Purification and characterization of an exo-β-1,4-glucanase from Ruminococcus flavefaciens FD-1. J. Bacteriol., **169**, 4581-4588 (1987)
[36] Grajek, W.: Temperature and pH optima of enzyme activities produced by cellulolytic thermophilic fungi in batch and solid-state culures. Biotechnol. Lett., **8**, 587-590 (1986)
[37] Huang, L.; Forsberg, C.W.: Isolation of cellodextrinase from Bacteroides succinogenes. Appl. Environ. Microbiol., **53**, 1034-1041 (1987)
[38] Huang, L.; Forsberg, C.W.: Purification and comparison of the periplasmic and extracellular forms of the cellodextrinase from Bacteroides succinogenes. Appl. Environ. Microbiol., **54**, 1488-1493 (1988)
[39] Varrot, A.; Hastrup, S.; Schulein, M.; Davies, G.J.: Crystal structure of the catalytic core domain of the family 6 cellobiohydrolase II, Cel6A, from Humicola insolens, at 1.92 A resolution. Biochem. J., **337**, 297-304 (1999)
[40] Divne, C.; Stahlberg, J.; Teeri, T.T.; Jones, T.A.: High-resolution crystal structures reveal how a cellulose chain is bound in the 50 A long tunnel of

cellobiohydrolase I from Trichoderma reesei. J. Mol. Biol., **275**, 309-325 (1998)
[41] Mattinen, M.L.; Kontteli, M.; Kerovuo, J.; Linder, M.; Annila, A.; Lindeberg, G.; Reinikainen, T.; Drakenberg, T.: Three-dimensional structures of three engineered cellulose-binding domains of cellobiohydrolase I from Trichderma reesei. Protein Sci., **6**, 294-303 (1997)
[42] Divne, C.; Stahlberg, J.; Reinikainen, T.; Ruohonen, L.; Pettersson, G.; Knowles J.K.; Teeri, T.T.; Jones, T.A.: The three-dimensional crystal structure of the catalytic core of cellobiohydrolase I from Trichoderma reesei. Science, **265**, 524-528 (1994)
[43] Witte, K.; Heitz, H.-J.; Wartenberg, A.: Isolation and purification of isoenzymes of cellobiolhydrolase I and II of Trichoderma reesei using LPLC methods. Acta Biotechnol., **10**, 41-48 (1990)
[44] Igarashi, K.; Samejima, M.; Eriksson, K.-E.L.: Cellobiose dehydrogenase enhances Phanerochaete chrysosporium cellobiohydrolase I activity by relieving product inhibition. Eur. J. Biochem., **253**, 101-106 (1998)
[45] Woodward, J.; Brown, J.P.; Evans, B.R.; Affholter, K.A.: Papain digestion of crude Trichoderma reesei cellulase: purification and properties of cellobiohydrolase I and II core proteins. Biotechnol. Appl. Biochem., **19**, 141-153 (1994)
[46] Kruus, K.; Andreacchi, A.; Wang, W.K.; Wu, J.H.D.: Product inhibition of the recombinant CelS, and exoglucanase component of the Clostridium thermocellum cellulosome. Appl. Microbiol. Biotechnol., **44**, 399-404 (1995)
[47] Linder, M.; Teeri, T.T.: The cellulose-binding domain of the major cellobiohydrolase of Trichoderma reesei exhibits true reversibility and a high exchange rate on crystalline cellulose. Proc. Natl. Acad. Sci. USA, **93**, 12251-12255 (1996)
[48] Singh, R.N.; Akimenko, V.K.: Isolation of a cellobiohydrolase of Clostridium thermocellum capable of degrading natural crystalline substrates. Biochem. Biophys. Res. Commun., **192**, 1123-1130 (1993)
[49] Zou, J.-y.; Kleywegt, G.J.; Stahlberg, J.; Driguez, H.; Nerinckx, W.; Claeyssens, M.; Koivula, A.; Teeri, T.T.; Jones, T.A.: Crystallographic evidence for substrate ring distortion and protein conformational changes during catalysis in cellobiohydrolase Cel6A from Trichoderma reesei. Structure, **7**, 1035-1045 (1999)
[50] Koivula, A.; Kinnari, T.; Harjunpaeae, V.; Ruohonen, L.; Teleman, A.; Drakenberg, T.; Rouvinen, J.; Jones, T.A.; Teeri, T.T.: Tryptophan 272: an essential determinant of crystalline cellulose degradation by Trichoderma reesei cellobiohydrolase Cel6A. FEBS Lett., **429**, 341-346 (1998)
[51] Bronnenmeier, K.; Rücknagel, K.P.; Staudenbauer, W.L.: Purification and properties of a novel type of exo-1,4-β-glucanase (avicelase II) from the cellulolytic thermophile Clostridium stercorarium. Eur. J. Biochem., **200**, 379-385 (1991)
[52] Kim, C.-H.; Kim, D.-S.: Purification and specificity of a specific endo-β-1,4-D-glucanase (avicelase II) resemling exo-cellobiohydrolase from Bacillus circulans. Enzyme Microb. Technol., **17**, 248-254 (1995)

[53] Singh, A.; Agrawal, A.K.; Abidi, A.B.; Darmwal, N.S.: Properties of exoglucanase from Aspergillus niger. J. Gen. Appl. Microbiol., **36**, 245-253 (1990)

[54] Ruttersmith, L.D.; Daniel, R.M.: Thermostable cellobiohydrolase from the thermophilic eubacterium Thermotoga sp. strain FjSS3-B.1. Purification and properties. Biochem. J., **277**, 887-890 (1991)

[55] Stahlberg, J.; Divne, C.; Koivula, A.; Piens, K.; Claeyssens, M.; Teeri, T.T.; Jones, T.A.: Activity studies and crystal structures of catalytically deficient mutants of cellobiohydrolase I from Trichoderma reesei. J. Mol. Biol., **264**, 337-349 (1996)

[56] Gotsche, S.; Dahl, M.K.: Purification and characterization of the phospho-α(1,1)glucosidase (TreA) of Bacillus subtilis 168. J. Bacteriol., **177**, 2721-2726 (1995)

[57] Morag, E.; Halevy, I.; Bayer, E.A.; Lamed, R.: Isolation and properties of a major cellobiolhydrolase from the cellulosome of Clostridium thermocellum. J. Bacteriol., **173**, 4155-4162 (1991)

Peptidoglycan β-N-acetylmuramidase 3.2.1.92

1 Nomenclature

EC number
3.2.1.92

Systematic name
peptidoglycan β-N-acetylmuramoylexohydrolase

Recommended name
peptidoglycan β-N-acetylmuramidase

Synonyms
β-2-acetamido-3-O-(D-1-carboxyethyl)-2-deoxy-D-glucoside acetamidodeoxy-glucohydrolase
exo-β-N-acetylmuramidase
exo-β-acetylmuramidase

CAS registry number
52219-03-1

2 Source Organism

<1> *Bacillus subtilis* (B) [1-3]

3 Reaction and Specificity

Catalyzed reaction
hydrolysis of terminal, non-reducing N-acetylmuramic residues

Reaction type
hydrolysis of O-glycosyl bond

Natural substrates and products
S Additional information <1> (hydrolysis of terminal, non-reducing N-acetylmuramic residues) [1-3]
P ?

Substrates and products
S 4-methylumbelliferyl-2-acetamido-3-O-(D-1-carboxyethyl)-2-deoxy-β-D-glucose + H_2O <1> [1, 2]
P 4-methylumbelliferone + 2-acetamido-3-O-(D-1-carboxyethyl)-2-deoxy-β-D-glucose (i.e. N-acetylmuramic acid)

S O-2-acetamido-2-deoxy-β-D-glucopyranosyl-(1-4)-2-acetamido-3-O-(D-1-carboxyethyl)-2-deoxy-D-glucose + H_2O <1> [1]
P 2-acetamido-2-deoxy-β-D-glucopyranose + 2-acetamido-3-O-(D-1-carboxyethyl)-2-deoxy-D-glucose
S O-[2-acetamido-3-O-(D-1-carboxyethyl)-2-deoxy-β-D-glucopyranosyl]-(1-4)-2-acetamido-2-deoxy-D-glucose + H_2O <1> [1]
P 2-acetamido-3-O-(D-1-carboxyethyl)-2-deoxy-β-D-glucose + 2-acetamido-2-deoxy-D-glucose

Inhibitors
2-acetamido-2-deoxy-D-galactose <1> [1, 3]
2-acetamido-2-deoxy-D-glucose <1> [1, 3]
2-acetamido-2-deoxy-D-mannose <1> [1, 3]
2-acetamido-3-O-(D-1-carboxyethyl)-2-deoxy-D-glucose <1> (<1> MurAc, 1 mM, activity 76,9% [1, 3]) [1, 3]

Specific activity (U/mg)
24.2 <1> [1]

K_m-Value (mM)
0.19 <1> (4-methylumbelliferyl-2-acetamido-3-O-(D-1-carboxyethyl)-2-deoxy-β-D-glucose) [1]
0.65 <1> (O-[2-acetamido-3-O-(D-1-carboxyethyl)-2-deoxy-β-D-glucopyranosyl]-(1-4)-2-acetamido-2-deoxy-D-glucose) [1]

pH-Optimum
5-5.2 <1> [3]

pH-Range
4-10 <1> [3]

4 Enzyme Structure

Molecular weight
90000 <1> (<1> estimated by SDS-PAGE, purified enzyme [1]) [1]

5 Isolation/Preparation/Mutation/Application

Source/tissue
culture medium <1> [1]

Purification
<1> [1]

6 Stability

pH-Stability

8 <1> (<1> most stable at [1]) [1]

References

[1] Del Rio, L.A.; Berkeley, R.C.W.: Exo-β-N-acetylmuramidase - a novel hexosaminidase. Production by Bacillus subtilis B, purification and characterization. Eur. J. Biochem., **65**, 3-12 (1976)
[2] Del Rio, L.A.; Berkeley, R.C.W.: A substrate for the fluorogenic assay of exo-β-N-acetylmuramidase: synthesis and purification of 4-methylumbelliferyl N-acetylmuramide. Anal. Biochem., **66**, 405-411 (1975)
[3] Del Rio, L.A.; Berkeley, R.C.W.; Brewer, S.J.; Roberts, S.E.: An enzyme from Bacillus subtilis B with exo-β-N-acetylmuramidase activity. FEBS Lett., **37**, 7-9 (1973)

α,α-Phosphotrehalase 3.2.1.93

1 Nomenclature

EC number
3.2.1.93

Systematic name
α,α-trehalose-6-phosphate phosphoglucohydrolase

Recommended name
α,α-phosphotrehalase

Synonyms
phosphotrehalase

CAS registry number
54576-93-1

2 Source Organism

<1> *Bradyrhizobium japonicum* (in symbiosis *Glycine max* [1]) [1]
<2> *Bacillus popilliae* [2]
<3> *Bacillus subtilis* [3]

3 Reaction and Specificity

Catalyzed reaction
α,α-trehalose 6-phosphate + H_2O = D-glucose + D-glucose 6-phosphate

Reaction type
hydrolysis of O-glycosyl bond

Natural substrates and products
S α,α-trehalose 6-phosphate + H_2O <1-3> [1-3]
P D-glucose + D-glucose 6-phosphate <1-3> [1-3]

Substrates and products
S α,α-trehalose 6-phosphate + H_2O <1, 2> [1, 2]
P D-glucose + D-glucose 6-phosphate <1, 2> [1, 2]

Inhibitors
fructose <3> [3]
glucose <3> [3]

glycerol <3> (<3> moderate [3]) [3]
mannitol <3> [3]
sucrose <3> (<3> moderate [3]) [3]

Specific activity (U/mg)
2.83 <2> [2]

K_m-Value (mM)
1.8 <2> (α,α-trehalose 6-phosphate) [2]

pH-Optimum
6.5-7 <2> [2]

Temperature optimum (°C)
30 <2> (<2> assay at [2]) [2]

5 Isolation/Preparation/Mutation/Application

Source/tissue
nodule <1> (<1> bacteroids [1]) [1]

Localization
cytoplasm <1, 3> [1, 3]
soluble <1> [1]

Purification
<2> [2]

References

[1] Salminen, S.O.; Streeter, J.G.: Enzymes of α,α-trehalose metabolism in soybean nodules. Plant Physiol., **81**, 538-541 (1986)
[2] Bhumiratana, A.; Anderson, R.L.; Costilow, R.N.: Trehalose metabolism by Bacillus popillae. J. Bacteriol., **119**, 484-493 (1974)
[3] Helfert, C.; Gotsche, S.; Dahl, M.K.: Cleavage of trehalose-phosphate in Bacillus subtilis is catalysed by a phospho-α(1,1)-glucosidase encoded by the treA gene. Mol. Microbiol., **16**, 111-120 (1995)

Glucan 1,6-α-isomaltosidase 3.2.1.94

1 Nomenclature

EC number
3.2.1.94

Systematic name
1,6-α-D-glucan isomaltohydrolase

Recommended name
glucan 1,6-α-isomaltosidase

Synonyms
G2-dextranase <1> [1]
IMD
dextranase, isomalto-
exo-isomalto-hydrolase
isomalto-dextranase
isomaltodextranase
isomaltohydrolase, exo

CAS registry number
56467-68-6

2 Source Organism

<1> *Arthrobacter globiformis* (T6 [1]) [1-12]
<2> *Actinomadura sp.* (strain R10 [4]) [4]

3 Reaction and Specificity

Catalyzed reaction
hydrolysis of 1,6-α-D-glucosidic linkages in polysaccharides, to remove successive isomaltose units from the
hydrolysis of 1,6-α-D-glucosidic linkages in polysaccharides, to remove successive isomaltose units from the non-reducing ends of the chains (hydrolysis of α-1,2-, α-1,3- and α-1,4-linkages in polysaccharides, to remove successive isomaltose units [7])

Reaction type
condensation
hydrolysis of O-glycosyl bond

Natural substrates and products

S glucooligosaccharides + H_2O <1, 2> [4]
P isomaltose + D-glucose <1, 2> [4]

Substrates and products

S α-linked gluco-oligosaccharide + H_2O <1> [7]
P isomaltose + D-glucose <1> [7]
S dextran + 1-butanol <1> [10]
P 1-butyl α-isomaltoside + H_2O [10]
S dextran + 1-propanol <1> [10]
P 1-propyl α-isomaltoside + H_2O [10]
S dextran + H_2O <1, 2> [4, 5, 7]
P α-isomaltose + O-α-D-Glcp-(1-3)-O-α-D-Glcp-(1-6)-D-Glc + O-α-D-Glcp-(1-4)-O-α-D-Glcp-(1-6)-D-Glc + O-α-D-Glcp-(1-2)-O-α-D-Glcp-(1-6)-D-Glc <1, 2> [4, 5, 7]
S dextran + ethanol <1> [10]
P ethyl α-isomaltoside + H_2O [10]
S dextran + methanol <1> [10]
P methyl α-isomaltoside + H_2O [10]
S dextran + sucrose <1> [10]
P 6G-α-isomaltosylsucrose + H_2O [10]
S dextran + trehalose <1> [10]
P 6-α-isomaltosyltrehalose + H_2O [10]
S isomaltohexatiol + H_2O <1> [7]
P isomaltose + Isomaltitol <1> [7]
S isomaltooligosaccharide + H_2O <1> (condensation of isomaltose to isomaltotetraose [8]) [7, 8]
P isomaltose + glucose <1> (<1> glucose not with isomaltotetraose and isomaltohexaose [7]) [7]
S isomaltotriitol + H_2O <1, 2> [4, 9]
P isomaltose + sorbitol <1> [9]
S isomaltotriose + H_2O <1, 2> [3, 4]
P α-isomaltose + D-glucose <1, 2> [3, 4]
S panose + H_2O <1, 2> [3, 4, 9]
P α-isomaltose <1, 2> [3, 4]
S panose + H_2O <1, 2> [3, 4, 9]
P isomaltose + D-glucose <1> [9]
S pullulan + H_2O <1> [1]
P isopanose <1> [1]

Inhibitors

Ag^+ <1> (<1> partial [2]) [1, 2]
Fe^{3+} <1> [1, 2]
Hg^{2+} <1> [1, 2]
$KMnO_4$ <1> [1, 2]
N-bromosuccinimide <1> [1, 2]

Specific activity (U/mg)
5 <2> [4]
15.3 <1> [9]
18.8 <1> [1, 2]

pH-Optimum
4 <1> [4]
4-4.4 <1> [9]
5 <2> [4]
5.3 <1> [1, 2]

pH-Range
2.5-7.5 <1> (<1> 50% of maximum activity at pH 2.5 and 10% at pH 7.5 [9]) [9]
3-7.5 <2> (<2> 10% of maximum activity at pH 3 and 20% at pH 7.5 [4]) [4]

Temperature optimum (°C)
65 <1> [1, 2]

Temperature range (°C)
30-70 <1> (<1> 20% of maximum activity at 30°C and 65% at 70°C [2]) [2]

4 Enzyme Structure

Subunits
? <1> (<1> x * 69000, SDS-PAGE [2]) [2]

5 Isolation/Preparation/Mutation/Application

Source/tissue
culture filtrate <1, 2> [1-5, 8-10]

Localization
extracellular <1, 2> [4, 9]

Purification
<1> [1, 2]
<2> [4]

Cloning
<1> (<1> expression in E. coli [11]) [11]

6 Stability

pH-Stability
2.8-8 <1> (<1> 4°C, 24 h [1, 2]) [1, 2]
4.1-8 <1> (<1> 45°C, 2 h [1]) [1]

Temperature stability
50 <1> (<1> stable below 50 [1, 2]) [1, 2]

70 <1> (<1> 20% activity after 10 min [1, 2]) [1, 2]
80 <1> (<1> complete inactivation after 10 min [1, 2]) [1, 2]

Oxidation stability

<1>, photooxidation with Rose Bengal [1]

Organic solvent stability

methanol <1> (<1> assay with 20% at 37°C for 4 min [12]) [12]

References

[1] Okada, G.; Takayanagi, T.; Miyahara, S.; Sawai, T.: An isomalto-dextranase accompanied by pullulanase activity from Arthrobacter globiformis T6. Agric. Biol. Chem., **52**, 829-836 (1988)
[2] Okada, G.; Takayanagi, T.; Sawai, T.: Improved purification and further characterization of an isomaltodextranase from Arthrobacter globiformis T6. Agric. Biol. Chem., **52**, 495-501 (1988)
[3] Takayanagi, T.; Okada, G.; Chiba, S.: Quantitative study of the anomeric forms of isomaltose and glucose produced by isomalto-dextranase and glucodextranase. Agric. Biol. Chem., **51**, 2337-2341 (1987)
[4] Sawai, T.; Ohara, S.; Ichimi, Y.; Okaji, S.; Hisada, K.; Fukaya, N.: Purification and some properties of the isomaltodextranase of Actinomadura strain R10 and comparison with that of Arthrobacter globiformis T6. Carbohydr. Res., **89**, 289-299 (1981)
[5] Sawai, T.; Tohyama, T.; Natsumi, T.: Hydrolysis of fourteen native dextrans by Arthrobacter isomaltodextranase and correlation with dextran structure. Carbohydr. Res., **66**, 195-205 (1978)
[6] Sawai, T.; Ukigai, Y.; Nawa, A.: Identification of an isomalto-dextranase producing bacterium, Arthrobacter globiformis. Agric. Biol. Chem., **40**, 1249-1250 (1976)
[7] Torii, M.; Sakakibara, K.; Misaki, A.; Sawai, T.: Degradation of α-linked D-gluco-oligosaccharides and dextrans by an isomalto-dextranase preparation from Arthrobacter globiformis T6. Biochem. Biophys. Res. Commun., **70**, 459-464 (1976)
[8] Sawai, T.; Niwa, Y.: Transisomaltosylation activity of a bacterial isomalto-dextranase. Agric. Biol. Chem., **39**, 1077-1083 (1975)
[9] Sawai, T.; Toriyama, K.; Yano, K.: A bacterial dextranase releasing only isomaltose from dextrans. J. Biochem., **75**, 105-112 (1974)
[10] Kim, Y.-K.; Tsumuraya, Y.; Sakano, Y.: Enzymatic preparation of novel non-reducing oligosaccharides having an isomaltosyl residue by using the transfer action of isomaltodextranase from Arthrobacter globiformis T6. Biosci. Biotechnol. Biochem., **59**, 1367-1369 (1995)
[11] Iwai, A.; Ito, H.; Mizuno, T.; Mori, H.; Matsui, H.; Honma, M.; Okada, G.; Chiba, S.: Molecular cloning and expression of an isomalto-dextranase gene from Arthrobacter globiformis T6. J.Bacteriol., **176**, 7730-7734 (1994)
[12] Takayanagi, T.; Kimura, A.; Okada, G.; Chiba, S.: Identification of ionizable groups essential to the activity of isomalto-dextranase from Arthrobacter globiformis. Biosci. Biotechnol. Biochem., **58**, 1995-1999 (1994)

Dextran 1,6-α-isomaltotriosidase 3.2.1.95

1 Nomenclature

EC number
3.2.1.95

Systematic name
1,6-α-D-glucan isomaltotriohydrolase

Recommended name
dextran 1,6-α-isomaltotriosidase

Synonyms
IMTD
exo-isomaltotriohydrolase
exoisomaltotriohydrolase
isomaltotrio-dextranase
isomaltotriohydrolase, exo-

CAS registry number
72561-11-6

2 Source Organism

<1> *Brevibacterium fuscum* (var. dextranlyticum [1, 2]; strain 0407 [2]) [1, 2]

3 Reaction and Specificity

Catalyzed reaction
hydrolysis of 1,6-α-D-glucosidic linkages in dextrans, to remove successive isomaltotriose units from the non-reducing ends of the chains

Reaction type
hydrolysis of O-glycosyl bond

Substrates and products
S Sephadex G-100 + H_2O <1> [1]
P ?
S Sephadex G-200 + H_2O <1> [1]
P ?
S dextran + H_2O <1> [1, 2]
P isomaltotriose <1> [1]

S isomaltodextrin + H_2O <1> (<1>, reduced [1]) [1]
P isomaltotriose <1> [1]

Inhibitors

$AgNO_3$ <1> (<1>, 2 mM, 97% inhibition [1]) [1]
$CuSO_4$ <1> (<1>, 2 mM, 76% inhibition [1]) [1]
$HgCl_2$ <1> (<1>, 2 mM, complete inhibition [1]) [1]
N-bromosuccinimide <1> (<1>, 2 mM, complete inhibition [1]) [1]
iodine <1> (<1>, 2 mM, complete inhibition [1]) [1]

Specific activity (U/mg)

21.5 <1> (<1>, recombinant enzyme [2]) [2]
Additional information <1> [1]

K_m-Value (mM)

Additional information <1>(<1>, K_m: 0.72 mg/ml for the recombinant enzyme and 0.26 mg/ml for the native enzyme [2]) [2]

pH-Optimum

7-7.5 <1> [1]
7.5 <1> (<1>, recombinant enzyme [2]) [2]

pH-Range

6.5-8 <1> (<1>, 85% of maximum activity at pH 6.5 and pH 8.0 [1]) [1]

4 Enzyme Structure

Subunits

? <1> (<1>, x * 68000, SDS-PAGE [2]) [2]

5 Isolation/Preparation/Mutation/Application

Purification

<1> (recombinant enzyme [2]) [1, 2]

Cloning

<1> (expression in E. coli [2]) [2]

6 Stability

pH-Stability

5.6-11 <1> (<1>, 37°C, 12 h, stable [1]) [1]

Temperature stability

50 <1> (<1>, pH 7.5, 30 min, stable up to [1]) [1]
60 <1> (<1>, pH 7.5, 30 min, 30% loss of activity [1]) [1]

References

[1] Sugiura, M.; Ito, A.; Yamaguchi, T.: Studies on dextranase. II. New exo-dextranase from Brevibacterium fuscum var. dextranlyticum. Biochim. Biophys. Acta, **350**, 61-70 (1974)
[2] Mizuno, T.; Mori, H.; Ito, H.; Matsui, H.; Kimura, A.; Chiba, S.: Molecular cloning of isomaltotrio-dextranase gene from Brevibacterium fuscum var. dextranlyticum strain 0407 and its expression in Escherichia coli. Biosci. Biotechnol. Biochem., **63**, 1582-1588 (1999)

Mannosyl-glycoprotein endo-β-N-acetylglucosaminidase 3.2.1.96

1 Nomenclature

EC number
3.2.1.96

Systematic name
glycopeptide-D-mannosyl-N^4-(N-acetyl-D-glucosaminyl)$_2$-asparagine 1,4-N-acetyl-β-glucosaminohydrolase

Recommended name
mannosyl-glycoprotein endo-β-N-acetylglucosaminidase

Synonyms
ENGase A <4> [55]
ENGase CI <12> [55]
ENGase CII <12> [55]
ENGase D <3> [55]
ENGase F <7> [55]
ENGase F1 <7> [55]
ENGase F2 <7> [55]
ENGase F3 <7> [55]
ENGase Fsp <4> [55]
ENGase H <8> [55]
ENGase L <8> [55]
ENGase Mx <28> [55]
ENGase PI <27> [55]
ENGase PII <27> [55]
ENGase St <23> [48, 55]
Endo-HO <13> [50]
acetylglucosaminidase, endo-β
di-N-acetylchitobiosyl β-N-acetylglucosaminidase
endo-BII <29> [58]
endo-BII <29> [58]
endo-GM <15> [20]
endo-LE <16> [20]
endo-M <5> [47]
endo-N-acetyl-β-D-glucosaminidase
endo-N-acetyl-β-glucosaminidase
endo-β-(1$\rightarrow$4)-N-acetylglucosaminidase
endo-β-GlcNAc-ase <5> [5]
endo-β-N-acetylglucosaminidase
endo-β-N-acetylglucosaminidase

endo-β-N-acetylglucosaminidase D <14> [19, 24, 38]
endo-β-N-acetylglucosaminidase F <7> [8, 26]
endo-β-N-acetylglucosaminidase H <8, 11> [9, 25, 27, 52]
endo-β-N-acetylglucosaminidase HS <1> [54]
endo-β-N-acetylglucosaminidase L <8> [39]
endo-β-N-acetylglucosaminidase M <5> [47]
endo-β-acetylglucosaminidase
endoglycosidase S
mannosyl-glycoprotein 1,4-N-acetamidodeoxy-β-D-glycohydrolase

CAS registry number
37278-88-9

2 Source Organism

<-2> no activity in *bovine* (cattle kidney [10]) [10]
<-1> no activity in *sheep* (kidney [10]) [10]
<1> *Homo sapiens* [1, 10, 43, 45, 54]
<2> *Staphylococcus simulans* (strain 22 [2]) [2]
<3> *Streptococcus pneumoniae* [3, 55]
<4> *Flavobacterium sp.* [4, 6, 11, 13, 55]
<5> *Mucor hiemalis* (f. hiemalis [47]) [5, 47]
<6> *Canavalia ensiformis* [7]
<7> *Flavobacterium meningosepticum* [8, 26, 55]
<8> *Streptomyces plicatus* (mutant enzymes D130N, D130E, D130A, E132Q, E132A, and D130N/E132Q [52]) [9, 16, 22, 39, 52, 55, 59]
<9> *Rattus norvegicus* [12, 15, 29, 40, 42, 45]
<10> *Aspergillus oryzae* [14]
<11> *Streptomyces griseus* [14, 25, 27, 28, 31, 32, 34]
<12> *Clostridium perfringens* [17, 33, 55]
<13> *Gallus gallus* (hen [35,50]) [18, 35, 50]
<14> *Diplococcus pneumoniae* [19, 24, 30, 31, 32, 34, 38]
<15> *Glycine max* [20]
<16> *Lycopersicon esculentum* [20]
<17> *Bacillus subtilis* (strain 168 [21]) [21]
<18> *Dictyostelium discoideum* [23]
<19> *Sus scrofa* (no activity detected in pig kidney [29]) [29]
<20> *Fiscus carica* [36, 37, 44]
<21> *Oryctolagus cuniculus* [41]
<22> *Acinetobacter sp.* [46, 55]
<23> *Stigmatella aurantiaca* (strain DW4 [48]) [48, 55]
<24> *Arthrobacter protophormiae* [49, 55]
<25> *Staphylococcus aureus* (mutant RUSAL2 [53]) [51, 53, 56]
<26> *Bacillus alvei* [55]
<27> *Pseudomonas sp.* [55, 57]
<28> *Myxococcus xanthus* [55]

<29> *Bacillus sp.* [58]
<30> *Pisum sativum* [60]

3 Reaction and Specificity

Catalyzed reaction

endohydrolysis of the di-N-acetylchitobiosyl unit in high-mannose glycopeptides and glycoproteins containing the-[Man(GlcNAc)$_2$]Asn-structure. One N-acetyl-D-glucosamine residue remains attached to the protein, the rest of the oligosaccharide is released intact

Reaction type

hydrolysis of O-glycosyl bond

Natural substrates and products

S Additional information <1, 4, 12, 13, 13, 25> (<4>, inducible by yeast extract [6]; <12>, constitutive enzyme [17]; <13>, probably responsible for hydrolysis of both fucose-containing and fucose-depleted substrates [35]; <1>, the enzyme is responsible for the formation of most oligosaccharides excreted in the urine of patients with congenital exoglycosidase deficiencies and also explains why large amounts of glycopeptides are excreted in the urine of fucosidosis patients [45]; <13>, in cytosol the enzyme contributes to the production of free oligosaccharides with one reducing end N-acetylglucosamine residue in cooperation with neutral α-mannosidase [50]; <25>, inhibits response of human lymphocytes to mitogens and interferes with production of antibodies in mice [51]; <3, 4, 7, 8, 12, 22, 24, 26, 27, 28>, the enzyme is probably devoted to exogeneous functions, such as degrading macromolecules for feeding purposes [55]; <25>, the enzyme functions as a virulence determinant which interferes with the host cellular immune response [56]) [6, 17, 35, 45, 50, 51, 56]

Substrates and products

S (GlcNAc)$_2$(Man)$_5$(GlcNAc)$_2$Asn + H_2O <9> [15]
P ?
S (GlcNAc)$_2$(Man)$_6$(GlcNAc)$_2$ + H_2O <11> [28]
P GlcNAc + (GlcNAc)1(Man)$_6$(GlcNAc)$_2$ <11> [28]
S (GlcNAc)$_3$ + H_2O <8> [39]
P GlcNAc + (GlcNAc)$_2$ <8> [39]
S (GlcNAc)$_4$ + H_2O <8> [39]
P (GlcNAc)$_2$ <8> [39]
S (Man)$_3$(GlcNAc)$_2$-N-Asn-dansyl + H_2O <27> (<27>, endo-β-N-acetylglucosaminidase PII, no activity with endo-β-N-acetylglucosaminidase PII [57]) [57]
P ?
S (Man)$_3$(GlcNAc)$_2$-N-dansyl-Asn + H_2O <20> [36]
P GlcNAc-N-dansyl-Asn + (Man)$_3$-GlcNAc <20> [36]

S $(Man)_3(GlcNAc)_2Asn + H_2O$ <20> (<20>, di-N-acetylchitobiosyl linkage is hydrolyzed faster than the linkage in the glycopeptides $(Man)_5(GlcNAc)_3Asn$ and $(Man)_6(GlcNAc)_2Asn$ by enzyme form F-I. No activity with the enzyme form F-II [36]) [36]
P $(Man)_3GlcNAc$ + GlcNAc-Asn
S $(Man)_3(GlcNAc)_2Asn$-Ac + H_2O <20> (<20>, di-N-acetylchitobiosyl linkage is hydrolyzed faster than the linkage in the glycopeptides $(Man)_5(GlcNAc)_3Asn$ and $(Man)_6(GlcNAc)_2Asn$ by enzyme form F-I. No activity with the enzyme form F-II [44]) [44]
P $(Man)_3GlcNAc$ + GlcNAc-Asn-Ac
S $(Man)_3(Xyl)(GlcNAc)_2$-pyridylamino + H_2O <15, 16> [20]
P ?
S $(Man)_5(GlcNAc)_2$-N-acetyl-Asn + H_2O <14, 20> [15, 19, 37, 44]
P $(Man)_5GlcNAc$ + GlcNAc-N-acetyl-Asn <14, 20> [19, 37]
S $(Man)_5(GlcNAc)_2$-N-dansyl-Asn + H_2O <20> [36]
P GlcNAc-N-dansyl-Asn + $(Man)_5$-GlcNAc <20> [36]
S $(Man)_5(GlcNAc)_2$-pyridylamino + H_2O <15, 16> [20]
P ?
S $(Man)_5(GlcNAc)_2Asn + H_2O$ <9, 14, 20> (<20>, di-N-acetylchitobiosyl linkage is hydrolyzed faster than the linkage in the glycopeptides $(Man)_5(GlcNAc)_3Asn$ and $(Man)_6(GlcNAc)_2Asn$ by enzyme form F-I [36]) [15, 19, 36]
P $(Man)_5GlcNAc$ + GlcNAc-Asn
S $(Man)_5(GlcNAc)_2Asn$-Ac + H_2O <20> (<20>, di-N-acetylchitobiosyl linkage is hydrolyzed faster than the linkage in the glycopeptides $(Man)_5(GlcNAc)_3Asn$ and $(Man)_6(GlcNAc)_2Asn$ by enzyme form F-I [44]) [44]
P $(Man)_5GlcNAc$ + GlcNAc-Asn-Ac
S $(Man)_5GlcNAc_2$-pyridylamino + H_2O <30> [60]
P ?
S $(Man)_6(GlcNAc)_2$-Asn + glucose <24> [49]
P $(Man)_6GlcNAcGlc$ + ? <24> [49]
S $(Man)_6(GlcNAc)_2$-pyridylamino + H_2O <15, 16> [20]
P ?
S $(Man)_6(GlcNAc)_2Asn + H_2O$ <1, 12, 20> (<20>, di-N-acetylchitobiosyl linkage is hydrolyzed faster than the linkage in the glycopeptides $(Man)_5(GlcNAc)_3Asn$ and $(Man)_6(GlcNAc)_2Asn$ by enzyme form F-I [36]) [1, 33, 36, 37]
P $(Man)_6GlcNAc$ + GlcNAc-Asn <20> [37]
S $(Man)_6(GlcNAc)_2Asn$-Ac + H_2O <20> (<20>, di-N-acetylchitobiosyl linkage is hydrolyzed faster than the linkage in the glycopeptides $(Man)_5(GlcNAc)_3Asn$ and $(Man)_6(GlcNAc)_2Asn$ by enzyme form F-I [44]) [44]
P $(Man)_6GlcNAc$ + GlcNAc-Asn-Ac
S $(Man)_6(GlcNAc)_2$-pyridylamino + H_2O <30> [60]
P ?
S $(Man)_7(GlcNAc)_2$-pyridylamino + H_2O <15, 16> [20]

P ?
S (Man)$_8$(GlcNAc)$_2$-pyridylamino + H_2O <15, 16> [20]
P ?
S (Man)$_9$(GlcNAc)$_2$-pyridylamino + H_2O <15, 16, 30> [20, 60]
P ?
S 4-methylumbelliferyl di-N-acetyl-β-chitobioside + H_2O <21> [41]
P ?
S Asn(GlcNAc)$_2$(Man)$_3$ + H_2O <8, 13> [16, 18, 35]
P ?
S Asn(GlcNAc)$_2$(Man)$_3$(Fuc) + H_2O <13> [18]
P ?
S Asn(GlcNAc)$_2$-(Man)$_5$ + H_2O <8, 11, 13, 14, 18> [16, 18, 23, 28, 30, 35]
P AsnGlcNAc + Man$_5$(GlcNAc) <8, 14> [16, 30]
S Asn(GlcNAc)$_2$-(Man)$_6$ + H_2O <8, 11, 13, 18> [16, 18, 23, 28, 35]
P AsnGlcNAc + (GlcNAc)-(Man)$_6$ <8, 11> [16, 28]
S Asn(GlcNAc)$_2$-(Man)$_6$(GlcNAc)$_2$ + H_2O <8, 11> [16, 28]
P AsnGlcNAc + (GlcNAC)$_2$(Man)$_6$(GlcNAc)$_2$ <8, 11> [16, 28]
S Asn(GlcNAc)$_4$(Man)$_6$ + H_2O <11> [28]
P ?
S Asn-Tyr-Asn(GlcNAc)$_2$(Man)$_3$(GlcNAc)$_2$-(Gal)2-Lys + H_2O <1> [54]
P Asn-Tyr-Asn(GlcNAc)Lys + (Gal)$_2$(GlcNAc)$_2$(Man)$_3$(GlcNAc) <1> [54]
S GlcNAc(Man)$_5$(GlcNAc)$_2$-pyridylamino <15, 16> [20]
P ?
S Man(GlcNAc)$_2$Asn + H_2O <8> [39]
P ManGlcNAc + GlcNAcAsn
S Man(GlcNAc)$_2$Asn-dansyl + H_2O <8> [39]
P ManGlcNAc + GlcNAcAsn-dansyl
S Manα1 $\rightarrow$ 6(Manα1 $\rightarrow$ 3)Manα1 $\rightarrow$ 6(Manα1 $\rightarrow$ 2Manα1 $\rightarrow$ 3)Manβ1 $\rightarrow$ 4GlcNAcβ1 $\rightarrow$ 4GlcNAcAsn + H_2O <12> (<12>, endo-β-N-acetylglucosaminidase CII [17, 33]) [17, 33]
P GlcNAc-]N-acetylAsn + ? <12> [17]
S Manα1 $\rightarrow$ 6(Manα1 $\rightarrow$ 3)Manβ1 $\rightarrow$ 4GlcNAcβ1 $\rightarrow$ 4(Fucα1 $\rightarrow$ 6)GlcNAc $\rightarrow$ N-acetylAsn + H_2O <12> (<12>, endo-β-N-acetylglucosaminidase CI [17, 33]) [17, 33]
P Manα1 $\rightarrow$ 6(Manα1 $\rightarrow$ 3)Manβ1 $\rightarrow$ 4GlcNAc + Fucα1-6GlcNAc-N-acetyl-Asn <12> [17]
S Ricinus lectin + H_2O <4> [11]
P ?
S Taka-amylase A glycopeptide + H_2O <10> (<10>, dansyl derivative of taka-amylase A glycopeptide [14]) [14]
P Additional information <10> (<10>, product contains N-acetylglucosaminitol and mannose in the molar ratio 1.0:5.2, but no N-acetylglucosamine or mannitol [14]) [14]
S acetyl-Asn(GlcNAc)$_2$(Man)$_6$ + H_2O <9, 19> [29]
P acetyl-AsnGlcNAc <9, 19> [29]
S asialo-N-acetyl-lactosaminic-type glycoasparagine + H_2O <9> [42]
P GlcNAc-Asn + ? <9> [42]

S dansyl-Asn-$(GlcNAc)_2(Man)_6$ + H_2O <4, 13> [13, 18, 35]
P dansyl-Asn-GlcNAc + ? <4, 13> [13, 18]
S dansyl-Asn-$(GlcNAc)_2(Man)_6(GlcNAc)_2$ + H_2O <11> [28]
P ?
S dansyl-Asn-$(GlcNAc)_4(Man)_6$ + H_2O <8, 11> [16, 28]
P dansyl-Asn-GlcNAc + ? <8> [16]
S dolichyl pyrophosphoryl oligosaccharides + H_2O <11> (<11>, both glucosylated and unglucosylated polymannose oligosaccharides are released by the enzyme through cleavage of the di-N-acetylchitobiose squence [25]) [25]
P ?
S fluoresceinyl 2-acetamido-2-deoxy-β-D-glucopyranoside + H_2O <21> [41]
P fluorescein + ? <21> [41]
S fluoresceinyl di-N-acetyl-β-chitobioside + H_2O <21> [41]
P fluorescein + ? <21> [41]
S glycopeptide + H_2O <1, 5, 7, 12, 13, 14, 18, 22, 23, 29> (<1>, asparagine oligosaccharides of transferrin and fetuin [54]; <1>, endo-β-N-acetylglucosaminidase 1: high mannose-type asparaginylglycopeptide is hydrolyzed efficiently, little activity towards complex-type glycopeptides. endo-β-N-acetylglucosaminidase 2: hydrolyzes both the high mannose-type and the complex-type oligosaccharide with chitobiosyl group at the reducing end and without the Asn [1]; <7>, hydrolysis of the following oligosaccharide structures: ovalbumin-high mannose, ovalbumin-hybrid containing GlcNAc β1,4-Man, ovalbumin-hybrid without GlcNAc β1,4-Man, ovomucoid hybrids, biantennary complex, triantennary complex and tetraantennary complex oligosaccharide structure [8]; <12>, two endo-β-N-acetylglucosaminidase CI and CII with different substrate specificity [17]; <11>, generally hydrolyzes glycopeptides of the mannose-N-acetylglucosamine unit [31]; <14>, generally hydrolyzes side-chain-free glycopeptides of the complex heteropolysaccharide unit [31]; <11>, release of Asn-GlcNAc-Fuc-containing glycopeptides from exoglycosidase-treated acidic IgM glycopeptides, limited capacity to hydrolyze ovalbumin glycopeptides larger than Asn$(GlcNAc)_2(Man)_5$ [34]; <14>, hydrolysis of Asn$(GlcNAc)_2(Man)_5$ and larger neutral oligosaccharides, no hydrolysis of exoglycosidase-treated acidic IgM glycopeptides [34]; <13>, glycopeptide from immunoglobulin M, (aa)x-Asn$(GlcNAC)_2(Man)_3(Fuc)$ [35]; <11, 12, 14, 18>, specificity [23, 24, 30, 31, 33, 34]; <1>, glycoasparagines derived from ovalbumin and asialoglycoasparagines derived from human serotransferrin [43]; <22>, high activity on dansyl transferrin glycopeptide, can also hydrolyze dansyl asialotransferrin glycopeptide and dansyl transferrin glycopeptide containing complex type sugar chains [46]; <5>, high activity on dansyl ovalbumin glycopeptide, can also act on dansyl asialotransferrin glycopeptide containing biantennary type sugar chains [47]; <23>, high activity on oligomannoside-type glycoasparagines and glycoproteins, can also hydrolyze hybrid-type and complex-type glycoasparagines [48]; <29>, enzyme endo-BI hydrolyzes the pyridylamino-trimannosyl core and pyridylamino-agalacto biantennary complex-type oli-

gosaccharides, enzyme form endo-BII acts specifically on the high-mannose-type oligosaccharides $(Man)_{5-9}(GlcNAc)_2$ present in ovomucoid in small amounts [58]) [1, 8, 17, 19, 23, 24, 30, 31, 33, 34, 35, 43, 46, 47, 48, 54, 58]

P ?

S glycoprotein + H_2O <1, 5, 7, 11, 14, 29, 30> (<29>, endo-BI hydrolyzes oligosaccharides of both hen ovalbumin and ovomucoid. Endo-BII can act only on oligosaccharides of hen ovalbumin and shows almost no activity towards hen ovomucoid [58]; <11, 14>, specificity [31]; <5>, hydrolysis of asparagine-linked oligosaccharides of glycoproteins at the di-N-acetylchitobiosyl moiety of complex oligosaccharides, high-mannose oligosaccharides and hybrid structure oligosachharides [5]; <7>, high-mannose and complex glycoproteins linked through asparagine to the protein backbone, cleavage occurs via hydrolysis of the glycosidic bond of the N,N'-diacetylchiotobiose core structure adjacent to asparagine. Glycoproteins of retrovirus, lymphocytic choriomeningitis virus, Pichinde virus, and HLA-A and HLA-B antigens are cleaved in presence of nonionic detergent. Ovalbumin, fetuin, bromelain, ovomucoid, α1-acid glycoprotein, immunoglobulin G and influenza virus hemagglutinin are susceptible only after reduction and alkylation or in presence of 1% 2-mercaptoethanol [26]; <11>, release of intact neutral oligosaccharides that is attached to the protein by a glycosyl asparagine bond from sulfitolyzed ovalbumin, bovine pancreatic deoxyribonuclease A, ribonuclease B, and invertase from Saccharomyces cerevisiae. Release from neutral chain only from proteins that contain both neutral and acidic oligosaccharides such as thyroglobulin, immunoglobulin M, and porcine ribonuclease [27]; <1>, cleaves the N,N'-diacetylchitobiose moiety of the sugar chain of human salivary α-amylase family, human transferrin and calf fetuin [54]) [5, 26, 27, 31, 54, 58, 60]

P ?

S ovalbumin + H_2O <1, 4> (<4>, native ovalbumin [4, 11, 13]; <4>, the enzyme hydrolyzes all five heterogenous ovalbumin glycopeptides, the rate of hydrolysis decreases as the size of the sugar moiety increases [11]) [4, 10, 11, 13]

P oligosaccharides + ? <1, 4> [10, 11]

S peptidoglycan + H_2O <2, 17> (<2>, no attack of intact cell walls of Staphylococcus simulans 22, hydrolysis of cell walls of Micrococcus luteus and soluble peptidoglycan chains of Staphylococcus simulans 22 [2]; <17>, Micrococcus luteus cell wall and Bacillus subtilis cell walls [21]; <21>, no hydrolyis of Micrococcus lysodeikticus cell walls [41]) [2, 21]

P ?

S ribonuclease B + H_2O <4> (<4>, bovine pancreatic ribonuclease B [13]) [11, 13]

P ?

S transferrin + H_2O <5> [5]

P complex oligosaccharides + ? <5> [5]

S yeast carboxypeptidase + H_2O <4> [11]

P ?

S yeast invertase + H_2O <4> [13]

P ?

S Additional information <1, 3, 6, 8, 9, 10, 11, 13, 14, 15, 16, 17, 27> (<9, 11, 13, 14, 15, 16, 18>, specificity [20, 23, 24, 25, 45, 50, 60]; <14>, endo-β-N-acetylglucosaminidase D does not require the presence of a free hydroxyl group at the C-4 position of the α-mannosyl residue of the trisaccharide glycan Manα → 3Manβ1 → 4GlcNAcβ1 → [24]; <3>, requires choline in the teichoic acid of the cell wall substrate for the catalytic activity [3]; <6>, the enzyme is specific for high mannose glycans, no hydrolysis of either hybrid or complex glycans [7]; <9>, completely inactive towards conjugate glycans [12]; <9>, specific for glycans with a terminal chitobiose residue [12]; <10>, high-mannose-type glycopeptides are rapidly hydrolyzed, hybrid-type glycopeptides are good substrates, complex-type glycopeptides as well as smaller glycopeptides prepared by partial glycosidase digestion of complex-type glycopeptides are resistant to the enzyme [14]; <8>, no hydrolysis of dansyl-Asn$(GlcNAc)_2$Man, the presence of fucose attached to the asparagine-proximal core glucosamine renders the di-N-acetylchitobiosyl moiety an inactive substrate. As the number of mannose residues is increased, the rate of hydrolysis increases to a maximum at five residues. Thereafter enzyme activity declines, but the rate is still substantial on mannosyl chains containing 54 residues [16]; <12>, two endo-β-N-acetylglucosaminidase CI and CII with different substrate specificity [17]; <14>, structural basis of the specificity of endo-β-N-acetylglucosaminidase [19]; <15, 16>, description of substrate specificity, enzyme barely hydrolyzes xylose-containing N-glycans $(Man)_3(Xyl)1(GlcNAc)_2$-pyridylamino and $(Man)_3(Fuc)(Xyl)(GlcNAc)_2$-pyridylamino [20]; <17>, no exo-activity [21]; <11>, the rate of hydrolysis of lipid-linked $(Glc)3(Man)_9(GlcNAc)_2$ is substantially slower than that of the same oligosaccharide attached to asparagine in a peptide sequence [25]; <14>, in the presence of exoglycosidases removing peripheral sugars, the enzyme acts on serum glycoproteins such as transferrin and fetuin [38]; <20>, enzyme type F-I hydrolyzes the tri-mannosyl derivatives di-N-acetylglycosaminylasparagine faster than the penta-mannosyl compounds and the hexamannosyl compounds. Enzyme type F-II hydrolyzes the penta-mannosyl derivatives and hexa-mannosyl derivatives, but not the tri-mannosyl compounds [44]; <9>, cleaves the N,N'-diacetylchitobiose moiety of Manα1 → 6(manα1 → 3)Manβ1 → 4R in which R represents either GlcNAc → Asn or N-acetylglucosamine. The enzyme can hardly act on the sugar chains with Fucα1 → 3 or 6GlcNAc → Asn or N-acetylglucosaminitol as their R residues. The sugar chains substituted at C-3 and C-6 positions of the Manα1 → 6 residue and at C-2 position of the Manα1 → 3 residue by other sugars are also cleaved [45]; <13>, the enzyme hydrolyzes oligomannose-type sugar chains faster than complex-type chains and hybrid-type chains. Sugar chains containing the manα1-2Manα1-3Manβ1-4GlcNAcβ1-GlcNAc structure are good substrates [50]; <1>, specific for complex type sugar chains and can release the sugar chains from native glycoproteins and glycopeptides regardless of the existence of a Fuc residue on the proximal GlcNAc of the N,N'-diacetylchitobiose core of their sugar chains [54]; <27>, endo-β-N-acetylglucosaminidase PI hydrolyzes high-mannose type oligosaccharides GP-IV and GP-V faster than the hybrid type FP-I, GP-II, and GP-III. Endo-β-N-acetylglucosaminidase PII hydrolyzes high mannose and hybrid type oligosaccharides [57]) [3, 7, 12, 14, 16, 19, 20, 21, 24, 25, 38, 45, 50, 54, 57]

P ?

Inhibitors

Ag^{2+} <5> (<5>, 1 mM, significant inactivation [47]) [47]
Co^{2+} <30> (<30>, 10 mM, 75% inhibition [60]) [60]
Cu^{2+} <15, 30> (<15>, 6 mM, 80% inhibition [20]; <30>, 10 mM, 75% inhibition [60]) [20, 60]
EDTA <12> (<12>, 10 mM, 60% inhibition of endo-β-N-acetylglucosaminidase CI, 22% inhibition of endo-β-N-acetylglucosaminidase CII [17]; <12>, slight inhibition [33]) [17, 33]
Fe^{3+} <23, 30> (<23>, 10 mM, about 50% inhibition [48]; <30>, 10 mM, 75% inhibition [60]) [48, 60]
Hg^{2+} <4, 5, 11, 13, 22, 24> (<22>, 1 mM $HgCl_2$, about 50% inhibition [46]; <5>, 1 mM, significant inactivation [47]; <4>, 1 mM $HgCl_2$, 85% loss of activity [13]; <11>, 1 mM $HgCl_2$, 60% loss of activity [28]) [13, 28, 46, 47, 49, 50]
N-acetylglucosamino(1 → 5)lactone <14> [30]
PCMB <5, 11, 27> (<27>, 1 mM, 84% inhibition [57]; <11>, 1 mM, 30% loss of activity [28]; <5>, strong [47]) [28, 47, 57]
SDS <7, 8, 11> (<11>, 0.1%, strong inhibition [25]; <8>, 0.2%, complete inactivation [39]) [25, 26, 39]
Zn^{2+} <15, 30> (<15>, 6 mM, complete inhibition [20]; <30>, 10 mM, 50% inhibition [60]) [20, 60]
acetate <1> (<1>, endo-β-N-acetylglucosaminidase 2 [1]) [1]
α-D-mannosides <14> [19]
choline <3> (<3>, 2%, activates [3]) [3]
citric acid <23> [48]
formamide <23> [48]
galactose <14> (<14>, 0.5 M, 23% inhibition [32]) [32]
glycoprotein <9> (<9>, glycoproteins present in the rat liver homogenate [15]) [15]
lipoteichoic acid <3, 17> (<3>, from Pneumococcus sp. [3]) [3, 21]
mannose <14, 11> (<11>, 0.5 mM 20% inhibition [32]; <14>, 0.5 M, 94% inhibition [32]) [19, 32]
methyl α-D-mannoside <14> (<14>, 0.05 M, 53% inhibition [32]; <11>, 0.5 M, 10% inhibition [32]) [32]
p-chloromercuriphenyl sulfonate <12> [17, 33]
p-nitrophenyl α-D-mannoside <14> (<14>, 0.01 M, 70% inhibition [32]; <14>, 0.01 M, no inhibition [32]) [32]

urea <23, 30> (<23>, 3 M [48]; <30>, 50% inhibition by 2 M, 65% inhibition at 4 M [60]) [48, 60]
yeast mannan <11> (<11>, 15 mg/ml, 62% inhibition [32]; <14>, no inhibition [32]) [32]

Activating compounds

2-mercaptoethanol <7> (<7>, ovalbumin, fetuin, bromelain, ovomucoid, α_1-acid glycoprotein, immunoglobulin G and influenza virus hemagglutinin are susceptible only after reduction and alkylation or in presence of 1% 2-mercaptoethanol [26]) [26]
L-Cys <10> (<10>, 1 mM, 18% stimulation [14]) [14]
Pep 5 <2> (<2>, cationic peptide antibiotic, activates hydrolysis of cell walls of Micrococcus luteus, no influence of hydrolysis of soluble peptidoglycan [2]) [2]
Triton X-100 <11> (<11>, maximal stimulation at 0.03%, or small amounts of other detergents [25]) [25]
choline <3> (<3>, 0.1%, activates [3]) [3]
dithiothreitol <1, 10> (<10>, 1 mM, 18% stimulation [14]; <1>, optimal activation at 10 mM [43]) [14, 43]
nisin <2> (<2>, activates hydrolysis of cell walls of Micrococcus luteus, no influence of hydrolysis of soluble peptidoglycan [2]) [2]
nonionic detergent <7> (<7>, glycoproteins of retrovirus, lymphocytic choriomeningitis virus; Pichinde virus, and HLA-A and HLA-B antigens are cleaved in presence of nonionic detergent [26]) [26]
poly(L-Lys) <2> (<2>, activates hydrolysis of cell walls of Micrococcus luteus, no influence of hydrolysis of soluble peptidoglycan [2]) [2]

Metals, ions

Li^+ <17> (<17>, maximal activation near 0.2 M LiCl with Micrococcus luteus cell walls as substrate, sharp optimal activation at 0.1 M with Bacillus subtilis cell walls as substrate [21]) [21]
Mg^{2+} <2, 17> (<2>, activates [2]; <17>, slight activation [21]) [2, 21]
Na^+ <2> (<2>, activates [2]) [2]

Turnover number (min^{-1})

Additional information <3, 17> [3, 21]

Specific activity (U/mg)

0.24 <13> [50]
0.299 <12> (<12>, endo-β-N-acetylglucosaminidase CI [33]) [33]
0.304 <12> (<12>, endo-β-N-acetylglucosaminidase CII [33]) [33]
0.752 <11> [31]
4.67 <14> [30]
12 <11> [28]
13.4 <14> [19]
13.4 <14> [38]
39.9 <4> [13]
400 <30> [60]

Additional information <1, 5, 6, 7, 8, 10, 12, 13, 15, 16, 17, 22, 23, 29> [1, 7, 8, 9, 14, 17, 18, 20, 21, 35, 39, 46, 47, 48, 54, 58]

K_m-Value (mM)

0.00625 <21> (4-methylumbelliferyl di-N-acetyl-β-chitobioside) [41]
0.025 <13> ((Man)$_6$(GlcNAc)$_2$-pyridylamidated) [50]
0.032 <12, 12> (Manα1 $\rightarrow$ 6(Manα1 $\rightarrow$ 3)Manβ1 $\rightarrow$ 4GlcNAcβ1 $\rightarrow$ 4(Fucα1 $\rightarrow$ 6)GlcNAc $\rightarrow$ N-acetylAsn, <12>, endo-β-N-acetylglucosaminidase CI [17,33]) [17, 33]
0.05 <12> (Manα1 $\rightarrow$ 6(Manα1 $\rightarrow$ 3)Manα1 $\rightarrow$ 6(Manα1 $\rightarrow$ 2Manα1 $\rightarrow$ 3)Manβ1 $\rightarrow$ 4GlcNAcβ1 $\rightarrow$ 4GlcNAcAsn, <12>, endo-β-N-acetylglucosaminidase CII [17, 33]) [17, 33]
0.0868 <10> (dansyl derivative of Taka-amylase A glycopeptide) [14]
0.2 <14> ((Man)$_5$(GlcNAc)$_2$Asn-acetyl) [19, 38]
0.25 <30> ((Man)$_6$(GlcNAc)$_2$-pyridylamino) [60]
0.25 <14> (Asn(GlcNAc)$_2$-(Man)$_5$) [30]
0.3 <4> (dansyl-Asn-(GlcNAc)$_2$(Man)$_6$) [13]
0.3 <8, 11> (dansyl-Asn-(GlcNAc)$_4$(Man)$_6$) [16, 28]
0.32 <30> ((Man)$_9$(GlcNAc)$_2$-pyridylamino) [60]
0.4 <30> ((Man)$_5$(GlcNAc)$_2$-pyridylamino) [60]
0.68 <22> (dansyl asialotransferrin glycopeptide) [46]
0.7 <8> (Man(GlcNAc)$_2$Asn-dansyl) [39]
2 <5> (dansyl asialotransferrin glycopeptide) [47]
2.56 <4> (ovalbumin) [4]
Additional information <3> [3]

pH-Optimum

3-3.5 <1, 22> (<1>, endo-β-N-acetylglucosaminidase 2 [1]) [1, 46, 55]
3-5 <23> [48]
3.5 <9> [12]
4 <23> [48]
4-4.5 <8> (<8>, hydrolysis of Man(GlcNAc)$_2$Asn-dansyl [39]; <8>, ENGase L [55]) [39, 55]
4-4.8 <10> (<10>, hydrolysis of dansyl derivative of Taka-amylase A glycopeptide [14]) [14]
4-5 <23, 28> [55]
4-6 <7> [8, 55]
4.5 <1, 17> (<1>, endo-β-N-acetylglucosaminidase 1 [1]; <17>, Bacillus subtilis cell walls [21]) [1, 21]
4.5-5.5 <17> (<17>, Micrococcus luteus cell walls as substrate, in 50 mM acetic acid/acetate buffer or succinate/NaOH buffer [21]) [21]
4.5-7 <18> [23]
5 <4, 17> (<4>, hydrolysis of ovalbumin [4]) [4, 21]
5-6 <4, 8, 11, 27> (<8>, hydrolysis of dansyl-Asn-(GlcNAc)$_4$(Man)$_6$ [16]; <11> hydrolysis of dansyl-Asn-(GlcNAc)$_4$(Man)$_6$ [28]; <27>, ENGase PI [55, 57]) [13, 16, 28, 55, 57]
5-9 <24> [49]
5-11 <24> [55]

5.4 <20> (<20>, hydrolysis of (Man)$_5$(GlcNAc)$_2$Asn-acetyl, enzyme type F-II [44]) [44]
5.5 <8, 13> (<13>, hydrolysis of dansyl-Asn(GlcNAc)$_2$-(Man)$_6$ [18, 35]; <13>, hydrolysis of peptide-Asn(GlcNAc)$_2$(Man)$_3$(Fuc) [18]; <8>, ENGase H [55]) [18, 35, 55]
5.5-6 <29> (<29>, enzyme form endo-BI and endo-BII [58]) [58]
5.9 <20> (<20>, hydrolysis of (Man)$_5$(GlcNAc)$_2$Asn-acetyl, enzyme type F-I [44]) [44]
6 <16> (<16>, hydrolysis of (Man)$_6$(GlcNAc)$_2$-pyridylamino [20]) [20]
6-6.5 <5, 12> (<12>, ENGase CI [55]) [47, 55]
6-8 <27> (<27>, ENGase PII [55,57]) [55, 57]
6.5 <1, 2, 3, 12, 13, 14, 16> (<2>, Micrococcus luteus cell walls as substrate [2]; <3>, in 20 mM phosphate buffer, at 30°C [3]; <12>, endo-β-N-acetylglucosaminidase CI, citrate-phosphate buffer [17]; <14>, hydrolysis of (Man)$_5$(GlcNAc)$_2$Asn-acetyl [19]; <15>, hydrolysis of (Man)$_6$(GlcNAc)$_2$-pyridylamino [20]; <14>, hydrolysis of Asn(GlcNAc)$_2$-(Man)$_5$ [30]; <12>, endo-β-N-acetylglucosaminidase CI [33]) [2, 3, 17, 19, 20, 30, 33, 43, 50, 55]
6.5-7 <26> [55]
7 <12, 19, 30> (<12>, endo-β-N-acetylglucosaminidase CII, in Tris buffer and citrate-phosphate buffer. endo-β-N-acetylglucosaminidase CI in Tris buffer [17]; <12>, endo-β-N-acetylglucosaminidase CI [33]; <12>, ENGase CII [55]; <30>, hydrolysis of (Man)$_6$(GlcNAc)$_2$-pyridylamino [60]) [17, 29, 33, 55, 60]

pH-Range

2.3-5.5 <9> (<9>, pH 2.3: about 20% of maximal activity, pH 5.5: about 25% of maximal activity [12]) [12]
2.5-4.5 <1> (<1>, about 70% of maximal activity at pH 2.5 and at pH 4.5, endo-β-N-acetylglucosaminidase 2 [1]) [1]
2.5-5 <22> (<22>, about 30% of maximal activity at pH 2.5 and at pH 5.0 [46]) [46]
2.5-8 <4> (<4>, pH 2.5: about 20% of maximal activity, pH 8.0: about 30% of maximal activity [13]) [13]
3-8 <23> (<23>, pH 3.0-5.0: optimum, pH 8.0: 60% of maximal activity [48]) [48]
3.3-6.3 <4> (<4>, pH 3.3: about 60% of maximal activity, pH 6.3: about 50% of maximal activity [4]) [4]
4.5-7 <1> (<1>, pH 4.5: about 60% of maximal activity, pH 7.0: about 50% of maximal activity, endo-β-N-acetylglucosaminidase 1 [1]) [1]
4.5-8 <18> (<18>, pH 4.5: about 75% of maximal activity, pH 8.0: about 40% of maximal activity [23]) [23]
4.5-8.5 <13> (<13>, pH 4.5: about 25% of maximal activity, pH 8.5: about 30% of maximal activity [50]) [50]
5-8.5 <30> (<30>, about 55% of maximal activity at pH 5.0 and at pH 8.5 [60]) [60]
5.3-8.2 <14> (<14>, about 50% of maximal activity at pH 5.3 and at pH 8.2 [30]) [30]

5.5-9.5 <13> (<13>, pH 5.5: maximal activity, pH 9.5: about 40% of maximal activity, hydrolysis of dansyl-Asn$(GlcNAc)_2(Man)_6$ [35]) [35]
6-7.5 <30> (<30>, 80% of maximal activity at pH 6.0 and at pH 7.5 [60]) [60]

Temperature optimum (°C)
35-40 <18> [23]
40 <15, 16, 30> (<15,16>, hydrolysis of $(Man)_6(GlcNAc)_2$-pyridylamino [20]) [20, 60]
45-50 <5> [47]
50 <22> [46]

Temperature range (°C)
25-60 <30> (<30>, 25°C: about 80% of maximal activity, 60°C: about 45% of maximal activity [60]) [60]

4 Enzyme Structure

Molecular weight
17500 <20> (<20>, enzyme type F¯II, gel filtration [44]) [44]
23500 <11> (<11>, gel filtration [28]) [28]
27000 <4, 8> (<4>, gel filtration [13]; <8>, equilibrium sedimentation analysis [16]) [13, 16]
27200 <11> (<11>, equilibrium sedimentation [28]) [28]
31000 <10> (<10>, gel filtration [14]) [14]
35000 <22> (<22>, gel filtration [46]) [46]
41670 <30> (<30>, MALDI-TOF mass spectrometry [60]) [60]
46000 <8> (<8>, gel filtration [39]) [39]
49500 <8> (<8>, equilibrium sedimentation [39]) [39]
52000 <20> (<20>, enzyme type F-I, gel filtration [44]) [44]
54000 <13> (<13>, gel filtration [50]) [50]
56000 <13> (<13>, gel filtration [18,35]) [18, 35]
62000 <15, 16> (<15, 16>, gel filtration [20]) [20]
95000 <5> (<5>, gel filtration [47]) [47]
150000 <29> (<29>, enzyme form endo-BII, non-denaturing PAGE [58]) [58]
190000 <17> (<17>, gel filtration [21]) [21]
210000 <29> (<29>, enzyme form endo-BI, non-denaturing PAGE [58]) [58]
280000 <14> (<14>, gel filtration [19, 38]) [19, 38]

Subunits
? <3, 4, 7, 8, 14, 23, 24, 27> (<8, 23>, x * 27000, SDS-PAGE [48,55]; <4>, x * 30000, SDS-PAGE [55]; <7>, x * 32000, ENGase F1, SDS-PAGE [55]; <7>, x * 33000, ENGase F3, SDS-PAGE [55]; <7>, x * 39500, ENGase F2, SDS-PAGE [55]; <3>, x * 64000, SDS-PAGE [3]; <24>, x * 72000, SDS-PAGE [49]; <27>, x * 94000, endo-β-N-acetylglucosaminidase PI, SDS-PAGE [57]; <27>, x * 106000, endo-β-N-acetylglucosaminidase PI, SDS-PAGE [57]; <14>, x * 150000, SDS-PAGE [19]) [3, 19, 48, 49, 55, 57]

dimer <5, 17> (<5>, 2 * 55000, SDS-PAGE [47]; <17>, 2 * 90000, SDS-PAGE [21]) [21, 47]
monomer <4, 8, 13, 22, 30> (<8>, 1 * 27000, SDS-PAGE [16]; <4>, 1 * 30000, SDS-PAGE [13]; <22>, 1 * 35000, SDS-PAGE [46]; <30>, 1 * 42000, SDS-PAGE [60]; <13>, 1 * 52000, SDS-PAGE in presence of 2-mercaptoethanol [50]; <8>, 1 * 52000-54000, SDS-PAGE [39]) [13, 16, 39, 46, 50, 60]

Posttranslational modification

no modification <9, 11> (<9, 11>, contains no carbohydrate [15, 28]) [15, 28]
proteolytic modification <8, 25> (<8>, a proenzyme or more probably a prepro enzyme may be the primary product of translation. The 42 or 44 residue leader sequence is partially removed by E. coli [22]; <25>, autolysin atl gene product is a bifunctional protein that has an amidase domain and an endo-β-N-acetylglucosaminidase domain which must undergo proteolytic processing to generate two extracellular lytic enzymes [53]) [22, 53]

5 Isolation/Preparation/Mutation/Application

Source/tissue

brain <1> [43]
culture medium <4, 5, 12, 14, 23, 24, 27, 29> (<24>, recombinant enzyme [49]) [5, 13, 17, 19, 30, 33, 47, 48, 49, 57, 58]
fruit <16> [20]
kidney <1, 9> [1, 10, 29, 40, 43]
latex <20> [36, 44]
liver <1, 9, 19> [12, 15, 29, 40, 42, 43, 45]
oviduct <13> [18, 35, 50]
saliva <1> (<1>, epithelial cells peeling from the oral cavity epithelium into saliva [54]) [54]
seed <15, 30> [20, 60]
serum <21> [41]
spleen <1, 9, 19> [29, 43]

Localization

cell membrane <1> [54]
cytoplasm <9> [15]
cytosol <9, 13> [40, 42, 50]
extracellular <4, 5, 8, 12, 14, 25> [5, 13, 17, 19, 51, 53, 59]
intracellular <8> (<8>, recombinant enzyme produced in E. coli appears both in the periplasmic space and in the cell [22]) [22]
lysosome <9> (<9>, no evidence for a lysosomal enzyme form [12]) [12]
periplasm <24> (<24>, recombinant enzyme [49]) [49]
periplasmic space <8> (<8>, recombinant enzyme produced in E. coli appears both in the periplasmic space and in the cell [22]) [22]
soluble <18> (<18>, associated with [23]) [23]

Purification

<1> (endo-β-N-acetylglucosaminidase 1 and 2 [1]) [1, 54]
<3> [3]
<4> [13]
<5> (partial [5]) [5, 47]
<6> [7]
<7> [8]
<8> [9, 16, 39]
<9> (partial [15]) [15]
<10> [14]
<11> [28, 31]
<12> (endo-β-N-acetylglucosaminidase CI and CII [17, 33]) [17, 33]
<13> [35, 50]
<14> [19, 30, 38]
<15> [20]
<16> [20]
<17> [21]
<18> (partial [23]) [23]
<19> [29]
<20> (partial [36, 37]; endo-β-N-acetylglucosaminidase F-I and F-II [36, 44]) [36, 37, 44]
<21> [41]
<22> [46]
<23> [48]
<27> (endo-β-N-acetylglucosaminidase PI and PII [57]) [57]
<29> (enzyme form endo-BI and endo-BII [58]) [58]
<30> [60]

Crystallization

<8> (wild-type enzyme and mutant enzymes D130N, D130E, D130A, E132Q, E132A, and D130N/E132Q [52]; at 1.9 A resolution [59]) [52, 59]

Cloning

<8> (expression in E. coli with 150fold improvement in yield [9]; expression in E. coli [22]) [9, 22]
<24> (expression in E. coli [49]) [49]
<25> (expression in E. coli [53]) [53]

Engineering

D130 <8> (<8>, 1.25-3.14% of the activity of the wild-type enzyme, depending on assay method [52]) [52]
D130A <8> (<8>, 0.04-0.1% of the activity of the wild-type enzyme [52]) [52]
D130N <8> (<8>, 0.1-0.21% of the activity of the wild-type enzyme, depending on assay method [52]) [52]
D130N/E132Q <8> (<8>, no actvity [52]) [52]
E132A <8> (<8>, no actvity [52]) [52]

E132D <8> (<8>, 0.03-0.05% of the activity of the wild-type enzyme [52]) [52]
E132Q <8> (<8>, less than 0.05% of the activity of the wild-type enzyme [52]) [52]

Application

analysis <3, 4, 5, 7, 8, 11, 12, 22, 24, 26, 27, 28> (<5, 27>, the enzyme will be an excellent tool for clarification of the structures and functions of complex oligosaccharides in glycoproteins [5, 57]; <11>, useful in metabolic studies dealing with oligosaccharide-lipid assembly and their involvement in the N-glycosylation of proteins [25]; <7>, study of glycans and protein backbones as discrete entities and for defining the nature of the glycan-protein interface [26]; <3, 4, 7, 8, 12, 22, 24, 26, 27, 28>, essential reagent for the investigation of the structure and functions of glycoproteins [55]; <8>, commonly used reagent in glycobiology research, including the characterization of oligosaccharides in glycoproteins [59]) [5, 25, 26, 55, 57, 59]
nutrition <29> (<29>, deglycosylation of hen ovomucoid, enzyme may be useful for the preparation of allergen-free egg products [58]) [58]

6 Stability

pH-Stability

3.2-8 <20> (<20>, room temperature, 24 h, enzyme type F-II retains 80-100% of its activity [44]) [44]
4-6 <10> (<10>, in the refrigerator, stable for at least 4 months [14]) [14]
4-7 <30> (<30>, stable [60]) [60]
4-8 <23> (<23>, 4°C, 48 h, stable [48]) [48]
4.5 <11> (<11>, 37°C, rapid inactivation below [28]) [28]
4.5-8.5 <8, 11> (<8,11>, 37°C, stable between pH 4.5 and pH 8.5, rapid loss of activity below pH 4.5 [16, 28]) [16, 28]
5-7 <4> (<4>, 4°C, 2 days, stable [13]) [13]
5-8 <20> (<20>, room temperature, 24 h, enzyme type F-I is stable, unstable below pH 5 [44]) [44]
6-7 <29> (<29>, 4°C, 10 min, stable [58]) [58]
6-8 <22> (<22>, 4°C, 5 d, stable [46]) [46]
7-8 <5> (<5>, 4°C, 3 days, stable [47]) [47]

Temperature stability

25-50 <30> (<30>, stable [60]) [60]
37 <8, 11, 13> (<8>, pH 4.5-8.5, protein concentration 0.0003 mg/ml, stable. Rapid loss of activity below pH 4.5 [16]; <11>, 48 h, stable [28]; <13>, 24 h, 30% loss of activity [35]) [16, 28, 35]
40 <5, 29> (<29>, 10 min, stable up to [58]; <5>, 10 min, 10 mM phosphate buffer, pH 7.0, stable up to [47]) [47, 58]
45 <4, 22, 27> (<4>, 10 min, stable up to [13]; <22>, 10 min, 10 mM phosphate buffer, pH 7.0, stable below [46]; <27>, 10 min, 5% loss of activity, endo-β-N-acetylglucosaminidase PII [57]) [13, 46, 57]

50 <10, 18, 27, 30> (<10>, pH 5.0, 0.1 M acetate buffer, 2 h, less than 10% loss of activity [14]; <18>, 70 min, 75% loss of activity [23]; <27>, 10 min, 60% loss of activity, endo-β-N-acetylglucosaminidase PII [57]; <30>, unstable [60]) [14, 23, 57, 60]
55 <5, 27> (<5>, 10 min, 10 mM phosphate buffer, pH 7.0, 70% loss of activity [47]; <27>, 10 min, stable, endo-β-N-acetylglucosaminidase I [57]) [47, 57]
60 <10, 24> (<10>, 10 min, more than 90% loss of activity [14]; <24>, stable up to [49]) [14, 49]
80 <4> (<4>, 10 min, 65% loss of activity [13]) [13]

General stability information

<5>, activity decreases on repeated freezing [47]
<8>, 20-30% loss of initial activity is lost when the protein concentration is decreased to less than 0.02-0.03 mg/ml [39]
<8>, retains nearly 30% of its activity in 1% SDS [16]
<8>, stable to incubation with 250fold excess of either Pronase, chymotrypsin, or trypsin w/w. Slow inactivation by 1000fold excess of Pronase during 12 h incubation [16]
<8>, stable to repeated freezing and thawing [16, 39]
<11>, pronase, trypsin, and chymotrypsin at 2.5 mg per ml, when incubated with 0.005-0.010 mg of enzyme per ml, has little or no effect after 5 h at 37°C. 5 mg/ml pronase inactivates slowly during a 12 h period [28]
<11>, stable to repeated freezing and thawing or lyophilization [28]
<13>, stable to freezing at -70°C [18, 35]
<17>, trypsin, 0.01 mg/ml, pH 7.2, 50 mM Tris/HCl, 70% inactivation after 15 min [21]
<7, 8>, stable to lyophilization [8, 39]

Storage stability

<5>, -20°C, pH 7.0, 3 months, stable [47]
<7>, 4°C, stable for months [8]
<8>, -20°C, stable for at least 2 years [16]
<8>, 4°C, stable for up to 6 months [39]
<10>, stable for at least 4 months in the refrigerator at pH 4.0-6.0 [14]
<13>, -10°C up to -20°C, inactivation [18]
<13>, 4°C, 6 months, 10-15% loss of activity [35]
<13>, 4°C, loss of activity within two to three days [50]
<13>, 4°C, with a drop of toluene, stable for at least 1 year [18]
<14>, -20°C, protein concentration more than 0.5 mg/ml, less than 30% loss of activity after 2 months [30]
<15, 16>, 0°C, stable for at least 1 month [20]

References

[1] DeGasperi, R.; Li, Y.-T.; Li, S.-C.: Presence of two endo-β-N-acetylglucosaminidases in human kidney. J. Biol. Chem., **264**, 9329-9334 (1989)

[2] Bierbaum, G.; Sahl, H.-G.: Influence of cationic peptides on the activity of the autolytic endo-β-N-acetylglucosaminidase of Staphylococcus simulans 22. FEMS Microbiol. Lett., **58**, 223-228 (1988)
[3] Garcia, P.; Garcia, J.L.; Garcia, E.; Lopez, R.: Purification and characterization of the autolytic glycosidase of Streptococcus pneumoniae. Biochem. Biophys. Res. Commun., **158**, 251-256 (1989)
[4] Kitabatake, N.; Ishida, A.; Yamamoto, K.; Tochikura, T.; Doi, E.: Deglycosylation of hen ovalbumin in native form by endo-β-N-acetylglucosaminidase. Agric. Biol. Chem., **52**, 2511-2516 (1988)
[5] Kadowaki, S.; Yamamoto, K.; Fujisaki, M.; Kumagai, H.; Tochikura, T.: A novel endo-β-N-acetylglucosaminidase acting on complex oligosaccharides of glycoproteins in a fungus. Agric. Biol. Chem., **52**, 2387-2389 (1988)
[6] Kadowaki, S.; Takegawa, K.; Yamamoto, K.; Kumagai, H.; Tochikura, T.: Effect of yeast on endo-β-N-acetylglucosaminidase production by a Flavobacterium sp.. Agric. Biol. Chem., **52**, 2105-2106 (1988)
[7] Yet, M.-G.; Wold, F.: Purification and characterization of two glycopeptide hydrolases from jack beans. J. Biol. Chem., **263**, 118-122 (1988)
[8] Tarentino, A.L.; Plummer, T.H.: Peptide-N^4-(N-acetyl-β-glucosaminyl) asparagine amidase and endo-β-N-acetylglucosaminidase from Flavobacterium meningosepticum. Methods Enzymol., **138**, 770-778 (1987)
[9] Trimble, R.B.; Trumbly, R.J.; Maley, F.: Endo-β-N-acetylglucosaminidase H from Streptomyces plicatus. Methods Enzymol., **138**, 763-770 (1987)
[10] Song, Z.; Li, S.-C.; Li, Y.-T.: Absence of endo-β-N-acetylglucosaminidase activity in the kidneys of sheep, cattle and pig. Biochem. J., **248**, 145-149 (1987)
[11] Yamamoto, K.; Takegawa, K.; Fan, J.; Kumagai, H.; Tochikura, T.: The release of oligosaccharides from glycoproteins by endo-β-N-acetylglucosaminidase of Flavobacterium sp.. J. Ferment. Technol., **64**, 397-403 (1986)
[12] Baussant, T.; Strecker, G.; Wieruszeski, J.-M.; Montreuil, J.; Michalski, J.-C.: Catabolism of glycoprotein glycans. Characterization of a lysosomal endo-N-acetyl-β-D-glucosaminidase specific for glycans with a terminal chitobiose residue. Eur. J. Biochem., **159**, 381-385 (1986)
[13] Yamamoto, K.; Kadowaki, S.; Takegawa, K.; Kumagai, H.; Tochikura, T.: Purification and characterization of endo-β-N-acetylglucosaminidase from a Flavobacterium sp.. Agric. Biol. Chem., **50**, 421-429 (1986)
[14] Hitomi, J.; Murakami, Y.; Saitoh, F.; Shigemitsu, N.; Yamaguchi, H.: Purification and characterization of endo-β-N-acetylglucosaminidase of Aspergillus oryzae. J. Biochem., **98**, 527-533 (1985)
[15] Lisman, J.J.W.; van der Wal, C.J.; Overdijk, B.: endo-N-Acetyl-β-D-glucosaminidase activity in rat liver. Studies on substrate specificity, enzyme inhibition, subcellular localization and partial purification. Biochem. J., **229**, 379-385 (1985)
[16] Tarentino, A.L.; Trimble, R.B.; Maley, F.: endo-β-N-Acetylglucosaminidase from Streptomyces plicatus. Methods Enzymol., **50 C**, 574-580 (1978)
[17] Kobata, A.: endo-β-N-Acetylglucosaminidases CI and CII from Clostridium perfringens. Methods Enzymol., **50**, 567-574 (1978)

[18] Tarentino, A.L.; Maley, F.: endo-β-N-Acetylglucosaminidase from hen oviduct. Methods Enzymol., **50**, 580-584 (1978)
[19] Muramatsu, T.: endo-β-N-Acetylglucosaminidase D from Diplococcus pneumoniae. Methods Enzymol., **50**, 555-559 (1978)
[20] Kimura, Y.; Tokuda, T.; Ohno, A.; Tanaka, H.; Ishiguro, Y.: Enzymatic properties of endo-β-N-acetylglucosaminidase from developing tomato fruits and soybean seeds: substrate specificity of plant origin endoglycosidase. Biochim. Biophys. Acta, **1381**, 27-36 (1998)
[21] Rogers, H.J.; Taylor, C.; Rayter, S.; Ward, J.B.: Purification and properties of autolytic endo-β-N-acetylglucosaminidase and the N-acetylmuramyl-L-alanine amidase from Bacillus subtilis strain 168. J. Gen. Microbiol., **130**, 2395-2402 (1984)
[22] Robbins, P.W.; Trimble, R.B.; Wirth, D.F.; Hering, C.; Maley, F.; Maley, G.F.; Das, R.; Gibson, B.W.; Royal, N.; Biemann, K.: Primary structure of the Streptomyces enzyme endo-β-N-acetylglucosaminidase H. J. Biol. Chem., **259**, 7577-7583 (1984)
[23] Freeze, H.H.; Etchison, J.R.: Presence of a nonlysosomal endo-β-N-acetylglucosaminidase in the cellular slime mold Dictyostelium discoideum. Arch. Biochem. Biophys., **232**, 414-421 (1984)
[24] Mizuochi, T.; Amano, J.; Kobata, A.: New evidence of the substrate specificity of endo-β-N-acetylglucosaminidase D. J. Biochem., **95**, 1209-1213 (1984)
[25] Chalifour, R.J.; Spiro, R.G.: Cleavage of dolichyl pyrophosphoryl oligosaccharides by endo-β-N-acetylglucosaminidase H: comparison of enzymatic and acid hydrolysis techniques for saccharide release. Arch. Biochem. Biophys., **229**, 386-394 (1984)
[26] Elder, J.H.; Alexander, S.: endo-β-N-acetylglucosaminidase F: endoglycosidase from Flavobacterium meningosepticum that cleaves both high-mannose and complex glycoproteins. Proc. Natl. Acad. Sci. USA, **79**, 4540-4544 (1982)
[27] Tarentino, A.L.; Plummer, T.H.; Maley, F.: The release of intact oligosaccharides from specific glycoproteins by endo-β-N-acetylglucosaminidase H. J. Biol. Chem., **249**, 818-824 (1974)
[28] Tarentino, A.L.; Maley, F.: Purification and properties of an endo-β-N-acetylglucosaminidase from Streptomyces griseus. J. Biol. Chem., **249**, 811-817 (1974)
[29] Nishigaki, M.; Muramatsu, T.; Kobata, A.: Endoglycosidases acting on carbohydrate moieties of glycoproteins: demonstration in mammalian tissue. Biochem. Biophys. Res. Commun., **59**, 638-645 (1974)
[30] Koide, N.; Muramatsu, T.: Endo-β-N-acetylglucosaminidase acting on carbohydrate moieties of glycoproteins. Purification and properties of the enzyme from Diplococcus pneumoniae. J. Biol. Chem., **249**, 4897-4904 (1974)
[31] Arakawa, M.; Muramatsu, T.: Endo-β-N-acetylglucosaminidases acting on the carbohydrate moieties of glycoproteins. The differential specificities of the enzymes from Streptomyces griseus and Diplococcus pneumoniae. J. Biochem., **76**, 307-317 (1974)

[32] Koide, N.; Muramatsu, T.: Specific inhibition of endo-β-N-acetylglucosaminidase D by mannose and simple mannosides. Biochem. Biophys. Res. Commun., **66**, 411-416 (1975)
[33] Ito, S.; Muramatsu, T.; Kobata, A.: Endo-β-N-acetylglucosaminidases acting on carbohydrate moieties of glycoproteins: purification and properties of the two enzymes with different specificities from Clostridium perfringens. Arch. Biochem. Biophys., **171**, 78-86 (1975)
[34] Tarentino, A.L.; Maley, F.: A comparison of the substrate specificities of endo-β-N-acetylglucosaminidases from Streptomyces griseus and Diplococcus pneumoniae. Biochem. Biophys. Res. Commun., **67**, 455-462 (1975)
[35] Tarentino, A.L.; Maley, F.: Purification and properties of an endo-β-N-acetylglucosaminidase from hen oviduct. J. Biol. Chem., **251**, 6537-6543 (1976)
[36] Chien, S.-F.; Weinburg, R.; Li, S.-C.; Li, Y.-T.: endo-β-N-Acetylglucosaminidase from fig latex. Biochem. Biophys. Res. Commun., **76**, 317-323 (1977)
[37] Ogata-Arakawa, M.; Muramatsu, T.; Kobata, A.: Partial purification and characterization of an engo-β-N-acetylglucosaminidase from fig extract. J. Biochem., **82**, 611-614 (1977)
[38] Muramatsu, T.; Koide, N.; Maeyama, Ken-ichi: Further studies on endo-β-N-acetylglucosaminidase D. J. Biochem., **83**, 363-370 (1978)
[39] Trimble, R.B.; Tarentino, A.L.; Evans, G.; Maley, F.: Purification and properties of endo-β-N-acetylglucosaminidase L from Streptomyces plicatus. J. Biol. Chem., **254**, 9708-9713 (1979)
[40] Pierce, R.J.; Spik, G.; Montreuil, J.: Cytosolic location of an endo-N-acetyl-β-D-glucosaminidase activity in rat liver and kidney. Biochem. J., **180**, 673-676 (1979)
[41] Delmotte, F.; Kieda, C.; Bouchard, M.; Monsigny, M.: Preparation and properties of an endo-β-N-acetylglucosaminidase from rabbit serum. Glycoconjugate Res., **2**, 881-884 (1979)
[42] Pierce, R.J.; Spik, G.; Montreuil, J.: Demonstration and cytosolic location of an endo-N-acetyl-β-D-glucosaminidase activity towards an asialo-N-acetyllactosaminic-type substrate in rat liver. Biochem. J., **185**, 261-264 (1980)
[43] Overdijk, B.; van der Kroef, W.M.J.; Lisman, J.J.W.; Pierce, R.J.; Montreuil, J.; Spik, G.: Demonstration and partial characterization of endo-N-acetyl-β-D-glucosaminidase in human tissues. FEBS Lett., **128**, 364-366 (1981)
[44] Li, S.-C.; Asakawa, M.; Hirabayashi, Y.; Li, Y.-T.: Isolation of two endo-β-N-acetylglucosaminidases from fig latex. Biochim. Biophys. Acta, **660**, 278-283 (1981)
[45] Tachibana, Y.; Yamashita, K.; Kobata, A.: Substrate specificity of mammalian endo-β-N-acetylglucosaminidase: study with the enzyme of rat liver. Arch. Biochem. Biophys., **214**, 199-210 (1982)
[46] Fujisaki, M.; Tsuchida, T.; Kadowaki, S.; Yamamoto, K.; Tochikura, T.: Purification and properties of Acinetobacter sp. Endo-β-N-acetylglucosaminidase acting on complex oligosaccharides of glycoproteins. J. Ferment. Bioeng., **71**, 242-247 (1991)
[47] Kadowaki, S.; Yamamoto, K.; Fujisaki, M.; Izumi, K.; Tochikura, T.; Yokoyama, T.: Purification and characterization of a novel fungal endo-β-N-acet-

ylglucosaminidase acting on complex oligosaccharides of glycoproteins. Agric. Biol. Chem., **54**, 97-106 (1990)
[48] Bourgerie, S.; Karamanos, Y.; Grard, T.; Julien, R.: Purification and characterization of an endo-N-acetyl-β-D-glucosaminidase from the cultured medium of Stigmatella aurantiaca DW4. J. Bacteriol., **176**, 6170-6174 (1994)
[49] Takegawa, K.; Yamabe, K.; Fujita, K.; Tabuchi, M.; Mita, M.; Izu, H.; Watanabe, A.; Asada, Y.; Sano, M.; Kondo, A.; Kato, I.; Iwahara, S.: Cloning, sequencing, and expression of Arthrobacter protophormiae endo-β-N-acetylglucosaminidase in Escherichia coli. Arch. Biochem. Biophys., **338**, 2-28 (1997)
[50] Kato, T.; Hatanaka, K.; Mega, T.; Hase, S.: Purification and characterization of endo-β-N-acetylglucosaminidase from hen oviduct. J. Biochem., **122**, 1167-1173 (1997)
[51] Valisena, S.; Varaldo, P.E.; Satta, G.: Staphylococcal endo-β-N-acetylglucosaminidase inhibits response of human lymphocytes to mitogens and interfers with production of antibodies in mice. J. Clin. Invest., **87**, 1969-1976 (1991)
[52] Rao, V.; Cui, T.; Guan, C.; van Roey, P.: Mutations of endo-β-N-acetylglucosaminidase H active site residues Asp130 and Glu132: activities and conformations. Protein Sci., **8**, 2338-2346 (1999)
[53] Oshida, T.; Sugai, M.; Komatsuzawa, H.; Hong, Y.-M.; Suginaka, H.; Tomasz, A.: A Staphylococcus aureus autolysin that has an N-acetylmuramoyl-L-alanine amidase domain and an endo-β-N-acetylglucosaminidase domain: cloning, sequence analysis, and characterization. Proc. Natl. Acad. Sci. USA, **92**, 285-289 (1995)
[54] Ito, K.; Okada, Y.; Ishida, K.; Minamiura, N.: Human salivary endo-β-N-acetylglucosaminidase HS specific for complex type sugar chains of glycoproteins. J. Biol. Chem., **268**, 16074-16081 (1993)
[55] Karamanos, Y.; Bourgerie, S.; Barreaud, J.-P.; Julien, R.: Are there biological functions for bacterial endo-N-acetyl-β-D-glucosaminidases?. Res. Microbiol., **146**, 437-443 (1995)
[56] Guardati, M.C.; Guzman, C.A.; LiPira, G.; Piatti, G.; Robbiati, F.; Pruzzo, C.: The use of monoclonal antibodies of Staphylococcus aureus endo-β-N-acetylglucosaminidase. FEMS Microbiol. Lett., **112**, 73-80 (1993)
[57] Takegawa, K.; Naitoh, T.; Iwahara, S.: Isolation of two endo-β-N-acetylglucosaminidases with different specificities from Pseudomonas sp.. Biochem. Int., **24**, 793-799 (1991)
[58] Yamamoto, K.; Tanaka, T.; Fujimori, K.; Kang, C.-S.; Ebihara, H.; Kanamori, J.; Kadowaki, S.; Tochikura, T.; Kumagai, H.: Characterization of Bacillus sp. endo-β-N-acetylglucosaminidase and its application to deglycosylation of hen ovomucoid. Biotechnol. Appl. Biochem., **28**, 235-242 (1998)
[59] Rao, V.; Guan, C.; van Roey, P.: Crystal structure of endo-β-N-acetylglucosaminidase H at 1.9 A resolution: active-site geometry and substrate recognition. Structure, **3**, 449-457 (1995)
[60] Kimura, Y.; Iwata, K.; Sumi, Y.; Takagi, S.: Purification and substrate specificity of an endo-β-N-acetylglucosaminidase from pea (Pisum sativum) seeds. Biosci. Biotechnol. Biochem., **60**, 228-232 (1996)

Glycopeptide α-N-acetylgalactosaminidase 3.2.1.97

1 Nomenclature

EC number
3.2.1.97

Systematic name
D-galactosyl-N-acetyl-α-D-galactosamine D-galactosyl-N-acetylgalactosamino-hydrolase

Recommended name
glycopeptide α-N-acetylgalactosaminidase

Synonyms
acetylgalactosaminidase, endo-α
endo-GalNAc-ase S <2> [15]
endo-α-N-acetylgalactosaminidase
endo-α-N-acetylgalactosaminidase S <2> [10, 15]
endo-α-acetylgalactosaminidase

CAS registry number
59793-96-3

2 Source Organism

<1> *Alcaligenes sp.* (strain F-1906 [1]) [1, 9, 12, 13]
<2> *Streptomyces sp.* (strain OH-11242 [2,10,15,16]) [2, 10, 15, 16]
<3> *Diplococcus pneumoniae* [3-7, 13, 18]
<4> *Bacillus lateropsorus* (strain SS-16 [8]) [8]
<5> *Bacillus sp.* [11]
<6> *Gliocladium virens* [14]
<7> *Mus musculus* [17]

3 Reaction and Specificity

Catalyzed reaction
hydrolysis of terminal D-galactosyl-N-acetyl-α-D-galactosaminidic residues from a variety of glycopeptides and glycoproteins

Reaction type
hydrolysis of O-glycosyl bond

Natural substrates and products

S Additional information <1, 2, 6> (<1, 2>, highly induced by porcine gastric mucin [1, 2]; <2>, enzyme production is induced by mucin, the induction is inhibited by glucose and other easily assimilable carbon sources, as well as by complex nitrogen sources, addition of palmitate, lambda-carrageenan or crude mucin to a mucin-based production medium enhances enzyme production [10]; <6>, induced when asialofetuin is the sole carbon source [14]) [1, 2, 10, 14]
P ?

Substrates and products

S Galβ(1-3)GalNAcα(1-)p-nitrophenyl + H_2O <5> [11]
P Galβ(1-3)GalNAc + *p*-nitrophenol
S Galβ(1-3)GalNAcα(1-)p-nitrophenyl + galactose <5> (<5>, transglycosylation [11]) [11]
P ?
S Galβ(1-3)GalNAcα(1-)p-nitrophenyl + glucose <5> (<5>, transglycosylation [11]) [11]
P ?
S Galβ(1-3)GalNAcα(1-)p-nitrophenyl + maltose <5> (<5>, transglycosylation [11]) [11]
P ?
S Galβ(1-3)GalNAcα(1-)p-nitrophenyl + mannitol <5> (<5>, transglycosylation [11]) [11]
P ?
S Galβ(1-3)GalNAcα(1-)p-nitrophenyl + mannose <5> (<5>, transglycosylation [11]) [11]
P ?
S Galβ(1-3)GalNAcα(1-)p-nitrophenyl + sorbitol <5> (<5>, transglycosylation [11]) [11]
P ?
S Galβ(1-3)GalNAcα(1-)p-nitrophenyl + sucrose <5> (<5>, transglycosylation [11]) [11]
P ?
S asialo κ-casein + H_2O <1, 4, 5> [1, 8, 11]
P Galβ(1-3)GalNAc + ?
S asialofetuin + H_2O <1, 2, 3, 4, 5, 6> (<2>, hydrolysis of the O-glycosidic linkage between GalNAc and Ser or Thr [16]) [1, 5, 8, 9, 11, 14, 16]
P Additional information <1, 2> (<1>, Galβ(1-3)GalNAc + ? [1,9]; <2>, Galβ(1-3)GalNAc + Galβ(1-3)(Galβ(1-4)GlcNAcβ(1-6))GalNAc [16]) [1, 9, 16]
S asialoglycophorin + H_2O <1, 4, 5> (<5>, asialoglycophorin A [11]) [1, 8, 11, 12]
P Galβ(1-3)GalNAc + ? <1> [12]
S dansyl-Ser-GalNAc-Gal + H_2O <1> [13]
P dansyl-Ser + GalNAc-Gal + H_2O <1> [13]
S dansyl-Thr-GalNAc-Gal + H_2O <1> [13]

P dansyl-Thr + GalNAc-Gal + H_2O <1> [13]
S fetuin + H_2O <2> (<2>, hydrolysis of the O-glycosidic linkage between GalNAc and Ser or Thr [15]) [15]
P ?
S glycopeptide + H_2O <3> (<3>, release of Galβ(1-3)GalNAc from glycopeptides possesing Ser or Thr O-glycosidic linkages [3, 5]; <3>, specific requirement for an unsubstituted galactose at the nonreducing terminus and an α-linkage between N-acetylgalactosamine and the aglycone [5]; <3>, glycopeptides from mouse myeloma, fetuin and pig submaxillary mucin [7]) [3, 5, 7]
P Galβ(1-3)GalNAc + ? <3> [7]
S glycoprotein + H_2O <1, 2, 3, 4, 7> (<1, 3>, release of Galβ(1-3)GalNAc from glycoproteins possesing Ser or Thr O-glycosidic linkages [1, 3, 6, 7]; <2>, liberates Galβ(1-3)4GalNAc and also other larger oligosaccharides by hydrolysis of O-glycosidic linkage between GalNAc and Ser or Thr [2]; <3>, specific requirement for an unsubstituted galactose in the nonreducing terminus and an α-linkage between N-acetylgalactosamine and the aglycone [5]; <3>, disialized human erythrocyte membrane glycoprotein, porcine submaxillarry mucin [6]; <4>, hydrolyzes the O-glycosidic linkage between α-N-acetylgalactosamine and the hydroxyl group of Ser or Thr residues in mucus and mucin-type glycoproteins [8]; <2>, purified pig gastric mucus glycoproteins [16]; <7>, digestion of nerve cell surface glycoproteins or their terminal N-acetylgalactosamines, no digestion of galactosamines associated with the collagenous ligands or perineuronal proteoglycan [17]; <3>, bovine submaxillary mucin, no hydrolysis of remodelled antifreeze glycoprotein containing NeuAcα2-3Galβ1-3GalNAc and Galβ1-3(NeuAcα2-6)GalNAc [18]) [1, 2, 3, 5, 6, 7, 8, 16, 17, 18]
P Galβ(1-3)4GalNAc + ? <1, 2, 3> (<3>, no oligosaccharides larger than trisaccharides are liberated from porcine submaxillarry mucin [6]) [1, 2, 5, 6]
S mucin + H_2O <2, 3> (<2>, release of Galβ(1-3)GalNAc from mucins possesing Ser or Thr O-glycosidic linkages [10]; <3>, desialylated bovine submaxillary mucin [18]) [10, 18]
P Galβ1-3GalNAc + ? <3> [18]
S Additional information <3> (<3>, no action on phenyl α-N-acetyl-D-galactosaminide, asialo ovine submaxillary mucin or monosialoganglioside [7]) [7]
P ?

Inhibitors

Ag^+ <4> (<4>, 5 mM, 35% inhibition [8]) [8]
C_2H_5HgCl <4> (<4>, 5 mM, 23% inhibition [8]) [8]
CH_3HgCl <4> (<4>, 5 mM, 18% inhibition [8]) [8]
Ca^{2+} <3, 4> (<3>, 2.5 mM $CaCl_2$, 3.3% inhibition [6]; <4>, 5 mM, 26% inhibition [8]) [6, 8]
Cd^{2+} <4> (<4>, 5 mM, 22% inhibition [8]) [8]
Co^{2+} <4> (<4>, 5 mM, 18% inhibition [8]) [8]

Cr^{3+} <4> (<4>, 5 mM, 58% inhibition [8]) [8]
Cu^{2+} <1, 4> (<1>, 2.5 mM, 26% inhibition [1]; <4>, 5 mM, 20% inhibition [8]) [1, 8]
Cys <1, 3> (<1>, 2.5 mM, 32% inhibition [1]; <3>, 2 mM, 12% inhibition [5]; <5>, no effect [11]) [1, 5]
DFP <4> (<4>, 1 mM, 48% inhibition [8]) [8]
EDTA <3> (<3>, 2.5 mM, 63% inhibition [3,6]; <3>, 2 mM, 40% inhibition [5]; <5>, no effect [11]) [3, 5, 6]
Fe^{2+} <5> (<5>, 1 mM [11]) [11]
Fe^{3+} <4> (<4>, 5 mM, 15% inhibition [8]) [8]
Hg^{2+} <1, 3, 4, 5> (<1>, 2.5 mM, complete inhibition [1]; <3>, 2 mM, complete inhibition [5]; <4>, 5 mM, 20% inhibition [8]) [1, 5, 8, 11]
Mg^{2+} <3, 4> (<3>, 2 mM, 28% inhibition [5]; <3>, 2.5 mM, 2.2% inhibition [6]; <4>, 5 mM, 17% inhibition [8]) [5, 6, 8]
Mn^{2+} <1, 3, 4> (<1>, 2.5 mM, 35% inhibition [1]; <3>, 2.5 mM, 44% inhibition [3, 6]; <3>, 2 mM, 29% inhibition [5]; <4>, 5 mM, 22% inhibition [8]) [1, 3, 5, 6, 8]
N-bromosuccinimide <4> (<4>, 1 mM, complete inhibition [8]) [8]
PCMB <4> (<4>, 1 mM, 26% inhibition [8]; <5>, no effect [11]) [8]
Zn^{2+} <3, 4> (<3>, 2.5 mM, 66% inhibition [3,6]; <4>, 5 mM, 9% inhibition [8]) [3, 6, 8]
galactonolactone <4> (<4>, 5 mM, 11% inhibition [8]) [8]
glucose <4> (<4>, 5 mM, 11% inhibition [8]) [8]
iodoacetate <4> (<4>, 5 mM, 17% inhibition [8]) [8]
p-chloromercuribenzenesulfonate <3> (<3>, 1 mM, 60% inhibition [3, 6]) [3, 6]

Metals, ions

Ca^{2+} <3> (<3>, at a level twice the EDTA concentration completely restores enzyme activity [3]) [3]
Mg^{2+} <3> (<3>, at a level twice the EDTA concentration completely restores enzyme activity [3]) [3]

Specific activity (U/mg)

2.18 <1> [1]
11.5 <3> [3]
30.1 <5> [11]
Additional information <2, 3> [6, 16]

K_m-Value (mM)

0.17 <1> (dansyl-Ser-GalNAc-Gal) [13]
0.32 <1> (asialoglycophorin) [12]
0.52 <4> (asialofetuin) [8]
0.78 <4> (asialo-kappa-casein) [8]
1 <3> (asialofetuin) [5]
1.43 <1> (dansyl-Thr-GalNAc-Gal) [13]
3.2 <1> (asialo κ-casein) [1, 13]
3.7 <1> (asialofetuin) [1, 13]

Additional information <3>(<3>, K_m for glycopeptides obtained by trypsin digestion of human erythrocyte membrane [6]) [3, 6]

pH-Optimum

4.5-5 <1> [1]
4.5-6 <1> (<1>, hydrolysis of dansyl-Thr-GalNAc-Gal [13]) [13]
4.5-7.5 <1> (<1>, hydrolysis of dansyl-Ser-GalNAc-Gal [13]) [13]
5 <5> [11]
5.5 <2> [16]
6 <3, 6> [3, 6, 14]
7.6 <3> [5]
8 <4> [8]

pH-Range

4-9 <1> (<1>, about 50% of maximal activity at pH 4.0 and at pH 9.0 [1]) [1]
5.5-7 <3> (<3>, about 50% of maximal activity at pH 5.5 and at pH 7.0 [3,6]) [3, 6]
6-8.8 <3> (<3>, pH 6.0: about 40% of maximal activity, pH 8.8: about 60% of maximal activity [5]) [5]
Additional information <2> [16]

Temperature optimum (°C)

35-37 <4> [8]
37 <6> [14]
40-45 <1> [1]

4 Enzyme Structure

Molecular weight

105000 <2> (<2>, gel filtration [16]) [16]
110000 <5> (<5>, gel filtration [11]) [11]
160000 <1, 3> (<1, 3>, gel filtration [1, 5]) [1, 5]
210000 <4> (<4>, gel filtration [8]) [8]

Subunits

? <3> (<3>, x * 190000, SDS-PAGE [4]) [4]
monomer <1, 5> (<5>, 1 * 110000, SDS-PAGE [11]; <1>, 1 * 160000, SDS-PAGE [1]) [1, 11]
tetramer <4> (<4>, 4 * 48000, SDS-PAGE [8]) [8]

5 Isolation/Preparation/Mutation/Application

Source/tissue

brain <7> (<7>, extracellular matrix [17]) [17]
culture filtrate <6> [14]
culture medium <1, 2, 3, 5, 6> [1, 2, 4, 6, 11, 14, 15, 16]

Localization

extracellular <7> [17]

Purification

<1> [1]
<2> (partial [16]) [16]
<3> (partial [6, 7]) [3-7]
<4> [8]
<5> [11]

Application

analysis <1, 3> (<3>, the enzyme is useful in the study of the structures of the carbohydrate moieties of various complex carbohydrates [3]; <3>, the enzyme is ideally suited to explore the distribution and function of mucin-type glycoproteins on normal and cancer cell surfaces [5]; <1>, the enzyme can release T-antigen, Galβ(1-3)GalNAc, a carcinoma-associated marker from human cells, enzyme is useful for detection and determination of T-antigen from human erythrocytes and native human gastric carcinoma cells [12]) [3, 5, 12]

synthesis <1> (<1>, production of Galβ(1-3)GalNAc from asialofetuin [9]) [9]

6 Stability

pH-Stability

3.5-9 <4> (<4>, 37°C, 1 h, stable [8]) [8]
4-10 <5> (<5>, 37°C, 30 min, stable [11]) [11]
4.5-6.5 <1> (<1>, 30°C, 1 h, stable [1]) [1]

Temperature stability

35 <1> (<1>, 10 min, stable below [1]) [1]
55 <1, 4, 5> (<1>, 10 min, complete inactivation [1]; <4>, 1 h, stable below [8]; <5>, 10 min, stable [11]) [1, 8, 11]
70 <4> (<4>, 1 h, complete inactivation [8]) [8]

Storage stability

<3>, -20°C, 2 months, stable [4]
<3>, -20°C, stable for 3 months [6]

References

[1] Fan, J.Q.; Kadowaki, S.; Yamamoto, K.; Kumagai, H.; Tochikura, T.: Purification and characterization of endo-α-N-acetylgalactosaminidase from Alcaligenes sp.. Agric. Biol. Chem., **52**, 1715-1723 (1988)

[2] Iwase, H.; Ishii, I.; Ishihara, K.; Tanaka, Y.; Omura, S.; Hotta, K.: Release of oligosaccharides possessing reducing-end N-acetylgalactosamine from mucus glycoprotein in Streptomyces sp. OH-11242 culture medium through

action of endo-type glycosidase. Biochem. Biophys. Res. Commun., **151**, 422-428 (1988)
[3] Kobata, A.; Takasaki, S.: endo-β-Galactosidase and endo-α-N-acetylgalactosaminidase from Diplococcus pneumoniae. Methods Enzymol., **50**, 560-567 (1978)
[4] Glasgow, L.R.; Paulson, J.C.; Hill, R.L.: Systematic purification of five glycosidases from Streptococcus (Diplococcus) pneumoniae. J. Biol. Chem., **252**, 8615-8623 (1977)
[5] Umemoto, J.; Bhavanandan, V.P.; Davidson, E.A.: Purification and properties of an endo-α-N-acetyl-D-galactosaminidase from Diplococcus pneumoniae. J. Biol. Chem., **252**, 8609-8614 (1977)
[6] Endo, Y.; Kobata, A.: Partial purification and characterization of an endo-α-N-acetylgalactosaminidase from the culture medium of Diplococcus pneumoniae. J. Biochem., **80**, 1-8 (1976)
[7] Bhavanandan, V.P.; Umemoto, J.; Davidson, E.A.: Characterization of an endo-α-N-acetyl galactosaminidase from Diplococcus pneumoniae. Biochem. Biophys. Res. Commun., **70**, 738-745 (1976)
[8] Yoshimoto, T.; Takahashi, E.; Tsuru, D.: Purification and enzymatic properties of an endo-α-N-acetylgalactosaminidase from Bacillus laterosporus SS-16. J. Ferment. Bioeng., **70**, 348-350 (1990)
[9] Fan, J.-Q.; Yamamoto, K.; Kumagai, H.; Tochikura, T.: Unique production of a disaccharide, Galβ1 → 3GalNAc, by endo-α-N-acetylgalactosaminidase of Alcaligenes sp.. J. Ferment. Bioeng., **72**, 92-95 (1991)
[10] Tanaka, Y.; Takahashi, Y.; Shinose, M.; Omura, S.; Karakasa, I.I.; Iwase, H.; Hotta, K.: Screening and fermentation of endo-α-N-acetylgalactosaminidase S, a mucin-hydrolyzing enzyme from Streptomyces acting on the GalNAc-O-Ser(Thr) linkage. J. Ferment. Bioeng., **85**, 381-387 (1998)
[11] Ashida, H.; Yamamoto, K.; Murata, T.; Usui, T.; Kumagai, H.: Characterization of endo-α-N-acetylgalactosaminidase from Bacillus sp. and syntheses of neo-oligosaccharides using its transglycosylation activity. Arch. Biochem. Biophys., **373**, 394-400 (2000)
[12] Fan, J.-Q.; Kanatani, K.; Andoh, N.; Yamamoto, K.; Kumagai, H.; Tochikura, T.: Release of T-antigen, a carcinoma marker from native human cells by endo-α-N-acetylgalactosaminidase of Alcaligenes sp.. Biochem. Biophys. Res. Commun., **172**, 341-347 (1990)
[13] Fan, J.-Q.; Yamamoto, K.; Matsumoto, Y.; Hirabayashi, Y.; Kumagai, H.; Tochikura, T.: Action of endo-α-N-acetylgalactosaminidase from Alcaligenes sp. on amino acid-O-glycans: comparison with the enzyme from Diplococcus pneumoniae. Biochem. Biophys. Res. Commun., **169**, 751-757 (1990)
[14] Prasad, V.P.; Devaraj, H.: Production of endo-N-acetyl-galactosaminidase from a mycoparasitic fungus Gliocladium virens. Indian J. Exp. Biol., **35**, 405-407 (1997)
[15] Ishii-Karakasa, I.; Iwase, H.; Hotta, K.: Structural determination of the O-linked sialyl oligosaccharides liberated from fetuin with endo-α-N-acetylgalactosaminidase-S by HPLC analysis and 600-MHz 1H-NMR spectroscopy. Eur. J. Biochem., **247**, 709-715 (1997)

[16] Ishii-Karakasa, I.; Iwase, H.; Hotta, K.; Tanaka, Y.; Omura, S.: Partial purification and characterization of an endo-α-N-acetylgalactosaminidase from the culture medium of Streptomyces sp. OH-11242. Biochem. J., **288**, 475-482 (1992)
[17] Murakami, T.; Ohtsuka, A.; Su, W.-D.; Taguchi, T.; Oohashi, T.; Murakami, T.; Abe, K.; Ninomiya, Y.: The extracellular matrix in the mouse brain: its reactions to endo-α-N-acetylgalactosaminidase and certain other enzymes. Arch. Histol. Cytol., **62**, 273-281 (1999)
[18] Brooks, M.M.; Savage, A.V.: The substrate specificity of the enzyme endo-α-N-acetyl-D-galactosaminidase from Diplococcus pneumonia. Glycoconjugate J., **14**, 183-190 (1997)

Glucan 1,4-α-maltohexaosidase 3.2.1.98

1 Nomenclature

EC number
3.2.1.98

Systematic name
1,4-α-D-glucan maltohexaohydrolase

Recommended name
glucan 1,4-α-maltohexaosidase

Synonyms
Aerobacter aerogenes amylase <1> [2]
G6-amylase <2> [10]
G6-forming amylase <1> [1, 2, 9]
G6-forming exo-α amylase <1> [2]
G6-producing amylase <1, 4> [7]
exo-maltohexao-hydrolase
exo-maltohexaohydrolase
exo-maltohexaose hydrolase
exomaltohexaohydrolase
maltohexaohydrolase, exo-
maltohexaose producing amylase <1> [4, 7]
maltohexaose-forming amylase <1, 2> [1, 6, 10]
maltohexaose-producing exo-amylase <1> [4]
Additional information (compare EC 3.2.1.3 glucan 1,4-α-glucosidase, which removes successive glucose residues, EC 3.2.1.2 β-amylase, which removes successive maltose residues, EC 3.2.1.116 glucan 1,4-α-maltotriohydrolase, which removes successive maltotriose units and EC 3.2.1.60 glucan 1,4-α-maltotetraohydrolase, which removes successive maltotetraose residues)

CAS registry number
72561-12-7

2 Source Organism

<1> *Klebsiella pneumoniae* (Aerobacter aerogenes; IFO-3321 [8, 9]; ultraviolet-induced mutant [1, 3, 5, 8, 9]; ultraviolet-induced mutant from Aerobacter aerogenes IFO-3321 [8]; native and immobilized, by irradiation method, enzyme [5]) [1, 2, 3, 4, 5, 7, 8, 9]

<2> *Bacillus sp. H-167* (alkalophilic bacterium, 3 enzymes: H-I-1, H-I-2, H-II [6]; alkalophilic bacterium: multiple forms of maltohexaose-forming amylase due to intracellular proteolysis of the enzyme [10]) [6, 10]
<3> *Bacillus circulans* (G-6 [11]; F-2 [12]) [11, 12]
<4> *Aerobacter cloacae* [7]

3 Reaction and Specificity

Catalyzed reaction

hydrolysis of 1,4-α-D-glucosidic linkages in amylaceous polysaccharides, to remove successive maltohexaose residues from the non-reducing chain ends (<1> mechanism of transglycosylation [1, 3]; <1> exo-α amylase, products have α-configuration [2, 4]; <1> reaction mechanism is exo [4, 7, 8, 9]; <2,3> α-amylase, product is linear oligosaccharide containing α-1,4-glucosidic linkages [6, 11])

Reaction type

hydrolysis of O-glycosyl bond

Substrates and products

S amylopectin + H_2O <1, 2, 4> [4, 6, 7, 8]
P maltohexaose <1, 2, 4> (<1> more than 35% maltohexaose [7]) [4, 6, 7, 8]
S amylopectin β-amylase limit dextrin + H_2O <1> [4, 8, 9]
P oligosaccharide <1> (<1> branched oligosaccharide: B6, B7 and B8 [4]; <1> branched oligosaccharides containing 5 to 10 glucose units, B5-B10 [8]) [4, 8, 9]
S amylose + H_2O <1, 2, 4> (<1> short-chain amylose, DP, average degree of polymerization, = 23 [1, 2, 3, 4, 5, 8]; <1> partially-oxidized potato amylose [2]; <1> potato amylose: about 57% of the rate of soluble starch [8]; <1> high molecular weight amylose: about 50% of the rate of short-chain amylose [4]; <1> reduced short-chain amylose: about 95% of the rate of short-chain amylose [4, 8]) [1, 2, 3, 4, 5, 6, 7, 8]
P maltohexaose <1, 2> (<1> main product, 40% maltohexaose [1, 2, 3]; <1> main product, about 38% maltohexaose [5]; <1> α-configuration [2]) [1, 2, 3, 4, 5, 6, 7, 8]
S glycogen + H_2O <1, 2> [4, 6, 8]
P maltohexaose <1, 2> [4, 6, 8]
S glycogen β-amylase limit dextrin + H_2O <1> (<1> poor substrate [4]) [4, 8, 9]
P oligosaccharide <1> (<1> branched oligosaccharide: B6, B7 and B8 [4]; <1> branched oligosaccharides containing 5 to 10 glucose units, B5-B10 [8]) [4, 8, 9]
S maltoheptaose + *p*-nitrophenyl α-glucoside <1> (<1> transglycosylation, mechanism [1]) [1]
P *p*-nitrophenyl α-maltoheptaoside + D-glucose <1> (<1> *p*-nitrophenyl α-maltoheptaoside useful substrate for assay of human α-amylase in serum and urine [1]) [1]

S maltohexaose + H_2O <1, 2> [2, 3, 5, 6]
P D-glucose + maltopentaose <1, 2> (<1, 2> small amounts [3, 6]) [3, 5, 6]
S maltohexaose + H_2O <1, 2> [2, 3, 5, 6]
P maltotriose <1, 2> (<1, 2> small amounts [3, 6]) [3, 5, 6]
S maltohexaose + H_2O <1, 2, 3> (<1, 3> poor substrate [4, 7, 8, 11]) [2, 3, 4, 5, 6, 7, 8, 11, 12]
P maltose + maltotetraose <1, 2, 3> (<1, 2, 3> main product [2, 3, 4, 6, 7, 8, 11, 12]) [2, 3, 4, 5, 6, 7, 8, 11, 12]
S maltooligosaccharide + H_2O <1> (<1> 6 glucose units and shorter [3]) [3]
P maltooligosaccharide <1> (<1> with less than 6 glucose units [3]) [3]
S maltooligosaccharide + H_2O <1> (<1> maltooligosaccharide larger than G6, maltohexaose [3, 7]) [2, 3, 7]
P maltohexaose <1> (<1> main product [3, 7]; <1> α-configuration [2]) [2, 3, 7]
S maltotetraose <1> (<1> transglycosylation, mechanism [3]) [3, 7]
P maltohexaose + maltose <1> (<1> main product [3]) [3, 7]
S starch + H_2O <1, 2, 3, 4> (<1, 2, 3> soluble starch [6, 8, 11]; <1> waxy-maize starch [7, 8]; <3> F-2, potato starch granules: 40%, corn starch: 83% and barley starch: 71% [12]; <3> F-2, gelatinized starch [12]; <1> blue starch: not a substrate [2]) [3, 4, 5, 6, 7, 8, 11, 12]
P maltohexaose <1, 2, 3, 4> (<1, 2, 3, 4> main product [4, 5, 6, 7, 8, 11, 12]; <1> more than 30% maltohexaose [4]; <2> 25-30% maltohexaose in early stage of hydrolysis, α-anomer product [6]; <3> G-6, about 30-34% maltohexaose [11]; <3> F-2, produces only maltohexaose in the early stage of the reaction [12]) [4, 5, 6, 7, 8, 11, 12]
S Additional information <1, 2, 3> (<1, 2, 3> products have α-configuration [2, 4, 6, 11]; <1> substrate and product specificity, overview [3, 4, 8]; <1> exo action pattern [4, 7, 8]; <1> action pattern on substrates of native and immobilized enzyme is almost the same [5]; <3> G-6, action pattern [11]; <1> α-cyclodextrin, β-cyclodextrin, pullulan or maltohexaitol: not a substrate [4]; <1> no reaction on pullulan [7]; <1> no reaction on α-, β- or γ-cyclodextrin, pullulan and maltose [8]; <2> pullulan, dextran or α-, β- or γ-cyclodextrin: not a substrate [6]; <3> G-6, α-1,6-glucosidic linkages in pullulan, amylopectin etc.: not a substrate [11]) [2, 3, 4, 5, 6, 7, 8, 11]
P ?

Inhibitors

1-propanol <1> (<1> transglycosylation, inactivation [1]) [1]
Ag^+ <1> (<1> extracellular enzyme, strong inhibition [8]) [8]
$AgNO_3$ <1, 3> (<1> native and immobilized enzyme, immobilized enzyme less than native enzyme [5]; <1> extracellular enzyme, strong inhibition [8]; <3> G-6 [11]) [4, 5, 8, 11]
$Al_2(SO_4)_3$ <1> (<1> immobilized enzyme, slight inactivation [5]) [5]
Ba^{2+} <1> (<1> very slight inactivation [4]; <1> immobilized enzyme [5]) [4, 5]
$BaCl_2$ <3> (<3> G-6, slight inactivation [11]) [11]

$CaCl_2$ <3> (<3> G-6 [11]) [11]
Co^{2+} <1, 3> (<1> very slight inactivation [4]; <3> G-6, strong inhibition [11]) [4, 11]
$CoCl_2$ <3> (<3> G-6, strong inhibition [11]) [11]
Cu^{2+} <1, 3> (<1> immobilized enzyme less than native enzyme [5]; <1> extracellular enzyme, strong inhibition [8]; <3> G-6, strong inhibition [11]) [4, 5, 8, 11]
$CuSO_4$ <1, 3> (<1> extracellular enzyme, strong inhibition [8]; <3> G-6, strong inhibition [11]) [8, 11]
Fe^{2+} <3> (<3> G-6, strong inhibition [11]) [11]
$FeCl_3$ <1> (<1> immobilized enzyme [5]) [5]
$FeSO_4$ <1, 3> (<1> native and immobilized enzyme [5]; <1> extracellular enzyme [8]; <3> G-6, strong inhibition [11]) [5, 8, 11]
Hg^{2+} <1, 3> (<1> native enzyme, completely [4, 5]; <1> immobilized enzyme, 61% inactivation [5]; <1> extracellular enzyme, completely [8]; <3> G-6, very strong inhibition [11]) [4, 5, 8, 11]
$HgCl_2$ <1, 3> (<1> native enzyme, completely [4, 5]; <1> immobilized enzyme, 61% inactivation [5]; <1> extracellular enzyme, completely [8]; <3> G-6, very strong inhibition [11]) [4, 5, 8, 11]
Mg^{2+} <1> (<1> slight inactivation [4]; <1> immobilized enzyme, slight inactivation [5]) [4, 5]
Mn^{2+} <1> (<1> native enzyme, slight inactivation [5]) [5]
$MnCl_2$ <3> (<3> G-6, slight inactivation [11]) [11]
$NiCl_2$ <3> (<3> G-6 [11]) [11]
Sn^{2+} <1> (<1> slight inactivation [4]) [4]
$SrCl_2$ <1, 3> (<1> immobilized enzyme [5]; <3> G-6 [11]) [5, 11]
Zn^{2+} <1, 3> (<1> native enzyme [5]; <1> extracellular enzyme, strong inhibition [8]; <3> G-6, strong inhibition [11]) [4, 5, 8, 11]
$ZnSO_4$ <1, 3> (<1> extracellular enzyme, strong inhibition [8]; <3> G-6, strong inhibition [11]) [8, 11]
iodoacetamide <1> (<1> slight inactivation [4, 8]; <1> native enzyme: 11% inhibition, immobilized enzyme: 4% inhibition [5]; <1> extracellular enzyme, 1 mM: 11% inactivation [8]) [4, 5, 8]
p-chloromercuribenzoate <1> (<1> slight inactivation [4]; <1> native enzyme: 46% inhibition, immobilized enzyme: 15% inhibition [5]; <1> extracellular enzyme, 1 mM: 46% inactivation [8]) [4, 5, 8]
Additional information <1> (<1> native enzyme compared with immobilized enzyme [5]) [5]

Activating compounds

cysteine <1> (<1> very slight activation [4]) [4]
ethanol <1> (<1> transglycosylation, activation [1]) [1]
methanol <1> (<1> aqueous methanol system, up to 40% methanol, increases transglycosylation [1]) [1]

Metals, ions

Ca^{2+} <1> (<1> native enzyme, activation [4, 5, 8]; <1> extracellular enzyme, slight activation [8]; <1> immobilized enzyme, no activation [5]) [4, 5, 8]

$CaCl_2$ <1> (<1> extracellular enzyme, slight activation [8]) [8]
Co^{2+} <1> (<1> extracellular enzyme, slight activation [8]) [8]
$CoCl_2$ <1> (<1> native and immobilized enzyme, very slight activation [5]; <1> extracellular enzyme, slight activation [8]) [5, 8]
$MgSO_4$ <3> (<3> G-6, slight activation [11]) [11]
Mn^{2+} <1> (<1> immobilized enzyme, activation [5]; <1> extracellular enzyme, slight activation [8]) [5, 8]
$MnSO_4$ <1> (<1> extracellular enzyme, slight activation [8]) [8]
$NiCl_2$ <1> (<1> immobilized enzyme, slight activation [5]) [5]
Sr^{2+} <1> (<1> activation [4]; <1> native enzyme, very slight activation [5]) [4, 5]
Zn^{2+} <1> (<1> immobilized enzyme, very slight activation [5]) [5]
Additional information <1> (<1> native enzyme compared with immobilized enzyme [5]) [5]

Specific activity (U/mg)
8.2 <1> (<1> cell-bound enzyme [4, 8]) [4, 8]
15.6 <1> [5]
18.4 <1> (<1> extracellular enzyme [8, 9]) [8, 9]
116 <2> (<2> H-I-2 [6]) [6]
124 <2> (<2> H-II [6]) [6]
147 <2> (<2> H-I-1 [6]) [6]
Additional information <3> (<3> G-6 [11]) [11]

K_m-Value (mM)
0.46 <1> (short-chain amylose) [3]
7.68 <1> (maltohexaose) [3]
10.02 <1> (maltopentaose) [3]
50 <1> (maltotetraose) [3]
Additional information <1, 3>(<1> kinetic data [3]; <1> native enzyme compared with immobilized enzyme, K_m-value of native and immobilized enzyme for short-chain amylose is the same, for soluble starch K_m-value of immobilized enzyme is larger than that of native enzyme [5]) [3, 5, 11]

pH-Optimum
6.8 <1> (<1> cell-bound enzyme [4, 8, 9]) [4, 8, 9]
7 <1> (<1> cell-bound enzyme [7]; <1> extracellular enzyme [8, 9]) [7, 8, 9]
8 <1, 3> (<1> transglycosylation [1]; <3> G-6 [11]) [1, 11]
10.5 <2> [10]
Additional information <1> (<1> native enzyme compared with immobilized enzyme, immobilized enzyme: optimum pH shifted slightly to acidic side [5]) [5]

pH-Range
3-9 <1> (<1> pH 3.0: 20% of maximum activity, pH 9.0: more than 60% of maximum activity [4]) [4]
5-8 <1> (<1> immobilized enzyme, more than 80% of maximum activity between 5.0 and 8.0 [5]) [5]
5-9 <1> (<1> transglycosylation [1]) [1]

5-10 <1, 3> (<1> extracellular enzyme, more than 20% activity at pH 5.0, more than 50% activity at pH 10.0 [8, 9]; <3> G-6, about 40% of activity at pH 5.0 and 10.0 [11]) [8, 9, 11]
5.5-8 <1> (<1> more than 80% of maximum activity at 5.5 and 8.0 [4]) [4]
6-8 <1> (<1> extracellular enzyme, more than 80% of maximum activity between pH 6.0 and 8.0 [8]) [8]
Additional information <1> (<1> native enzyme compared with immobilized enzyme [5]) [5]

Temperature optimum (°C)

35 <1> (<1> transglycosylation [1]) [1]
45 <1> (<1> cell-bound enzyme [7]) [7]
50 <1> (<1> cell-bound enzyme [4]) [4]
52 <1> (<1> immobilized enzyme [5]; <1> extracellular enzyme [8, 9]) [5, 8, 9]
60 <1, 2, 3> (<1> cell-bound enzyme [8, 9]; <3> G-6 [11]) [8, 9, 10, 11]
Additional information <1> (<1> native enzyme compared with immobilized enzyme [5]) [5]

Temperature range (°C)

20-80 <1> (<1> 20°C: about 30% of maximum activity, 80°C: more than 20% of maximum activity [4]) [4]
45 <1> (<1> transglycosylation, below, considerable inactivation at [1]) [1]
Additional information <1> (<1> native enzyme compared with immobilized enzyme [5]) [5]

4 Enzyme Structure

Molecular weight

48000 <1> (<1> cell-bound and extracellular enzyme, gel filtration [8, 9]) [8, 9]
54000 <1> (<1> gel filtration [4]) [4]
59000 <2> (<2> H-I-2, SDS-PAGE [6]) [6]
65000 <1> (<1> extracellular enzyme, SDS-PAGE [8, 9]) [8, 9]
67000 <1> (<1> cell-bound enzyme, SDS-PAGE [8, 9]) [8, 9]
73000 <2> (<2> H-I-1, SDS-PAGE [6]) [6]
76000 <3> (<3> G-6, gel filtration [11]) [11]
80000 <2> (<2> H-II, SDS-PAGE [6]) [6]
100000 <2> (<2> molecular weight of largest G6-amylase produced in E. coli HB101, SDS-PAGE [10]; <2> multiple forms G6-amylase having molecular masses: 100000, 90000, 80000, 73000, and 60000, due to intracellular proteolysis of the enzyme [10]) [10]
102600 <2> (<2> mature G6-amylase, before proteolytic processing at the COOH-terminal side of enzyme, calculated from nucleotide sequence [10]) [10]

Additional information <2> (<2> multiple forms G6-amylase having molecular masses: 100000, 90000, 80000, 73000, and 60000, due to intracellular proteolysis of the enzyme [10]) [10]

Subunits

monomer <1> (<1> extracellular enzyme, 1 * 65000, SDS-PAGE [8]) [8]

5 Isolation/Preparation/Mutation/Application

Source/tissue

culture filtrate <1, 2, 3> [6, 8, 9, 11, 12]

Localization

cell-bound <1> [4, 7, 8, 9]
extracellular <1, 3> (<1> ultraviolet induced mutant [1, 3, 5, 8, 9]; <1> extracellular enzyme has the same characteristics as the cell-bound enzyme: enzyme is secreted extracellulary due to mutation of parent strain of Klebsiella pneumoniae [3, 8, 9]) [1, 2, 3, 5, 8, 9, 11, 12]

Purification

<1> (partial [5, 7]) [2, 3, 4, 5, 7, 8, 9]
<2> [6, 10]
<3> (G-6 [11]) [11]

Cloning

<2> (expressed and multiform G6-amylase produced in E. coli HB101 [10]) [10]

Application

synthesis <1, 2> (<1> enzymatic reaction, transglycosylation, provides practical technique for industrial production of *p*-nitrophenyl α-maltoheptaoside, useful substrate for assay of human α-amylase in serum and urine [1]; <1> continuous production of maltohexaose on large scale using the immobilized exomaltohexaohydrolase [5]; <2> very advantageous for obtaining pure maltohexaose [6]) [1, 5, 6]

6 Stability

pH-Stability

3 <1> (<1> 70% loss of activity [4]; <1> immobilized enzyme, 93% of activity retains [5]; <1> extracellular enzyme, most of activity lost below pH 3.0 [8]) [4, 5, 8]
4 <1> (<1> transglycosylation, complete inactivation [1]) [1]
4-11 <1> (<1> immobilized enzyme, stable between [5]) [5]
5-10 <1, 3> (<1> extracellular enzyme, more than 80% of activity retained [8, 9]; <3> G-6, stable between [11]) [8, 9, 11]

6-9 <1> (<1> cell-bound enzyme: 20% loss of activity, 60 min, 40°C [4, 8, 9]) [4, 8, 9]
10 <1> (<1> 50% loss of activity [4]) [4]
11 <1> (<1> extracellular enzyme, most of activity lost above pH 11.0 [8]) [8]
11.5 <1> (<1> immobilized enzyme, 85% of activity retained [5]) [5]
Additional information <1> (<1> pH stability is improved by immobilization [5]; <1> extracellular enzyme, relatively wide pH stability [8, 9]; <1> pH stability: extracellular enzyme is a little more stable compared with the cell-bound enzyme [8, 9]) [5, 8, 9]

Temperature stability

20-45 <1> (<1> cell-bound enzyme, stable between, 1 h [7]) [7]
40 <1> (<1> completely stable up to, with Ca^{2+} [4]) [4]
45 <1> (<1> transglycosylation, considerabe inactivation [1]) [1]
50 <1, 3> (<1> 15 min at 50°C: 80% of activity retains, with Ca^{2+} [4]; <1> immobilized enzyme, completely stable up to [5]; <1> cell-bound enzyme, 50% of activity lost at 50°C [7]; <1> cell-bound and extracellular enzyme, stable below 50°C, thermostability improved by addition of Ca^{2+} [8, 9]; <3> G-6, stable below, Ca^{2+} not very effective for protecting the enzyme from heat [11]) [4, 5, 7, 8, 9, 11]
55 <1> (<1> extracellular enzyme, stable up to 55°C in the presence of 2 mM $CaCl_2$ [8]) [8]
60 <1> (<1> all activity lost after 15 min [4]) [4]
75 <1> (<1> immobilized enzyme, 15 min at 75°C: 40% of activity retained [5]) [5]
Additional information <1> (<1> enzyme more thermostable in the presence of Ca^{2+} [4, 8, 9]; <1> thermostability is improved by immobilization [5]) [4, 5, 8, 9]

Organic solvent stability

ethanol <1> (<1> stable and activation [1]) [1]
methanol <1> (<1> stable up to 40%, at 50% methanol: inactivation of transglycosylation and hydrolysis [1]) [1]

General stability information

<1>, Ca^{2+} stabilizes, enzyme is more thermostable in the presence of Ca^{2+} [4, 8, 9]
<1>, immobilized enzyme is very stable [5]
<1>, immobilized enzyme stable to repeated use of 20 cycles [5]
<1>, substrate has protective effect on enzyme stability [4]

References

[1] Ogawa, K.; Uejima, O.; Nakakuki, T.; Usui, T.; Kainuma, K.: Enzymatic synthesis of *p*-nitrophenyl α-maltoheptaoside by transglycosylation of maltohexaose-forming amylase. Agric. Biol. Chem., **54**, 581-586 (1990)

[2] Nakakuki, T.; Azuma, K.; Kainuma, K.: Action patterns of various exo-amylases and the anomeric configurations of their products. Carbohydr. Res., **128**, 297-310 (1984)
[3] Monma, M.; Nakakuki, T.; Kainuma, K.: Formation and hydrolysis of maltohexaose by an extracellular exo-maltohexaohydrolase. Agric. Biol. Chem., **47**, 1769-1774 (1983)
[4] Kainuma, K.; Wako, K.; Kobayashi, S.; Nogami, A.; Suzuki, S.: Purification and some properties of a novel maltohexaose-producing exo-amylase from Aerobacter aerogenes. Biochim. Biophys. Acta, **410**, 333-346 (1975)
[5] Nakakuki, T.; Hayashi, T.; Monma, M.; Kawashima, K.; Kainuma, K.: Immobilization of the exo-maltohexaohydrolase by the irradiation method. Biotechnol. Bioeng., **25**, 1095-1107 (1983)
[6] Hayashi, T.; Akiba, T.; Horikoshi, K.: Production and purification of new maltohexaose-forming amylases from alkalophilic Bacillus sp. H-167. Agric. Biol. Chem., **52**, 443-448 (1988)
[7] Kainuma, K.; Kobayashi, S.; Ito, T.; Suzuki, S.: Isolation and action pattern of maltohexaose producing amylase from Aerobacter aerogenes. FEBS Lett., **26**, 281-285 (1972)
[8] Nakakuki, T.; Azuma, K.; Monma, M.; Kainuma, K.: Purification and some properties of an extracellular exo-maltohexaohydrolase from an Aerobacter aerogenes mutant. J. Jpn. Soc. Starch Sci., **29**, 188-197 (1982)
[9] Nakakuki, T.; Kainuma, K: Exo-maltohexaohydrolase. Recent development of studies on the exo-maltohexaohydrolase. J. Jpn. Soc. Starch Sci., **29**, 138-144 (1982)
[10] Shirokizawa, O.; Akiba, T.; Horikoshi, K.: Nucleotide sequence of the G6-amylase gene from alkalophilic Bacillus sp. H-167. FEMS Microbiol. Lett., **70**, 131-135 (1990)
[11] Takasaki, Y.: Production of maltohexaose by α-amylase from Bacillus circulans G-6. Agric. Biol. Chem., **46**, 1539-1547 (1982)
[12] Taniguchi, H.; Odashima, F.; Igarashi, M.; Maruyama, Y.; Nakamura, M.: Characterization of a potato starch-digesting bacterium and its production of amylase. Agric. Biol. Chem., **46**, 2107-2115 (1982)

Arabinan endo-1,5-α-L-arabinosidase 3.2.1.99

1 Nomenclature

EC number
3.2.1.99

Systematic name
1,5-α-L-arabinan 1,5-α-L-arabinanohydrolase

Recommended name
arabinan endo-1,5-α-L-arabinosidase

Synonyms
1,5-α-arabinanase A <6> (<6> exhibits endo- and exo-mode of action [17]) [17]
ABN <1, 4> [7, 20]
ABN A <1> [20]
ABNA <1, 5> [9, 15]
PPase-C <2> (<2> IFO 3134 [19]) [19]
araban 5-α-L-arabinohydrolase
arabinanase A <6> (<6> exhibits endo- and exo-mode of action [16, 17]) [16, 17]
arabinase, endo-1,5-α-L-
arbA <6> (<6> exhibits endo- and exo-mode of action [16, 17]) [16, 17]
endo-(1 → 5)-α-L-arabinanase
endo-1,5-α-L-arabinanase
endo-1,5-α-L-arabinase
endo-1,5-α-L-arabinase A <1> [20]
endo-1,5-α-arabinanase
endo-ABA <1> [10, 11]
endo-L-arabinanase
endo-α-(1 → 5)-L-arabinanase
endo-α-1,5-arabanase
endo-ara <1> [14]
endo-arabanase
endo-arabinanase
endo-arabinase
endoarabinase
protopectinase C <2> (<2> IFO 3134 [18, 19]) [18, 19]
protopectinase-C <2> (<2> IFO 3134 [19]) [19]

CAS registry number
75432-96-1

2 Source Organism

<1> *Aspergillus niger* (creA mutant, mutant relieved of carbon repression [9]; wild-type N400 and mutant N572, lacks xylulose kinase activity [20]; wild-type N402 and mutant N572, xylulose kinase negative mutant [21]; N400 [22]) [1, 7, 8, 9, 10, 11, 12, 13, 14, 20, 21, 22, 23]
<2> *Bacillus subtilis* (F-11 [2, 3, 5, 6]; IFO 3022 [2, 4, 19]; IFO 3134 [18, 19]) [2, 3, 4, 5, 6, 18, 19]
<3> *Clostridium felsineum var. sikokianum* [2]
<4> *Aspergillus aculeatus* [7, 8]
<5> *Aspergillus nidulans* (wild-type WG096 and cre mutant, carbon catabolite derepressed mutant [15]; wild-type WG096 [20, 23]) [15, 20, 23]
<6> *Pseudomonas fluorescens* (subspecies cellulosa [16, 17]; arabinanase A, exhibits endo- and exo-mode of action [16, 17]) [16, 17]

3 Reaction and Specificity

Catalyzed reaction

endohydrolysis of 1,5-α-arabinofuranosidic linkages in 1,5-arabinans (also acts slowly on beet-arabinan; <1, 2, 5> regulation, induction and carbon catabolite repression of endo-1,5-α-arabinanase [9, 15, 18, 20, 21, 22, 23])

Reaction type

hydrolysis of O-glycosyl bond

Natural substrates and products

S 1,5-α-L-arabinan + H_2O <2> [3]
S Additional information <2> (<2> plays a role in degradation of plant tissues, macerating enzyme [4]) [4]

Substrates and products

S 1,5-α-L-arabinan + H_2O <5> [15, 23]
P L-arabinan oligomers <5> [15]
S 1,5-α-L-arabinan + H_2O <1, 4> (<1, 4> best substrate, endo mechanism [1, 8, 11, 20]; <1> best substrate: linear 1,5-α-L-haze arabinan [11]) [1, 7, 8, 11, 12, 13, 14, 20]
P α-L-arabinopyranosyl-(1-5)-O-α-L-arabinopyranose + α-L-Ara-(1-5)-O-α-L-Ara-(1-5)-O-α-L-Ara <1, 4> (<1, 4> major final end-products, upon exhaustive degradation of substrate some monomers, L-arabinose, are produced [1, 7]; <1, 4> action patterns of initial hydrolysis products [7, 8]; <1,4> product formation, overview [8]; <1> product formation, overview: oligoarabinosides with degree of polymerization between 2 and 5 are the major products, but only trace amounts of free arabinose [12]) [1, 7, 8, 12, 20]
S 1,5-α-L-arabinan + H_2O <2> (<2> F-11 and IFO 3022: best substrate, endo mechanism [2]; <2> F-11 best substrate [3]; <2> IFO 3022, best substrate [4]) [2, 3, 4]

P L-arabinose + α-L-arabinopyranosyl-(1-5)-O-α-L-arabinopyranose + α-L-Ara-(1-5)-O-α-L-Ara-(1-5)-O-α-L-Ara + α-L-Ara-(1-5)-O-α-L-Ara-(1-5)-O-α-L-Ara-(1-5)-O-α-L-Ara + α-L-Ara-(1-5)-O-α-L-Ara-(1-5)-O-α-L-Ara-(1-5)-O-α-L-Ara-(1-5)-O-α-L-Ara <2> (<2> IFO 3022 [4]) [4]

S 1,5-α-L-arabinan + H_2O <2> (<2> F-11 and IFO 3022: best substrate, endo mechanism [2]; <2> F-11 best substrate [3]; <2> IFO 3134, linear 1,5-α-L-arabinan [18]) [2, 3, 5, 18, 19]

P α-L-arabinopyranosyl-(1-5)-O-α-L-arabinopyranose + L-arabinose <2> (<2> end-product [2,3]; <2> F-11: releases arabinose, arabinobiose, arabinotriose and higher oligosaccharides during hydrolysis, end-product are arabinose and arabinobiose [3]) [2, 3, 19]

S 1,5-α-L-arabinan + H_2O <6> (<6> linear arabinan [16, 17]) [16, 17]

P α-L-Ara-(1-5)-O-α-L-Ara-(1-5)-O-α-L-Ara <6> (<6> main product [16, 17]) [16, 17]

S CM-cellulose + H_2O <4> (<4> poor substrate [8]) [8]

P ?

S L-arabinan + H_2O <1, 2, 3, 4, 5, 6> (<1> beet and apple juice ultrafiltration retentate, UFR, arabinan [1]; <1, 2, 3, 4> random hydrolysis of L-arabinan [2, 3, 7, 18, 20]; <1> debranched and branched sugar beet arabinan [11]; <1, 2, 5> sugar beet pulp [13, 18, 19, 20, 22, 23]; <5> apple and sugar beet pulp [23]; <1> debranched arabinan [13]; <1, 4> linear and branched arabinan [8]; <1, 4> apple juice ultrafiltration retentate arabinan, branched sugar-beet arabinan, linear apple arabinan [8]; <2> beet arabinan [2, 3, 4, 5, 6]; <2> sugar-beet arabinan in commercial apple juice concentrate [5]; <2> branched and partially debranched arabinan [6]; <6> linear arabinan [16]; <6> branched sugar beet arabinan containing α-1,3-arabinose side chains: not a substrate [16]) [1, 2, 3, 4, 5, 6, 7, 8, 9, 11, 13, 16, 17, 18, 19, 20, 22, 23]

P L-arabino-oligosaccharides <1, 2, 4, 6> (<1> end-product: L-arabinose disaccharide + L-arabinose trisaccharide, upon exhaustive degradation of substrate some monomers, L-arabinose, are produced [1, 20]; <1, 4> end-product: low molecular weight oligomeric products [8]; <6> main product: L-arabinotriose [16, 17]) [1, 2, 6, 7, 8, 16, 17, 20]

S L-arabino-oligosaccharide + H_2O <1, 4, 6> (<1> linear 1,5-α-L-arabino-oligosaccharide [10]; <1> degree of polymerization 2-8, DP2-8 [10]) [7, 8, 10, 16]

P L-arabino-oligosaccharide <1, 4, 6> (<1, 4> product formation [8, 10]; <1, 4> small oligomeric products [8]; <1> L-arabinose + L-arabinobiose: end-products after exhaustive enzymatic degradation [10]; <1> composition of end-products depends largely on substrate and enzyme concentrations, and on incubation times [10]; <6> main product: L-arabinotriose [16]) [7, 8, 10, 16]

S α-L-Ara-(1-5)-O-α-L-Ara-(1-5)-O-α-L-Ara + H_2O <1, 6> (<1> DP3 [10]; <1, 6> poor substrate [1, 16]; <1> very poor substrate [8]; <1> mild hydrolysis: no significant degradation, exhaustive enzymatic hydrolysis: extremely low degradation [10]) [1, 8, 10, 16]

P L-arabinose + α-L-arabinopyranosyl-(1-5)-O-α-L-arabinopyranose

S α-L-Ara-(1-5)-O-α-L-Ara-(1-5)-O-α-L-Ara-(1-5)-O-α-L-Ara + H_2O <6> [16]

P α-L-Ara-(1-5)-O-α-L-Ara-(1-5)-O-α-L-Ara + L-arabinose <6> [16]

S α-L-Ara-(1-5)-O-α-L-Ara-(1-5)-O-α-L-Ara-(1-5)-O-α-L-Ara + H_2O <1> (<1> DP4 [10]; <1> poor substrate [1]) [1, 10]

P L-arabinose + α-L-arabinopyranosyl-(1-5)-O-α-L-arabinopyranose <1> (<1> end-products after exhaustive enzymatic degradation [10]; <1> product formation [10]; <1> intermediate enzymatic hydrolysis: L-arabinobiose and much less L-arabinose + L-arabinotriose [10]; <1> composition of end-products depends largely on substrate and enzyme concentrations, and on incubation times [10]) [10]

S α-L-Ara-(1-5)-O-α-L-Ara-(1-5)-O-α-L-Ara-(1-5)-O-α-L-Ara-(1-5)-O-α-L-Ara + H_2O <6> [16]

P α-L-Ara-(1-5)-O-α-L-Ara-(1-5)-O-α-L-Ara + α-L-arabinopyranosyl-(1-5)-O-α-L-arabinopyranose <6> [16]

S α-L-Ara-(1-5)-O-α-L-Ara-(1-5)-O-α-L-Ara-(1-5)-O-α-L-Ara-(1-5)-O-α-L-Ara + H_2O <1> (<1> DP5 [10]) [1, 10]

P L-arabinose + α-L-arabinopyranosyl-(1-5)-O-α-L-arabinopyranose <1> (<1> end-products after exhaustive enzymatic degradation [10]; <1> product formation [10]; <1> intermediate enzymatic hydrolysis: L-arabinobiose + L-arabinotriose [10]; <1> composition of end-products depends largely on substrate and enzyme concentrations, and on incubation times [10]) [10]

S α-L-Ara-(1-5)-O-α-L-Ara-(1-5)-O-α-L-Ara-(1-5)-O-α-L-Ara-(1-5)-O-α-L-Ara-(1-5)-O-α-L-Ara + H_2O <6> [16]

P α-L-Ara-(1-5)-O-α-L-Ara-(1-5)-O-α-L-Ara <6> [16]

S α-L-Ara-(1-5)-O-α-L-Ara-(1-5)-O-α-L-Ara-(1-5)-O-α-L-Ara-(1-5)-O-α-L-Ara-(1-5)-O-α-L-Ara + H_2O <1> (<1> DP6 [10]) [10]

P L-arabinose + α-L-arabinopyranosyl-(1-5)-O-α-L-arabinopyranose <1> (<1> end-products after exhaustive enzymatic degradation [10]; <1> product formation [10]; <1> intermediate enzymatic hydrolysis: L-arabinobiose + L-arabinotetraose and L-arabinotriose [10]; <1> composition of end-products depends largely on substrate and enzyme concentrations, and on incubation times [10]) [10]

S α-L-Ara-(1-5)-O-α-L-Ara-(1-5)-O-α-L-Ara-(1-5)-O-α-L-Ara-(1-5)-O-α-L-Ara-(1-5)-O-α-L-Ara-(1-5)-O-α-L-Ara + H_2O <6> [16]

P α-L-Ara-(1-5)-O-α-L-Ara-(1-5)-O-α-L-Ara + α-L-Ara-(1-5)-O-α-L-Ara-(1-5)-O-α-L-Ara-(1-5)-O-α-L-Ara <6> [16]

S α-L-Ara-(1-5)-O-α-L-Ara-(1-5)-O-α-L-Ara-(1-5)-O-α-L-Ara-(1-5)-O-α-L-Ara-(1-5)-O-α-L-Ara-(1-5)-O-α-L-Ara + H_2O <1> (<1> DP7 [10]) [10]

P L-arabinose + α-L-arabinopyranosyl-(1-5)-O-α-L-arabinopyranose <1> (<1> end-products after exhaustive enzymatic degradation [10]; <1> product formation [10]; <1> intermediate enzymatic hydrolysis: L-arabinotriose + L-arabinotetraose [10]; <1> composition of end-products depends largely on substrate and enzyme concentrations, and on incubation times [10]) [10]

S α-L-Ara-(1-5)-O-α-L-Ara-(1-5)-O-α-L-Ara-(1-5)-O-α-L-Ara-(1-5)-O-α-L-Ara-(1-5)-O-α-L-Ara-(1-5)-O-α-L-Ara-(1-5)-O-α-L-Ara + H_2O <6> [16]
P α-L-Ara-(1-5)-O-α-L-Ara-(1-5)-O-α-L-Ara + α-L-Ara-(1-5)-O-α-L-Ara-(1-5)-O-α-L-Ara-(1-5)-O-α-L-Ara-(1-5)-O-α-L-Ara <6> [16]
S α-L-Ara-(1-5)-O-α-L-Ara-(1-5)-O-α-L-Ara-(1-5)-O-α-L-Ara-(1-5)-O-α-L-Ara-(1-5)-O-α-L-Ara-(1-5)-O-α-L-Ara-(1-5)-O-α-L-Ara + H_2O <1> (<1> DP8 [10]) [10]
P L-arabinose + α-L-arabinopyranosyl-(1-5)-O-α-L-arabinopyranose <1> (<1> end-products after exhaustive enzymatic degradation [10]; <1> product formation [10]; <1> intermediate enzymatic hydrolysis: L-arabinotriose + L-arabinopentaose and L-arabinobiose + L-arabinohexaose and L-arabinotetraose [10]; <1> composition of end-products depends largely on substrate and enzyme concentrations, and on incubation times [10]) [10]
S α-L-arabinopyranosyl-(1-5)-O-α-L-arabinopyranose + H_2O <1, 4> (<1> DP2 [10]; <1, 4> at very low rate, with high enzyme concentration [7]; <1> no or extremely low degradation [10]; <1> not a substrate [1]) [7, 10]
P L-arabinose <1, 4> [7]
S arabinogalactan + H_2O <1, 4, 5> (<1, 4> potato arabinogalactan [8, 14]; <1> for optimal breakdown of potato arabinogalactan combined action of endo-arabinase and endo-galactanase is required [14]; <5> very poor substrate [23]; <4> onion arabinogalactan: not a substrate [8]) [8, 14, 23]
P ?
S carboxymethylarabinan + H_2O <6> [16]
P ?
S cell wall polysaccharide + H_2O <2> (<2> F-11, apple [5]) [5]
P arabinose + galactose + rhamnose + fucose <2> (<2> F-11, 34.3% arabinose + 10% galactose + 2.6% rhamnose/fucose [5]) [5]
S cell wall polysaccharide + H_2O <2> (<2> F-11, sycamore [2, 6]; <2> F-11, primary cell wall polysaccharides [6]) [2, 5, 6, 19]
P L-arabinan <2> [2]
S feruloylated oligosaccharide + H_2O <1> (<1> from sugar beet pulp, SBP [13]; <1> endo-arabinanase alone: very poor substrate, endo-arabinanase and α-L-arabinofuranosidase together: good substrate, best: combination endo-arabinanase, α-L-arabinofuranosidase and cinnamoyl esterase [13]) [13]
P ferulic acid + ? <1> (<1> free and esterified ferulic acid [13]; <1> efficient releasing free ferulic acid from sugar beet pulp requires strong synergy involving three enzymes: endo-arabinanase, α-L-arabinofuranosidase and cinnamoyl esterase [13]) [13]
S galactan + H_2O <1> (<1> potato galactan: poor substrate, citrus galactan: very poor substrate [1]; <1> very poor substrate [11]; <6> not a substrate [16]) [1, 11]
P ?

S protopectin + H_2O <2> (<2> IFO 3134, protopectin of sugar-beet and apple pulp [18]; <2> IFO 3134, protopectin of sugar beet pulp and various citrus fruit peels, but much less on that from lemon peel [19]) [18, 19]
P pectin <2> (<2> IFO 3134, soluble pectin [18]; <2> IFO 3134, highly polymerized pectin [19]) [18, 19]
S Additional information <1, 2, 4, 5, 6> (<1, 4> substrate specificity, overview [1, 7, 8, 10, 11]; <1, 4> product specificity, overview [7, 8, 10]; <1> composition of end-products depends largely on substrate and enzyme concentrations, and on incubation times [10]; <2> F-11, substrate specificity [6]; <1, 4> substrate binding site, action pattern, subsite mapping [7]; <2> IFO 3022: potato disc macerating activity, able to loosen potato tissue [2, 4]; <1> inactive towards synthetic arabinosides, *p*-nitrophenyl-α-L-arabinoside, arabinose disaccharide, arabinogalactan, stractan and xylan from oat spelts [1]; <2> inactive towards phenyl and *p*-nitrophenyl α-L-arabinofuranoside, arabinoxylan, arabinogalactan, and gum arabic [2]; <2> F-11: inactive on phenyl α-L-arabinofuranoside, *p*-nitrophenyl α-D-galactopyranoside, *p*-nitrophenyl β-D-galactopyranoside, arabinoxylan, arabinogalactan and gum arabic [3]; <2> IFO 3022: inactive on *p*-nitrophenyl α-L-arabinofuranoside, *p*-nitrophenyl α-D-galactopyranoside, *p*-nitrophenyl β-D-galactopyranoside, arabinogalactan, carboxymethyl cellulose, filter paper, pectin and polygalacturonic acid [4]; <2> F-11, inactive on potato galactan [5]; <1, 4> no bi-substrate reactions such as condensation or transglycosylation [7]; <1, 4> inactive on arabinobiitol [7]; <4> inactive on *p*-nitrophenylarabinofuranoside, onion arabinogalactan, stractan, wheat arabinoxylan, oat spelts arabinoxylan, avicel, polygalacturonic acid [8]; <6> inactive on cellulose, avicel, amorphous cellulose, carboxymethylcellulose, xylan, galactan, mannan, 4-nitrophenyl α-L-arabinofuranoside, 4-nitrophenyl α-L-arabinopyranoside, 4-nitrophenyl β-xyloside and 2,4-dinitrophenyl β-xylobioside [16]; <2> IFO 3134, inactive on polygalacturonic acid and soluble pectin [19]; <5> inactive on *p*-nitrophenyl-α-L-arabinofuranoside [23]) [1, 2, 3, 4, 5, 6, 7, 8, 10, 11, 16, 19, 23]
P ?

Inhibitors

2-mercaptoethanol <2> (<2> IFO 3134, slight inhibition [19]) [19]
$AgNO_3$ <2> (<2> IFO 3134 [19]) [19]
CH_3COONH_2 <2> (<2> IFO 3134, complete inhibition of enzyme production [18]) [18]
CH_3COONa <2> (<2> IFO 3134, inhibition of enzyme production [18]) [18]
$CdCl_2$ <2> (<2> IFO 3134 [19]) [19]
$Fe_2(SO_4)_3$ <2> (<2> IFO 3134 [19]) [19]
Fe^{2+} <2> (<2> IFO 3022 [4]) [4]
Fe^{3+} <2> (<2> IFO 3022 [4]) [4]
Hg^{2+} <1, 2> (<2> F-11 [3]; <2> IFO 3022, completely [4]; <1> completely [11]; <2> IFO 3134, completely [19]) [3, 4, 11, 19]
$HgCl_2$ <1, 2> (<1> completely [11]; <2> IFO 3134, completely [19]) [11, 19]
$MgSO_4$ <1> (<1> very slight inhibition [11]) [11]

Mn^{2+} <2> (<2> IFO 3134, completely [19]) [19]
$MnCl_2$ <2> (<2> IFO 3134, completely [19]) [19]
NaN_3 <2> (<2> IFO 3134, slight inhibition [19]) [19]
NaNO2 <2> (<2> IFO 3134, strong inhibition [19]) [19]
$PbCl_2$ <2> (<2> IFO 3134 [19]) [19]
Sn^{2+} <2> (<2> IFO 3134, strong inhibition [19]) [19]
$SnCl_2$ <2> (<2> IFO 3134, strong inhibition [19]) [19]
Zn^{2+} <1> [11]
$ZnSO_4$ <1> [11]
branched arabinofuranosides <1, 4> [8]
iodoacetamide <1> (<1> very slight inhibition [11]) [11]
o-phenanthroline <2> (<2> IFO 3134, slight inhibition [19]) [19]
p-chloromercuribenzoate <2> (<2> IFO 3134, slight inhibition [19]) [19]
Additional information <1, 2> (<2> IFO 3022: no inhibition by L-arabonic-γ-lactone, D-and L-galactonic-γ-lactones [4]; <1> no inhibition by $CaCl_2$, EDTA, L-arabono-γ-lactone, L-arabinose, D-arabitol [11]; <2> IFO 3134, inhibition of enzyme production [18]) [4, 11, 18, 19]

Activating compounds

L-arabinose <1, 5> (<1,5> induction of endo-arabinase, addition of D-glucose prevents this induction [9, 15]) [9, 15]
L-arabitol <1, 5> (<1, 5> induction of endo-arabinase, addition of D-glucose prevents this induction [9, 15, 20, 21, 23]) [9, 15, 20, 21, 23]
apple pulp <5> (<5> induction of endo-arabinase and activation [23]) [23]
sugar beet pulp <1, 5> (<1, 5> induction of endo-arabinase and activation [22, 23]) [22, 23]

Metals, ions

$(NH_4)_2HPO_4$ <2> (<2> IFO 3134, stimulation of enzyme production and activity [18]) [18]
$(NH_4)H_2PO_4$ <2> (<2> IFO 3134, stimulation of enzyme production and activity [18]) [18]
$BaCl_2$ <2> (<2> IFO 3134, slight activation [19]) [19]
$CaCl_2$ <2> (<2> IFO 3134, slight activation [19]) [19]
$CoCl_2$ <2> (<2> IFO 3134, very slight activation [19]) [19]
Hg_2Cl_2 <2> (<2> IFO 3134, slight activation [19]) [19]
K_2HPO_4 <2> (<2> IFO 3134, stimulation of enzyme production and activity [18]) [18]
KH_2PO_4 <2> (<2> IFO 3134, stimulation of enzyme production and activity [18]) [18]
$MgCl_2$ <2> (<2> IFO 3134, activation [19]) [19]
Na_2HPO_4 <2> (<2> IFO 3134, stimulation of enzyme production and activity [18]) [18]
NaH_2PO_4 <2> (<2> IFO 3134, stimulation of enzyme production and activity [18]) [18]
Additional information <2> (<2> F-11, no divalent cation requirement [6]; <2> IFO 3134, stimulation of enzyme production by several phosphates, but not by chloride, acetate or sulfate [18]) [6, 18]

Specific activity (U/mg)

0.36 <1> (<1> apple juice ultrafiltration retentate, UFR, arabinan [1]) [1]
1.2 <1> (<1> best substrate: 1,5-α-L-arabinan [1]) [1]
5.8 <2> (<2> IFO 3134 [19]) [19]
7.78 <2> (<2> IFO 3022 [2, 4]) [2, 4]
14 <1> (<1> substrate: branched sugar beet arabinan [11]) [11]
46.5 <2> (<2> F-11 [5]) [5]
48.8 <2> (<2> F-11 [3]) [3]
81.1 <1> (<1> substrate: debranched sugar beet arabinan [11]) [11]
90.2 <1> (<1> best substrate: linear 1,5-α-L-arabinan [11]) [11]
429 <2> (<2> F-11 [6]) [6]
Additional information <6> [16]

K_m-Value (mM)

0.052 <1> (1,5-α-L-arabinan, <1> CM-linear 1,5-α-L-arabinan [23]) [23]
0.336 <5> (1,5-α-L-arabinan, <5> CM-linear 1,5-α-L-arabinan [23]) [23]
0.388 <1> (α-L-Ara-(1-5)-O-α-L-Ara-(1-5)-O-α-L-Ara-(1-5)-O-α-L-Ara-(1-5)-O-α-L-Ara-(1-5)-O-α-L-Ara-(1-5)-O-α-L-Ara-(1-5)-O-α-L-Ara) [7]
0.592 <4> (α-L-Ara-(1-5)-O-α-L-Ara-(1-5)-O-α-L-Ara-(1-5)-O-α-L-Ara-(1-5)-O-α-L-Ara-(1-5)-O-α-L-Ara-(1-5)-O-α-L-Ara-(1-5)-O-α-L-Ara) [7]
4.2 <1> (1,5-α-L-arabinan) [1]
9.3 <1> (L-arabinan, <1> apple juice ultrafiltration retentate, UFR, arabinan, linearised [1]) [1]
26 <1> (L-arabinan, <1> apple juice ultrafiltration retentate, UFR, arabinan [1]) [1]
Additional information <1, 4, 5, 6>(<1, 4> kinetic data and mechanism [7]; <1, 5> kinetic data [11, 22, 23]) [1, 7, 11, 16, 22, 23]

pH-Optimum

4.5-5 <1> [11]
4.5-5.5 <1> [8]
4.6 <1> (<1> maximum, pH optimum is quite broad [22]) [22, 23]
4.8 <1> (<1> maximum [11]) [11]
5 <1> [1]
5.5 <4, 5> [8, 23]
6 <2> (<2> F-11 [2, 3, 5, 6]; <2> IFO 3022 [2, 4]; <2> IFO 3134 [18]; <2> IFO 3134, at 37°C [19]) [2, 3, 4, 5, 6, 18, 19]
8 <2> (<2> IFO 3134, at 60°C [19]) [19]

pH-Range

3-9 <2> (<2> IFO 3022 [4]) [4]
4-9 <2> (<2> F-11 [3, 6]) [3, 6]
4-10 <2> (<2> IFO 3134 [19]) [19]

Temperature optimum (°C)

50 <1> [1]
51 <1> [22, 23]
55 <1> [11]

60 <2> (<2> IFO 3134 [18, 19]) [18, 19]
68 <5> [23]

Temperature range (°C)
20-80 <2> (<2> IFO 3134, activity between [19]) [19]

4 Enzyme Structure

Molecular weight
27000 <2> (<2> 3134, gel filtration on Toyopearl HW-55S [19]) [19]
30000 <2> (<2> IFO 3134, SDS-PAGE [18, 19]) [18, 19]
32000 <2> (<2> F-11, SDS-PAGE [2, 6]) [2, 6]
32500 <1> (<1> calculated from nucleotide sequence [20]) [20]
33000 <2> (<2> IFO 3022, SDS-PAGE [2, 4]) [2, 4]
33900 <1> (<1> MALDI-TOF-MS [11]) [11]
34000 <6> (<6> SDS-PAGE [16]) [16, 17]
35000 <1> (<1> SDS-PAGE [1]) [1]
35710 <2> (<2> IFO 3134, calculated from nucleotide sequence [18]) [18]
36000 <2> (<2> F-11, gel filtration [6]) [6]
39440 <6> (<6> calculated from nucleotide sequence [16]) [16]
40000 <5> (<5> SDS-PAGE [23]) [23]
41000 <1> (<1> gel filtration [11]) [11]
42500 <1> (<1> SDS-PAGE [11]) [11]
43000 <1> (<1> SDS-PAGE [20, 22, 23]) [20, 22, 23]
45000 <1, 4> (<1,4> SDS-PAGE [8]) [8]

Subunits
monomer <2> (<2> F-11, 1 * 32000, SDS-PAGE [6]; <2> IFO 3134, 1 * 30000, SDS-PAGE [19]) [6, 19]

Posttranslational modification
glycoprotein <1> (<1> O-linked glycosylation: approximately 10000 Da [20]) [1, 11, 20]

5 Isolation/Preparation/Mutation/Application

Source/tissue
culture medium <1, 2, 5> (<2> F-11 [3, 6]; <2> IFO 3022 [4]) [3, 4, 6, 20, 21, 22, 23]

Localization
extracellular <1, 2, 5, 6> (<2> F-11 [6]) [6, 15, 16, 20, 21, 22, 23]

Purification
<1> (by fast protein liquid chromatography, FPLC [12]; wild-type N400 [20, 22]) [1, 7, 10, 11, 12, 20, 22]
<2> (F-11 [2, 3, 5, 6]; IFO 3022 [2, 4]; IFO 3134 [19]) [2, 3, 4, 5, 6, 19]

<4> [7, 8]
<5> [23]
<6> (recombinant arabinanase [16]) [16]

Crystallization
<6> (resolution limit of 3.5 A [17]; preliminary X-ray analysis [17]) [17]

Cloning
<1> (cDNA clone, overexpression in Aspergillus niger and Aspergillus nidulans [20]) [20]
<2> (IFO 3134 [18]) [18]
<6> (expressed in E. coli [16]; hyperexpression in E. coli [16]) [16]

Application
analysis <1, 2> (<2> useful in structural analysis of primary cell wall polysaccharides [6, 10, 11, 12, 18]; <1> structural analysis of plant cell pectic polysaccharides and arabinans [12, 18]; <2> useful for studies on structure of protopectin [19]) [6, 10, 11, 12, 18, 19]
nutrition <1, 2, 4> (<2> role in juice clarification [5]; <1, 4> commercial applications in fruit and vegetable processing, particularly in apple juice manufacture where their use prevents haze formation in pectinase-treated apple pulp by depolymerizing debranched arabinan, linear α-1,5-arabinan, formed by the action of α-L-arabinofuranosidases during this process [7, 8]; <2> useful for pectin extraction from citrus peel from fruit-juice industries and sugar-beet pulp from sugar factories [18]; <2> useful for pectin production from sugar beet pulp [19]) [5, 7, 8, 11, 18, 19]
synthesis <2> (<2> applied to the refinement of cotton fiber, enzyme is able to release the cotton fiber coating, yielding product of high quality but with lower amounts of wastes [18]) [18]

6 Stability

pH-Stability
3.5-8 <1> (<1> 85% activity retained between, 67 h at 25°C [11]) [11]
5-9 <2> (<2> IFO 3134, stable between at 50°C: 30 min [19]) [19]
5.5-6.3 <4> (<4> stable between [8]) [8]
7-10 <2> (<2> F-11, stable between [3]) [3]
9 <2> (<2> F-11, most stable at [3]) [3]

Temperature stability
40 <1> (<1> long incubation times, 26.5 h, at 40°C: 90% activity retained [11]) [11]
45 <1> (<1> up to 45°C: more than 95% activity retained for 3 h [11]) [11]
50 <1> (<1> up to, full retention of activity, sharp drop in activity beyond 50°C [1]) [1]
60 <2> (<2> IFO 3134, up to, retains more than 80% of activity after 10 min [19]) [19]

General stability information

<2>, IFO 3134, EDTA stabilizes during purification [19]

<2>, IFO 3134, enzyme stable against protease secreted by the microorganism even during 2 days cultivation [18]

Storage stability

<1>, -20°C, 10 mM sodium acetate or piperazine buffer, pH 5.0 dispensed in small tubes [1]

References

[1] Rombouts, F.M.; Voragen, A.G.J.; Searle-van Leeuwen, M.F.; Geraeds, C.C.J.M.; Schols, H.A.; Pilnik, W.: The arabinanases of Aspergillus niger - purification and characterization of two α-L-arabinofuranosidases and an endo-1,5-α-L-arabinanase. Carbohydr. Polym., **9**, 25-47 (1988)

[2] Kaji, A.: L-Arabinosidases. Adv. Carbohydr. Chem. Biochem., **42**, 383-394 (1984)

[3] Kaji, A.; Saheki, T.: Endo-arabinanase from Bacillus subtilis F-11. Biochim. Biophys. Acta, **410**, 354-360 (1975)

[4] Yoshihara, O.; Kaji, A.: An endo-1,5-α-L-arabinase, which can disintegrate potato tissue. Agric. Biol. Chem., **47**, 1935-1940 (1983)

[5] Voragen, A.-G.-J.; Geerst, F.; Pilnik, W.: Hemi-cellulases in enzymic fruit processing. Util. Enzymes Technol. Aliment., Symp. Int. (Dupuy P., ed.), Tech. Doc. Lavoisier Paris, -fr., 497-502 (1982)

[6] Weinstein, L.; Albersheim, P.: Structure of plant cell walls. IX. Purification and partial characterization of a wall-degrading endo-arabanase and an arabinosidase from Bacillus subtilis. Plant Physiol., **63**, 425-432 (1979)

[7] Pitson, S.M.; Voragen, A.G.J.; Vincken, J.-P.; Beldman, G.: Action patterns and mapping of the substrate-binding regions of endo-(1 → 5)-α-L-arabinanases from Aspergillus niger and Aspergillus aculeatus. Carbohydr. Res., **303**, 207-218 (1997)

[8] Beldman, G.; Searle-van Leeuwen, M.J.F.; De Ruiter, G.A.; Siliha, H.A.; Voragen, A.G.J.: Degradation of arabinans by arabinanases from Aspergillus aculeatus and Aspergillus niger. Carbohydr. Polym., **20**, 159-168 (1993)

[9] Ruijter, G.J.G.; Vanhanen, S.A.; Gielkens, M.M.C.; van de Vondervoort, P.J.I.; Visser, J.: Isolation of Aspergillus niger creA mutants and effects of the mutations on expression of arabinases and L-arabinose catabolic enzymes. Microbiology, **143**, 2991-2998 (1997)

[10] Dunkel, M.P.H.; Amado, R.: Analysis of endo-(1 → 5)-α-L-arabinanase degradation patterns of linear (1 → 5)-α-L-arabino-oligosaccharides by high-performance anion-exchange chromatography with pulsed amperometric detection. Carbohydr. Res., **268**, 151-158 (1995)

[11] Dunkel, M.P.H.; Amado, R.: Purification and physico-chemical properties of an endo-1,5-α-L-arabinanase (EC 3.2.1.99) isolated from an Aspergillus niger pectinase preparation. Carbohydr. Polym., **24**, 247-263 (1994)

[12] Lerouge, P.; O'Neill, M.A.; Darvill, A.G.; Albersheim, P.: The purification of commercially available endo-α-L-arabinanases and α-L-arabinosidase for use in the structural analysis of pectic polysaccharides. Carbohydr. Res., **243**, 373-378 (1993)
[13] Kroon, P.A.; Williamson, G.: Release of ferulic acid from sugar-beet pulp by using arabinanase, arabinofuranosidase and an esterase from Aspergillus niger. Biotechnol. Appl. Biochem., **23**, 263-267 (1996)
[14] van de Vis, J.W.; Searle-van Leeuwen, M.J.F.; Siliha, H.A.; Kormelink, F.J.M.; Voragen, A.G.J.: Purification and characterization of endo-1,4-β-D-galactanase from Aspergillus niger and Aspergillus aculeatus: use in combination with arabinanases from Aspergills niger in enzymic conversion of potato arabinogalactan. Carbohydr. Polym., **16**, 167-187 (1991)
[15] van der Veen, P.; Arst, H.N.;Jr.; Flipphi, M.J.A.; Visser, J.: Extracellular arabinases in Aspergillus nidulans: the effect of different cre mutations on enzyme levels. Arch. Microbiol., **162**, 433-440 (1994)
[16] McKie, V.A.; Black, G.W.; Millward-Sadler, S.J.; Hazlewood, G.P.; Laurie, J.I.; Gilbert, H.J.: Arabinanase A from Pseudomonas fluorescens subsp. cellulosa exhibits both an endo- and an exo-mode of action. Biochem. J., **323**, 547-555 (1997)
[17] Scott, M.; Pickersgill, R.W.; Hazlewood, G.P.; Gilbert, H.J.; Harris, G.W.: Crystallization and preliminary X-ray analysis of arabinanase A from Pseudomonas fluorescens subspecies cellulosa. Acta Crystallogr. Sect. D, **D55**, 544-546 (1999)
[18] Sakamoto, T.; Yamada, M.; Kawasaki, H.; Sakai, T.: Molecular cloning and nucleotide sequence of an endo-1,5-α-L-arabinase gene from Bacillus subtilis. Eur. J. Biochem., **245**, 708-714 (1997)
[19] Sakai, T.; Sakamoto, T.: Purification and some properties of a protopectin-solubilizing enzyme that has potent activity on sugar beet protopectin. Agric. Biol. Chem., **54**, 879-889 (1990)
[20] Flipphi, M.J.A.; Panneman, H.; van der Veen, P.; Visser, J.; de Graaff, L.H.: Molecular cloning, expression and structure of the endo-1,5-α-L-arabinase gene of Aspergillus niger. Appl. Microbiol. Biotechnol., **40**, 318-326 (1993)
[21] van der Veen, P.; Flipphi, M.J.A.; Voragen, A.G.J.; Visser, J.: Induction of extracellular arabinases on monomeric substrates in Aspergillus niger. Arch. Microbiol., **159**, 66-71 (1993)
[22] van der Veen, P.; Flipphi, M.J.A.; Voragen, A.G.J.; Visser, J.: Induction, purification and characterisation of arabinases produced by Aspergillus niger. Arch. Microbiol., **157**, 23-28 (1991)
[23] Ramon, D.; van der Veen, P.; Visser, J.: Arabinan degrading enzymes from Aspergillus nidulans: induction and purification. FEMS Microbiol. Lett., **113**, 15-22 (1993)

Mannan 1,4-β-mannobiosidase 3.2.1.100

1 Nomenclature

EC number
3.2.1.100

Systematic name
1,4-β-D-mannan mannobiohydrolase

Recommended name
mannan 1,4-β-mannobiosidase

Synonyms
exo-1,4-β-mannobiohydrolase
exo-β-mannanase
exo-β-mannanase
mannobiohydrolase, exo-1,4-β-

CAS registry number
81811-49-6

2 Source Organism

<1> *Aeromonas hydrophila* (strain F-25, ssp. anaerogenes [1]) [1]

3 Reaction and Specificity

Catalyzed reaction
hydrolysis of 1,4-β-D-mannosidic linkages in 1,4-β-D-mannans, to remove successive mannobiose residues from non-reducing chain ends

Reaction type
hydrolysis of O-glycosyl bond
transglycosylation [1]

Substrates and products
S codium mannan + H_2O <1> [1]
P mannobiose <1> (<1> sole product [1]) [1]
S coffee mannan + H_2O <1> [1]
P mannobiose <1> (<1> sole product [1]) [1]
S mannobiosylmannitol + H_2O <1> [1]
P mannobiose + mannitol <1> [1]

S mannopentaose + H_2O <1> [1]
P mannobiose + mannose <1> [1]
S mannotetraose <1> (<1> transglycosylation [1]) [1]
P mannohexaose <1> [1]
S mannotetraose + H_2O <1> [1]
P mannobiose <1> [1]
S mannotetraosylmannitol + H_2O <1> [1]
P mannobiose + mannosylmannitol <1> [1]
S mannotriose + H_2O <1> [1]
P mannobiose + mannose <1> [1]
S mannotriose + mannose <1> (<1> transglycosylation [1]) [1]
P mannopentaose <1> [1]
S Additional information <1> (<1> not: mannobiose, *p*-nitrophenyl-β-D-mannoside, konjac glucomannan, guar gum galactomannan [1]) [1]
P ?

Inhibitors

Ag^+ <1> (<1> 1 mM, complete inhibition [1]) [1]
Cu^{2+} <1> (<1> 1 mM, complete inhibition [1]) [1]
EDTA <1> (<1> 1 mM, 30-50% inhibition [1]) [1]
Fe^{3+} <1> (<1> 1 mM, 30-50% inhibition [1]) [1]
Hg^{2+} <1> (<1> 1 mM, complete inhibition [1]) [1]
Pb^{2+} <1> (<1> 1 mM, 30-50% inhibition [1]) [1]
Zn^{2+} <1> (<1> 1 mM, 30-50% inhibition [1]) [1]
p-chloromercuribenzoate <1> (<1> 1 mM, 30-50% inhibition [1]) [1]

Specific activity (U/mg)

86.5 <1> [1]

K_m-Value (mM)

0.13 <1> (mannopentaose) [1]
0.24 <1> (mannotetraose) [1]
0.51 <1> (mannobiose) [1]

pH-Optimum

6 <1> [1]

Temperature optimum (°C)

37 <1> (<1> assay at [1]) [1]

4 Enzyme Structure

Molecular weight

64000 <1> (<1> gel filtration [1]) [1]

5 Isolation/Preparation/Mutation/Application

Source/tissue
culture fluid <1> [1]

Localization
extracellular <1> [1]

Purification
<1> (F-25 [1]) [1]

6 Stability

pH-Stability
5-8.5 <1> (20°C, 20 h [1]) [1]

Temperature stability
45 <1> (<1> 15 min, stable below [1]) [1]
55 <1> (<1> 15 min, complete loss of activity [1]) [1]

References

[1] Araki, T.; Kitamikado, M.: Purification and characterization of a novel exo-β-mannanase from Aeromonas sp. F-25. J. Biochem., **91**, 1181-1186 (1982)

Mannan endo-1,6-α-mannosidase 3.2.1.101

1 Nomenclature

EC number
3.2.1.101

Systematic name
1,6-α-D-mannan-mannanohydrolase

Recommended name
mannan endo-1,6-α-mannosidase

Synonyms
endo-1,6-α-mannanase
endo-α(1$\rightarrow$6)mannanase
endo-α-1$\rightarrow$6-D-mannanase
endo-α1$\rightarrow$6-D-mannase
endo-α1$\rightarrow$6-mannanase
endoM <1> [2]
mannanase, endo-1,6-α-

CAS registry number
58500-56-4
98148-86-8

2 Source Organism

<1> *Bacillus circulans* (Bacillus sp. TN-31: probably Bacillus circulans [1]) [1, 2, 3]

3 Reaction and Specificity

Catalyzed reaction
random hydrolysis of 1,6-α-D-mannosidic linkages in unbranched 1,6-mannans (<1> enzyme hydrolyzes α-1,6-D-mannosidic and not β-1,6-D-mannosidic linkages [1])

Reaction type
hydrolysis of O-glycosyl bond

Substrates and products

S α-1,6-D-mannan + H_2O <1> (<1> unbranched, with more than 80% α-1,6-D-mannosidic linkages, best substrate [1]; <1> Saccharomyces cerevisiae X2180-1A backbone mannan [1]; Saccharomyces cerevisiae X2180-1A-5 mannan [1, 3]) [1, 2, 3]

P mannose + α-1,6-mannobiose <1> (<1> end-product, with intermediate formation of α-1,6-mannooligosaccharides of various sizes [1]) [1]

S α-1,6-mannohexaose + H_2O <1> [1]

P mannose + α-1,6-mannobiose <1> (<1> end-product, with intermediate formation of mannose, mannobiose and mannotriose, the latter apparently converts more slowly to the smaller products [1]) [1]

S α-1,6-mannooctaose + H_2O <1> (<1> largest substrate [1]) [1]

P mannose + α-1,6-mannobiose <1> (<1> end-product, with intermediate formation of mannose, mannobiose and mannotriose, the latter apparently converts more slowly to the smaller products [1]) [1]

S α-1,6-mannooligosaccharides + H_2O <1> (<1> smallest substrate: α-1,6-mannotriose, largest substrate: α-1,6-mannooctaose [1]; <1> unbranched α-1,6 outer chain of N-linked oligosaccharides [2]) [1, 2]

P mannose + α-1,6-mannobiose <1> (<1> end-product, with intermediate formation of mannose, mannobiose and mannotriose, the latter apparently converts more slowly to the smaller products [1]) [1]

S α-1,6-mannopentaose + H_2O <1> [1]

P mannose + α-1,6-mannobiose <1> (<1> end-product, with intermediate formation of mannose, mannobiose and mannotriose, the latter apparently converts more slowly to the smaller products [1]) [1]

S α-1,6-mannosylated pro-α-factor + H_2O <1> [2]

P ?

S α-1,6-mannotetraose + H_2O <1> [1]

P mannose + α-1,6-mannobiose <1> (<1> end-product, with intermediate formation of mannose, mannobiose and mannotriose, the latter apparently converts more slowly to the smaller products [1]) [1]

S α-1,6-mannotriose + H_2O <1> (<1> smallest substrate [1]) [1]

P mannose + α-1,6-mannobiose <1> [1]

S reduced mannohexaose + H_2O <1> (<1> poor substrate [1]) [1]

P ?

S reduced mannooctaose + H_2O <1> (<1> poor substrate [1]) [1]

P ?

S reduced mannopentaose + H_2O <1> (<1> poor substrate [1]) [1]

P ?

S reduced mannotetraose + H_2O <1> (<1> poor substrate [1]) [1]

P ?

S Additional information <1> (<1> substrate specificity [1]; <1> enzyme hydrolyzes α-1,6-D-mannosidic and not β-1,6-D-mannosidic linkages [1]; <1> no activity on branched α-mannan, α-1,6-mannans that contain side chains, little or no activity with glucans: amylose, dextran, cellulose, laminarin, α-1,6-mannobiose, reduced mannobiose, reduced mannotriose, no activity with methyl α-D-mannoside or *p*-nitrophenyl α-D-mannoside, no activity with β-1,4-D-mannan, methyl α-D-mannopyranoside or *p*-nitrophenyl α-D-mannopyranoside [1]; <1> no activity with α-1,2-mannosylated pro-α-factor [2]) [1, 2, 3]

P ?

Inhibitors

Cu^{2+} <1> [1]
EDTA <1> (<1> 70% inhibition, 90% of original activity restored with 10 mM Ca^{2+} [1]) [1]
Fe^{2+} <1> [1]
Mg^{2+} <1> (<1> very slight inhibition [1]) [1]
N-ethylmaleimide <1> (<1> 2 mM: 50% inhibition [1]) [1]
diisopropyl fluorophosphate <1> [1]
ethanol <1> (<1> 10% ethanol: inactivation [1]) [1]
iodoacetamide <1> (<1> 2 mM: 50% inhibition [1]) [1]
mannan <1> (<1> Saccharomyces cerevisiae wild type mannan, 20% inhibition [1]) [1]
p-chloromercuribenzoate <1> (<1> 1 mM: complete inhibition, 0.001 mM: 50% inhibition [1]) [1]
reduced α-1,6-mannohexaose <1> (<1> 54% inhibition, also serves as a substrate [1]) [1]
reduced α-1,6-mannotriose <1> (<1> competitive inhibitor, 92% inhibition [1]) [1]

Activating compounds

N-ethylmaleimide <1> (<1> 1 mM: slight activation [1]) [1]
ethanol <1> (<1> 1% v/v ethanol: slight activation [1]) [1]
iodoacetamide <1> (<1> 1 mM: slight activation [1]) [1]
propanol <1> (<1> slight activation [1]) [1]

Metals, ions

Ca^{2+} <1> (<1> activation, required for full activity [1]) [1]
Co^{2+} <1> (<1> very slight activation [1]) [1]
Mn^{2+} <1> (<1> activation [1]) [1]

Turnover number (min^{-1})

440 <1> (α-1,6-D-mannan) [1]

Specific activity (U/mg)

0.45 <1> [2]
1.87 <1> [1]

K_m-Value (mM)

0.0044 <1> (α-1,6-D-mannan) [1]
0.072 <1> (α-1,6-mannohexaose) [1]
0.12 <1> (α-1,6-mannopentaose) [1]
0.33 <1> (α-1,6-mannotetraose) [1]
1 <1> (α-1,6-mannotriose) [1]
4 <1> (α-1,6-mannobiose) [1]
Additional information <1> (<1> kinetic data [1]) [1]

pH-Optimum
5.5-7 <1> [1, 3]

pH-Range
5-7.5 <1> [1, 3]

Temperature optimum (°C)
50 <1> (<1> pH 5.5 [1]) [1, 3]

4 Enzyme Structure

Molecular weight
125000 <1> (<1> gel filtration and SDS-PAGE [1]) [1]
131000 <1> (<1> sedimentation equilibrium [1]) [1]

Subunits
monomer <1> (<1>, 1 * 125000, SDS-PAGE [1]) [1]

5 Isolation/Preparation/Mutation/Application

Source/tissue
culture filtrate <1> [1]

Localization
extracellular <1> [1, 2, 3]

Purification
<1> (partial [2]) [1, 2]

Application
analysis <1> (<1> used to determine the structure of mannan-protein linkage region [1, 3]; <1> used to define kinetics of mannosylation of pro-α-factor, pro-αf, to distinguish α-1,6-mannosylated and α-1,2-mannosylated forms of glycoproteins [2]) [1, 2, 3]

6 Stability

Temperature stability
55 <1> (<1> stable at, in the presence of substrate, for at least 3 h [1]) [1]
60-65 <1> (<1> 5 min, denatured [1]) [1]
Additional information <1> (<1> unusually heat-stable [1]) [1]

Organic solvent stability
ethanol <1> (<1> 1% v/v ethanol: stable and slight activation, effect decreases at higher concentrations, at 10% ethanol: inactivation [1]) [1]
propanol <1> (<1> stable and slight activation [1]) [1]

Storage stability

<1>, -10°C, crude enzyme solution, 0.05 M citrate-phosphate buffer, pH 5.5, 0.001 mM $CaCl_2$ [3]
<1>, 4°C, 0.1 M citrate-phosphate buffer, pH 6.0, or 0.1 M potassium phosphate buffer, pH 6.8, 1 mM $CaCl_2$, stable for at least 1 month [1]
<1>, 4°C, 1 ml aliquots, 0.1 M citrate-phosphate buffer, pH 6.0, or 0.1 M potassium phosphate buffer, pH 6.8, 1 mM $CaCl_2$, 1 year, 80% original activity retained [2]
<1>, 4°C, crude enzyme solution, 0.1 M citrate-phosphate buffer, pH 5.5, more than 1 month, without appreciable loss of activity [3]
<1>, frozen and lyophilized, crude enzyme solution, more than 1 month, without appreciable loss of activity [3]

References

[1] Nakajima, T.; Maitra, S.K.; Ballou, C.E.: An endo-$\alpha 1 \rightarrow 6$-D-mannanase from a soil bacterium. Purification, properties, and mode of action. J. Biol. Chem., **251**, 174-181 (1976)
[2] Brigance, W.T.; Barlowe, C.; Graham, T.R.: Organization of the yeast golgi complex into at least four functionally distinct compartments. Mol. Biol. Cell, **11**, 171-182 (2000)
[3] Nakajima, T.; Ballou, C.E.: Structure of the linkage region between the polysaccharide and protein parts of Saccharomyces cerevisiae mannan. J. Biol. Chem., **249**, 7685-7694 (1974)

Blood group-substance endo-1,4-β-galactosidase 3.2.1.102

1 Nomenclature

EC number
3.2.1.102

Systematic name
blood group-substance 1,4-β-D-galactanohydrolase

Recommended name
blood group-substance endo-1,4-β-galactosidase

Synonyms
endo-β-galactosidase
endo-β-galactosidase DII <1> [4]
galactosidase, endo-β-

CAS registry number
52720-51-1

2 Source Organism

<1> *Diplococcus pneumoniae* [1, 4, 5]
<2> *Escherichia freundii* [2, 3, 6, 8, 9]
<3> *Bacteroides fragilis* [7, 8]

3 Reaction and Specificity

Catalyzed reaction
endohydrolysis of 1,4-β-D-galactosidic linkages in blood group A and B substances

Reaction type
hydrolysis of O-glycosyl bond

Substrates and products
S GalNAcα1-3(Fucα1-2)Galβ1-4Glc + H_2O <1> [1, 5]
P ?
S GalNAcα1-3(Fucα1-2)Galβ1-4GlcNAcβ1-3-hexene-1,2,5,6-tetrol + H_2O <1> [1, 5]
P ?

S Galα1-3(Fucα1-2)Galβ1-4(Fucα1-3)GlcNAcβ1-3-hexene-1,2,5,6-tetrol + H_2O <1> [1, 5]

P ?

S Galα1-3(Fucα1-2)Galβ1-4GlcNAcβ1-3-hexene-1,2,5,6-tetrol + H_2O <1> [1, 5]

P ?

S blood group O glycopeptide + H_2O <3> (<3>, desialylated human blood group O erythrocyte glycopeptide [7]) [7]

P GlcNAcβ1-3Gal.ol + Galβ1-4GlcNAcβ1-3Gal.ol + Fucα1-2Galβ1-4GlcNAcβ1-3Gal.ol + GlcNAcβ1-6[GlcNAcβ1-3]Galβ1-4GlcNAcβ1-3Gal.ol + Galβ1-4GlcNAcβ1-6[GlcNAcβ1-3]Galβ1-4GlcNAcβ1-3Gal.ol + GlcNAcβ1-6[Galβ1-4GlcNAcβ1-3]Galβ1-4GlcNAcβ1-3Gal.ol + Galβ1-4GlcNAcβ1-6[Galβ1-4GlcNAcβ1-3]Galβ1-4GlcNAcβ1-3Gal.ol + Fucα1-2Galβ1-4GlcNAcβ1-6[GlcNAcβ1-3]Galβ1-4GlcNAcβ1-3Gal.ol + Fucα1-2Galβ1-4GlcNAcβ1-6[Galβ1-4GlcNAcβ1-3]Galβ1-4GlcNAcβ1-3Gal.ol + Fucα1-2Galβ1-4GlcNAcβ1-6[Fucα1-2Galβ1-4GlcNAcβ1-3]Galβ1-4GlcNAcβ1-3Gal.ol <3> (<3>, characterization of the oligosaccharides [7]) [7]

S blood group substance A + H_2O <1> (<1>, degradation of blood type A determinant [5]) [5]

P GalNAcα_1-3(Fucα_1-2)Gal + Galα_1-3(Fucα_1-2)Gal + ? <1> [5]

S blood group substance B + H_2O <1> (<1>, degradation of blood type B determinant [5]) [5]

P ?

S lactosaminoglycan + H_2O <1> [4]

P small oligosaccharides + residual glycopeptide <1> [4]

S Additional information <1-3> (<1>, acts specifically on the oligosaccharides with blood group A and B antigenic determinants composed of type 2 chains: the A determinant present in erythrocyte glycoproteins is completely hydrolyzed [1]; <2>, treatment of human erythrocytes with the enzyme converts protein 3, the anion transporter of the erythrocyte, to a more compact staining band [2]; <2>, band 3 glycopeptide of cord and i variant adult human erythrocytes are hydrolyzed to mostly small oligosaccharides by endo-β-galactosidase from Escherichia freundii, whereas that of normal adult produces a number of oligosaccharides with various sizes which is caused by branched structures [3]; <1>, hydrolyzes linear N-acetyllactosamine repeating structures in glycolipids and glycopeptides to release oligosaccharides. Branched lactosaminyl glycolipids and H-antigenic glycolipids are resistant to endo-β-galactosidase DII, even when linear structure is present at the inner part. The enzyme releases mostly small oligosaccharides from substrates, stepwise release from the nonreducing terminal. No degradation of keratan sulfate [4]; <2>, directly modifies specific cell surface glycoconjugates: abolishment of Ii-antigenic activity, decrease of the reactivity with anti-ABH blood group and anti-paragloboside antibodies, a slight but distinct increase of the reactivity with anti-lacto-N-triosylceramide antibodies, no change of the reactivity with anti-globoside antibodies [6]; <2, 3>, hydrolyzes linear but not branched oligosaccharide domains of glycolipids of the neolacto series.

At an optimal concentration of detergent the endo-β-galactosidase susceptibility of the GlcNAcβ1-3Galβ1-4Glc sequence near the ceramide moiety of the branched glycolipids is similar to that of the corresponding sequence in the linear glycolipids [8]) [1-4, 6, 8]

P ?

Inhibitors

Ag_2SO_4 <1> (<1>, 3 mM, 82% inhibition [1, 5]) [1, 5]
Cys <1> (<1>, 3.3 mM, 15 min at 37°C, 52% inhibition [1, 5]) [1, 5]
$HgCl_2$ <1> (<1>, 3 mM, 98.6% inhibition [1, 5]) [1, 5]
PCMB <1> (<1>, 3.3 mM, 15 min at 37°C, 88% inhibition [1, 5]) [1, 5]

Specific activity (U/mg)

0.162 <1> [1, 5]

K_m-Value (mM)

0.068 <1> (GalNAcα1-3(Fucα1-2)Galβ1-4GlcNAcβ1-3-hexene-1,2,5,6-tetrol) [1, 5]
0.13 <1> (GalNAcα1-3(Fucα1-2)Galβ1-4Glc) [1, 5]
0.36 <1> (Galα1-3(Fucα1-2)Galβ1-4GlcNAcβ1-3-hexene-1,2,5,6-tetrol) [1, 5]
0.67 <1> (Galα1-3(Fucα1-2)Galβ1-4(Fucα1-3)GlcNAcβ1-3-hexene-1,2,5,6-tetrol) [1, 5]

pH-Optimum

6 <1> [1, 5]

pH-Range

5-7.5 <1> (<1>, pH 5.0: about 20% of maximal activity, pH 7.5: about 25% of maximal activity [5]) [5]

5 Isolation/Preparation/Mutation/Application

Source/tissue

culture fluid <1> [1]

Purification

<1> [1, 5]

Application

analysis <2> (<2> endo-β-galactosidase-Griffonia-simplicifolia-agglutinin-II-staining procedure into carbohydrate histochemistry of human tissues. The procedure may provide a powerful tool for analysing the backbone structures carrying the blood-group-related antigens in tissue sections [9]) [9]

References

[1] Kobata, A.; Takasaki, S.: endo-β-Galactosidase and endo-α-N-acetylgalactosaminidase from Diplococcus pneumoniae. Methods Enzymol., **50**, 560-570 (1978)
[2] Mueller, T.J.; Li, Y.T.; Morrison, M.: Effect of endo-β-galactosidase on intact human erythrocytes. J. Biol. Chem., **254**, 8103-8106 (1979)
[3] Fukuda, M.; Fukuda, M.N.; Hakomori, S.I.: Developmental change and genetic defect in the carbohydrate structure of band 3 glycoprotein of human erythrocyte membrane. J. Biol. Chem., **254**, 3700-3703 (1979)
[4] Fukuda, M.N.: Isolation and characterization of a new endo-β-galactosidase from Diplococcus pneumoniae. Biochemistry, **24**, 2154-2163 (1985)
[5] Takasaki, S.; Kobata, A.: Purification and characterization of an endo-β-galactosidase produced by Diplococcus pneumoniae. J. Biol. Chem., **251**, 3603-3609 (1976)
[6] Fukuda, M.N.; Fukuda, M.; Hakomori, S.-i.: Cell surface modification by endo-β-galactosidase. Change of blood group activities and release of oligosaccharides from glycoproteins and glycosphingolipids of human erythrocytes. J. Biol. Chem., **254**, 5458-5465 (1979)
[7] Scudder, P.; Lawson, A.M.; Hounsell, E.F.; Carruthers, R.A.; Childs, R.A.; Feizi, T.: Characterisation of oligosaccharides released from human-blood-group O erythrocyte glycopeptides by the endo-β-galactosidase of Bacteroides fragilis. A study of the enzyme susceptibility of branched poly(N-acetyllactosamine) structures. Eur. J. Biochem., **168**, 585-593 (1987)
[8] Scudder, P.; Hanfland, P.; Uemura, K.-i.; Feizi, T.: Endo-β-D-galactosidases of Bacteroides fragilis and Escherichia freundii hydrolyzes linear but not branched oligosaccharide domains of glycolipids of the neolacto series. J. Biol. Chem., **259**, 6586-6592 (1984)
[9] Ito, N.; Kawahara, S.; Hirano, Y.; Morimura, Y.; Nakajima, K.; Uchida, K.; Hirota, T.: Localization of blood-group-related linear poly-N-acetyllactosamine structure in different human tissues by Griffonia simplicifolia agglutinin-II staining following endo-β-galactosidase digestion. Histochem. J., **26**, 327-336 (1994)

Keratan-sulfate endo-1,4-β-galactosidase 3.2.1.103

1 Nomenclature

EC number
3.2.1.103

Systematic name
keratan-sulfate 1,4-β-D-galactanohydrolase

Recommended name
keratan-sulfate endo-1,4-β-galactosidase

Synonyms
endo-β-galactosidase
galactosidase, keratosulfate endo- β-
keratan sulfate endogalactosidase
keratanase

CAS registry number
55072-01-0

2 Source Organism

<1> *Bacteroides fragilis* [1, 11-13]
<2> *Flavobacterium keratolyticus* [2, 8, 9, 14]
<3> *Pseudomonas sp.* (IFO-13309 [3]) [3, 14, 16]
<4> *Escherichia freundii* [4, 6, 7, 10, 14, 15]
<5> *Escherichia coli* [5]

3 Reaction and Specificity

Catalyzed reaction
endohydrolysis of 1,4-β-D-galactosidic linkages in keratan sulfate

Reaction type
hydrolysis of O-glycosyl bond

Natural substrates and products
S ? <2, 4> (<4>, induced with high efficiency in culture medium containing Smith-degraded gastric mucin [6]; <2>, the enzyme is produced without induction by keratan sulfate [8, 9]) [6, 8, 9]
P ?

Substrates and products

S A1-glycolipid + H_2O <4> [6]
P ?
S Ab-glycolipid + H_2O <4> [6]
P ?
S Fucα1-2Galβ1-3GlcNAcβ1-3Galβ1-4Glc + H_2O <2> [9]
P ?
S Fucα1-2Galβ1-3GlcNAcβ1-3Galβ1-4Glc-Cer + H_2O <4> [7, 10]
P Fucα1-2Galβ1-3GlcNAcβ1-3Gal + Glc-Cer <4> [7, 10]
S Gal(β1-4)GlcNAc(β1-3)Gal(β1-4)sorbitol + H_2O <4> [15]
P Gal(β1-4)GlcNAc(β1-3)Gal + sorbitol <4> [15]
S Galα1-3Galβ1-4GlcNAcβ1-3Galβ1-4Glc-Cer + H_2O <2, 4> (<3>, no activity [14]) [7, 10, 14]
P ?
S Galβ1-3GlcNAc(6-2αNeuAc)β1-3Galβ1-4Glc + H_2O <2> [9]
P ?
S Galβ1-3GlcNAcβ1-3Galβ1-4Glc + H_2O <1, 2> [9, 11]
P ?
S Galβ1-3[Fucα1-4]GlcNAcβ1-3Galβ1-4Glc + H_2O <1, 2> [9, 11]
P ?
S Galβ1-3[Fucα1-4]GlcNAcβ1-3Galβ1-4Glc-Cer + H_2O <4> [7, 10]
P Galβ1-3[Fucα1-4]GlcNAcβ1-3Gal + Glc-Cer <4> [7, 10]
S Galβ1-4GlcNAcβ1-3Galβ1-3Galβ1-4Glc-Cer + H_2O <2, 4> [7, 9, 10]
P Galβ1-4GlcNAcβ1-3Galβ1-3Gal + Glc-Cer <4> [7, 10]
S Galβ1-4GlcNAcβ1-3Galβ1-4Glc + H_2O <1, 2> [9, 11]
P ?
S H1-glycolipid + H_2O <4> [6]
P ?
S NeuAcα2-3Galβ1-3GlcNAc(6-2αNeuAc)β1-3Galβ1-4Glc + H_2O <2> [9]
P ?
S NeuAcα2-3Galβ1-4GlcNAcβ1-3Galβ1-4Glc-Cer + H_2O <2, 4> [7, 9, 14]
P NeuAcα2-3Galβ1-4GlcNAcβ1-3Gal + Glc-Cer <2, 4> (<3>, no activity [14]) [7, 10, 14]
S NeuAcα2-3Galβ1-4GlcNAcβ1-4Galβ1-3GlcNAcβ1-3Galβ1-4Glc-Cer + H_2O <2, 4> (<3>, no activity [14]) [14]
P NeuAcα2-3Galβ1-4GlcNAcβ1-4Galβ1-3GlcNAcβ1-3Gal + Glc-Cer <2, 4> [14]
S NeuAcα2-6Galβ1-4GlcNAcβ1-3Galβ1-4Glc + H_2O <2> [9]
P ?
S erythroglycan + H_2O <2> [9]
P ?
S glycosphingolipids + H_2O <5> (<5>, lacto-N-glycosylceramide series having the common structure R-GlcAcβ1-3Galβ1-4Glc(or GlcNAc) are hydrolyzed at Galβ1-4Glc (or GlcNAC) linkages. β-Galactosyl linkages in globo series or ganglio series of glycolipids are not hydrolyzed [5]) [5]
P ?

S keratan sulfate + H_2O <1-4> (<3>, corneal keratan sulfate. Compared with corneal keratan sulfate, keratan sulfates from human nucleus pulposus and shark cartilage are attacked at lower rates with a resultant production of oligosaccharides of relative large size [3]; <2>, from nasal carilage [8, 9]; <1>, from bovine cornea [11-13]; <2-4>, from teleost skin [14]; <3>, keratan sulfate chains from bovine articular cartilage. Presence of fucose on N-acetylglucosamine residues in keratan sulfates protects both of the adjacent unsulfated galactose residues from keratanase cleavage [16]) [2, 3, 6, 7-9, 11, 12, 14-16]

P oligosaccharides <1-4> (<3>, differing in size and sulfate content [3]; <4>, 6-O-sulfo-GlcNAcβ1-3Gal is the major product [7]; <1>, GlcNAcβ1-3Gal + sulfated disaccharides, trisaccharides, tetrasaccharides, hexasaccharides, octasaccharides and decasaccharides based on the sequence (-3/4GlcNAcβ1-3Galβ1-), having 1, 3, 5, 7 and 9 sulfate groups [12]; <1>, more than 75% of the oligosaccharides are sulfated unbranched poly(N-acetyllactosamine) sequences, (-3/4GlcNAcβ1-3Galβ1-)n. SO3-β1-3Gal accounts for almost 35% of the oligosaccharide material while 40% consists of four oligosaccharides, unbranched tetrasaccharides, hexasaccharides, octasaccharides and decasaccharides of poly(N-acetyllactosamine)type, having 3, 5, 7 and 9 sulfate residues respectively. A sulfate residue is attached to the C_6 position of each N-acetylglucosamine and each internal galactose residue of the four oligosaccharides [13]; <2-4>, 6-O-sulfo-GlcNAcβ1-3Gal is one of the major products [14]; <2,4>, GlcNAcβ1-3Gal [14]; <3>, no production of GlcNAcβ1-3Gal [14]; <4>, 2-acetamido-2-deoxy-6-O-sulfo-β-D-Gal-(1-3)-D-Gal + 2-acetamido-2-deoxy-6-O-sulfo-β-D-Glc-(1-3)-6-O-sulfoD-Gal-(1-4)-2-acetamido-2-deoxy-6-O-sulfo-β-D-Glc-(1-3)-D-Gal [15]; <3>, GlcNAc(6S)β1-3Galβ1-4(Fucα1-3)GlcNAc(6S)-β1-3Galβ1-4GlcNAc(6S)β1-3Gal-ol + GlcNAc(6S)β1-3Galβ1-4(Fucα-3)-GlcNAc(6S)β1-3Galβ1-4GlcNAc(6S)β1-6(Galβ1-33)GalNAc-ol [16]) [3, 7, 12-16]

S lacto-N-fucopentaose + H_2O <4> [6]

P ?

S lacto-N-fucopentaose II + H_2O <4> [6]

P ?

S lacto-N-fucopentaose III + H_2O <4> [6]

P ?

S lacto-N-neotetraosylceramide + H_2O <1> [11]

P ?

S lacto-N-tetraitol + H_2O <4> [4]

P sorbitol + Galβ(1-3)GlcNAcβ(1-3)Gal <4> [4]

S lacto-N-tetraose + H_2O <4> [4, 6]

P glucose + Galβ(1-3)GlcNAcβ(1-3)Gal <4> [4]

S lacto-N-tetraose + H_2O <4> [6]

P ?

S mucin + H_2O <4> (<4>, desialyzed pig colonic mucin [4]) [4]

P GlcNAcβ(1-3)Gal + GlcNAc-6Sβ(1-3)Gal + resistant polymer <4> [4]

S neolacto-type glycosphingolipid + H_2O <2, 4> (<2, 4>, hydrolysis of the internal β1,4-galactosyl linkage of various neolacto-type glycosphingolipids to produce glucosylceramides [14]) [14]
P glucosylceramide + ? <2, 4> [14]
S paragloboside + H_2O <4> [6]
P ?
S polyglycosylceramides + H_2O <4> (<4>, in absence of sodium taurodeoxycholate conversion into at least three shorter-chain glycosphingolipids, in presence of sodium taurodeoxycholate conversion into glucosylceramide [10]) [10]
P ? <4> (<4>, in absence of sodium taurodeoxycholate conversion into at least three shorter-chain glycosphingolipids, in presence of sodium taurodeoxycholate conversion into glucosylceramide [10]) [10]
S sialosylparagloboside + H_2O <4> [6]
P ?
S Additional information <1-3, 5> (<1>, no activity with Galβ1-4GlcNAcβ1-6Galβ1-4GlcNAcβ1-6(Galβ1-3)GalNAc-ol [1]; <3>, no hydrolysis of phenyl β-D-galactoside [3]; <5>, sialosyl substitition at the terminal galactosyl residue of lacto-N-neotetraosyl structure of paragloboside greatly enhances the hydrolyzability of the internal Galβ1-4Glc linkage, thus producing a high yield of sialotetrasaccharide (AcNeuα2-3Galβ1-4GlcNAcβ-3Gal). The Galβ1-4GlcNAc linkage located in the middle of the repeating Galβ1-4GlcNAc(or Glc) unit seen in H_2-glycolipid or Ab-glycolipid, (R-Galβ1-4GlcNAcβ1-3Galβ1-4GlcNAcβ1-3Galβ1-4Glc-Cer) is preferentially hyydrolyzed relative to the Galβ1-4Glc linkage directly attached to ceramide. The branched structure as is found in H3-glycolipid greatly reduces the hydrolyzability [5]; <2>, hydrolyzes glycoconjugates containing N-acetyllactosamine repeating units [9]; <1>, hydrolyzes the type 2 sequence Galβ1-4GlcNAcβ1-3Galβ1-4Glc at about twice the rate of the type 1 isomer, Galβ1-3GlcNAcβ1-3Galβ1-4Glc [11]; <4>, AB active blood group substance can be hydrolyzed only after Smith degradation [15]) [1, 3, 5, 9, 11, 15]
P ?

Inhibitors

Ag^+ <2, 4> (<4>, 1 mM, complete inhibition [7, 10]) [7-10]
Cu^{2+} <1, 2, 4> (<4>, 1 mM, 30% inhibition [7, 10]; <1>, 1 mM, partial inhibition [11]) [7-11]
Fe^{2+} <1> (<1>, 1 mM, partial inhibition [11]) [11]
Galβ1-4GlcNAcβ1-3Gal <1> (<1>, when assayed with lacto-N-neotetraosylceramide [11]) [11]
GlcNAcβ1-3Gal <1> (<1>, when assayed with lacto-N-neotetraosylceramide [11]) [11]
Hg^{2+} <2, 4> (<4>, 1 mM, complete inhibition [7, 10]) [7-10]
PCMB <1, 2, 4> (<4>, 0.25 mM, complete inhibition [7, 10]; <1>, 1 mM, complete inhibition [11]) [7-11]

Zn^{2+} <1, 4> (<4>, 1 mM, 30% inhibition [7, 10]; <1>, 1 mM, partial inhibition [11]) [7, 10, 11]

Activating compounds

acetate <3> (<3>, slight activation [3]) [3]
maleate <3> (<3>, slight activation [3]) [3]

Specific activity (U/mg)

3.11 <4> [15]
44 <2> [8]
62.8 <4> [6]
156 <1> [11]
Additional information <3, 4> [3, 7, 10]

K_m-Value (mM)

0.3 <4> (paragloboside) [6]
0.35 <4> (keratan sulfate) [6]
3.91 <4> (lacto-N-tetraose) [6]

pH-Optimum

5.5-5.8 <4> (<4>, hydrolysis of keratan sulfate from nasal cartilage or Galα_1-3Galβ1-4GlcNAcβ1-3Galβ1-4Glc-Cer [7]) [7, 10]
5.7 <1> [11]
6 <2> (<2>, hydrolysis of keratan sulfate [8, 9]) [8, 9]
7.2-7.4 <3> [3]

Temperature optimum (°C)

37 <3> [3]
55 <1> [11]

Temperature range (°C)

30-45 <3> (<3>, 30°C: about 80% of maximal activity, 45°C: about 50% of maximal activity [3]) [3]

4 Enzyme Structure

Molecular weight

28000 <4> (<4>, gel filtration [7, 10]) [7, 10]
30000 <1, 2> (<1, 2>, gel filtration [8, 9, 11]) [8, 9, 11]

Subunits

? <2> (<2>, x * 41000, mature recombinant enzyme, SDS-PAGE [2]; <2>, x * 43000, native enzyme, SDS-PAGE [2]) [2]

5 Isolation/Preparation/Mutation/Application

Source/tissue

culture medium <2, 4> [7, 8, 10]

Purification

<1> [11]
<2> [2, 8]
<3> [3]
<4> [6, 7, 15]

Cloning

<2> (expression in E. coli [2]) [2]

Application

analysis <1, 3, 4> (<1>, the enzyme can be used to distinguish between GlcNAcβ1-6Gal and GlcNAcβ1-3Gal units within linear backbone sequences of all known types of oligo-(N-acetyllactosamino)glycans [1]; <3>, the enzyme can be used for studies on structural variability of keratan sulfates of different origin [3]; <4>, useful for studying the structure of glycoproteins and oligosaccharides [4, 7, 10]; <4>, enzyme can be useful in studying the precise structures of keratan sulfates, related glycoproteins and oligosaccharides [15]) [1, 3, 4, 7, 10, 15]

6 Stability

pH-Stability

3.5-9 <1> (<1>, 25°C, 16 h, stable in presence of 0.2 mg of albumin per ml [11]) [11]
4 <2, 4> (<4>, 17 h at room temperature, rapid loss of activity below [7, 10]; <2>, rapid loss of activity below [9]) [7, 9, 10]
4.5-5.5 <4> (<4>, 17 h at room temperature, retains 80 to 100% activity [7]) [7, 10]
5-10 <2> (<2>, stable [9]) [9]
6 <4> (<4>, 17 h at room temperature, rapid loss of activity above [7]) [7, 10]
11 <2> (<2>, rapid loss of activity above [9]) [9]

Temperature stability

37 <1, 4> (<4>, 1 h, 50% loss of activity, bovine serum albumin, 0.4 mg/ml, prevents inactivation [7, 10]; <1>, 0.003 mg/ml protein, 40% loss of activity after 30 min [11]) [7, 10, 11]

General stability information

<1>, Ca^{2+} stabilizes [11]

Storage stability

<1>, -20°C, 0.044 mg/ml protein, stable for at least 12 months [11]
<1>, 4°C, stable for at least 2-3 months at a protein concentration of 0.044 mg/ml protein, 20% loss of activity in 3 h at a protein concentration of 0.003 mg/ml [11]
<2>, 20°C, stable for at least 5 months [9]
<3>, in an ice bath, stable for 1 year [3]

<4>, -20°C, stable for several months [6]
<4>, 0°C, stable for several months [7, 10]

References

[1] Maaheimo, H.; Penttilä, L.; Renkonen, O.: Enzyme-aided construction of medium-sized alditols of complete O-linked saccharides. The constructed hexasaccharide alditol Galβ1-4GlcNAcβ1-6Galβ1-4GlcNAcβ1-6(Galβ1-3)-GalNAc-ol resists the action of endo-β-galactosidase from Bacteroides fragilis. FEBS Lett., **349**, 55-59 (1994)
[2] Leng, L.; Zhu, A.; Zhang, Z.; Hurst, R.; Goldstein, J.: Cloning, functional expression and purification of endo-β-galactosidase from Flavobacterium keratolyticus. Gene, **222**, 187-194 (1998)
[3] Nakazawa, K.; Suzuki, S.: Purification of Keratan sulfate-endogalactosidase and its action on keratan sulfates of different origin. J. Biol. Chem., **250**, 912-917 (1975)
[4] Fukuda, M.; Matsumara, G.: Endo-β-galactosidase of Escherichia freundii. Hydrolysis of pig colonic mucin and milk oligosaccharides by endoglycosidic action. Biochem. Biophys. Res. Commun., **64**, 465-471 (1975)
[5] Fukuda, M.; Watanabe, K.; Hakomori, S.I.: Release of oligosaccharides from various glycosphingolipids by endo-β-galactosidase. J. Biol. Chem., **253**, 6814-6819 (1978)
[6] Fukuda, M.: Purification and characterization of endo-β-galactosidase from Escherichia freundii induced by hog gastric mucin. J. Biol. Chem., **256**, 3900-3905 (1981)
[7] Nakagawa, H.; Yamada, T.; Chien, J.-L.; Gardas, A.; Kitamikado, M.; Li, S.C.; Li, Y.T.: Isolation and characterization of an endo-β-galactosidase from a new strain of Escherichia freundii. J. Biol. Chem., **255**, 5955-5959 (1980)
[8] Kitamikado, M.; Ito, M.; Li, Y.T.: Isolation and characterization of a keratan sulfate-degrading endo-β-galactosidase from Flavobacterium keratolyticus. J. Biol. Chem., **256**, 3906-3909 (1981)
[9] Kitamikado, M.; Ito, M.; Li, Y.-T.: Endo-β-galactosidase from Flavobacterium keratolyticus. Methods Enzymol., **83**, 619-625 (1982)
[10] Li, Y.T.; Nakagawa, H.; Kitamikado, M.; Li, S.C.: Endo-β-galactosidase from Escherichia freundii. Methods Enzymol., **83**, 610-619 (1982)
[11] Scudder, P.; Uemura, K.-i.; Dolby, J.; Fukuda, M.N.; Feizi, T.: Isolation and characterization of an endo-β-galactosidase from Bacteroides fragilis. Biochem. J., **213**, 484-494 (1983)
[12] Hounsell, E.F.; Feeney, J.; Scudder, P.; Tang, P.W.; Feizi, T.: ^{1}H-NMR studies at 500 MHz of a neutral disaccharide and sulphated di-, tetra-, hexa- and larger oligosaccharides obtained by endo-β-galactosidase treatment of keratan sulphate. Eur. J. Biochem., **157**, 375-384 (1986)
[13] Scudder, P.; Tang, P.W.; Hounsell, E.F.; Lawson, A.M.; Mehemet, H.; Feizi, T.: Isolation and characterization of sulphated oligosaccharides released from bovine corneal keratan sulphate by the action of endo-β-galactosidase. Eur. J. Biochem., **157**, 365-373 (1986)

[14] Ito, M.; Hirabayashi, Y.; Yamagata, T.: Substrate specificity of endo-β-galactosidases from Flavobacterium keratolyticus and Escherichia freundii is different from that of Pseudomonas sp.. J. Biochem., **100**, 773-780 (1986)
[15] Fukuda, M.N.; Matsumura, G.: Endo-β-galactosidase of Escherichia freundii. Purification and endoglycosidic action on keratan sulfates, oligosaccharides, and blood group active glycoprotein. J. Biol. Chem., **251**, 6218-6225 (1976)
[16] Tai, G.-H.; Huckerby, T.N.; Nieduszynski, A.: N.M.R. spectroscopic studies of fucose-containing oligosaccharides derived from keratanase digestion of articular cartilage keratan sulphates. Influence of fucose residues on keratanase cleavage. Biochem. J., **291**, 889-894 (1993)

Steryl-β-glucosidase 3.2.1.104

1 Nomenclature

EC number
3.2.1.104

Systematic name
cholesteryl-β-D-glucoside glucohydrolase

Recommended name
steryl-β-glucosidase

Synonyms
SG hydrolase <1> [2]
glucosidase, hydroxy steroid β-
steryl β-D-glucoside hydrolase
steryl glucoside hydrolase

CAS registry number
69494-88-8

2 Source Organism

<1> *Sinapis alba* [1-3]

3 Reaction and Specificity

Catalyzed reaction
cholesteryl-β-glucoside + H_2O = D-glucose + cholesterol

Reaction type
hydrolysis of O-glycosyl bond

Substrates and products
S cholesteryl β-D-glucoside + H_2O <1> [1, 2]
P D-glucose + cholesterol
S *p*-nitrophenyl β-D-glucoside + H_2O <1> [1]
P D-glucose + *p*-nitrophenol
S sitosteryl β-D-glucoside + H_2O <1> [1]
P D-glucose + sitosterol

Inhibitors
2-lyso-phosphatidylcholine <1> [2]

2-lyso-phosphatidylethanolamine <1> [2]
5α-cholestanol <1> (<1>, hydrolysis of cholesteryl glucoside [1]) [1]
campesterol <1> (<1>, hydrolysis of cholesteryl glucoside [1]) [1]
cholesterol <1> (<1>, hydrolysis of cholesteryl glucoside [1]) [1]
ethanol <1> (<1>, 39% inhibition at a concentration of 50%, hydrolysis of *p*-nitrophenyl β-D-glucoside [1]) [1]
gluconolactone <1> (<1>, 50% inhibition at 4 mM, hydrolysis of cholesteryl β-D-glucoside. 50% inhibition at 2 mM, hydrolysis of *p*-nitrophenyl β-D-glucoside [1]) [1]
monogalactosyldiacylglycerol <1> [2]
phosphatidylcholine <1> [2]
phosphatidylethanolamine <1> [2]
sitosterol <1> (<1>, hydrolysis of cholesteryl glucoside [1]) [1]

Activating compounds

cholesteryl acetate <1> (<1>, 150% stimulation at 0.1 mM [2]) [2]
dithiothreitol <1> (<1>, 45% stimulation at 0.1 mM, hydrolysis of cholesteryl β-D-glucoside. No effect on hydrolysis of *p*-nitrophenyl β-D-glucoside [1]) [1]
ethanol <1> (<1>, 64% stimulation at a concentration of 20%, hydrolysis of cholesteryl β-D-glucoside. 65% stimulation at a concentration of 20%, hydrolysis of *p*-nitrophenyl β-D-glucoside [1]) [1]
methyl stearate <1> (<1>, 150% stimulation at 0.1 mM [2]) [2]
phytol <1> (<1>, 200% stimulation at 0.002 mM [2]) [2]
tripalmitoylglycerol <1> (<1>, 145% stimulation at 0.1 mM [2]) [2]

Specific activity (U/mg)

0.3621 <1> [1]

pH-Optimum

5.2-5.6 <1> (<1>, hydrolysis of cholesteryl β-D-glucoside or *p*-nitrophenyl β-D-glucoside [1]) [1]

Temperature optimum (°C)

30 <1> (<1>, hydrolysis of cholesteryl β-D-glucoside [1]) [1]
60 <1> (<1>, hydrolysis of *p*-nitrophenyl β-D-glucoside [1]) [1]

4 Enzyme Structure

Molecular weight

65000 <1> (<1>, gel filtration [1]) [1]

5 Isolation/Preparation/Mutation/Application

Source/tissue

seedling <1> (<1>, changes in activity during germination, distribution between cotyledons and roots [3]) [1-3]

Localization

cell membrane <1> [1]

Purification

<1> (partial [1]) [1]

References

[1] Kalinowska, M.; Wojciechowski, Z.A.: Purification and some properties of steryl β-D-glucoside hydrolase from Sinapis alba seedlings. Phytochemistry, **17**, 1533-1537 (1978)

[2] Kalinowska, M.; Wojciechowski, Z.A.: Modulation of activities of steryl glucoside hydrolase and UDPG:sterol glucosyltransferase from Sinapis alba by detergent and lipids. Phytochemistry, **25**, 45-49 (1986)

[3] Kalinowska, M.; Wojciechowski, Z.A.: Sterol conjugate interconversion during germination of white mustard (Sinapis alba). Phytochemistry, **23**, 2485-2488 (1984)

Strictosidine β-glucosidase 3.2.1.105

1 Nomenclature

EC number
3.2.1.105

Systematic name
strictosidine β-D-glucohydrolase

Recommended name
strictosidine β-glucosidase

Synonyms
SG
SGD
glucosidase, strictosidine β-
strictosidine β-D-glucosidase
strictosidine glucosidase

CAS registry number
73379-57-4

2 Source Organism

<1> *Tabernaemontana divaricata* (cell line 59 and cell line 6B [1]) [1, 4]
<2> *Catharanthus roseus* [2-6]
<3> *Strychnos mellodora* [3]
<4> *Amsonia salicifolia* [6]
<5> *Catharanthus ovalis* [6]
<6> *Catharanthus pusilus* [6]
<7> *Catharanthus trichophyllus* [6]
<8> *Rauwolfia vomitoria* (weak activity [6]) [6]
<9> *Rhazya orientalis* (weak activity [6]) [6]
<10> *Vinca major* (weak activity [6]) [6]

3 Reaction and Specificity

Catalyzed reaction
strictosidine + H_2O = D-glucose + strictosidine aglycone

Reaction type
hydrolysis of O-glycosyl bond

Natural substrates and products

S strictosidine <2> (<2>, the enzyme may play an important role in steering the monoterpenoid indole alkaloid biosynthesis in a specific direction [2]; <2>, enzyme in terpenoid indole alkaloid biosynthesis [5]) [2, 5, 6]
P ?

Substrates and products

S dolichantoside + H_2O <3> [3]
P ?
S palicoside + H_2O <2, 3> [3]
P ?
S strictosidine + H_2O <1-3> [1-5]
P D-glucose + strictosidine aglycone <1> (<1>, the major intermediate after rearrangement of the strictosidine aglycone is thought to be 21-hydroxyajmalicine [1]) [1]
S vincoside + H_2O <2> (<2>, hydrolyzed at less than 10% of strictosidine [6]) [6]
P ?
S Additional information <1> (<1>, no activity with secologanin, strictosamide or *p*-nitrophenyl-β-D-glucopyranoside [1]) [1]
P ?

Inhibitors

Cu^{2+} <2> (<2>, 1 mM [2]) [2]
D-gluconic acid δ-lactone <1, 2> (<1>, some inhibition at a concentration of 10 mM and higher [1]; <2>, enzyme I is much less sensitive than enzyme II [6]) [1, 6]
serpentine <2> (<2>, 1 mM [2]) [2]
strictosidine <2> (<2>, enzyme I shows substrate inhibition at 1 mM, enzyme II not [6]) [6]
tryptamine <2> (<2>, 2.5 mM, 15% inhibition of enzyme I and II [6]) [6]

Specific activity (U/mg)

Additional information <1, 2> [1, 2]

K_m-Value (mM)

0.01-0.018 <2> (strictosidine) [3]
0.018 <2> (strictosidine) [2]
0.1 <2> (strictosidine, <2>, enzyme II [6]) [6]
0.12 <3> (strictosidine) [3]
0.2 <2> (strictosidine, <2>, enzyme I [6]) [6]
0.25-0.33 <1> (strictosidine, <1>, enzyme form SGIII [1]) [1]
0.335 <3> (palicoside) [3]
0.564 <3> (dolichantoside) [3]
1 <1> (strictosidine, <1>, above, enzyme form SGII [1]) [1]
1.13 <2> (dolichantoside) [3]
7.015 <2> (palicoside) [3]

pH-Optimum

4.5-5 <1> (<1>, enzyme form SGI, a second pH-optimum at pH 7.0 [1]) [1]
6-6.4 <2> (<2>, enzyme I and II [6]) [6]
6-8.5 <2> [2]
7 <1> (<1>, enzyme form SGI, a second pH-optimum at pH 4.5-5.0 [1]) [1]

Temperature optimum (°C)

30 <2> [6]

4 Enzyme Structure

Molecular weight

22000 <1> (<1>, enzyme form SGII, non-denaturing PAGE [1]) [1]
31000 <1> (<1>, enzyme form SGIII, non-denaturing PAGE [1]) [1]
187000 <1> (<1>, enzyme form SGII, non-denaturing PAGE [1]) [1]
190000 <1> (<1>, enzyme form SGII, non-denaturing PAGE [1]) [1]
200000 <3> (<3>, non-denaturing PAGE [3]) [3]
230000 <2> (<2>, enzyme I, gel filtration [6]) [6]
250000 <2> (<2>, enzyme form I, non-denaturing PAGE [5]) [5]
450000 <2> (<2>, above, enzyme II, gel filtration [6]) [6]
500000 <2> (<2>, enzyme form II, non-denaturing PAGE [5]) [5]
630000 <2> (<2>, enzyme form III, non-denaturing PAGE [5]) [5]
670000 <1> (<1>, above, enzyme form SGI, gel filtration [1]) [1]

Subunits

? <2, 3> (<2>, x * 63000, SDS-PAGE [2, 5]; <3>, x * 80000, SDS-PAGE [3]) [2, 3, 5]

5 Isolation/Preparation/Mutation/Application

Source/tissue

cell suspension culture <1, 2> [1, 2]
leaf <3> [3]
root <3> (<3>, bark [3]) [3]
stembark <3> [3]

Localization

endoplasmic reticulum <2> (<2>, most likely associated with endoplasmic reticulum [5]) [5]
vacuole <1, 2> (<1, 2>, inside the vacuole, outside of the tonoplast [4]) [4]

Purification

<1> (partial, three enzyme forms: SGI, SGII and SGIII [1]) [1]
<2> (3 different enzyme forms [2, 5]) [2, 5]

Cloning

<2> (expression in Saccharomyces cerevisiae [5]) [5]

6 Stability

Temperature stability

50 <2> (<2>, stable at [2]) [2]

General stability information

<1>, highly increased sensitivity for freezing and thawing after separation over Superose 6 [1]

Storage stability

<1>, -80°C, stable for 6 months [1]
<1>, 4°C, 24 h, stable [1]
<2>, -80°C, 50% loss of activity [2]

References

[1] Luijendijk, T.J.C.; Nowak, A.; Verpoorte, R.: Strictosidine glucosidase from suspension cultured cells of Tabernaemontana divaricata. Phytochemistry, **41**, 1451-1456 (1996)

[2] Luijendijk, T.J.C.; Stevens, L.H.; Verpoorte, R.: Purification and characterization of strictosidine β-D-glucosidase from Catharanthus roseus cell suspension cultures. Plant Physiol. Biochem., **36**, 419-425 (1998)

[3] Brandt, V.; Geerlings, A.; Tits, M.; Delaude, C.; van der Heijden, R.; Verpoorte, R.; Angenot, L.: New strictosidine β-glucosidase from Strychnos mellodora. Plant Physiol. Biochem., **38**, 187-192 (2000)

[4] Stevens, L.H.; Blom, T.J.M.; Verpoorte, R.: Subcellular localization of tryptophan decarboxylase, strictosidine synthase and strictosidine glucosidase in suspension cultured cells of Catharanthus roseus and Tabernaemontana divaricata. Plant Cell Rep., **12**, 573-576 (1993)

[5] Geerlings, A.; Martinez-Lozano Ibanez, M.; Memlink, J.; van der Heijden, R.; Verpoorte, R.: Molecular cloning and analysis of strictosidine β-D-glucosidase, an enzyme in terpenoid indole alkaloid biosynthesis in Catharanthus roseus. J. Biol. Chem., **275**, 3051-3056 (2000)

[6] Hemscheid, T.; Zenk, M.H.: Glucosidases involved in indole alkaloid biosynthesis of Catharanthus cell culture. FEBS Lett., **110**, 187-191 (1992)

Mannosyl-oligosaccharide glucosidase 3.2.1.106

1 Nomenclature

EC number
3.2.1.106

Systematic name
mannosyl-oligosaccharide glucohydrolase

Recommended name
mannosyl-oligosaccharide glucosidase

Synonyms
(Glc3)-glucosidase
Cwh41p <1> (<1> calcofluor white hypersensitivity [9, 20, 21]; <1> CWH41 gene encodes glucosidase I [19, 20]; <1> CWH41 gene encodes a type-II transmembrane protein, Cwh41p [9]; <1> Cwh41p is yeast glucosidase I [20]) [9, 19, 20, 21]
Glc I
Glc3 oligosaccharide glucosidase
Glc3-OS-glucosidase
Glc3-glucosidase
Glc3-oligosaccharide glucosidase
$Glc_3Man_9GlcNAc_2$ oligosaccharide glucosidase
$Glc_3Man_9NAc_2$ oligosaccharide glucosidase
α-glucosidase I
glucosidase I
glucosidase, mannosyloligosaccharide
glucosidase-1
processing α-glucosidase I
processing exoglucosidase I
processing glucosidase I
trimming enzyme I
trimming glucosidase I

CAS registry number
78413-07-7

2 Source Organism

<1> *Saccharomyces cerevisiae* (X-2180 [2]) [1, 2, 9, 19, 20, 21]

<2> *Bos taurus* (calf [1, 3, 6, 15, 24, 25]; cow [22]; lactating cow [22]) [1, 3, 6, 10, 15, 22, 24, 25]
<3> *Rattus norvegicus* [4, 5, 11, 16, 22, 23]
<4> *Entamoeba histolytica* (strain HM1:IMSS [7]) [7]
<5> *mammalia* (generally present in mammalian tissues [8]) [8, 9]
<6> *Homo sapiens* [8, 17, 18, 24]
<7> *yeast* [10]
<8> *Sus scrofa* [11, 17, 24]
<9> *Mus musculus* (female C57BL/6 [12]) [12, 22]
<10> *Gallus gallus* (hen) [13]
<11> *Vigna radiata* (mung bean [14]) [14]
<12> *Penicillium funiculosum* (QM474 [16]) [16]
<13> *Cavia porcellus* [22]

3 Reaction and Specificity

Catalyzed reaction

exohydrolysis of the non-reducing terminal glucose residues in the mannosyl-oligosaccharide $Glc_3Man_9GlcNAc_2$ (also acts, more slowly, on the corresponding glycolipids and glycopeptides; <2, 3, 11> glucosidase I is a specific α-(1,2)-glucosidase [5, 14, 15]; <2,7> mechanism [10]; <2, 7> stereochemistry, enzyme is an inverting glycosidase, operates with inversion of anomeric configuration [10])

Reaction type

hydrolysis of O-glycosyl bond

Natural substrates and products

S $Glc_3Man_9GlcNAc_2 + H_2O$ <1-13> (<1, 2, 3, 4, 5, 6, 7, 8, 10, 11, 12> hydrolyses specifically terminal α1,2-linked glucose residue [1, 2, 3, 4, 6, 7, 8, 9, 10, 13, 14, 15, 16, 17, 18, 19, 20, 22, 23, 24, 25]; <1, 2, 3, 5, 6, 7, 8, 11> specific involvement in the N-linked oligosaccharide-processing pathway [1, 2, 3, 4, 8, 10, 11, 14, 17, 18, 19, 20, 22, 23, 24, 25]; <1, 2, 3, 5, 6, 7, 8, 9, 10, 11> involved in first step of processing of oligosaccharides after transfer from dolichyl diphosphate to proteins [2, 3, 8, 10, 11, 12, 13, 14, 17, 18, 23, 24, 25]; <2, 3, 5, 6, 7, 8, 9, 10, 11> involved in processing of asparagine-linked oligosaccharides [3, 4, 5, 6, 8, 10, 11, 12, 13, 14, 17, 22, 24, 25]; <3, 9, 10> catalyzes initial step in post-glycosylation processing of cellular glycoprotein [5, 12, 13]) [1-25]

S Additional information <1, 2, 3, 5, 6, 7, 8, 9, 10, 11, 12> (<1, 2, 5, 6, 8, 9, 10, 11> involved in the formation of high mannose and complex glycoproteins [1, 2, 3, 8, 12, 13, 14, 15, 17, 18, 19, 22]; <1, 2, 3, 5, 6, 7, 8, 9, 10, 11, 12> involved in first step of processing of oligosaccharides after transfer from dolichyl diphosphate to proteins [2, 3, 8, 10, 11, 12, 13, 14, 15, 16, 17, 18, 19, 20, 22, 23, 24, 25]; <2, 3, 5, 6, 7, 8, 9, 10, 11, 12> involved in processing of asparagine-linked oligosaccharides [3, 4, 5, 6, 8, 10, 11, 12, 13, 14, 15, 16, 17, 18, 22, 23, 24, 25]; <2, 3, 9, 10> catalyzes initial step in post-glycosylation processing of cellular glycoprotein [5, 12, 13, 22]; <5, 6> role in the folding of newly formed glycoprotein and in endoplasmic reticulum quality control [8]; <2> glucosidase I is a control enzyme for the regulation of glycoprotein processing [15, 22]; <5, 6> processing α-glucosidases in the endoplasmic reticulum act in concert with two lectin chaperones, membrane-bound calnexin and its soluble homolog calreticulin, and with the enzyme UDP-glucose:glycoprotein glucosyltransferase to facilitate glycoprotein folding [8]; <5> essential for maturation of glycoproteins containing N-glycans in mammalian cells, α-glucosidase I has been conserved through eukaryotic evolution from yeast to mammals [8]; <1, 5> trimming of the glucose residues is identical in Saccharomyces cerevisiae and in mammalian cells [9]; <1> indirect role in cell wall β1,6-glucan biosynthesis and/or in β1,6-glucan insertion into the cell wall [9, 19]; <2, 7> significant role in protein folding by allowing the binding of glycoproteins to the endoplasmic reticulum chaperones calnexin and calreticulin [10]; <11> first enzyme of processing of N-linked oligosaccharides in plants, plant glucosidase I might be analogous to glucosidase I of animal cells [14]; <1> cell wall 1,6-β-glucan synthesis depends on endoplasmic reticulum glucosidase I and II, and the molecular chaperone BiP/Kar2p in a synergistic manner [19, 20, 21]; <2> play a crucial role in the overall regulation of glycoprotein biosynthesis [22]; <3> critical role in overall biosynthesis and regulation of protein N-glycosylation [23]; <8> glucose processing is essential for pro-receptor transport and maturation [24]) [1, 2, 3, 4, 5, 6, 8, 9, 10, 11, 12, 13, 14, 15, 16, 17, 18, 19, 20, 21, 22, 23, 24, 25]

Substrates and products

S 4-methylumbelliferyl-α-D-glucoside + H_2O <4> (<2, 8> not a substrate [22, 24]) [7]

P 4-methylumbelliferone + D-glucose

S $Glc_3Man_5GlcNAc$ + H_2O <2, 3> (<2, 3> hydrolyses specifically terminal α1,2-linked glucose residue [4, 25]; <3> 11% activity compared with $Glc_3Man_9GlcNAc$ [4]; <2> hydrolyzed at a rate similar to that of natural substrate $Glc_3Man_9GlcNAc_2$ [25]) [4, 25]

P D-glucose + $Glc_2Man_5GlcNAc$ <3> [4]

S $Glc_3Man_6GlcNAc$ + H_2O <2, 3> (<2, 3> hydrolyses specifically terminal α1,2-linked glucose residue [4, 25]; <3> 15% activity compared with $Glc_3Man_9GlcNAc$ [4]; <2> slowly hydrolyzed [25]) [4, 25]

P D-glucose + $Glc_2Man_6GlcNAc$ <2, 3> [4, 25]

S $Glc_3Man_7GlcNAc$ + H_2O <2, 3> (<2, 3> hydrolyses specifically terminal α1,2-linked glucose residue [4, 25]; <3> 30% activity compared with $Glc_3Man_9GlcNAc$ [4]; <2> slowly hydrolyzed [25]) [4, 25]

P D-glucose + $Glc_2Man_7GlcNAc$ <2, 3> [4, 25]

S $Glc_3Man_8GlcNAc$ + H_2O <2, 3> (<2, 3> hydrolyses specifically terminal α1,2-linked glucose residue [4, 25]; <3> 85% activity compared with $Glc_3Man_9GlcNAc$ [4]; <2> slowly hydrolyzed [25]) [4, 25]

P D-glucose + $Glc_2Man_8GlcNAc$ <2, 3> [4, 25]

S $Glc_3Man_9GlcNAc$ + H_2O <2, 3, 11> (<2, 3, 11> hydrolyses specifically terminal α1,2-linked glucose residue [4, 5, 6, 14]; <3> very good substrate: activity is highest against free oligosaccharide which contains a single N-acetylglucosamine at the reducing end and is lowered about 10% by the presence of a second N-acetylglucosamine at the reducing end [4]) [4, 5, 6, 14]

P D-glucose + Glc$_2$Man$_9$GlcNAc <2, 3, 11> [4, 5, 6, 14]

S Glc$_3$Man$_9$GlcNAc$_2$ + H_2O <1-13> (<1, 2, 3, 4, 5, 6, 7, 8, 10, 11, 12> hydrolyses specifically terminal α1,2-linked glucose residue [1, 2, 3, 4, 6, 7, 8, 9, 10, 13, 14, 15, 16, 17, 18, 19, 20, 22, 23, 24, 25]) [1-25]

P D-glucose + Glc$_2$Man$_9$GlcNAc$_2$ <1-13> [1-25]

S Glc$_3$Man$_9$GlcNAc$_2$-lipid + H_2O <2, 3, 10, 11, 12> (<3, 10, 11, 12> hydrolyses specifically terminal α1,2-linked glucose residue [4, 5, 13, 14, 16, 25]; <3> lower activity than towards free oligosaccharide substrates [4]; <10> activity with oligosaccharide-lipid 2.5 times higher than with glycosylated α-lactalbumin and 3 times higher than with free oligosaccharide [13]; <2> no significant difference in enzyme activity towards free, peptide-bound, or lipid-linked oligosaccharide, lipid- and peptide-linked oligosaccharide are hydrolyzed at a rate only 10% lower than free oligosaccharide [15]; <2> slowly hydrolyzed, free and peptide-bound Glc$_3$Man$_9$GlcNAc$_2$ oligosaccharides are hydrolyzed at similar rates, both being about threefold better substrates than the lipid-linked structure [25]) [4, 5, 13, 14, 15, 16, 25]

P D-glucose + Glc$_2$Man$_9$GlcNAc$_2$-lipid <2, 3, 10, 11> [4, 5, 13, 14, 15, 16, 25]

S Glc$_3$Man$_9$GlcNAc$_2$-peptide + H_2O <2, 3, 10, 11> (<3, 10, 11> hydrolyses specifically terminal α1,2-linked glucose residue [4, 5, 13, 14, 23, 25]; <3> lower activity than towards free oligosaccharide substrates [4]; <10> Glc-containing oligosaccharide attached to S-carboxymethylated α-lactalbumin [13]; <2> no significant difference in enzyme activity towards free, peptide-bound, or lipid-linked oligosaccharide, lipid- and peptide-linked oligosaccharide are hydrolyzed at a rate only 10% lower than free oligosaccharide [15]; <2> free and peptide-bound Glc$_3$Man$_9$GlcNAc$_2$ oligosaccharides are hydrolyzed at similar rates, both being about threefold better substrates than the lipid-linked structure [25]) [4, 5, 13, 14, 15, 23, 25]

P D-glucose + Glc$_2$Man$_9$GlcNAc$_2$-peptide <2, 3, 10, 11> [4, 5, 13, 14, 15, 25]

S α-D-Glc-(1-2)-α-D-Glc-(1-3)-α-D-Glc-O$(CH_2)_8$-$COOCH_3$ + H_2O <1, 2, 7> (<1> synthetic trisaccharide [9, 10, 20]) [9, 10, 20]

P β-D-glucose + α-D-Glc-(1-3)-α-D-Glc-O$(CH_2)_8$-$COOCH_3$ <1, 2, 7> (<2, 7> formation of α-glucose is delayed with respect to β-glucose: enzyme operates with inversion of anomeric configuration [10]) [9, 10]

S Additional information <1, 2, 3, 4, 5, 6, 7, 8, 10, 11> (also acts, more slowly, on the corresponding glycolipids and glycopeptides; <2, 3, 8> substrate specificities [4, 24, 25]; <3> comparison of free and linked oligosaccharides as substrates: activity is lower against linked substrates than against free substrates [4]; <11> substrate and product specificity [14]; <10> substrate selectivity [13]; <2> structure of the Glc$_3$ branch rather than the more distant domains of the oligosaccharide molecule are controlling specificity [25]; <1> very slightly active with Glc$_2$Man$_9$GlcNAc$_2$ [2]; <1, 2, 6, 8, 11> Glc$_2$Man$_9$GlcNAc$_2$: not a substrate [1, 3, 9, 14, 17, 22,

24, 25]; <1, 2, 6, 8, 11> $Glc1Man_9GlcNAc_2$: not a substrate [2, 3, 9, 14, 17, 22, 24, 25]; <1, 2, 5, 6, 8> aryl-α-D-glucosides: not a substrate [1, 3, 8, 17, 20]; <1, 2, 4, 6, 8, 11> *p*-nitrophenyl-α-D-glucopyranoside: not a substrate [1, 2, 3, 7, 9, 14, 22, 24]; <2,8> *p*-nitrophenyl-β-D-glucopyranoside: not a substrate [22, 24]; <2, 7> *p*-nitrophenyl-glucosides: no substrates [10, 25]; <8> methylumbelliferyl-β-glucosides: no substrates [24]; <10> *p*-NO_2-phenyl-α-glucoside and *p*-NO_2-phenyl-β-glucoside: no substrates [13]) [1, 2, 3, 4, 7, 8, 9, 10, 13, 14, 17, 20, 22, 24, 25]

P ?

Inhibitors

1-deoxymannojirimycin <2> (<2> very slight inhibition, Ki = 0.19 mM [25]) [25]

1-deoxynojirimycin <1, 2, 3, 4, 5, 6, 8, 9, 11> (<1, 2, 5, 6, 8, 11> strong inhibition [1, 3, 9, 14, 15, 17, 22, 24, 25]; <4> 50% inhibition [7]; <8> 0.01 mM: 50% inhibition [11]; <2> 0.01 mM: 80% inhibition [15]; <8> strong inhibiton, Ki = 0.0021 mM [24]; <2> strong inhibition, K_i = 0.0001 mM [25]) [1, 3, 7, 8, 9, 11, 12, 14, 15, 17, 19, 20, 22, 24, 25]

2,5-dihydroxymethyl-3,4-dihydroxypyrrolidine <11> (<11> plant pyrrolidine alkaloid, DMDP, 0.04 mM: 50% inhibition [14]) [14]

Ag^+ <2> (<2> potent inhibitor, inhibition reversed by adding an excess of dithiothreitol [22]) [22]

Co^{2+} <2, 11> (<11> quite inhibitory [14]; <2> potent inhibitor, inhibition reversed by adding an excess of dithiothreitol [22]) [14, 22]

Cu^{2+} <2, 8, 11> (<11> quite inhibitory [14]; <2> potent inhibitor, inhibition reversed by adding an excess of dithiothreitol [22]; <8> partial inhibition [24]) [14, 22, 24]

D-glucose <3> (<3> 5 mM: 25% inhibition [5]; <2> no significant inhibition, 10 mM: 13% inhibition [22]) [5]

D-mannose <3> (<3> 1 mM: 4% inhibition, 25 mM: 60% inhibition [4]; <3> 5 mM: 25% inhibition [5]; <2> 10 mM: no inhibition [22]) [4, 5]

Hg^{2+} <2> (<2> potent inhibitor, inhibition reversed by adding an excess of dithiothreitol [22]) [22]

$Man_9GlcNAc$ <3> (<3> 1 mM thyroglobulin-derived $Man_9GlcNAc$: 22% inhibition [4]) [4]

Mg^{2+} <4> (<4> slight inhibition [7]) [7]

Mn^{2+} <4> (<4> slight inhibition [7]) [7]

N,N-dimethyl-1-deoxynojirimycin <2, 6, 8> (<2, 6, 8> strong inhibition [3, 17, 24, 25]; <6, 8> specific inhibitor [17]; <8> strong inhibition, K_i = 0.0005 mM [24]; <2> strong inhibition, K_i = 0.0004 mM [25]) [3, 17, 24, 25]

N-(5-carboxypentyl)-1-deoxynojirimycin <1, 2, 8> (<1, 2, 8> strong inhibition [1, 24, 25]; <2> slight inhibition [3]; <8> strong specific inhibition, K_i = 0.00045 mM [24]; <2> strong inhibition, K_i = 0.0005 mM [25]) [1, 3, 22, 24, 25]

N-5-carboxypentyl-1-deoxymannojirimycin <2> (<2> very slight inhibition, K_i = 0.1 mM [25]) [25]

N-butyl-1-deoxynojirimycin <2> (<2> strong inhibition, K_i = 0.00009 mM [25]) [25]
N-decanoyl-1-deoxynojirimycin <2> (<2> very slight inhibition, K_i = 0.07 mM [25]) [25]
N-decyl-1-deoxynojirimycin <2> (<2> strong inhibition, K_i = 0.0004 mM [25]) [25]
N-dodecyl-1-deoxynojirimycin <2> (<2> slight inhibition [3]) [3]
N-ethylmaleimide <3> (<3> inhibition when membranes are disrupted with Lubrol PX or saponin, inhibition abolished completely in the presence of dithiothreitol [23]) [23]
N-hexyl-1-deoxynojirimycin <2> (<2> strong inhibition, K_i = 0.00013 mM [25]) [25]
N-methyl-1-deoxymannojirimycin <2> (<2> very slight inhibition, K_i = 0.017 mM [25]) [25]
N-methyl-1-deoxynojirimycin <1, 2, 9> (<1, 2> strong inhibition [1, 3, 25]; <2> strong inhibition, K_i = 0.00007 mM [25]) [1, 3, 12, 25]
SDS <4> (<4> strong inhibition [7]) [7]
Tricine <11> [14]
Tris <1, 11> (<11> 20 mM: 50% inhibition [14]) [9, 14]
Tris/maleate <1> (<1> 10 mM: 60% inhibition [2]) [2]
Zn^{2+} <11> (<11> quite inhibitory [14]; <2> potent inhibitor, inhibition reversed by adding an excess of dithiothreitol [22]) [14, 22]
castanospermine <1, 4, 5, 11> (<4> 15% inhibition [7]; <11> plant indolizidine alkaloid, strong inhibition [14]) [7, 8, 9, 14, 19, 20]
chloride <11> (<11> inhibitory above 0.3 M, 0.5 M: 30% inhibition [14]) [14]
ethanolamine <11> [14]
glycerol <11> (<11> 2% or higher: 25% or more inhibition [14]) [14]
heavy-metal ions <2> (<2> potent inhibitors, inhibition reversed by adding an excess of dithiothreitol [22]) [22]
kojibiose <1, 2, 8, 11> (<1, 2, 8, 11> α-1,2-linked glucose disaccharide [1, 14, 22, 24]; <11> 0.5 mM: 50% inhibition [14]; <2> 1 mM: 76% inhibition [22]; <8> K_i = 0.095 mM [24]) [1, 9, 14, 22, 24]
nojirimycin <2> (<2> potent inhibitor [3]; <2> strong inhibition, 0.1 mM: 40% inhibition, 0.16 mM: 50% inhibition [15]) [3, 15]
octyl-β-glucoside <1, 4> (<1> 10 mM: 90% inhibition [2]) [2, 7]
p-aminophenyl-β-thioglucoside <1> (<1> 100 mM: 73% inhibition [2]) [2]
p-nitrophenyl-α-D-glucoside <2, 3> (<3> 5 mM [5]; <2> 0.0005 mM: 19% inhibition [22]) [5, 22]
p-nitrophenyl-β-D-glucoside <2, 3> (<3> 5 mM [5]; <2> 0.0005 mM: 14% inhibition [22]) [5, 22]
phosphate <11> (<11> inhibitory above 0.3 M, 0.5 M: 30% inhibition [14]) [14]
sulfate <11> (<11> 0.1-0.2 M: 15-20% inhibition, somewhate more inhibitory at higher concentrations [14]) [14]
trypsin <3> (<3> up to 40% inhibition in the presence of Lubrol PX, without Lubrol PX: no inhibition [23]) [23]

Additional information <1, 2, 3, 5, 6, 8, 9, 10, 11> (<2> inhibition of glucosidase I as a function of pH, pH dependence of inhibition, the inhibiting species is the cationic form of the inhibitor [25]; <1, 8> no inhibition by 1 mM EDTA or 1 mM *o*-phenanthroline [1, 24]; <1> no inhibition by dithiothreitol, 2-mercaptoethanol, Mg^{2+}, Mn^{2+}, EDTA, potassium phosphate [2]; <2, 3, 11> no inhibition by EDTA [5, 14, 22, 25]; <2> no inhibition by α,α'-bipyridyl and *o*-phenanthroline [22]; <5> no inhibition by bromoconduritol [8]; <9> inhibitors of processing glucosidase enhances anti CD3-induced interleukin-2 production and cause an accumulation of glucose-containing high-mannose-type oligosaccharides and suppress the secretion of interleukin-4, interferon-γ and interleukin-5. Processing inhibitors inhibit the N-linked oligosaccharide processing of cytokines, inhibition of processing enzyme glucosidase I induces a decreased secretion of cytokines with N-linked oligosaccharides [12]; <10> no inhibition by methyl-β-D-glucoside, sophorose laminaribiose, cellobiose, gentiobiose and methyl-α-D-glucoside [13]; <10> no inhibition by α- or β-linked glucose disaccharides [13]; <11> no inhibition by swainsonine, deoxymannojirimycin or 1,4-dideoxy-1,4-imino-D-mannitol [14]; <11> no inhibition by nigerose, maltose, isomaltose at 2 mM [14]; <11> no inhibition by α-1,3-, α-1,4-, and α-1,6-linked glucose disaccharides [14]; <2> no inhibition by conduritol B epoxide, bromoconduritols A and B, acarbose at 1 mM [15]; <6, 8> no inhibition by 1-deoxymannojirimycin [17]; <2> no significant inhibition by nigerose, Glc(α1-3)Glc disaccharide, maltose, gentibiose, cellobiose, sucrose [22]) [1, 2, 5, 8, 12, 13, 14, 15, 17, 22, 24, 25]

Activating compounds

Brij 35 <4> (<4> at 0.12%: 98% activation [7]) [7]
Lubrol PX <2, 3> (<2, 3> activation [22, 23]) [22, 23]
NP-40 <10> (<10> crude enzyme: 3 to 10fold activation [13]) [13]
Triton X-100 <1, 2, 3, 4, 8, 11> (<1> about 45% activation of membrane-bound activity, maximum stimulation of membrane-bound activity at 0.5% Triton X-100, no activation of soluble activity [2, 9]; <2> several-fold activation [3, 15]; <2> activation, maximal at 0.5% Triton X-100 [15]; <3> activation, required in order to measure glucosidase activity [5]; <3> activation, maximal activity in the presence of 5-10 mg/ml Triton X-100 [11]; <4> 77% activation [7]; <8> activation, maximal activity in the presence of 0.5-2 mg/ml Triton X-100 [11]; <11> 0.2% Triton X-100: 4 to 5fold activation, glucosidase I releases in a soluble form by treating microsomal particles with 1.5% Triton X-100, not an absolute requirement for Triton X-100 [14]) [2, 3, 5, 7, 9, 11, 14, 15, 21, 22]
Additional information <2> (<2> no significant activation by phospholipids [22]) [22]

Metals, ions

phosphate <2, 10> (<10> activation [13]; <2> activation up to 25 mM [22]) [13, 22]
Additional information <1, 2, 3, 4, 5, 6, 8, 11> (<1, 2, 4, 6, 8> metal ions not required [1, 7, 17, 22, 24, 25]; <1, 3, 5, 11> no requirement for divalent cations [2, 5, 8, 9, 14]) [1, 2, 5, 7, 8, 9, 14, 17, 22, 24, 25]

Specific activity (U/mg)

2.6 <1> (<1> soluble activity [2]) [2]
5.5 <1> (<1> membrane-bound activity, with Triton X-100 [2]) [2]
Additional information <1, 2, 7, 8, 10, 11> [1, 2, 3, 6, 10, 13, 14, 22, 24]

K_m-Value (mM)

0.036 <4> (4-methylumbelliferyl-α-D-glucoside) [7]
Additional information <2, 8>(<2, 8> kinetic data [24, 25]) [24, 25]

pH-Optimum

6 <4> [7]
6.2 <2> [3, 15]
6.3-6.5 <11> [14]
6.4 <8> [17, 24]
6.5-7.5 <3> (<3> broad pH optimum between [5]) [5]
6.5-8 <10> (<10> broad pH optimum between [13]) [13]
6.6-7 <2> [22]
6.7 <2> (<2> calf [1, 25]; <2> sharp maximum close to [25]) [1, 25]
6.8 <1> (<1> pH optimum close to [1]; <1> rapid decrease in activity on acidic side [2]; <1> assay at [20]) [1, 2, 9, 20]
Additional information <5, 10> (<5, 10> near neutral pH optimum [8, 13]) [8, 13]

pH-Range

5.5-6.8 <11> (<11> activity greatly decreases at pH values below 5.5 or above 6.8 [14]) [14]
5.6-7.3 <8> (<8> half-maximal activity at, steep decrease in activity on acidic side [24]) [24]
5.8-7.4 <2> (<2> half-maximal activity [25]) [25]
6-7.5 <2> (<2> half-maximal activity at [22]) [22]
6-8.5 <1> (<1> pH 6.0: about 35% of maximal activity, pH 8.5: about 15% of maximal activity, pH 5.8: no activity [2]) [2]
6.2-7.2 <1> (<1> half-maximal activity at [1]) [1]

Temperature optimum (°C)

37 <1, 2, 3, 8> (<1, 2, 8> assay at [1, 20, 22, 24, 25]) [1, 4, 5, 20, 22, 24, 25]
47 <4> [7]

4 Enzyme Structure

Molecular weight

85000 <3, 8> (<3, 8> SDS-PAGE [23, 24]) [23, 24]
92000 <6> (<6> calculated from nucleotide sequence [8, 17]) [8, 17]
93000 <6, 8> (<6, 8> enzyme of 95000 Da degraded by endoglycosidase H, SDS-PAGE [17]) [17]
95000 <1, 6, 8> (<1> proteolytically generated soluble form of the enzyme, SDS-PAGE [9]; <6> cDNA cloned, expressed in COS 1 cells, SDS-PAGE [17, 18]; <6, 8> SDS-PAGE [17]) [9, 17, 18]

97000 <1> (<1> calculated from nucleotide sequence [20]) [20]
105000 <1> (<1> Cwh41p, SDS-PAGE with tunicamycin, which specifically inhibits N-glycosylation [21]) [21]
107000 <1> (<1> cloned CWH41 gene encodes a type-II transmembrane protein, Cwh41p, of 107 kDa, SDS-PAGE [9, 21]) [9, 21]
150000-170000 <2> (<2> calf, gel filtration [25]) [25]
316000 <11> (<11> gel fitration [14]) [14]
320000-330000 <2> (<2> gel filtration [22]) [22]
320000-350000 <2> (<2> calf, gel filtration [3, 25]) [3, 25]

Subunits

? <1, 2, 3> (<1>, x * 95000, SDS-PAGE, reducing conditions [1,9]; <2>, x * 83000, SDS-PAGE, reducing conditions [1]; <3, 8>, x * 85000, SDS-PAGE, reducing conditions [23, 24]) [1, 9, 23, 24]
dimer <2> (<2> calf, 2 * 83000, formation of tetramers and dimers, SDS-PAGE [25]) [25]
tetramer <2> (<2> calf, 4 * 85000, SDS-PAGE [3]; <2>, 4 * 85000, SDS-PAGE reducing conditions [22]; <2> calf, 4 * 83000, formation of tetramers and dimers, SDS-PAGE [25]) [3, 22, 25]

Posttranslational modification

glycoprotein <1, 2, 3, 6, 8, 11> (<1, 2, 3, 6, 8, 11> with high-mannose oligosaccharides [1, 14, 17, 18, 23, 24, 25]; <6,8> single N-glycosylation site, N-glycosylation about 2-3 kDa [17]; <1> N-glycoprotein, four potential N-linked glycosylation sites, N-glycosylation about 2 kDa [21]; <3> N-glycoprotein, N-glycosylation about 2 kDa [23]; <8> N-glycoprotein, one N-linked high-mannose sugar chain of less than 3 kDA [24]) [1, 9, 14, 17, 18, 20, 21, 23, 24, 25]

5 Isolation/Preparation/Mutation/Application

Source/tissue

brain <8> [11]
heart <2, 13> [22]
hippocampus <6> [17]
intestine <2> (<2> calf [24]) [24]
kidney <2, 3> (<2> calf [24]) [11, 24]
liver <2, 3, 6, 8, 9, 13> (<2> calf [1, 3, 15, 24, 25]) [1, 3, 4, 5, 6, 15, 16, 17, 22, 24, 25]
mammary gland <2, 3, 9, 13> (<2> lactating mammary glands [10, 22]; <3> mammary microsomes [23]) [10, 22, 23]
oviduct <10> [13]
seedling <11> [14]
spleen <9> [12]
trophozoite <4> [7]

Localization

endoplasmic reticulum <1, 2, 3, 5, 6, 8, 10, 11> (<3> cisternal surface of endoplasmic reticulum [4]; <3> rough and smooth endoplasmic reticulum [5]; <1, 3, 5, 6, 8> endoplasmic reticulum-resident glycoprotein [8, 9, 17, 21, 23, 24]; <3, 6, 8> catalytic site/active centre located at lumenal side of endoplasmic reticulum [11, 17, 23]; <3, 8> endoplasmic reticulum vesicles [11]; <2, 3, 8, 10> rough endoplasmic reticulum [13, 15, 23, 24]) [3, 4, 5, 8, 9, 11, 13, 14, 15, 17, 18, 19, 20, 21, 22, 23, 24]

membrane <1, 2, 3, 4, 5, 6, 8, 10, 11> (<1> X-2180, equally distributed between particulate and supernatant fractions [2]; <1, 2, 3, 4, 5, 6, 8, 10, 11> membrane-bound [3, 6, 7, 8, 11, 13, 14, 15, 17, 20, 21, 23, 24]; <3> microsomal membranes [4, 5]; <3> integral membrane protein [5]; <1, 5, 6, 8> type-II transmembrane protein [8, 9, 17, 18, 19, 20, 21]; <3> transmembrane, endoplasmic reticular N-linked glycoprotein with distinct domains within its structure. It contains distinct membrane anchoring and luminally oriented catalytic domains, the enzyme has a cytoplasmically exposed tail [23]; <3> membrane orientation [23]) [2, 3, 4, 5, 6, 7, 8, 9, 11, 13, 14, 15, 17, 18, 19, 20, 21, 23, 24, 25]

microsome <1, 2, 3, 6, 8, 10, 11> (<3> localized in rough and smooth microsomes, located on the cisternal surface of microsomal vesicles [5]; <3, 8> microsomal vesicles [11]) [1, 3, 4, 5, 10, 11, 13, 14, 15, 16, 17, 22, 23, 24, 25]

Additional information <1, 2, 3, 6, 8, 10> (<1> X-2180, equally distributed between particulate and supernatant fractions [2, 9]; <3> subcellular distribution and orientation [5, 23]; <3, 8> topological studies [11]; <1, 3, 6, 8, 10> subcellular localization [13, 17, 21, 23]; <2> specific antibodies against glucosidase I allow subcellular localization of glucosidase I [22]; <3> membrane orientation [23]; <3> transmembrane, endoplasmic reticular N-linked glycoprotein with distinct domains within its structure. It contains distinct membrane anchoring and luminally oriented catalytic domains, the enzyme has a cytoplasmically exposed tail [23]) [2, 5, 8, 9, 11, 13, 17, 21, 22, 23]

Purification

<1> (partial, soluble glucosidase I [2, 9]; partial [9]) [1, 2, 9]
<2> (calf [3, 25]; partial, calf [6]; lactating mammary gland [10, 22]) [3, 6, 10, 22, 25]
<3> [16, 23]
<7> [10]
<8> [17, 24]
<10> [13]
<11> [14]

Cloning

<1> (CWH41 gene, encodes yeast glucosidase I [9, 19]) [9, 19, 21]
<6> (cDNA, overexpression in COS 1 cells [17, 18]) [8, 17, 18]

Application

analysis <1, 3, 8, 12> (<1> contributes to knowledge of N-glycan biosynthesis [9]; <3, 8> using the processing glucosidase activity as latency marker for

topological studies with microsomal vesicles from nongluconeogenic tissues in addition to brain and cultured mammalian cells lacking glucose 6-phosphate phosphatase, to assess the intactness of microsomal vesicles [11]; <3, 12> structural analysis of lipid-linked oligosaccharide precursor of asparaginyl oligosaccharides [16]) [9, 11, 16]
pharmacology <2, 5, 6, 7> (<2, 5, 6, 7> glucosidase I may be a target for antiviral and antitumor drugs, glucosidase I inhibitors are potentially useful as antiviral, antitumor and antimetastatic agents [8, 9, 10]; <2, 7> enzyme is an inhibitor target for antiviral agents that interfere with the formation of essential glycoproteins required in viral assembly, secretion and infectivity [10]) [8, 9, 10]

6 Stability

pH-Stability
6.5-7 <11> (<11> most stable at [14]) [14]

Temperature stability
25 <2> (<2> 50% activity lost after 24 h, in buffer containing 1% poly(ethylene glycol) 4000: 75% activity retains after 72 h [22]) [22]
27 <11> (<11> 40 h: 50% activity retained [14]) [14]
37 <11> (<11> most activity lost during storage [14]) [14]

General stability information
<1>, bovine serum albumin, 1-8 mg/ml, enhances stability during assay [2]
<1>, quite stable at all stages of purification [2]
<1>, stable to lyophilization without dialysis after step 3 of purification [2]
<1>, unstable to dialysis against, 10 mM potassium phosphate buffer, pH 6.8, containing 0.002% Hibitane, using either dialysis bags or diafiltration [2]
<2>, 1% poly(ethylene glycol) 4000 stabilizes [22]
<8>, native form of glucosidase I is unusually resistant to non-specific proteolysis, partial digestion with trypsin requires rather high concentrations of the protease, at high, that is equimolar, concentrations of trypsin enzyme is cleaved into a defined and enzymatically active mixture of protein fragments with molecular mass of 69 kDa, 45 kDa and 29 kDa [24]
<8>, resistant to trypsin or subtilisin in sealed microsomal vesicles, in unsealed vesicles by the addition of 2 mg/ml Triton X-100, inactivation within 5 min [11]

Storage stability
<1>, -26°C, lyophilized, stable for at least 1 month [2]
<2>, -20°C, several months, without loss of activity [22]
<2>, 2-4°C, 0.8% Lubrol PX, 200 mM phosphate, pH 6.8, 3 weeks, 50% enzyme activity retained, activity lost rapidly on the removal of detergent [3]
<2>, 25°C, buffer containing 1% poly(ethylene glycol) 4000, 72 h, 75% activity retained [22]
<2>, 4°C, 1 week, without loss of activity [22]

<2>, 4°C, presence of detergent, 0.8% Lubrol PX, $t_{1/2}$: 20 days [3]
<3>, -70°C, 200 mM potassium phosphate buffer, pH 6.8, 0.8% Lubrol PX [23]
<8>, 4°C, stable for several weeks [24]
<11>, -18°C, 10% glycerol, few weeks, no loss of activity [14]
<11>, 0°C, very slight loss of activity [14]
<11>, 27°C, 40 h, more than 50% activity retained [14]
<11>, 37°C, most of activity lost during storage [14]

References

[1] Bause E.; Erkens, R.; Schweden, J.; Jaenicke, L.: Purification and characterization of trimming glucosidase I from Saccharomyces cerevisiae. FEBS Lett., **206**, 208-212 (1986)

[2] Kilker, R.D. Jr.; Saunier, B.; Tkacz, J.S.; Herscovics, A.: Partial purification from Saccharomyces cerevisiae of a soluble glucosidase which removes the terminal glucose from the oligosaccharide $Glc_3Man_9GlcNAc_2$. J. Biol. Chem., **256**, 5299-5303 (1981)

[3] Hettkamp, H.; Legler, G.; Bause, E.: Purification by affinity chromatography of glucosidase I, an endoplasmic reticulum hydrolase involved in the processing of asparagine-linked oligosaccharides. Eur. J. Biochem., **142**, 85-90 (1984)

[4] Grinna, L.S.; Robbins, P.W.: Substrate specificities of rat liver microsomal glucosidases which process glycoproteins. J. Biol. Chem., **255**, 2255-2258 (1980)

[5] Grinna, L.S.; Robbins, P.W.: Glycoprotein biosynthesis. Rat liver microsomal glucosidases which process oligosaccharides. J. Biol. Chem., **254**, 8814-8818 (1979)

[6] Michael, J.M.; Kornfeld, S.: Partial purification and characterization of the glucosidases involved in the processing of asparagine-linked oligosaccharides. Arch. Biochem. Biophys., **199**, 249-258 (1980)

[7] Zamarripa-Morales, S.; Villagomez-Castro, J.C.; Calvo-Mendez, C.; Flores-Carreon, A.; Lopez-Romero, E.: Entamoeba histolytica: identification and properties of membrane-bound and soluble α-glucosidases. Exp. Parasitol., **93**, 109-115 (1999)

[8] Herscovics, A.: Importance of glycosidases in mammalian glycoprotein biosynthesis - review. Biochim. Biophys. Acta, **1473**, 96-107 (1999)

[9] Herscovics, A.: Processing glycosidases of Saccharomyces cerevisiae - review. Biochim. Biophys. Acta, **1426**, 275-285 (1999)

[10] Palcic, M.M.; Scaman, C.H.; Otter, A.; Szpacenko, A.; Romaniouk, A.; Li, Y.X.; Vijay, I.K.: Processing α-glucosidase I is an inverting glycosidase. Glycoconjugate J., **16**, 351-355 (1999)

[11] Rush, J.S.; Waechter, C.J.: Topological studies on the enzymes catalyzing the biosynthesis of Glc-P-dolichol and the triglucosyl cap of $Glc_3Man_9GlcNAc_2$-P-P-dolichol in microsomal vesicles from pig brain:

use of the processing glucosidases I/II as latency markers. Glycobiology, **8**, 1207-1213 (1998)
[12] Kosuge, T.; Tamura, T.; Nariuchi, H.; Toyoshima, S.: Effect of inhibitors of glycopotein processing on cytokine secretion and production in anti CD3-stimulated T cells. Biol. Pharm. Bull., **23**, 1-5 (2000)
[13] Elting, J.J.; Chen, W.W.; Lennarz, W.J.: Characterization of a glucosidase involved in an initial step in the processing of oligosaccharide chains. J. Biol. Chem., **255**, 2325-2331 (1980)
[14] Szumilo, T.; Kaushal, G.P.; Elbein, A.D.: Purification and properties of glucosidase I from mung bean seedlings. Arch. Biochem. Biophys., **247**, 261-271 (1986)
[15] Hettkamp, H.; Bause, E.; Legler, G.: Inhibition by nojirimycin and 1-deoxynojirimycin of microsomal glucosidases from calf liver acting on the glycoprotein oligosaccharides Glc1-3$Man_9GlcNAc_2$. Biosci. Rep., **2**, 899-906 (1982)
[16] Liu, T.; Stetson, B.; Turco, S.J.; Hubbard, S.C.; Robbins, P.W.: Arrangement of glucose residues in the lipid-linked oligosaccharide precursor of asparaginyl oligosaccharides. J. Biol. Chem., **254**, 4554-4559 (1979)
[17] Kalz-Füller, B.; Bieberich, E.; Bause, E.: Cloning and expression of glucosidase I from human hippocampus. Eur. J. Biochem., **231**, 344-351 (1995)
[18] Kalz-Füller, B.; Heidrich-Kaul, C.; Nöthen, M; Bause, E.; Schwanitz, G.: Localization of the human glucosidase I gene to chromosome 2p12-p13 by fluorescence in situ hybridization and PCR analysis of somatic cell hybrids. Genomics, **34**, 442-443 (1996)
[19] Simons, J.F.; Ebersold, M.; Helenius, A.: Cell wall 1,6-β-glucan synthesis in Saccharomyces cerevisiae depends on ER glucosidases I and II, and the molecular chaperone BiP/Kar2p. EMBO J., **17**, 396-405 (1998)
[20] Romero, P.A.; Dijkgraaf, G.J.P.; Shahinian, S.; Herscovics, A.; Bussey, H.: The yeast CWH41 gene encodes glucosidase I. Glycobiology, **7**, 997-1004 (1997)
[21] Jiang, B.; Sheraton, J.; Ram, A.F.J.; Dijkgraaf, G.J.P.; Klis, F.M.; Bussey, H.: CWH41 encodes a novel endoplasmic reticulum membrane N-glycoprotein involved in β1,6-glucan assembly. J. Bacteriol., **178**, 1162-1171 (1996)
[22] Shailubhai, K.; Pratta, M.A.; Vijay, I.K.: Purification and characterization of glucosidase I involved in N-linked glycoprotein processing in bovine mammary gland. Biochem. J., **247**, 555-562 (1987)
[23] Shailubhai, K.; Pukazhenthi, B.S.; Saxena, E.S.; Varma, G.M.; Vijay, I.K.: Glucosidase I, a transmembrane endoplasmic reticular glycoprotein with a luminal catalytic domain. J. Biol. Chem., **266**, 16587-16593 (1991)
[24] Bause, E.; Schweden, J.; Gross, A.; Orthen, B.: Purification and characterization of trimming glucosidase I from pig liver. Eur. J. Biochem., **183**, 661-669 (1989)
[25] Schweden, J.; Borgmann, C.; Legler, G.; Bause, E.: Characterization of calf liver glucosidase I and its inhibition by basic sugar analogs. Arch. Biochem. Biophys., **248**, 335-340 (1986)

Protein-glucosylgalactosylhydroxylysine glucosidase 3.2.1.107

1 Nomenclature

EC number
3.2.1.107

Systematic name
protein-α-D-glucosyl-1,2-β-D-galactosyl-L-hydroxylysine glucohydrolase

Recommended name
protein-glucosylgalactosylhydroxylysine glucosidase

Synonyms
2-O-α-D-glucopyranosyl-5-O-β-D-galactopyranosylhydroxy-L-lysine glucohydrolase
2-O-α-D-glucopyranosyl-O-β-D-galactopyranosyl-hydroxylysine glucohydrolase
GGHG
glucosidase, glycosylgalactosylhydroxylysine
glycosyl-galactosyl-hydroxylysine glucosidase
glycosylgalactosylhydroxylysine glucosidase

CAS registry number
72829-45-9

2 Source Organism

<1> *Gallus gallus* (White Leghorns [1]) [1]
<2> *Rattus norvegicus* (Sprague-Dawley [2]; male albino rat Charles River CD strain [3]; female, normale and diabetic mal Wistar rats [4]) [2-4]

3 Reaction and Specificity

Catalyzed reaction
protein α-D-glucosyl-1,2-β-D-galactosyl-L-hydroxylysine + H_2O = D-glucose + protein β-D-galactosyl-L-hydroxylysine

Reaction type
hydrolysis of O-glycosyl bond

Substrates and products

S 2-O-α-D-glucopyranosyl-O-β-D-galactopyranosylhydroxylysine + H_2O <1, 2> (<1> N-acetylated [1]) [1-3]
P α-D-glucose + β-D-galactopyranosyl-L-hydroxylysine <1> [1]
S basement membrane <1> [1]
P D-glucose + ? <1> [1]
S isomaltose + H_2O <2> [3]
P ?
S kojibiose + H_2O <2> [3]
P α-D-glucose
S maltose + H_2O <2> [3]
P α-D-glucose
S Additional information <2> [4]
P ?

Inhibitors

Cu^{2+} <1, 2> (<1> 1 mM [1]) [1, 2]
D-glucono-1,5-lactone <2> (<2> 10 mM complete inhibition [3]) [3]
NaCl <2> (<2> 0.8 mM 62% inhibition [3]) [3]
Zn^{2+} <1, 2> (<1> 1 mM [1]) [1, 2]
glucosamine <1, 2> (<2> competitive [2]) [1, 2]
iodoacetic acid <1> [1]
p-chloromercuribenzoate <1, 2> (<2> 4 micromol complete inhibition [2]; <2> 1 mM complete inhibition [3]) [1-3]

Specific activity (U/mg)

8.3 <2> [2]
18.7 <1> [1]
Additional information <2> (<2> highest activity in the kidneys of neonatal rats and decreases with age [3]) [3]

K_m-Value (mM)

2.2 <1> (2-O-α-D-glucopyranosyl-O-β-D-galactopyranosylhydroxylysine) [1]
5.7 <2> (2-O-α-D-glucopyranosyl-O-β-D-galactopyranosylhydroxylysine) [3]
5.9 <2> (2-O-α-D-glucopyranosyl-O-β-D-galactopyranosylhydroxylysine) [2]

pH-Optimum

4.4-4.7 <2> [3]
5.3 <1> [1]
5.8 <2> [2]

pH-Range

3.9-5.8 <2> (<2> pH 3.9 and pH 5.8, 50% of maximal activity [3]) [3]

4 Enzyme Structure

Molecular weight

65000 <1> (<1> gel filtration [1]) [1]

75000 <1> (<1> SDS-PAGE [1]) [1]
85000 <2> (<2> SDS-PAGE [2]) [2]

5 Isolation/Preparation/Mutation/Application

Source/tissue

brain <2> [2]
cortex <2> [3, 4]
embryo <1> [1]
heart <2> [2]
intestine <2> [2]
kidney <2> [2-4]
liver <2> [2]
lung <2> [2]
spleen <2> [2, 4]
stomach <2> [2]
testis <2> [2]

Purification

<1> [1]
<2> [2, 3]

6 Stability

General stability information

<2>, 15 mg/l bovine serum albumin stabilize [2]

References

[1] Hamazaki, H., Hotta, K.: Purification and characterization of an α-glucosidase specific for hydroxylysine-linked disaccharide of collagen. J. Biol. Chem., **254**, 9682-9687 (1979)
[2] Hamazaki, H., Hotta, K.: Enzymatic hydrolysis of disaccharide unit of collagen. Isolation of 2-O-α-D-glucopyranosyl-O-β-D-galactopyranosyl-hydroxylysine glucohydrolase from rat spleens. Eur. J. Biochem., **111**, 587-591 (1980)
[3] Sternberg, M., Spiro, R.G.: Studies on the catabolism of the hydroxylysine-linked disaccharide units of basement membranes and collagens. Isolation and characterization of a rat kidney α-glucosidase of high specificity. J. Biol. Chem., **254**, 10329-10336 (1979)
[4] A. M. Grigorova-Boros, Tacnet, F., R.W. Fischer, Winterhalter, K. H., Sternberg, M.: Studies on the catabolism of glycated haemoglobin. Glycated globin is not a substrate for the glycosyl-galactosyl-hydroxylysine glucosidase but for a peptidase activity present in rat kidney and spleen. Amino Acids, **4**, 63-72 (1993)

Lactase 3.2.1.108

1 Nomenclature

EC number
3.2.1.108

Systematic name
lactose galactohydrolase

Recommended name
lactase

Synonyms
lactase-phlorizin hydrolase
lactase/phlorizin hydrolase

CAS registry number
9031-11-2

2 Source Organism

<1> *Rattus norvegicus* [1, 5]
<2> *Homo sapiens* [1, 3, 4]
<3> *monkey* (<3>, lactase-phlorizin hydrolase complex [2]) [2]
<4> *Sus scrofa* [3]

3 Reaction and Specificity

Catalyzed reaction
lactose + H_2O = D-glucose + D-galactose

Reaction type
hydrolysis of O-glycosyl bond

Natural substrates and products
S Additional information <1> (<1>, total lactase activity increases to reach maximal levels around weaning, and remains high subsequently, enzyme activity appears to be regulated during development predominantly by transcription mechanism, while alterations during lactation, and along the proximal to distal gradient, are the result of other control mechanism [5]) [5]
P ?

Substrates and products

S β-methyl glucoside + H_2O <3> (<3>, 7.3% of the activity with lactose [2]) [2]
P methanol + β-D-glucose
S cellobiose + H_2O <2, 3, 4> (<3>, 16% of the activity with lactose [2]; <4>, 25% of the activity with lactose [3]; <2>, 14% of the activity with lactose [3, 4]) [2, 3, 4]
P β-D-glucose
S cellotetraose + H_2O <2, 4> (<4>, 14% of the activity with lactose [3]; <2>, 9% of the activity with lactose [3]) [3]
P β-D-glucose
S cellotriose + H_2O <2, 4> (<4>, 9% of the activity with lactose [3]; <2>, 14% of the activity with lactose [3]) [3]
P β-D-glucose
S cellulose + H_2O <2, 4> (<2,4>, 0.1% of the activity with lactose [3]) [3, 4]
P β-D-glucose
S gentiobiose + H_2O <2> (<2>, 0.7% of the activity with lactose [4]) [4]
P β-D-glucose
S lactose + H_2O <1-4> [1-5]
P D-glucose + D-galactose <1-4> (<3>, most active substrate [2]) [1-5]
S *o*-nitrophenyl-β-galactoside + H_2O <2, 3> (<3>, 15.4% of the activity with lactose [2]; <2>, 17% of the activity with lactose [4]) [2, 4]
P *o*-nitrophenol + β-D-galactose
S *o*-nitrophenyl-β-glucoside + H_2O <3> (<3>, 13% of the activity with lactose [2]) [2]
P *o*-nitrophenol + β-D-glucose
S *p*-nitrophenyl-β-galactoside + H_2O <2, 3> (<2>, 13% of the activity with lactose [4]; <3>, 9.3% of the activity with lactose [2]) [2, 4]
P *p*-nitrophenol + β-D-galactose
S *p*-nitrophenyl-β-glucoside + H_2O <2, 3> (<3>, 12.7% of the activity with lactose [2]; <2>, 10% of the activity with lactose [4]) [2, 4]
P *p*-nitrophenol + β-D-glucose
S phenyl-β-galactoside + H_2O <2> (<2>, 0.1% of the activity with lactose [4]) [4]
P phenol + β-D-galactose
S phlorizin + H_2O <1, 2, 3> (<3>, at 2.3% of the activity with lactose [2]; <2>, 22% of the activity with lactose [4]) [1, 2, 3, 4]
P phloretin + D-glucose <1> [1]
S salicin + H_2O <3> (<3>, 1.6% of the activity with lactose [2]) [2]
P ?

Inhibitors

Tris <2, 4> (<2, 4>, competitive [3, 4]) [3, 4]
cerebroside <3> (<3>, bovine brain cerebroside [2]) [2]
phlorizin <2, 3, 4> (<3, 4>, competitive inhibitor of lactose hydrolysis [2,3]) [2, 3, 4]

Specific activity (U/mg)

16.1 <3> (<3>, lactase [2]) [2]
18 <4> [3]
22 <1> [1]
32 <2> [4]
450 <3> (<3>, phlorizin hydrolase [2]) [2]

K_m-Value (mM)

0.4 <3> (phlorizin) [2]
0.44 <2> (phlorizin) [4]
2.78 <3> (cellobiose) [2]
4.4 <2> (cellobiose) [4]
5.1 <4> (lactose) [3]
21 <1, 2> (lactose) [1, 4]
21.8 <3> (lactose) [2]

pH-Optimum

5.6-6.2 <4> [3]
5.8-6 <2> (<2>, hydrolysis of lactose [4]) [4]

pH-Range

4-10 <4> (<4>, pH 4.0: about 60% of maximal activity, pH 10.0: about 50% of maximal activity [3]) [3]
5-7 <2> (<2>, pH 5.0: about 70% of maximal activity, pH 7.0: about 65% of maximal activity [4]) [4]

4 Enzyme Structure

Molecular weight

280000 <2> (<2>, hydrophilic enzyme form, gel filtration [4]) [4]
320000 <2, 4> (<4>, gel filtration [3]; <2>, amphiphilic enzyme form, gel filtration [4]) [3, 4]

Subunits

dimer <2, 4> (<2, 4>, 2 * 160000, SDS-PAGE [3, 4]) [3, 4]

5 Isolation/Preparation/Mutation/Application

Source/tissue

small intestine <1-4> (<1>, enzyme activity appears to be regulated during development predominantly by transcription mechanism, while alterations during lactation, and along the proximal to distal gradient, are the result of other control mechanism [5]) [1-5]

Purification

<1> [1]
<3> [2, 4]
<4> [3]

Cloning

<1> [5]

6 Stability

pH-Stability

5.5 <1> (<1>, 30% loss of activity after 0.5 min, 55% loss of activity after 120 min [1]) [1]

Temperature stability

37 <1> (<1>, about 30% loss of lactase activity after 30 min [1]) [1]
46 <1> (<1>, 35 min, about 50% loss of phlorizin hydrolase activity. About 70% loss of lactase activity after 25 min [1]) [1]
55 <3> (<3>, lactase and phlorizin hydrolase activity are identically heat inactivated [2]) [2]

General stability information

<2>, repeated freezing and thawing, activity declines fast [4]

Storage stability

<1>, -20°C, stable for several months [1]
<2>, -20°C, stable for several months [4]
<2>, 4°C, stable for several weeks [4]

References

[1] Schlegel-Haueter, S.; Hore, P.; Kerry, K.R.; Semenza, G.: The preparation of lactase and glucoamylase of rat small intestine. Biochim. Biophys. Acta, **258**, 506-519 (1972)
[2] Ramaswamy, S.; Radhakrishnan, A.N.: Lactase-phlorizin hydrolase complex from monkey small intestine. Purification, properties and evidence for two catalytic sites. Biochim. Biophys. Acta, **403**, 446-455 (1975)
[3] Skovbjerg, H.; Noren, O.; Sjoestroem, H.; Danielsen, E. M.; Enevoldsen, B.S.: Further characterization of intestinal lactase/phlorizin hydrolase. Biochim. Biophys. Acta, **707**, 89-97 (1982)
[4] Skovbjerg, H.; Sjoestroem, H.; Noren, O.: Purification and characterisation of amphiphilic lactase/phlorizin hydrolase from human small intestine. Eur. J. Biochem., **114**, 653-661 (1981)
[5] Bueller, H.A.; Kothe, M.J.C.; Goldman, D.A.; Grubman, S.A.; Sasak, W.V.; Matsudaira, P.T.; Montgomery, R.K.; Grand, R.J.: Coordinate expression of lactate-phlorizin hydrolase mRNA and enzyme levels in rat intestine during development. J. Biol. Chem., **265**, 6978-6953 (1990)

Endogalactosaminidase 3.2.1.109

1 Nomenclature

EC number
3.2.1.109

Systematic name
galactosylaminoglycan glycanohydrolase

Recommended name
endogalactosaminidase

CAS registry number
59088-26-5

2 Source Organism

<1> *Streptomyces griseus* [1]
<2> *Pseudomonas sp.* (881 [2]) [2]

3 Reaction and Specificity

Catalyzed reaction
endohydrolysis of 1,4-α-D-galactosaminidic linkages in poly(D-galactosamine)galactosaminidic linkages in poly(D-galactosamine)

Reaction type
hydrolysis of O-glycosyl bond

Natural substrates and products
S polygalactosaminoglycan + H_2O <1> (<1> high molecular weight substrate obtained from Neurospora [1]) [1]
P ?

Substrates and products
S oligogalactosaminoglycan + H_2O <1> (<1> endohydrolysis of 1,4-α-galactosaminidic linkages [1]) [1]
P ?
S polygalactosaminoglycan + H_2O <1, 2> (<1> high molecular weight substrate obtained from Neurospora [1]; <2> α-1,4-linkage [2]) [1, 2]

P galactosamine + GalNα1-4GalN + GalNα1-4GalNα1-4GalN + GalNα1-4GalNα1-4GalNα1-4GalN + α1-4GalNα1-4GalNα1-4GalNα1-4GalNα1-4GalN <2> [2]

S Additional information <1> (<1> sporeling detaching activity, not: N-acetyl-oligogalactosaminoglycan and chitosan [1]) [1]

P ?

Inhibitors

Cu^{2+} <1> [1]
EDTA <1> [1]

K_m-Value (mM)

Additional information <2>(<2> 0.05% polygalactosaminoglycan [2]) [2]

pH-Optimum

5 <2> [2]
5.4 <1> (<1> sporeling-detaching activity [1]) [1]

pH-Range

3-8 <1> (<1> very low activity at pH 3 and 8 [1]) [1]

Temperature optimum (°C)

35 <1> (<1> assay at [1]) [1]
50 <2> [2]

4 Enzyme Structure

Molecular weight

31000 <2> (<2> gel filtration, SDS-PAGE [2]) [2]

Subunits

monomer <2> (<2> 1 * 31000, SDS-PAGE [2]) [2]

5 Isolation/Preparation/Mutation/Application

Source/tissue

culture filtrate <1, 2> [1, 2]

Localization

extracellular <1, 2> [1, 2]

Purification

<1> [1]
<2> [2]

6 Stability

pH-Stability

2 <1> (<1> 20°C, inactivated within a few min [1]) [1]
4-7 <2> (<2> citrate-phosphate buffer [2]) [2]
7 <1> (<1> quite stable at neutrality [1]) [1]

Temperature stability

20 <1> (<1> pH 2, inactivated within a few min [1]) [1]
70 <2> (<2> 15 min, 60% loss of activity [2]) [2]
100 <1> (<1> half-life: 3 min [1]) [1]
Additional information <1> (<1> heat stability increases at less acid pH [1]) [1]

General stability information

<1>, repeated freezing and thawing, no loss of activity [1]

Storage stability

<1>, -20°C, 8 months, distilled water, stable to repeated freezing and thawing [1]

References

[1] Reissig, J.L; Lai, H.-H.; Glasgow, J.E.: An endogalactosaminidase from Streptomyces griseus. Can. J. Biochem., **53**, 1237-1249 (1975)
[2] Tamura, J.-i.; Takagi, H.; Kadowaki, K.: Purification and some properties of the endo α-1,4 polygalactosaminidase from Pseudomonas sp.. Agric. Biol. Chem., **52**, 2475-2484 (1988)

Mucinaminylserine mucinaminidase 3.2.1.110

1 Nomenclature

EC number
3.2.1.110

Systematic name
D-galactosyl-3-(N-acetyl-β-D-galactosaminyl)-L-serine mucinaminohydrolase

Recommended name
mucinaminylserine mucinaminidase

Synonyms
EC 3.2.1.97 (formerly)
acetylgalactosaminidase, endo-α-
endo-α-N-acetyl-D-galactosaminidase
endo-α-N-acetylgalactosaminidase
endo-α-acetylgalactosaminidase

CAS registry number
59793-96-3

2 Source Organism

<1> *Diplococcus pneumoniae* [1-4]

3 Reaction and Specificity

Catalyzed reaction
D-galactosyl-3-(N-acetyl-β-D-galactosaminyl)-L-serine + H_2O = D-galactosyl-3-N-acetyl-β-D-galactosamine + L-serine

Reaction type
hydrolysis of O-glycosyl bonds

Substrates and products
S Galβ(1-3)GalNAcα(1-)OC_6H_4-*o*-NO_2 + H_2O <1> [2]
P Galβ(1-3)GalNAc + *o*-nitrophenol
S Galβ(1-3)GalNAcα(1-)OC_6H_4-*p*-NO_2 + H_2O <1> [2]
P Galβ(1-3)GalNAc + *p*-nitrophenol
S Galβ(1-3)GalNAcα(1-)OC_6H_5 + H_2O <1> [2]
P Galβ(1-3)GalNAc + phenol

S glycopeptide + H_2O <1> (<1>, liberates Galβ(1-3)GalNAc from glycopeptides with Galβ(1-3)GalNAcα(1-)Ser and Thr moieties [1]; <1>, glycopeptides from mouse melanoma, fetuin and pig submaxillary mucin [4]) [1, 4]
P Galβ(1-3)GalNAc + ?
S glycoprotein + H_2O <1> (<1>, liberates Galβ(1-3)GalNAc from glycoproteins with Galβ(1-3)GalNAcα(1-)Ser and Thr moieties [1]; <1>, hydrolyzes the O-glycosidic linkage between α-N-acetyl-D-galactosamine and serine or threonine in mucins and mucin-type glycoproteins [3]) [1, 3]
P Galβ(1-3)GalNAc + ?
S Additional information <1> (<1>, no activity when the aglycone is methanol [2]; <1>, no action on phenyl α-N-acetyl-D-galactosaminide, asialo ovine submaxillary mucin or monosialoganglioside [4]) [2, 4]
P ?

Inhibitors

L-Cys <1> (<1>, 2 mM, 12% inhibition [2]) [2]
EDTA <1> (<1>, 2.5 mM, 62.9% inhibition [1]; <1>, 2 mM, 40% inhibition [3]) [1, 3]
Hg^{2+} <1> (<1>, 2 mM, complete inhibition [3]) [3]
Mg^{2+} <1> (<1>, 2 mM, 28% inhibition [3]) [3]
Mn^{2+} <1> (<1>, 2.5 mM $MnCl_2$, 44.3% inhibition [1]; <1>, 2 mM, 29% inhibition [3]) [1, 3]
PCMB <1> (<1>, 1 mM, 60.2% inhibition [1]) [1]
$ZnCl_2$ <1> (<1>, 2.5 mM, 65.5% inhibition [1]) [1]
galactose <1> [3]
p-nitrophenyl-2-acetamido-2-deoxy-3-O-β-D-galactopyranosyl-β-D-galactopyranoside <1> [2]

Metals, ions

K^+ <1> (<1>, 2 mM, slight activation [3]) [3]
Na^+ <1> (<1>, 2 mM, slight activation [3]) [3]

Specific activity (U/mg)

Additional information <1> [1]

K_m-Value (mM)

0.2 <1> (glycoprotein, <1>, glycoprotein from human erythrocyte membrane, K_m-value determined on the basis of the amount of Galβ(1-3)GalNAc residues [1]) [1]
0.23 <1> (Galβ(1-3)GalNAcα(1-)OC_6H_4-*o*-NO_2) [2]
0.25 <1> (Galβ(1-3)GalNAcα(1-)OC_6H_4-*p*-NO_2) [2]
Additional information <1> [3]

pH-Optimum

5.5-7 <1> [2]
6 <1> [1]
7.6 <1> [3]

pH-Range

6-9 <1> (<1>, pH 6.0: about 45% of maximal activity, pH 9.0: about 60% of maximal activity [3]) [3]

4 Enzyme Structure

Molecular weight

160000 <1> (<1>, gel filtration [3]) [3]

5 Isolation/Preparation/Mutation/Application

Source/tissue

culture medium <1> [1, 2]

Localization

extracellular <1> [1]

Purification

<1> (partial [1, 4]) [1, 3, 4]

References

[1] Endo, Y.; Kobata, A.: Partial purification and characterization of an endo-α-N-acetylgalactosaminidase from the culture medium of Diplococcus pneumoniae. J. Biochem., **80**, 1-8 (1976)

[2] Umemoto, J.; Matta, K.L.; Barlow, J.J.; Bhavanandan, V.P.: Action of endo-α-N-acetyl-D-galactosaminidase on synthetic glycosides including chromogenic substrates. Anal. Biochem., **91**, 186-193 (1978)

[3] Umemoto, J.; Bhavanandan, V.P.; Davidson, E.A.: Purification and properties of an endo-α-N-acetyl-D-galactosaminidase from Diplococcus pneumoniae. J. Biol. Chem., **252**, 8609-8614 (1977)

[4] Bhavanandan, P.; Umemoto, J.; Davidson, E.A.: Characterization of an endo-α-N-acetyl galactosaminidase. Biochem. Biophys. Res. Commun., **70**, 738-745 (1976)

1,3-α-L-Fucosidase 3.2.1.111

1 Nomenclature

EC number
3.2.1.111

Systematic name
3-α-L-fucosyl-N-acetylglucosaminyl-glycoprotein fucohydrolase

Recommended name
1,3-α-L-fucosidase

Synonyms
almond emulsin fucosidase I
fucosidase, 1,3-α-L-
Additional information (cf. EC 3.2.1.51, not identical with EC 3.2.1.63)

CAS registry number
83061-50-1

2 Source Organism

<1> *Prunus dulcis* (almond) [1-3]

3 Reaction and Specificity

Catalyzed reaction
hydrolysis of 1,3-linkages between α-L-fucose and N-acetylglucosamine residues in glycoproteins

Reaction type
hydrolysis of O-glycosyl bond

Substrates and products
S asialoorosomucoid + H_2O <1> (specificially hydrolyzes fucose in α-1,3-linkage to N-acetylglucosamine of Asn-linked oligosaccharide chain branches) [2]
P ?
S lacto-N-difucohexaitol + H_2O <1> (poor substrate) [1]
P ?
S lacto-N-fucopentaitol + H_2O <1> (lacto-N-fucopentaitol II or III, no substrate: lacto-N-fucopentaitol I, <1> [1]) [1, 3]

P ?

S lactoferrin + H_2O <1> (hydrolyzes only α-1,3-linkages between fucose and N-acetylglucosamine) [2]

P ?

S Additional information <1> (no hydrolysis of fucose α-1,6 or α-1,2-linkages to N-acetylglucosamine, as in IgG glycopeptides or 2'-fucosyllactitol, <1> [1]; no substrates are α_2-macroglobulin (MW 72600), bovine IgG glycopeptides or milk oligosaccharide 2'-fucosyllactose, <1> [2]) [1, 2]

P ?

Inhibitors
Hg^{2+} <1> (strong) [2]
L-fucose <1> [2]
malonate <1> (above 20 mM) [2]

Metals, ions
Ca^{2+} <1> (slight activation) [2]
Mg^{2+} <1> (slight activation) [2]
Additional information <1> (no activation by Mn^{2+}) [2]

Specific activity (U/mg)
0.00146 <1> [1]
0.0645 <1> [2]
Additional information <1> [3]

K_m-Value (mM)
0.095 <1> (lacto-N-fucopentaitol III) [1]
0.101 <1> (lacto-N-fucopentaitol II) [1]

pH-Optimum
5.3 <1> [2]
5.5 <1> [1]

pH-Range
4-6.8 <1> (about half-maximal activity at pH 4 and 6.8) [1]
4.3-5.8 <1> (about half-maximal activity at pH 4.3 and about 80% of maximal activity at pH 5.8) [2]

Temperature optimum (°C)
37 <1> (assay at) [2, 3]

5 Isolation/Preparation/Mutation/Application

Source/tissue
emulsin <1> (commercially available, <1> [2]) [1-3]

Purification
<1> (partial [1-3]; affinity chromatography on Blue Sepharose [2]) [1-3]

6 Stability

General stability information

<1>, decreasing ionic strength inactivates [2]
<1>, freeze-thawing results in 30% loss of initial activity [2]
<1>, major loss of activity by ion-exchange chromatography [1]

Storage stability

<1>, 4°C, less than 10% loss of activity within 6 months [2]
<1>, frozen, retains more than 90% of activity after 1 month [3]
<1>, storage of lyophilized enzyme preparation results in 60% loss of activity within 1 month [3]

References

[1] Ogata-Arakawa, M.; Muramatsu, T.; Kobata, A.: α-L-Fucosidases from almond emulsin: characterization of the two enzymes with different specificities. Arch. Biochem. Biophys., **181**, 353-358 (1977)
[2] Imbert, M.J.; Glasgow, L.R.; Pizzo, S.V.: Purification of an almond emulsin fucosidase on Cibacron blue-Sepharose and demonstration of its activity toward fucose-containing glycoproteins. J. Biol. Chem., **257**, 8205-8210 (1982)
[3] Yoshima, H.; Takasaki, S.; Ito-Mega, S.; Kobata, A.: Purification of almond emulsin α-L-fucosidase I by affinity chromatography. Arch. Biochem. Biophys., **194**, 394-398 (1979)

2-Deoxyglucosidase 3.2.1.112

1 Nomenclature

EC number
3.2.1.112

Systematic name
2-deoxy-α-D-glucoside deoxyglucohydrolase

Recommended name
2-deoxyglucosidase

Synonyms
2-deoxy-α-D-glucosidase
glucosidase, 2-deoxy-α

CAS registry number
92480-05-2

2 Source Organism

<1> *Rattus norvegicus* (Sprague Dawley) [1]

3 Reaction and Specificity

Catalyzed reaction
a 2-deoxy-α-D-glucoside + H_2O = an alcohol + 2-deoxy-D-glucose

Reaction type
hydrolysis of O-glycosyl bond

Substrates and products
S 4-nitrophenyl 2-deoxy-α-D-glucopyranoside + H_2O <1> [1]
P 4-nitrophenol + 2-deoxy-D-glucose

Inhibitors
D-mannono-1,5-lactone <1> (4.5 mM: complete inhibition) [1]
Additional information <1> (L-mannono-1,5-lactone, D-glucono-1,5-lactone, D-galactono-1,4-lactone and L-galactono-1,4-lactone, all up to 4.5 mM, 2-deoxyribose, 2-deoxy-D-glucose, 2-deoxycoformycin, 1-amino-2-deoxyglucose, up to 5.5 mM, are no inhibitors) [1]

Specific activity (U/mg)
0.035 <1> [1]

K_m-Value (mM)
2 <1> (4-nitrophenyl 2-deoxy-α-D-glucopyranoside) [1]

pH-Optimum
7.5 <1> (broad optimum) [1]

pH-Range
4-10 <1> (30% of maximal activity at pH 4.0, 80% of maximal activity at pH 10.0) [1]

5 Isolation/Preparation/Mutation/Application

Source/tissue
HTC cell line <1> [1]
leukemia, P388 cell line <1> [1]
liver <1> [1]

Purification
<1> (liver, P388 leukemia cells, HTC cells, all partial) [1]

6 Stability

Temperature stability
50 <1> (5 min, stable) [1]
60 <1> (5 min, 98% loss of activity) [1]

Storage stability
<1>, -20°C, partially purified enzyme, stable up to 18 months [1]

References

[1] Canellakis, Z.N.; Bondy, P.K.; May, J.A.; Myers-Robfogel, M.K.; Sartorelli, A.C.: Identification of a glycosidase activity with apparent specificity for 2-deoxy-D-glucose in glycosidic linkage. Eur. J. Biochem., **143**, 159-163 (1984)

Mannosyl-oligosaccharide 1,2-α-mannosidase 3.2.1.113

1 Nomenclature

EC number
3.2.1.113

Systematic name
1,2-α-mannosyl-oligosaccharide α-D-mannohydrolase

Recommended name
mannosyl-oligosaccharide 1,2-α-mannosidase

Synonyms
(α1,2)-mannosidase-I
1,2-α-mannosidase
Man$_9$-mannosidase
α-1,2-mannosidase
exo-α-1,2-mannanase
glycoprotein processing mannosidase I
mannose-9 processing α-mannosidase
mannosidase 1A
mannosidase 1B
mannosidase I
mannosidase, exo-1,2-α-
Additional information <7> (not identical with aryl-α-mannosidase) [6]

CAS registry number
9068-25-1

2 Source Organism

<1> *Rattus norvegicus* (male [1, 2]; Sprague-Dawley [1]; Wistar [2]) [1, 2, 7, 16, 19, 21]
<2> *Oryctolagus cuniculus* [3, 15, 17]
<3> *Bos taurus* (calf) [10]
<4> *Homo sapiens* [5]
<5> *Sus scrofa* [14, 20]
<6> *Canis familiaris* [21]
<7> *Vigna radiata var. radiata* [6, 9, 13]
<8> *Ricinus communis* (castor bean) [18]
<9> *Plants* [19, 21]
<10> *Spodoptera frugiperda* (IPLB-SF-21AE, baculovirus-infected) [22]

<11> *Aspergillus oryzae* [4, 8]
<12> *Aspergillus saitoi* [11, 12]
<13> *Saccharomyces cerevisiae* [24]
<14> *Bacillus sp.* (M-90) [23]

3 Reaction and Specificity

Catalyzed reaction

hydrolysis of the terminal 1,2-linked α-D-mannose residues in the oligo-mannose oligosaccharide $Man_9(GlcNAc)_2$

Reaction type

hydrolysis of O-glycosyl bond

Natural substrates and products

S Additional information <1-4> (reaction during Asn-linked oligosaccharide biosynthesis, <1> [1]; involved in synthesis, <2> [3] and processing, <4> [5] of Asn-linked complex-type oligosaccharides of mammalian glycoproteins, <2, 4> [3, 5]; involved in glycoprotein processing, <1, 3, 4> [5, 7, 10]) [1, 3, 5, 7, 10]

Substrates and products

S $(Glc)(Man)_9(GlcNAc)_2 + H_2O$ <3> (removal of one α1,2-linked mannose) [10]
P ?
S $(Glc)_3(Man)_9(GlcNAc)_2 + H_2O$ <3> (removal of one α1,2-linked mannose) [10]
P ?
S (Manα1-2)nManα1-6(Manα1-3)Manα1-6(Manα1-3)Manβ1-4GlcNAcβ1-4GlcNAc + H_2O <12> [12]
P Manα1-6(Manα1-3)Manα1-6(Manα1-3)Manβ1-4GlcNAcβ1-4GlcNAc + mannose <12> [12]
S 2-O-α-D-Mannopyranosyl-D-mannose + H_2O <11> [4]
P D-mannose <11> [4]
S GlcNAc-$(Man)_5$GlcNAc + H_2O <1> (poor substrate) [2]
P ?
S High mannose chains of thyroglobulin and phytohemagglutinin-P <2> (thyroglobulin: 70% of α1,2-mannose residues accessible) [15]
P ?
S Manα(1-2)Manα(1-3)Manβ(1-4)GlcNAc + H_2O <12> [12]
P ?
S Manα(1-2)Manα(1-6)(Manα(1-2)Manα(1-3))Manα(1-6)(Manα(1-2)Manα(1-2)Manα(1-3)Manβ(1-4)GlcNAcβ(1-4)GlcNAc-hexapeptide <5> (substrate specificity not influenced by peptide moiety, removes 3 mannose residues) [20]
P $(Man)_6(GlcNAc)_2$-hexapeptide + mannose <5> [20]

S Manα(1-2)Manα(1-6)(Manα(1-2)Manα(1-3)Manα(1-6)(Manα(1-2) Manα(1-2)Manα(1-3)Manβ(1-4)GlcNAc <1, 5, 7, 13, 14> (i.e. (Man)$_9$ GlcNAc, specific for α-1,2-linked mannose residues, <1, 14> [1, 23]; best substrate, <1> [2]; removes all 4 α1,2-linked mannosyl residues, <1, 7> [7, 13]; the terminal mannose on the middle antenna appears to be the most susceptible residue, <1> [16]) [1, 2, 7, 9, 13, 16, 20, 21, 23, 24]

P (Man)$_5$GlcNAc + mannose <1, 5, 7, 14> (and (Man)$_6$GlcNAc, <1> [1]; after long-term incubation, <7> [9]; (Man)$_8$GlcNAc, (Man)$_7$GlcNAc and (Man)$_6$GlcNAc after 4 h, (Man)$_8$GlcNAc, (Man)$_7$GlcNAc, (Man)$_6$GlcNAc and (Man)$_5$GlcNAc after 12 h incubation, <7> [9]) [1, 7, 9, 13, 20, 23]

S Manα(1-2)Manα(1-6)(Manα(1-2)Manα(1-3)Manα(1-6)(Manα(1-2) Manα(1-2)Manα(1-3)Manβ(1-4)GlcNAcβ(1-4)GlcNAc <1, 3-5, 11> (i.e. (Man)$_9$ (GlcNAc)$_2$ <4> [5]; highly active on α(1-2) linked mannooligosaccharides, <5, 11> [4, 14]; removes 3 of the 4 α1,2-linked mannosyl residues, <3, 5> [10, 14, 20]; intact chitobiose core affects Man$_9$-mannosidase specificity: reduction or removal of terminal N-acetylglucosamine residue increases hydrolytic susceptibility of the fourth α1,2-mannosyl-linkage, <5> [20]; enzyme from endoplasmic reticulum removes only 1 or 2 α1,2-linked mannose residues, <1> [21]) [4, 5, 10, 14, 20, 21]

P (Man)$_6$(GlcNAc)$_2$ + mannose <3, 5> [10, 20]

S Manα(1-2)Manα(1-6)(Manα(1-3))Manα(1-6)(Manα(1-2)Manα(1-3)) Manβ(1-4)GlcNA <1> (hydrolysis at 39% (mannosidase IB) or 67% (mannosidase IA) the rate of (Man)$_9$GlcNAc hydrolysis, <1> [2]) [1, 2, 16]

P ?

S Manα(1-2)Manα(1-6)(Manα(1-3))Manα(1-6)(Manα(1-2)Manα(1-3)) Manβ(1-4)GlcNAcβ(1-4)GlcNAc <3, 5> [10, 14]

P ?

S Manα(1-2)Manα(1-6)(Manα(1-3))Manα(1-6)(Manα(1-2Manα(1-2)) Manα(1-3))Manβ(1-4)GlcNAc <1> [1, 16]

P ?

S Manα(1-2)Manα(1-6)(Manα(1-3))Manα(1-6)(Manα(1-2Manα(1-2)) Manα(1-3))Manβ(1-4)GlcNAcβ(1-4)GlcNAc <3, 5> [10, 14]

P ?

S Manα(1-2)Manα(1-6)(Manα(1-3))Manα(1-6)(Manα(1-2Manα(1-2)) Manα(1-3))Manβ(1-4)GlcNAcβ(1-4)GlcNAc-peptide <1> [1]

P (Man)$_5$(GlcNAc)-peptide + mannose <1> [1]

S Manα(1-6)(Manα(1-3))Manα(1-6)(Manα(1-2)Manα(1-3))Manβ(1-4) GlcNAc <5, 8, 10, 12> (preferred substrate, <10> [22]; as in Taka-amylase A, <12> [11]; very poor substrate, <5> [14, 20]; pyridylamino derivative of Manα1-6(Manα1-3)Manα1-6(Manα1-2Man α1-3)ManβGlcNAc2, <8> [18]; not, <3> [10]) [11, 14, 18, 20, 22]

P (Man)$_5$(GlcNAc)$_2$ + mannose <8, 10> (i.e. pyridylamino derivative of (Manα1-3)Manβ1-4GlcNAcβ1-4GlcNAc, <8> [18]) [18, 22]

S Manα(1-6)(Manα(1-3))Manα(1-6)(Manα(1-2)Manα(1-3))Manβ(1-4) GlcNAcβ(1-4)GlcNAc <4, 11> (i.e. GPIV, <4,11> [4, 5]) [4, 5]

P (Man)$_5$(GlcNAc)$_2$-Asn + mannose <4, 11> (i.e. GPI, <4> [5]; major product formed, 3.5 mol mannose/mol GP IV per h, <4> [5]) [4, 5]

S Manα(1-6)(Manα(1-3))Manα(1-6)(Manα(1-2)Manα(1-3))Manβ(1-4) GlcNAcβ(1-4)GlcNAc-Asn <1> (hydrolysis at 18-25% the rate of $(Man)_9$ GlcNAc hydrolysis, <1> [2]) [1, 2, 16]
P ?
S Manα(1-6)(Manα(1-3))Manα(1-6)(Manα(1-3))Manβ(1-4)GlcNAc <1> (without α1,2-linked mannose residues: poor substrate, <1> [1, 2]) [1, 2]
P $(Man)_4$GlcNAc + $(Man)_3$GlcNAc + mannose <1> [2]
S Manα(1-6)(Manα(1-3))Manα(1-6)(Manα(1-3))Manβ(1-4)GlcNAcβ(1-4) GlcNAc <2, 7> (i.e. Manα1-2Manα1-2-Manα1-3(Manα1-6)ManβGlcNAc-βGlcNAc, <2, 7> [3, 6, 15]; or heptasaccharide I, <7> [6]; without α1,2-linked mannose residues: no substrate, <1> [16]) [3, 6, 15]
P $(Man)_4(GlcNAc)_2$ + $(Man)_3(GlcNAc)_2$ + mannose <7> [6]
S pyridylamino derivative of $(Man)_9(GlcNAc)_2$ + H_2O <8> [18]
P pyridylamino derivative of $(Man)_5(GlcNAc)_2$ + mannose <8> (after 120 h incubation, via specific intermediates: overview) [18]
S α-1,2-linked mannotetraose + H_2O <14> [23]
P ?
S α-D-Man-(1-2)-D-Man + H_2O <11> [8]
P D-Mannose <11> [8]
S baker's yeast mannan + H_2O <11, 12, 14> (releases 9.2% of total mannose, <11> [4]; α-1,2-linked side-chains, <14> [23]) [4, 8, 11, 23]
P ?
S mannobiose + H_2O <12> [11]
P D-mannose <12> [11]
S methyl-2-O-α-D-mannopyranosyl-α-D-mannopyranoside + H_2O <2> [15, 17]
P ?
S ovalbumin glycopeptide + H_2O <2, 10> (glycopeptide IV, <2> [15]) [15, 22]
P ?
S *p*-nitrophenyl-α-D-mannoside + H_2O <1> (mannosidase IB: poor substrate, mannosidase IA: no substrate, <1> [2]; not, <1-3, 5, 7, 8, 10-12, 14> [4, 6, 10, 12, 14-16, 18, 20-23]) [2]
P ?
S Additional information <1, 3-5, 7, 8, 11, 12, 14> (mannosidase IA and B are quite similar in substrate specificity, <1> [2]; pig liver enzyme substrate specificity resembles that of calf liver enzyme, <5> [14]; no hydrolysis of α1,6-linkages, <7, 12> [6, 12] or α1,3-linkages, <12> [12]; no substrates are 3-O-α- and 6-O-α-D-mannobioses, <11> [4]; $(Man)_5$ $(GlcNAc)_2$-Asn (i.e. GPI, <4> [5]), <4,11> [4, 5]; GP III, <4> [5]; $(Man)_4$-$(GlcNAc)_2$, <8> [18]; 4-methylumbelliferyl-α-mannoside, <5> [14, 20]; 1,2-α-D-mannobiitol, <14> [23]; little or no activity with 4-methylumbelliferyl-α-mannoside, aryl-α-D-mannosides, <3> [10]) [2, 4-6, 10, 12, 14, 18, 20, 23]
P ?

Inhibitors

1,4-dideoxy-1,4-imino-D-mannitol <1> (in vitro and in vivo, <1> [7]; not, <1, 7, 9> [9, 13, 19]) [7]
1-deoxymannojirimycin <1-3, 5, 7, 8, 10> (mannose analog, in vivo and in vitro, <1,2> [17, 19]; reversible (not by Ca^{2+}, <2> [17]), <1> [16]; rat liver enzyme, <1> [19]; strong, <5> [20]; not (ER-enzyme, <1> [21]), <1, 14> [21, 23]) [9, 10, 13-22]
1-cyclohexyl-3-(2-morpholinyl-4-ethyl)carbodiimide/glycine methylester <12> [11]
Ag^{2+} <11> (not, <12> [11]) [4]
BAPTA <2> (i.e. 1,2-bis(2-aminophenoxy) ethane N,N,N',N'-tetraacetic acid) [15]
Ba^{2+} <11> (10 mM) [4]
Cd^{2+} <8, 11> (strong, <8> [18]; 10 mM, activation at 1 mM, <11> [4]) [4, 18]
Co^{2+} <1, 2, 11, 12> (weak, <1> [1]) [1, 4, 11, 15-17]
Cu^{2+} <1, 3, 5, 7, 8, 11, 12> (strong, <7, 8> [6, 18]; 10 mM, <11> [4]) [1, 4, 6, 10, 11, 14, 16, 18]
D-mannono-γ-lactone <11> [4]
D-mannonolactam amidrazone <1> (mannosidase inhibitor, MDCK-cells: in vivo and in vitro) [21]
EDTA <1-3, 7, 8, 11-14> (weak, <1, 12> [1, 11]; divalent cations, <13> [24]; Ca^{2+}, <1, 2, 7> [6, 9, 13, 15, 21]; Mg^{2+} (slightly, <1> [21]; not, <2,7> [9, 15]), <3> [10]; Sr^{2+} (slightly), <1> [21]; Zn^{2+} (not, <1, 2> [15, 21]), <3> [10] or Mn^{2+} (less efficiently, <3> [10]; not, <1,2> [15, 21]) restores, <1-3, 7, 13> [6, 9, 10, 13, 15, 21, 24]; not Cu^{2+}, <1> [21]; Co^{2+}, <1, 2> [15, 21]; Ni^{2+} or Ba^{2+}, <2> [15]) [1, 4, 6, 9-11, 15, 16, 18, 21, 23, 24]
EGTA <2> (i.e. ethylene glycol-bis[β-aminoethylether]-N,N,N',N'-tetraacetic acid) [15]
Fe^{2+} <11, 12> [4, 11]
Fe^{3+} <8, 11> (strong, <8> [18]) [4, 18]
Hg^{2+} <8, 11> (strong) [4, 18]
Mn^{2+} <2, 12> [11, 15, 17]
N-5-carboxypentyldeoxymannojirimycin <3, 5> (strong, <5> [20]) [10, 20]
N-methyl-deoxymannojirimycin <1, 3, 5> [10, 14, 19]
Na^{+} <11> (1 mM, weak, not at 10 mM) [4]
Ni^{2+} <3> [10]
PCMB <2> (Ca^{2+} protects, <2> [15]; not, <12> [11]) [15]
Sn^{2+} <8> (strong) [18]
Ti^{4+} <8> (strong) [18]
Tris <1, 2, 8> (in vivo and in vitro, <2> [17]) [16-18]
Tris/maleate <1> [1]
Zn^{2+} <2, 7, 8, 11> (strong, <7> [6]; reversible by Ca^{2+}, <2> [17]) [4, 6, 15, 17, 18]
α1,2-linked oligosaccharides <7> [9, 13]
α1,3-linked oligosaccharides <7> [9, 13]
basic sugar analogs of mannose <5> (strong) [14]

iodoacetic acid <11, 12> (monoiodoacetic acid, 1 mM, weak, <12> [11]) [4, 11]
kifunensine <1, 9> (plant and animal Golgi mannosidase I, <1, 9> [19]; mannosidase I, not ER-enzyme from rat liver, <1> [21]) [19, 21]
mannosamine <7> (not, <12> [11]) [9, 13]
mannosyl-mannose disaccharide <7> (in decreasing order of efficiency: α1,2, α 1,3 and α1,6-mannosylmannose) [6]
methyl-α-mannoside <12> (not, <11> [4]) [11]
p-chloromercuriphenylsulfonic acid <1> [16]
phosphatidylinositol <2> ($CaCl_2$ or NaCl restores) [15]
Additional information <1, 2, 5-12> (mannosidase IA and B are quite similar in response to inhibitors, <1> [2]; no inhibition by L-cysteine, <11> [4]; mannose, α1,6-linked mannose oligosaccharide, castanospermine, deoxynojirimycin, 2,5-dihydroxymethyl-3,4-dihydroxypyrrolidine, <7> [9, 13]; 4-nitrophenyl-α-mannoside, mannitol, L-mannono-1,4-lactone, PMSF, <12> [11]; Ca^{2+}, <1> [16]; Mg^{2+}, <1,2> [15-17]; swainsonine (little or not, <5> [14]), <1, 2, 5, 7-10, 14> [9, 13-15, 18-20, 22, 23]; mannostatin A, <1, 6> [19, 21] or N,N-dimethyl-1-deoxymannojirimycin (little or not, <5> [14]), <5> [14, 20]) [2, 4, 9, 11, 13-22]

Activating compounds

Triton X-100 <1, 2, 7> (requirement, <1> [16]; activation (together with zwitterionic phospholipids, <2> [3]), <2,7> [3, 9, 13]; 0.1%, <7> [9, 13]; 0.12%, no activation at 0.07% and below, <1> [16]) [3, 9, 13, 16]
phosphatidylcholines <2> (activation, with acyl-chains of different lengths, above C-4) [3]
phosphatidylethanolamine <2> (activation, in the presence of 3.5 mM Triton X-100) [3]
zwitterionic phospholipids <2> (activation, required for solubilized and purified enzyme, independent of acyl chain length or degree of saturation, an ordered lipid structure of either micelles or bilayers, mixed micelles together with Triton X-100) [3]
Additional information <2> (no activation by negatively charged phospholipids, dibutyryl-phosphocholine) [3]

Metals, ions

Ca^{2+} <2, 5, 7, 8, 10> (requirement, <2> [17]; activation, <2, 5, 7, 8, 10> [6, 9, 13-15, 18, 22]; most effective cation, <5> [14]; slight, <8> [18]; 5 mM, <7> [13]; microsomal enzyme, <7> [6]; mechanism, <2> [17]; not, <14> [23]) [6, 9, 13-15, 17, 18, 22]
Cd^{2+} <11> (activation, 1 mM, inhibits at 10 mM) [4]
Co^{2+} <7> (slight activation) [9]
Cu^{2+} <7> (slight activation) [9]
Fe^{2+} <7> (slight activation) [9]
Mg^{2+} <5, 7> (activation, <5, 7> [13, 14]; less effective than Ca^{2+}, <5> [14]; slight, <7> [13]; not, <11> [4]) [13, 14]
Mn^{2+} <7> (slight activation) [9, 13]

Zn^{2+} <5> (activation, less effective than Ca^{2+}) [14]
divalent cations <3> (requirement, not, <12> [12]) [10]

Specific activity (U/mg)
0.00000036 <8> ($(Man)_6$-GlcNAc as substrate) [18]
12 <11> [8]
34600 <12> [12]
Additional information <1-3, 5, 7> [2, 9, 10, 13-16]

K_m-Value (mM)
0.1 <1> ($(Man)_8$GlcNAc) [1]
0.26 <8> ($(Man)_6$GlcNAc) [18]
0.3 <13> ($(Man)_9$GlcNAc) [24]
0.55 <2> (GPIV from ovalbumin) [15]
0.57 <11> (2-O-α-D-mannopyranosyl-D-mannose) [4]
0.6 <2> (methyl-mannopyranosyl-mannopyranoside) [17]
0.67 <12> (mannobiose) [11]
1.25 <12> (Manα1-2Manα1-3Manβ1-4GlcNAc) [12]
Additional information <5> (kinetic constants of various substrate derivatives) [20]

pH-Optimum
4.9-5.3 <11> (2-O-α-D-mannopyranosyl-D-mannose as substrate) [4]
5 <1, 12> (Man(α1-2)Man as substrate, <1> [21]) [12, 21]
5-6 <2> (in the presence of 10 mM $CaCl_2$, <2> [15]) [15, 17]
5.5-6 <7, 8> [6, 13, 18]
5.5-6.5 <2> (in the presence of egg yolk lyso-phosphatidylcholine) [3]
6 <1, 5, 7, 10> (mannosidase IA and B, <1> [2]) [2, 9, 14, 16, 22]
6-6.2 <3> [10]
6-6.5 <1> ($(Man)_8(GlcNAc)_2$-peptide as substrate) [1]
6.5 <2> (in the presence of soybean phosphatidylinositol) [3]
7 <14> [23]
Additional information <14> (pI: 3.6) [23]

pH-Range
4-5.8 <11> (about half-maximal activity at pH 4 and 5.8, 0.1 M acetate buffer) [4]
4.8-6.8 <7> (about half-maximal activity at pH 4.8 and 6.8) [9]
5.2-6.7 <5> (about half-maximal activity at pH 5.2 and 6.7) [14]
5.2-7.3 <2> (about half-maximal activity at pH 5.2 and 7.3, in the presence of egg yolk lyso-phosphatidylcholine) [3]
5.3-6.6 <10> (about half-maximal activity at pH 5.3 and 6.6) [22]
5.5-7 <1, 3> (about half-maximal activity at pH 5.5 and 7, <1, 3> [1, 10]; $(Man)_8(GlcNAc)_2$-peptide as substrate, <1> [1]) [1, 10]
5.5-7.5 <2> (about half-maximal activity at pH 5.5 and 7.5, in the presence of soybean phosphatidylinositol) [3]

Temperature optimum (°C)
30 <12> (assay at) [11]

37 <1-5, 7, 10-12> (assay at) [1-6, 10, 12-17, 22]
40 <8> (assay at) [18]

4 Enzyme Structure

Molecular weight
39000 <7> (cytosolic enzyme, gel filtration) [6]
40000 <2> (gel filtration) [15]
49000 <11> (gel filtration) [4]
49200 <2> (sedimentation analysis) [15]
65000 <5> (detected in freshly prepared crude microsomal extracts using polyclonal antibodies, during purification the native enzyme, MW 65000, loses a membrane-spanning domain without losing its catalytic activity) [14]
230000 <1> (PAGE) [16]
380000 <14> [23]
460000 <7> (mung bean, microsomal enzyme, gel filtration, presumably in a micelle or aggregate containing detergent, phospholipid or other protein) [6]
Additional information <2, 12> (amino acid composition) [11, 15]

Subunits
? <3, 10, 12> (x * 56000, bovine, SDS-PAGE [10]; x * 63000, Spodoptera frugiperda, SDS-PAGE [22]; x * 64000, Aspergillus saitoi, SDS-PAGE [11]) [10, 11, 22]
dimer <14> (2 * 190000, Bacillus sp.) [23]
monomer <2, 5> (1 * 49000, pig, SDS-PAGE [14, 20]; 1 * 52000, rabbit, SDS-PAGE, under reducing and non-reducing conditions [15]) [14, 15, 20]
tetramer <1> (4 * 57000-58000, rat, SDS-PAGE) [16]

Posttranslational modification
glycoprotein <1, 2, 12> (19.4% neutral and 3.9% amino sugars, <12> [11]; 0.88% hexoses, <1> [16]) [11, 15, 16]

5 Isolation/Preparation/Mutation/Application

Source/tissue
Taka-diastase <11> (enzyme-product from Sankyo Co., Ltd.) [4, 8]
adenocarcinoma, colon, liver <4> [5]
cell suspension culture <6, 10> (MDCK-cells, <6> [21]) [21, 22]
cotyledon <8> [18]
culture filtrate <12> [11]
culture medium <12> (commercially available under the name of Morushin) [12]
culture supernatant <14> [23]
hypocotyl <7> [6]
kidney, Madin-Darby canine cell <1> (influenza virus-infected) [21]

liver <1-3, 5, 6> [1-3, 7, 10, 14-17, 19-21]
mycelium <12> (commercially available under the name of Morushin) [12]
seedling <7, 8> [6, 9, 18]

Localization

Golgi membrane <1, 9, 10> (cis-Golgi, <1, 9> [19]) [1, 2, 16, 19, 22]
cytosol <1, 7> (2 enzymes: a membrane-bound microsomal and a soluble cytoplasmic one, <7> [6]; rat liver, <1> [21]) [6, 21]
endoplasmic reticulum <1, 3, 5> (rat liver, <1> [21]) [10, 20, 21]
membrane-bound <1-7, 9> (2 enzymes: a membrane-bound microsomal and a soluble cytoplasmic one, <7> [6]) [1-3, 5, 6, 9, 10, 13-17, 21]
microsome <1-7, 9> (2 enzymes: a membrane-bound microsomal and a soluble cytoplasmic one, <7> [6]; trans-membrane protein with cytosolic domain, <5> [14]) [3, 5, 6, 9, 10, 13-15, 17, 21]
soluble <1, 7> (2 enzymes: a membrane-bound microsomal and a soluble cytoplasmic one, <7> [6]; rat liver, <1> [21]) [6, 21]

Purification

<1> (partial [1, 2, 7]; mannosidase IA [16]; enzyme from endoplasmic reticulum [21]) [1, 2, 7, 16, 21]
<2> (partial, solubilization requires non-ionic detergents) [3]
<3> [10]
<4> (partial) [5]
<5> (affinity chromatography on immobilized N-5-carboxypentyl-1-deoxymannojirimycin) [14]
<6> (partial) [21]
<7> (partial [6]; affinity chromatography on mannan and mannosamine-Sepharose [9]) [6, 9, 13]
<8> (partial) [18]
<10> [22]
<11> (affinity chromatography with baker's yeast mannan gel [8]) [4, 8]
<12> [12]
<13> (recombinant enzyme) [24]

6 Stability

pH-Stability

5 <2> (above, stable) [15]

Temperature stability

37 <1> (24 h, in 0.1 M sodium acetate buffer, pH 5.8, 30% loss of activity) [2]
50 <1, 11> (90 min, about 60-70% loss of activity, <1> [1]; 1 h, in 0.1 M acetate buffer, pH 5, 10% loss of activity, <11> [4]; mannosidase IB less stable than IA, <1> [2]) [1, 2, 4]
Additional information <1> (mannosidase IA and B differ in thermostability) [2]

General stability information

<1>, mannosidase IA and B differ in sensitivity to storage in liver cytoplasmic extracts, no interconversion between mannosidase IA and IB during cellulose phosphate chromatography [2]
<5>, enzyme loses a membrane-spanning domain during purification which does not alter its catalytic properties [20]
<5>, native enzyme protein is highly susceptible to proteolytic cleavage, not by glucopeptidase F [14]

Storage stability

<1>, 0-4°C, in 10 mM potassium phosphate buffer, pH 7.2, 0.5% Triton X-100, 4-6 months [16]
<1>, 2-4°C, in 10 mM potassium phosphate buffer, pH 7.2, 1% Triton X-100, 20% loss within 4 weeks [2]
<1>, 4°C, 1 mg protein/ml, in 50 mM phosphate buffer, pH 6.5, 5 mM $MgCl_2$, 0.1% Triton X-100, 30 days [1]
<2>, -20°C, in 10 mM HEPES/NaOH, pH 7.5, 2 mM $MgCl_2$, 0.1% Nonidet-P40, 0.2 M NaCl, less than 10% loss of activity within 3 months [15]
<7>, 0-4°C, in buffer with 10% glycerol, purified enzyme, a few days, partially purified enzyme, at least 2 weeks [9]
<7>, 4°C, in buffer with 10% glycerol, a few days, at early stage of purification at least 2 weeks [13]
<11>, 4°C, pH 4.5-6, at least 6 months [4]
<12>, -20°C, in 10 mM sodium acetate buffer, pH 5, with or without 1 mM $CaCl_2$, at least 4 months [12]
<12>, 4°C, in 10 mM sodium acetate buffer, pH 5, with or without 1 mM $CaCl_2$, at least 4 months [12]

References

[1] Tabas, I.; Kornfeld, S.: Purification and characterization of a rat liver Golgi α-mannosidase capable of processing asparagine-linked oligosaccharides. J. Biol. Chem., **254**, 11655-11663 (1979)
[2] Tulsiani, D.R.P.; Hubbard, S.C.; Robbins, P.W.; Touster, O.: α-D-Mannosidases of rat liver Golgi membranes. Mannosidase II is the $GlcNAcMan_5$-cleaving enzyme in glycoprotein biosynthesis and mannosidases IA and IB are the enzymes converting Man_9 precursors to Man_5 intermediates. J. Biol. Chem., **253**, 3660-3668 (1982)
[3] Forsee, W.T.; Springfield, J.D.; Schutzbach, J.S.: Effect of phospholipids on α-1,2-mannosidase activity. J. Biol. Chem., **257**, 9963-9967 (1982)
[4] Yamamoto, K.; Hitomi, J.; Kobatake, K.; Yamaguchi, H.: Purification and characterization of 1,2-α-mannosidase of Aspergillus oryzae. J. Biochem., **91**, 1971-1979 (1982)
[5] Chandrasekaran, E.V.; Savila, M.; Nixon, D.; Mendicino, J.: Purification and properties of α-D-mannose:β-1,2-N-acetylglucosaminyl-transferases and α-

D-mannosidases from human adenocarcinoma. Cancer Res., **44**, 4059-4068 (1984)
[6] Forsee, W.T.: Characterization of microsomal and cytosolic α-1,2-mannosidases from mung bean hypocotyls. Arch. Biochem. Biophys., **242**, 48-57 (1985)
[7] Palamarczyk, G.; Mitchell, M.; Smith, P.W.; Fleet, G.W.; Elbein, A.D.: 1,4-Dideoxy-1,4-imino-D-mannitol inhibits glycoprotein processing and mannosidase. Arch. Biochem. Biophys., **243**, 35-45 (1985)
[8] Tanimoto, K.; Nishimoto, T.; Saitoh, F.; Yamaguchi, H.: Improved purification of Aspergillus oryzae 1,2-α-mannosidase by using mannan gel affinity chromatography. J. Biochem., **99**, 601-604 (1986)
[9] Szumilo, T.; Kaushal, G.P.; Hori, H.; Elbein, A.D.: Purification and properties of a glycoprotein processing α-mannosidase from mung bean seedlings. Plant Physiol., **81**, 383-389 (1986)
[10] Schweden, J.; Legler, G.; Bause, E.: Purification and characterization of a neutral processing mannosidase from calf liver acting on $(Man)_9(GlcNAc)_2$ oligosaccharides. Eur. J. Biochem., **157**, 563-570 (1986)
[11] Shigematsu, Y.; Tsukahara, K.; Tanaka, T.; Takeuchi, M.; Ichishima, E.: Molecular and enzymic properties of 1,2-α-D-mannosidase from Aspergillus saitoi. Curr. Microbiol., **13**, 43-46 (1986)
[12] Kobata, A.; Amano, J.: α-Mannosidases I and II from Aspergillus saitoi. Methods Enzymol., **138**, 779-785 (1987)
[13] Kaushal, G.P.; Elbein, A.D.: Glycoprotein processing enzymes of plants. Methods Enzymol., **179**, 452-475 (1989)
[14] Schweden, J.; Bause, E.: Characterization of trimming Man_9-mannosidase from pig liver. Purification of a catalytically active fragment and evidence for the transmembrane nature of the intact 65 kDa enzyme. Biochem. J., **264**, 347-355 (1989)
[15] Forsee, W.T.; Palmer, C.F.; Schutzbach, J.S.: Purification and characterization of an α-1,2-mannosidase involved in processing asparagine-linked oligosaccharides. J. Biol. Chem., **264**, 3869-3876 (1989)
[16] Tulsiani, D.R.P.; Touster, O.: Mannosidase IA from rat liver Golgi membranes. Methods Enzymol., **179**, 446-451 (1989)
[17] Schutzbach, J.S.; Forsee, W.T.: Calcium ion activation of rabbit liver α 1,2-mannosidase. J. Biol. Chem., **265**, 2546-2549 (1990)
[18] Kimura, Y.; Yamaguchi, O.; Suehisa, H.; Tagaki, S.: In vitro hydrolysis of oligomannose-type sugar chains by an α-1,2-mannosidase from microsomes of developing castor bean cotyledons. Biochim. Biophys. Acta, **1075**, 6-11 (1991)
[19] Elbein, A.D.: Glycosidase inhibitors: inhibitors of N-linked oligosaccharide processing. FASEB J., **5**, 3055-3063 (1991)
[20] Bause, E.; Breuer, W.; Schweden, J.; Roeser, R.; Geyer, R.: Effect of substrate structure on the activity of Man_9-mannosidase from pig liver involved in N-linked oligosaccharide processing. Eur. J. Biochem., **208**, 451-457 (1992)
[21] Pan, Y.-T.; Kaushal, G.P.; Papandreou, G.; Ganem, B.; Elbein, A.D.: D-Mannonolactam amidrazone. A new mannosidase inhibitor that also inhibits

the endoplasmic reticulum or cytoplasmic α-mannosidase. J. Biol. Chem., **267**, 8313-8318 (1992)
[22] Ren, J.; Bretthauer, R.K.; Castellino, F.J.: Purification and properties of a Golgi-derived (α 1,2)-mannosidase-I from baculovirus-infected lepidopteran insect cells (IPLB-SF21AE) with preferential activity toward mannose$_6$-N-acetylglucosamine$_2$. Biochemistry, **34**, 2489-2495 (1995)
[23] Maruyama, Y.; Nakajima, T.; Ichishima, E.: A 1,2-α-D-mannosidase from a Bacillus sp.: purification, characterization, and mode of action. Carbohydr. Res., **251**, 89-98 (1994)
[24] Lipari, F.; Herscovics, A.: Production, purification and characterization of recombinant yeast processing α 1,2-mannosidase. Glycobiology, **4**, 697-702 (1994)

Mannosyl-oligosaccharide 1,3-1,6-α-mannosidase 3.2.1.114

1 Nomenclature

EC number
3.2.1.114

Systematic name
1,3-(1,6-)mannosyl-oligosaccharide α-D-mannohydrolase

Recommended name
mannosyl-oligosaccharide 1,3-1,6-α-mannosidase

Synonyms
GlcNAc transferase I-dependent α1,3[α1,6]mannosidase
Golgi α-mannosidase II
α-D-mannosidase II
α-mannosidase II
α1-3,6-mannosidase
mannosidase II
mannosidase, exo-1,3-1,6-α-

CAS registry number
82047-77-6

2 Source Organism

<1> *Rattus norvegicus* (male [3, 4, 17]; Wistar [3, 4]; Sprague-Dawley [17]) [1-4, 6, 8, 10, 15-17, 19]
<2> *Homo sapiens* [5, 16]
<3> *Canis familiaris* [7, 14, 20]
<4> *Mus musculus* [16, 18]
<5> *Drosophila melanogaster* [18]
<6> *Vigna radiata* (mung bean) [11, 13, 14]
<7> *Plants* [19]
<8> *Aspergillus saitoi* [9, 12]

3 Reaction and Specificity

Catalyzed reaction
hydrolysis of the terminal 1,3- and 1,6-linked α-D-mannose residues in the mannosyl oligosaccharide $Man_5(GlcNAc)_3$

Reaction type

hydrolysis of O-glycosyl bond

Natural substrates and products

S GlcNAc$(Man)_5$-species + H_2O <1, 2, 4> (involved in complex-type glycoprotein synthesis, <1, 2> [3-5]; by completing mannosyl trimming process, <1> [3, 4]; final step in N-glycan maturation pathway, <1, 2, 4> [16]) [3-5, 16, 17]

Substrates and products

S $(Man)_3$GlcNAc + H_2O <6> (poor substrate) [11, 13]

P ?

S $(Man)_5$GlcNAc + H_2O <1> (very poor substrate, <1> [4]; not, <6> [11, 13]) [4]

P Man_4GlcNAc + mannose <1> [4]

S $(Man)_6$GlcNAc + H_2O <1> (rate at 2% of GlcNAc$(Man)_5$GlcNAc hydrolysis) [4]

P ?

S $(Man)_7$GlcNAc + H_2O <1> (rate at 3% of GlcNAc$(Man)_5$GlcNAc hydrolysis) [4]

P ?

S $(Man)_8$GlcNAc + H_2O <1> (rate at 5% of GlcNAc$(Man)_5$GlcNAc hydrolysis) [4]

P ?

S $(Man)_9$GlcNAc + H_2O <1> (rate at 6.4% of GlcNAc$(Man)_5$GlcNAc hydrolysis, <1> [4]; not, <1, 6> [3, 11, 13]) [4]

P ?

S 4-methylumbelliferyl-α-D-mannoside + H_2O <1, 2, 4, 6> [8, 11, 16]

P ?

S 4-nitrophenyl α-mannoside + H_2O <1, 3, 6> (rate at 24% of GlcNAc $(Man)_5$GlcNAc hydrolysis, <1> [4]; not, <8> [9, 12]) [1, 3, 4, 6, 8, 11, 14, 15, 17]

P 4-nitrophenol + mannose <1, 3, 6> [1, 3, 4, 6, 14]

S D-Manα(1-2)Manα(1-3)Manβ(1-4)GlcNAc + H_2O <8> (reduced with sodium borate, poor substrate) [9, 12]

P ?

S D-Manα(1-3)Manβ(1-4)GlcNAc + H_2O <8> (sodium borate reduced, best substrate) [9, 12]

P D-Manβ(1-4)GlcNAc + mannose <8> (reduced with sodium borate) [9]

S D-Manα(1-6)(Manα(1-3))Manα(1-6)(GlcNAcβ(1-2)Manα(1-3))Manβ(1-4) GlcNAcβ(1-4)GlcNAc-Asn + H_2O <2> (i.e. glycopeptide III, derived from ovalbumin, no hydrolysis of the innermost α-1,6-linked or the α-1,2-linked mannose) [5]

P D-Manα(1-6)(GlcNAcβ(1-2)Manα(1-3))Manβ(1-4)GlcNAcβ(1-4)GlcNAc-Asn + mannose <2> (i.e. glycopeptide II) [5]

S D-Manα(1-6)Manβ(1-4)GlcNAcβ(1-4)GlcNAc + H_2O <8> (sodium borate reduced, at 10% the rate of Manα(1-3)Manβ(1-4)GlcNAc hydrolysis) [9]

P D-Manβ(1-4)GlcNAcβ(1-4)GlcNAc + mannose <8> (reduced with sodium borate) [9]

S GlcNAc$(Man)_5$GlcNAc + H_2O <1, 3, 6> (preferred substrate, <1> [4]; strict substrate specificity, <6> [13]; removes the exposed α-1,3- and α-1,6-mannosyl residues, <1, 6> [3, 11]; with preference for the α-1,6-linkage, <6> [11]) [3, 4, 11, 13, 14, 17, 19, 20]

P GlcNAc$(Man)_3$GlcNAc + mannose <1, 6> (i.e. GlcNAcManα(1-3)(Manα(1-6))Manβ(1-4)GlcNAc, <6> [11]; final product, via GlcNAc$(Man)_4$GlcNAc, <6> [11, 13]) [3, 11, 13, 19]

S GlcNAcβ(1-2)Manα(1-6)(Manα(1-3))Manβ(1-4)GlcNAcβ(1-4)GlcNAc + H_2O <8> (reduced with sodium borate, hydrolysis at equal rate as Man_3 $(GlcNAc)_2$, galactosyl and sialyl-galactosyl derivatives are substrates, too) [9]

P ?

S GlcNAcβ(1-6)(GlcNAcβ(1-6))Manα(1-6)(Manα(1-3))Manβ(1-4) GlcNAcβ(1-4)GlcNAc + H_2O <8> (reduced with sodium borate) [9]

P ?

S Manα(1-3)(Manα(1-6))Manα(1-6)(Manα(1-3))Manβ(1-4)GlcNAcβ(1-4) GlcNAc <8> (reduced with sodium borate, <8> [9]; not, <6> [11]) [9]

P Manα(1-3)(Manα(1-6))D-Manα(1-6)Manβ(1-4)GlcNAcβ(1-4)GlcNAc + mannose <8> (reduced with sodium borate) [9]

S Manα-1,6(Manα-1,3)Manβ-1,4GlcNAcβ-1,4GlcNAc <8> (reduced with sodium borate, cleaves 1,6-linkage after 1,3-linkage, <8> [12]) [9, 12]

P D-Manα(1-6)Manβ(1-4)GlcNAcβ(1-4)GlcNAc + mannose <8> (reduced with sodium borate) [9]

S R-Manα(1-6)(Manα(1-3))Manβ(1-4)GlcNAc + H_2O <8> (not Manα(1-6) (R-Manα(1-3))Manβ(1-4)GlcNAc) [12]

P ?

S Additional information <1, 6, 8> (hydrolysis mechanism of branched oligosaccharides, <8> [9]; no substrates are GlcNAc$(Man)_3$GlcNAc, <1> [3]; D-Manα(1-6) (Manα(1-3))Manα(1-6)(GlcNAcβ(1-4))[GlcNAcβ(1-4) (GlcNAcβ(1-2))Manα(1-3)]Manβ(1-4)GlcNAcβ(1-4)GlcNAc (reduced with sodium borate), <8> [9]; $Glc_3(Man)_9$GlcNAc, <6> [11, 13]) [3, 9, 11, 13]

P ?

Inhibitors

1,4-dideoxy-1,4-imino-D-mannitol <1, 3, 6> (less effective than swainsonine, <6> [11]) [7, 11, 13, 19]
Cu^{2+} <1, 6, 8> (strong, <1> [1]) [1, 4, 9-11, 13]
D-mannonolactam amidrazone <3> (broad spectrum mannosidase inhibitor) [20]
Fe^{2+} <1> (strong) [1, 4]
Fe^{3+} <1> (weak) [4]
α-1,6-linked mannose trisaccharide <6> [11]
antibodies to purified rat enzyme <1, 2, 4> (mouse) [16]
mannose <1> [1]

mannostatin A <1, 3, 6> (cyclopentanol structure containing thiomethyl and amino functional groups, in vivo and in vitro) [14, 19, 20]
p-chloromercuriphenylsulfonate <1> [1, 4]
swainsonine <1, 2, 4, 6> (i.e. (1S,2R,8R,8aR)-trihydroxy indolizidine, <6> [11]; alkaloid, strong, <1> [3]; kinetics, <1> [6]; in vivo and in vitro, <1> [6]; partially reversible depending on inhibitor concentration, <1> [6]) [3, 6, 11, 13, 16, 17, 19]
Additional information <1, 3, 6> (no inhibition by chelating agents, <1> [1]; Mn^{2+}, Mg^{2+}, <1, 6> [1, 11]; Co^{2+}, Zn^{2+}, <1,6> [4, 11]; Tris-Cl, IAA, α-methylmannoside, <1> [4]; Tris-maleate, <1> [4, 10]; EDTA, <1, 6, 8> [1, 4, 9, 11]; 1,10-phenanthroline, 2,2'-dipyridyl, glucose, galactose, <1> [1]; EGTA, 2-epicastanospermine, 2-episwainsonine, trisaccharides of mannose residues linked in α-1,2-linkages, tetrasaccharides with nonreducing mannose linked in α-1,3-linkages to an α-1,2-linked trisaccharide, <6> [11]; castanospermine, deoxynojirimycin, deoxymannojirimycin, <6> [13]; N-acetylmannostatin A, <1, 3, 6> [14, 19]) [1, 4, 9-11, 13, 14, 19]

Cofactors/prosthetic groups
ATP <1> (activation) [1]
CMP <1> (activation) [1]

Activating compounds
2-mercaptoethanol <1> (activtion) [1]
DTT <1> (activation) [1]
Triton X-100 <6> (activation, <6> [11, 13]; of solubilized enzyme, <6> [11]; 0.15-0.2%, <6> [13]) [11, 13]
bovine serum albumin <1> (activation) [1]
fucose <1> (activation) [1]
Additional information <1> (no activation by glucose, galactose, AMP, ADP, GMP, GDP, GTP, CDP, CTP, UMP, UDP, UTP, cAMP or cGMP) [1]

Metals, ions
Ca^{2+} <8> (activation, 1 mM, <8> [9, 12]; not, <6> [11]) [9, 12]
Co^{2+} <1> (activation, <1> [1, 4]; not, <6> [11]) [1, 4]
Zn^{2+} <1> (activation, <1> [1, 4]; not, <6,8> [9, 11]) [1, 4]
Additional information <1, 6, 8> (no activation by Mg^{2+}, Mn^{2+}, <1, 6, 8> [1, 9, 11]; Na^{+}, <1> [1]; Ba^{2+}, <8> [9]) [1, 9, 11]

Specific activity (U/mg)
1.425 <1> (4-nitrophenyl α-D-mannoside as substrate) [4]
3.1 <1> [1]
7.03 <1> (rat, chymotrypsin-cleaved enzyme) [17]
25.3 <1> (rat, intact enzyme) [17]
52.9 <1, 2, 4> (mouse) [16]
38000 <8> [9, 12]
Additional information <6> [11, 13]

K_m-Value (mM)

0.2 <2> (D-Manα(1-6)(Manα(1-3))Manα(1-6)(GlcNAcβ(1-2)Manα(1-3)) Manβ(1-4)GlcNAcβ(1-4)GlcNAc-Asn, , i.e. glycopeptide III) [5]
1.25 <8> (Man3$(GlcNAc)_2$, sodium borate reduced) [9, 12]
40 <1> (4-nitrophenyl mannoside) [17]

pH-Optimum

5 <8> [9, 12]
5.5 <1, 2, 4> (4-nitrophenyl mannoside as substrate, <1> [4]; mouse, <1, 2, 4> [16]) [1, 4, 16]
5.8 <1> (Man_8GlcNAc as substrate) [4]
5.8-6.4 <6> [13]
6 <6> [11]
Additional information <1> (pI: 5.8) [1]

pH-Range

3.5-6 <8> (more than 70% of maximal activity at pH 3.5 and about half-maximal activity at pH 6) [9]
4.2-6 <1> (about half-maximal activity at pH 4.2 and about 80% of maximal activity at pH 6) [1]
4.8-7 <6> (about half-maximal activity at pH 4.8 and 7) [11]

Temperature optimum (°C)

37 <1, 2, 6, 8> (assay at) [1, 3-6, 9, 10, 12, 13]

4 Enzyme Structure

Molecular weight

110000 <1> (rat, sucrose density gradient centrifugation) [1]
230000 <1> (rat, chymotrypsin-cleaved enzyme, gel filtration) [17]
285000 <1> (rat, intact enzyme, gel filtration in the presence of detergent) [17]
295000 <1> (rat, gel filtration) [1]

Subunits

? <1, 4, 6> (x * 75000-80000, rat, SDS-PAGE [1]; x * 110000 + x * 14000, rat, catalytic domain and membrane-anchoring domain, generated by mild chymotrypsin digest, SDS-PAGE [8]; x * 124000, rat, SDS-PAGE [8, 10]; x * 125000, mung bean, SDS-PAGE [11]; x * 128000, mung bean, SDS-PAGE [13]; x * 131000, mouse, predicted from nucleotide sequence [16]) [1, 8, 10, 11, 13, 16]
dimer <1> (2 * 124000, rat, intact enzyme, SDS-PAGE [17]; 2 * 128000, rat, SDS-PAGE [15]) [15, 17]

5 Isolation/Preparation/Mutation/Application

Source/tissue

NRK-cell <1> (i.e. normal kidney cells) [15]
adenocarcinoma, colon, liver <2> [5]
embryo <5> [18]
kidney, Madin-Darby canine cell <6> (i.e. MDCK-cells, influenza virus-infected) [7, 14, 20]
liver <1, 2, 4> [1-4, 6, 8, 10, 15-17, 19]
mycelium <8> (freeze-dried mixture of mycelium and culture medium, commercially available as Morushin) [9, 12]
seedling <6> [11, 13, 14]
Additional information <1> (rat tissue distribution) [16]

Localization

Golgi membrane <1> (topology study: integral membrane protein with lumenal orientation, consists of hydrophobic membrane anchoring domain and hydrophilic catalytic domain, <1> [8]; transmembrane protein with short cytoplasmic tail, single transmembrane domain and large catalytic domain, rat, <1> [16]) [1-4, 6, 8, 15-17]
membrane-bound <1, 2, 4, 6> (topology study: integral membrane protein with lumenal orientation, consists of hydrophobic membrane anchoring domain and hydrophilic catalytic domain, <1> [8]) [1-4, 6, 8, 11, 13, 15-17]
microsome <1, 2, 4, 6> (mouse, <4> [16]) [5, 11, 13, 15, 16]

Purification

<1> (intact and soluble, chymotrypsin-cleaved enzyme [17]) [1, 4, 8, 10, 17]
<2> (partial) [5]
<6> [11, 13]

Cloning

<2, 4, 5> (mouse [16]; human (partial) [16]; Drosophila melanogaster [18]) [16, 18]

6 Stability

pH-Stability

4 <1> (1 h, inactivation at 37°C, Zn^{2+} partially stabilizes) [4]
4-7 <1> (unstable at 37°C, no activity below pH 4) [1]
6 <1, 6> (1 h, 30% loss of activity at 37°C, <1> [4]; below, loss of activity within 1 week, <6> [11]) [4, 11]
6.5 <6> (below, loss of activity) [13]
7-9 <1> (stable at 37°C) [1]
8 <1> (1 h, stable at 37°C) [4]

Temperature stability

37 <1> (24 h, in 0.1 M sodium acetate buffer, pH 5.8, 30-50% loss of activity, <1> [4]; stable at pH 7-9, unstable at acidic pH-values, Zn^{2+} increases stability, <1> [1]) [1, 4]
40 <1> (and below, 15 min stable, rapid inactivation above) [1]
50 <1> (3 h, 30% loss of activity, 4-nitrophenyl mannoside or $(Man)_9$GlcNAc as substrate) [4]

General stability information

<1>, 4-nitrophenyl α-D-mannoside stabilizes [1]
<1>, enzyme retains full activity after incubation with chymotrypsin or other proteases that cleave the enzyme into catalytic and membrane-anchoring domains, a decrease in activity occurs with a corresponding cleavage of the catalytic subunit to a MW 105000 species [8]
<1>, freeze-thawing greatly enhances the rate of cleavage of the intact enzyme [17]
<1>, requires presence of detergent, 0.1% Triton X-100, to maintain enzyme in soluble form [1]
<1>, ultrafiltration, even in the presence of PMSF at 0°C, results in a slow conversion of the intact enzyme to the MW 110000 degradation product, the catalytic domain, which itself is extremely stable [17]
<6>, glycerol stabilizes [11, 13]

Storage stability

<1>, 4°C, intact enzyme, several weeks [17]
<1>, 5°C, in 0.1 M potassium phosphate buffer, pH 7.2, Triton X-100, 20% loss of activity within 30 days [1]
<1>, frozen, at least 2 months [1]
<1>, stable to repeated freeze-thawing cycles [1]
<6>, 0-4°C, in 10 mM HEPES buffer, pH 7.1, 0.1% Triton X-100, 0.5 mM PMSF, 0.2 mM DTT, 10% glycerol, at least 2 weeks [13]
<6>, 0-4°C, in 10 mM HEPES buffer, pH 7.2, 10% glycerol, several days [11]
<6>, frozen, in 10 mM HEPES buffer, pH 7.1, 0.1% Triton X-100, 0.5 mM PMSF, 0.2 mM DTT, 20% glycerol, at least 2 months [11]
<8>, -20°C, 4 months [9]
<8>, -20°C, in 10 mM sodium acetate buffer, pH 5, with or without $CaCl_2$, 4 months [12]
<8>, 4°C, 4 months [9]
<8>, 4°C, in 10 mM sodium acetate buffer, pH 5, with or without $CaCl_2$, 4 months [12]

References

[1] Tulsiani, D.R.P.; Opheim, D.J.; Touster, O.: Purification and characterization of α-D-mannosidase from rat liver Golgi membranes. J. Biol. Chem., **252**, 3227-3233 (1977)

[2] Harpaz, N.; Schachter, H.: Control of glycoprotein synthesis. Processing of asparagine-linked oligosaccharides by one or more rat liver Golgi α-D-mannosidases dependent on the prior action of UDP-N-acetylglucosamine: α-D-mannoside β 2-N-acetylglucosaminyltransferase I. J. Biol. Chem., **255**, 4894-4902 (1980)
[3] Tulsiani, D.R.P.; Harris, T.M.; Touster, O.: Swainsonine inhibits the biosynthesis of complex glycoproteins by inhibition of Golgi mannosidase II. J. Biol. Chem., **257**, 7936-7939 (1982)
[4] Tulsiani, D.R.P.; Hubbard, S.C.; Robbins, P.W.; Touster, O.: α-D-Mannosidases of rat liver Golgi membranes. Mannosidase II is the $GlcNAcMAN_5$-cleaving enzyme in glycoprotein biosynthesis and mannosidases Ia and IB are the enzymes converting Man_9 precursors to Man_5 intermediates. J. Biol. Chem., **257**, 3660-3668 (1982)
[5] Chandrasekaran, E.V.; Savila, M.; Nixon, D.; Mendicino, J.: Purification and properties of α-D-mannose:β-1,2-N-acetylglucosaminyl-transferases and α-D-mannosidases from human adenocarcinoma. Cancer Res., **44**, 4059-4068 (1984)
[6] Tulsiani, D.R.P.; Broquist, H.P.; Touster, O.: Marked differences in the swainsonine inhibition of rat liver lysosomal α-D-mannosidase, rat liver Golgi mannosidase II, and jack bean α-D-mannosidase. Arch. Biochem. Biophys., **236**, 427-434 (1985)
[7] Palamarczyk, G.; Mitchell, M.; Smith, P.W.; Fleet, G.W.; Elbein, A.D.: 1,4-Dideoxy-1,4-imino-D-mannitol inhibits glycoprotein processing and mannosidase. Arch. Biochem. Biophys., **243**, 35-45 (1985)
[8] Moremen, K.W.; Touster, O.: Topology of mannosidase II in rat liver Golgi membranes and release of the catalytic domain by selective proteolysis. J. Biol. Chem., **261**, 10945-10951 (1986)
[9] Amano, J.; Kobata, A.: Purification and characterization of a novel α-mannosidase from Aspergillus saitoi. J. Biochem., **99**, 1645-1654 (1986)
[10] Moremen, K.W.; Touster, O.: Biosynthesis and modification of Golgi mannosidase II in HeLa and 3T3 cells. J. Biol. Chem., **260**, 6654-6662 (1985)
[11] Kaushal, G.P.; Szumilo, T.; Pastuszak, I.; Elbein, A.D.: Purification to homogeneity and properties of mannosidase II from mung bean seedlings. Biochemistry, **29**, 2168-2176 (1990)
[12] Kobata, A.; Amano, J.: α-Mannosidases I and II from Aspergillus saitoi. Methods Enzymol., **138**, 779-785 (1987)
[13] Kaushal, G.P.; Elbein, A.D.: Glycoprotein processing enzymes of plants. Methods Enzymol., **179**, 452-475 (1989)
[14] Tropea, J.E.; Kaushal, G.P.; Pastuszak, I.; Mitchell, M.; Aoyagi, T.; Molyneux, R.J.; Elbein, A.D.: Mannostatin A, a new glycoprotein-processing inhibitor. Biochemistry, **29**, 10062-10069 (1990)
[15] Baron, M.D.; Garoff, H.: Mannosidase II and the 135-kDa Golgi-specific antigen recognized monoclonal antibody 53FC3 are the same dimeric protein. J. Biol. Chem., **265**, 19928-19931 (1990)
[16] Moremen, K.W.; Robbins, P.W.: Isolation, characterization, and expression of cDNAs encoding murine α-mannosidase II, a Golgi enzyme that controls

conversion of high mannose to complex N-glycans. J. Cell Biol., **115**, 1521-1534 (1991)
[17] Moremen, K.W.; Touster, O.; Robbins, P.W.: Novel purification of the catalytic domain of Golgi α-mannosidase II. Characterization and comparison with the intact enzyme. J. Biol. Chem., **266**, 16876-16885 (1991)
[18] Foster, J.M.; Yudkin, B.; Lockyer, A.E.; Roberts, D.B.: Cloning and sequence analysis of GmII, a Drosophila melanogaster homologue of the cDNA encoding murine Golgi α-mannosidase II. Gene, **154**, 183-186 (1995)
[19] Elbein, A.D.: Glycosidase inhibitors: inhibitors of N-linked oligosaccharide processing. FASEB J., **5**, 3055-3063 (1991)
[20] Pan, Y.-T.; Kaushal, G.P.; Papandreou, G.; Ganem, B.; Elbein, A.D.: D-Mannonolactam amidrazone. A new mannosidase inhibitor that also inhibits the endoplasmic reticulum or cytoplasmic α-mannosidase. J. Biol. Chem., **267**, 8313-8318 (1992)

Branched-dextran exo-1,2-α-glucosidase 3.2.1.115

1 Nomenclature

EC number
3.2.1.115

Systematic name
1,2-α-D-glucosyl-branched-dextran 2-glucohydrolase

Recommended name
branched-dextran exo-1,2-α-glucosidase

Synonyms
dextran α-1,2 debranching enzyme
glucosidase, dextran 1,2-α

CAS registry number
72840-94-9

2 Source Organism

<1> *Flavobacterium sp.* (strain M-73) [1]

3 Reaction and Specificity

Catalyzed reaction
hydrolysis of 1,2-α-D-glucosidic linkages at the branched points of dextrans and related polysaccharides, producing free D-glucose

Reaction type
hydrolysis of O-glycosyl bond

Natural substrates and products
S Additional information <1> (hydrolysis of α-1,2-glucosidic linkages at branching points) [1]

Substrates and products
S clinical dextran + H_2O <1> (major linkages: α-1,6-, α-1,3-, minor activity) [1]
P glucose <1> (hydrolysis of 1,2-α-D-glucosidic linkages at the branched points, producing free D-glucose) [1]
S mutan IG-1 + H_2O <1> (major linkage: α-1,3-, α-1,6-, small activity) [1]

P glucose <1> (hydrolysis of 1,2-α-D-glucosidic linkages at the branched points, producing free D-glucose) [1]
S soluble dextran + H_2O <1> (B-1299, major linkages: α-1,6-, α-1,3-, α-1,2-, highest activity) [1]
P glucose <1> (hydrolysis of 1,2-α-D-glucosidic linkages at the branched points, producing free D-glucose) [1]
S Additional information <1> (no activity with α-1,3-glucan or starch) [1]
P ?

Specific activity (U/mg)
66 <1> [1]

pH-Optimum
5.6 <1> (assay at) [1]

Temperature optimum (°C)
40 <1> (assay at) [1]

4 Enzyme Structure

Subunits
? <1> (x * 125000, Flavobacterium sp., SDS-PAGE) [1]

5 Isolation/Preparation/Mutation/Application

Purification
<1> (strain M-73, partial) [1]

References

[1] Mitsuishi,Y.; Kobayashi, M.; Matsuda, K.: Dextran α-1,2 debranching enzyme from Flavobacterium sp. M-73: its production and purification. Agric. Biol. Chem., **43**, 2283-2290 (1979)

Glucan 1,4-α-maltotriohydrolase 3.2.1.116

1 Nomenclature

EC number
3.2.1.116

Systematic name
1,4-α-D-glucan maltotriohydrolase

Recommended name
glucan 1,4-α-maltotriohydrolase

Synonyms
exo-maltotriohydrolase
maltotriohydrolase
maltotriohydrolase, exo
Additional information (cf. EC 3.2.1.2, EC 3.2.1.60 and EC 3.2.1.98)

CAS registry number
91273-84-6

2 Source Organism

<1> *Streptomyces griseus* [1]
<2> *Microbacterium sp.* [2]
<3> *Natronococcus sp.* (strain Ah-36) [3]

3 Reaction and Specificity

Catalyzed reaction
hydrolysis of 1,4-α-D-glucosidic linkages in amylaceous polysaccharides so as to remove successive maltotriose residues from the nonreducing chain ends

Reaction type
hydrolysis of O-glycosyl bond

Substrates and products
S amylopectin + H_2O <3> [3]
P maltotriose <3> (+ small amounts of maltose and glucose, α-configuration) [3]
S amylose + H_2O <3> [3]

P maltotriose <3> (+ small amounts of maltose and glucose, α-configuration) [3]
S γ-cyclodextrin + H_2O <3> (not: α- or β-cyclodextrin) [3]
P ?
S glycogen + H_2O <3> [3]
P maltotriose <3> (+ small amounts of maltose and glucose, α-configuration) [3]
S maltooligosaccharide + H_2O <3> (ranging from maltotetraose to maltoheptaose, <3> [3]; not: maltotriose, maltose) [3]
P ?
S short-chain amylose + H_2O <1> [1]
P maltotriose <1> (prolonged action leads to gradual hydrolysis of maltotriose to glucose and maltose) [1]
S starch + H_2O <2, 3> (liquified, <2> [2]; soluble, <3> [3]) [2, 3]
P maltotriose + maltose + glucose + higher oligosaccharides <2> (in amounts of 55-60%, 10-20%, 1-3%, 20-40%, respectively) [2]

Inhibitors
Ca^{2+} <3> [3]
Hg^{2+} <3> (no strong inhibition, 10 mM: 75% loss of activity) [3]
N-bromosuccinimide <3> [3]
Ni^{2+} <3> [3]
SDS <3> [3]
$ZnCl_2$ <3> [3]
α-cyclodextrin, β-cyclodextrin <3, 3> [3, 3]
urea <3> [3]

Metals, ions
CsCl <3> (can partially substitute for NaCl at higher concentration) [3]
KCl <3> (can partially substitute for NaCl at higher concentration) [3]
NaCl <3> (2.5 M, required, <3> [3]) [3]
RbCl <3> (can partially substitute for NaCl at higher concentration) [3]

Specific activity (U/mg)
Additional information <2, 3> [2, 3]

K_m-Value (mM)
6.8 <3> (γ-cyclodextrin, K_m-value shown as glucose equivalent) [3]
19 <3> (amylose, DP, 17, K_m-value shown as glucose equivalent) [3]
25 <3> (starch, soluble, K_m-value shown as glucose equivalent) [3]
42 <3> (glycogen, from oyster, K_m-value shown as glucose equivalent) [3]
49 <3> (amylopektin, K_m-value shown as glucose equivalent) [3]

pH-Optimum
6.5-7 <2> [2]
8.7 <3> [3]

pH-Range
5.8-8.5 <2> (pH 5.8: about 90% of activity maximum, pH 8.5: about 30% of activity maximum) [2]

Temperature optimum (°C)
50 <2> [2]
55 <3> [3]

Temperature range (°C)
25-55 <2> (about 50% of activity maximum at 25°C and 55°C) [2]

4 Enzyme Structure

Molecular weight
40000 <2> (Microbacterium sp., gel filtration) [2]
76000 <3> (Natronococcus sp., gel filtration) [3]

Subunits
? <2> (x * 25500, Microbacterium sp., SDS-PAGE) [2]
monomer <2, 3> (1 * 74000, Natronococcus sp., SDS-PAGE [3]) [2, 3]

5 Isolation/Preparation/Mutation/Application

Purification
<2> [2]
<3> (strain Ah-36) [3]

6 Stability

pH-Stability
4.2 <3> (30 min, 50°C, 80% loss of activity) [3]
5 <3> (30 min, 50°C, 44% loss of activity) [3]
5-9 <2> (stable) [2]
6-8.6 <3> (50°C, 30 min, stable) [3]
10.5 <3> (30 min, 50°C, 30% loss of activity) [3]

Temperature stability
40 <2> (unstable above) [2]
50 <3> (30 min, pH 4.2: 80% loss of activity, pH 5.0: 44% loss of activity, pH 6.0-8.6: stable, pH 10.5: 30% loss of activity) [3]
55 <3> (pH 8.0, 30 min, 50% loss of activity without stabilizer, no loss of activity in presence of 0.5% soluble starch) [3]
Additional information <2, 3> (Ca^{2+} protects against heat inactivation, <2> [2]; Ca^{2+} does not enhance thermal stability, <3> [3]; soluble starch stabilizes against thermal inactivation, <3> [3]) [2, 3]

General stability information
<2>, Ca^{2+} protects against heat inactivation [2]
<3>, Ca^{2+} does not enhance thermal stability [3]
<3>, Soluble starch stabilizes against thermal inactivation [3]

<3>, activity irreversibly lost at low ionic strength [3]
<3>, does not lose activity after dialysis [3]

References

[1] Nakakuki, T.; Azuma, K.; Kainuma, K.: Action patterns of various exo-amylases and the anomeric configurations of their products. Carbohydr. Res., **128**, 297-310 (1984)
[2] Takasaki, Y.; Kitajima, M.; Tsuruta, T.; Nonoguchi, M.; Hayashi, S.; Imada, K.: Studies on enzymic production of oligosaccharides. Part VI. Maltotriose-producing amylase from Microbacterium imperiale. Agric. Biol. Chem., **55**, 687-692 (1991)
[3] Kobayashi, T.; Kanai, H.; Hayashi, T.; Akiba, T.; Akaboshi, R.; Horikoshi, K.: Haloalkaliphilic maltotriose-forming α-amylase from the Archaebacterium Natronococcus sp. strain Ah-36. J. Bacteriol., **174**, 3439-3444 (1992)

Amygdalin β-glucosidase 3.2.1.117

1 Nomenclature

EC number
3.2.1.117

Systematic name
amygdalin β-D-glucohydrolase

Recommended name
amygdalin β-glucosidase

Synonyms
amygdalase
amygdalin glucosidase
amygdalin hydrolase
amygdalinase
glucosidase, amygdalin
Additional information (cf. EC 3.2.1.21)

CAS registry number
51683-43-3

2 Source Organism

<1> *Prunus serotina* (Ehrh., black cherry) [1-3, 5]
<2> *Prunus domestica* (plum, var. Stanley) [4]

3 Reaction and Specificity

Catalyzed reaction
(R)-amygdalin + H_2O = (R)-prunasin + D-glucose

Reaction type
hydrolysis of O-glycosyl bond

Natural substrates and products
S (R)-amygdalin + H_2O <1> (involved in catabolism of cyanogenic glycosides, <1> [1]; initial reaction in sequential degradation of amygdalin to HCN, <1> [2, 3]; cf. EC 3.2.1.118) [1-3]

Substrates and products

S (R)-amygdalin + H_2O <1, 2> (highly specific, <1> [1]) [1-5]
P (R)-prunasin + D-glucose <1> [1-3]
S 2-nitrophenyl-β-D-glucopyranoside + H_2O <1> (reaction at 57% the rate of amygdalin hydrolysis) [1]
P 2-nitrophenol + D-glucose
S 4-methylumbelliferyl-β-D-glucoside + H_2O <1> [3]
P 4-methylumbelliferone + D-glucose
S 4-nitrophenyl-β-D-galactopyranoside + H_2O <1> (poor substrate) [1]
P 4-nitrophenol + D-galactose
S 4-nitrophenyl-β-D-glucopyranoside + H_2O <1> (reaction at 46% the rate of amygdalin hydrolysis) [1]
P 4-nitrophenol + D-glucose
S indoxyl-β-D-glucoside + H_2O <2> (i.e. indican) [4]
P indoxyl + D-glucose <2> (insoluble) [4]
S Additional information <1> (very little activity (1% the rate of amygdalin hydrolysis and below) with 2-nitrophenyl-β-D-xylopyranoside, maltose or 4-nitrophenyl-β-D-xylopyranoside. No substrates are 4-nitrophenyl-α-D-glucopyranoside, (R)-prunasin, linamarin, β-gentiobiose, 4-nitrophenyl-α-glucopyranoside, β-phenyl-D-glucoside, salicin, β-methyl-D-glucoside, sucrose, lactose or cellobiose) [1]
P ?

Specific activity (U/mg)

6.5 <1> (cotyledon extract) [2]
9.9 <1> (embryonic axes) [2]
450 <1> (isozyme II') [3]
483 <1> (isozyme I' or II) [3]
587 <1> (isozyme I) [3]

K_m-Value (mM)

1.65 <1> ((R)-amygdalin, isozyme II') [3]
1.88 <1> ((R)-amygdalin, isozyme II) [3]
2.01 <1> ((R)-amygdalin, isozyme I) [3]
2.2 <1> ((R)-amygdalin, isozyme I') [3]

pH-Optimum

4.5-5.5 <1> [3]
5.5 <1> [1]
Additional information <1, 2> (pI: 5.6-6, <2> [4]; pI: 5.8 (isozyme II'), pI: 6 (isozyme II), pI: 6.2 (isozyme I), pI: 6.4 (isozyme I'), <1> [3]) [3, 4]

pH-Range

5-7 <1> (more than 50% of maximal rates at pH 5 and 7) [1]

Temperature optimum (°C)

30 <1> (assay at) [1]

4 Enzyme Structure

Molecular weight

52000 <1> (*Prunus serotina*, FPLC gel filtration) [3]

Subunits

monomer <1, 2> (1 * 60000, *Prunus serotina*, SDS-PAGE [3]; *Prunus domestica*, SDS-PAGE [4]) [3, 4]

Posttranslational modification

glycoprotein <1, 2> [3, 4]

5 Isolation/Preparation/Mutation/Application

Source/tissue

cotyledon <1, 2> [2-5]
procambium <1, 2> [2, 4, 5]
seed <1, 2> [1-4]
seedling <1> (of different stages of development) [5]
Additional information <1, 2> (tissue distribution, <1, 2> [2, 4, 5]; not in cotyledonary storage parenchyma, bundle sheath cells or endosperm, <1, 2> [2, 4]) [2, 4, 5]

Localization

protein body <1, 2> [2, 4, 5]

Purification

<1> (partial [1]; 4 isozymes: I, I', II, II', differing in pI-value and N-terminal amino acid sequence [3]) [1, 3]
<2> (to near homogeneity) [4]

References

[1] Kuroki, G.; Lizotte, P.A.; Poulton, J.E.: Catabolism of (R)-amygdalin and (R)-vicianin by partially purified β-glycosidases from Prunus serotina Ehrh. and Davallia trichomanoides. Z. Naturforsch. C, **39c**, 232-239 (1983)

[2] Swain, E.; Li, C.P.; Poulton, J.E.: Tissue and subcellular localization of enzymes catabolizing (R)-amygdalin in mature Prunus serotina seeds. Plant Physiol., **100**, 291-300 (1992)

[3] Li, C.P.; Swain, E.; Poulton, J.E.: Prunus serotina amygdalin hydrolase and prunasin hydrolase. Purification, N-terminal sequencing, and antibody production. Plant Physiol., **100**, 282-290 (1992)

[4] Poulton, J.E.; Li, C.P.: Tissue level compartmentation of (R)-amygdalin and amygdalin hydrolase prevents large-scale cyanogenesis in undamaged Prunus seeds. Plant Physiol., **104**, 29-35 (1994)

[5] Swain, E.; Poulton, J.E.: Utilization of amygdalin during seedling development of Prunus serotina. Plant Physiol., **106**, 437-445 (1994)

Prunasin β-glucosidase 3.2.1.118

1 Nomenclature

EC number
3.2.1.118

Systematic name
prunasin β-D-glucohydrolase

Recommended name
prunasin β-glucosidase

Synonyms
hydrolase, prunasin
prunasin hydrolase
Additional information (cf. EC 3.2.1.21)

CAS registry number
9023-41-0

2 Source Organism

<1> *Prunus serotina* (Ehrh., black cherry) [1-3, 5, 6]
<2> *Prunus domestica* (plum, var. Stanley) [4]

3 Reaction and Specificity

Catalyzed reaction
(R)-prunasin + H_2O = mandelonitrile + D-glucose

Reaction type
hydrolysis of O-glycosyl bond

Natural substrates and products
S (R)-prunasin + H_2O <1> (involved in catabolism of cyanogenic glycosides, <1> [1]; participates in sequential degradation of (R)-amygdalin to HCN, <1> [2, 3, 6]; cf. EC 3.2.1.117) [1-3, 6]

Substrates and products
S (R)-prunasin + H_2O <1, 2> (highly specific, <1> [1]; preferred substrate, <1> [6]) [1-6]
P mandelonitrile + D-glucose <1> [1-3, 6]
S (S)-sambunigrin + H_2O <1> (i.e. epimer of (R)-prunasin, poor substrate) [6]

P ?
S 2-nitrophenyl-β-D-glucopyranoside + H_2O <1> (reaction at 79% the rate of prunasin, <1> [1]) [1, 6]
P 2-nitrophenol + D-glucose
S 4-nitrophenyl-β-D-galactopyranoside + H_2O <1> (very poor substrate, <1> [1]) [1, 6]
P 4-nitrophenol + D-galactose
S 4-nitrophenyl-β-D-glucopyranoside + H_2O <1> (reaction at 18% the rate of prunasin, <1> [1]) [1, 6]
P 4-nitrophenol + D-glucose
S Additional information <1> (very little activity (1% of prunasin hydrolysis and below) with maltose, 4-nitrophenyl-β-D-xylopyranoside, cellobiose or sucrose, <1> [1]; no substrates are (R)-amygdalin, linamarin, <1> [1, 6]; gentiobiose, 4-nitrophenyl-α-D-glucopyranoside, β-phenyl-D-glucoside, β-methyl-D-glucoside, lactose, <1> [1]; (neo)linustatin or (S)-dhurrin, <1> [6]) [1, 6]
P ?
S indoxyl-β-D-glucoside + H_2O <2> (i.e. indican) [4]
P indoxyl + D-glucose <2> (insoluble) [4]
S salicin + H_2O <1> (poor substrate) [1]
P ?

Inhibitors

2-mercaptoethanol <1> (weak) [6]
DTT <1> (weak) [6]
castanospermine <1> (indolizidine alkaloid) [6]
iodoacetamide <1> (weak) [6]
iodoacetic acid <1> (weak) [6]
mandelamide <1> (weak) [6]
mandelic acid <1> (weak) [6]
p-chloromercuriphenylsulfonate <1> (weak) [6]
p-hydroxymandelic acid <1> (weak) [6]
Additional information <1> (little or no inhibition by Cu^{2+}, Mg^{2+}, Zn^{2+}, Pb^{2+}, Fe^{2+}, Fe^{3+}, Ag^{2+}, no inhibition by diethyldithiocarbamate, 1,10-phenanthroline, 2,2'-dipyridyl or EDTA) [6]

Metals, ions

Additional information <1> (no metal ion requirement) [6]

Specific activity (U/mg)

2.8 <1> (embryonic axes) [2]
4.5 <1> (cotyledon extract) [2]
4933 <1> [3]

K_m-Value (mM)

1.35 <1> ((R)-prunasin, isozyme IIb) [3]
1.73 <1> ((R)-prunasin, isozyme I) [3]
2.3 <1> ((R)-prunasin, isozyme IIa) [3]

pH-Optimum
5 <1> [6]
6.5 <1> [1]

pH-Range
4-7 <1> (about half-maximal activity at pH 4 and 7) [6]
5-7 <1> (more than 50% of maximal rates at pH 5 and 7) [1]

Temperature optimum (°C)
30 <1> (assay at) [1]

4 Enzyme Structure

Molecular weight
68000 <1> (Prunus serotina, isozymes I and IIb, PAGE [3]; gel filtration [6]) [3, 6]
140000 <1> (Prunus serotina, isozyme IIa, PAGE [3]; gel filtration [6]) [3, 6]

Subunits
dimer <1> (2 * 69500, Prunus serotina, isozyme IIa, SDS-PAGE) [6]
monomer <1, 2> (1 * 60000, Prunus domestica, SDS-PAGE [4]; 1 * 69500, Prunus serotina, isozymes I and IIb, SDS-PAGE [6]) [4, 6]

Posttranslational modification
glycoprotein <1, 2> [3, 4, 6]

5 Isolation/Preparation/Mutation/Application

Source/tissue
cotyledon <1, 2> [2-5]
procambium <1, 2> [2, 4, 5]
seed <1, 2> [1-6]
seedling <1> (of different stages of development) [5]
Additional information <1, 2> (tissue distribution, <1, 2> [2, 4, 5]; not in cotyledonary parenchyma, bundle sheath cells, <2> [4] or endosperm, <1, 2> [2, 4]) [2, 4, 5]

Localization
protein body <1, 2> [2, 4, 5]

Purification
<1> (partial [1]; 3 isozymes: I, IIa and IIb [3]) [1, 3]
<2> (to near homogeneity) [4]

6 Stability

Storage stability

<1>, 4°C, concentrated enzyme preparation with 0.04% NaN_3 retains 90% of activity after 2 months [6]

References

[1] Kuroki, G.; Lizotte, P.A.; Poulton, J.E.: Catabolism of (R)-amygdalin and (R)-vicianin by partially purified β-glycosidases from Prunus serotina Ehrh. and Davallia trichomanoides. Z. Naturforsch. C, **39c**, 232-239 (1983)
[2] Swain, E.; Li, C.P.; Poulton, J.E.: Tissue and subcellular localization of enzymes catabolizing (R)-amygdalin in mature Prunus serotina seeds. Plant Physiol., **100**, 291-300 (1992)
[3] Li, C.P.; Swain, E.; Poulton, J.E.: Prunus serotina amygdalin hydrolase and prunasin hydrolase. Purification, N-terminal sequencing, and antibody production. Plant Physiol., **100**, 282-290 (1992)
[4] Poulton, J.E.; Li, C.P.: Tissue level compartmentation of (R)-amygdalin and amygdalin hydrolase prevents large-scale cyanogenesis in undamaged Prunus seeds. Plant Physiol., **104**, 29-35 (1994)
[5] Swain, E.; Poulton, J.E.: Utilization of amygdalin during seedling development of Prunus serotina. Plant Physiol., **106**, 437-445 (1994)
[6] Kuroki, G.; Poulton, J.E.: Isolation and characterization of multiple forms of prunasin hydrolase from black cherry (Prunus serotina Ehrh.) seeds. Arch. Biochem. Biophys., **255**, 19-26 (1987)

Vicianin β-glucosidase 3.2.1.119

1 Nomenclature

EC number
3.2.1.119

Systematic name
(R)-vicianin β-D-glucohydrolase

Recommended name
vicianin β-glucosidase

Synonyms
hydrolase, vicianin
vicianin hydrolase

CAS registry number
91608-93-4

2 Source Organism

<1> *Davallia trichomanoides* (Squirrel's foot fern, var. Blume [2]) [1, 2]

3 Reaction and Specificity

Catalyzed reaction
(R)-vicianin + H_2O = mandelonitrile + vicianose

Reaction type
hydrolysis of O-glycosyl bond

Natural substrates and products
S (R)-vicianin + H_2O <1> (involved in catabolism of cyanogenic glycosides) [2]

Substrates and products
S (R)-amygdalin + H_2O <1> (reaction at 27%, <1> [2] or 16%, <1> [1] the rate of vicianin hydrolysis) [1, 2]
P mandelonitrile + β-gentiobiose <1> [1]
S (R)-prunasin + H_2O <1> (reaction at 14%, <1> [2] or 12%, <1> [1] the rate of vicianin hydrolysis) [1, 2]
P mandelonitrile + D-glucose <1> [1]

S (R)-vicianin + H_2O <1> (i.e. β-vicianoside of (R)-mandelonitrile, strict requirement for (R)-epimers, <1> [2]; cyanogenic disaccharides preferred over monosaccharides, <1> [2]) [1, 2]
P mandelonitrile + vicianose <1> [1, 2]
S 2-nitrophenyl-β-D-glucopyranoside + H_2O <1> (poor substrate) [1]
P 2-nitrophenol + D-glucose
S 2-nitrophenyl-β-D-xylopyranoside + H_2O <1> (poor substrate) [1]
P 2-nitrophenol + D-xylose
S 4-nitrophenyl-β-D-glucopyranoside + H_2O <1> (poor substrate) [1, 2]
P 4-nitrophenol + D-glucose
S 4-nitrophenyl-β-D-xylopyranoside + H_2O <1> (poor substrate) [1]
P 4-nitrophenol + D-xylose
S Additional information <1> (no substrates are lactose, 2-nitrophenyl-β-D-galactopyranoside, β-methyl-D-glucoside, β-gentiobiose, linamarin, cellobiose, 4-nitrophenyl-α-D-glucopyranoside, salicin, phenyl-β-D-glucoside, maltose or sucrose) [1]
P ?

Inhibitors

Ag^{2+} <1> (4-nitrophenyl-β-D-glucopyranoside as substrate) [2]
Cu^{2+} <1> (4-nitrophenyl-β-D-glucopyranoside as substrate) [2]
Fe^{2+} <1> (weak, 4-nitrophenyl-β-D-glucopyranoside as substrate) [2]
Hg^{2+} <1> (4-nitrophenyl-β-D-glucopyranoside as substrate) [2]
castanospermine <1> (indolizidine alkaloid, strong, (R)-prunazin as substrate) [2]
diethyldithiocarbamate <1> (weak, 4-nitrophenyl-β-D-glucopyranoside as substrate) [2]
glucono-1,5-lactone <1> (strong, (R)-prunazin as substrate) [2]
p-chloromercuriphenylsulfonate <1> (strong, 4-nitrophenyl-β-D-glucopyranoside as substrate) [2]
Additional information <1> (less than 10% inhibition by Ca^{2+}, Zn^{2+}, Fe^{3+}, Co^{2+}, Cd^{2+}, Pb^{2+}, no inhibition by EDTA, 2,2'-dipyridyl, 1,10-phenanthroline, 2-mercaptoethanol, DTT, IAA, iodoacetic acid or NEM) [2]

K_m-Value (mM)

3.2 <1> ((R)-prunasin) [2]
3.39 <1> (4-nitrophenyl-β-D-glucopyranoside) [2]
4.9 <1> (vicianin) [2]

pH-Optimum

5 <1> (4-nitrophenyl-β-D-glucopyranoside) [2]
5.5 <1> ((R)-vicianin) [2]
6 <1> ((R)-prunasin, <1> [2]) [1, 2]
Additional information <1> (pI: 4.6-4.7) [2]

Temperature optimum (°C)

30 <1> (assay at) [1, 2]

4 Enzyme Structure

Molecular weight

340000 <1> (Davallia trichomanoides, gel filtration) [2]

Subunits

? <1> (x * 56000 + x * 49000 + x * 32000, Davallia trichomanoides, SDS-PAGE) [2]

Posttranslational modification

Additional information <1> (no glycoprotein) [2]

5 Isolation/Preparation/Mutation/Application

Source/tissue

fiddle head <1> [2]
frond <1> (young) [1, 2]

Purification

<1> (partial [1]) [1, 2]

References

[1] Kuroki, G.; Lizotte, P.A.; Poulton, J.E.: Catabolism of (R)-amygdalin and (R)-vicianin by partially purified β-glycosidases from Prunus serotina Ehrh. and Davallia trichomanoides. Z. Naturforsch. C, **39c**, 232-239 (1983)

[2] Lizotte, P.A.; Poulton, J.E.: Catabolism of cyanogenic glycosides by purified vicianin hydrolase from squirrel's foot fern (Davallia trichomanoides Blume). Plant Physiol., **86**, 322-324 (1988)

Oligoxyloglucan β-glycosidase 3.2.1.120

1 Nomenclature

EC number
3.2.1.120

Systematic name
oligoxyloglucan xyloglucohydrolase

Recommended name
oligoxyloglucan β-glycosidase

Synonyms
hydrolase, oligoxyloglucan
isoprimeverose-producing oligoxyloglucan hydrolase
oligoxyloglucan hydrolase

CAS registry number
97162-80-6

2 Source Organism

<1> *Aspergillus oryzae* [1]

3 Reaction and Specificity

Catalyzed reaction
hydrolysis of 1,4-β-D-glucosidic links in oligoxyloglucans so as to remove successive isoprimeverose residues from the non-reducing chain ends (i.e. α-xylo-1,6-β-D-glucosyl-)

Reaction type
hydrolysis of O-glycosyl bond

Substrates and products
S xyloglucan fragment oligosaccharide + H_2O <1> (highly specific for) [1]
P isoprimeverose <1> (isoprimeverose is split off from the nonreducing end of the backbone of the substrate) [1]

Inhibitors
Cu^{2+} <1> [1]
Fe^{2+} <1> [1]
Fe^{3+} <1> [1]

Hg^{2+} <1> [1]
SDS <1> [1]

Specific activity (U/mg)
38.42 <1> [1]

pH-Optimum
5 <1> [1]

Temperature optimum (°C)
60 <1> [1]

4 Enzyme Structure

Molecular weight
230000 <1> (Aspergillus oryzae, gel filtration) [1]

Subunits
dimer <1> (2 * 115000, Aspergillus oryzae, SDS-PAGE) [1]

5 Isolation/Preparation/Mutation/Application

Source/tissue
commercial preparation <1> [1]

Purification
<1> [1]

6 Stability

pH-Stability
5-7 <1> (45°C, 10 min, stable) [1]

Temperature stability
50 <1> (stable up to) [1]

References

[1] Kato, Y.; Matsushita, J.; Kubodera, T.; Matsuda, K.: A novel enzyme producing isoprimeverose from oligoxyloglucans of Aspergillus oryzae. J. Biochem., **97**, 801-810 (1985)

Polymannuronate hydrolase 3.2.1.121

1 Nomenclature

EC number
3.2.1.121

Systematic name
poly(mannuronide) mannuronohydrolase

Recommended name
polymannuronate hydrolase

Synonyms
depolymerase, polymannuronate
polymannuronic acid polymerase

CAS registry number
99283-64-4

2 Source Organism

<1> *Pseudomonas aeruginosa* [1]

3 Reaction and Specificity

Catalyzed reaction
endohydrolysis of the D-mannuronide linkages of polymannuronate

Reaction type
hydrolysis of O-glycosyl bond

Natural substrates and products
S Additional information <1> (involved in degradation of polymannuronic acid exopolysaccharide from Pseudomonas aeruginosa, strain PM1) [1]

Substrates and products
S alginic acid + H_2O <1> (little activity) [1]
P ?
S polymannuronic acid exopolysaccharide + H_2O <1> (from Pseudomonas aeruginosa, strain PM1) [1]
P ? <1> (endohydrolysis of the D-mannuronide linkages of polymannuronic acid exopolysaccharide from Pseudomonas aeruginosa, strain PM1) [1]

Inhibitors

$CaCl_2$ <1> (17 mM: inhibition, at 5 mM slight stimulation) [1]
EDTA <1> (0.5 mM: 35% inhibition) [1]
KCl <1> (50 mM: 50% inhibition, at 3 mM stimulation) [1]
$MgCl_2$ <1> (9 mM: inhibition, at 3 mM slight stimulation) [1]
$MgSO_4$ <1> (14 mM: inhibition, at 3 mM slight stimulation) [1]
NaCl <1> (50 mM: 40% inhibition, at 3 mM stimulation) [1]

Metals, ions

$CaCl_2$ <1> (5 mM: slight stimulation, at 17 mM inhibition) [1]
KCl <1> (3 mM: stimulation, more than 10 mM: inhibition) [1]
$MgCl_2$ <1> (3 mM: slight stimulation, at 9 mM inhibition) [1]
$MgSO_4$ <1> (3 mM: slight stimulation, at 14 mM inhibition) [1]
NaCl <1> (3 mM: stimulation, more than 10 mM: inhibition) [1]

K_m-Value (mM)

0.096 <1> (polymannuronic acid exopolysaccharide, from Pseudomonas aeruginosa, strain PM1) [1]

pH-Optimum

8 <1> [1]

Temperature optimum (°C)

37 <1> (assay at) [1]

4 Enzyme Structure

Subunits

? <1> (x * 53000 (major band), x * 23000 (minor band), Pseudomonas aeruginosa, SDS-PAGE) [1]

5 Isolation/Preparation/Mutation/Application

Localization

periplasm <1> (or associated with outer cell membrane) [1]

Purification

<1> (partial) [1]

References

[1] Dunne, M.W.; Buckmire, F.L.A.: Partial purification and characterization of a polymannuronic acid depolymerase produced by a mucoid strain of Pseudomonas aeruginosa isolated from a patient with cystic fibrosis. Appl. Environ. Microbiol., **50**, 562-567 (1985)

Maltose-6′-phosphate glucosidase 3.2.1.122

1 Nomenclature

EC number
3.2.1.122

Systematic name
maltose-6'-phosphate 6-phosphoglucohydrolase

Recommended name
maltose-6'-phosphate glucosidase

Synonyms
phospho-α-glucosidase

CAS registry number
98445-08-0

2 Source Organism

<1> *Streptococcus mutans* (OMZ 176) [1]

3 Reaction and Specificity

Catalyzed reaction
maltose 6'-phosphate + H_2O = D-glucose 6-phosphate + D-glucose

Reaction type
hydrolysis of O-glycosyl bond

Natural substrates and products
S maltose 6'-phosphate + H_2O <1> (maltose-inducible enzyme) [1]

Substrates and products
S maltose 6'-phosphate + H_2O <1> [1]
P D-glucose 6-phosphate + D-glucose <1> [1]

Inhibitors
maltotriitol <1> [1]

Temperature optimum (°C)
37 <1> (assay at) [1]

References

[1] Würsch, P.; Koellreutter, B.: Maltotriitol inhibition of maltose metabolism in Streptococcus mutans via maltose transport, amylomaltase and phospho-α-glucosidase activities. Caries Res., **19**, 439-449 (1985)

Endoglycosylceramidase 3.2.1.123

1 Nomenclature

EC number
3.2.1.123

Systematic name
oligoglycosylglucosylceramide glycohydrolase

Recommended name
endoglycosylceramidase

Synonyms
EGCase
ceramidase, endoglycosyl-
endoglycoceramidase
glycosyl-N-acetyl-sphingosine 1,1-β-D-glucanohydrolase
Additional information (cf. EC 3.2.1.62)

CAS registry number
105503-61-5

2 Source Organism

<1> *Rhodococcus sp.* (G-74-2 [1]; mutant strain M750 [5]; M-777 [6]; 2 molecular species: I [5] and II [5, 7]) [1, 4-7]
<2> *Corynebacterium sp.* (A-809 [3]) [2, 3]

3 Reaction and Specificity

Catalyzed reaction
oligoglycosylglucosylceramide + H_2O = oligoglycosylglucose + ceramide

Reaction type
hydrolysis of O-glycosyl bond

Natural substrates and products
S Additional information <2> (transglycosylation reaction may be useful in the synthesis of neoglycoconjugates) [3]

Substrates and products
S Fucα(1-3)GalNAcα(1-3)(Fucα(1-3))Galβ(1-4)Glcβ(1-1)ceramide + H_2O <1> (84% of the activity with galactosylcerebroside) [1]

P Fucα(1-3)GalNAcα(1-3)(Fucα(1-3))Galβ(1-4)Glc + ceramide <1> [1]

S GalNAcβ(1-4)(NeuAcα(2-3))Galβ(1-4)Glcβ(1-1)ceramide + H_2O <1> (94% of the activity with lactosylceramide) [1]

P GalNAcβ(1-4)(NeuAcα(2-3))Galβ(1-4)Glc + ceramide <1> [1]

S Galβ(1-3)GalNAcβ(1-4)(NeuAcα(2-3))Galβ(1-4)Glcβ(1-1)ceramide + H_2O <1, 2> (i.e. GM1, as active as lactosylceramide, <1> [1]) [1, 2, 4, 5]

P Galβ(1-3)GalNAcβ(1-4)(NeuAcα(2-3))Galβ(1-4)Glc + ceramide <1> [1]

S Galβ(1-3)GalNAcβ(1-4)Galβ(1-4)Glcβ(1-4)Glcβ(1-1)ceramide + H_2O <1> (as active as lactosylceramide) [1]

P Galβ(1-3)GalNAcβ(1-4)Galβ(1-4)Glcβ(1-4)Glc + ceramide <1> [1]

S Galβ(1-4)Glcβ(1-1)ceramide + H_2O <1> (i.e. lactosylceramide) [1]

P Galβ(1-4)Glc + ceramide <1> [1]

S Galβ(1-6)Galβ(1-6)Galβ(1-1)ceramide + H_2O <1> (i.e. neogalactosylceramide, 77% of the activity with lactosylceramide) [1]

P Galβ(1-6)Galβ(1-6)Gal + ceramide <1> [1]

S GlcNAcβ(1-2)Manα(1-3)Manβ(1-4)Glcβ(1-1)ceramide + H_2O <1> (35% of the activity with lactosylceramide) [1]

P GlcNAcβ(1-2)Manα(1-3)Manβ(1-4)Glc + ceramide <1> [1]

S Manα(1-3)Manβ(1-4)Glcβ(1-1)ceramide + H_2O <1> (88% of the activity with lactosylceramide) [1]

P Manα(1-3)Manβ(1-4)Glc + ceramide <1> [1]

S Manβ(1-4)Glcβ(1-1)ceramide + H_2O <1> (93% of the activity with lactosylceramide) [1]

P Manβ(1-4)Glc + ceramide <1> [1]

S NeuAcα(2-3)Galβ(1-3)GalNAcβ(1-4)(NeuAcα(2-3))Galβ(1-4)Glcβ(1-1) ceramide + H_2O <1> (93% of the activity with lactosylceramide) [1]

P NeuAcα(2-3)Galβ(1-3)GalNAcβ(1-4)(NeuAcα(2-3))Galβ(1-4)Glc + ceramide <1> [1]

S NeuAcα(2-3)Galβ(1-4)GlcNAcβ(1-3)Galβ(1-4)Glcβ(1-1)ceramide + H_2O <1, 2> (i.e. sialylparagloboside, as active as lactosylceramide, <1> [1]) [1, 2]

P NeuAcα(2-3)Galβ(1-4)GlcNAcβ(1-3)Galβ(1-4)Glc + ceramide <1> [1]

S NeuAcα(2-3)Galβ(1-4)Glcβ(1-1)ceramide + H_2O <1> (93% of the activity with lactosylceramide) [1]

P NeuAcα(2-3)Galβ(1-4)Glc + ceramide <1> [1]

S NeuAcα(2-8)NeuAcα(2-3)Galβ(1-3)GalNAcβ(1-4)(NeuAcα(2-3))Galβ(1-4) Glcβ(1-1)ceramide + H_2O <1> (81% of the activity with lactosylceramide) [1]

P NeuAcα(2-8)NeuAcα(2-3)Galβ(1-3)GalNAcβ(1-4)(NeuAcα(2-3))Galβ(1-4) Glc + ceramide <1> [1]

S glycosphingolipid + H_2O <1, 2> (GM2, <1> [4]; GM3, <1> [4]; neutral and acidic in presence of 0.1% Triton X-100, lacto-type, <1, 2> [2, 5] and ganglio-type, <1, 2> [2, 5] glycosphingolipids readily hydrolyzed, globo-type (slowly, <2> [2]; strongly resistant to hydrolysis by endoglycoceramidase II in comparison with endoglycoceramidase I, <1> [5]), <1, 2> [2, 5] glycosphingolipids, overview, <2> [2]; cell-surface glycosphingolipids of erythrocytes, horse-erythrocyte cell-surface GM3 and guinea pig erythrocyte cell-surface Gg3Cer, <1> [7]) [2, 4, 5, 7]

P ?

S Additional information <1, 2> (hardly attacks linkages between monosaccharides and ceramides (cerebrosides) or between oligosaccharides and diacylglycerol (glycoglycerolipid), <1> [1]; not: N-linked glycoproteins, <1> [1]; O-linked glycoproteins, <1> [1]; gala-type glycosphingolipids, <1> [5]; cerebrosides, <1> [5]; sulfatides, <1> [5]; glycoglycerolipids, <1> [5]; sphingomyelins, <1> [5]; enzyme has transglycosylation activity (1-hexanol is best acceptor for GM1), <2> [3]) [1, 3, 5]

P ?

Inhibitors

Ag^{+} <1> (1 mM, 63% inhibition of endoglycoceramidase I, 94% inhibition of endoglycoceramidase II) [5]
Brij 58 <1> [5]
Cu^{2+} <1> [1, 5]
Hg^{2+} <1, 2> [1, 2, 5]
Tween 20 <1> [5]
Zn^{2+} <1, 2> (1 mM, 20% inhibition of endoglycoceramidase I, 48% inhibition of endoglycoceramidase II, <1> [5]) [1, 2, 5]

Activating compounds

Nonidet P-40 <1, 2> (activates) [2, 5]
Triton X-100 <1, 2> (activates, <1,2> [2, 5]; maximal concentration: 0.2%, <2> [2]; 0.3-0.4% w/v (endoglycoceramidase I), <1> [5]; 0.2-0.4% (endoglycoceramidase II), <1> [5]) [2, 5]
activator protein <1> (purified from culture supernatant of Rhodococcus sp. M-777, enhances the hydrolysis of glycosphingolipids in vitro and on the cell-surface by EGCase in absence of detergent) [4, 6, 7]
detergent <1> (required) [4, 6]
sodium cholate <1> [5]
sodium taurodeoxycholate <1> (activates endoglycoceramidase I and II, inhibition of endoglycoceramidase I above 0.4% w/v) [5]
taurodeoxycholate <1> (0.5 mg/ml, 2.5times increase of activity) [1]

Specific activity (U/mg)

Additional information <1> [1, 5]

K_m-Value (mM)

0.044 <1> (horse erythrocyte cell-surface GM3) [7]
0.047 <1> (guinea pig erythrocyte cell-surface Gg3Cer) [7]
0.1 <1> (asialo GM1, in presence of sodium taurodeoxycholate, endoglycoceramidase I) [5]
0.15 <2> (Galβ(1-3)GalNAcβ(1-4)(NeuAcα(2-3))Galβ(1-4)Glcβ(1-1)ceramide) [2]
0.21 <1> (asialo GM1, in presence of sodium taurodeoxycholate, endoglycoceramidase II) [5]

0.23 <1> (asialo GM1, in presence of Triton X-100, endoglycoceramidase I) [5]
0.5 <1> (asialo GM1, in presence of Triton X-100, endoglycoceramidase II) [5]

pH-Optimum
5 <1> (endoglycoceramidase I, substrate: GM1) [5]
5.5 <1> (endoglycoceramidase II, substrate: GM1) [5, 7]
5.5-6.5 <1> [1]

pH-Range
Additional information <1> [5]

Temperature optimum (°C)
6 <1> [1]

4 Enzyme Structure

Molecular weight
32700 <1> (Rhodococcus sp., endoglycoceramidase I, HPLC gel filtration) [5]
43000 <1> (Rhodococcus sp., endoglycoceramidase II, HPLC gel filtration) [5]
55900 <1> (Rhodococcus sp., endoglycoceramidase I, PAGE in absence and presence of SDS) [5]
58900 <1> (Rhodococcus sp., endoglycoceramidase II, PAGE in absence and presence of SDS) [5]
250000-450000 <2> (Corynebacterium sp., HPLC gel filtration) [2]

Subunits
? <2> (x * 65000, Corynebacterium sp., SDS-PAGE) [2]

5 Isolation/Preparation/Mutation/Application

Source/tissue
cell membrane <2> [3]
culture medium <1> [5]
culture supernatant <1, 2> [1, 2]

Localization
cell membrane <2> (bound, <2> [2]) [2, 3]
extracellular <2> [2]
intracellular <2> [2]

Purification
<1> (2 molecular species: I and II [5]) [1, 5]
<2> [2, 3]

6 Stability

pH-Stability

3 <1> (37°C, 60 min, 50.4% of endoglycoceramidase I activity and 27.5% of endoglycoceramidase II activity lost) [5]
4-8 <1> (37°C, 60 min, endoglycoceramidase I, stable) [5]
4.5-8 <2> (37°C, 30 min, stable) [2]
5-8.5 <1> (37°C, 60 min, endoglycoceramidase II, stable) [5]
5-9 <1> (stable) [1]

Temperature stability

45 <2> (pH 7.0, 10 min, stable up to) [2]
60 <2> (pH 7.0, 10 min, complete inactivation) [2]

General stability information

<1>, unstable at protein concentration below 0.2 mg/ml or without Triton X-100 [5]
<1>, unstable to repeated freezing and thawing [5]
<2>, stable to repeated freezing and thawing [2]

Storage stability

<1>, 0°C, protein concentration greater than 0.2 mg/ml, 50 mM sodium acetate buffer at pH 6.0 or 20 mM phosphate buffer at pH 7.2, 0.1-1.0% Triton X-100, stable for several months [5]
<1>, stored frozen, stable for at least 6 months [5]
<2>, -20°C, stable for at least 1 year [2]

References

[1] Ito, M.; Yamagata, T.: A novel glycosphingolipid-degrading enzyme cleaves the linkage between the oligosaccharide and ceramide of neutral and acidic glycosphingolipids [published erratum appears in J Biol Chem 1987 Feb 5;262(4):1926]. J. Biol. Chem., **261**, 14278-14282 (1986)
[2] Ashida, H.; Yamamoto, K.; Kumagai, H.; Tochikura, T.: Purification and characterization of membrane-bound endoglycoceramidase from Corynebacterium sp. Eur. J. Biochem., **205**, 729-735 (1992)
[3] Ashida, H.; Tsuji, Y.; Yamamoto, K.; Kumagai, H.; Tochikura, T.: Transglycosylation activity of endoglycoceramidase from Corynebacterium sp. Arch. Biochem. Biophys., **305**, 559-562 (1993)
[4] Ito, M.; Ikegami, Y.; Yamagata, T.: Activator proteins for glycosphingolipid hydrolysis by endoglycoceramidases. Elucidation of biological functions of cell-surface glycosphingolipids in situ by endoglycoceramidases made possible using these activator proteins. J. Biol. Chem., **266**, 7919-7926 (1991)
[5] Ito, M.; Yamagata, T.: Purification and characterization of glycosphingolipid-specific endoglycosidases (endoglycoceramidases) from a mutant strain of Rhodococcus sp. Evidence for three molecular species of endoglycoceramidase with different specificities. J. Biol. Chem., **264**, 9510-9519 (1989)

[6] Ito, M.; Ikegami, Y.; Omori, A.; Yamagata, T.: Conversion of endoglycoceramidase-activator II by trypsin to the 27.9 kDa polypeptide possessing full activity: purification of activator for endoglycoceramidase by trypsin treatment followed by trypsin-inhibitor agarose column application. J. Biochem., **110**, 328-332 (1991)
[7] Ito, M.; Ikegami, Y.; Yamagata, T.: Kinetics of endoglycoceramidase action toward cell-surface glycosphingolipids of erythrocytes. Eur. J. Biochem., **218**, 645-649 (1993)

3-Deoxy-2-octulosonidase 3.2.1.124

1 Nomenclature

EC number
3.2.1.124

Systematic name
capsular-polysaccharide 3-deoxy-D-manno-2-octulosonohydrolase

Recommended name
3-deoxy-2-octulosonidase

Synonyms
hydrolase, 2-keto-3-deoxyoctonate
octulofuranosylono hydrolase
octulopyranosylonohydrolase
octulosylono hydrolase

CAS registry number
103171-48-8

2 Source Organism

<1> *Coliphage PHI20* (growing on E.coli K13, K20 or K23) [1]
<2> *Coliphage PHI95* [2]

3 Reaction and Specificity

Catalyzed reaction
endohydrolysis of the β-ketopyranosidic linkages of 3-deoxy-D-manno-2-octulosonate in capsular polysaccharides

Reaction type
hydrolysis of O-glycosyl bond

Natural substrates and products
S 3-deoxy-D-manno-2-octulosonic acid glycoside + H_2O <1> (catalyzes depolymerization of 3-deoxy-2-octulosonide containing capsular polysaccharides from E. coli K13, K20 and K23) [1]

Substrates and products
S 3-deoxy-D-manno-2-octulosonic acid glycoside + H_2O <1, 2> (i.e. KDO, native or deacetylated polysaccharides are substrates, <2> [2]; specific for

K95-antigen, <2> [2]; cleaves β-octylfuranosidonic linkages of K95-glycan, <2> [2]) [1, 2]

P tetrasaccharide of two repeating units <1> (i.e. main product, β-Ribf1$\rightarrow$7β-KDOp2$\rightarrow$3β-Ribf1$\rightarrow$7KDO) [1]

pH-Optimum

6.5 <2> (assay at) [2]

Temperature optimum (°C)

30 <2> (assay at) [2]
37 <1> (assay at) [1]

References

[1] Altmann, F.; Kwiatkowski, B.; Stirm, S.; März, L.; Unger, F. M.: A bacteriophage-associated glycanase cleaving β-pyranosidic linkages of 3-deoxy-D-manno-2-octulosonic acid (KDO). Biochem. Biophys. Res. Commun., **136**, 329-335 (1986)

[2] Altmann, F.; März, L.; Stirm, S.; Unger, F.M.: Two additional bacteriophage-associated glycan hydrolases cleaving ketosidic bonds of 3-deoxy-D-manno-octulosonic acid in capsular polysaccharides of Escherichia coli. FEBS Lett., **221**, 145-149 (1987)

Raucaffricine β-glucosidase 3.2.1.125

1 Nomenclature

EC number
3.2.1.125

Systematic name
raucaffricine β-D-glucohydrolase

Recommended name
raucaffricine β-glucosidase

Synonyms
glucosidase, raucaffricine β-
raucaffricine β-D-glucosidase
raucaffricine glucosidase

CAS registry number
102925-37-1

2 Source Organism

<1> *Rauwolfia serpentina* [1]
<2> *Rauwolfia caffra* [1]
<3> *Rauwolfia verticillata* [1]
<4> *Rauwolfia mannii* [1]

3 Reaction and Specificity

Catalyzed reaction
raucaffricine + H_2O = vomilenine + D-glucose

Reaction type
hydrolysis of O-glycosyl bond

Substrates and products
S 1,2,19,20-tetrahydroraucaffricine + H_2O <1> (87% of activity compared to raucaffricine) [1]
P ?
S 1,2-dihydro-1-methylraucaffricine + H_2O <1> (36% of activity compared to raucaffricine) [1]
P ?

S 1,2-dihydroraucaffricine + H_2O <1> (98% of activity compared to raucaffricine) [1]
P ?
S 17-O-deacetyl-1,2-dihydroraucaffricine + H_2O <1> (95% of activity compared to raucaffricine) [1]
P ?
S raucaffricine + H_2O <1> [1]
P vomilenine + D-glucose <1> [1]

Inhibitors

fructose <1> (0.95 M: 30% inhibition) [1]
glucose <1> (0.8 M: complete inhibition) [1]
Additional information <1> (glucono-1,5-lactone up to 1 M, phenylmethylsulfonyl fluoride up to 80 mM, EDTA up to 80 mM, iodoacetamide up to 80 mM, strictosidine or amygdalin up to 50 mM, mannitol up to 1 M are no inhibitors) [1]

Specific activity (U/mg)

22.5 <1> (raucaffricine) [1]

K_m-Value (mM)

1.4 <1> (raucaffricine, 1,2-dihydro-1-methylraucaffricine) [1]
1.5 <1> (1,2-dihydroraucaffricine) [1]
1.7 <1> (17-O-deacetyl-1,2-dihydroraucaffricine) [1]
2 <1> (1,2,19,20-tetrahydroraucaffricine) [1]

pH-Optimum

5.1 <1> [1]

pH-Range

4.2-6 <1> (half-maximal activity at pH 4.2 and pH 6.0) [1]

Temperature optimum (°C)

38 <1> [1]

4 Enzyme Structure

Molecular weight

66600 <1> (Rauwolfia serpentina, gel chromatography) [1]

5 Isolation/Preparation/Mutation/Application

Purification

<1> (partial) [1]

6 Stability

Temperature stability

10-55 <1> (50% of maximal activity after 30 min) [1]

Storage stability

<1>, 4°C, crude enzyme, 30% loss of activity within 6 months [1]

References

[1] Schuebel, H.; Stoekigt, J.; Feicht, R.; Simon, H.: Partial purification and characterization of raucaffricine β-D-glucosidase from plant cell-suspension cultures of Rauwolfia serpentina Benth. Helv. Chim. Acta, **69**, 538-547 (1986)

Coniferin β-glucosidase 3.2.1.126

1 Nomenclature

EC number
3.2.1.126

Systematic name
coniferin β-D-glucosidase

Recommended name
coniferin β-glucosidase

Synonyms
coniferin-hydrolyzing β-glucosidase
glucosidase, coniferin β-

CAS registry number
83869-30-1

2 Source Organism

<1> *Cicer arietinum* (chickpea) [1]
<2> *Petroselinum hortense* (parsley) [2]
<3> *Triticum aestivum* (wheat) [2]
<4> *Picea abies* (spruce) [3]
<5> *Pinus banksiana* [4]

3 Reaction and Specificity

Catalyzed reaction
coniferin + H_2O = coniferol + D-glucose

Reaction type
hydrolysis of O-glycosyl bond

Natural substrates and products
S coniferin + H_2O <4> (plant cell wall enzyme, involved in lignin biosynthesis) [3]

Substrates and products
S 2-nitrophenyl β-D-galactoside + H_2O <1, 4> (reaction at 16% (isozyme I) or 13% (isozyme II) the rate of 2-nitrophenyl β-D-glucoside hydrolysis, <4> [3]) [1, 3]

P 2-nitrophenol + D-galactose

S 2-nitrophenyl β-D-glucoside + H_2O <1, 4> (best substrate, <4> [3]) [1, 3]

P 2-nitrophenol + D-glucose

S 2-nitrophenyl β-D-xyloside + H_2O <4> (poor substrate) [3]

P 2-nitrophenol + D-xylose

S 4-coumaryl β-D-glucoside + H_2O <4> (reaction at 25% (isozyme II) or 4% (isozyme I) the rate of 2-nitrophenyl β-D-glucoside hydrolysis) [3]

P ? + D-glucose

S 4-methylumbelliferyl β-D-glucosaminide + H_2O <1> (poor substrate) [1]

P 4-methylumbelliferone + D-glucosamine

S 4-methylumbelliferyl β-D-glucoside + H_2O <1, 4> (reaction at 79% (isozyme II) or 51% (isozyme I) the rate of 2-nitrophenyl β-D-glucoside hydrolysis, <4> [3]) [1, 3]

P 4-methylumbelliferone + D-glucose

S 4-nitrophenyl β-D-galactoside + H_2O <1> (poor substrate, not, <4> [3]) [1]

P 4-nitrophenol + D-galactose

S 4-nitrophenyl β-D-glucoside + H_2O <1, 4> (reaction at 56% (isozyme II) or 8% (isozyme I) the rate of 2-nitrophenyl β-D-glucoside hydrolysis, <4> [3]) [1, 3]

P 4-nitrophenol + D-glucose

S coniferin + H_2O <1-5> (i.e. coniferyl alcohol β-D-glucoside, (E)-coniferin, <5> [4]; reaction at 52% (isozyme I) or 38% (isozyme II) the rate of 2-nitrophenyl β-D-glucoside hydrolysis, <4> [3]) [1-4]

P coniferol + D-glucose <1, 5> (i.e. trans-coniferyl alcohol, <5> [4]) [1, 4]

S ferulic acid *p*-β-glucoside + H_2O <1> [1]

P ferulic acid + D-glucose

S feruloyl glucoside + H_2O <4> (poor substrate) [3]

P ferulic acid + glucose

S picein + H_2O <1, 4> (reaction at 50% (isozyme II) or 12% (isozyme I) the rate of 2-nitrophenyl β-D-glucoside hydrolysis, <4> [3]) [1, 3]

P ?

S salicin + H_2O <1, 4> (reaction at 25% (isozyme II) or 2% (isozyme I) the rate of 2-nitrophenyl β-D-glucoside hydrolysis, <4> [3]; poor substrate, <1> [1]) [1, 3]

P ?

S syringin + H_2O <1, 4> (poor substrate, <1> [1]; reaction at about 30% the rate of 2-nitrophenyl β-D-glucoside hydrolysis, <4> [3]) [1, 3]

P ?

S Additional information <1> (very poor substrates are arbutin or indican, no substrates are α-glucosides, 4-methylumbelliferyl β-cellobioside, 4-methylumbelliferyl β-glucuronide, quercetin 3-rhamnosyl glucoside, quercetin 3-rhamnoside or cellobiose) [1]

P ?

Specific activity (U/mg)

2.7 <1> [1]

4.5 <4> (hypocotyls) [3]
Additional information <5> [4]

K_m-Value (mM)

0.3 <1> (formononetin 7-β-glucoside, 2-nitrophenyl β-glucoside) [1]
0.38 <4> (coniferin, roots) [3]
0.4 <1> (4-methylumbelliferyl β-D-glucoside, ferulic acid p-β-glucoside, 4-nitrophenyl β-D-glucoside) [1]
0.6 <1, 4> (picein, <1> [1]; 2-nitrophenyl β-D-glucoside, 2-nitrophenyl β-D-galactoside, hypocotyls, <4> [3]) [1, 3]
0.7 <4> (4-nitrophenyl β-D-glucoside, hypocotyls) [3]
0.8 <1> (coniferin) [1]
0.9 <1> (4-methylumbelliferyl β-D-glucosaminide) [1]
1 <4> (4-methylumbelliferyl β-D-glucoside, hypocotyls) [3]
1.3 <4> (coniferin, hypocotyls) [3]
1.8 <4> (syringin, hypocotyls) [3]
3 <1> (2-nitrophenyl β-D-galactoside) [1]
5 <1> (syringin, 4-nitrophenyl β-D-glucoside) [1]
6.3 <4> (picein, hypocotyls) [3]

pH-Optimum

4.5-5.5 <4> (hypocotyls) [3]
5 <1> [1]
Additional information <4, 5> (pI: 3.8, <5> [4]; pI: 10 (isozyme I), pI: 10.3 (isozyme II), <4> [3]) [3, 4]

pH-Range

3.5-6.5 <4> (about half-maximal activity at pH 3.5 and 6.5, hypocotyls) [3]
4-7.5 <1> (about 20% of maximal activity at pH 4 and 7.5) [1]

Temperature optimum (°C)

30 <1, 4, 5> (assay at) [1, 3, 4]
40-45 <4> [3]

4 Enzyme Structure

Molecular weight

58570 <4> (Picea abies, sedimentation equilibrium method) [3]
90000 <5> (Pinus banksiana, 2 isozymes, MW 90000 and 110000) [4]
110000 <1, 5> (Cicer arietinum, gel filtration [1]; Pinus banksiana, 2 isozymes, MW 90000 and 110000 [4]) [1, 4]

Subunits

dimer <1> (1 * 43000 + 1 * 63000, Cicer arietinum, SDS-PAGE) [1]
monomer <4> (1 * 57000, Picea abies, SDS-PAGE) [3]

Posttranslational modification

glycoprotein <5> [4]

5 Isolation/Preparation/Mutation/Application

Source/tissue

cell suspension culture <1-3> [1, 2]
hypocotyl <4> [3]
root <4> [3]
seedling <4> (tissue distribution) [3]
xylem <5> (lignifying) [4]

Localization

cell wall <1, 4> [1, 3]
cytosol <5> [4]
soluble <5> [4]

Purification

<1> (to near homogeneity) [1]
<4> (hypocotyls) [3]
<5> (2 isozymes) [4]

6 Stability

General stability information

<1>, 0.5 M NaCl stabilizes, removal inactivates [1]

Storage stability

<1>, 4°C, in the presence of 0.5 M NaCl, several months [1]
<4>, -20°C, crude cell wall preparation, frozen with liquid nitrogen, at least 3 months [3]
<4>, 4°C, in 0.025 M McIlvaine buffer, pH 5, 0.1 M NaCl, 0.02% NaN_3, less than 10% loss of activity within 15 days [3]

References

[1] Hösel, W.; Surholt, E.; Borgmann, E.: Characterization of β-glucosidase isoenzymes possibly involved in lignification from chick pea (Cicer arietinum L.) cell suspension cultures. Eur. J. Biochem., **84**, 487-492 (1978)
[2] Hösel, W.; Fiedler-Preiss, A.; Borgmann, E.: Relationship of coniferin β-glucosidase to lignification in various plant cell suspension cultures. Plant Cell Tissue Organ Cult., **1**, 137-148 (1982)
[3] Marcinowski, S.; Grisebach, H.: Enzymology of lignification. Cell-wall bound β-glucosidase for coniferin from spruce (Picea abies) seedlings. Eur. J. Biochem., **87**, 37-44 (1978)
[4] Leinhos, V.; Udagama-Randeniya, P.V.; Savidge, R.A.: Purification of an acidic coniferin-hydrolysing β-glucosidase from developing xylem of Pinus banksiana. Phytochemistry, **37**, 311-315 (1994)

1,6-α-L-Fucosidase 3.2.1.127

1 Nomenclature

EC number
3.2.1.127

Systematic name
1,6-L-fucosyl-N-acetyl-D-glucosaminylglycopeptide fucohydrolase

Recommended name
1,6-α-L-fucosidase

Synonyms
α-L-fucosidase
fucosidase, 1,6-α-L-

CAS registry number
102925-35-9

2 Source Organism

<1> *Aspergillus niger* [1]

3 Reaction and Specificity

Catalyzed reaction
hydrolysis of 1,6-linkages between α-L-fucose and N-acetyl-D-glucosamine in glycopeptides such as immunoglobulin G glycopeptide and fucosyl-asialo-agalacto-fetuin

Reaction type
hydrolysis of O-glycosyl bond

Substrates and products
S α-1,6-fucosyl asialo-agalacto-fetuin + H_2O <1> [1]
P α-L-fucose + asialo-agalacto-fetuin (α-1,6-fucosyl linkage is cleaved)
S immunoglobulin G glycopeptide + H_2O <1> (bovine) [1]
P fragments of immunoglobuline G glycopeptide <1> (α-1,6-fucosyl linkage is cleaved) [1]
S Additional information <1> (α-1,2-, α-1,3- or α-1,4-α-L-fucosides are not cleaved, α-L-fucosyl asialo-fetuin is no substrate) [1]
P ?

Specific activity (U/mg)
0.000000606 <1> [1]

K_m-Value (mM)
0.00000277 <1> (α-L-fucosyl asialo-agalacto-fetuin) [1]

pH-Optimum
4.5 <1> [1]

Temperature optimum (°C)
37 <1> (assay at) [1]

4 Enzyme Structure

Molecular weight
180000 <1> (Aspergillus niger, gel filtration) [1]

5 Isolation/Preparation/Mutation/Application

Purification
<1> (partial) [1]

References

[1] Yazawa, S.; Madiyalakan, R.; Chawda, R.P.; Matta, K.L.: α-L-Fucosidase from Aspergillus niger: demonstration of a novel α-L-(1–6)-fucosidase acting on glycopeptides. Biochem. Biophys. Res. Commun., **136**, 563-569 (1986)

Glycyrrhizinate β-glucuronidase 3.2.1.128

1 Nomenclature

EC number
3.2.1.128

Systematic name
glycyrrhizinate glucuronosylhydrolase

Recommended name
glycyrrhizinate β-glucuronidase

Synonyms
glycyrrhizin hydrolase
glycyrrhizinic acid hydrolase
hydrolase, glycyrrhizin β-

CAS registry number
102484-56-0
108891-59-4

2 Source Organism

<1> *Aspergillus niger* (GRM 3 [1]) [1, 2]
<2> *Eubacterium sp.* [3]

3 Reaction and Specificity

Catalyzed reaction
glycyrrhizinate + H_2O = 1,2-β-D-glucuronosyl-D-glucuronate + glycyrrhetinate

Reaction type
hydrolysis of O-glycosyl bond

Natural substrates and products
S glycyrrhizinic acid + H_2O <1> (specific for hydrolysis of triterpenoid glycoside from roots of Glycyrrhiza sp.) [1]

Substrates and products
S chenodeoxycholic acid 3-O-β-D-glucuronide + H_2O <1> [2]
P chenodeoxycholic acid + glucuronic acid <1> [2]
S chikusetsusaponin-IVa + H_2O <1> (from Panax japonicus) [2]

P β-D-glucopyranosyl oleanolate + D-glucuronic acid <1> [2]
S chikusetsusaponin-V + H_2O <1> (i.e. ginsenoside-Ro, from Panax japonicus) [2]
P β-D-glucopyranosyl oleanolate + 2-O-β-D-glucopyranosyl D-glucuronic acid <1> [2]
S glycyrrhizinic acid + H_2O <1, 2> (a pentacyclic triterpene saponin from liquorice roots, best substrate, <1> [1]; substrate specificity, <1> [2]; selectively hydrolyzes the 3-O-β-D-glucuronide linkage of glycyrrhinic acid, <1> [2]) [1-3]
P glycyrrhetinic acid + D-glucuronyl β-1,2-D-glucuronic acid <1> [1]
S hemsloside Ma1 + H_2O <1> (from Hemsleya macrosperma C.Y.Wu) [2]
P β-D-glucopyranosyl oleanolate + 3-O-α-L-arabinopyranosyl D-glucuronic acid <1> [2]
S hemsloside Ma2 + H_2O <1> (from Hemsleya macrosperma C.Y.Wu) [2]
P β-D-glucopyranosyl oleanolate + 2-O-β-D-glucopyranosyl-(3-O-α-L-arabinopyranosyl) α-D-glucuronic acid <1> [2]
S pseudo-ginsenoside-RT1 + H_2O <1> (from Panax pseudoginseng ssp. himalaicus Hara) [2]
P β-D-glucopyranosyl oleanolate + 2-O-β-D-xylopyranosyl D-glucuronic acid <1> [2]
S Additional information <1> (no substrates are 4-nitrophenyl β-D-glucuronide, phenolphthalein β-D-glucuronide, <1> [1]; chikusetsusaponin-IV, stipuleanosides R1 and 2 (from Panax stipuleanatus H.T. Tsai and K.M. Feng), chikusetsusaponin-IVa ethyl or chikusetsusaponin-V methyl ester and β-D-glucuronides of small-size aliphatic alcohols (e.g. methanol, propanol or cyclohexanol), 5-α-cholestan-3β-ol, l-menthol or d-menthol) [2]
P ?

Specific activity (U/mg)

Additional information <1> [1]

pH-Optimum

4.1-4.5 <1> [1]

pH-Range

3-4.8 <1> (about half-maximal activity at pH 3 and 4.8) [1]

Temperature optimum (°C)

45 <1> (at pH 5) [1]

Temperature range (°C)

30-53 <1> (about 63% of maximal activity at 30°C and about half-maximal activity at 53°C, pH 5) [1]

4 Enzyme Structure

Molecular weight

150000 <1> (Aspergillus niger, gel filtration) [1]

5 Isolation/Preparation/Mutation/Application

Localization

cytosol <2> [3]

Purification

<1> [1]
<2> (partial) [3]

6 Stability

pH-Stability

3 <1> (about 50% of activity after 1 h at 40°C) [1]
5-6.5 <1> (at least 1 h stable at 40°C) [1]
7.5 <1> (about 50% of activity after 1 h at 40°C) [1]

Temperature stability

20-45 <1> (at least 30 min stable at pH 5) [1]

References

[1] Muro, T.; Kuramoto, T.; Imoto, K.; Okada, S.: Purification and some properties of glycyrrhizinic acid hydrolase from Aspergillus niger GRM3. Agric. Biol. Chem., **50**, 687-692 (1986)
[2] Sasaki, Y.; Morita, T.; Kuramoto, T.; Mizutani, K.; Ikeda, R.; Tanaka, O.: Substrate specificity of glycyrrhizinic acid hydrolase. Agric. Biol. Chem., **52**, 207-210 (1988)
[3] Akao, Ta.; Akao, Te.; Kobashi, K.: Glycyrrhizin β-D-glucuronidase of Eubacterium sp. from human intestinal flora. Chem. Pharm. Bull., **35**, 705-710 (1987)

Endo-α-sialidase 3.2.1.129

1 Nomenclature

EC number
3.2.1.129

Systematic name
poly(sialoside) α-2,8-sialosylhydrolase

Recommended name
endo-α-sialidase

Synonyms
endo-N
endo-N-acetylneuraminidase
endoneuraminidase
endosialidase
neuraminidase, endo-
poly(α-2,8-sialoside) α-2,8-sialosylhydrolase
poly(α-2,8-sialosyl) endo-N-acetylneuraminidase
Additional information (an exo-α-sialidase activity is listed as EC 3.2.1.18, see also EC 3.2.1.138)

CAS registry number
91195-87-8

2 Source Organism

<1> *Bacteriophage* (K1F phage [6]; Phi1.2 [3]; PK1A [2, 4]; PK1E [2]; coliphage E [5]; host strain: E. coli K1 (IH3088) [2]; E. coli K235 [3]; E. coli U9/41 [4]) [1-6]

3 Reaction and Specificity

Catalyzed reaction
endohydrolysis of α-2,8-sialosyl linkages in oligo or poly(sialic) acids

Reaction type
hydrolysis of O-glycosyl bond

Substrates and products

S poly(sialic) acids or oligo(sialic) acids containing α-2,8-linked N-acetylneuraminic acid + H_2O <1> (brain glycoproteins, <1> [4]; enzyme requires the simultaneous presence adjacent to the site of cleavage a minimum of 3 sialic acid residues on the distal side and a minimum of 5 sialic acid residues on the proximal side, <1> [4]; catalyzes the hydrolysis of both α-2,8-linked poly(N-acetyl-D-neuraminic acid) and poly(N-glycoloyl-D-neuraminic acid), <1> [1]; rapid hydrolysis of both the O-acetylated and non-O-acetylted forms of the K1 antigen, an α-2,8-linked homopolymer of N-acetylneuraminic acid, and of Meningococcus B antigen, <1> [5]; limited hydrolysis of the E. coli K92 antigen, an N-acetylneuraminic acid homopolymer containing alternating α-2,8-, and α-2,9-linkages, <1> [5]; no release of α-2,3-, α-2,6-, or α-2,9-linked sialic residues from a variety of other substrates, <1> [5]; bacterial and neural membrane glycoconjugates, <1> [6]; polysialic acid capsule isolated from E. coli N67 or E. coli K1, <1> [6]; oligosialic acid from E. coli K1, <1> [6]; the minimum requirement for cleavage is α-2,8-$(NeuAc)_5$, <1> [6]) [1-6]

P fragments of poly(sialic) acid or oligo(sialic) acids <1> [1-6]

Inhibitors

Ca^{2+} <1> [5]

Specific activity (U/mg)

0.95 <1> [5, 6]

K_m-Value (mM)

0.0066 <1> (polysialic acid, capsule isolated from E. coli N67, soluble enzyme) [6]
0.051 <1> (polysialic acid, capsule isolated from E. coli K1, soluble enzyme) [6]
0.071 <1> (polysialic acid, capsule isolated from E. coli K1, phage-bound enzyme) [6]
1.2 <1> (oligosialic acid, isolated from E. coli K1, phage-bound enzyme) [6]
1.6 <1> (oligosialic acid, from E. coli K1, soluble enzyme) [6]
7.41 <1> (K1 antigen, an α-2,8-linked homopolymer of N-acetylneuraminic acid) [5]

pH-Optimum

5.2-5.5 <1> [5]

4 Enzyme Structure

Molecular weight

208000 <1> (coliphage E, enzyme complex, SDS-PAGE at 37°C) [5]
328000 <1> (K1F phage, gel filtration) [6]

Subunits

? <1> (coliphage E, enzyme complex dissociated by SDS at 100°C yields 2 polypeptides with apparent MW of 74000 and 38500) [5]
trimer <1> (3 * 105000, K1F phage, SDS-PAGE) [6]

5 Isolation/Preparation/Mutation/Application

Source/tissue

phage <1> (lysate) [6]

Purification

<1> (coliphage E [5]; K1F phage [6]) [5, 6]

References

[1] Kitajima, K.; Inoue, S.; Inoue, Y.; Troy, F.A.: Use of a bacteriophage-derived endo-N-acetylneuraminidase and an equine antipolysialyl antibody to characterize the polysialyl residues in salmonid fish egg polysialoglycoproteins. Substrate and immunospecificity studies. J. Biol. Chem., **263**, 18269-18276 (1988)

[2] Pelkonen, S.; Pelkonen, J.; Finne, J.: Common cleavage pattern of polysialic acid by bacteriophage endosialidases of different properties and origins. J. Virol., **63**, 4409-4416 (1989)

[3] Kwiatowski, B.; Boschek, B.; Thiele, H.; Stirm, S.: Endo-N-acetylneuraminidase associated with bacteriophage particles. J. Virol., **43**, 697-704 (1982)

[4] Finne, J.; Mäkelä, P.H.: Cleavage of the polysialosyl units of brain glycoproteins by a bacteriophage endosialidase. Involvement of a long oligosaccharide segment in molecular interactions of polysialic acid. J. Biol. Chem., **260**, 1265-1270 (1985)

[5] Tomlinson, S.; Taylor, P.W.: Neuraminidase associated with coliphage E that specifically depolymerizes the Escherichia coli K1 capsular polysaccharide. J. Virol., **55**, 374-378 (1985)

[6] Hallenbeck, P.C.; Vimr, E.R.; Yu, F.; Bassler, B.; Troy, F.A.: Purification and properties of a bacteriophage-induced endo-N-acetylneuraminidase specific for poly-α-2,8-sialosyl carbohydrate units. J. Biol. Chem., **262**, 3553-3561 (1987)

Glycoprotein endo-α-1,2-mannosidase 3.2.1.130

1 Nomenclature

EC number
3.2.1.130

Systematic name
glycoprotein glucosylmannohydrolase

Recommended name
glycoprotein endo-α-1,2-mannosidase

Synonyms
endo-α-D-mannosidase
endo-α-mannosidase
endomannosidase
glucosyl mannosidase
glucosylmannosidase
mannosidase, endo-α-
mannosidase, glucosyl-

CAS registry number
108022-16-8
125858-79-9

2 Source Organism

<1> *Rattus norvegicus* (male [1-3]; Wistar [2]; CD-strain [1, 3]) [1-3]

3 Reaction and Specificity

Catalyzed reaction
hydrolysis of the terminal α-D-glucosyl-(1,3)-D-mannosyl unit from the $GlcMan_9(GlcNAc)_2$ oligosaccharide component of the glycoprotein produced in the Golgi membrane

Reaction type
hydrolysis of O-glycosyl bond

Natural substrates and products

S $GlcMan_9GlcNAc + H_2O$ <1> (involved in N-linked oligosaccharide processing, <1> [1]; accomplishes deglucosylation of glycoproteins with N-linked carbohydrate units, <1> [3]) [1, 3]

Substrates and products

S $Glc_2Man_9GlcNAc + H_2O$ <1> (at 14% the rate of $GlcMan_9GlcNAc$ hydrolysis) [2]

P ?

S $Glc_3Man_9GlcNAc + H_2O$ <1> (at 11% the rate of $GlcMan_9GlcNAc$ hydrolysis) [2]

P ?

S $GlcMan_7GlcNAc + H_2O$ <1> (at 71% the rate of $GlcMan_9GlcNAc$ hydrolysis) [2]

P ?

S $GlcMan_8GlcNAc + H_2O$ <1> (at 89% the rate of $GlcMan_9GlcNAc$ hydrolysis) [2]

P ?

S $GlcMan_9GlcNAc + H_2O$ <1> (best substrate, <1> [2]) [1-3]

P $Man_8GlcNAc$ + glucosyl α1,3-mannose <1> [1-3]

Inhibitors

glucosyl α1-3(1-deoxy)mannojirimycin <1> [3]

Additional information <1> (no inhibition by 1-deoxynojirimycin or EDTA) [1]

Activating compounds

Triton X-100 <1> (requirement, membrane-bound enzyme) [2]

Specific activity (U/mg)

Additional information <1> (10855000 cpm of glucosyl α1,3-mannose/h and mg protein, <1> [3]) [1-3]

pH-Optimum

6.2-6.8 <1> [3]

7 <1> [1]

pH-Range

5.8-7.7 <1> (about half-maximal activity at pH 5.8 and 7.7) [2]

6-8.5 <1> (about half-maximal activity at pH 6 and 8.5) [1]

Temperature optimum (°C)

37 <1> (assay at) [1-3]

4 Enzyme Structure

Molecular weight

380000 <1> (rat, solubilized enzyme, HPLC gel filtration) [2]

560000 <1> (rat, gel filtration) [3]

Subunits

multimer <1> (x * 60000 + x * 56000, rat, SDS-PAGE) [3]

5 Isolation/Preparation/Mutation/Application

Source/tissue

brain <1> [2]
liver <1> [1, 3]

Localization

Golgi apparatus <1> [1, 3]
membrane-bound <1> [1-3]
microsome <1> (intrinsic membrane component) [2]
Additional information <1> (subcellular distribution) [1]

Purification

<1> (partial (solubilized with Triton X-100/deoxycholate, less effectively with Triton X-100 alone or Triton X-100/NaCl) [2]; ligand affinity chromatography on glucosyl α1,3-O-$(CH_2)_8$-CONH-agarose [3]) [2, 3]

6 Stability

Storage stability

<1>, -70°C, in isolated Golgi membranes, 30% loss of activity within 40 days, with several freeze-thawing cycles [1]
<1>, -80°C, in intact Golgi membranes, 10% loss of activity within 6 months [3]
<1>, -80°C, purified enzyme, at 0.03 mg protein/ml pH 6.8 buffer, 50% loss of activity within 6 months, increased stability at 0.5 mg protein/ml, bovine serum albumin stabilizes [3]

References

[1] Lubas, W.A.; Spiro, R.G.: Golgi endo-α-D-mannosidase from rat liver, a novel N-linked carbohydrate unit processing enzyme. J. Biol. Chem., **262**, 3775-3781 (1987)
[2] Tulsiani, D.R.P.; Coleman, V.D.; Touster, O.: Asparagine-linked glycoprotein biosynthesis in rat brain: identification of glucosidase I, glucosidase II, and and endomannosidase (glucosyl mannosidase). Arch. Biochem. Biophys., **277**, 114-121 (1990)
[3] Hiraizumi, S.; Spohr, U.; Spiro, R.G.: Ligand affinity chromatographic purification of rat liver Golgi endomannosidase. J. Biol. Chem., **269**, 4697-4700 (1994)

Xylan α-1,2-glucuronosidase 3.2.1.131

1 Nomenclature

EC number
3.2.1.131

Systematic name
xylan α-D-1,2-(4-O-methyl)glucuronohydrolase

Recommended name
xylan α-1,2-glucuronosidase

Synonyms
α-(1$\rightarrow$2)-glucuronidase
glucuronidase, 1,2-α-

CAS registry number
114921-73-2

2 Source Organism

<1> *Tyromyces palustris* (highest active producer among Trichoderma and 5 basidiomycete species) [1]
<2> *Trichoderma aureoviride* [1]
<3> *Trichoderma hamatum* [1]
<4> *Trichoderma harzianum* [1]
<5> *Trichoderma koningii* [1]
<6> *Trichoderma longibrachiatum* [1]
<7> *Trichoderma viride* [1]
<8> *Trichoderma sp.* (No.3, No.4, No.5) [1]
<9> *Agaricus bisporus* (307) [1]
<10> *Laetiporus sulphueus var. miniatus* [1]
<11> *Pleurotus ostreatus* [1]
<12> *Polyporus versicolor* [1]

3 Reaction and Specificity

Catalyzed reaction
hydrolysis of α-D-1,2-(4-O-methyl)glucuronosyl links in the main chain of hardwood xylans

Reaction type

hydrolysis of O-glycosyl bond

Natural substrates and products

S xylan + H_2O <1-12> (liberates 4-O-methyl-D-glucuronic acid from the main chain of the hardwood xylan) [1]

Substrates and products

S xylan + H_2O <1-12> (liberates 4-O-methyl-D-glucuronic acid from the main chain of the hardwood xylan) [1]

P 4-O-methyl-D-glucuronic acid <1-12> (liberates 4-O-methyl-D-glucuronic acid from the main chain of the hardwood xylan) [1]

6 Stability

Storage stability

<1>, labile even if frozen [1]

References

[1] Ishihara, M.; Shimizu, K.: α-(1→2)-glucuronidase in the enzymatic saccharification of hardwood xylan. I. Screening of a-glucuronidase producing fungi. Mokuzai Gakkaishi, **34**, 58-64 (1988)

Chitosanase 3.2.1.132

1 Nomenclature

EC number
3.2.1.132

Systematic name
chitosan N-acetylglucosaminohydrolase

Recommended name
chitosanase

CAS registry number
51570-20-8

2 Source Organism

<1> *Penicillium islandicum* [1]
<2> *Bacillus sp.* (No. 7-M [2, 21]; R-4 [4]; PI-7S [10]) [2, 4, 10, 21]
<3> *Streptomyces griseus* [3, 16]
<4> *Streptomyces sp.* (No. 6 [5]; N174 [12, 22]) [5, 12, 22]
<5> *Rhodotorula gracilis* (CFR-1) [6]
<6> *Bacillus circulans* (MH-K1) [7, 8]
<7> *Mucor rouxii* [9]
<8> *Triticum aestivum* [11]
<9> *Hordeum vulgare* [11]
<10> *Pisum sativum* [11]
<11> *Cucumis sativus* [11]
<12> *Onoclea sensibilis* [11]
<13> *Campanula rotundifolia* [11]
<14> *Petroselinum crispum* [11]
<15> *Setaria italica* [11]
<16> *Solanum melongena* [11]
<17> *Capsicum annuum* [11]
<18> *Citrus medica* [11]
<19> *Thuja occidentalis* [11]
<20> *Amycolatopsis sp.* (CsO-2) [13]
<21> *Bacillus megaterium* [14]
<22> *Nocardia orientalis* [15]
<23> *Citrus sinensis* [16, 23, 24]
<24> *Bacillus pumilus* (Bn-262) [17]
<25> *Enterobacter sp.* [18]

<26> *Fusarium solani* (f. sp. phaseoli SUF386) [19]
<27> *Pseudomonas sp.* (H-14) [20]
<28> *Macrotermes estherae* (workers, soldiers, males, winged females) [25]

3 Reaction and Specificity

Catalyzed reaction
endohydrolysis of β-1,4-linkages between N-acetyl-D-glucosamine and D-glucosamine residues in a partly acetylated chitosan

Reaction type
hydrolysis of O-glycosyl bond

Natural substrates and products
S Additional information <1, 5> (inducible enzyme, <1> [1]; enzyme may be involved in the budding process of the yeast, <5> [6]) [1, 6]

Substrates and products
S N-acetyl-chitooligosaccharides + H_2O <25> [18]
P ?
S carboxymethylcellulose + H_2O <2, 3> (slow cleavage, <2> [2]; not, <1, 2, 4, 6, 20> [1, 5, 7, 10, 12, 13]) [2, 3]
P ?
S carboxymethylchitosan + H_2O <3, 6> [3, 7]
P ?
S chitopentaose + H_2O <26> [19]
P ?
S chitosan + H_2O <1-28> (polymers with 30% to 60% acetylation, <1> [1]; deacetylation degree from 70% to 100%, <20,26> [13, 19]; polymers containing equal proportions of acetylated and non-acetylated sugars are optimal for chitosanase activity, <1> [1]; most susceptible substrate is: 80% deacetylated chitosan, <6> [7]; 100% deacetylated chitosan, <20> [13]; chitohexaose and chitoheptaose, <22> [15]; substrates with less than 40% deacetylation are not affected by the enzyme, <6> [7]; the smallest of the substrates is a tetramer of glucosamine, <4, 6, 22> [7, 12, 15]; hydrolysis of: soluble chitosan, <2, 3> [2, 3]; glycol chitosan, <2, 3, 6, 26> [2-4, 7, 19]; colloidal chitosan (80% deacetylated, not 100% deacetylated, <25> [18]), <3, 25> [3, 18]; chitosan extracted from Mucor rouxii cell walls, <7> [9]; enzyme A is highly specific for chitosan, enzyme B and C possess comparable specific activities towards chitosan, chitin and cellulose, <21> [14]; enzyme is capable of cleaving between glucosamine and glucosamine or N-acetylglucosamine, but not cleaving between N-acetylglucosamine and glucosamine, <22> [15]; specificity, <24> [17]; high substrate specificity for highly deacetylated chitosan, <27> [20]; enzyme is absolutely specific towards the GlcN-GlcN bonds in partially N-acetylated chitosan and at least three GlcN residues are necessary to hydrolyze chitosan, <2> [21]) [1-25]

P Additional information <1, 2, 4-7, 21, 22, 24, 25, 27> (the products depend on the degree of acetylation of the polymer, <1> [1]; major oligomeric product from 30% acetylated chitosan is a trimer, <1> [1]; with 60% acetylated chitosan as substrate a dimer is also found, <1> [1]; glucosamine oligomers with the degree of polymerization from 2 to 8 are produced during hydrolysis of soluble chitosan, <2> [2]; diglucosamine and triglucosamine, <4> [5]; chitosan of an average MW 36000 is reduced by the enzymatic catalysis to nearly one-fourth this size, <5> [6]; mixture of dimer and trimer of glucosamine, <4, 6, 22, 24, 27> [7, 12, 15, 17, 20]; mixture of dimer, trimer and higher molecular weight oligomers of glucosamine, <7> [9]; glucosamine oligomers, predominantly dimers and trimers, <21> [14]; $GlcNAc_{4-6}$ and colloidal chitin are hydrolyzed to $GlcNAc_2$, <25> [18]; the main cleavage site with $GlcNAc_{3-6}$ is the second linkage from the non-reducing end, based on the pattern of pNp-$GlcNAc_{2-5}$, <25> [18]; colloidal chitosan is hydrolyzed to $GlcNAc_2$ and to similar partially N-acetylated chitooligosaccharides, <25> [18]) [1, 2, 5-7, 9, 12, 14, 15, 17, 18, 20]

S colloidal chitin + H_2O <23, 25> [18, 23, 24]

P ?

S glycol chitin + H_2O <25> [18]

P ?

S Additional information <1-4, 6, 20> (slow cleavage of: pentamer of N-acetylglucosamine, glucosamine oligomers, <1> [1]; degrades glucosamine tetramer to dimer and pentamer and to dimer and trimer, <20> [13] no activity: chitin (i.e. 100% acetylated chitosan), <1, 2, 4, 6, 20> [1, 2, 5, 7, 10, 12, 13]; trimers or tetramers of N-acetylglucosamine, carboxymethylcellulose, dextran, galactan, polygalacturonic acid, laminaran, mucoran, pectin, pullulan, salicin, xylan, 4-nitrophenyl-β-D-acetylglucosamine, 4-nitrophenyl-β-D-glucosamine, <1> [1]; glycol chitin, <2, 3> [2, 3]; colloidal chitin, <1-3> [1-3]; powdered chitosan, <2> [2]; cellulose, <2> [2]; methylcellulose, <2> [2]; glucosamine dimer and trimer, <20> [13]) [1-3, 5, 7, 10, 12, 13]

P ?

Inhibitors

Ag^+ <2, 3> [3, 10]
Cu^{2+} <1, 3, 5-7> ($CuSO_4$, <5> [6]) [1, 3, 6, 7, 9]
Fe^{2+} <2, 3> [3, 4]
Hg^{2+} <1-3, 5-7, 20> ($HgCl_2$, <5> [6]) [1, 3, 4, 6, 7, 9, 10, 13]
L-cysteine <2> [4]
Mn^{2+} <7> [9]
N-bromosuccinimide <5, 20> [6, 13]
Ni^{2+} <2, 6> [4, 7]
PCMB <2, 5, 6, 20> [4, 6, 7, 10, 13]
Pb^{2+} <7> [9]
Sn^{2+} <7> [9]
Zn^{2+} <2, 6> [4, 7]

acetate <7> (enzyme A, enzyme B is not affected) [9]
chitosan <4, 7> (substrate inhibition, enzyme B, <7> [9]; soluble chitosan, above 0.5 g/l, <4> [5]) [5, 9]
monoiodoacetate <2> [4]
reduced glutathione <2> [4]
sulfhydryl reagents <2> (glutathione or L-cysteine restores activity) [4]
tetramers of N-acetylglucosamine <1> [1]
trimers of N-acetylglucosamine <1> [1]

Activating compounds
chelating agents <2> (especially EDTA, activate) [4]

Metals, ions
Ca^{2+} <7> (activates enzyme B) [9]
K^+ <7> (activates enzyme A) [9]
$MnCl_2$ <5> (slight stimulation) [6]

Specific activity (U/mg)
2.76 <3> [3]
319 <2> [2]
Additional information <1, 4, 5, 20-23, 25, 27> [1, 6, 12-15, 18, 20, 23]

K_m-Value (mM)
Additional information <1-7, 21> (1.4 mg/ml: 5% or 30% acetylated chitosan, <1> [1]; K_m: 3.3% soluble chitosan, <2> [2]; 2.0% glycol chitosan, <2> [2]; 0.21% glycol chitosan, <3> [3]; 0.18% carboxymethylcellulose, <3> [3]; 0.688 g/l chitosan, at pH 5.5, <4> [5]; 0.83 mg/ml chitosan, <5> [6]; 0.63 mg chitosan/ml, <6> [7]; 0.088 mg/ml chitosan, <4> [12]; 0.8 mg/ml chitosan, <21> [14]) [1-3, 5-7, 9, 12, 14]

pH-Optimum
4 <27> [20]
4.5-6 <1, 4> (with 30% or 5% acetylated chitosan, <1> [1]; at 25°C, <4> [5]) [1, 5]
4.9 <28> (enzyme from males) [25]
5 <7, 23> [9, 24]
5-6 <22> [15]
5.3 <28> (enzyme from winged females) [25]
5.5 <4> [12]
5.6 <2, 26> [4, 19]
5.8 <28> (enzyme from soldiers) [25]
6.2 <28> (enzyme from workers) [25]
6.5 <3, 6> (carboxymethylcellulose hydrolysis, <3> [3]) [3, 7]
8 <3> (chitosanase activity) [3]

pH-Range
4-6 <4> [12]
4-6.5 <1> (rapid drop in activity below pH 4.0 and above pH 6.5) [1]
4-8 <6> (about 20% of maximal activity at pH 4 and 8) [7]
4.5-6.5 <22> (more than 80% of activity maximum at pH 4.5 and 6.5) [15]

Temperature optimum (°C)

40 <2, 26> [4, 19]
45 <1> [1]
50 <2, 6, 7, 21, 22, 25> (at pH 6.5, <6> [7]; enzyme B, <7> [9]; enzyme A, <21> [14]) [2, 7, 9, 14, 15, 18]
55 <7, 20> (enzyme A, <7> [9]; pH 5.3, <20> [13]) [9, 13]
60 <4> [5]
65 <4> [12]

4 Enzyme Structure

Molecular weight

10000 <3> (Streptomyces griseus, gel filtration) [3]
25000 <2> (Bacillus sp. PI-7S, gel filtration) [10]
27000 <6, 20> (Bacillus circulans, gel filtration [7]; Amycolatopsis sp. CsO-2, gel filtration [13]) [7, 13]
29000 <4> (Streptomyces sp. No. 6, gel filtration) [5]
29020 <6> (Bacillus circulans, deduced from nucleotide sequence) [8]
30000 <1, 2> (Penicillium islandicum, gel filtration [1]; Bacillus sp. (No. 7-M), gel filtration [2]) [1, 2]
31000 <2, 28> (Bacillus sp. R-4, gel filtration [4]; Macrotermes estherae, enzyme from males and winged females [25]) [4, 25]
35000 <28> (Macrotermes estherae, enzyme from soldiers) [25]
40000 <28> (Macrotermes estherae, enzyme from workers) [25]
63000 <7> (Mucor rouxii, enzyme B, gel filtration) [9]
79000 <7> (Mucor rouxii, enzyme A, gel filtration) [9]

Subunits

? <3, 8, 10-13, 15, 16, 19, 21, 23, 25-27> (x * 35000, Streptomyces griseus, SDS-PAGE [3]; x * 18000 + x * 21000, Onoclea sensibilis, SDS-PAGE without reducing agent [11]; x * 19000 + x * 22000, Thuja occidentalis, SDS-PAGE without reducing agent [11]; x * 10000 + x * 16000 + x * 18000 + x * 20000, Cucumis sativus, SDS-PAGE without reducing agent [11]; x * 23000 + x * 19000, Pisum sativum, SDS-PAGE without reducing agent [11]; x * 23000 + x * 22000 + x * 19000, Solanum melongena, SDS-PAGE without reducing agent [11]; x * 18000 + x * 19000, Triticum aestivum, SDS-PAGE without reducing agent [11]; x * 18000 + x * 22000, Setaria italica, SDS-PAGE without reducing agent [11]; x * 19000, Campanula rotundifolia, SDS-PAGE without reducing agent [11]; x * 43000, Bacillus megaterium P1, enzyme A, SDS-PAGE [14]; x * 39500, Bacillus megaterium P1, enzyme B, SDS-PAGE [14]; x * 22000, Bacillus megaterium P1, enzyme C, SDS-PAGE [14]; x * 50000, Enterobacter sp., SDS-PAGE [18]; Fusarium solani f. sp. phaseoli, SDS-PAGE [19]; x * 35000, Pseudomonas sp. H-14, SDS-PAGE [20]; x * 30500, Citrus sinensis, SDS-PAGE [23]) [3, 11, 14, 18-20, 23]
monomer <1, 2, 4, 6, 20> (Penicillium islandicum [1]; 1 * 41000, Bacillus sp. (No. 7-M), SDS-PAGE [2]; 1 * 26000, Streptomyces sp., gel filtration [5]; 1 *

27000, Amycolatopsis sp. CsO-2, SDS-PAGE [13]; 1 * 32000, Bacillus circulans, SDS-PAGE [7]; 1 * 31000, Bacillus circulans, SDS-PAGE [8]; 1 * 43000, Bacillus sp. PI$^-$7S, SDS-PAGE [10]) [1, 2, 5, 7, 8, 10, 13]

Posttranslational modification

glycoprotein <6> [7]
Additional information <1, 4> (carbohydrate content is less than 0.5%, <1> [1]; no glucosamine or other carbohydrate, <4> [5]) [1, 5]

5 Isolation/Preparation/Mutation/Application

Source/tissue

callus <3, 23> [16, 23]
culture filtrate <22, 26> [15, 19]
culture medium <2, 6> [7, 10]
fruit <11> (best source of Cucumis sativus enzyme) [11]
midgut secretion <28> [25]
seed <8-10> (best source of enzyme from Triticum aestivum, Hordeum vulgare, Pisum sativum) [11]

Localization

extracellular <1, 4-6, 20> [1, 5-7, 13]

Purification

<1> [1]
<2> (No. 7-M [2]; R-4 [4]; PI-7S [10]) [2, 4, 10]
<3> [3]
<4> [5, 12]
<5> [6]
<6> [7, 8]
<7> (2 forms: A and B) [9]
<20> [13]
<21> (3 enzymes: A, B and C) [14]
<22> [15]
<23> (2 isoforms: A1-CF1 and A1-CF2) [23]
<25> [18]
<26> (f. sp. phaseoli) [19]
<27> [20]

Crystallization

<4> [22]

Cloning

<6> [8]

6 Stability

pH-Stability

2.5-5.5 <7> (22°C, 4 h, enzyme B stable) [9]
3.3-7.4 <2> (37°C, 60 min, stable) [10]
3.5-7 <7> (2°C, 4 h, enzyme A stable) [9]
3.5-8.5 <1> (quite stable in this range, optimum stability between pH 3.5 and 5.5) [1]
4-7 <27> (stable) [20]
4-9 <6> (stable in this range) [7]
4.5-6 <4, 20> (37°C, 4 h, stable, <4> [12]; stable, <20> [13]) [12, 13]
4.5-6.5 <21> (up to 45°C, stable) [14]
4.5-7.5 <2> (40°C, 3 h, stable, highest stability at pH 5.6) [4]
5-6.5 <22> (stable) [15]
6-8 <3> (37°C, stable) [3]

Temperature stability

0-40 <6> (stable in this range) [7]
0-50 <20> (stable) [13]
20-50 <25> (stable) [18]
30 <7> (pH 5.0, 30 min, stable) [9]
37 <1, 3, 4> (denaturation in 1-2 h, stabilized by 1.0 mM Ca^{2+}, <1> [1]; pH 6.0-8.0, stable, unstable at elevated temperatures, <3> [3]; stable, <4> [12]) [1, 3, 12]
40 <4, 7, 22> (pH 5.0, 30 min, stable, <7> [9]; rapid decrease of activity above, <4> [12]; 15 min, stable up to, <22> [15]) [9, 12, 15]
45 <2> (pH 6.8, 15 min, stable below) [10]
50 <2, 20> (stable below, pH 6.0, 15 min, <2> [2]; stable, <2> [4]; 60 min, stable, <20> [13]) [2, 4, 13]
60 <2, 7, 20, 23> (pH 6.0, 15 min, 40% loss of activity, <2> [2]; 30 min, pH 5.0, complete inactivation, <7> [9]; 60 min, 40% loss of activity, <20> [13]; stable up to, <23> [24]) [2, 9, 13, 24]
70 <2, 20> (pH 6.0, 15 min, complete loss of activity, <2> [2]; 15 min, 30% loss of activity, <2> [4]; 60 min, 70% loss of activity, <20> [13]) [2, 4, 13]
80 <20, 22> (60 min, 80% loss of activity, <20> [13]; 15 min, 55% loss of activity, <22> [15]) [13, 15]
100 <20> (10 min, 50% loss of activity) [13]

Storage stability

<21>, -20°C, 50% glycerol, stable for at least 6 months [14]

References

[1] Fenton, D.M.; Everleigh, D.E.: Purification and mode of action of a chitosanase from Penicillium islandicum. J. Gen. Microbiol., **126**, 151-165 (1981)

[2] Uchida, Y.; Ohtakara, A.: Chitosanase from Bacillus species. Methods Enzymol., **161**, 501-505 (1988)
[3] Ohtakara, A.: Chitosanase from Streptomyces griseus. Methods Enzymol., **161**, 505-510 (1988)
[4] Tominaga, Y.; Tsujisaka, Y.: Purification and some enzymatic properties of the chitosanase from Bacillus R-4 which lyses Rhizopus cell walls. Biochim. Biophys. Acta, **410**, 145-155 (1975)
[5] Price, J.S.; Storck, R.: Production, purification, and characterization of an extracellular chitosanase from Streptomyces. J. Bacteriol., **124**, 1574-1585 (1975)
[6] Somashekar, D.; Joseph, R.: Partial purification and properties of a novel chitosanase secreted by Rhodotorula gracilis. Lett. Appl. Microbiol., **14**, 1-4 (1992)
[7] Yabuki, M.; Uchiyama, A.; Suzuki, K.; Ando, A.; Fujii, T.: Purification and properties of chitosanase from Bacillus circulans MH-K1. J. Gen. Appl. Microbiol., **34**, 255-270 (1988)
[8] Ando, A.; Noguchi, K.; Yanagi, M.; Shinoyama, H.; Kagawa, Y.; Hirata, H.; Yabuchi, M.; Fujii, T.: Primary structure of chitosanase produced by Bacillus circulans MH-K1. J. Gen. Appl. Microbiol., **38**, 135-144 (1992)
[9] Alfonso, C.; Martinez, M.J.; Reyes, F.: Purification and properties of two endochitosanases from Mucor rouxii implicated in its cell wall degradation. FEMS Microbiol. Lett., **95**, 187-194 (1992)
[10] Seino, H.; Tsukuda, K.; Shimasue, Y.: Properties and action pattern of a chitosanase from Bacillus sp. PI-7S. Agric. Biol. Chem., **55**, 2421-2423 (1991)
[11] Ouakfaoui, S.E.; Asselin, A.: Multiple forms of chitosanase activities. Phytochemistry, **31**, 1513-1518 (1992)
[12] Boucher, I.; Dupuy, A.; Vidal, P.; Neugebauer, W.A.; Brzezinski, R.: Purification and characterization of a chitosanase from Streptomyces N174. Appl. Microbiol. Biotechnol., **38**, 188-193 (1992)
[13] Okajima, S.; Ando, A.; Shinoyama, H.; Fujii, T.: Purification and characterization of an extracellular chitosanase produced by Amycolatopsis sp. CsO-2. J. Ferment. Bioeng., **77**, 617-620 (1994)
[14] Pelletier, A.; Sygusch, J.: Purification and characterization of three chitosanase activities from Bacillus megaterium P1. Appl. Environ. Microbiol., **56**, 844-848 (1990)
[15] Sakai, K.; Katsumi, R.; Isobe, A.; Nanjo, F.: Purification and hydrolytic action of a chitosanase from Nocardia orientalis. Biochim. Biophys. Acta, **1079**, 65-72 (1991)
[16] Osswald, W.F.; McDonald, R.E.; Niedz, R.P.; Shapiro, J.P.; Mayer, R.T.: Quantitative fluorometric analysis of plant and microbial chitosanases. Anal. Biochem., **204**, 40-46 (1992)
[17] Fukamizo, T.; Ohkawa, T.; Ikeda, Y.; Goto, S.: Specificity of chitosanase from Bacillus pumilus. Biochim. Biophys. Acta, **1205**, 183-188 (1994)
[18] Yamasaki, Y.; Hayashi, I.; Ohta, Y.; Nakagawa, T.; Kawamukai, M.; Matsuda, H.: Purification and mode of action of chitosanolytic enzyme from Enterobacter sp. G-1. Biosci. Biotechnol. Biochem., **57**, 444-449 (1993)

[19] Shimosaka, M.; Nogawa, M.; Ohno, Y.; Ozaki, M.: Chitosanase from the plant pathogenic fungus, Fusarium solani f. sp. phaseoli. Purification and some properties. Biosci. Biotechnol. Biochem., **57**, 231-235 (1993)
[20] Yoshihara, K.; Hosokawa, J.; Kubo, T.; Nishiyama, M.; Koba, Y.: Purification and properties of a chitosanase from Pseudomonas sp. H-14. Biosci. Biotechnol. Biochem., **56**, 972-973 (1992)
[21] Izume, M.; Nagae, S.; Kawagishi, H.; Mitsutomi, M.; Ohtakara, A.: Action pattern of Bacillus sp. no. 7-M chitosanase on partially N-acetylated chitosan. Biosci. Biotechnol. Biochem., **56**, 448-453 (1992)
[22] Marcotte, E.; Hart, P.J.; Boucher, I.; Brzezinski, R.; Robertus, J.D.: Crystallization of a chitosanase from Streptomyces N174. J. Mol. Biol., **232**, 995-996 (1993)
[23] Osswald, W.F.; Shapiro, J.P.; McDonald, R.E.; Niedz, R.P.; Mayer, R.T.: Some citrus chitinases also possess chitosanase activities. Experientia, **49**, 888-892 (1993)
[24] Osswald, W.F.; Shapiro, J.P.; Doostdiar, H.; McDonald, R.E.; Niedz, R.P.; Nairn, C.J.; Hearn, C.J.; Mayer, R.T.: Identification and characterization of acidic hydrolases with chitinase and chitosanase activities from sweet orange callus tissue. Plant Cell Physiol., **35**, 811-820 (1994)
[25] Aruchami, M.; Gowri, N.; Rajulu, G.S.: Detection and characterization of chitosanase from termites. Chitin Chitosan, Proceeding Int. Conf., 2nd (Hirano, S., Tokura, S., eds.) Jpn. Soc. Chitin Chitosan, Tottori, Japan, 131-141 (1982)

Glucan 1,4-α-maltohydrolase 3.2.1.133

1 Nomenclature

EC number
3.2.1.133

Systematic name
1,4-α-D-glucan α-maltohydrolase

Recommended name
glucan 1,4-α-maltohydrolase

Synonyms
maltohydrolase, glucan 1,4-α-
Additional information (cf. EC 3.2.1.2)

CAS registry number
160611-47-2

2 Source Organism

<1> *Bacillus stearothermophilus* (expression in Bacillus subtilis) [1, 2]

3 Reaction and Specificity

Catalyzed reaction
hydrolysis of 1,4-α-D-glucosidic linkages in polysaccharides so as to remove successive α-maltose units from the nonreducing ends of the chains

Reaction type
hydrolysis of O-glycosyl bond

Substrates and products
S α-cyclodextrin + H_2O <1> [1]
P maltose + glucose <1> (ratio 10:1) [1]
S β-cyclodextrin + H_2O <1> [1]
P maltose + glucose <1> (ratio 3:1) [1]
S starch + H_2O <1> [1]
P α-maltose <1> [1]

Inhibitors
$CoCl_2$ <1> [1]
$CuCl_2$ <1> [1]

$HgCl_2$ <1> [1]
$MnCl_2$ <1> [1]
$ZnCl_2$ <1> [1]
Additional information <1> (not: PCMB, Schardinger dextrins) [1]

pH-Optimum
5.3 <1> [1]

pH-Range
3.5-7.3 <1> (3.5: about 65% of activity maximum, 7.3: about 35% of activity maximum) [1]

Temperature optimum (°C)
60 <1> [1]

Temperature range (°C)
30-80 <1> (30°C: about 20% of activity maximum, 80°C: about 60% of activity maximum) [1]

4 Enzyme Structure

Subunits
? <1> (x * 70000, Bacillus stearothermophilus, SDS-PAGE) [1]

5 Isolation/Preparation/Mutation/Application

Cloning
<1> [1, 2]

6 Stability

Temperature stability
60 <1> (pH 5.5, 60 min, stable) [1]
70 <1> (pH 5.5, 60 min, about 25% loss of activity) [1]

References

[1] Outtrup, H.; Norman, B.E.: Properties and application of a thermostable maltogenic amylase produced by a strain of Bacillus modified by recombinant-DNA techniques. Starch Staerke, **36**, 405-411 (1984)
[2] Diderichsen, B.; Christiansen, L.: Cloning of a maltogenic α-amylase from Bacillus stearothermophilus. FEMS Microbiol. Lett., **56**, 53-60 (1988)

Difructose-anhydride synthase 3.2.1.134

1 Nomenclature

EC number
3.2.1.134

Systematic name
bis-D-fructose-2',1:2,1'-dianhydride fructohydrolase

Recommended name
difructose-anhydride synthase

Synonyms
hydrolase, inulobiose

CAS registry number
121479-55-8

2 Source Organism

<1> *Aspergillus fumigatus* [1, 2]

3 Reaction and Specificity

Catalyzed reaction
bis-D-fructose 2',1:2,1'-dianhydride + H_2O = inulobiose

Reaction type
hydrolysis of O-glycosyl bond

Natural substrates and products
S bis-D-fructose 2',1:2,1' dianhydride + H_2O <1> (might act as a regulator for a piecemeal use of inulin as a carbon source) [1]

Substrates and products
S inulobiose <1> (r, <1> [1]) [1, 2]
P bis-D-fructose 2',1:2,1' dianhydride + H_2O <1> [1, 2]

Inhibitors
Co^{2+} <1> [1]
Cu^{2+} <1> [1]
DTT <1> (slight) [1]
L-cysteine <1> [1]
Zn^{2+} <1> [1]

K_m-Value (mM)

241 <1> (bis-D-fructose 2',1:2,1' dianhydride) [1]
377 <1> (inulobiose) [1]

pH-Optimum

6 <1> [1]

Temperature optimum (°C)

48 <1> [1]

5 Isolation/Preparation/Mutation/Application

Source/tissue

mycelium <1> [1, 2]

Purification

<1> [2]

6 Stability

pH-Stability

6-8.6 <1> (37°C) [1]

Temperature stability

50 <1> (pH 6.0, stable up to) [1]

Storage stability

<1>, under toluene in the refrigerator at pH 6.0, stable for several months [1]

References

[1] Matsuyama, T.; Tanaka, K.: On the enzyme of Aspergillus fumigatus producing difructose anhydride I from inulobiose. Agric. Biol. Chem., **53**, 831-832 (1989)
[2] Matsuyama, T.; Tanaka, K.; Mashiko, M.; Kanamoto, M.: Enzymic formation of di-D-fructose 1,2; 2,1 dianhydride from inulobiose by Aspergillus fumigatus. J. Biochem., **92**, 1325-1328 (1982)

Neopullulanase 3.2.1.135

1 Nomenclature

EC number
3.2.1.135

Systematic name
pullulan 4-D-glucanohydrolase (panose-forming)

Recommended name
neopullulanase

Synonyms
pullulanase II
pullulanase, neo-
Additional information (cf. EC 3.2.1.41 and EC 3.2.1.57)

CAS registry number
119632-58-5

2 Source Organism

<1> *Bacillus stearothermophilus* (expression in Bacillus subtilis [2]) [1, 2, 4]
<2> *Bacillus thetaiotaomicron* (95-1) [3]
<3> *Bacillus subtilis* (NA-1) [5, 6]

3 Reaction and Specificity

Catalyzed reaction
hydrolysis of pullulan to panose (6-α-D-glucosylmaltose)

Reaction type
hydrolysis of O-glycosyl bond

Substrates and products
S Additional information <1-3> (enzyme hydrolyzes not only α-1,4-glucosidic linkages but also specific α-1,6-glucosidic linkages of several branched oligosaccharides, <1> [1]; only hydrolyzes a small amount of starch, <1> [2]; enzyme also cleaves α-1,4-bonds in amylose and in oligosaccharides of maltotriose through maltoheptaose in chain length, <2> [3]; enzyme is able to cleave β-cyclodextrin, <2> [3]; analysis of the active centre, one active center participates in the dual activity toward α-1,4-,

and α-1,6-glucosidic linkages, <1> [4]; 4 reactions are catalyzed by the enzyme: 1. hydrolysis of α-1,4-glucosidic linkage, 2. hydrolysis of α-1,6-glucosidic linkage, 3. transglycosylation to form α-1,4-glucosidic linkage, 4. transglycosylation to form α-1,6-glucosidic linkage, <3> [5]) [1-5]

P ?

S pullulan + H_2O <1-3> [1-6]

P panose <1-3> (+ maltose + glucose (in a molar ratio of 3:1:1, <1> [1]; small amount, <1> [2]), <1> [1, 2]; in the first step, the enzyme hydrolyzes only α-1,4-glucosidic linkages on the nonreducing side of α-1,6-linkages of pullulan and produces panose and several intermediate products composed of some panose units. In the second step, taking 62-O-α-(63-O-α-glucosyl-maltotriosyl)-maltose as an example of one of the intermediate products, the enzyme hydrolyzes either α-1,4-, or α-1,6-linkages and produces panose or 63-O-α-glucosyl-maltotriose plus maltose, respectively. In the third step, the α-1,4-linkage of the 63-O-α-glucosyl-maltotriose is hydrolyzed by the enzyme, and glucose and another panose are produced, <1> [1]) [1-6]

Inhibitors

6-azido-6-deoxypullulan <3> [6]
6-chloro-6-deoxypullulan <3> [6]

Specific activity (U/mg)

Additional information <1, 2> [2, 3]

pH-Optimum

6 <1> [2]

pH-Range

5-7 <1> (5: about 25% of activity maximum, 7: about 40% of activity maximum) [2]

Temperature optimum (°C)

40 <3> (assay at) [5]
60-65 <1> [2]

Temperature range (°C)

40-80 <1> (40°C: about 50% of activity maximum, 80°C: about 25% of activity maximum) [2]

4 Enzyme Structure

Molecular weight

62000 <1> (Bacillus stearothermophilus expressed in Bacillus subtilis) [2]
70000 <2> (Bacillus thetaiotaomicron 95-1, gel filtration) [3]

Subunits

monomer <2> (1 * 70000, Bacillus thetaiotaomicron 95-1, SDS-PAGE) [3]

5 Isolation/Preparation/Mutation/Application

Purification

<1> [2]
<2> [3]

Cloning

<1> [2]

6 Stability

pH-Stability

6-9 <1> (60 min, stable) [2]

Temperature stability

60 <1> (60 min, about 10% loss of activity) [2]
70 <1> (60 min, about 80% loss of activity) [2]
Additional information <1> (EDTA stabilizes against thermal inactivation) [2]

Storage stability

<2>, -20°C, 20% glycerol, stable [3]

References

[1] Imanaka, T.; Kuriki, T.: Pattern of action of Bacillus stearothermophilus neopullulanase on pullulan. J. Bacteriol., **171**, 369-374 (1989)
[2] Kuriki, T.; Okada, S.; Imanaka, T.: New type of pullulanase from Bacillus stearothermophilus and molecular cloning and expression of the gene in Bacillus subtilis. J. Bacteriol., **170**, 1554-1559 (1988)
[3] Smith, K.A.; Salyers, A.A.: Characterization of a neopullulanase and an α-glucosidase from Bacteroides thetaiotaomicron 95-1. J. Bacteriol., **173**, 2962-2968 (1991)
[4] Kuriki, T.; Takata, H.; Okada, S.; Imanaka, T.: Analysis of the active center of Bacillus stearothermophilus neopullulanase. J. Bacteriol., **173**, 6147-6152 (1991)
[5] Takata, H.; Kuriki, T.; Okada, S.; Takesada, Y.; Iizuka, M.; Minamiura, N.; Imanaka, T.: Action of neopullulanase. Neopullulanase catalyzes both hydrolysis and transglycosylation at α-(1–4)- and α-(1–6)-glucosidic linkages. J. Biol. Chem., **267**, 18447-18452 (1992)
[6] Ball, D.H.; Wiley, B.J.; Reese, E.T.: Effect of substitution at C-6 on the susceptibility of pullulan to pullulanases. Enzymatic degradation of modified pullulans. Can. J. Microbiol., **38**, 324-327 (1992)

Glucuronoarabinoxylan endo-1,4-β-xylanase 3.2.1.136

1 Nomenclature

EC number
3.2.1.136

Systematic name
glucuronoarabinoxylan 1,4-β-D-xylanohydrolase

Recommended name
glucuronoarabinoxylan endo-1,4-β-xylanase

Synonyms
endoarabinoxylanase
feraxan endoxylanase
feraxanase
glucuronoxylan xylanohydrolase
glucuronoxylan xylohydrolase
glucuronoxylanase
xylanase, glucuronoarabinoxylan endo-1,4-β-

CAS registry number
123609-77-8

2 Source Organism

<1> *Bacillus subtilis* [1-3]

3 Reaction and Specificity

Catalyzed reaction
endohydrolysis of 1,4-β-D-xylosyl links in some glucuronoarabinoxylans

Reaction type
hydrolysis of O-glycosyl bond

Natural substrates and products
S glucuronoarabinoxylans + H_2O <1> (involved in heteroxylan metabolism) [1, 2]

Substrates and products
S Vigna angularis xylan + H_2O <1> (i.e. heteroxylan with glucuronic acid side chain) [2]

P ?

S feraxan + H_2O <1> (i.e. feruloylated glucuronoarabinoxylans, highly specific endo-type hydrolase, <1> [1]; hydrolyzes feraxan both in intact maize coleophile cell wall and maize arabinoxylans extracted from cell walls with alkaline, <1> [1]; selectively dissociates feraxan from maize cell wall, <1> [3]) [1, 3]

P ?

S maize xylan + H_2O <1> (i.e. heteroxylan with glucuronic acid side chain, enzyme recognizes glucuronic acid side chains along the xylan main chain and mediates hydrolysis of β-1,4-xylosyl linkages of adjacent unsubstituted xylosyl residues, endo-type hydrolase, possible recognition sites) [2]

P ?

S Additional information <1> (no substrates are xylan derivatives whose glucuronic acid residues have been converted to glucose, β-1,4-xylan, <1> [2]; larch arabino-1,4-β-D-xylan, Rhodymenia 1,3-1,4-β-D-xylan, <1> [1, 3]; cellulose, carboxymethyl cellulose, lichenan, oat or barley β-glucan, laminarin, dextran, starch, citrus or maize pectin, polyuronides, arabinan, galactan, arabinogalactan, yeast mannan, galactomannan, <1> [1]) [1-3]

P Additional information <1> (products from hydrolysis of extracted maize arabinoxylan: major linkages in smaller fractions are terminal arabinofuranosyl, 5-linked arabinofuranosyl, 4-linked xylopyranoslyl, 3,4-linked xylopyranosyl and terminal glucuronosyl residues, <1> [1]; structural analysis of products of maize xylan: sugar composition and glycosidic linkage composition of xylan fragments, <1> [2]; no products are xylohomooligomers, <1> [2] or mono-, di-, tri-, or tetramers of xylose and/or arabinose, <1> [1]) [1, 2]

K_m-Value (mM)

Additional information <1>(K_m-values: 5 mg/ml (Vigna angularis xylan, derivative with reduced hemiacetal at reducing end), 6.9 mg/ml (maize xylan derivative with reduced hemiacetal at reducing end)) [2]

pH-Optimum

6.5-7 <1> [1, 3]

Temperature optimum (°C)

35 <1> (assay at) [2]
37 <1> (assay at) [1]

4 Enzyme Structure

Molecular weight

45000 <1> (Bacillus subtilis) [3]

Subunits

? <1> (x * 45000, Bacillus subtilis, SDS-PAGE) [1]

5 Isolation/Preparation/Mutation/Application

Source/tissue
industrial amylase preparation <1> (Novo Ban L-120, <1> [1, 2]) [1-3]

Purification
<1> (to near homogeneity) [3]

References

[1] Nishitani, K.; Nevins, D.J.: Enzymic analysis of feruloylated arabinoxylans (Feraxan) derived from Zea mays cell walls. I. Purification of novel enzymes capable of dissociating Feraxan fragments from Zea mays coleoptile cell wall. Plant Physiol., **87**, 883-890 (1988)
[2] Nishitani, K.; Nevins, D.J.: Glucuronoxylan xylanohydrolase. A unique xylanase with the requirement for appendant glucuronosyl units. J. Biol. Chem., **266**, 6539-6543 (1991)
[3] Nevins, D.J.; Nishitani, K.: Purification and characterization of Bacillus feraxanase, a highly specific enzyme for hydrolysis of complex polysaccharides. PCT Int. Appl. WO9001059 A1, (1990)

Mannan exo-1,2-1,6-α-mannosidase 3.2.1.137

1 Nomenclature

EC number
3.2.1.137

Systematic name
1,2-1,6-α-D-mannan D-mannohydrolase

Recommended name
mannan exo-1,2-1,6-α-mannosidase

Synonyms
mannosidase, exo-1,2-1,6-α

CAS registry number
123175-72-4

2 Source Organism

<1> *Cellulomonas sp.* (production of the enzyme strongly induced by yeast extract, 2 different MW forms) [1]

3 Reaction and Specificity

Catalyzed reaction
hydrolysis of 1,2-α-D- and 1,6-α-D-linkages in yeast mannan, releasing D-mannose

Reaction type
hydrolysis of O-glycosyl bond

Natural substrates and products
S mannan + H_2O <1> (hydrolysis of 1,2-α-D-, and 1,6-α-D-linkages in yeast mannan, releasing D-mannose) [1]

Substrates and products
S mannan <1> (hydrolysis of 1,2-α-D-, and 1,6-α-D-linkages in yeast mannan, releasing D-mannose, hydrolysis rate of α-1,3-linkage is very low) [1]
P D-mannose <1> [1]
S Additional information <1> (only slight activity towards 4-nitrophenyl-α-D-mannoside and no hydrolysis of O-α-D-mannopyranosyl-1,2-D-mannitol) [1]
P ?

Inhibitors
Cu^{2+} <1> [1]
EDTA <1> [1]
Hg^{2+} <1> [1]
PCMB <1> [1]
Zn^{2+} <1> [1]

Specific activity (U/mg)
12.8 <1> [1]

K_m-Value (mM)
2.5 <1> (hydrolysis of α-1,2-linkage in yeast mannan) [1]
3.2 <1> (hydrolysis of α-1,6-linkage in yeast mannan) [1]
200 <1> (hydrolysis of α-1,3-linkage in yeast mannan) [1]

pH-Optimum
6.5-8 <1> [1]

pH-Range
5-9 <1> (5: about 20% of activity maximum, 9: about 15% of activity maximum) [1]

Temperature optimum (°C)
37 <1> (assay at) [1]
50 <1> [1]

4 Enzyme Structure

Molecular weight
450000 <1> (Cellulomonas sp., gel filtration) [1]

5 Isolation/Preparation/Mutation/Application

Purification
<1> (high molecular weight form) [1]

6 Stability

pH-Stability
5-9.5 <1> (4°C, 48 h, stable) [1]

Temperature stability
40 <1> (10 min, stable) [1]
50 <1> (10 min, 50% loss of activity) [1]

Storage stability
<1>, -20°C [1]

References

[1] Takegawa, K.; Miki, S.; Jikibara, T.; Iwahara, S.: Purification and characterization of exo-a-D-mannosidase from a Cellulomonas sp.. Biochim. Biophys. Acta, **991**, 431-437 (1989)

Anhydrosialidase 3.2.1.138

1 Nomenclature

EC number
3.2.1.138

Systematic name
sialglycoconjugate N-acylneuraminylhydrolase (2,7-cyclizing)

Recommended name
anhydrosialidase

Synonyms
anhydroneuraminidase
neuraminidase, anhydro-
sialidase L
Additional information (cf. EC 3.2.1.18 and EC 3.2.1.129)

CAS registry number
157857-11-9

2 Source Organism

<1> *Macrobdella decora* [1, 2]

3 Reaction and Specificity

Catalyzed reaction
hydrolysis of α-sialosyl linkages in N-acetylneuraminic acid glycosides, releasing 2,7-anhydro-α-N-acetylneuraminic acid

Reaction type
hydrolysis of O-glycosyl bond

Substrates and products
S 4-methylumbelliferyl-α-N-acetylneuraminic acid glycoproteins or ganglioside + H_2O <1> (only sialoglycoconjugates, not free α-N-acetylneuraminic acid, <1> [1]; strict specificity towards the hydrolysis of NeuAcα(2-3)Gal-linkages, <1> [2]) [1, 2]
P 2,7-anhydro-α-N-acetylneuraminic acid <1> [1, 2]

Inhibitors
Additional information <1> (not: 2-deoxy-2,3-dehydroacetylneuraminic acid, EDTA, Ca^{2+}, Mg^{2+}, Mn^{2+}) [2]

Specific activity (U/mg)
Additional information <1> [2]

K_m-Value (mM)
0.68 <1> (4-methylumbelliferyl-α-N-acetylneuraminic acid) [2]

pH-Optimum
5.5-7 <1> [2]

Temperature optimum (°C)
37 <1> (assay at) [2]

4 Enzyme Structure

Subunits
? <1> (x * 84000, Macrobdella decora, SDS-PAGE) [2]

5 Isolation/Preparation/Mutation/Application

Purification
<1> (partial [1]) [1, 2]

6 Stability

General stability information
<1>, 50% loss of activity after one freeze and thaw cycle, in presence of 10% glycerol stable to several freeze and thaw cycles [2]

Storage stability
<1>, -20°C, 10% glycerol, stable for more than 4 months [2]

References

[1] Li, Y.-T.; Nakagawa, H.; Ross, S.A.; Hansson, G.C.; Li, S.-C.: A novel sialidase which releases 2,7-anhydro-α-N-acetylneuraminic acid from sialoglycoconjugates. J. Biol. Chem., **265**, 21629-21633 (1990)
[2] Chou, M.-Y.; Li, S.-C.; Kiso, M.; Hasegawa, A.; Li, Y.-T.: Purification and characterization of sialidase L, a NeuAc α 2→3Gal-specific sialidase. J. Biol. Chem., **269**, 18821-18826 (1994)

α-Glucosiduronase 3.2.1.139

1 Nomenclature

EC number
3.2.1.139

Systematic name
α-D-glucosiduronate glucuronohydrolase

Recommended name
α-glucosiduronase

Synonyms
(4-O-methyl)-α-glucuronidase
4-O-methylglucuronidase
aryl α-glucuronidase
PNP-GAase
α-(4-O-methyl)-D-glucuronidase
α-D-glucuronidase
α-glucuronidase
glucuronidase, α-
p-nitrophenyl α-D-glucuronide-hydrolyzing enzyme

CAS registry number
37259-81-7

2 Source Organism

<-2> no activity in *Neocallimastix patriciarum* [5]
<-1> *no activity on Neocallimastix frontalis* [5]
<1> *Aspergillus niger* (strain 5-16 [1]) [1, 6]
<2> *Agaricus bisporus* [2]
<3> *Trichoderma reesei* (strain RUT C-30 [10]) [2, 10]
<4> *Aspergillus awamori* [2, 6]
<5> *Pleurotus ostreatus* [2]
<6> *Streptomyces olivochromogenes* [2]
<7> *Dactylium dendroides* [2]
<8> *Tyromyces palustris* [2]
<9> *Polyporus versicolor* [2]
<10> *Laetiporus sulphureus* [2]
<11> *Thermoascus aurantiacus* [2, 8]
<12> *Bacteroides sp.* (strain J-37 [3]) [3]

<13> *Clostridium stercorarium* [4]
<14> *Thermoanaerobacterium saccharolyticum* [4]
<15> *Piromonas communis* [5]
<16> *Phanerochaete chrysosporium* [6]
<17> *Thermotoga maritima* (strain MSB8 [11]) [11]
<18> *Aspergillus phoenicis* [6]
<19> *Helix pomatia* [7, 12]
<20> *Thermoanaerobacterium sp.* (strain JW/SL-YS485 [9]) [9]

3 Reaction and Specificity

Catalyzed reaction

an α-D-glucuronoside + H_2O = an alcohol + D-glucuronate

Reaction type

hydrolysis of O-glycosyl bond

Natural substrates and products

S Additional information <2, 11, 20> (<2>, enzyme is involved in hydrolysis of wood xylans [2]; <11>, the three enzymes, xylanase, α-D-glucuronidase, and β-glucosidase, together are responsible for complete degradation of larchwood xylan to xylose and 4-O-methyl-α-D-glucuronic acid [8]; <20>, important enzyme for the utilization of substituted xylans [9]; <20>, enzyme is induced by growth on beech wood or birch wood [9]) [2, 8, 9]

Substrates and products

S 18β-glycyrrhetinic acid β-D-glucuronyl α-D-glucuronic acid + H_2O <12> [3]
P ?
S 2-O-α-D-glucopyranosyluronic acid-D-xylose + H_2O <19> [12]
P α-D-glucopyranosyluronic acid + D-xylose
S 2-O-α-D-glucuronosyl-D-xylose + H_2O <19> [7]
P D-glucuronate + D-xylose
S 3-O-α-D-glucuronosyl-D-xylose + H_2O <19> [7]
P D-glucuronate + D-xylose
S 4-O-α-D-glucuronosyl-D-xylose + H_2O <19> [7]
P D-glucuronate + D-xylose
S 4-O-methyl-glucoronosyl-xyloheptaose + H_2O <11> [8]
P 4-O-methyl-glucoronate + xyloheptaose
S 4-O-methyl-glucuronosyl-xylan + H_2O <11, 20> (<11>, at 52% of the activity with 4-O-methyl-glucuronosyl-xylobiose [2]) [2, 9]
P 4-O-methyl-glucoronate + xylan
S 4-O-methyl-glucuronosyl-xylitol + H_2O <2, 5, 8-11> [2]
P 4-O-methyl-glucuronate + xylitol
S 4-O-methyl-glucuronosyl-xylobiose + H_2O <2, 3, 11, 15, 16, 20> [2, 5, 6, 8-10]

P 4-O-methyl-glucuronate + xylobiose
S 4-O-methyl-glucuronosyl-xylohexaose + H_2O <2, 11> (<2>, at 53% of the activity with 4-O-methyl-glucuronosyl-xylobiose [2]) [2, 8]
P 4-O-methyl-glucuronate + xylohexaose
S 4-O-methyl-glucuronosyl-xylopentaose + H_2O <2, 11> (<2>, at 50% of the activity with 4-O-methyl-glucuronosyl-xylobiose [2]) [2, 8]
P 4-O-methyl-glucuronate + xylopentaose
S 4-O-methyl-glucuronosyl-xylose + H_2O <3, 11, 15, 19> (<11>, 96% of the activity with 4-O-methyl-glucuronosyl-xylobiose [2]; <2, 16>, no activity [2, 6]) [1, 2, 5, 8, 10, 12]
P 4-O-methyl-glucuronate + xylose
S 4-O-methyl-glucuronosyl-xylotetraose + H_2O <2, 3, 11, 20> (<2, 11>, at 96% of the activity with 4-O-methyl-glucuronosyl-xylobiose [2]) [2, 8-10]
P 4-O-methyl-glucuronate + xylotetraose
S 4-O-methyl-glucuronosyl-xylotriose + H_2O <1-3, 11, 13-16, 20> (<2>, at 99% of the activity with 4-O-methyl-glucuronosyl-xylobiose [2]; <11>, at 96% of the activity 4-O-methyl-glucuronosyl-xylobiose [2]) [1, 2, 4-6, 8-10]
P 4-O-methyl-glucuronate + xylotriose
S O-α-D-glucopyranosyluronic acid-α-D-glucopyranosiduronic acid + H_2O <19> [12]
P α-D-glucopyranosyluronic acid + α-D-glucopyranosiduronic acid
S O-α-glucuronosyl α-D-glucuronide + H_2O <19> [7]
P D-glucuronate + D-glucuronide
S acetylated glucuronoxylan + H_2O <3> [10]
P ?
S benzyl 4-O-α-D-glucuronosyl-β-D-glucoside + H_2O <19> [7]
P ?
S β-D-glucuronyl α-D-glucuronic acid + H_2O <12> [3]
P D-glucuronate + α-D-glucuronic acid
S glucuronosyl-xylotriose + H_2O <1> [1]
P glucuronate + xylotriose
S *p*-nitrophenyl α-D-glucuronide + H_2O <19> [7, 12]
P *p*-nitrophenol + D-glucuronide
S *p*-nitrophenyl-β-D-glucuronide + H_2O <12> (<12>, 3% of the activity with 18β-glycyrrhetinic acid β-D-glucuronyl α-D-glucuronic acid) [3]
P *p*-nitrophenol + D-glucuronide
S Additional information <3, 12-14> (<12>, no activity with baicalin [3]; <13, 14>, no activity with *p*-nitrophenyl-α-D-glucuronoside [4]; <3>, enzyme prefers low-molecular-weight xylooligomers as substrates. The enzyme acts almost exclusively on the bond between the terminal xylose at the nonreducing end of the xylose chain and the methyl glucuronic acid attached to it [10]) [3, 4, 10]
P ?

Inhibitors

Ag^+ <1, 17> (<3>, 1 mM, strong inhibition) [1, 11]

Al^{3+} <1> [1]
Cd^{2+} <17> (<17>, 1 mM, strong inhibition) [11]
Cu^{2+} <17> (<17>, 1 mM, strong inhibition) [11]
$CuCl_2$ <13, 14> [4]
Fe^{2+} <1> [1]
Fe^{3+} <1> [1]
Hg^{2+} <1, 17> (<17>, 1 mM, strong inhibition) [1, 11]
N-bromosuccinimide <1, 19> [1, 7]
PCMB <19> [7]
Pb^{2+} <17> (<17>, 1 mM, strong inhibition) [11]
SDS <1, 19> [1, 7]
Sn^{2+} <1> [1]
Zn^{2+} <17> (<17>, 1 mM, strong inhibition) [11]
$ZnCl_2$ <13> (<14>, no inhibition) [4]
monoiodoacetic acid <1, 19> [1, 7]
p-hydroxymercuribenzoate <17> [11]

Metals, ions
$BaCl_2$ <19> (<19>, slight activation) [7]
$CaCl_2$ <19> (<19>, slight activation) [7]
$FeCl_3$ <19> (<19>, slight activation) [7]

Turnover number (min^{-1})
Additional information <20> [9]

Specific activity (U/mg)
1.647 <13> [4]
3.98 <11> [8]
4.5 <16> [6]
8.4 <20> [9]
9.6 <14> [4]
Additional information <1, 3, 19> [1, 7, 10]

K_m-Value (mM)
0.12 <12> (18β-glycyrrhetinic acid β-D-glucuronyl α-D-glucuronic acid) [3]
0.13 <19> (*p*-nitrophenyl α-D-glucopyranosyluronic acid) [7, 12]
0.145 <11> (4-O-methyl-glucuronosyl-xylotriose) [8]
0.33 <19> (2-O-α-D-glucopyranosyluronic acid-D-xylose) [12]
0.36 <19> (O-α-D-glucopyranosyluronic acid-α-D-glucopyranosiduronic acid) [12]
0.37 <1> (4-O-methyl-glucuronosyl-xylotriose, enzyme form CM-I, <1>) [1]
0.47 <1> (4-O-methyl-glucuronosyl-xylotriose, enzyme form CM-II, <1>) [1]
0.76 <20> (4-O-methyl-glucuronosyl-xylobiose, <20>) [9]
0.77 <1> (glucuronosyl-xylotriose, enzyme form CM-I, <1>) [1]
0.82 <1> (glucuronosyl-xylotriose, enzyme form CM-II, <1>) [1]
0.95 <17> (4-O-methyl-glucuronosyl-xylobiose, <17>) [11]
17.6 <19> (2-O-(4-methyl-α-D-glucopyranosyluronic acid)-D-xylose, <19>) [12]

pH-Optimum

3 <19> [7]
3.3-3.8 <2> [2]
3.5 <16> [6]
4.5 <11> [2, 8]
4.5-6 <3> [10]
4.8 <1> [1]
5 <3> [2]
5.4 <20> [9]
5.5 <15> [5]
6 <13, 14> [4]
6-7 <12> [3]
6.3 <17> [11]

pH-Range

2-4 <19> (<19>, pH 2: about 30% of maximal activity, pH 4: about 70% of maximal activity) [7]
3-5 <16> (<16>, pH 3.0: about 80% of maximal activity, pH 5.0: about 50% of maximal activity) [6]
3-5.8 <11> (<11>, about 45% of maximal activity at pH 3 and at pH 5.8) [8]
3.3-7 <3> (<3>, pH 3.3: about 60% of maximal activity, pH 7: about 50% of maximal activity) [10]
4-5.6 <1> (<1>, about 50% of maximal activity at pH 4.5 and 5.6) [1]
4.8-6.3 <15> (<15>, pH 5.8: about 80% of maximal activity, pH 6.3: about 50% of maximal activity) [5]
5-6.2 <20> (<20>, pH 5: about 85% of maximal activity, pH 6.2: about 65% of maximal activity) [9]
5-8 <13, 14, 17> (<13, 14>, at least 40% of maximal activity at pH-values ranging from 5 to 8 [4]; <17>, pH 5: about 35% of maximal activity, pH 8: about 45% of maximal activity [11]) [4, 11]

Temperature optimum (°C)

50 <14-16> (<16> at pH 3.5 [6]) [4-7]
60 <1, 3, 13, 20> [1, 4, 9, 10]
65 <11> [8]
85 <17> [11]

Temperature range (°C)

30-60 <19> (<19>, about 40% of maximal activity at 30°C and at 60°C) [7]
40-75 <11> (<11>, 40°C: about 35% of maximal activity, 75°C: about 30% of maximal activity) [8]
45-65 <20> (<20>, 45°C: about 80% of maximal activity, 65°C: about 85% of maximal activity) [9]
60-100 <17> (<17>, about 45% of maximal activity at 60°C and at 100°C) [11]

4 Enzyme Structure

Molecular weight

80000 <12> [3]
100000 <3> [2]
100000-300000 <15> (<15>, MW ranging from 100000 to 300000, the fraction with the highest activity has a MW of 103000, gel filtration) [5]
118000 <11, 14> (<11>, gel filtration [8]; <14>, gel filtration [4]) [4, 8]
124000 <13> (<13>, gel filtration) [4]
130000 <20> (<20>, gel filtration, native PAGE) [9]
150000 <1> (<1>, gel filtration) [1]
160000 <2> (<2>, gel filtration) [2]
180000 <19> (<19>, gel filtration) [7, 12]

Subunits

? <3, 19> (<3>, x * 91000, SDS-PAGE [10]; <19>, x * 97000, SDS-PAGE [7,12]) [7, 10, 12]
dimer <13, 14, 20> (<14>, 2 * 71000, SDS-PAGE [4]; <13>, 1 * 72000 + 1 * 76000, SDS-PAGE [4]; <20>, 2 * 74000, SDS-PAGE [9]) [4, 9]
monomer <1, 11> (<11>, 1 * 107000, SDS-PAGE [8]; <1>, 1 * 130000, SDS-PAGE [1]) [1, 8]

5 Isolation/Preparation/Mutation/Application

Source/tissue

culture filtrate <11> [8]
culture medium <15> (<15>, of the enzyme grown in cultures containing filter paper, barley straw, birchwood xylan or birchwood sawdust as carbon source) [5]

Localization

cell-associated <20> [9]
extracellular <15, 16> [5, 6]
intracellular <1, 13, 14> [1, 4]

Purification

<1> (2 enzym forms: CM-I and CM-II) [1]
<3> [10]
<11> [8]
<12> (partial) [3]
<13> [4]
<14> [4]
<16> [6]
<17> (recombinant enzyme) [11]
<19> [7]
<20> [9]

Cloning

<17> (expression in E. coli) [11]

6 Stability

pH-Stability

3-7 <19> (<19>, 30°C, 2 h, stable) [7]
3.5 <16> (<16>, 50°C, in absence of substrate, inactivated in 2 h) [6]
4.5-7 <1> (<1>, 50°C, 1 h, stable) [1]
4.8-5.5 <3> (<3>, 40°C, 24 h, stable) [10]
5.5 <15, 16> (<15>, 50°C, absence of substrate, 15 min, 79% loss of activity [5]; <15>, 40°C, absence of substrate, 20% loss of activity [5]; <16>, 50°C, in absence of substrate, 58% loss of activity in 24 h [6]) [5, 6]
6.2-7.9 <20> (<20>, 60°C, stable for 1 h) [9]
6.6 <3> (<3>, 40°C, 24 h, 22% loss of activity) [10]

Temperature stability

40 <3, 15> (<15>, pH 5.5, in absence of substrate, 15 min, 20% loss of activity [5]; <3>, pH 6.2-7.9, stable for 1 h [10]) [5, 10]
50 <1, 3, 11, 13-16, 19> (<1>, pH 4.5-7.0, 1 h, stable [1]; <13>, pH 6.0, 4 h, stable [4]; <15>, pH 5.5, absence of substrate, 15 min, 79% loss of activity [5]; <16>, pH 3.5, in absence of substrate, inactivation in 2 h [6]; <16>, pH 5.0, in absence of substrate, 58% loss of activity in 24 h [6]; <19>, stable up to [7]; <11>, in absence of substrate, 6 h, 15% loss of activity [8]; <3>, pH 4.8, 24 h, 50% loss of activity [10]) [1, 4-8, 10]
55-75 <17> (<17>, partial inactivation, but thereafter the residual activity remains nearly constant for several days) [11]
60 <11, 13, 14, 20> (<13>, half-life: 14 h [4]; <14>, half-life: 2-3 h [4]; <11>, in absence of substrate, 6 h, 58% loss of activity [8]; <11>, in absence of substrate, $t_{1/2}$: 192 min [8]; <20>, pH 6.2-7.9, stable for 1 h [9]) [4, 8, 9]
62 <20> (<20>, half-life: 1 h) [9]
70 <11> (<11>, in absence of substrate, $t_{1/2}$: 28 min) [8]
Additional information <11> (<11>, 1 mg/ml of substrate protects against heat inactivation) [8]

Storage stability

<16>, -18°C, 4 months, complete inactivation [6]

References

[1] Uchida, H.; Nanri, T.; Kawabata, Y.; Kusakabe, I.; Murakami, K.: Purification and characterization of intracellular α-glucuronidase from Aspergillus niger 5-16. Biosci. Biotechnol. Biochem., **56**, 1608-1615 (1992)
[2] Puls, J.: α-Glucuronidases in the hydrolysis of wood xylans. Xylans and Xylanases (Visser, J., et al., Eds.) Elsevier, Amsterdam, 213-224 (1992)

[3] Kim, D.-H.; Jang, H.-S.; Lee, S.-W.: Bacteroides J-37, a human intestinal bacterium, produces α-glucuronidase. Biol. Pharm. Bull., **20**, 834-837 (1997)
[4] Bronnenmeier, K.; Meissner, H.; Stocker, S.; Staudenbauer, W.L.: α-D-Glucuronidases from the xylanolytic thermophiles Clostridium stercorarium and Thermoanaerobacterium saccharolyticum. Microbiology, **141**, 2033-2040 (1995)
[5] Wood, T.M.; Wilson, C.A.: α-(4-O-Methyl)-D-glucuronidase activity produced by rumen anaerobic fungus Piromonas communis a study of selected properties. Appl. Microbiol. Biotechnol., **43**, 893-900 (1995)
[6] Castanares, A.; Hay, A.J.; Gordon, A.H.; McCrae, S.I.; Wood, T.M.: D-Xylan-degrading enzyme system for the fungus Phanerochaete chrysosporium: isolation and partial characterisation of an α(4-O-methy)-D-glucuronidase. J. Biotechnol., **43**, 183-194 (1995)
[7] Kawabata, Y.; Ono, K.; Gama, Y.; Yoshida, S.; Kobayashi, H.; Kusakabe, I.: Purification and characterization of α-glucuronidase from snail acetone powder. Biosci. Biotechnol. Biochem., **59**, 1086-1090 (1995)
[8] Khandke, K.M.; Vithayathil, P.J.; Murthy, S.K.: Purification and characterization of an α-D-glucuronidase from a thermophilic fungus, Thermoascus aurantiacus. Arch. Biochem. Biophys., **274**, 511-517 (1989)
[9] Shao, W.; Obi, S.K.C.; Puls, J.; Wiegel, J.: Purification and characterization of the α-glucuronidase from Thermoanaerobacterium sp. strain JW/SL-YS485, an important enzyme for the utilization of substituted xylans. Appl. Environ. Microbiol., **61**, 1077-1081 (1995)
[10] Siika-aho, M.; Tenkanen, M.; Buchert, J.; Puls, J.; Viikari, L.: An α-glucuronidase from Trichoderma reesei RUT C-30. Enzyme Microb. Technol., **16**, 813-819 (1994)
[11] Ruile, P.; Winterhalter, C.; Liebl, W.: Isolation and analysis of a gene encoding α-glucuronidase, an enzyme with a novel primary structure involved in the breakdown of xylan. Mol. Microbiol., **23**, 267-279 (1997)
[12] Kawabata, Y.; Kuno, A.; Gama, Y.; Kusakabe, I.: Substrate specificity of α-glucuronidase isolated from snail acetone powder. Biosci. Biotechnol. Biochem., **60**, 524-525 (1996)

Lacto-N-biosidase 3.2.1.140

1 Nomenclature

EC number
3.2.1.140

Systematic name
oligosaccharide lacto-N-biosylhydrolase

Recommended name
lacto-N-biosidase

Synonyms
lacto-N-biosidase
lacto-N-biosidase (Streptomyces strain 142)

CAS registry number
146359-52-6
183511-11-7 (lacto-N-biosidase (Streptomyces strain 142))

2 Source Organism

<1> *Streptomyces sp.* (strain 142 [1,2]) [1, 2]

3 Reaction and Specificity

Catalyzed reaction
β-D-Gal-(1→3)-β-D-GlcNAc-(1→3)-β-D-Gal-(1→4)-D-Glc + H_2O = β-D-Gal-(1→3)-D-GlcNAc + β-D-Gal-(1→4)-D-Glc

Reaction type
hydrolysis of O-glycosyl bond

Substrates and products
- **S** lacto-N-tetraose + H_2O <1> [2]
- **P** lacto-N-biose
- **S** *p*-nitrophenyl-lacto-N-bioside + H_2O <1> [1, 2]
- **P** *p*-nitrophenol + lacto-N-bioside <1> [1]
- **S** pyridylamine-oligosaccharides which have type 1 chains at the nonreducing terminus + H_2O <1> (<1>, no reaction with pyridylamine-oligosaccharides which contain only type 2 chains. No reaction with oligosacchar-

ides in which the non-reducing terminal Gal or the penultimate GlcNAc are replaced by fucose or sialic acid are [1]) [1, 2]

P lacto-N-biose + ? <1> (<1>, from N-acetyllactosamine type of triantennary sugar chain with type 1 structure and N-acetyllactosamine type of biantennary sugar chain is produced [2]) [2]

S Additional information <1> (<1>, no reaction with lacto-N-neotetraose, lacto-N-triose, sialyl lacto-N-tetraose, lacto-N-fucopentaose I, II, or III, asialo-GM1 tetrasaccharide, poly-N-acetyllactosamine [2]) [2]

P ?

Inhibitors

Fe^{3+} <1> [1]
guanidine hydrochloride <1> [1]
Hg^{2+} <1> [1]
lacto-N-biose <1> [1]

Specific activity (U/mg)

18.97 <1> [1]

K_m-Value (mM)

0.0068 <1> (Galβ1-3GlcNAcβ1-3Galβ1-4Glc-pyridylamine, <1>) [1]
Additional information <1> [1]

pH-Optimum

4.5 <1> (<1>, *p*-nitrophenyl-β-lacto-N-bioside) [1]
5.5 <1> (<1>, Galβ1-3GlcNAcβ1-3Galβ1-4Glc-pyridylamine [1]) [1, 2]

Temperature optimum (°C)

50 <1> [2]
60 <1> (<1>, *p*-nitrophenyl-β-lacto-N-bioside) [1]

4 Enzyme Structure

Molecular weight

46000 <1> (<1>, gel filtration) [1]

Subunits

monomer <1> (<1>, 1 * 60000, SDS-PAGE) [1]

5 Isolation/Preparation/Mutation/Application

Source/tissue

culture medium <1> [1]

Purification

<1> [1]

Application

analysis <1> (the enzyme may be useful in identifying type 1 structure in glycoconjugates and to distinguish type 1 oligosaccharides from type 2 oligosaccharides) [2]

6 Stability

pH-Stability

4-7 <1> (<1>, 4°C, 16 h, stable) [2]
4-10 <1> (<1>, 4°C, 16 h, stable) [1]

Temperature stability

45 <1> (<1>, 30 min, stable) [1]
50 <1> (<1>, 30 min, 20% loss of activity) [2]

General stability information

<1>, dilution without Brij58 causes 30-50% inactivation [1]
<1>, stabilized by 0.05% Brij58, 0.05% Nonidet P-40 or 0.01% bovine serum albumin [1]

Storage stability

<1>, 4°C or -20°C, protein concentration 0.01 mg/ml or 0.00005 mg/ml, 0.05% Brij 58, stable for at least 6 months [1]

References

[1] Sano, M.; Hayakawa, K.; Kato, I.: Purification and characterization of an enzyme releasing lacto-N-biose from oligosaccharides with type 1 chain. J. Biol. Chem., **268**, 18560-18566 (1993)
[2] Sano, M.; Hayakawa, K.; Kato, I.: An enzyme releasing lacto-N-biose from oligosaccharides. Proc. Natl. Acad. Sci. USA, **89**, 8512-8516 (1992)

4-α-D-{(1→4)-α-D-Glucano}trehalose trehalohydrolase 3.2.1.141

1 Nomenclature

EC number

3.2.1.141

Systematic name

4-α-D-{(1→4)-α-D-glucano}trehalose glucanohydrolase (trehalose-producing)

Recommended name

4-α-D-{(1→4)-α-D-glucano}trehalose trehalohydrolase

Synonyms

MTHase
malto-oligosyltrehalose trehalohydrolase
maltooligosyl trehalose trehalohydrolase (Rhizobium strain M-11 clone pMBTU1 gene treZ reduced)
maltooligosyltrehalose trehalohydrolase
maltooligosyltrehalose trehalosidase
trehalosidase, maltooligosyltrehalose
trehalosidase, maltooligosyltrehalose (Rhizobium strain M-11 clone pBMTU1 gene treZ reduced)

CAS registry number

170780-50-4
176431-02-0 (trehalosidase, maltooligosyltrehalose (Rhizobium strain M-11 clone pBMTU1 gene treZ reduced) /maltooligosyl trehalose trehalohydrolase (Rhizobium strain M-11 clone pMBTU1 gene treZ reduced))

2 Source Organism

<1> *Sulfolobus acidocaldarius* (ATCC 33909) [1, 5]
<2> *Arthrobacter sp.* (strain O36 [2]) [2-4]

3 Reaction and Specificity

Catalyzed reaction

hydrolysis of α-(1→4)-D-glucosidic linkage in 4-α-{(1→4)-α-D-glucanosyl}n trehalose to yield trehalose and α-(14)-D-glucan

Reaction type

hydrolysis of O-glycosyl bond

Natural substrates and products

S Additional information <2> (<2>, the enzyme is involved in the biosynthesis of trehalose from maltooligosaccharides) [3]

Substrates and products

S glucosyl trehalose + H_2O <1> [1]
P glucose + trehalose <1> [1]
S maltopentaosyltrehalose + H_2O <1, 2> [1, 2]
P maltopentaose + trehalose <1> [1]
S maltosyl trehalose + H_2O <1, 2> [1, 2]
P maltose + trehalose <1> [1]
S maltotetraosyl trehalose + H_2O <1, 2> [1, 2]
P maltotetraose + trehalose <1> [1]
S maltotriosyl trehalose + H_2O <1, 2> [1-3]
P maltotriose + trehalose <1> [1]

Inhibitors

Cu^{2+} <1, 2> [1, 2]
Fe^{2+} <1, 2> (<1>, slight [1]) [1, 2]
Hg^{2+} <1, 2> [1, 2]
Pb^{2+} <1> (<1>, slight) [1]
Zn^{2+} <1, 2> (<1>, slight [1]) [1, 2]

Specific activity (U/mg)

497 <2> [2]
730 <1> [1]

K_m-Value (mM)

2.7 <1> (maltotriosyl trehalose, <1>) [1]
3.7 <1> (maltotetraosyl trehalose, <1>) [1]
4.2 <2> (maltopentaosyl trehalose, <2>) [2]
4.6 <2> (maltotriosyl trehalose, <2>) [2]
4.9 <1> (maltopentaosyl trehalose, <1>) [1]
5.5 <2> (maltosyl trehalose, <2>) [2]
7 <2> (maltotetraosyl trehalose, <2>) [2]
16.7 <1> (maltosyl trehalose, <1>) [1]

pH-Optimum

5.5-6 <1> [1]
6.5 <2> [2]

pH-Range

5-7 <1> (<1>, pH 5: about 80% of maximal activity, pH 7: about 65% of maximal activity) [1]
5-8 <2> (<2>, pH 5: about 50% of maximal activity, pH 8: about 40% of maximal activity) [2]

Temperature optimum (°C)
45 <2> [2]
75 <1> [1]

Temperature range (°C)
30-55 <2> (<2>, 30°C: about 75% of maximal activity, 55°C: about 40% of maximal activity) [2]
40-90 <1> (<1>, 40°C: about 45% of maximal activity, 90°C: about 65% of maximal activity) [1]

4 Enzyme Structure

Subunits
? <1> (<1>, x * 59000, SDS-PAGE) [1]

5 Isolation/Preparation/Mutation/Application

Localization
extracellular <2> [3]

Purification
<1> [1]
<2> (partial [3]) [2, 3]

Application
synthesis <1> (<1>, production of trehalose from starch) [5]
synthesis <2> (<2> production of trehalose from starch. Trehalose is utilized as a stabilizer for dried or frozen food, in cosmetics and in medicines as a drug additive) [4]

6 Stability

pH-Stability
5-10 <2> (<2>, 4°C, 24 h, stable) [2]
5.5-6 <1> (<1>, 4°C, 24 h, stable) [1]

Temperature stability
45 <2> (<2>, 60 min, stable up to) [2]
85 <1> (<1>, 60 min, stable below) [1]

References

[1] Nakada, T.; Ikegami, S.; Chaen, H.; Kubota, M.; Fukuda, S.; Sugimoto, T.; Kurimoto, M.; Tsujisaka, Y.: Purification and characterization of thermostable maltooligosyl trehalose trehalohydrolase from the thermoacidophilic

archaebacterium Sulfolobus acidocaldarius. Biosci. Biotechnol. Biochem., **60**, 267-270 (1996)

[2] Nakada, T.; Maruta, K.; Mitsuzumi, H.; Kubota, M.; Chaen, H.; Sugimoto, T.; Kurimoto, M.; Tsujisaka, Y.: Purification and characterization of a novel enzyme, maltooligosyl trehalose trehalohydrolase, from Arthrobacter sp. Q36. Biosci. Biotechnol. Biochem., **59**, 2215-2218 (1995)

[3] Maruta, K.; Nakada, T.; Kubota, M.; Chaen, H.; Sugimoto, T.; Kurimoto, M.; Tsujisaka, Y.: Formation of trehalose from maltooligosaccharides by a novel enzymatic system. Biosci. Biotechnol. Biochem., **59**, 1829-1834 (1995)

[4] Takahashi, N.; Kawaguchi, K.: Novel enzymes, maltooligosyl trehalose synthase and maltooligosyl trehalose trehalohydrolase and their application to the production of trehalose from starch. Trends Glycosci. Glycotechnol., **8**, 369-370 (1996)

[5] Mukai, K.; Tabuchi, A.; Nakada, T.; Shibuya, T.; Chaen, H.; Fukuda, S.; Kurimoto, M.; Tsujisaka, Y.: Production of trehalose from starch by thermostable enzymes from Sulfolobus acidocaldarius. Starch, **44**, 26-30 (1997)

Limit dextrinase 3.2.1.142

1 Nomenclature

EC number
3.2.1.142

Systematic name
dextrin α-1,6-glucanohydrolase

Recommended name
limit dextrinase

Synonyms
pullulanase

CAS registry number
9032-15-9

2 Source Organism

<1> *Hordeum vulgare* (germinating barley [1]) [1, 2]
<2> *Pisum sativum* (ungerminated peas [3]) [2, 3]
<3> *Sorghum vulgare* (malted sorghum [4]) [2, 4]
<4> *Avena sativa* (ungerminated oats [5]) [2, 5]
<5> *Oryza sativa* (ungerminated rice [5]) [2, 5]
<6> *Vicia faba* (broad bean [6]) [2, 6]

3 Reaction and Specificity

Catalyzed reaction
hyrolysis of (1,6)-α-D-glucosidic linkages in α- and β-limit dextrins of amylopectin and glycogen, and in amylopectin and pullulan

Reaction type
hyrolysis of O-glycosyl bonds

Natural substrates and products
S polysaccharide + H_2O <1-6> (<2> natural substrates appear to be α-limit dextrins [3]) [1-6]
P oligosaccharides <1-6> [1-6]

Substrates and products

S polysaccharide + H_2O <1-6> (<1-6> oligosaccharides containing maltosyl or larger side chains are hydrolysed as well as amylopectin and amylopectin β-limit dextrin, oligo- and poly-saccharides containing single D-glucosyl groups attached by an α-(1,6) linkage to the rest of the molecule are not hydrolysed, glycogen is not a substrate [1-6]) [1-6]

P oligosaccharides <1-6> [1-6]

Inhibitors

6,3-α-D-glucosylmaltotriose <5> [5]
6,3-α-D-maltosylmaltotetraose <5> [5]
6,3-α-D-maltosylmaltotriose <5> [5]
6,3-α-D-maltotriosylmaltotetraose <5> [5]
6,3-α-D-maltotriosylmaltotriose <5> [5]
cyclohexa-amylose <2> (<2> competitive inhibition [3]) [3]
maltose <5> [5]
oligo-maltoses <5> [5]

Specific activity (U/mg)

0.6 <5> [5]
0.8 <4> [5]
1.04 <6> [6]
1.08 <2> [3]
1.08 <3> [4]

K_m-Value (mM)

0.16 <6> (6-α-maltotriosylmaltotetraose) [6]
0.2 <4> (6-α-maltosylmaltotetraose) [5]
0.2 <4> (6-α-maltosylmaltotriose) [5]
0.2 <4> (6-α-maltotriosylmaltotetraose) [5]
0.2 <4> (6-α-maltotriosylmaltotriose) [5]
0.4 <3> (pullulan) [4]
0.6 <5> (6-α-maltotriosylmaltotetraose) [5]
0.6 <6> (amylopectin β-limit dextrin) [6]
1.5 <3> (6-α-maltotriosylmaltotetraose) [4]
1.5 <3> (amylopectin β-limit dextrin) [4]
3.24 <2> (amylopectin β-limit dextrin) [3]
26 <3> (glycogen β-limit dextrin) [4]

pH-Optimum

5 <4, 5> [5]
5-5.4 <3> (<3> pullulan as substrate [4]) [4]
5.8-6.2 <2> (<2> with both pullulan and amylopectin β-limit dextrin as substrates [3]) [3]
6.4-6.8 <6> (<6> pullulan and amylopectin as substrates, maximum activity occurrs over a relatively broad range [6]) [6]

4 Enzyme Structure

Molecular weight

80000 <4> (<4> preliminary studies, gel filtration [5]) [5]
80000 <6> (<6> gel filtration [6]) [6]
90000 <3> (<3> gel filtration [4]) [4]
99000 <1> (<1> calculated from amino acid sequence [1]) [1]
105000 <1> (<1> SDS-PAGE [1]) [1]
180000 <2> (<2> gel filtration [3]) [3]

Subunits

monomer <1> [1]

5 Isolation/Preparation/Mutation/Application

Purification

<2> ($(NH_4)_2SO_4$ precipitation, acetone precipitation, affinity chromatography [3]) [3]
<3> ($(NH_4)_2SO_4$ precipitation, continuous electrophoresis, starch-celite, Sephadex G-150 [4]) [4]
<4-6> ($(NH_4)_2SO_4$ precipitation, acetone precipitation, DEAE-cellulose, Sephadex G-150 [5,6]) [5, 6]

6 Stability

Temperature stability

60 <3> (<3> 60% inactivation [4]) [4]

References

[1] Kristensen, M.; Lok, F.; Planchot, V.; Svendsen, I.; Leah, R.; Svensson, B.: Isolation and characterization of the gene encoding the starch debranching enzyme limit dextrinase from germinating barley. Biochim. Biophys. Acta, **1431**, 538-546 (1999)
[2] Manners, D.J.: Observation on the specificity and nomenclature of starch debranching enzymes. J. Appl. Glycosci., **44**, 83-85 (1997)
[3] Yellowlees, D.: Purification and characterization of limit dextrinase from Pisum sativum L.. Carbohydr. Res., **83**, 109-118 (1980)
[4] Hardie, D.G.; Manners, D.J.; Yellowlees, D.: The limit dextrinase from malted sorghum (Sorghum vulgare). Carbohydr. Res., **50**, 75-85 (1976)
[5] Dunn, G.; Manners, D.J.: The limit dextrinase from ungerminated oats (Avena sativa L.) and ungerminated rice (Oryza sativa L.). Carbohydr. Res., **39**, 283-293 (1975)
[6] Gordon, R.W.; Manners, D.J.; Stark, J.R.: The limit dextrinase of the broad bean (Vicia faba L.). Carbohydr. Res., **42**, 125-134 (1975)

Poly(ADP-ribose) glycohydrolase 3.2.1.143

1 Nomenclature

EC number
3.2.1.143

Recommended name
poly(ADP-ribose) glycohydrolase

Synonyms
ADP-ribose glycohydrolase
Genbank AB019366-derived protein GI 6518480
Genbank U78975-derived protein GI 2062407
PARG <1> [1, 3]
glycohydrolase, poly(adenosine diphosphoribose)
glycohydrolase, poly(adenosine diphosphoribose) (Rattus norvegicus strain BUF gene Parg)
glycohydrolase, poly(adenosine diphosphoribose) (cattle clone 4/5)
poly(ADP-ribose) glycohydrolase (Arabidopsis thaliana gene At2g31860)
poly(ADP-ribose) glycohydrolase (Arabidopsis thaliana gene At2g31870)
poly(ADP-ribose) glycohydrolase (cattly clone 4/5)
poly(ADP-ribosyl) glycohydrolase
poly(adenosine diphosphoribose) glycohydrolase
poly(adenosine diphosphoribose) glycosidase

CAS registry number
190209-88-2
253766-41-5
253766-42-6
253767-74-7
9068-16-0

2 Source Organism

<1> *Bos taurus* (calf [3, 4, 8, 9, 10]) [1, 3, 4, 8, 9, 10, 14]
<2> *Cavia porcellus* [2, 11]
<3> *Rattus norvegicus* [5, 6]
<4> *Homo sapiens* [7, 12]
<5> *Sus scrofa* [13, 15]

3 Reaction and Specificity

Catalyzed reaction

hydrolyses poly(ADP-ribose) at glycosidic (1'→2') linkage of ribose-ribose bond to produce free ADP-ribose

Reaction type

hydrolysis of O-glycosyl bonds

Natural substrates and products

S poly(ADP-ribose)$_n$ + H_2O <1, 4> (<1>, the enzyme is responsible for the catabolism of poly(ADP-ribose), the enzyme is a crucial determinant of polymer metabolism which is known to be implicated in DNA repair and other cellular processes [3]; <1>, the enzyme may regulate functionally the chain length of poly(ADP-ribose) which plays a role in DNA synthesis or in the structure of chromatin [4,9]; <1>, regulates differentially the levels of large and small poly(ADP-ribose) in the cell [10]; <4>, poly-(ADP-ribose) glycohydrolase II may be involved in extranuclear de(ADP-ribosyl)n-ation, but not in membrane de-mono(ADP-ribosyl)ation [12]) [3, 4, 9, 10, 12]

P ?

Substrates and products

S poly(ADP-ribose)$_n$ + H_2O <1-5> (<1>, n > 20, degradation proceeds in a biphasic as well as bimodal manner [10]) [1-15]

P ADP-ribose <1, 2, 3, 4, 5> (<1>, + oligo(ADP-ribose) [8]) [2, 5, 8, 9, 12, 15]

Inhibitors

1-naphthylamine <5> (<5>, 1 mM, 45% inhibition [13]) [13]
2':3'-cyclic NADP$^+$ <1> [10]
2-naphthalene sulfonic acid <5> (<5>, 5 mM, 26% inhibition [13]) [13]
2-naphthylamine <5> (<5>, 1 mM, 50% inhibition [13]) [13]
4-amino-3-hydroxynaphthalene 1-sulfonic acid <5> (<5>, 5 m, 79% inhibition [13]) [13]
8-bromoadenosine 3',5'-monophosphate <5> (<5>, 5 mM, 13% inhibition [13]) [13]
8-methylaminoadenosine 3',5'-monophosphate <5> (<5>, 5 mM, 14% inhibition [13]) [13]
ADP <1> (<1>, 10 mM, 39% inhibition [9]) [4, 9]
ADP-ribose <1, 2, 4, 5> (<2>, inhibits poly(ADP-ribose) glycohydrolase II more strongly than poly(ADP-ribose) glycohydrolase I [2]; <1>, 1.1 mM, 50% inhibition [9]; <1>, 5 mM, 55% inhibition [10]) [2, 4, 9, 10, 11, 12, 13]
AMP <1> (<1>, 10 mM, 64% inhibition [9]) [4, 9]
ATP <1> (<1>, 10 mM, 23% inhibition [9]) [4, 9]
$CaCl_2$ <2, 5> (<2>, 1.5 mM, 50% inhibition. 96% inhibition at 10 mM [11]; <5>, 5 mM 86% inhibition [13]) [11, 13]

DNA <1, 2> (<1, 2>, denatured [10, 11]; <1,2>, no inhibition by native DNA [10, 11]; <5>, single-stranded DNA, no inhibition by double-stranded DNA [13]) [10, 11]
$HgCl_2$ <1> [9]
KCl <2, 4, 5> (<2>, 100 mM, 50% inhibition, poly(ADP-ribose) glycohydrolase II [2]; <2, 5>, 200 mM or more [11,15]) [2, 11, 12, 15]
$MgCl_2$ <5> (<5>, 10 mM, 18% inhibition [13]) [13]
N^1-naphthylethanenediamine <5> (<5>, 5 mM, 70% inhibition [13]) [13]
N^6,$O^{2'}$-dibutyryl adenosine cyclic 3':5'-monophosphate <1> (<1>, 10 mM, 11% inhibition [9]) [9]
N^6-monobutyryl adenosine cyclic 3':5'-monophosphate <1> (<1>, 10 mM, 50% inhibition [9]) [9]
NAD^+ <1> (<1>, 10 mM, 40% inhibition [9]) [4, 9]
NaCl <2, 5> (<2>, 50-100 mM, poly(ADP-ribose) glycohydrolase II [2]; <2, 5>, 200 mM or more [11, 15]; <5>, 400 mM, 75% inhibition [13]) [2, 11, 13, 15]
RNA <1> (<1>, calf liver RNA [10]) [10]
SDS <1> (<1>, 0.01%, 96% inhibition [10]) [10]
acetone <1> (<1>, 20%, 95% inhibition [10]) [10]
adenine(5'-9)ribofuranoside <5> (<5>, 5 mM, 24% inhibition [13]) [13]
adenosine 3':5'-cyclic monophosphate <1, 2> (<1>, 0.3 mM, 50% inhibition [9]) [4, 9, 10, 11]
adenosine 5'-carboxylic acid <5> (<5>, 5 mM, 26% inhibition [13]) [13]
adenosine cyclic 2':3'-monophosphate <1> (<1>, 10 mM, 23% inhibition [9]) [4, 9]
adenosine diphosphate (hydroxymethyl)pyrrolidinediol <1> (<1>, partial noncompetitive inhibition [14]) [14]
cAMP <2, 4, 5> (<2>, inhibits poly(ADP-ribose) glycohydrolase II more strongly than poly(ADP-ribose) glycohydrolase I [2]) [2, 12, 13]
cyclic dAMP <5> (<5>, 5 mM, 34% inhibition [13]) [13]
diadenosine 5',5"→p1,p4-tetraphosphate <2> [11]
ethanol <1> (<1>, 20%, 91% inhibition [10]) [10]
gallic acid <4> (<4>, 0.1 mg/ml, 9% inhibition [7]) [7]
galloylgallic acid <4> (<4>, 0.1 mg/ml, 11% inhibition [7]) [7]
glucuronic acid <4> (<4>, 0.1 mg/ml, 14% inhibition [7]) [7]
guanosine cyclic 3':5'-monophosphate <1> (<1>, 10 mM, 42% inhibition [9]) [4, 9]
histone <1, 2, 3, 5> (<1>, lysine-rich histone of calf thymus [4]; <3>, due to noncovalent interactions of the protein with free ADP-ribose polymers [5]; <1>, histone f2a, f2b and f3. Inhibition by histone f2a is reversed by DNA [9]; histone H1, H2A, H2B, H3 and H4 [10]; <5>, histone H1, H2A, H2B and H3 [13]) [4, 5, 9, 10, 11, 13]
naphthylamine <5> [13]
nuclear matrix proteins <3> (<3>, due to non-covalent interactions of the protein with free ADP-ribose polymers [5]) [5]
p-chloromercuribenzenesulfonate <1> (<1>, 0.01 mM [9]) [4, 9]

poly(A) <1, 3, 5> (<3>, inhibitory effect is eliminated when 250 mM KCl is present in the reaction mixture, inhibition is markedly diminished after hybridization with polyT [6]) [6, 10, 13]
poly(G) <3> (<3>, inhibitory effect is eliminated when 250 mM KCl is present in the reaction mixture [6]) [6]
poly(I) <3, 5> (<3>, inhibitory effect is eliminated when 250 mM KCl is present in the reaction mixture, inhibition is markedly diminished after hybridization with polyC [6]) [6, 13]
poly(L-Arg) <5> [13]
poly(L-Lys) <1, 5> [9, 13]
poly(L-Orn) <5> [13]
polyvinylsulfate <5> [13]
protamine <1, 5> [9, 13]
tannin <4> (<4>, 0.01 mg/ml, 89% inhibition, competitive with respect to poly(ADP-ribose) [7]) [7]

Activating compounds

2-mercaptoethanol <1, 2, 5> (<1>, enhances activity [4]; <1>, 10 mM, required for maximal activity [9]; <2>, 10 mM, 3fold stimulation [11]; <5>, 10 mM, stimulates by 70% [13]) [4, 9, 11, 13]
ADP <1> (<1>, 5 mM, slight activation [10]) [10]
ADP-glucose <1> (<1>, 5 mM, slight activation [10]) [10]
AMP <1> (<1>, 5 mM, slight activation [10]) [10]
ATP <1> (<1>, 5 mM, slight activation [10]) [10]
NADH <1> (<1>, 5 mM, slight activation [10]) [10]
$NADP^+$ <1> (<1>, 5 mM, slight activation [10]) [10]
NADPH <1> (<1>, 5 mM, slight activation [10]) [10]
adenosine 5'-tetraphosphate <1> (<1>, 5 mM, slight activation [10]) [10]
dithiothreitol <1, 2, 5> (<1>, enhances activity [4]; <1>, 10 mM, required for maximal activity [9]; <2>, 10 mM, 3fold stimulation [11]; <5>, 10 mM, stimulates by 70% [13]) [4, 9, 11, 13]
p1,p5-di(adenosine-5')pentaphosphate <1> (<1>, 5 mM, slight activation [10]) [10]

Metals, ions

KCl <1, 2> (<2>, 50-100 mM, stimulates poly(ADP-ribose) glycohydrolase I [2]; <1>, 100 mM, slight activation [10]; <2>, 50 mM, 1.4-1.5fold stimulation [11]; <5>, 50-100 mM, stimulates [15]) [2, 10, 11, 15]
NaCl <1, 2, 5> (<2>, 50-100 mM, stimulates poly(ADP-ribose) glycohydrolase I [2]; <1>, 100 mM, slight activation [10]; <2>, 50 mM, 1.4-1.5fold stimulation [11]; <5>, 50-100 mM, stimulates [15]) [2, 10, 11, 15]
sodium phosphate <1> (<1>, 100 mM, slight activation [10]) [10]

Turnover number (min^{-1})

5400 <1> (poly(ADP-ribose)) [10]

Specific activity (U/mg)

0.182 <1> [9]
13.4 <2> (<2>, poly(ADP-ribose) glycohydrolase II [2]) [2]

13.8 <5> [13]
14 <4> [12]
32.1 <5> [15]
32.5 <2> [11]
69.95 <1> [3]
85 <1> [10]

K_m-Value (mM)

0.0003 <1> (poly(ADP-ribose)) [10]
0.00058 <1, 2> ((ADP-ribose)$_n$, <2>, average chain length n = 27, poly(ADP-ribose) glycohydrolase I [2]) [2, 9]
0.0011 <2> ((ADP-ribose)$_n$, <2>, average chain length n = 21, poly(ADP-ribose) glycohydrolase I [2]) [2]
0.0018 <5> (poly(ADP-ribose), <5>, as monomer units [13]) [13]
0.002 <2> ((ADP-ribose)$_n$, <2>, average chain length n = 15, poly(ADP-ribose) glycohydrolase I [2]) [2]
0.0023 <2> ((ADP-ribose)$_n$) [11]
0.0031 <2> ((ADP-ribose)$_n$, <2>, average chain length n = 27, poly(ADP-ribose) glycohydrolase II [2]) [2]
0.004 <2> ((ADP-ribose)$_n$, <2>, average chain length n = 21, poly(ADP-ribose) glycohydrolase II [2]) [2]
0.0045 <2> ((ADP-ribose)$_n$, <2>, average chain length n = 6.7, poly(ADP-ribose) glycohydrolase I [2]) [2]
0.0048 <2> ((ADP-ribose)$_n$, <2>, average chain length n = 15, poly(ADP-ribose) glycohydrolase II [2]) [2]
0.0054 <5> ((ADP-ribose)15) [15]
0.0058 <4> ((ADP-ribose)15) [12]
0.0066 <2> ((ADP-ribose)$_n$, <2>, average chain length n = 6.7, poly(ADP-ribose) glycohydrolase II [2]) [2]

pH-Optimum

6.8-7 <2> [11]
7 <1, 2> (<2>, poly(ADP-ribose) glycohydrolase I [2]) [2, 4]
7-7.5 <1> [10]
7.2 <5> [15]
7.4 <2, 4> (<2>, poly(ADP-ribose) glycohydrolase II [2]) [2, 12]
7.5 <1, 5> (<1>, activity is less in Tris-HCl buffer than in sodium phosphate buffer or N-2-hydroxyethylpiperazine-N-ethanesulfonic acid-NaOH buffer [9]) [9, 13]

pH-Range

6-8 <1, 2> (<1>, about 45% of maximal activity at pH 6.0 and 8.0 [10]; <2>, about 55% of maximal activity at pH 6.0 and 8.0 [11]) [10, 11]
6-8.5 <5> (<5>, pH 6.0: about 50% of maximal activity, pH 8.5: about 35% of maximal activity [15]) [15]
6-9 <5> (<5>, the enzyme is active between pH 6 and 9 [13]) [13]

Temperature optimum (°C)

50 <1> [10]

Temperature range (°C)

30-50 <1> (<1>, 30°C: about 25% of maximal activity, 55°C: about 50% of maximal activity, no activity at 60°C [10]) [10]

4 Enzyme Structure

Molecular weight

48000 <1> (<1>, gel filtration [9]) [9]
51000 <5> (<5>, gel filtration [15]) [15]
53000 <1> (<1>, sucrose density gradient centrifugation [9]) [9]
56000 <4> (<4>, gel filtration [12]) [12]
57000 <2> (<2>, poly(ADP-ribose) glycohydrolase II, gel filtration [2]) [2]
59000 <5> (<5>, gel filtration [13]) [13]
72000 <2> (<2>, gel filtration [11]) [11]

Subunits

? <1> (<1>, x * 59000, SDS-PAGE [1]; <1>, x * 60000, poly(ADP-ribose) glycohydrolase G1, G2 and G3, SDS-PAGE [3]; <1>, x * 74000, poly(ADP-ribose) glycohydrolase G4, SDS-PAGE [3]) [1, 3]
monomer <2, 4, 5> (<5>, 1 * 58000, SDS-PAGE [15]; <4>, 1 * 59000, SDS-PAGE [12]; <2>, 1 * 59500, poly(ADP-ribose) glycohydrolase II, SDS-PAGE [2]; <2>, 1 * 75500, SDS-PAGE [11]) [2, 11, 12, 15]

5 Isolation/Preparation/Mutation/Application

Source/tissue

erythrocyte <4> [12]
kidney <1> [1]
liver <2, 3> [2, 5, 11]
placenta <4> [7]
testis <3, 5> [6, 15]
thymus <1, 5> [1, 3, 4, 8, 9, 10, 13, 14]

Localization

cytoplasm <2> (<2>, poly(ADP-ribose) glycohydrolase II [2]) [2]
nucleus <1, 2, 3, 5> (<2>, poly(ADP-ribose) glycohydrolase I [2]) [2, 5, 8, 10, 15]
Additional information <1> (<1>, contains a putative bipartite nuclear location signal [1]) [1]

Purification

<1> (4 poly(ADP-ribose) glycohydrolase isoforms: G1, G2, G3, G4 [3]) [1, 3, 9]
<2> (cytoplasmic enzyme [2]) [2, 11]
<3> (partial [5]) [5]
<4> (poly(ADP-ribose) glycohydrolase II [12]) [12]
<5> [13, 15]

Cloning

<1> (expression in E. coli [1]) [1]

6 Stability

pH-Stability

5-9 <1> (<1>, 37°C, 10 min, stable [10]) [10]
6-9 <2> (<2>, optimal stability [2]) [2]

Temperature stability

30 <1> (<1>, 10 min, stable [10]) [10]
45 <1> (<1>, 10 min, about 10% loss of activity [10]) [10]
50 <1> (<1>, 10 min, complete inactivation [10]) [10]
Additional information <1> (<1>, poly(ADP-ribose) protects the enzyme from thermal inactivation [10]) [10]

General stability information

<1>, poly(ADP-ribose) protects the enzyme from thermal inactivation [10]

Storage stability

<1>, -20°C, retains 90-100% of its activity for at least 1.5 years [10]
<1>, -80°C, stable for 3 months [9]
<1>, 4°C, 20% loss of activity after 1 month, addition of 0.1 mg/ml bovine serum albumin or 0.1 M NaCl to the diluted enzyme solution stabilizes the enzyme activity [9]
<2>, -20°C, stable for at least 6 months [2]

References

[1] Lin, W.; Ame, J.-C.; Aboul-Ela, N.; Jacobson, E.L.; Jacobson, M.K.: Isolation and characterization of the cDNA encoding bovine poly(ADP-ribose) glycohydrolase. J. Biol. Chem., **272**, 11895-11901 (1997)
[2] Maruta, H.; Inageda, K.; Aoki, T.; Nishina, H.; Tanuma, S.-i.: Characterization of two forms of poly(ADP-ribose) glycohydrolase in guniea pig liver. Biochemistry, **30**, 5907-5912 (1991)
[3] Brochu, G.; Shah, G.M.; Poirier, G.G.: Purification of poly(ADP-ribose) glycohydrolase detection of its isoforms by a zymogram following one- or two-dimensional electrophoresis. Anal. Biochem., **218**, 265-272 (1994)
[4] Sugimura, T.; Yamada, M.; Miwa, M.; Matsushima, T.; Hidaka, T.; Nagao, M.; Inui, N.; Takayama, S.: Properties of poly(adenosine diphosphate ribose) polymerase, poly(adenosine diphosphate ribose) glycohydrolase and poly-(adenosine diphosphate ribose). Biochem. Soc. Trans., **1**, 642-644 (1973)
[5] Pacheco-Rodriguez, G.; Alvarez-Gonzalez, R.: Measurement of poly(ADP-ribose) glycohydrolase activity by high resolution polyacrylamide gel electrophoresis: specific inhibition by histones and nuclear matrix proteins. Mol. Cell. Biochem., **193**, 13-18 (1999)

[6] Concha, I.I.; Koide, S.S.; Burzio, L.O.: Characteristics of the inhibition of poly(ADP-ribose) glycohydrolase by homopolypurines. Biochem. Int., **16**, 397-403 (1988)
[7] Tanuma, S.-i.; Sakagami, H.; Endo, H.: Inhibitory effect of tannin on poly(-ADP-ribose) glycohydrolase. Biochem. Int., **18**, 701-708 (1989)
[8] Miwa, M.; Sugimura, T.: Splitting of the ribose-ribose linkage of poly(adenosine diphosphate-ribose) by a calf thymus extract. J. Biol. Chem., **246**, 6362-6364 (1971)
[9] Miwa, M.; Tanaka, M.; Matsushima, T.; Sugimura, T.: Purification and properties of a glycohydrolase from calf thymus splitting ribose-ribose linkages of poly(adenosine diphosphate ribose). J. Biol. Chem., **249**, 3475-3482 (1974)
[10] Hatakeyama, K.; Nemoto, Y.; Ueda, K.; Hayaishi, O.: Purification and characterization of poly(ADP-ribose) glycohydrolase. Different modes of action on large and small poly(ADP-ribose). J. Biol. Chem., **261**, 14902-14911 (1986)
[11] Tanuma, S.-i.; Kawashima, K.; Endo, H.: Purification and properties of an $(ADP\text{-}ribose)_n$ glycohydrolase from guinea pig liver nuclei. J. Biol. Chem., **261**, 965-969 (1986)
[12] Tanuma, S.-i.; Endo, H.: Purification and characterization of an $(ADP\text{-}ribose)_n$ glycohydrolase from human erythrocytes. Eur. J. Biochem., **191**, 57-63 (1990)
[13] Tavassoli, M.; Tavassoli, M.H.; Shall, S.: Isolation and purification of poly(-ADP-ribose) glycohydrolase from pig thymus. Eur. J. Biochem., **135**, 449-455 (1983)
[14] Slama, J.T.; Aboul-Ela, N.; Jacobson, M.K.: Mechanism of inhibition of poly(ADP-ribose) glycohydrolase by adenosine diphosphate (hydroxymethyl)pyrrolidinediol. J. Med. Chem., **38**, 4332-4336 (1995)
[15] Abe, H.; Tanuma, S.-i.: Properties of poly(ADP-ribose) glycohydrolase purified from pig testis nuclei. Arch. Biochem. Biophys., **336**, 139-146 (1996)

3-Deoxyoctulosonase 3.2.1.144

1 Nomenclature

EC number
3.2.1.144

Systematic name
3-deoxyoctulosonyl-lipopolysaccharide hydrolase

Recommended name
3-deoxyoctulosonase

Synonyms
α-Kdo-ase <1> (<1> enzyme releases α linked 3-deoxy-D-manno-octulosonic acid (KDO) from lipopolysaccharides [1]) [1]

CAS registry number
199128-67-1

2 Source Organism

<1> *Crassostrea virginica* (oyster [1]) [1]

3 Reaction and Specificity

Catalyzed reaction
3-deoxyoctulosonyl-lipopolysaccharide + H_2O = 3-deoxyoctulosonic acid + lipopolysaccharide

Reaction type
hydrolysis of O-glycosyl bond

Natural substrates and products
S re-lipopolysaccharide + H_2O <1> [1]
P 3-deoxy-D-manno-octulosonic acid + lipopolysaccharide <1> [1]

Substrates and products
S 4-methylumbelliferyl 3-deoxy-α-D-manno-2-octulopyranosidonic acid + H_2O <1> [1]
P 3-deoxy-D-manno-octulosonic acid + 4-methylumbelliferone <1> [1]

Specific activity (U/mg)
1433 <1> [1]

pH-Optimum

4.5 <1> (<1> pH-optimum for natural substrate re-lipopolysaccharide, optimum for 4-methylumbelliferyl 3-deoxy-α-D-manno-2-octulopyranosidonic acid at pH 3 [1]) [1]

5 Isolation/Preparation/Mutation/Application

Source/tissue

hepatopancreas <1> [1]

Purification

<1> ($(NH_4)_2SO_4$ precipitation, Sephacryl S200, SP-Fractogel, Con A-Sepharose, octyl-Sepharose [1]) [1]

6 Stability

pH-Stability

3-8 <1> [1]

References

[1] Li, Y.; Wang, L.; Pavlovat, N.V.; Li, S.; Lee, Y.C.: α-KDOase activity in oyster and synthesis of α- and β-4-methylumbelliferyl ketosides of 3-deoxy-D-manno-octulosonic acid (KDO). J. Biol. Chem., **272**, 26419-26424 (1997)

Galactan 1,3-β-galactosidase 3.2.1.145

1 Nomenclature

EC number
3.2.1.145

Systematic name
galactan (1→3)-β-D-galactosidase

Recommended name
galactan 1,3-β-galactosidase

Synonyms
arabinogalactan endo-1,3-β-galactosidase

CAS registry number
62213-17-6

2 Source Organism

<1> *Irpex lacteus* (Polyporus tulipiferae [1]) [1]

3 Reaction and Specificity

Catalyzed reaction
hydrolysis of terminal, non-reducing β-D-galactose residues in (1→3)-β-D-galactopyranans

Reaction type
hydrolysis of O-glycosyl bond

Natural substrates and products
S β-(1-3)-D-galactan + H_2O <1> [1]
P β-(1-3)-D-galactan + D-galactose <1> [1]

Substrates and products
S β-(1-3)-D-galactan + H_2O <1> [1]
P β-(1-3)-D-galactan + D-galactose <1> [1]
S β-(1-3)-D-galactotriose + H_2O <1> [1]
P β-(1-3)-D-galactobiose + D-galactose
S β-(1-3)-galactobiose + H_2O <1> [1]
P D-galactose <1> [1]

S galactooligosaccharide + H_2O <1> (<1> only β-(1→3) oligomers act as substrates [1]) [1]

P galactooligosaccharide + D-galactose <1> [1]

Specific activity (U/mg)

87.8 <1> [1]

K_m-Value (mM)

19.9 <1> (β-(1-3)-D-galactotriose) [1]
31.6 <1> (β-(1-3)-D-galactobiose) [1]

pH-Optimum

4.6 <1> (<1> enzyme action on β-(1-3)-D-galactan [1]) [1]

Temperature range (°C)

30-48 <1> (<1> retains full activity up to 48°C, completely inactivated at 60°C [1]) [1]

4 Enzyme Structure

Molecular weight

42000 <1> (<1> gel filtration [1]) [1]
51000 <1> (<1> native PAGE [1]) [1]

Subunits

monomer <1> (<1> 1 * 51000, SDS-PAGE [1]) [1]

5 Isolation/Preparation/Mutation/Application

Purification

<1> (DEAE-cellulose, CM-Toyopearl, hydroylapatite, electrofocusing, purified from commercial driselase [1]) [1]

6 Stability

pH-Stability

3.5-7.5 <1> (<1> after incubation for 24 h at 4°C [1]) [1]

Temperature stability

30-48 <1> (<1> retains full activity up to 48°C, completely inactivated at 60°C [1]) [1]

References

[1] Tsumuraya, Y.; Mochizuki, N.; Hashimoto, Y.; Kovacß, P.: Purification of an exo-β-(1→3)-D-galactanase of Irpex lacteus (Polyporus tulipiferae) and its action on arabinogalactan-proteins. J. Biol. Chem., **265**, 7207-7215 (1990)

β-Galactofuranosidase 3.2.1.146

1 Nomenclature

EC number
3.2.1.146

Systematic name
β-D-galactofuranoside hydrolase

Recommended name
β-galactofuranosidase

Synonyms
β-D-galactofuranosidase
exo-β-D-galactofuranosidase
exo-β-galactofuranosidase

CAS registry number
52357-57-0

2 Source Organism

<1> *Aspergillus niger* (enzyme is induced in the presence of galactofuranoside containing glycoconjugates [1,5]) [1, 5]
<2> *Aspergillus fumigatus* [1, 5]
<3> *Aspergillus nidulans* [1, 5]
<4> *Aspergillus parasiticus* [1, 5]
<5> *Aspergillus tamari* [1, 5]
<6> *Aspergillus versicolor* [1, 5]
<7> *Penicillium charlesii* (enzyme synthesis is blocked by 2-deoxy-D-glucose [9]) [8, 9]
<8> *Penicillium fellutanum* [2, 3, 6]
<9> *Penicillium expansum* [5]
<10> *Penicillium funiculosum* [5]
<11> *Penicillium implicatum* [5]
<12> *Penicillium islandicum* [5]
<13> *Penicillium palitans* [5]
<14> *Penicillium rubrum* [5]
<15> *Penicillium simplissimum* [5]
<16> *Penicillium tardum* [5]
<17> *Penicillium viridicatum* [5]
<18> *Penicillium verrucosum* [5]

<19> *Trichoderma harzianum* [4]
<20> *Helminthosporium maydis* [7]
<21> *Helminthosporium carbonum* [7]
<22> *Helminthosporium turcicum* [7]
<23> *Helminthosporium victoriae* [7]
<24> *Helminthosporium saccharii* [7]

3 Reaction and Specificity

Catalyzed reaction

hydrolysis of terminal non-reducing β-D-galactofuranosides, releasing galactose

Reaction type

hydrolysis of O-glycosyl bond

Natural substrates and products

S peptidophosphogalactomannan + H_2O <7> [8]
P peptidophosphogalactomannan + D-galactose <7> [8]

Substrates and products

S Helminthosporium sacchari toxin + H_2O <24> (<24> enzyme cleaves only the terminal unit of galactose [7]) [7]
P Helminthosporium sacchari toxin + D-galactose [7]
S β-D-(1→5) galactofuranose tetramer + H_2O <19> [4]
P β-D-(1→5) galactofuranose trimer + D-galactose <19> [4]
S methyl-β-D-galactofuranoside + H_2O <7> [8]
P methanol + D-galactose <7> [8]
S *p*-nitrophenyl β-D-galactofuranoside + H_2O <1-18> (<8> 4-nitrophenyl β-D-fucofuranoside is not hydrolyzed [2]) [1, 2, 3, 5, 6, 8, 9]
P *p*-nitrophenol + D-galactose <1-18> [1, 3, 5, 6, 8, 9]

Inhibitors

D-galactono-1,4-lactone <1, 2> (<1> competitive inhibition, 50% inhibition in the presence of 1 mM $ZnCl_2$, $CaCl_2$, $NiCl_2$ or $AgNO_3$ [1]) [1, 3]

Specific activity (U/mg)

1 <19> (<19> extracellular polysaccharides of Penicillium digitatum as substrate [4]) [4]
6.77 <7> (<7> with 70% galactose in substrate peptidophosphogalactomannan [8]) [8]
26.3 <1> [1]

K_m-Value (mM)

0.31 <8> (*p*-nitrophenyl β-D-galactofuranoside) [3]
3.87 <1> (*p*-nitrophenyl β-D-galactofuranoside) [1]

pH-Optimum

3-4 <1> (<1> very little activity above pH 6.0 [1]) [1]

3-4 <7> (<7> sharp drop above pH 5, no activity above pH 7 [8]) [8]
3-6 <8> (<8> enzyme from Penicillium fellutanum, sharp drop at pH 8.0 or above [3]) [3]
4-4.5 <19> (<19> activity drops above pH 5, no activity at pH 7 or higher [4]) [4]

Temperature optimum (°C)

35-40 <19> [4]
47 <7> [8]

4 Enzyme Structure

Molecular weight

35000 <19> (<19> SDS-PAGE [4]) [4]
55000 <1> (<1> non-denaturing PAGE [1]) [1]
70000 <8> (<8> SDS-PAGE [3]) [3]
90000 <1> (<1> SDS-PAGE [1]) [1]
150000 <8> (<8> non-denaturing PAGE [3]) [3]

Posttranslational modification

glycoprotein <1, 8> (<8> stains with Schiff-reagent, binds to Concanavalin A-Sepharose, 17% carbohydrate content [3]) [1, 3]

5 Isolation/Preparation/Mutation/Application

Purification

<1> (dialysed culture filtrate, Sepharose-Q, NeobarAQ-1, Neobar AQ-2, Superose-12, GNA-agarose [1]) [1]
<7> (affinity chromatography on Sepharose 4B-peptidophosphogalactomannan [8]) [8]
<8> (culture filtrate of Penicillium fellutanum, DEAE-Sepharose, 4-aminophenyl 1-thio-β-D-galactofuranoside-Sepharose [3]) [3]
<19> (purified from Glucanex, a crude enzyme preparation isolated from Trichoderma harzianum [4]) [4]

6 Stability

pH-Stability

3-10 <1> [1]

Storage stability

<7>, 25°C, 66 mM acetate buffer pH 4, 1 week, 20% loss in activity [8]

References

[1] Wallis, G.L.F.; Hemming, F.W.; Peberdy, J.F.: An extracellular β-galactofuranosidase from Aspergillus niger and its use as a tool for glycoconjugate analysis. Biochim. Biophys. Acta, **1525**, 19-28 (2001)

[2] Chiocconi, A.; Marino, C.; de Lederkremer, R.M.: Synthesis of 4-nitrophenyl β-D-fucofuranoside and β-D-fucofuranosyl-(1→3)-D-mannopyranose: modified substrates for studies on catalytic requirements of β-D-galactofuranosidase. Carbohydr. Res., **323**, 7-13 (2000)

[3] Miletti, L.C.; Marino, C.; Marino, K.; de Lederkremer, R.M.; Colli, W.; Alves, M.J.M.: Immobilized 4-aminophenyl 1-thio-β-D-galactofuranoside as a matrix for affinity purification of an exo-β-D-galactofuranosidase. Carbohydr. Res., **320**, 176-182 (1999)

[4] van Bruggen van der Lugt, A.W.; Kamphuis, H.J.; de Ruiter, G.A.; Mischnik, P.; van Boom, J.H.; Rombouts, F.M.: New structural features of the antigenic extracellular polysaccharides of Penicillium and Aspergillus species revealed with exo-β-D-galactofuranosidase. J. Bacteriol., **174**, 6096-6102 (1992)

[5] Cousin, M.A.; Notermans, S.; Hoogerhout, P; van Boom, J.H.: Detection of β-galactofuranosidase production by Penicillium and Aspergillus species using 4-nitrophenyl β-D-galactofuranosidase. J. Appl. Bacteriol., **66**, 311-317 (1989)

[6] Varela, O.; Marino, C.; de Lederkremer, R.M.: Synthesis of *p*-nitrophenyl β-D-galactofuranoside. A convenient substrate for β-galactofuranosidase. Carbohydr. Res., **155**, 247-251 (1986)

[7] Livingston, R.S.; Scheffer, R.P.: Conversion of Helminthosporium sacchari toxin to toxoids by β-galactofuranosidase from Helminthosporium. Plant Physiol., **72**, 530-534 (1983)

[8] Rietschel-Berst, M.; Jentoft, N.H.; Rick, P.D.; Pletcher, C.; Fang, F.; Gander, J.E.: Extracellular exo-β-galactofuranosidase from Penicillium charlesii. J. Biol. Chem., **252**, 3219-3226 (1977)

[9] Gander, J.E; Fang, F.: Influence of 2-deoxy-D-glucose on galactofuranosidase and peptidophosphogalactomannan synthesis in Penicillium charlesii. Biochem. Biophys. Res. Commun., **58**, 368-374 (1974)

Thioglucosidase 3.2.1.147

1 Nomenclature

EC number
3.2.1.147

Systematic name
thioglucoside glucohydrolase

Recommended name
thioglucosidase

Synonyms
EC 3.2.3.1 (formerly)
β-thioglucosidase
β-thioglucoside glucohydrolase
glucosidase, thio-
glucosinolase
myrosin
myrosinase
sinigrase
sinigrinase

CAS registry number
9025-38-1

2 Source Organism

<1> *Brassica napus* (i.e. oilseed rape [39]; cv. panter [9]; i.e. rape [16]; cv. Zephyr [22]; 4 isoenzymes: RA, RB, RC and RD [24]; cv. Turret [30]; cv. Bronowski [22, 30]; L. cv. Bienvenu [37]; myrosinase M65, M70 and M75 [48]) [1, 4, 7, 9, 16, 22, 24, 25, 30, 37, 39, 42, 43, 44, 45, 48]
<2> *Lepidium sativum* (i.e. cress [16]) [2, 11, 15, 16, 39, 43, 47]
<3> *Sinapis alba* (i.e. white mustard [16]; 3 isoenzymes: SA, SB and SC [24]) [3, 7, 8, 10, 11, 12, 13, 16, 22, 24, 30, 31, 38, 39, 49, 50, 52]
<4> *Brassica juncea* (i.e. yellow mustard [18]; i.e. leaf mustard [46]) [5, 6, 18, 19, 35, 39, 46]
<5> *Brassica nigra* [7]
<6> *Crambe abyssinicy* [8, 12, 39]
<7> *Aspergillus sydowi* (slight activity [26]; IFO 4284 [32]) [8, 10, 26, 32, 33, 34, 39]

<8> *Brassica rapa* (ssp. rapifera [8]; ssp. rapifera, i.e. turnip [12]; cv. Echo [30]; var. chinensis [39]; subsp. rapa [51]) [8, 12, 30, 39, 51]
<9> *Brassica chinensis* (i.e. chineses mustard [46]) [11, 16, 39, 46]
<10> *Brassica oleracea* (L. var. capitata L. [46]; var. capitata L. f. alba DC i.e. white cabbage [12]; var. gemmifera i.e. Brussel sprouts [12]; var. bullata subvar. gemmifera, i.e. Brussel sprouts [39]; L. var. alboglabra, i.e. Chineses kale [46]; L. var. italica Plenca, i.e. broccoli [46]; L. var. botrytis, i.e. cauliflower [46]) [11, 12, 39, 46]
<11> *Rhaphanus sativus* (i.e. radish [16, 41, 46]) [11, 16, 39, 40, 41, 46]
<12> *Armoracia rusticana* (i.e. horseradish [12]) [12, 39]
<13> *Iberis amara* (i.e. wild candytuft [16]; var. coronaria, i.e. candytuft [16]) [16]
<14> *Wasabia japonica* [17, 39]
<15> *Enterobacter cloacae* (strain 506 [21]) [20, 21, 39]
<16> *Aspergillus niger* (AKU 3302 [23]) [23, 26, 39]
<17> *Carica papaya* [28]
<18> *Brevicoryne brassicae* (i.e. cabbage aphid [36]) [36, 39]
<19> *Paracolobactrum aerogenoides* [39]
<20> *Lipaphis erisimi* [39]
<21> *Cheiranthus cheiri* [39]
<22> *Arabis alpina* [39]
<23> *Tropaeolum majus* [39]
<24> *Iberis umbellata* [39]
<25> *Sinapis sp.* (myrosinase F-IA, D-IB, F-IIA and F-IIB [29]) [14, 27, 29, 36]
<26> *Brassica pekinensis* (i.e. Chineses cabbage [46]) [46]
<27> *Brassica caulorapa* (i.e. kohlrabi [46]) [46]

3 Reaction and Specificity

Catalyzed reaction

a thioglucoside + H_2O = a thiol + a sugar (<2>, the mechanism involves a first step acid-base catalyzed reaction with the formation of the glycosyl enzyme intermediate and the benzyl isothiocyanate obtained by the enzyme-independent Lossen rearrangement of the transient thiohydroximate-O-sulfonate, a sulfate anion and a proton are also released [47])

Reaction type

hydrolysis of O-β-glucosyl bond
hydrolysis of O-glycosyl bond
hydrolysis of S-glycosyl bond

Natural substrates and products

S 2-hydroxybut-3-enylglucosinolate + H_2O <1> [9]
P ?
S Additional information <11, 15> (<15>, induced by addition of 0.01% sinigrin and 6% mustard extract [21]; <11>, the enzyme is normally physically segregated from the glucosinolates, but when the cells are damaged,

e.g. during food preparation, mastication or injury by predators such as insects, the enzyme is released and catalyzes their hydrolysis of glucosinolates [40]) [21, 40]

P ?

Substrates and products

S 2-hydroxybut-3-enylglucosinolate + H_2O <1, 3> (<3>, (S)-2-hydroxy-3-butenylglucosinolate, i.e. epi-progoitrin [8]) [4, 8, 9]

P Additional information <1, 3> (<1>, in absence of ascorbate 5-vinyloxazolidine-2-thione is the only product, whereas in the presence of added ascorbate 1-cyano-2-hydroxybut-3-ene is also formed [4]; <3>, a mixture of products which includes 1-cyano-2-hydroxy-3-butene, (R)-5-vinyloxazolidine-2-thione, D-glucose, HSO_4^- and elemental sulfur is formed without epithiospecifier protein at pH 5.9. These products as well as erythro-1-cyano-2-hydroxy-3,4-epithiobutanes and threo-1-cyano-2-hydroxy-3,4-epithiobutanes are formed by combination of the enzyme and epithiospecifier protein from various sources [8]; <1>, 5-vinyloxazolidine-2-thione is the sole product above pH 5.4, reaching a maximum at pH 8.0, whilst 1-cyano-2-hydroxybut-3-ene is the main product at low pH reaching a maximum at pH 3.4 [9]) [4, 8, 9]

S 2-phenethylglucosinolate + H_2O <2, 25> [14, 15]

P Additional information <2, 25> (<2>, in presence of ascorbic acid, isothiocyanate and nitrile are obtained but no thiocyanate [15]; <25>, nitrile formation is favoured at lower pH levels, the ratio of nitrile to isothiocyanate is directly related to the hydrogen ion concentration of the medium [14]) [14, 15]

S 3-deoxyglucotropaeolin + H_2O <2> [47]

P ?

S 4-deoxyglucotropaeolin + H_2O <2> [47]

P ?

S 6-deoxyglucotropaeolin + H_2O <2> [47]

P ?

S allylglucosinolate + H_2O <1-3, 6, 7, 8, 10, 12, 25> [9, 10, 12, 14, 15]

P Additional information <2, 3, 7, 25> (<3, 7>, in presence of epithiospecifier protein 1-cyano-2,3-epithiopropane and allyl isothiocyanate are formed, in absence of epithiospecifier protein only allyl isothiocyanate is formed [10]; <3, 6, 8, 10, 12>, conversion to 1-cyano-2,3-epithiopropane in presence of epithiospecifier protein, conversion to allyl cyanide in absence of epithiospecifier protein [12];<2>, in presence of ascorbic acid, isothiocyanate and nitrile are obtained but no thiocyanate [15]; <25>, nitrile formation is favoured at lower pH levels, the ratio of nitrile to isothiocyanate is directly related to the hydrogen ion concentration of the medium [14]) [10, 12, 14, 15]

S benzylglucosinolate + H_2O <1, 2> [15, 37]

P Additional information (<2>, in presence of ascorbic acid, isothiocyanate and nitrile are obtained but no thiocyanate, in absence of ascorbic acid, benzyl thiocyanate is an additional product [15]) [15]

S glucoapparin + H_2O <1, 3> [24]
P ?
S glucobrassicin + H_2O <1> [37]
P ?
S glucocheirolin + H_2O <1, 3> [24]
P ?
S glucosinalbin + H_2O <1, 3, 11> [24, 41]
P ?
S glucotropaeolin + H_2O <1, 3> [24]
P ?
S glucotropaeolin + H_2O <2> [47]
P ?
S neo-glucobrassicin + H_2O <1> [37]
P ?
S *p*-hydroxybenzylglucosinolate + H_2O <1> [45]
P ?
S *p*-nitrophenyl β-D-glucopyranoside + H_2O <3, 7, 11, 15> [2, 20, 33, 40, 50]
P ?
S phenylethylglucosinolate + H_2O <1> [37]
P ?
S progoitrin + H_2O <1, 3, 6, 11, 18> [24, 36, 37, 39, 41]
P Additional information <6> (<6>, mainly oxazolidine-2-thione in the absence of epithiospecifier protein and mainly epithionitrile in the presence of epithiospecifier protein [39]) [39]
S sinigrin + H_2O <1-27> (<1>, 2-propenylglucosinolate [1]) [1-52]
P glucose + isothiocyanate <1, 4> [1, 6]
S Additional information <1, 3> (<3>, hydrolysis of mustard oil glucosides to isothiocyanate, glucose and sulfate [13]; <1,3>, hydrolysis of rapeseed meal yields 1-cyano-2-hydroxy-3-butene, 5-vinyloxazolidine-2-thione, 3-butenylisothiocyanate and 4-pentenylisothiocyanate [22]; <3>, the enzyme does not catalyze a transglycosylation reaction either with alcohols or with other suitable glycosyl acceptors [50]; <3>, enzyme retains the anomeric configuration at the cleavage point, the catalytic residue is a nucleophilic glutamate [52]) [13, 22, 50, 52]
P ?

Inhibitors

(2R,5R)-dihydroxymethyl-(3R,4R)-dihydroxypyrrolidine <18, 25> (<18>, inhibition of hydrolysis of sinigrin and progoitrin at pH 5 and at pH 7 [36]) [36]

1,10-phenanthroline <14, 16> (<16>, 1 mM, 26% inhibition [23]) [17, 23]

1,4-dideoxy-1,4-imino-D-arabinitol <18> (<18>, inhibition of hydrolysis of progoitrin at pH 5 in citrate buffer and at pH 7 in phosphate buffer, inhibition of hydrolysis of sinigrin at pH 7 in phosphate buffer, no hydrolysis of progoitrin and sinigrin at pH 5 in acetate buffer [36]) [36]

1-deoxynojirimycin <18> (<18>, inhibition of hydrolysis of progoitrin at pH 5 in citrate buffer and at pH 7 in phosphate buffer, inhibition of hydrolysis of sinigrin at pH 7 in phosphate buffer, no hydrolysis of progoitrin and sinigrin at pH 5 in acetate buffer [36]) [36]
2-deoxy-2-fluoroglucotropaeolin <1> [45]
2-deoxy-2-fluoroglucotropaeolin <3> (<3>, inhibition occurs via the accumulation of a long-life glucosyl-enzyme intermediate [52]) [52]
2-deoxyglucotropaeolin <2> [47]
2-methoxy-5-nitrotropone <4, 14> (<4>, 1 mM, strong [18]) [17, 18]
5,5'-dithiobis(2-nitrobenzoic acid) <4, 14, 25> [17, 18, 27]
Ag^+ <11> (<11>, 1 mM $AgNO_3$, 79% inhibition [41]) [41]
Ba^{2+} <8> [51]
Ca^{2+} <8, 11, 15> (<15>, 1 mM, $CaCl_2$, 14% inhibition [20]; <11>, 1 mM $CaSO_4$, 55% inhibition [41]) [20, 41, 51]
Co^{2+} <11, 15> (<15>, 1 mM, $CoCl_2$, 27% inhibition [20]; <11>, 1 mM $CoNO_3$, 66% inhibition [41]) [20, 41]
Cu^+ <15> (<15>, 1 mM, CuCl, 56% inhibition [20]) [20]
Cu^{2+} <1, 7, 8, 11, 14, 15> (<1>, strong [4]; <14>, 1 mM $CuCl_2$, complete inhibition [17]; <15>, 1 mM, $CuCl_2$, 78% inhibition [20]; <7>, 1 mM $CuCl_2$, 25% inhibition [34]; <11>, 1 mM $CuSO_4$, 30% inhibition [41]) [4, 17, 20, 34, 41, 51]
DTT <4> (<4>, strong inhibition of formation of allyl isothiocyanate in presence of Fe^{2+} at pH 6.5 and at pH 5.0, little influence on glucose production from sinigrin [5]) [5]
EDTA <11, 14, 15, 16> (<16>, 1 mM, 20% inhibition [23]) [17, 20, 23, 41]
Fe^{2+} <1, 4, 7, 11, 15> (<1>, slightly suppresses reaction but causes a significant effect by directing degradation of 2-hydroxybut-3-enylglucosinolate to 1-cyano-2-hydroxy-3-ene rather than to 5-vinyloxazolidine-2-thione [4]; <4>, at pH 4.5 and 5.5 the formation of isothiocyanate is strongly inhibited, depressed effect at pH 6.5, no effect at pH 7.5. No inhibition of glucose liberation [6]; <15>, 1 mM, $FeCl_2$, 68% inhibition [20]; <7>, 1 mM $FeSO_4$, 42% inhibition [34]; <11>, 1 mM $FeSO_4$, 61% inhibition [41]) [4, 6, 20, 34, 41]
Fe^{3+} <14, 15, 16> (<14>, 1 mM $FeCl_3$, 16% inhibition [17]; <15>, 1 mM, $FeCl_3$, 18% inhibition [20]; <16>, 1 mM $FeCl_3$, 26% inhibition [23]) [17, 20, 23]
$HgCl_2$ <7, 14, 15, 16> (<14>, 1 mM $HgCl_2$, 93% inhibition [17]; <15>, 1 mM, $HgCl_2$, 87% inhibition [20]; <16>, 1 mM $HgCl_2$, 95% inhibition [23]; <7>, 1 mM $HgCl_2$, 89% inhibition [34]) [17, 20, 23, 34]
K_2SO_4 <11> (<11>, 1 mM, complete inhibition [41]) [41]
KNO_3 <4> (<4>, activates in absence of ascorbate, inhibits in presence of 1 mM ascorbate [19]) [19]
L-Cys <4, 11> (<4>, 5 mM, and 2.5 mM Fe^{2+}, pH 5.0 or 6.5, strong inhibition of formation of allyl isothiocyanate [5]) [5, 41]
Mg^{2+} <11, 15, 16> (<15>, 1 mM, $MgCl_2$, 22% inhibition [20]; <16>, 1 mM $MgCl_2$, 23% inhibition [23]; <11>, 1 mM $MgSO_4$, 93% inhibition [41]) [20, 23, 41]

NaBr <4, 16> (<4>, activates in absence of ascorbate, inhibits in presence of 1 mM ascorbate [19]; <16>, 1 mM, 16% inhibition [23]) [19, 23]

NaCl <2> (<2>, 0.5 M, 30% inhibition, 2 M, complete inhibition [47]) [47]

$NaNO_3$ <4> (<4>, activates in absence of ascorbate, inhibits in presence of 1 mM ascorbate [19]) [19]

Ni^{2+} <15> (<15>, 1 mM, $NiCl_2$, 30% inhibition [20]) [20]

PCMB <11, 16, 25> (<16>, 0.06 mM, 91% inhibition [23]) [23, 27, 41]

Pb^{2+} <8, 11> (<11>, 1 mM Pb-acetate, 60% inhibition [41]) [41, 51]

SO_4^{2-} <11> [40]

Sn^{2+} <16> (<16>, 1 mM, $SnCl_2$, 66% inhibition [23]) [23]

Sr^{2+} <15> (<15>, 1 mM, $SrCl_2$, 21% inhibition [20]) [20]

Zn^{2+} <11, 15, 16> (<15>, 1 mM, $ZnCl_2$, 45% inhibition [20]; <16>, 1 mM $ZnCl_2$, 28% inhibition [23]; <11>, 1 mM $ZnSO_4$, 37% inhibition [41]) [20, 23, 41]

alexine <18> (<18>, inhibition of hydrolysis of progoitrin at pH 5 and at pH 7, inhibition of hydrolysis of sinigrin at pH 7, no inhibition of hydrolysis of sinigrin at pH 5 [36]) [36]

amygdalin <7, 15, 16> (<16>, 0.02 M, 14% inhibition [23]; <7>, 0.1 mM, 98% inhibition [34]) [20, 23, 34]

arbutin <4, 7, 15, 16> (<16>, 0.1 M, 21% inhibition [23]; <7>, 0.1 M, 27% inhibition [34]; <4>, 0.1 M, 7% inhibition in absence of ascorbate, 26% inhibition in presence of ascorbate [35]) [20, 23, 34, 35]

ascorbic acid <11, 15> (<15>, 0.1 mM L-ascorbic acid, 88% inhibition [20]) [20, 41]

castanospermine <1, 2, 18, 25> (<2>, 0.3 mM, 50% inhibition, competitive [2]) [2, 36, 45]

cysteamine <4> (<4>, strong inhibition of formation of allyl isothiocyanate in presence of Fe^{2+} at pH 6.5 and at pH 5.0, little influence on glucose production from sinigrin [5]) [5]

δ-gluconolactone <3, 7, 15, 16> (<16>, 1 mM, 62% inhibition [23]; <7>, 10 mM, 68% inhibition [34]; <3>, poor noncompetitive inhibitor [50]) [20, 23, 34, 50]

diisopropyl fluorophosphate <16> (<16>, 1 mM, 18% inhibition [23]) [23]

fluorodinitrobenzene <25> (<25>, rate of inhibition is accelerated by 1 mM ascorbic acid [27]) [27]

fructose <4, 15> (<4>, 1.0 M, 16% inhibition in presence of ascorbate [35]) [20, 35]

galactose <4, 15> (<4>, 1.0 M, 22% inhibition in presence of ascorbate [35]) [20, 35]

glucose <4, 11, 15> (<4>, 1.0 M, 22% inhibition in presence of ascorbate, competitive. No inhibition in absence of ascorbate [35]; <11>, very weak inhibition [40]) [20, 35, 40]

maltose <4, 15> (<4>, 1.0 M, 33% inhibition in absence of ascorbate, 6% inhibition in presence of ascorbate [35]) [20, 35]

mannose <4, 15> (<4>, 1.0 M, 13% inhibition in presence of ascorbate [35]) [20, 35]

methyl-α-D-glucopyranoside <7, 15> (<7>, 0.1 M, 25% inhibition [34]) [20, 34]
methyl-β-D-glucopyranoside <7, 15> (<7>, 0.1 M, 21% inhibition [34]) [20, 34]
monochlorotrifluoro-*p*-benzoquinone <25> [27]
p-diazabenzenesulfonic acid <25> [27]
p-mercuribenzoate <4, 14, 25> (<4>, 0.1 mM, strong inhibition [18]) [17, 18, 27]
p-nitrophenyl-β-D-glucopyranoside <7, 15> (<7>, 0.05 M, 31% inhibition [34]) [20, 34]
phenyl-β-D-glucopyranoside <7, 15> (<7>, 0.1 M, 52% inhibition [34]) [20, 34]
salicin <4, 7, 15, 16> (<16>, 0.1 M, 15% inhibition [23]; <7>, 0.1 M, 22% inhibition [34]; <4>, 0.1 M, 1% inhibition in presence of ascorbate, 6% inhibition in absence of ascorbate [35]) [20, 23, 34, 35]
sinigrin <7> (<7>, competitive inhibition of hydrolysis of *p*-nitrophenyl β-glucoside [33]) [33]
sorbitol <15> [20]
sucrose <15> [20]
thiobenzoate <4> (<4>, strong inhibition of formation of allyl isothiocyanate in presence of Fe^{2+} at pH 6.5 and at pH 5.0 [5]) [5]
thiomalate <4> (<4>, strong inhibition of formation of allyl isothiocyanate in presence of Fe^{2+} at pH 6.5 and at pH 5.0 [5]) [5]
thiophenol <4> (<4>, strong inhibition of formation of allyl isothiocyanate in presence of Fe^{2+} at pH 6.5 and at pH 5.0, little influence on glucose production from sinigrin [5]) [5]
trinitrobenzenesulfonic acid <14, 25> (<25>, rate of inhibition is accelerated by 1 mM ascorbic acid [27]) [17, 27]
xylose <4> (<4>, 1.0 M, 13% inhibition [35]) [35]

Activating compounds

2-mercaptoethanol <11, 15> (<15>, 1 mM, activates [20]; <11>, 1 mM, slight activation [41]) [20, 41]
D-araboascorbate <14> (<14>, lower activation than with ascorbic acid [17]) [17]
EDTA <8> (<8>, 0.04 M, 70% activation [51]) [51]
ascorbic acid <1, 2, 3, 4, 11, 14> (<1, 3, 11, 14>, activates [22, 37, 39, 40, 45, 50]; <3>, 1 mM, activation of isoenzyme SA, SB, very low activation of isoenzyme SC [24]; <3>, 1 mM, activation of isoenzyme RA, RB and RC [24]; <4>, activates, enzyme is conformationally changed, no activation by analogs [18]; <2>, activates, enzyme is still active in absence of ascorbic acid although to much lesser extent, in this circumstances benzyl thiocyanate is an additional product in hydrolysis of benzylglucosinolate [15]; <1>, maximal activation at 0.3-0.5 mM [1]; <1>, maximal activation at 1.57 mM [4]; <14>, active only in presence of L-ascorbic acid, maximal activation at 2 mM [17]; <3>, considerable activation in the pH range 5.7-7.5, maximal activation at 0.375 mM [30]; <1>, maximal activation for myrosinase I, 50fold

at 0.3 mM. Maximal activation for myrosinase II, 106fold between 0.3 and 0.5 mM [37]; <3>, maximal activation at 0.7 mM [39]; <3>, relatively unresponsible to ascorbic acid [22]) [1, 4, 15, 17, 18, 22, 24, 30, 37, 39, 40, 45, 50]
ascorbyl palmitate <14> (<14>, lower activation than with ascorbic acid [17]) [17]
ascorbyl stearate <14> (<14>, lower activation than with ascorbic acid [17]) [17]
epithiospecifier protein <3, 6-8, 10, 12> (<3, 6-8, 10, 12> ESP, is a small protein of molecular weight 30 to 40 kDa, which co-occurs with myrosinase. ESP does not have thioglucosidase activity, but interacts with myrosinase to promote the transfer of sulfur from the S-glucose moiety of terminally unsaturated glucosinolates to the alkenyl moiety, resulting in the formation of epithionitriles. The presence of ferrous ions is essential for ESP function. [8, 10, 12, 39]; <3>, a mixture of products which includes 1-cyano-2-hydroxy-3-butene, (R)-5-vinyloxazolidine-2-thione, D-glucose, HSO_4^- and elemental sulfur is formed from (S)-2-hydroxy-3-butenylglucosinolate without epithiospecifier protein at pH 5.9. These products as well as erythro- and threo-1-cyano-2-hydroxy-3,4-epithiobutanes are formed by combination of the enzyme and epithiospecifier protein from various sources [8]; <3, 7>, in presence of epithiospecifier protein 1-cyano-2,3-epithiopropane and allyl isothiocyanate are formed, in absence of epithiospecifier protein only allyl isothiocyanate is formed [10]; <6>, non-specific requirement for epithiospecifier protein [12]; <6>, interacts with myrosinase to promote sulfur transfer from the S-glucose moiety to the terminal alkenyl moiety. Degradation of progoitrin produces mainly oxazolidine-2-thione in the absence of epithiospecifier protein and mainly epithionitrile in the presence of epithiospecifier protein [39]) [8, 10, 12, 39]

Metals, ions

$(NH_4)_2SO_4$ <4> (<4>, activation is stronger in presence of 1 mM ascorbate [19]) [19]
Ba^{2+} <4> (<4>, activation is stronger in presence of 1 mM ascorbate [19]) [19]
Ca^{2+} <4, 14> (<14>, 1 mM $CaCl_2$, activates [17]; <4>, slight activation in presence of 1 mM ascorbate, no effect in absence of ascorbate [19]) [17, 19]
Co^{2+} <7, 14, 16> (<14>, 1 mM $CoCl_2$, activates [17]; <7, 16>, 1 mM, stimulates [23, 34]) [17, 23, 34]
Cu^+ <16> (<16>, 1 mM, stimulates [23]) [23]
Cu^{2+} <16> (<16>, 1 mM, stimulates [23]) [23]
K^+ <4, 14> (<14>, 1 mM KCl_2, activates [17]; <4>, K_2SO_4, activation is stronger in presence of 1 mM ascorbate [19]) [17, 19]
KNO_3 <4> (<4>, activates in absence of ascorbate, inhibits in presence of 1 mM ascorbate [19]) [19]
Li^+ <15> (<15>, 1 mM $LiCl_2$, activates [20]) [20]
Mg^{2+} <4, 7, 14> (<14>, 1 mM $MgCl_2$, slight activation [17]; <4>, activation is stronger in presence of 1 mM ascorbate [19]; <7>, 1 mM, slight activation [34]) [17, 19, 34]

Mn^{2+} <14, 16> (<14>, 1 mM $MnCl_2$, activates [17]; <16>, 1 mM, stimulates [23]) [17, 23]
Na^{+} <4, 14> (<14>, 1 mM $NaCl_2$, slight activation [17]; <4>, Na_2SO_4, activation is stronger in presence of 1 mM ascorbate [19]) [17, 19]
NaBr <4> (<4>, activates in absence of ascorbate, inhibits in presence of 1 mM ascorbate [19]) [19]
$NaNO_3$ <4> (<4>, activates in absence of ascorbate, inhibits in presence of 1 mM ascorbate [19]) [19]
Ni^{2+} <14> (<14>, 1 mM $NiCl_2$, activates [17]) [17]
Sn^{2+} <14> (<14>, 1 mM $SnCl_2$, activates [17]) [17]
Sr^{2+} <14> (<14>, 1 mM $SrCl_2$, activates [17]) [17]
Zn^{2+} <3, 7, 14> (<14>, 1 mM $ZnCl_2$, activates [17]; <7>, 1 mM $ZnCl_2$, slight activation [34]; <3>, dimeric enzyme is stabilized by a Zn^{2+}-ion bound on a twofold axis. The Zn^{2+} has a tetrahedral coordination, with angles between 97° and 118°, though four residues: His56 and Asp70, and their symmetry-related equivalents [49]) [17, 34, 49]

Turnover number (min^{-1})

96 <2> (3-deoxyglucotropaeolin) [47]
126.6 <11> (sinigrin, <11>, in absence of ascorbate [40]) [40]
186 <2> (4-deoxyglucotropaeolin) [47]
564 <2> (6-deoxyglucotropaeolin) [47]
3936 <2> (glucotropaeolin) [47]
17220 <11> (sinigrin, <11>, in presence of 0.5 mM ascorbate [40]) [40]

Specific activity (U/mg)

0.58 <16> (<16>, hydrolysis of *p*-nitrophenyl *β*-glucoside [23]) [23]
1.31 <2> [2]
1.916 <16> (<16>, hydrolysis of sinigrin [23]) [23]
5.26 <7> (<7>, hydrolysis of *p*-nitrophenyl *β*-glucoside [33]) [33]
10.07 <7> [32]
11.5 <15> [20]
25.33 <1> (<1>, enzyme form Cc [1]) [1]
28.94 <1> (<1>, enzyme form Ca [1]) [1]
32.11 <1> (<1>, enzyme form Cb [1]) [1]
60 <3> (<3>, myrosinase C [31]) [31]
111 <7> (<7>, hydrolysis of sinigrin [33]) [33]
114 <1> (<1>, myrosinase I [37]) [37]
164 <11> [40]
283 <1> (<1>, myrosinase II [37]) [37]
Additional information <3, 11, 14> (<3>, direct ultraviolet spectrophotometric assay [13]) [13, 17, 41]

K_m-Value (mM)

0.023 <11> (sinigrin, <11>, in absence of ascorbate [40]) [40]
0.03 <1> (sinigrin, <1>, isoenzyme RC, without ascorbic acid [24]) [24]
0.033 <2> (4-deoxyglucotropaeolin) [47]
0.045 <8> (sinigrin) [51]

0.05 <1> (sinigrin, <1>, isoenzyme RA, without ascorbic acid [24]) [24]

0.06 <1, 3> (sinigrin, <3>, isoenzyme SB, without ascorbic acid [24]; <1>, isoenzyme RB, without ascorbic acid [24]) [24]

0.068 <2> (6-deoxyglucotropaeolin) [47]

0.07 <1> (sinigrin) [1]

0.075 <2> (glucotropaeolin) [47]

0.094 <4> (sinigrin, <4>, in presence of 0.05 M NaCl [19]) [19]

0.1 <7> (*p*-nitrophenyl-*β*-D-glucoside) [33]

0.115 <2> (sinigrin) [2]

0.156 <3> (sinigrin, <3>, at pH 7.0, at 30°C [13]) [13]

0.17 <3> (sinigrin, <3>, isoenzyme SC, without ascorbic acid [24]) [24]

0.18 <4> (sinigrin, <4>, in presence of 0.5 M NaCl [19]; <4>, in absence of ascorbate [35]) [19, 35]

0.25 <3, 11> (sinigrin, <3>, isoenzyme SA, in presence of 1 mM ascorbic acid [24]; <11>, in presence of 0.5 mM ascorbate [40]) [24, 40]

0.26 <4> (sinigrin, <4>, in presence of 0.05 M NaCl and 1 mM ascorbate [19]) [19]

0.3 <3> (sinigrin, <3>, isoenzyme SB, in presence of 1 mM ascorbic acid [24]; <1>, isoenzyme RC in presence of 1 mM ascorbic acid [24]) [24]

0.37 <15> (sinigrin) [20]

0.4 <1, 3> (sinigrin, <3>, isoenzyme SC, in presence of 1 mM ascorbic acid [24]; <1>, isoenzyme RA and RB, in presence of 1 mM ascorbic acid [24]) [24]

0.42 <3> (sinigrin, <3>, with H_2O or 6.25 M methanol as solvent [50]) [50]

0.47 <11, 14> (sinigrin) [17, 41]

0.54 <3> (sinigrin, <3>, with acetonitrile as solvent [50]) [50]

0.57 <3> (sinigrin, <3>, with 12.5 M methanol as solvent [50]) [50]

0.57 <3> (sinigrin, <3>, with 6.25 M ethanol as solvent [50]) [50]

0.64 <4> (sinigrin, <4>, in presence of 2.5 M NaCl [19]) [19]

0.71 <15> (*p*-nitrophenyl *β*-D-glucopyranoside) [20]

0.87 <3> (sinigrin, <3>, with dioxane as solvent [50]) [50]

0.92 <4> (sinigrin, <4>, in presence of 0.5 M NaCl and 1 mM ascorbate [19]) [19]

0.93 <4> (sinigrin, <4>, in presence of 1 mM ascorbate [35]) [35]

1.5 <16> (*p*-nitrophenyl-*β*-D-glucopyranoside) [23]

2.78 <4> (sinigrin, <4>, in presence of 2.5 M NaCl and 1 mM ascorbate [19]) [19]

3.3 <16> (sinigrin) [23]

3.6 <7> (sinigrin) [32]

4 <3> (*p*-nitrophenyl-*β*-D-glucopyranoside, <3>, with dioxane as solvent [50]) [50]

12 <3> (*p*-nitrophenyl-*β*-D-glucopyranoside, <3>, with 12.5 M methanol as solvent [50]) [50]

14 <11> (*p*-nitrophenyl-*β*-D-glucopyranoside, <11>, in absence of ascorbate [40]) [40]

17 <3> (*p*-nitrophenyl-*β*-D-glucopyranoside, <3>, with acetonitrile as solvent [50]) [50]

30 <3> (*p*-nitrophenyl-β-D-glucopyranoside, <3>, with 6.25 M methanol as solvent [50]) [50]
49 <3> (*p*-nitrophenyl-β-D-glucopyranoside, <3>, with 0.2 M methanol as solvent [50]) [50]
51 <3> (*p*-nitrophenyl-β-D-glucopyranoside, <3>, with 6.25 M ethanol as solvent [50]) [50]
61 <3> (*p*-nitrophenyl-β-D-glucopyranoside, <3>, with water as solvent [50]) [50]

pH-Optimum

4 <3> (<3>, isoenzyme SA [24]) [24]
4.3 <1> (<1>, isoenzyme RC [24]) [24]
4.4 <1> (<1>, and a second isoenzyme with pH-optimum 8.0 [9]) [9]
4.5-5.5 <3> (<3>, isoenzyme SB [24]) [24]
4.5-7.8 <3> (<3>, hydrolysis of sinigrin, in citrate and phosphate buffer [13]) [13]
5 <1> (<1>, isoenzyme RA and RB [24]) [24]
5-6 <1> (<1>, myrosinase I and II [37]) [37]
5.2-5.5 <1> [1]
5.5 <2, 3, 8> (<2>, hydrolysis of sinigrin [2]; <3>, isoenzyme SC [24]) [2, 24, 51]
5.8 <3> [39]
6-6.5 <11> (<11>, at 37°C [41]) [41]
6-7 <1> (<1>, recombinant enzyme [45]) [45]
6.2 <16> [23]
6.5 <2> (<2>, hydrolysis of *p*-nitrophenyl-β-D-glucopyranoside [2]) [2, 47]
6.5-7 <14> [17]
6.8 <15> [20]
7 <7> [32]
8 <1, 10> (<1>, and a second isoenzyme with pH-optimum 8.0 [9]) [9, 46]

pH-Range

3-8 <2> (<2>, pH 3.0: about 45% of maximal activity, pH 8.0: about 55% of maximal activity [47]) [47]
3-9 <10> (<10>, pH 3.0: about 60% of maximal activity, pH 9.0: 80-90% of maximal activity [46]) [46]
4-7 <1> (<1>, about 35% of maximal activity of enzyme form Cc and Cb, about 45% of maximal activity of enzyme form Ca at pH 4.0 and at pH 7.0 [1]) [1]
4-9 <14> (<14>, pH 4.0: about 45% of maximal activity, pH 9.0: about 40% of maximal activity [17]) [17]
4.2-10 <7> (<7>, pH 4.2: about 40% of maximal activity, pH 10.0: about 55% of maximal activity [32]) [32]
4.5-7 <8> (<8>, pH 4.5: about 40% of maximal activity, pH 7.0: about 70% of maximal activity [51]) [51]
4.5-9 <1> (<1>, pH 4.5: about 55% of maximal activity, pH 9.0: about 80% of maximal activity, recombinant enzyme [45]) [45]

5-7.5 <16> (<16>, about 35% of maximal activity at pH 5.0 and at pH 7.5 [23]) [23]
5.2-6.5 <11> (<11>, pH 5.2: about 50% of maximal activity, pH 6.0-6.5: optimum [41]) [41]
5.5-7.5 <15> (<15>, pH 5.5: about 40% of maximal activity, pH 7.5: about 60% of maximal activity [20]) [20]
6-7.2 <1> (<1>, activity in presence of ascorbic acid is greater in sodium phosphate buffer than in citric acid /Na_2HPO_4 buffer [30]) [30]

Temperature optimum (°C)

34 <16> [23]
35 <4> (<4>, in presence of L-ascorbic acid [18]) [18]
36-60 <1> (<1>, recombinant enzyme [45]) [45]
37 <8, 11, 14> [17, 41, 51]
55 <4> (<4>, in absence of L-ascorbic acid [18]) [18]
60 <1, 3> (<1>, isoenzyme RA, RB, RC and RD [24]; <3>, isoenzyme SA, SB and SC [24]) [24, 39]
70 <1> [1]

Temperature range (°C)

20-65 <1> (<1>, 20°C: about 85% of maximal activity, 65°C: about 45% of maximal activity, recombinant enzyme [45]) [45]
25-50 <4> (<4>, about 45% of maximal activity at 25°C and at 50°C, in presence of 1 mM ascorbic acid [18]) [18]
30-50 <14> (<14>, about 40% of maximal activity at 30°C and at 50°C [17]) [17]
30-70 <10> (<10>, 30°C, about 60% of maximal activity of the enzyme from white cabbage, about 45% of maximal activity of the enzyme from red cabbage, 70°C: about 75% of maximal activity of the enzyme from red cabbage, about 50% of maximal activity of the enzyme from white cabbage [46]) [46]
40-80 <1> (<1>, 40°C: about 55% of maximal activity of enzyme form Ca, about 50% of maximal activity of enzyme form Cb, about 60% of maximal activity of enzyme form Cc, 80°C: about 55% of maximal activity of enzyme form Ca, about 60% of maximal activity of enzyme form Cb, about 60% of maximal activity of enzyme form Cc [1]) [1]

4 Enzyme Structure

Molecular weight

61000 <15> (<15>, gel filtration [20]) [20]
62000 <8> (<8>, gel filtration [51]) [51]
120000 <11> (<11>, gel filtration [40]) [40]
124000 <25> (<25>, myrosinase F-IIB, equilibrium sedimentation [29]) [29]
125000 <25> (<25>, myrosinase F-IIB, gel filtration [29]) [29]
130000 <2> (<2>, gel filtration [2]) [2]
133000-135000 <1> (<1>, gel filtration, ultracentrifugation [25]) [25]
140000 <1> (<1>, M75 myrosinase, gel filtration [48]) [48]

150000 <25> (<25>, myrosinase F-IB, gel filtration [29]) [29]
150000-151000 <3> (<3>, gel filtration, equilibrium sedimentation [31]) [31]
152000 <25> (<25>, myrosinase F-IB, equilibrium sedimentation [29]) [29]
154000 <1> (<1>, non-denaturing PAGE [1]) [1]
156000 <1> (<1>, myrosinase I, gel filtration [37]) [37]
188000 <1> (<1>, myrosinase II, gel filtration [37]) [37]
580000 <14> (<14>, gel filtration [17]) [17]

Subunits

? <1, 25> (<25>, x * 30000, enzyme consists of at least 4 subunits, myrosinase F-IIB, SDS-PAGE [29]; <25>, x * 40000, enzyme consists of at least 4 subunits, myrosinase F-IA, F-IB and F-IIA, SDS-PAGE [29]; <1>, x * 65000, myrosinase B, SDS-PAGE [44]; <1>, x * 70000, myrosinase C, SDS-PAGE [44]; <1>, x * 75000, myrosinase A, SDS-PAGE [44]) [44, 29]
dimer <1, 2, 3, 11> (<11>, 1 * 61000 + 1 * 62000, SDS-PAGE [40]; <2>, 1 * 62000 + 1 * 65000, SDS-PAGE [2]; <1>, 2 * 65000, SDS-PAGE [25]; <3>, 2 * 66000, SDS-PAGE [31]; <1>, 2 * 75000, myrosinase I, SDS-PAGE [37]; <1>, 2 * 75000, M75 myrosinase, SDS-PAGE [48]; <1>, 2 * 77000, SDS-PAGE [1]) [1, 2, 25, 31, 37, 40, 48]
dodecamer <14> (<14>, 12 * 45000-47000, SDS-PAGE [17]) [17]
trimer <1> (<1>, 3 * 62000, myrosinase II, SDS-PAGE [37]) [37]

Posttranslational modification

glycoprotein <1, 2, 3, 25> (<1>, contains fucose, mannose and N-acetylglucosamine. Enzyme form Ca contains 18.9% carbohydrate, enzyme form Cb contains 11.7% carbohydrate, enzyme form Cc contains 9.6% carbohydrate [1]; <2>, both subunits contain carbohydrate [2]; <1>, myrosinase C1 contains 9.3% carbohydrate, myrosinase C2 contains 15.2% carbohydrate, myrosinase C3 contains 17.4% carbohydrate [25]; <3>, contains 18% carbohydrate: 4% glucosamine, 12% hexose and 2% pentose [31]; <1>, N-linked glycosylation sites, myrosinase I contains 18% carbohydrate [43]; <1>, size differences between the different myrosinases are mainly due to differences in glycosylation [44]; <3>, contains 13000 Da carbohydrate per dimer, the carbohydrate is required to maintain molecular stability and solubility in the dehydrated environment of the seed [49]; <25>, hexose content of enzyme F-IA is 15.8%, hexose content of enzyme F-IB is 17.8%, hexose content of enzyme F-IIA is 22.5%, hexose content of enzyme F-IIB is 8.6% [29]) [1, 2, 25, 29, 31, 43, 44]
proteolytic modification <1> (<1>, signal peptide [43]) [43]

5 Isolation/Preparation/Mutation/Application

Source/tissue

cotyledon <1, 3> [39, 44]
culture medium <7> [32]
embryo axis <1> [44]
fruit <17> (<17>, no activity found in latex [28]) [28]

leaf <8, 9, 10, 11, 13> (<9, 13>, very low activity [16]; <10>, activity is 4-5 times higher in the outer leaves than in the stalk, the inner leaves and the centre of the Brussels sprouts [39]) [12, 16, 39, 41, 51]
muscle <18> (<18>, of head and thorax [36]) [36]
mycelium <16> [23, 26]
myrosin cells <1, 3, 11> (<1>, of the embryo axis and the cotyledons [44]) [39, 44]
root <2, 3, 9, 11, 12> (<1-3, 9, 11>, seedling roots, highest activity [16]; <9, 11>, highest activity 1 day after germination [16]; <1-3>, highest activity 2 days after germination [16]) [11, 12, 16, 40, 41]
seed <1-4, 6, 8, 11, 17> (<17>, found in sarcotestae but not in endosperms [28]; <1>, Myrl encodes seed myrosinase [43]) [1, 3, 6, 9, 12, 13, 15, 19, 24, 28, 38, 39, 41, 43, 47, 49, 50]
seedling <1, 2, 3, 11> (<1, 3>, cotyledon [39]; <3>, hypocotyl [39]; <3>, leaf [39]; <3>, root [39]; <3>, myrosinase gene MYR2 is transcribed in all organs of the seedling, myrosinase gene Myrl shows no transcription in seedling [39]) [2, 37, 39, 40]
stem <9, 13> [16]
vegetative cells <1> (<1>, Myr2 encodes vegetative type myrosinase [43]) [43]

Localization

extracellular <12> [39]
intracellular <16> [23, 26]
membrane <12> (<12>, associated with the cytoplasmic side of internal membranes [39]) [39]
myrosin grains <3> (<3>, exclusively localized to the interior of the myrosin grains [38]) [38]
protoplast <2, 3, 9-11> (<2, 3, 9-11>, the specific activity of the protoplast is less than that of the intact root tissue [11]) [11]

Purification

<1> (three different forms of enzyme: Ca, Cb and Cc [1]; at least 2 isoenzymes [9]; myrosinase C1, C2 and C3 [25]; myrosinase I and II [37]; <1>, complexes with myrosinase and myrosinase-binding protein [44]) [1, 7, 9, 25, 37, 44, 48]
<2> [2]
<3> (myrosinase A, B and C [31]) [3, 7, 31]
<5> [7]
<7> [32, 33]
<11> (2 isoenzymes [41]) [40, 41]
<14> [17]
<15> [20]
<25> [29]

Crystallization

<3> (native myrosinase and stable glycosyl-enzyme intermediate [49]) [49]

Cloning

<1> (myrosinase is encoded by two different families of genes: MA and MB. The MA type myrosinases are encoded by approximately 4 genes, the MB type myrosinases are encoded by more than 10 genes [42]; myrosinases are encoded by a multigene family consisting of two subgroups. Cloning and sequencing of the two nuclear genes, Myr1.Mn1 and Myr2.Mn1, representing each of these two subgroups [43]; expression of myrosinase MYR1 in Saccharomyces cerevisiae. The recombinant enzyme is enzymatically active which shows that the enzyme, which in plant is complex bound with myrosinase-binding proteins and myrosinase-associated proteins, is functional in its free form. Expression of the enzyme is transient and the growth of the yeast cells is significantly reduced during the period of expression of myrosinase [45]) [42, 43, 44, 45]

6 Stability

pH-Stability

4 <14> (<14>, 5°C, 41 h, about 60% loss of activity [17]) [17]
5-7 <15> (<15>, below 40°C, 24 h, stable [20]) [20]
5.5-8.5 <7> (<7>, below 45°C, stable [32]) [32]
6 <1, 3, 14> (<14>, 5°C, 41 h, about 10% loss of activity [17]; <1, 3>, maximal stability [24]) [17, 24]
6-9 <10> (<10>, 30°C, 1 h, stable [46]) [46]
7 <14, 16> (<14>, 5°C, 41 h, stable [17]; <16>, 5°C, 24 h, about 20% loss of activity [23]) [17, 23]
7.6-8 <16> (<16>, 5°C, 24 h, stable [23]) [23]
9 <16> (<16>, 5°C, 24 h, about 55% loss of activity [23]) [23]
10 <14> (<14>, 5°C, 41 h, about 25% loss of activity [17]) [17]
11 <14> (<14>, 5°C, 41 h, about 30% loss of activity [17]) [17]

Temperature stability

30 <8, 11, 14, 15, 16> (<14>, pH 7.0, 20 min, stable [17]; <15>, pH 6.8, 30 min, stable [20]; <16>, pH 7.8, 20 min, stable [23]; <11>, 90 min, stable [41]; <8>, 60 min, 24% loss of activity [51]) [17, 20, 23, 41, 51]
35 <16> (<16>, pH 7.8, 20 min, about 40% loss of activity [23]) [23]
40 <10, 11, 14, 15, 16> (<14>, pH 7.0, 20 min, about 25% loss of activity [17]; <15>, pH 6.8, 30 min, about 20% loss of activity [20]; <16>, pH 7.8, 20 min, complete inactivation [23]; <7>, pH 7.0, 15 min, stable [32]; <11>, 45 min, 18% loss of activity [41]; <10>, 30 min, 30 min, 5% loss of activity of the enzyme from red cabbage, 15% loss of activity of the enzyme from white cabbage [46]) [17, 20, 23, 32, 41, 46]
50 <7, 10, 14, 15> (<14>, pH 7.0, 20 min, about 60% loss of activity [17]; <15>, pH 6.8, 30 min, about 50% loss of activity [20]; <7>, pH 7.0, 15 min, complete loss of activity [32]; <10>, 30 min, stable [46]) [17, 20, 32, 46]

60 <10, 15> (<15>, pH 6.8, 30 min, about 90% loss of activity [20]; <10>, 30 min, enzyme from red cabbage is stable, enzyme from white cabbage loses about 60% of its activity [46]) [20, 46]
70 <10, 14> (<14>, pH 7.0, 20 min, about 80% loss of activity [17]; <10>, 30 min, 90% loss of activity [46]) [17, 46]

Oxidation stability

<25>, photooxidation by methylene blue [27]

General stability information

<3>, good static and operational stability after immobilization on γ-alumina [3]
<16>, unstable after extraction from mycelium, considerably inactivated during dialysis and precipitation process with ammonium sulfate, stabilized by 10 mM 2-mercaptoethanol and 1 mM ascorbic acid [26]

References

[1] Bones, A.; Slupphaug, G.: Purification, characterization and partial amino acid sequencing of β-thioglucosidase from Brassica napus. J. Plant Physiol., **134**, 722-729 (1989)
[2] Durham, P.L.; Poulton, J.E.: Effect of castanospermine and related polyhydroxyalkaloids on purified myrosinase from Lepidium sativum seedlings. Plant Physiol., **90**, 48-52 (1989)
[3] Iori, R.; Leoni, O.; Palmieri, S.: Immobilization of myrosinase (thioglucoside glycohydrolase EC 3.2.3.1). Biotechnol. Lett., **10**, 575-578 (1988)
[4] MacLeod, A.J.; Rossiter, J.T.: Degradation of 2-hydroxybut-3-enylglucosinolate (progoitrin). Phytochemistry, **26**, 669-673 (1987)
[5] Uda, Y.; Kurata, T.; Arakawa, N.: Effects of thiol compounds on the formation of nitriles from glucosinolates. Agric. Biol. Chem., **50**, 2741-2746 (1986)
[6] Uda, Y.; Kurata, T.; Arakawa, N.: Effects of pH and ferrous ion on the degradation of glucosinolates by myrosinase. Agric. Biol. Chem., **50**, 2735-2740 (1986)
[7] Buchwaldt, L.; Larsen, L.M.; Ploeger, A., Sorensen, H.: Fast polymer liquid chromatography isolation and characterization of plant myrosinase, β-thioglucoside glucohydrolase, isoenzymes. J. Chromatogr., **363**, 71-80 (1986)
[8] Petroski, R.J.: Stereoselectivity of the interaction of thioglucoside glucohydrolase and epithiospecifier protein from various sources. Plant Sci., **44**, 85-88 (1986)
[9] MacLeod, A.J.; Rossiter, J.T.: Isolation and examination of thioglucoside glucohydrolase from seeds of Brassica napus. Phytochemistry, **25**, 1047-1051 (1986)
[10] Petroski, R.J.; Kwolek, W.F.: Interaction of a fungal thioglucoside glycohydrolase and cruciferous plant epithiospecifier protein to form 1-cyanoe-

pithioalkanes: implication of an allosteric mechanism. Phytochemistry, **24**, 213-216 (1985)
[11] Iversen, T.-H.; Myhre, S.; Evjen, K.; Baggerud, C.: Morphology and myrosinase activity in root protoplast of Brassicaceae. Z. Pflanzenphysiol., **112**, 391-401 (1983)
[12] Petroski, R.J.; Tookey, H.L.: Interactions of thioglucoside glucohydrolase and epithiospecifier protein of cruciferous plants to form 1-cyanoepithioalkanes. Phytochemistry, **21**, 1903-1905 (1982)
[13] Palmieri, S.; Leoni, O.; Iori, R.: A steady-state kinetics study of myrosinase with direct ultraviolet spectrophotometric assay. Anal. Biochem., **123**, 320-324 (1982)
[14] Gil, V.; MacLeod, A.J.: The effects of pH on glucosinolate degradation by a thioglucoside glucohydrolase preparation. Phytochemistry, **19**, 2547-2551 (1980)
[15] Gil, V.; MacLeod, A.J.: Effects of a Lepidium sativum enzyme preparation on the degradation of glucosinolates. Phytochemistry, **19**, 2071-2076 (1980)
[16] Iversen, T.-H.; Baggerud, C.: Myrosinase activity in differentiated and undifferentiated plants of brassicaceae. Z. Pflanzenphysiol., **97**, 399-407 (1980)
[17] Ohtsuru, M.; Kawatani, H.: Studies on the myrosinase from Wasabia japonica: purification and some properties of Wasabi myrosinase. Agric. Biol. Chem., **43**, 2249-2255 (1979)
[18] Ohtsuru, M.; Hata, T.: The interaction of L-ascorbic acid with the active center of myrosinase. Biochim. Biophys. Acta, **567**, 384-391 (1979)
[19] Tsuruo, I.; Hata, T.: Studies on the myrosinase in mustard seed. Part III. On the effect of neutral salts. Agric. Biol. Chem., **32**, 479-483 (1968)
[20] Tani, N.; Ohtsuru, M.; Hata, T.: Purification and general characteristics of bacterial myrosinase produced by Enterobacter cloacae. Agric. Biol. Chem., **38**, 1623-1630 (1974)
[21] Tani, N.; Ohtsuru, M.; Hata, T.: Isolation of myrosinase producing microorganism. Agric. Biol. Chem., **38**, 1617-1622 (1974)
[22] Srivastava, V.K.; Hill, D.C.: Glucosinolate hydrolytic products given by Sinapis alba, and Brassica napus thioglucosidases. Phytochemistry, **13**, 1043-1046 (1974)
[23] Ohtsuru, M.; Hata, T.: General characterization of the intracellular myrosinase from Aspergillus niger. Agric. Biol. Chem., **37**, 2543-2548 (1973)
[24] Bjoerkman, R.; Loennerdal, B.: Studies on myrosinases. III. Enzymatic properties of myrosinases from Sinapis alba and Brassica napus seeds. Biochim. Biophys. Acta, **327**, 121-131 (1973)
[25] Loennerdal, B.; Janson, J.-C.: Studies on myrosinases. II. Purification and characterization of a myrosinase from rapeseed (Brassica napus L.). Biochim. Biophys. Acta, **315**, 421-429 (1973)
[26] Ohtsuru, M.; Tsuruo, I.; Hata, T.: The production and stability of intracellular myrosinase from Aspergillus niger. Agric. Biol. Chem., **37**, 967-971 (1973)
[27] Ohtsuru, M.; Hata, T.: Studies on the functional groups of plant myrosinase. Agric. Biol. Chem., **37**, 269-275 (1973)

[28] Tang, C.-S.: Localization of benzyl glucosinolate and thioglucosidase in Carica papaya fruit. Phytochemistry, **12**, 769-773 (1973)
[29] Ohtsuru, M.; Hata, T.: Molecular properties of multiple forms of plant myrosinase. Agric. Biol. Chem., **36**, 2495-2503 (1972)
[30] Henderson, H.M.; McEwen, T.J.: Effect of ascorbic acid on thioglucosidases from different crucifers. Phytochemistry, **11**, 3127-3133 (1972)
[31] Bjoerkman, R., Janson, J.-C.: Studies on myrosinases. I. Purification and characterization of a myrosinase from white mustard seed (Sinapis alba, L.). Biochim. Biophys. Acta, **276**, 508-518 (1972)
[32] Ohtsuru, M.; Tsuruo, I.; Hata, T.: Studies on fungous myrosinase. Part I. Production, purification and some characteristics. Agric. Biol. Chem., **33**, 1309-1314 (1969)
[33] Ohtsuru, M.; Tsuruo, I.; Hata, T.: Studies on fungous myrosinase. Part III. On the β-glucosidase activity of fungous myrosinase and the relationship of fungous and plant myrosinases to β-glucosidases. Agric. Biol. Chem., **33**, 1320-1325 (1969)
[34] Ohtsuru, M.; Tsuruo, I.; Hata, T.: Studies on fungous myrosinase. Part II. Effects of various reagents on its enzymatic activities. Agric. Biol. Chem., **33**, 1315-1319 (1969)
[35] Tsuruo, I.; Hata, T.: Studies on myrosinase in mustard seed. Part IV. Sugar and glucosides as competitive inhibitors. Agric. Biol. Chem., **32**, 1420-1424 (1968)
[36] Scofield, A.M.; Rossiter, J.T.; Witham, P.; Kite, G.C.; Nash, R.J.; Fellows, E.: Inhibition of thioglucosidase-catalysed glucosinolate hydrolysis by castanospermine and related alkaloids. Phytochemistry, **29**, 107-109 (1990)
[37] James, D.C.; Rossiter, J.T.: Development and characteristics of myrosinase in Brassica napus during early seedling growth. Physiol. Plant., **82**, 163-170 (1991)
[38] Hoeglund, A.-S.; Lenman, M.; Rask, L.: Myrosinase is localized to the interior of myrosin grains and is not associated to the surrounding tonoplast membrane. Plant Sci., **85**, 165-170 (1992)
[39] Bones, A.M.; Rossiter, J.T.: The myrosinase-glucosinolate system, its organisation and biochemistry. Physiol. Plant., **97**, 194-208 (1996)
[40] Shikita, M.; Fahey, J.W.; Golden, R.R.; Holtzclaw, W.D.; Talalay, P.: An unusual case of uncompetitive activation by ascorbic acid: purification and kinetic properties of a myrosinase from Raphanus sativus seedlings. Biochem. J., **341**, 725-732 (1999)
[41] Jwanny, E.W.; El-Sayed, S.T.; Rashad, M.M.; Mahmoud, A.E.; Abdallah, N.M.: Myrosinase from roots of Raphanus sativus. Phytochemistry, **39**, 1301-1303 (1995)
[42] Lenman, M.; Falk, A.; Xue, J.; Rask, L.: Characterization of a Brassica napus myrosinase pseudogene: myrosinases are members of the BGA family of β-glycosidases. Plant Mol. Biol., **21**, 463-474 (1993)
[43] Thangstad, O.P.; Winge, P.; Husebye, H.; Bones, A.: The myrosinase (thioglucoside glucohydrolase) gene family in brassicaceae. Plant Mol. Biol., **23**, 511-524 (1993)

[44] Falk, A.; Ek, B.; Rask, L.: Characterization of a new myrosinase in Brassica napus. Plant Mol. Biol., **27**, 863-874 (1995)
[45] Chen, S.; Halkier, B.A.: Functional expression and characterization of the myrosinase MYR1 from Brassica napus in Saccharomyces cerevisiae. Protein Expr. Purif., **17**, 414-420 (1999)
[46] Yen, G.-C.; Wei, Q.-K.: Myrosinase activity and total glucosinolate content of cruciferous vegetables, and some properties of cabbage myrosinase in Taiwan. J. Sci. Food Agric., **61**, 471-475 (1993)
[47] Iori, R.; Rollin, P.; Streicher, H.; Thiem, J.; Palmieri, S.: The myrosinase-glucosinolate interaction mechanism studied using some synthetic competitive inhibitors. FEBS Lett., **385**, 87-90 (1996)
[48] Lenman, M.; Roedin, J.; Josefsson, L.-G.; Rask, L.: Immunological characterization of rapeseed myrosinase. Eur. J. Biochem., **194**, 747-753 (1990)
[49] Burmeister, W.P.; Cottaz, S.; Driguez, H.; Iori, R.; Palmieri, S.; Henrissat, B.: The crystal structures of Sinapis alba myrosinase and a covalent glycosyl-enzyme intermediate provide insight into the substrate recognition and active-site machinery of an S-glycosidase. Structure, **5**, 663-675 (1997)
[50] Botti, M.G.; Taylor, M.G.; Botting, N.P.: Studies on the mechanism of myrosinase. Investigation of the effect of glycosyl acceptors on enzymatic activity. J. Biol. Chem., **270**, 20530-20535 (1995)
[51] El-Sayed, S.T.; Jwanny, E.W.: Glycosidases of turnip leaf tissues. Appl. Biochem. Biotechnol., **49**, 11-22 (1994)
[52] Cottaz, S.; Henrissat, B.; Driguez, H.: Mechanism-based inhibition and stereochemistry of glucosinolate hydrolysis by myrosinase. Biochemistry, **35**, 15256-15259 (1996)

Ribosylhomocysteinase 3.2.1.148

1 Nomenclature

EC number
3.2.1.148

Systematic name
S-ribosyl-L-homocysteine ribohydrolase

Recommended name
ribosylhomocysteinase

Synonyms
EC 3.3.1.3 (formerly)
S-ribosylhomocysteine hydrolase
S-ribosylhomocysteine-cleavage enzyme
homocysteinase, ribosyl-

2 Source Organism

<-20> no activity in *Arthrobacter globiformis* [4]
<-19> no activity in *Alcaligenes faecalis* [4]
<-18> no activity in *Acinetobacter calcoaceticus* [4]
<-17> no activity in *Peudomonas aeruginosa* [4]
<-16> no activity in *Pseudomonas putida* [4]
<-15> no activity in *Mycobacterium avium* [4]
<-14> no activity in *Nocardia asteroides* [4]
<-13> no activity in *Streptomyces hygroscopicus* [4]
<-12> no activity in *Micromonospora faeni* [4]
<-11> no activity in *Streptoverticillium kentuchense* [4]
<-10> no activity in *Streptosporangium roseum* [4]
<-9> no activity in *Micropolyspora angiospora* [4]
<-8> no activity in *Microellobosporia violacea* [4]
<-7> no activity in *Saccharomyces cerevisiae* [1]
<-6> no activity in *Candida cerevisiae* [1]
<-5> no activity in *Candida utilis* [1]
<-4> no activity in *Phaseolus vulgaris* [1]
<-3> no activity in *Zea mays* [1]
<-1> no activity in *Hordeum vulgare* [1]
<1> *Alcaligenes faecalis* [1]
<2> *Citrobacter freundii* [1]
<3> *Enterobacter aerogenes* [1]

<4> *Escherichia coli* (ATCC 14948 [4]) [1, 2, 3, 4]
<5> *Proteus vulgaris* [1]
<6> *Bacillus cereus* [1]
<7> *Sarcina lutea* (low activity [1]) [1]
<8> *Staphylococcus aureus* (low activity [1]) [1]
<9> *Proteus inconstans* (IFO 12930 [4]) [4]
<10> *Citrobacter intermedius* (IFO 13539 [4]) [4]
<11> *Salmonella typhimurium* (AKU 90 [4]) [4]
<12> *Bacillus subtilis* (IFO 3007 [4]) [4]

3 Reaction and Specificity

Catalyzed reaction

H_2O + S-ribosyl-L-homocysteine = D-ribose + L-homocysteine

Substrates and products

S H_2O + S-ribosyl-L-homocysteine <1-12> [1-4]
P D-ribose + L-homocysteine <4> [3]

K_m-Value (mM)

1.94 <4> (S-ribosyl-L-homocysteine) [2]

pH-Optimum

7.5 <1-8> [1]
7.6 <4> [2]
7.9-8.1 <4> [2]
8 <4> [3]

pH-Range

7-8.6 <4> (<4>, no activity below pH 7.0 and above pH 8.6 [2]) [2]
7.5-9 <4> (<4>, pH 7.5: about 35% of maximal activity, pH 9.0: about 25% of maximal activity [3]) [3]

5 Isolation/Preparation/Mutation/Application

Purification

<4> [2]

6 Stability

General stability information

<4>, Tris-HCl buffer, 0.1 M, pH 7.5, provides the best conditions for stability [2]
<4>, partially purified enzyme after ethyl alcohol fractionation is extremely unstable and loses 95% of its activity within 24 h at 4°C [2]

Storage stability

<4>, 2°C, retains about 75% of ist activity after 48 h, retains 50-60% of its activity after 1 week [3]

References

[1] Walker, R.D.; Duerre, J.A.: S Adenosylhomocysteine metabolism in various species. Can. J. Biochem., **53**, 312-319 (1975)
[2] Miller, C.H.; Duerre, J.A.: S-Ribosylhomocysteine cleavage enzyme from Escherichia coli. J. Biol. Chem., **243**, 92-97 (1968)
[3] Duerre, J.A.; Miller, C.H.: Cleavage of S-ribosyl-L-homocysteine by extract from Escherichia coli. J. Bacteriol., **91**, 1210-1217 (1966)
[4] Shimizu, S.; Shiozaki, S.; Ohshiro, T.; Yamada, H.: Occurence of S-adenosylhomocysteine hydrolase in prokaryote cells. Charactzerization if the enzyme from Alcaligenes faecalis and role of the enzyme in the activated methyl cycle. Eur. J. Biochem., **141**, 385-392 (1984)

β-Primeverosidase 3.2.1.149

1 Nomenclature

EC number
3.2.1.149

Systematic name
6-O-(β-D-xylopyranosyl)-β-D-glucopyranoside 6-O-(β-D-xylosyl)-β-D-glucohydrolase

Recommended name
β-primeverosidase

Synonyms
4-methylumbelliferyl-β-D-glucopyranoside:β-glucosidase
arbutinase
ary β-glucosidase
β-D-glucosidase
β-D-glucoside glucohydrolase
β-glucosidase
β-glucoside hydrolase
β-primeverosidase
p-nitrophenyl β-glucosidase
primeverosidase
vicianase

CAS registry number
9001-22-3

2 Source Organism

<1> *Camellia sinensis* (var. assamica [1]; var. sinensis cv. Shuixian [1, 2]; var. sinensis cv. Yabukita [1, 3, 4]) [1-4]

3 Reaction and Specificity

Catalyzed reaction
a 6-O-(β-D-xylopyranosyl)-β-D-glucopyranoside + H_2O = 6-O-(β-D-xylopyranosyl)-β-D-glucopyranose + an alcohol (The enzyme is responsible for the formation of the alcoholic aroma in oolong and black tea. In addition to β-primeverosides, i.e. 6-O-(β-D-xylopyranosyl)-β-D-glucopyranosides, it also

hydrolyses 6-O-(β-D-apiofuranosyl)-β-D-glucopyranosides and, less rapidly, β-vicianosides and 6-O-(α-L-arabinofuranosyl)-β-D-glucopyranosides, but not β-glucosides. Geranyl-, linaloyl-, benzyl- and *p*-nitrophenol glycosides are all hydrolysed)

Reaction type

hydrolysis of O-glycosyl bond

Natural substrates and products

S (R)-linalyl β-vicianoside + H_2O <1> [1]
P linalol + 6-O-α-L-arabinopyranosyl-β-D-glucopyranoside <1> [1]
S 2-phenylethyl 6-O-(β-D-xylopyranosyl)-β-D-glucopyranoside + H_2O <1> (i.e. 2-phenylethyl β-D-primeveroside) [3, 4]
P 2-phenylethanol + 6-O-(β-D-xylopyranosyl)-β-D-glucopyranose <1> [3, 4]
S 2-phenylethyl β-vicianoside + H_2O <1> [4]
P 2-phenylethanol + 6-O-α-L-arabinopyranosyl-β-D-glucopyranoside <1> [4]
S benzyl 6-O-(β-D-xylopyranosyl)-β-D-glucopyranoside + H_2O <1> (i.e. 2-benzyl β-D-primeveroside) [3]
P benzyl alcohol + 6-O-(β-D-xylopyranosyl)-β-D-glucopyranose <1> [3]
S eugenyl β-primeveroside + H_2O <1> (i.e. eugenyl 6-O-β-D-xylopyranosyl-β-D-glucopyranose) [1]
P eugenol + primeverose <1> [2]
S geranyl 6-O-(β-D-xylopyranosyl)-β-D-glucopyranoside + H_2O <1> (i.e. geranyl β-D-primeveroside) [3]
P geraniol + 6-O-(β-D-xylopyranosyl)-β-D-glucopyranose <1> [3]
S geranyl 6-O-α-L-arabinofuranosyl-β-D-glucopyranoside + H_2O <1> [1]
P geraniol + 6-O-α-L-arabinofuranosyl-β-D-glucopyranoside <1> [1]
S linalyl 6-O-(β-D-xylopyranosyl)-β-D-glucopyranoside + H_2O <1> [3]
P linalol + 6-O-(β-D-xylopyranosyl)-β-D-glucopyranose <1> [3]

Substrates and products

S (R)-linalyl β-vicianoside + H_2O <1> [1]
P linalol + 6-O-α-L-arabinopyranosyl-β-D-glucopyranoside <1> [1]
S 2-phenylethyl 6-O-(β-D-xylopyranosyl)-β-D-glucopyranoside + H_2O <1> (i.e. 2-phenylethyl β-D-primeveroside) [3, 4]
P 2-phenylethanol + 6-O-(β-D-xylopyranosyl)-β-D-glucopyranose <1> [3, 4]
S 2-phenylethyl β-acuminoside + H_2O <1> (i.e. 2-phenylethyl 6-O-β-D-apiofuranosyl-β-D-glucopyranoside) [4]
P 2-phenylethanol + 6-O-β-D-apiofuranosyl-β-D-glucopyranoside <1> [4]
S 2-phenylethyl β-vicianoside + H_2O <1> [4]
P 2-phenylethanol + 6-O-α-L-arabinopyranosyl-β-D-glucopyranoside <1> [4]
S 2-phenylethyl-β-D-glucopyranoside + H_2O <1> [4]
P 2-phenylethanol + β-D-glucopyranose
S 4-nitrophenyl β-D-glucopyranoside + H_2O <1> [4]

P 4-nitrophenol + β-D-glucopyranose <1> [4]
S 4-nitrophenyl β-primeveroside + H_2O <1> (i.e. 4-nitrophenyl-6-O-β-D-xylopyranosyl-β-D-glucopyranose) [1]
P 4-nitrophenol + primeverose <1> [1]
S benzyl 6-O-(β-D-xylopyranosyl)-β-D-glucopyranoside + H_2O <1> (i.e. 2-benzyl β-D-primeveroside) [3]
P benzyl alcohol + 6-O-(β-D-xylopyranosyl)-β-D-glucopyranose <1> [3]
S benzyl 6-O-α-L-arabinofuranosyl-β-D-glucopyranoside + H_2O <1> [1]
P benzyl alcohol + 6-O-α-L-arabinofuranosyl-β-D-glucopyranoside
S eugenyl β-primeveroside + H_2O <1> (i.e. eugenyl 6-O-β-D-xylopyranosyl-β-D-glucopyranose) [1]
P eugenol + primeverose <1> [2]
S geranyl 6-O-(β-D-xylopyranosyl)-β-D-glucopyranoside + H_2O <1> (i.e. geranyl β-D-primeveroside) [3]
P geraniol + 6-O-(β-D-xylopyranosyl)-β-D-glucopyranose <1> [3]
S geranyl 6-O-α-L-arabinofuranosyl-β-D-glucopyranoside + H_2O <1> [1]
P geraniol + 6-O-α-L-arabinofuranosyl-β-D-glucopyranoside <1> [1]
S linalyl 6-O-(β-D-xylopyranosyl)-β-D-glucopyranoside + H_2O <1> [3]
P linalool + 6-O-(β-D-xylopyranosyl)-β-D-glucopyranose <1> [3]

Turnover number (min^{-1})
351 <1> (2-phenylethyl β-vicianoside) [4]
2622 <1> (2-phenylethyl β-vicianoside) [4]

Specific activity (U/mg)
0.04 <1> (<1> glucosidase I [3]) [3]
0.15 <1> (<1> glucosidase II [3]) [3]
0.27 <1> (<1> glucosidase IIa [3]) [3]
0.44 <1> (<1> glucosidase IIb [3]) [3]
0.9-0.99 <1> [1, 2]
1.61 <1> (<1> glucosidase III [3]) [3]

K_m-Value (mM)
2 <1> (2-phenylethyl β-vicianoside) [4]
14.9 <1> (2-phenylethyl β-vicianoside) [4]

pH-Optimum
4 <1> [1, 2]
5 <1> (<1> var. sinensis cv. Yabukita [3]) [3]

pH-Range
3-6 <1> (<1> less than 50% of maximal activity above and below [2]) [2]

Temperature optimum (°C)
45 <1> [1, 2]
50 <1> (<1> var. sinensis cv. Yabukita [3]) [3]

Temperature range (°C)
30-60 <1> (<1> less than 50% of maximal activity above and below [2]) [2]

4 Enzyme Structure

Molecular weight

60170 <1> (<1> var. sinensis cv. Shuixian, MALDI-TOF-MS [1]) [1]
60310 <1> (<1> var. assamica, MALDI-TOFMS [1]) [1]
60480 <1> (<1> var. sinensis cv. Yabukita, MALDI-TOF-MS [1]) [1]
61000 <1> (<1> SDS-PAGE [1]) [1]

5 Isolation/Preparation/Mutation/Application

Source/tissue

leaf <1> (<1> fresh, non-fermented [1,2]) [1-4]

Purification

<1> (var. assamica [1]; var. sinensis cv. Shuixian [2]) [1, 2]

6 Stability

pH-Stability

3-8 <1> [3]
4-9 <1> [2]

Temperature stability

45 <1> (<1> stable below [3]) [3]
55 <1> (<1> 1 h, 50% loss of activity [2]) [2]

References

[1] Ijima, Y.; Ogawa, K.; Watanabe, N.; Usui, T.; Ohnishi-Kameyama, M.; Nagata, T.; Sakata, K.: Characterization of β-primeverosidase, being concerned with alcoholic aroma formation in tea leaves to be processed into black tea, and preliminary observations on its substrate specificity. J. Agric. Food Chem., **46**, 1712-1718 (1998)

[2] Ogawa, K.; Ijima, Y.; Guo, W.; Watanabe, N.; Usui, T.; Dong, S.; Tong, Q.; Sakata, K.: Purification of a β-primeverosidase concerned with alcoholic aroma formation in tea leaves (cv. Shuxian) to be processed to oolong tea. J. Agric. Food Chem., **45**, 877-882 (1997)

[3] Guo, W.; Ogawa, K.; Yamauchi, K.; Watanabe, N.; Usui, T.; Luo, S.; Sakata, K.: Studies on the aroma formation mechanism in oolong tea. Part V. isolation and characterization of a β-primeverosidase concerned with alcoholic aroma formation in tea leaves. Biosci. Biotechnol. Biochem., **60**, 1810-1814 (1996)

[4] Ma, S.-J.; Mizutani, M.; Hiratake, J.; Hayashi, K.; Yagi, K.; Watanabe, N.; Sakata, K.: Substrate specificity of β-primeverosidase, a key enzyme in aroma formation during oolong tea and black tea manufacturing. Biosci. Biotechnol. Biochem., **65**, 2719-2729 (2001)

Be the first to know

with the new online notification service

Springer Alert

You decide how we keep you up to date on new publications:

- Select a specialist field within a subject area
- Take your pick from various information formats
- Choose how often you'd like to be informed

And receive customised information to suit your needs

Register now

http://www.springer.de/alert

and then you are one click away from a world of chemistry information!

Come and visit Springer's Chemistry Online Library

http://www.springer.de/chem

Springer

008818_001x_1c

Printing: Mercedes-Druck, Berlin
Binding: Stein+Lehmann, Berlin